Crustal Evolution of Southern Africa

A.J. Tankard M.P.A. Jackson
K.A. Eriksson D.K. Hobday
D.R. Hunter W.E.L. Minter

Crustal Evolution of Southern Africa

3.8 Billion Years of Earth History

With a contribution by S.C. Eriksson

With 182 Figures

Springer-Verlag
New York Heidelberg Berlin

A.J. TANKARD Petro-Canada, Calgary, Alberta T2P 3E3 Canada
M.P.A. JACKSON Bureau of Economic Geology, The University of Texas at Austin, Austin, Texas, 78712 U.S.A.
K.A. ERIKSSON Department of Geological Sciences, Virginia Polytechnic Institute and State University, Blacksburg, Virginia 24061 U.S.A.
D.K. HOBDAY Department of Geology and Geophysics, University of Sydney, Sydney, N.S.W. 2006 Australia
D.R. HUNTER Department of Geology and Mineralogy, University of Natal, Pietermaritzburg 3200 South Africa
W.E.L. MINTER Anglo American Corporation of South Africa, Welkom 9460 South Africa

On the front cover. An artistic rendering of the Stage 3 photograph shown on page 218.

Library of Congress Cataloging in Publication Data

Main entry under title:
Crustal evolution of southern Africa.
Bibliography: p.
Includes index.
1. Earth—Crust. 2. Geology—Africa, Southern.
I. Tankard, A.J.
QE511.C72 551.1′3′0968 81-9413
AACR2

Printed in the United States of America.

9 8 7 6 5 4 3 2 1

ISBN 0-387-90608-8 Springer-Verlag New York Heidelberg Berlin
ISBN 3-540-90608-8 Springer-Verlag Berlin Heidelberg New York

To
Alex. L. du Toit

Foreword

Syntheses of the geology of major areas of the Earth's crust are increasingly needed in order that the features of, and the problems associated with, the secular evolution of the continents can be understood by a wide audience. Southern Africa is fortunate in having a remarkable variety of geological environments developed without many breaks over 3.8 Ga, and many of the rock groups are household names throughout the geological world. In one respect the geology of Southern Africa is particularly important: cratonization clearly began as early as 3.0 Ga ago, in contrast to about 2.5 Ga in most other continental areas such as North America. This book documents very well the remarkable change in tectonic conditions that took place between the Early and Mid-Precambrian; we have here evidence of the very earliest development of rigid lithospheric plates.

This book is a tribute to the multitudes of scientists who have worked out the geology of Southern Africa over many years and decades. Whatever their discipline, each provided a step in the construction of this fascinating story of 3.8 Ga of crustal development. In the book the reader will find a detailed review of the factual data, together with a balanced account of interpretative models without the indulgence of undue speculation. One of its attractions is its multidisciplinary approach which provides a stimulating challenge to the reader. All the important features of earth history are here: greenstone belts, granulite-gneiss belts, intra-cratonic basins, miogeoclinal troughs, abortive rifts, igneous intrusions, rifted continental margins, calc-alkaline arcs, Himalayan-type collisions, and changing climates and geography. This is an overview of the geological development of one of the key segments of the Earth's crust which took place in a well-defined sequence and which clearly illustrates the changes that occurred between the Archean and the Cenozoic.

BRIAN F. WINDLEY
University of Leicester
England

Preface

For various reasons the geology of southern Africa has considerable international appeal. Some of the world's oldest crust, dated at 3.8 billion years, is preserved in the Limpopo Valley, and the course of crustal development can be traced through a virtually complete Precambrian record. The great antiquity and the unique development and preservation of several Archean and Proterozoic stratigraphic sequences is complemented by a complete record of Phanerozoic geologic history. Well-known igneous suites include the Bushveld Complex, komatiites, and alkaline and kimberlitic rocks. The earliest unequivocable traces of life are in stromatolitic limestone greater than $\sim$ 3.5 billion years old in the Fort Victoria greenstone belt of Zimbabwe. Other important records of evolutionary history include the largest and most diverse assemblages of mammal-like reptilian faunas in the Paleozoic Karoo basin and classic localities of early hominid remains in the continental interior. The region is also endowed with great mineral wealth, including type localities such as the gold and uranium of the Witwatersrand; the diamondiferous kimberlites and beach gravels; platinum, chromium, and vanadium in the Bushveld Complex; and vast accumulations of stratiform manganese. In many cases these deposits represent the principal global concentrations of vital resources.

The broad stratigraphy of southern Africa is well established (see for example A.L. Du Toit, 1954; Haughton, 1969; Truswell, 1977), but this established hierarchical order does not address the dynamics of its component parts. The basic facts and the broad tabular stratigraphic model have not changed, but the way we view the evidence has. The impetus for this book lies in the considerable number of process-related studies that have been undertaken in the past decade. It is our aim to incorporate these recent studies within the established stratigraphic framework to produce a dynamic account of the geology of southern Africa. Our approach to the southern African rock record is primarily interpretative and, where possible, the sedimentary, igneous, structural, and metamorphic events are considered as integral components of basin evolution. The scope of this study ranges from the analysis of individual formations and stratigraphic sequences to an understanding of large-scale phenomena such as crustal evolution and the stratigraphic setting of southern Africa as the hub of the Gondwana supercontinent.

This book is written for advanced undergraduates, graduate students, and professional geologists worldwide. Familiarity with the crustal processes, mineral deposits, and fossil history of the southern African "treasure chest" enables a deeper understanding of global geology through the study of some of the most famous and chronologically continuous rocks in the world.

Anthony Tankard
Martin Jackson

Acknowledgments

By its very nature this text encompasses almost the entire spectrum of geology, extending beyond the experience of only six authors. Initially, the task of preparing an up-to-date text of this breadth was daunting. However, one of the pleasures of authorship has been the unselfish cooperation of friends and colleagues. This is partly reflected in the numerous "personal communications" cited.

We are particularly grateful to the following people for their generous investment of time and expertise: Nic Beukes (Chuniespoort and Ghaap Groups), John Bristow (Karoo volcanism), Gerard Germs (Nama Group), Henno Martin (Damara Supergroup), Izak Rust (Table Mountain Group), Noel Tyler (Ventersdorp volcanism), and Johan (J.N.J.) Visser (Karoo Supergroup). Beryl Tankard assisted with the compilation of the index and bibliography.

We would also like to thank those who provided data or critically read sections of the manuscript for this book: Jay Barton, Gavin Birch, Tim Broderick, Stuart Buck, Andrew Button, Gene Cameron, Robin Cleverly, Tom Clifford, Dave Cornell, Mike Coward, Richard Dingle, Allan Donaldson, Schalk du Toit, Marc Edwards, Pat Eriksson, John Ferguson, Burg Flemming, Rod Fripp, Ingo Hälbich, Anton Hales, Chris Hawkesworth, Brett Hendey, Norton Hiller, Nick Hotton, Roger Jacob, Karl Kasch, Fred Keller, Roger Key, Herbert Klinger, Alfred Kröner, Mike Leith, Brian Lock, Johan Loock, Roddie MacLennan, John McCarthy, Ian McLachlan, Peter Matthews, Andrew Miall, Tim Partridge, Hubertus Porada, Des Pretorius, Dave Reid, Dairne Rowsell, Ted Saggerson, Dan Schultze, Russell Shone, Bill Siesser, Norman Smith, Willo Stear, Meiring Strydom, John Sutton, Ron Tavener-Smith, Hannes Theron, Brian Turner, Jan van Bever Donker, Willem Verwoerd, Victor von Brunn, John Wakefield, Mike Watkeys, Janet Watson, Alan Wilson, Henk Winter, and George Zelt.

Barbara Hartmann drafted all the maps and figures apart from Figures 10-7, 10-14, and 12-5, which were prepared by Cedric Hunter. In addition we wish to thank those who have supported us in other ways, including Vic Goodwin, Lucille Harrell, Barbara Dudgeon, Leslie leRoux, Deborah Love, Barbara Rimbault, Johan Ross, David Stephens, and Ginger Zeikus.

Preparation of this book was aided by the much-appreciated support of the Anglo American Corporation of South Africa, the Bureau of Economic Geology (The University of Texas at Austin), the geology departments at the University of Natal (Pietermaritzburg), University of Tennessee (Knoxville), University of Texas (Dallas), and the South African Museum (Cape Town).

Finally, we would like to record our appreciation to Beryl, Jo, Susan, Eugenia, Val, and Pam for their strong support and for their good-natured tolerance of our reclusion.

A.J. Tankard M.P.A. Jackson
K.A. Eriksson D.K. Hobday
D.R. Hunter W.E.L. Minter

Contents

Crustal Evolution of Southern Africa

Chapter 1

Tectonic Framework

This book synthesizes the geologic evolution of southern Africa, a subcontinent comprising the Republic of South Africa, Namibia (formerly known as South West Africa), Zimbabwe (formerly known as Rhodesia), Botswana, Lesotho, and Swaziland. In terms of mineral wealth, geologic diversity, and degree of documentation, the geologic center of gravity of this region is unquestionably the Republic of South Africa. Accordingly most of this volume concerns itself with that country. Nevertheless, where tectonic zones, sedimentary basins, or igneous provinces are shared by more than one state we have ignored political boundaries in order to provide more comprehensive coverage.

The southern African subcontinent comprises several Early Precambrian to Cambrian structural provinces overlain by relatively undeformed cover sequences whose ages vary from Late Archean to Cenozoic. Stripped of their cover, these structural provinces are shown in Figure 1-1. In some cases boundaries have been precisely delineated according to selected criteria (Table 1-1), but the choice of these criteria is somewhat arbitrary. For instance, geochronologic, metamorphic, or structural discontinuities may not coincide exactly in places; furthermore, successive phases of deformation may have created structural discontinuities of different type or location.

1.1. Cratons, Mobile Belts, and Structural Provinces

The terms "mobile belt" and "craton" have commonly been used to describe the Precambrian

Table 1–1. Boundaries of Tectonic Provinces in Southern Africa

Tectonic Provinces	Definition of Common Boundaries[a]
Limpopo–Zimbabwe	1. Northern margin of D4 Tuli–Sabi shear belt
	2. "Orthopyroxene-in" isograd of Limpopo metamorphism
Limpopo–Kaapvaal	3. "Orthoamphibole-in" isograd of Limpopo metamorphism
Natal–Kaapvaal	4. D3 Natal thrust belt; contact between paraautochthonous Ntingwe Formation (Kaapvaal) and allochthonous Mfongosi Metamorphic Suite (Natal)
Namaqua–Kaapvaal	5. Doornberg fault
	6. Western margin of Draghoender and Skalkseput granitoids
	7. Axial trace of D2 Orange River synform
Namaqua–Namibia	8. Contact between Naisib River Igneous Suite (Namaqua) and intruding or overlying Sinclair Group (Namibia)
Gariep–Namaqua	9. Eastern margin of Gariep Group or penetratively reworked Namaqua basement
	10. Contact between allochthonous Spencer Bay Formation (Gariep–Damara) and reworked Namaqua basement
Damara–Namibia	11. Southern boundary thrust zone separating allochthonous Swakop Group (Damara) from Namibia basement
Damara–Congo	12. Northern known limit of Otavi Group or Mulden Group (Damara) in subsurface

[a]Reasons for selection of boundaries and data sources are given in Chapters 3, 8, and 9. Extent of boundaries is shown in Figure 1–1.

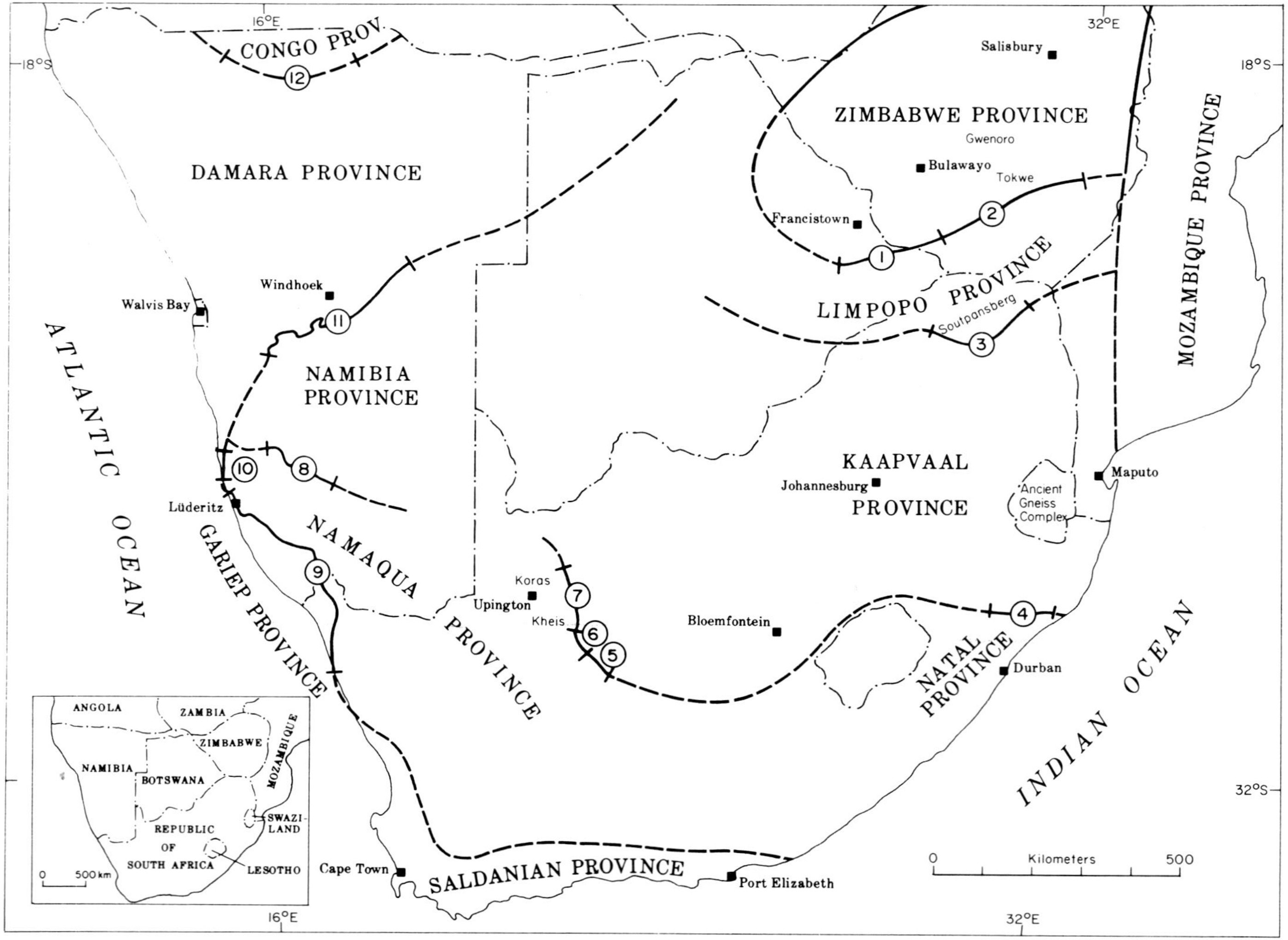

Figure 1-1. Tectonic provinces of southern Africa. See Table 1-1 for details of numbered boundaries. Boundaries obscured by cover are shown as dashed lines; their positions have been estimated by interpolation, extrapolation, consideration of the regional gravity field, and postulated crosscutting of one province by another, based on known age relations.

tectonic units that make up the structural framework of southern Africa. The concept of "mobile belts" and "cratons" was undoubtedly valuable in drawing attention to the processes that characterized their evolution. Nevertheless the burgeoning of new data from both types of structural province necessitates a reevaluation of the use of these terms.

"Mobile belts" in the southern African context were defined as "younger, linear, metamorphic belts which tend to surround the ancient cratonic nucleii of shield areas and which are characterized by high-grade metamorphism, granitization and often by transcurrent dislocation" (Anhaeusser *et al.*, 1969). However, the most recent age determinations from the Limpopo "mobile belt" have yielded dates that are more than 200 Ma (1 Ma = 10^6 years) older than those so far reported from the adjacent Zimbabwe* and Kaapvaal "cratons" (Barton *et al.*, 1978). The use of the qualifying adjective "younger" is therefore no longer appropriate. The term "belt" implies a linear tract with a high length to breadth ratio. Although the Limpopo "mobile belt" is linear, the geometry of the Namaqua and Natal terranes of high-grade gneisses is not beltlike.

Pretorius (1974) introduced the concept of crustal provinces with respect to the Namaqua gneiss terrane, independently of a simultaneous proposal by Blignault *et al.* (1974). Pretorius stressed that the Namaqua gneiss terrane "might well have acted as a mobile belt between 2000 and 1000 m.y. ago, but subsequent to the latter date after which it attained a stable state and deformed in a brittle manner, it assumed the role of a craton." An earlier stage of cratonization was reached by the Limpopo gneiss terrane, which had attained sufficient stability by 1.8 Ga (1 Ga = 10^9 years) ago for the Soutpansberg cover sequence to accumulate. Cratonization is the final stage in the life span of a "mobile belt" and all "mobile belts" are destined to become "cratons."

"Cratons" are now known to include extensive areas of high-grade gneisses (e.g., Tokwe and Gwenoro gneisses in Zimbabwe; Ancient Gneiss Complex in Swaziland; Fig. 1-1) that have the structural and metamorphic characteristics of "mobile belt" terranes, apart from their smaller size, which is a consequence of intrusion by granitoids or mantling by cover rocks. The presence of these gneisses indicates that parts of these cratons initially evolved in a manner analogous to "mobile belts." Whether a region is referred to as a "mobile belt" or a "craton" therefore depends upon its evolutionary stage.

The term "mobile belt" implies deformation and metamorphism under conditions of high heat flow. However, there is no evidence for fundamentally different metamorphic facies series indicating the former existence of higher geotherms in "mobile belts" than in Precambrian "cratons." Rather "mobile belts" mark the sites of much greater uplift than the surrounding areas and so contain metamorphic rocks of higher metamorphic grade at the present level of exposure.

For all these reasons the use of the nongenetic term "province" is preferred here, as has been customary in North America (e.g., Stockwell, 1969; Price and Douglas, 1972).

Kröner and Blignault (1976) defined a tectonic province as "a geographic region that is characterized by a combination of such parameters as lithology, structure, metamorphism, and predominant radiometric age differing significantly from those of adjacent areas." The sum total of these parameters should be used to define a structural or tectonic province because an individual parameter taken in isolation may apply equally to a given structural province and its neighbor.

The boundaries between structural provinces can seldom be unequivocally defined. In the case of the Natal–Kaapvaal boundary or the Damara–Namibia boundary, prominent thrust zones representing fundamental structural discontinuities provide obvious tectonic boundaries (Fig. 1-1). However, there is no simple criterion where broad transitional zones separate structural terranes, such as the Kaapvaal and Namaqua Provinces. Pretorius (1974) concluded from a study of the gravity field that these terranes formed parts of the same crustal fragment during the Archean and Early Proterozoic. Subsequent uplift of the Namaqua Province relative to the Kaapvaal Province after 1.8 Ga continued until deposition of the Koras Group at 1.2 Ga (Pretorius, 1974; Stowe, 1979). Between these provinces is a zone where the regional metamorphic pattern is transitional, but where the structural and radiometric imprints are more characteristic of the Namaqua Province. Because of its gradational character,

*In 1980 under a new constitution Rhodesia became known as Zimbabwe. The tectonic unit forming the core of the country is conventionally known as the Rhodesian Craton, but it is unlikely that this name will continue to be used by the Zimbabwe Geological Survey (as was the case with the former Tanganyika Shield of East Africa). The name applied in this volume is the Zimbabwe Province.

some authors have designated this transitional area as a separate zone, known as the Kheis domain, with equal rank to the Kaapvaal and Namaqua Provinces (Vajner, 1974a; Kröner and Blignault, 1976; Botha and Grobler, 1979). In this volume the Kheis domain is regarded as a tectonic zone within the Namaqua Province analogous to tectonic zones within the Limpopo, Natal, and Damara Provinces.

The boundaries between the Limpopo and adjoining provinces are also gradational, and the lack of known structural or stratigraphic discontinuities has led to the adoption of metamorphic isograds for the northern and southern boundaries in Zimbabwe and South Africa (Table 1-1). However, the pervasive granulite metamorphic imprint found along the northern margin of the province in Zimbabwe is apparently absent in Botswana, and a structural zone known as the Tuli–Sabi shear belt has been selected as the northern boundary in the western part of the Limpopo Province in Botswana (Table 1-1). The definitions of boundaries thus differ from province to province and within a province itself (Fig. 1-1).

1.2. Gravity Field and Crustal Structure

The structural framework of the southern African shield is a combination of ancient tectonic components and displacements and neotectonic crustal warping. In a synthesis of gravity data Pretorius (1979) suggested that the shield consists of a central core in Zimbabwe surrounded by a platform that has been deformed by relatively recent concentric and radial flexures. The structural framework is formed by first-order swells and sags within which are second-order concentric and radial upwarps and downwarps and third-order antiforms and synforms. The swells generally mark areas of greater age, lower metamorphic grade, and more-negative gravity field; the sags tend to coincide with areas of younger age, higher metamorphic grade, and more-positive gravity field. "Cratons" are located within swells and "mobile belts," within sags. Although structural provinces such as the Limpopo are located within first-order sags, their boundaries are parallel and adjacent to concentric second-order upwarps.

1.3. Evolutionary Stages in the Southern African Crust

The geologic history of South Africa stretches far back in time to a limit presently fixed at approximately 3.8 billion years. Most of this extensive history is Precambrian. Table 1-2 is a summary of this history for reference throughout this volume. Chronometric divisions within the Precambrian are still debated; in the interests of clearer communication we have followed the most recent suggestions of the International Union of Geological Sciences (IUGS) Subcommission on Stratigraphy (Sims, 1980; see also Harrison and Peterman, 1980) rather than the chronostratigraphic subdivisions proposed by the South African Committee for Stratigraphy (SACS) (Kent and Hugo, 1978). The chronometric divisions proposed are shown below:

IUGS		SACS	
>3500 Ma	Early Archean		
3500–2900 Ma	Middle Archean	3750–2870 Ma	Swazian
2900–2500 Ma	Late Archean	2870–2630 Ma	Randian
2500–1600 Ma	Early Proterozoic	2630–2070 Ma	Vaalian
1600–970 Ma	Middle Proterozoic	2070–1080 Ma	Mokolian
970–570 Ma	Late Proterozoic	1080–570 Ma	Namibian

Some of these chronologic boundaries, particularly that separating the Archean from the Proterozoic, areknown to be highly diachronous (Rankama, 1970). All Rb-Sr ages quoted in this volume are based on the decay constant $^{87}Rb{:}\lambda = 1.42 \times 10^{-11}$/year.

Because the Precambrian time scale is based almost entirely on radiometric age determinations, the terms listed above serve to characterize both time units and their associated rock units (Douglas, 1980). The modifiers "lower" and "upper" are generally reserved for their conventional strati-

Table 1–2. Simplified Chronology of the Crustal Evolution of Southern Africa

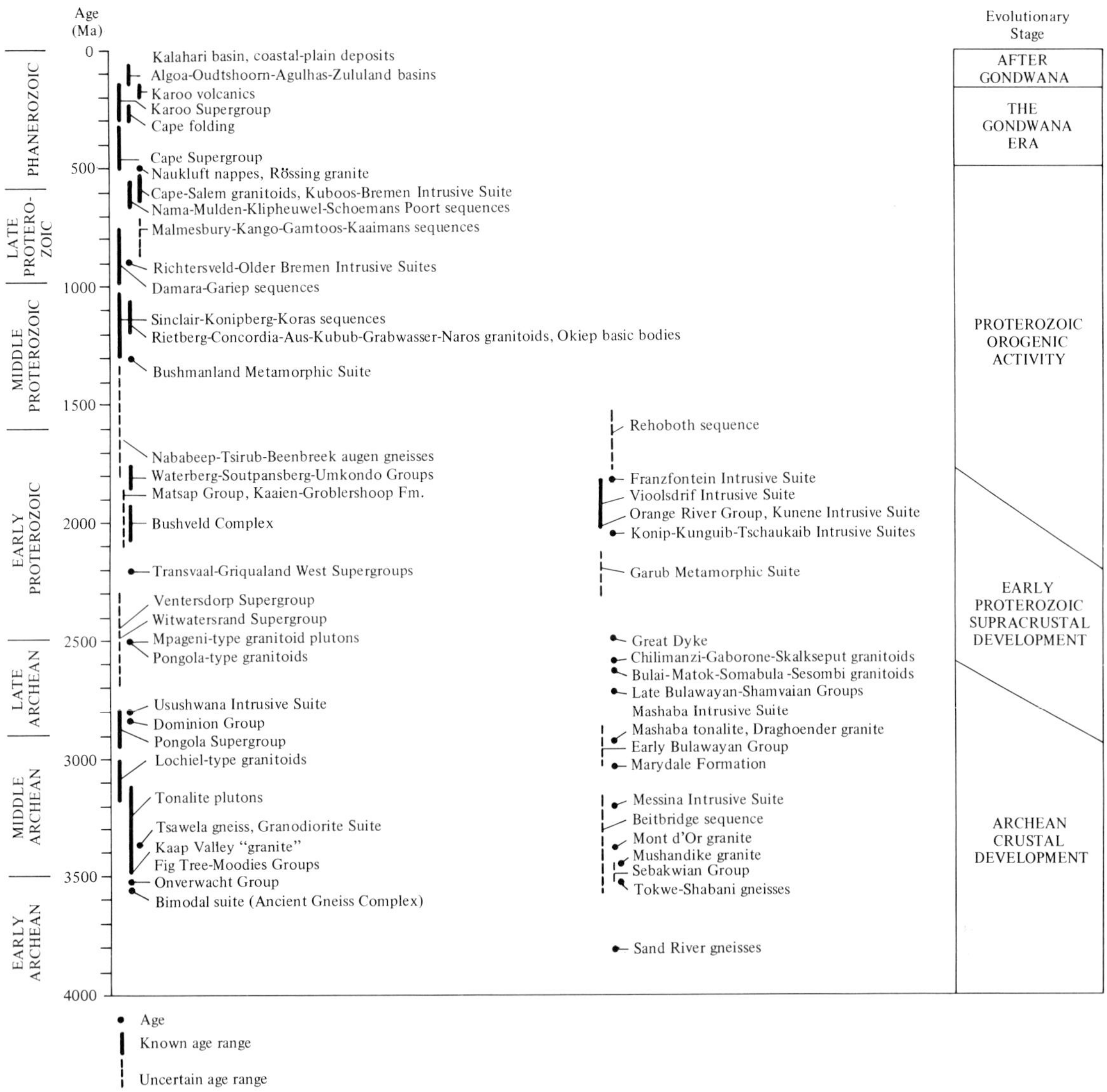

graphic use in Late Precambrian and Phanerozoic series. In general the lithostratigraphic nomenclature applied here to high-grade metamorphic or igneous rocks follows that prepared for the American Commission on Stratigraphic Nomenclature (Sohl, 1977), except in the case of entrenched names such as the Bushveld Complex or the Great Dyke.

An overview of the course of geologic events in southern Africa indicates that the crust has passed through a well-defined sequence of evolutionary stages which are emphasized in the layout of this book.

(1) A period of Archean crustal development gave rise to crystalline massifs represented by the Kaapvaal, Limpopo, and Zimbabwe Provinces.

(2) This ancient basement was buried beneath a largely sedimentary cover, represented by the Pongola Supergroup, the Witwatersrand triad, the Transvaal–Griqualand West Supergroups, and the Waterberg–Soutpansberg–Matsap red beds, and was punctured by massive injections of basic magma in the form of the Great Dyke and Bushveld Complex during the Early Proterozoic.

(3) In several Proterozoic orogens in the southern and western parts of the subcontinent older crystalline rocks and their supracrustal cover

were reworked tectonically, geosynclinal deposits accumulated, and massive granitoid intrusions were emplaced by partial melting of older crust and by additions from the mantle. Geosynclinal opening and closing of the proto-South Atlantic is recorded along the west coast.

(4) The Paleozoic Gondwana era ushered in a period of aborted rifting and unparalleled continental sedimentation throughout the supercontinent of Gondwana, of which southern Africa formed the hub.

(5) Mesozoic fragmentation of Gondwana was preceded by continental rift volcanism and the injection of diamondiferous kimberlites, carbonatites, and other alkaline intrusions. Late Mesozoic and Cenozoic sedimentation was restricted to the newly formed margins of the stable subcontinent and depressed areas of the interior.

1.4. Stage 1: Archean Crustal Development

Figure 1-2 summarizes the presently known geologic development at the end of the Archean. Basement in the Kaapvaal Province consists of massive and foliated granitoids and gneisses with deformed greenstone relicts of basic and ultrabasic volcanic and sedimentary sequences, of which the Swaziland Supergroup is the most famous. The basal Onverwacht Group of the Swaziland Supergroup has yielded a Nd-Sm age of 3.51 Ga, which is close to the age of 3.55 Ga obtained from the bimodal assemblage of interlayered tonalitic–trondhjemitic gneisses and amphibolites of the bimodal suite (Ancient Gneiss Complex) in Swaziland. There is as yet no unequivocal field evidence to demonstrate the age relations of the gneisses and greenstones. Other volcanic sequences lithologically similar to the Swaziland Supergroup are preserved within the Kaapvaal Province, but lack of geochronologic data precludes chronostratigraphic correlation. Small plutons of tonalite and trondhjemite were intruded close to the margins of the greenstone relicts at 3.2 Ga. Widespread granitoid intrusion at about 3.0 Ga in the southeast and east was followed by a similar event between 2.7 Ga and 2.6 Ga in the northern and northwestern parts of the Kaapvaal Province. Potassic granites, emplaced as sheetlike bodies and smaller plutons, were the dominant rock types.

The earliest evidence of cratonization in the Kaapvaal Province is seen in the accumulation of the epicontinental Pongola Supergroup between 3.0 Ga and 2.8 Ga. Shelf sedimentation and volcanism in this basin were terminated by intrusion of the Usushwana Intrusive Suite, a layered, dominantly gabbroic body consisting of two northwesterly striking dikes linked by a sheetlike mass, and by emplacement of the 2.7–2.6 Ga granitoid sheets and plutons.

Farther north in the Limpopo Province a supracrustal sequence at least 3.8 Ga old had been deformed and metamorphosed to form a sialic basement, the Sand River gneisses, at the time of the Onverwacht extrusion at 3.5 Ga. By this stage granitoid gneissic crust, together with supracrustal rocks of the Sebakwian Group containing ~3.5 Ga old stromatolites, had also formed in at least the southern and central parts of the Zimbabwe Province. Younger supracrustal sequences deposited in the precratonic stage are widespread in the central part of the Limpopo Province (3.6–3.2-Ga Beitbridge sequence) and in the Zimbabwe Province (2.9-Ga Early Bulawayan and 2.7-Ga Late Bulawayan–Shamvaian Groups. Emplacement of basic dikes and layered basic–ultrabasic intrusions took place at 3.2 Ga in the Limpopo Province (Messina Intrusive Suite) and 2.7 Ga in the Zimbabwe Province (e.g., Mashaba Intrusive Suite and Mashaba-Chibi dike swarm).

Metamorphism and deformation of these units followed, concentrated around 2.7–2.6 Ga, with concomitant syntectonic and posttectonic intrusion of widespread granitoids in both the Limpopo and the Zimbabwe Provinces. The process of cratonization in the Zimbabwe Province and the northern Limpopo Province culminated in the emplacement of the 500-km-long Great Dyke and its satellite dikes at 2.5 Ga at the end of the Archean.

1.5. Stage 2: Early Proterozoic Supracrustal Development

The Archean–Proterozoic boundary is conventionally drawn at the time of cratonic stabilization, which resulted from a decrease in heat flow and an increase in the thickness of continental crust. This process of cratonization enabled ensialic supra-

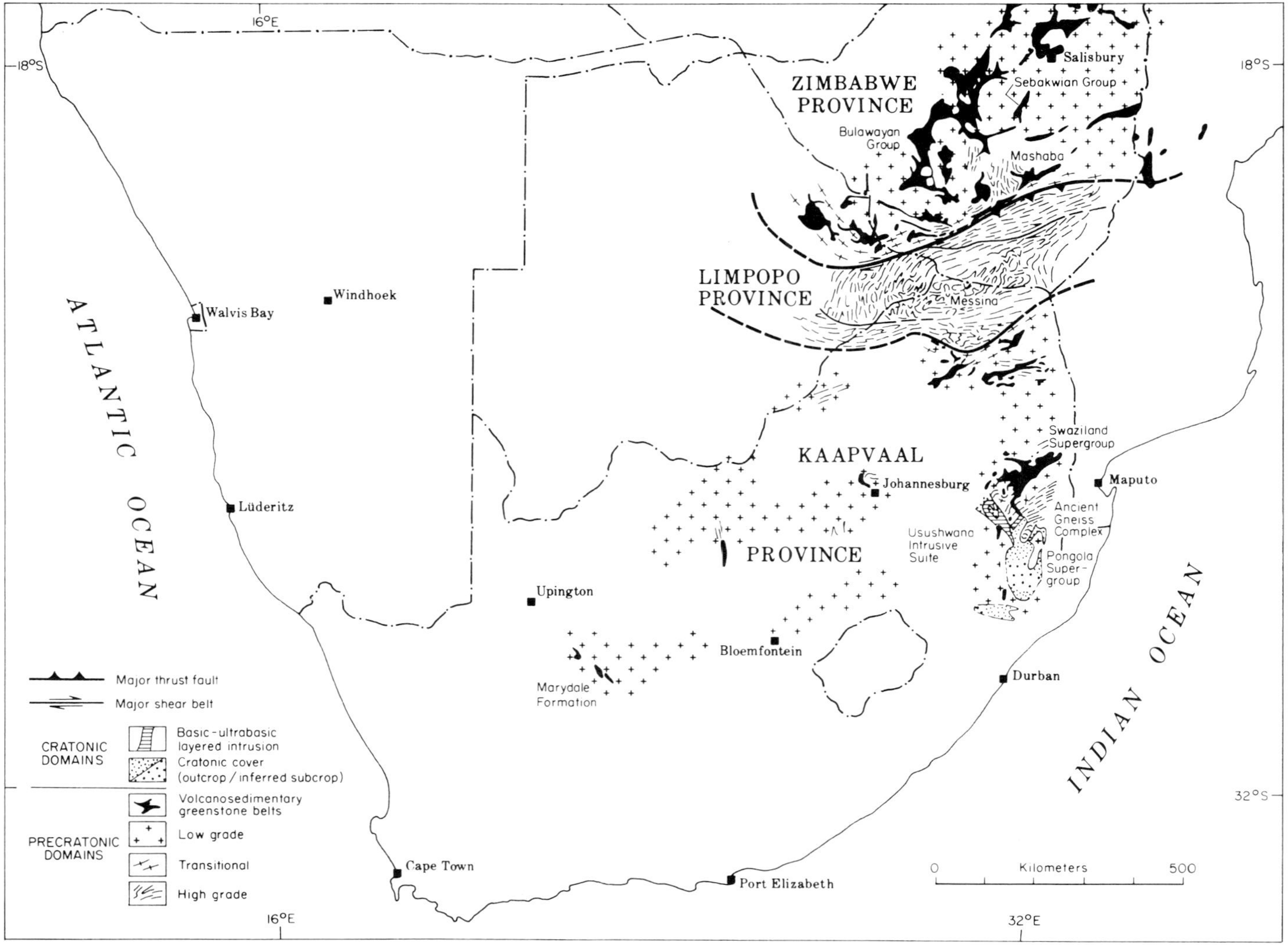

Figure 1-2. Nonpalinspastic map showing Archean crustal development from 3800 Ma to 2600 Ma. The Pongola Supergroup is a Proterozoic-type succession deposited during the Late Archean.

crustal successions to accumulate and be preserved as cratonic cover that is typically only gently deformed and metamorphosed to low or very low grade. The Archean–Proterozoic boundary, however, is highly diachronous on a global and continental scale and ranges in age from 1.7 Ga to 3.0 Ga (Rankama, 1970). The Pongola Supergroup can be regarded as an epicontinental succession of Proterozoic character that accumulated during the Late Archean, a time when the Kaapvaal Province was only partly cratonized.

The Late Archean and Early Proterozoic crustal history in southern Africa is characterized by a considerable overlap of granitoid intrusion and supracrustal processes from 3.0 Ga to 1.8 Ga (Hunter, 1974a,b). Supracrustal sequences in southern Africa developed on continental crust consisting of the Kaapvaal, Limpopo, and Zimbabwe Provinces (Fig. 1-3); some of these sequences may have extended southward into the Namaqua Province, where they were subsequently metamorphosed or destroyed in the Early Proterozoic. The cratonic cover includes the Pongola, Witwatersrand, Ventersdorp, and Transvaal–Griqualand West Supergroups and the Waterberg–Soutpansberg–Umkondo–Matsap Groups (Fig. 1-3). The Bushveld Complex represents a massive injection of basic magma into the Transvaal Supergroup during the Early Proterozoic.

Many of South Africa's strategic mineral resources are present in these Early Proterozoic successions, including gold and uranium in the Witwatersrand and basal Ventersdorp Supergroups; iron and manganese in the Griqualand West Supergroup; and chromium, vanadium, and platinum in the Bushveld Complex.

The Late Archean and Early Proterozoic successions in southern Africa accumulated in response to the progressive northward migration of depositional axes; the ancient Pongola Supergroup is confined to the southeastern margin of the Kaapvaal Province, whereas the much younger Soutpansberg Group and its correlatives are most prevalent along the northern and western margin of the Kaapvaal Province and in the Limpopo Province. This migration of depositional axes has been related to a complementary northward migration of loci of granitoid emplacement (Hunter, 1974b) resulting in a diachronous Archean–Proterozoic boundary from southeast to northwest.

The cratonic sequences display a cyclic pattern of basin development, all having commenced with, and frequently terminated by, volcanism. Fluvial conglomerates and arkoses commonly underlie or are intercalated with the basal volcanic rocks. Between the volcanic rocks, thick sedimentary intervals consist of marine quartz arenites and shales and fluvial conglomerates and arkoses. These lithologies occur in varying proportion in the different stratigraphic units. Nonterrigenous sediments are present only in the Transvaal–Griqualand West Supergroups. Sedimentation took place during periods of submergence; major unconformities represent episodes of continental emergence. Emergence and submergence of continents have been attributed by Sloss and Speed (1974) to mantle processes. They envisaged that trapping of melt beneath continents resulted in expansion of the asthenosphere and emergence; conversely, lateral migration of melt from beneath the continents to the suboceanic asthenosphere produced continental subsidence. Another explanation, particularly applicable to southern Africa, is that deflation of the swollen asthenosphere was accomplished by an upward release of magma. In many of the Early Proterozoic sedimentary basins, seas transgressed over the subsiding continental crust, whereas regional regressions preceded the terminal volcanic events (Hunter, 1974b).

1.6. Stage 3: Proterozoic Orogenic Activity

After a billion years of comparative stability marked by cyclic epeirogeny during the Late Archean and Early Proterozoic, intense orogenic activity disturbed vast areas in the southern and western parts of the subcontinent. This episode of crustal instability persisted until the start of the Phanerozoic and can be conveniently divided on a chronologic and geographic basis into: (1) Early to Middle Proterozoic tectonism that gave rise to the Namaqua and Natal Provinces (Fig. 1-4); and (2) a long chain of Late Proterozoic geosynclines along the present southwestern coast and Namibian interior, including the Damara, Gariep, and Saldanian Provinces, collectively termed the Pan African Geosynclines (Fig. 1-5).

The two groups of orogens were both characterized by abundant syntectonic partial melting and intrusion of granitoids with inferred remobilization

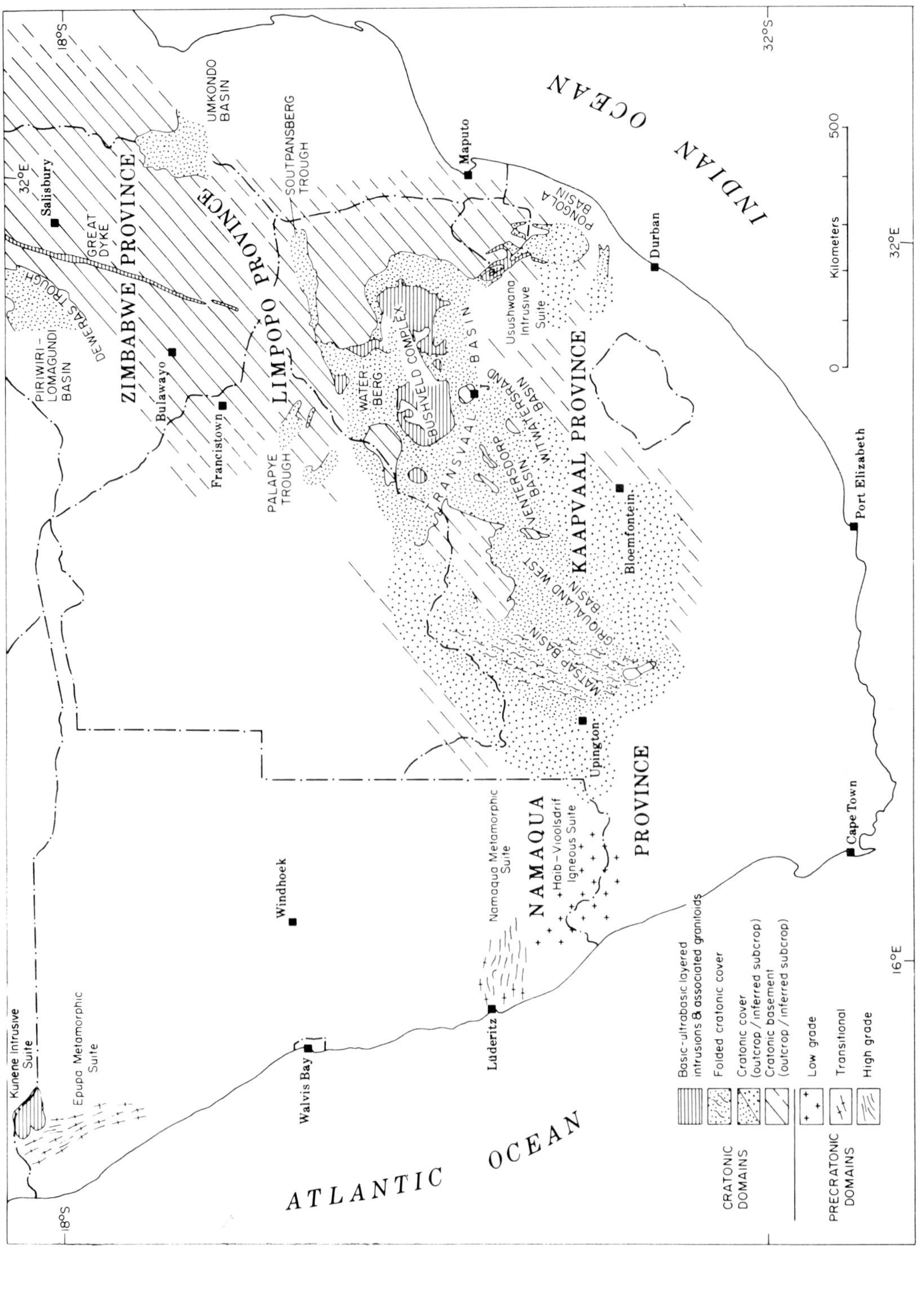

Figure 1-3. Nonpalinspastic map showing Late Archean and Early Proterozoic supracrustal development from 2900 Ma to 1800 Ma.

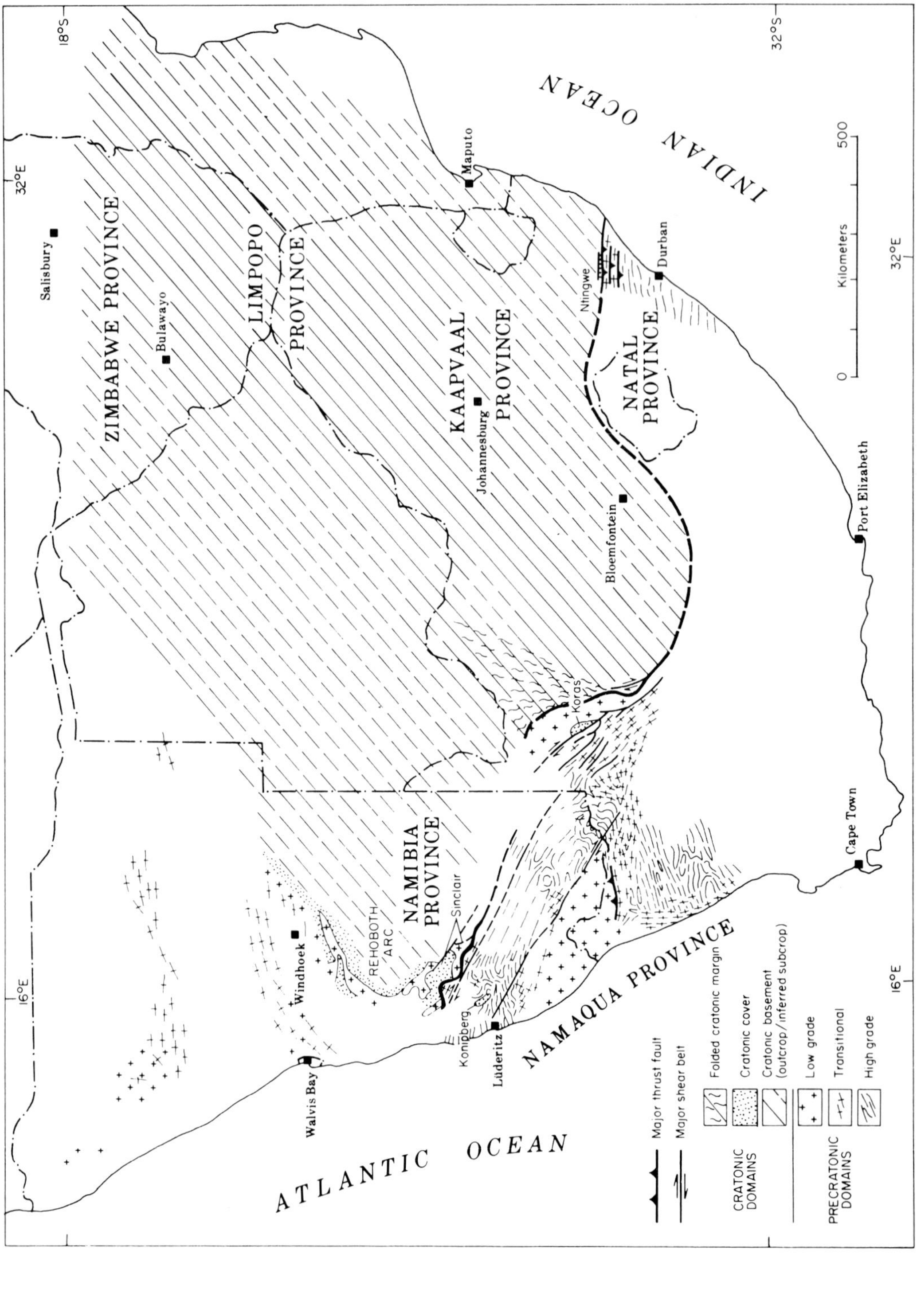

Figure 1-4. Nonpalinspastic map showing Early-Middle Proterozoic orogeny form 2100 Ma to 1000 Ma.

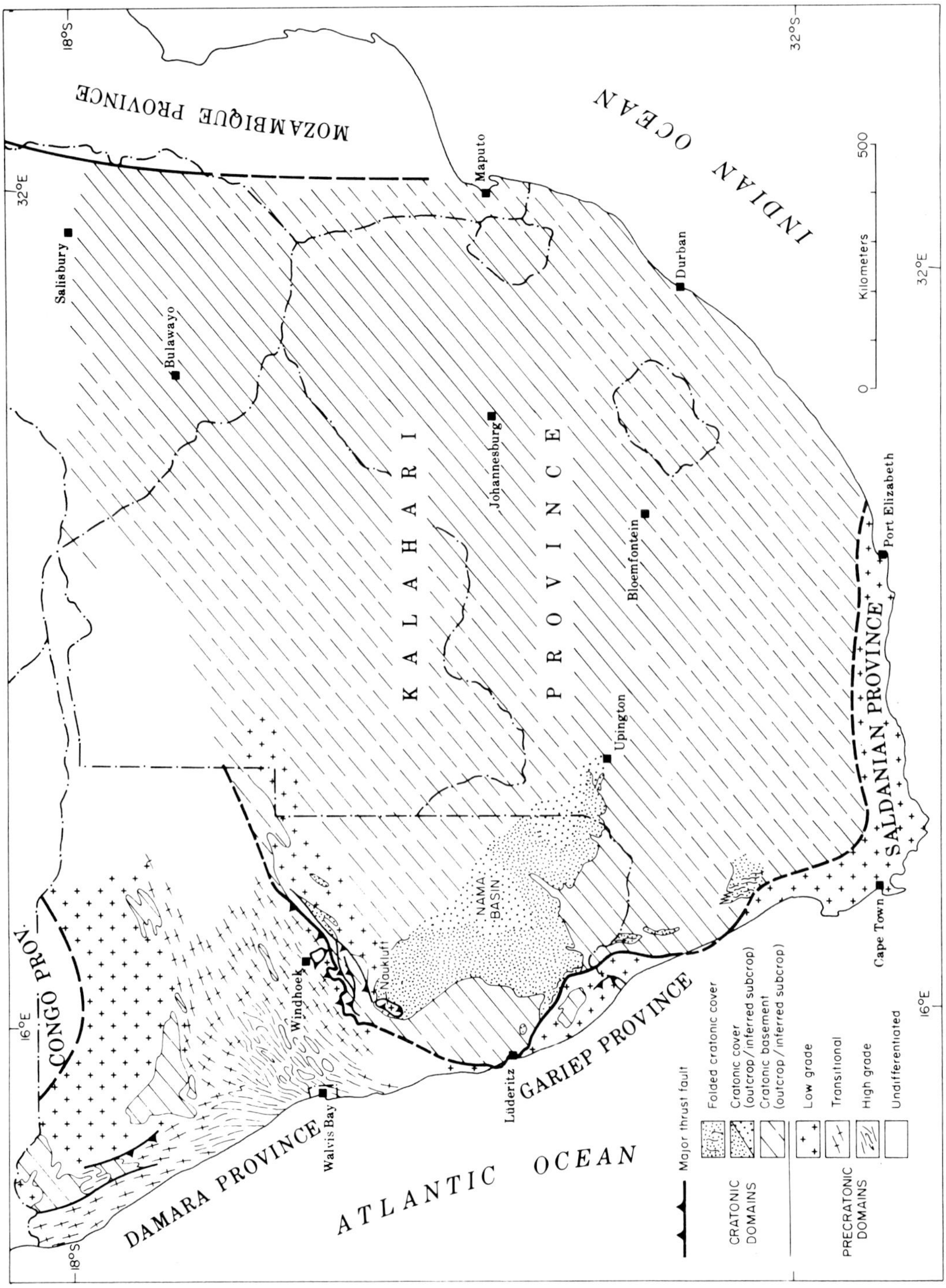

Figure 1-5. Nonpalinspastic map showing Late Proterozoic geosynclines and the ensialic Nama basin from 1000 Ma to 500 Ma.

of sialic basement. They also resemble each other in that spectacular thrust zones formed along the northern margin of the Natal Province and along the southern boundary of the Damara Province.

However, the Early to Middle Proterozoic orogens differed from the Late Proterozoic orogens in a number of important respects. First, recognizable basement beneath the older orogens is extremely rare, whereas it is common in the Damara and eastern Gariep Provinces. Second, the supracrustal successions in the Namaqua and Natal Provinces are thin and have been correlated with Early Proterozoic cratonic successions in the Kaapvaal Province, such as the Matsap and Griqualand West sequences; syntectonic deposits are extremely rare. In contrast, all the Pan African orogens contain thick geosynclinal deposits that accumulated during the early stages of trough formation and deformation in the Late Proterozoic; in the miogeosynclinal parts of the Damara and Gariep Provinces the supracrustal successions have been preserved to a large extent by their rigid crystalline basements.

In the Natal Province metamorphosed oceanic crust has been allochthonously transported northward as a series of nappe sheets onto the Kaapvaal foreland. Continental convergence may have induced this orogeny. In the case of the Namaqua Province the cover sequences range from the 3.0-Ga Marydale greenstone belt in the east (Fig. 1-2), through >2.1-Ga shelf sequences in the northwest (Fig. 1-3), possibly coeval with the Griqualand West Supergroup, to 1.3-Ga volcanosedimentary stratabound Pb-Zn ore deposits in the central area (Fig. 1-4). Tectonism ceased near the margins of the Namaqua Province before 1.2 Ga, and the Namaqua gneisses acted as basement to volcanosedimentary successions of this age known as the Sinclair and Koras Groups (Fig. 1-4). Tectonism in the more deeply buried central parts continued until about 1.0 Ga and deformed the syntectonic 1.2-Ga Konipberg sequence. Coeval basement reworking along the coast at the close of Namaqua tectonism set the stage for Pan African geosynclinal development along similar trends. Cratonization fused the Namaqua and Natal Provinces to the Kaapvaal and Namibia Provinces; the aggregate craton is known as the Kalahari Province (Fig. 1-5).

Rifting of continental crust was followed by opening of the proto-South Atlantic Ocean during the Late Proterozoic at about 900 Ma. This allowed the accumultion of clastic wedges along a passive continental margin fed by detritus from the rising Namaqua massif. The rifts largely coincided with the present southern and western coast of South Africa and Namibia, but they also extended intracontinentally through Namibia toward the continental interior, thus forming a triple junction between the diverging Kalahari, Congo, and South American plates. The extent of ocean opening along the intracontinental plate junction is currently being debated.

Subsequent plate convergence, starting some 700 Ma ago and persisting into the Cambrian Period, is most frequently suggested as the cause of the metamorphism and deformation in these geosynclines. The miogeosynclinal parts of the Damara and Gariep Provinces were protected from orogeny by the underlying cratons of the Congo and Kalahari Provinces. In contrast the southern margin of the Damara Eugeosyncline and the eastern margin of the Gariep Eugeosyncline were thrust toward the Kalahari foreland over distances of at least 50 km in the case of the Naukluft nappe complex. Molasse deposits accumulated on the northern and southern flanks of the Damara Province. The Mozambique Province underwent Pan African basement reworking during the Late Proterozoic and Early Paleozoic, but no syntectonic deposits are recognized.

Continental crust appears to have been present north and south of the poorly exposed Saldanian Province; this has also been deduced in the case of the younger Cape trough, which formed over the Saldanian basins as a result of rifting during the Early Paleozoic along Pan African structural trends.

1.7. Stage 4: The Gondwana Era

During the Early Paleozoic southern Africa lay at the heart of Gondwana, bounded in the west by South America, in the south by the Falkland Plateau, and to the east by Antarctica. Abortive rifting around the southern and eastern fringe of the Kalahari Province resulted in accumulation of continental and marine clastic successions, known as the Cape Supergroup, in elongate troughs in the southern Cape and Natal (Fig. 1-6). Similar successions are preserved in South America and the Falkland Islands (Du Toit, 1927, 1937).

Up to 8 km of sediment accumulated in the Cape basin. The lower 4 km of quartz arenites, mud-

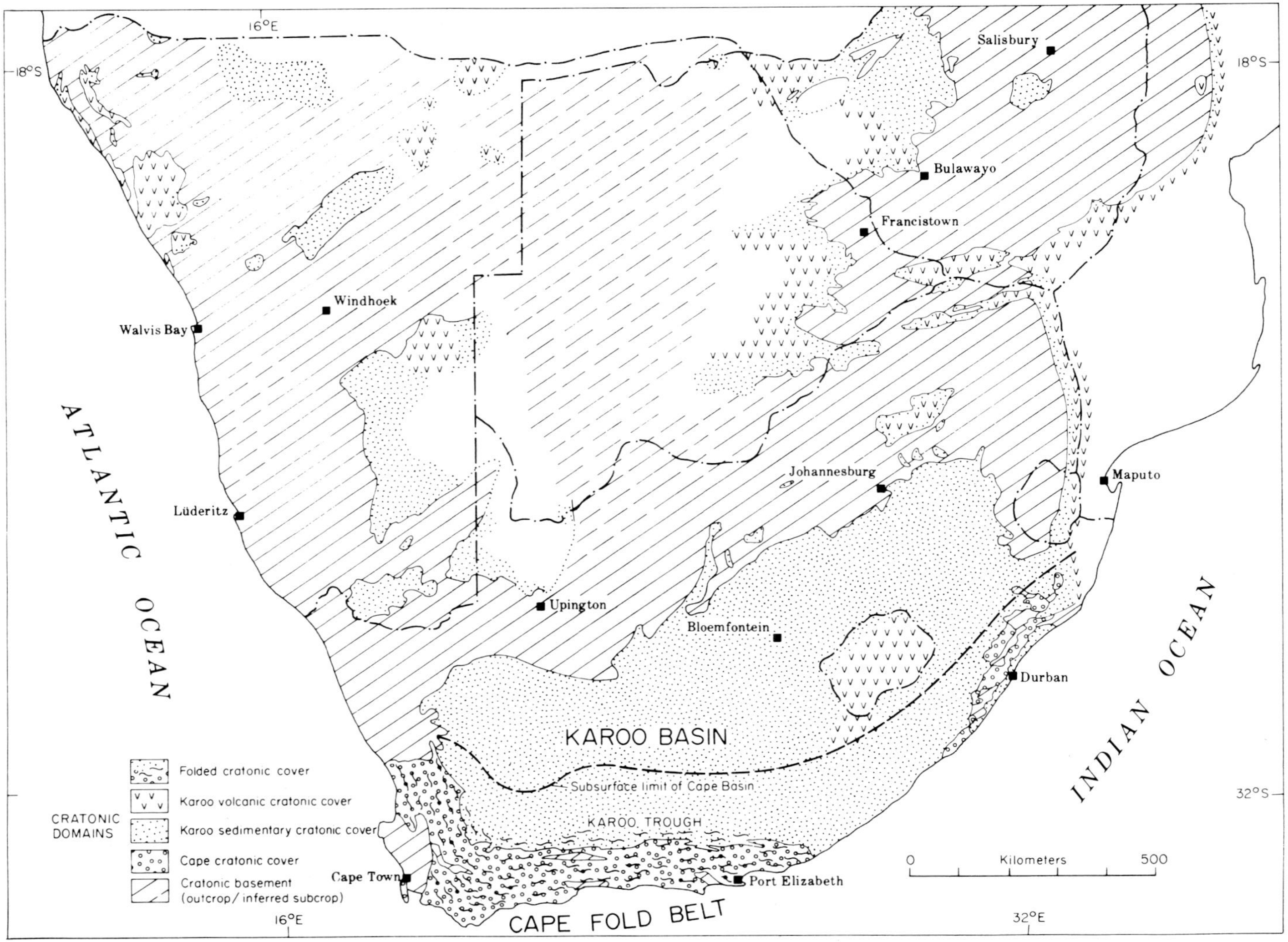

Figure 1-6. Nonpalinspastic map showing the Cape and Karoo successions of Gondwana from 500 Ma to 140 Ma.

stones, and conglomerates in the Table Mountain Group record terrestrial and shallow-marine environments and intermittent northward transgression of the Cape sea during the Ordovician and Early Devonian. Prolonged periods of tectonic and eustatic stability are reflected in quartz arenites up to 2100 m thick, representing one of the greatest known accumulations of quartz sand (Visser, 1974). These deposits are succeeded by destructive deltaic and shallow-marine shelf deposits of the Bokkeveld and Witteberg Groups.

The Natal embayment developed along a trend parallel to the Pan African Mozambique Province to the north (compare Figs. 1-5 and 1-6). Proximal coarse alluvial sediments were deposited at the rugged northern end of the embayment, which opened southward into a tide-dominated marine reentrant where considerable thicknesses of marine quartz sands accumulated.

By the Carboniferous Period the Cape basin lay on the periphery of an extensive Gondwana ice sheet which migrated progressively southeastward in response to drift of the supercontinent across the southern polar regions. As the ice sheets melted, a temporary marine incursion from the west was followed by formation of the extensive Ecca sea during the Permian.

Subsidence of a large intracratonic basin such as the Karoo is an enigma (Bally and Snelson, 1980). Possible mechanisms, none of which is satisfactory in itself, could have involved subcrustal erosion or asthenospheric deflation, mantle phase changes, or a lag in isostatic rebound after melting of the Dwyka ice sheets. Sedimentary loading could have been a contributing factor in the southern part of the basin where the Karoo trough was located before it migrated northward and lost its identity. A subordinate trough in Natal represented reactivation of the Early Paleozoic zone of rifting. On the craton a stable to gradually downwarped platform prevailed.

Fluvial and shoal-water deltaic environments dominated Ecca and Beaufort sedimentation, apart from short-lived turbidite deposition in the early Karoo trough in the south. Basement topography played a dominant role in controlling facies distributions in the north. Important coal reserves are associated with paraglacial, fluvial, and deltaic deposits in the northern Karoo basin. Potentially significant epigenetic uranium deposits have been discovered in the southern Karoo basin.

The effects of tectonic shortening of the Cape and Karoo successions in the Cape Fold Belt were first manifest during Beaufort deposition in the Triassic Period (Fig. 1-6). The paradox of a collision-type fold belt 1000 km in the interior of Gondwana is possibly accounted for by flat-plate subduction (Lock, 1980). Oceanic crust, subducted at a low angle at the supercontinental margin, became indirectly coupled to overriding continental crust (Falkland Plateau) and thereby transmitted the stresses of convergence into the interior of the supercontinent; the Cape Fold Belt developed along the Late Proterozoic structural trends over the zone of "unpeeling" and steep northward subduction of the oceanic plate.

Major clastic wedges extended northward from the rising mountain chain into a foredeep where fluvial sediments completely filled lacustrine remnants of the formerly extensive Ecca sea. Progressively drier climates prevailed during the closing phases of Karoo sedimentation. Broad alluvial flats gave way to eolian dune fields, playa lakes, and ephemeral streams. Finally vast outpourings of basaltic and rhyolitic lavas heralded the end of Karoo deposition and the close of the Gondwana era.

1.8. Stage 5: After Gondwana

Prior to Mesozoic rifting and dispersion of Gondwana the Falkland Plateau lay between the African and Antarctic plates. During subsequent divergence of these plates and opening of the South Atlantic Ocean, the Falkland Plateau on the tip of the South American plate sheared westward from southern Africa along a transform boundary. The continental margins of southern Africa reflect these styles of fragmentation (Fig. 1-7): the western and eastern margins are passive, whereas the southern margin lies along the Agulhas–Falkland fracture zone initiated by transform faulting (Francheteau and Le Pichon, 1972; Rabinowitz and LaBrecque, 1979). Many of the subcontinent's kimberlite diatremes were emplaced at this stage of continental upwarp and nearby sea-floor spreading.

Regional stresses accompanying transform movement along the southern continental margin generated numerous fault-controlled basins onshore and offshore (Fig. 1-7). Alluvial fans developed in response to active boundary faulting, but the first marine incursion was delayed until

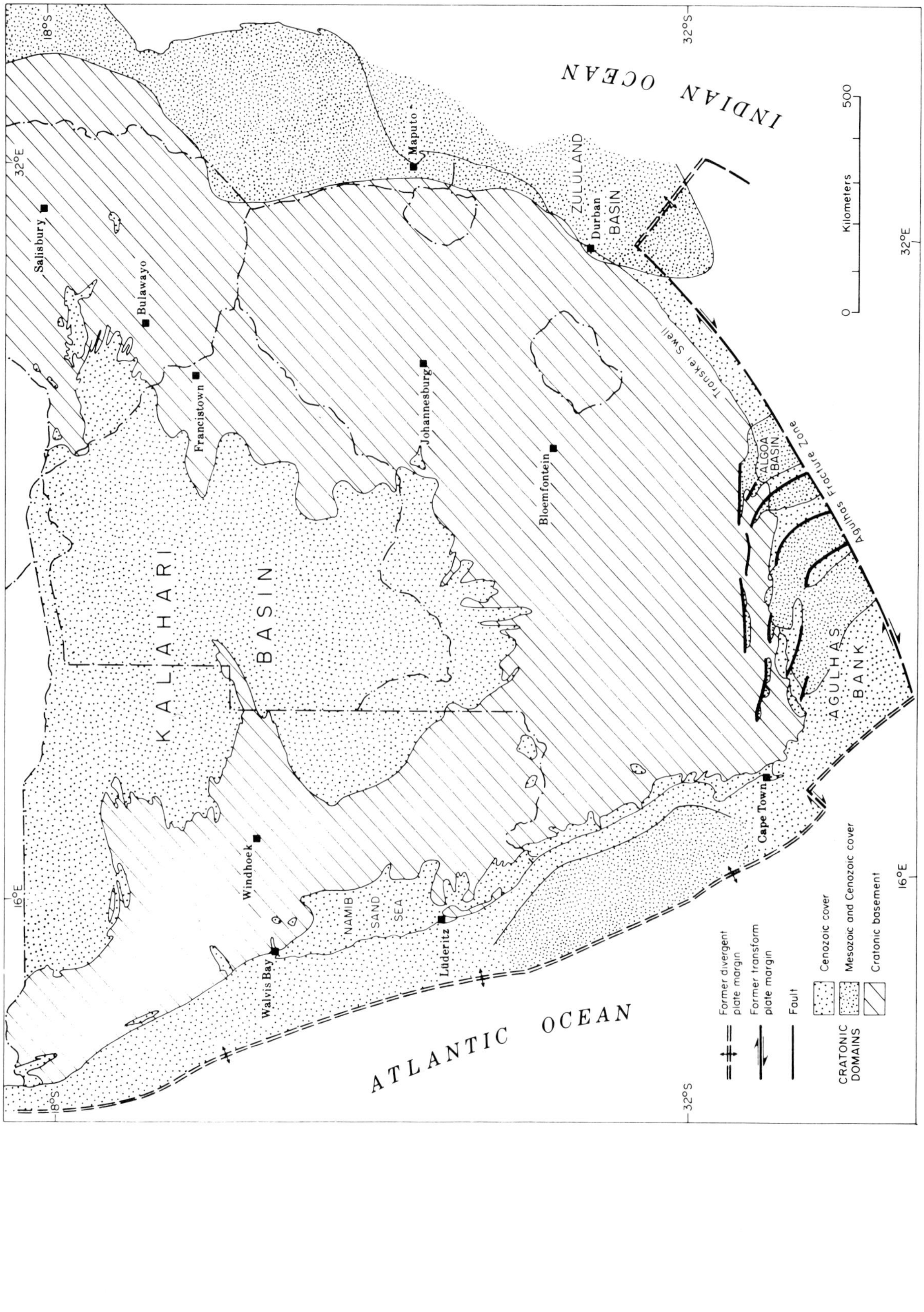

Figure 1-7. Partly palinspastic map showing sedimentary basins around the continental margins and in the continental interior after the breakup of Gondwana 150 Ma ago.

the Late Jurassic Epoch, when Africa and Antarctica finally separated. The basins are bounded in the south beneath the Agulhas Bank by a fracture ridge, which was finally buried by sediments toward the end of the Cretaceous Period.

In contrast, the Zululand basin is characterized by north–south tensional faulting related to rifting between east and west Gondwana. Coarse fluvial deposition along the basin's western margin was a response to the steep gradients and high basement relief produced by faulting. The Indian Ocean first transgressed in the Barremian (Early Cretaceous Epoch). An anomalous region of positive relief in the southeastern part of the subcontinent, the Transkei swell (Simpson and Dingle, 1973), kept faunas of the Zululand and Algoa–Agulhas basins separate until the Coniacian.

The main depositories of marine sedimentation in the Cenozoic Era coincide with those of the Mesozoic Era and record several transgressive cycles which correlate with eustatic fluctuations. Depositional history was complicated locally by seaward tilting and epeirogenic uplift. The continental interior was dominated by the internal drainage system of the Kalahari basin and the Orange–Vaal drainage basin, where there is evidence of epeirogenic control over sedimentation.

STAGE 1

Archean Crustal Evolution

The geologic record begins with Early Archean continental crust in the Central Zone of the Limpopo Province. Among the oldest rocks in the world, the 3786-Ma layered granodioritic–dioritic Sand River gneisses are cut by a 3560-Ma deformed metatholeiitic dike in the Sand River near Messina, South Africa.

Chapter 2

Granite–Greenstone Terrane: Kaapvaal Province

The exposed Archean rocks of the Kaapvaal Province consist predominantly of various granitoids with subordinate gneisses and relicts of volcanosedimentary greenstone belts (Swaziland Supergroup and equivalents). These are overlain by the Pongola Supergroup, a sequence of volcanic and sedimentary rocks that accumulated in a generally shallow-water, ensialic basin. Granitoids, basic layered intrusions, and syenites were intruded at various times throughout the Archean. The Pongola Supergroup rests nonconformably on quartz monzonites and granodiorites emplaced during a widespread granitic event at about 3.0 Ga. This event effectively divides the evolutionary history of the Kaapvaal Province into periods older and younger than 3.0 Ga. It is still debated whether a further broad subdivision of the Archean can be made, separating an earlier (>3.5 Ga) development of sialic crust from the 3.5–3.3-Ga volcanosedimentary event (Swaziland Supergroup) (Table 2-1).

Only 14 percent of the Kaapvaal basement (3.0 Ga and older) is exposed. Most of the basement comprises granitoids and gneisses; greenstone belts constitute less than 10 percent (Anhaeusser, 1976a). Much of this granitoid terrane has not been mapped in detail, most studies having been undertaken in the eastern Transvaal and Swaziland (Fig. 2-1).

2.1. The Early Gneiss Terranes

Inliers of layered granitoid gneisses crop out at a number of localities within the Archean basement of the Kaapvaal Province; for example, north of Johannesburg, near the Murchison greenstone belt, and northwest of Pretoria (Fig. 2-1). Few of these areas have been mapped in detail and little is known of their petrologic characteristics, ages, or regional extent. The central area of Swaziland is an exception, where approximately 2500 km^2 (Fig. 2-2) are underlain by various gneisses, collectively known as the Ancient Gneiss Complex (Hunter, 1968, 1970).

Ancient Gneiss Complex of Swaziland

Lithology and field relations. The following lithologic units have been recognized in the Ancient Gneiss Complex and are listed in order of decreasing age:

(1) Bimodal gneiss suite, age 3555 Ma (Barton *et al.*, 1980)
(2) Migmatitic gneisses, age not known
(3) Dwalile Metamorphic Suite, age not known but probably pre-3320 Ma
(4) Biotite–hornblende tonalite gneiss (Tsawela gneiss), age 3321 Ma (Davies and Allsopp, 1976)
(5) Metaanorthosite and metagabbro (Mponono Intrusive Suite), age not known
(6) Lenses of homogeneous, medium-grained quartz monzonite, age 3150 Ma (Davies and Allsopp, 1976)
(7) Mkhondo Valley Metamorphic Suite, age not known

The dominant rock type is the bimodal gneiss suite, which constitutes about 80 percent of the total area of the Ancient Gneiss Complex in Swaziland. The Dwalile Metamorphic Suite is preserved as small remnants in the bimodal suite in southwestern Swaziland (Fig. 2-2B) where the Tsawela gneiss has intruded both these units. The Mponono Intrusive Suite is now represented by discontinuous outcrops along the Mponono Valley (Fig. 2-2B). The migmatitic gneisses are con-

Table 2–1. Chronologic Summary of the Archean in Swaziland and the Eastern Transvaal[a]

	Mhlosheni-type granite plutons				
	Mpageni-type granite plutons				
$\sim$ 2.6 Ga	Kwetta-type granite plutons				
	Pongola granite				
	Cunning Moor tonalite				
$\sim$ 2.9 Ga	Usushwana Intrusive Suite				
	Pongola Supergroup	Mozaan Group Nsuze Group			
	Dalmein-type granodioritic plutons				
$\sim$ 3.0 Ga	Hebron granodiorite				
	Lochiel granite				
	Nelspruit porphyritic granite and migmatites			Ancient Gneiss Complex (Swaziland)	
				Mkhondo Valley Metamorphic Suite (?) (stratigraphic position uncertain: age > 2.6 Ga)	
$\sim$ 3.1 Ga	Bosmanskop syenite				
$\sim$ 3.2 Ga	Leucotonalitic plutons		$\sim$ 3.15 Ga	Quartz monzonite (minor intrusions)	
3.3 Ga	Kaap Valley "granite"		$\sim$ 3.3 Ga	Granodiorite Suite	
				Mponono Intrusive Suite	
				Tsawela gneiss	pre-3.3 Ga
	Swaziland Supergroup	Moodies Group		Dwalile Metamorphic Suite	
		Fig Tree Group		Migmatitic gneiss (stratigraphic position uncertain)	
3.5 Ga		Onverwacht Group	$\sim$ 3.5 Ga	Bimodal gneiss suite	

[a]No relative age sequence is implied with respect to granitoid intrusions aged 3.2–3.3 Ga, 3.0 Ga, and 2.6 Ga.

fined to an area east of Mbabane, about 300 km^2 in extent, and apparently grade into the bimodal gneiss suite (Fig. 2-2A). The Mkhondo Valley Metamorphic Suite underlies an area of about 300 km^2 in southern Swaziland. Younger granitoids and a major fault zone preclude the establishment of field relationships between the Mkhondo Valley Metamorphic Suite and other lithlogic units of the Ancient Gneiss Complex.

The correlation of the Dwalile and Mkhondo Valley Suites is uncertain. The former is lithologically similar to the Onverwacht Group, with which a possible correlation is suggested by its pre-3320 Ma age. If so, these supracrustal rocks assume considerable importance in determining the relationship between the Onverwacht Group and the bimodal gneiss suite. Relations between the bimodal suite and the Dwalile supracrustal rocks have been obscured by repeated high strains, but the common occurrence of the supracrustal rocks in the cores of synforms suggests that these rocks are likely to have overlain the bimodal suite at the time of their deposition.

The Mkhondo Valley Metamorphic Suite is intruded by 2.6-Ga granites, but no other geochronologic data are available to fix its stratigraphic position more precisely.

The bimodal gneiss suite consists of leucocratic gneisses, dominantly of trondhjemitic or tonalitic composition, that are complexly interlayered with plagioclase amphibolites. The lighter colored siliceous layers are medium grained and commonly display layering because of variations in the proportions of dark minerals, mainly biotite (0–10 percent by volume). Plagioclase (An_{25}), the most abundant mineral in these gneisses, constitutes up to 55 volume percent of the leucocratic layers.

Individual amphibolite layers vary considerably in thickness from a few centimeters up to 500 m. The interlayered amphibolites are medium grained and comprise plagioclase ($An_{35\text{-}45}$) and hornblende with minor quartz. Some amphibolites consist entirely of hornblende, whereas others have microscopic layers of diopside. Garnet is typically absent from the amphibolites but is locally prominent in both amphibolite and siliceous layers east of Mankayane. The amphibolites have acted competently during the intense polyphase deformation of the bimodal suite and are commonly boudinaged. These strain features

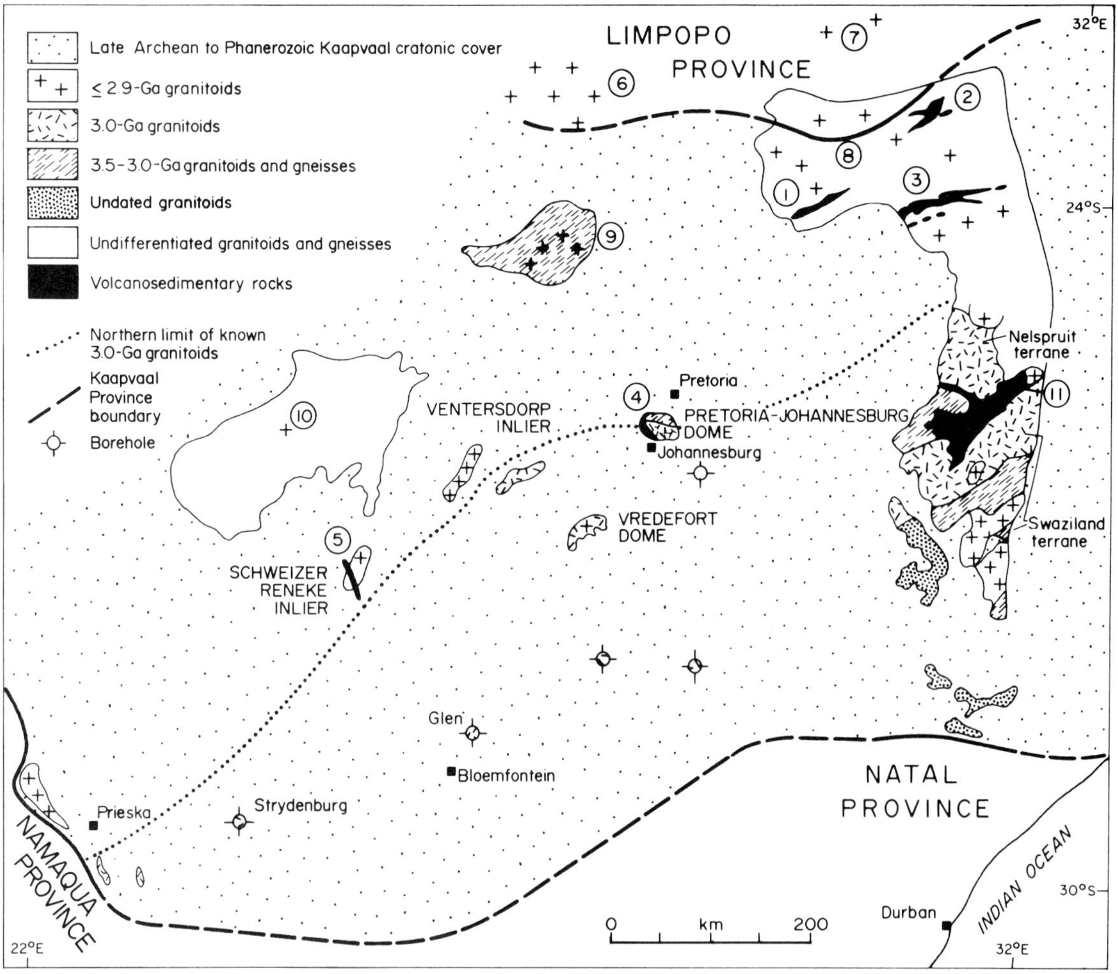

Figure 2-1. Inliers of Archean volcanosedimentary greenstone belts and granitoids in the Kaapvaal Province. (1) Pietersburg belt. (2) Sutherland belt. (3) Murchison belt. (4) Muldersdrif belt. (5) Amalia belt. (6) Mahalapye area, Botswana; radiometric date may indicate age of reworking of older crust. (7) Granitoids in Limpopo Province. (8) Granitoids have yielded widespread Rb-Sr and U-Pb ages between 2.5 Ga and 2.7 Ga; layered gneisses are probably pre-3.0 Ga. (9) Makoppa Dome: granitoids have yielded ages of about 2.6 Ga; layered gneisses are probably pre-3.0 Ga; distribution of various granitoids not mapped. (10) Two radiometric ages reported; remainder of granitoid terrane not dated but is pre-2.3 Ga. (11) Barberton belt. Vredefort Dome: a minimum U-Pb age of 2.5 Ga on zircon. Undated granitoids in eastern Kaapvaal Province have not been mapped, but some are known to be pre-Pongola in age (>2.9 Ga). Massive granites intruded into various granitoid gneisses are found over a wide area of the Kaapvaal Province, but the figure shows only the locations of massive granitoids that have been radiometrically dated. Data from Burger and Coertze (1973), Burger and Walraven (1979).

have been interpreted by some as the result of intrusion of the trondhjemitic and tonalitic gneisses into the amphibolites.

Hornblende anorthosites (the Mponono Intrusive Suite) crop out within the bimodal suite along the Mponono River in western Swaziland (Fig. 2-2B). These rocks consist of aggregates of plagioclase up to 10 cm in diameter, set in a matrix of hornblende and even-grained metagabbro. The hornblende-rich parts are more highly strained than the plagioclase aggregates. The relationship of the anorthositic rocks to the bimodal suite is uncertain, but they appear to represent an original sheetlike intrusion that has suffered intense deformation by folding after being tectonically intersliced with the bimodal suite by imbricate thrusting. The Mponono anorthosites appear to have suffered deformation that predates the

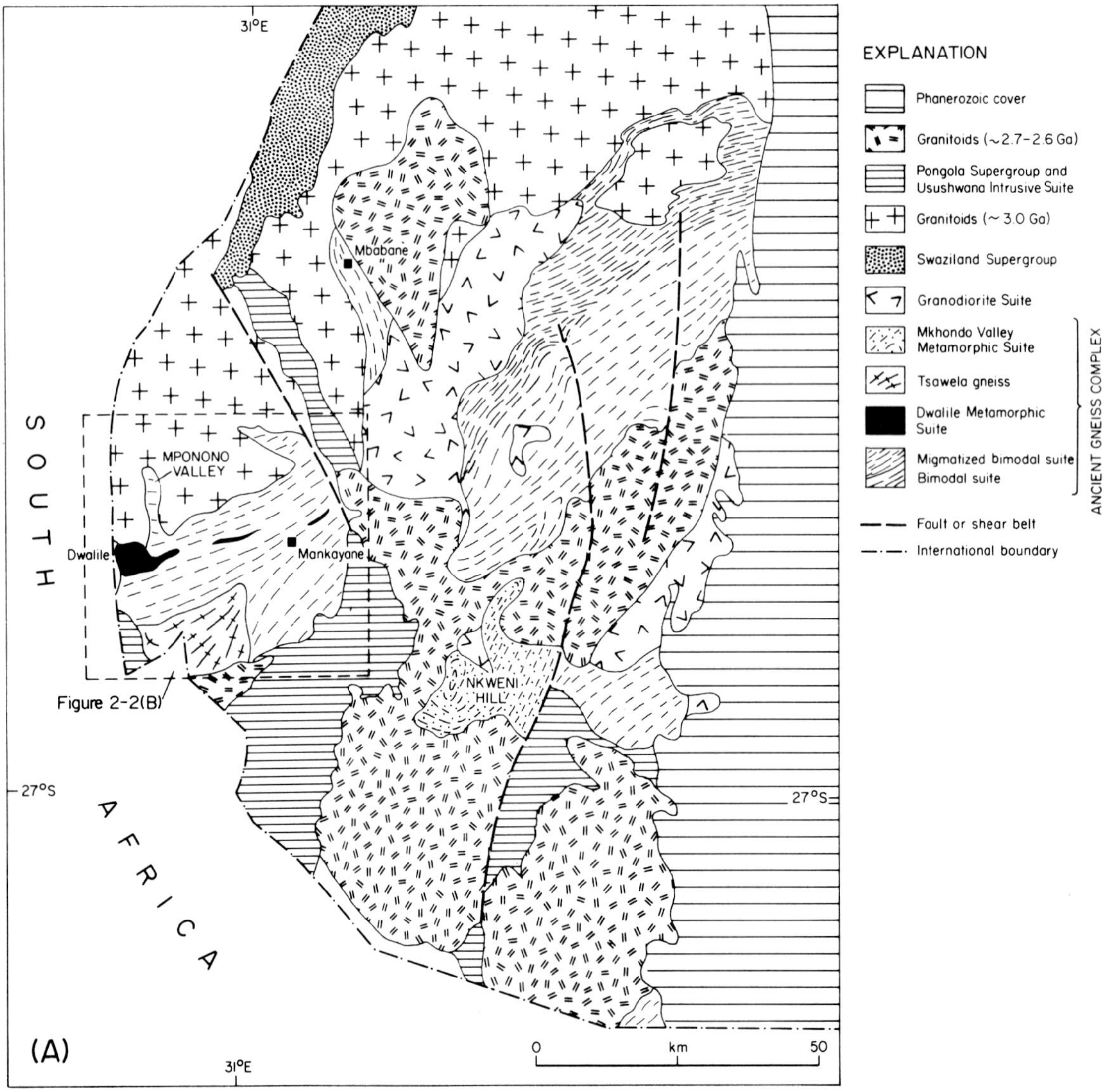

Figure 2-2. (A) Simplified geologic map of part of the gneiss terrane of Swaziland showing the distribution of the Ancient Gneiss Complex.

homogeneous quartz monzonite intrusions dated at 3.15 Ga, but the anorthosites intrude the biotite–hornblende tonalite gneiss. Accordingly the Mponono Intrusive Suite is likely to be 3.2–3.3 Ga old and hence very similar in age to the Messina Intrusive Suite (Chapter 3).

Narrow, deformed amphibolite dikes in the bimodal suite suggest that this suite suffered at least one period of brittle deformation following an early period of metamorphism and folding.

The biotite–hornblende tonalite gneiss (Tsawela gneiss) is coarser grained and more homogeneous than the gneisses of the bimodal suite. It occupies a discrete area in southwest Swaziland and, on structural and geochemical grounds, is regarded as a calc–alkaline intrusion into the bimodal suite, both units subsequently being tightly folded about an east-northeasterly striking axial plane. The biotite–hornblende tonalite gneiss consists of plagioclase (An_{25-30}), brown biotite, green hornblende, and quartz, the mafic minerals forming up to 15 percent by volume of the rock.

Gneisses that display less well-defined layering but that contain abundant quartz–feldspar veins crop out north of Manzini in central Swaziland in a zone approximately 12 km wide. These nebulitic

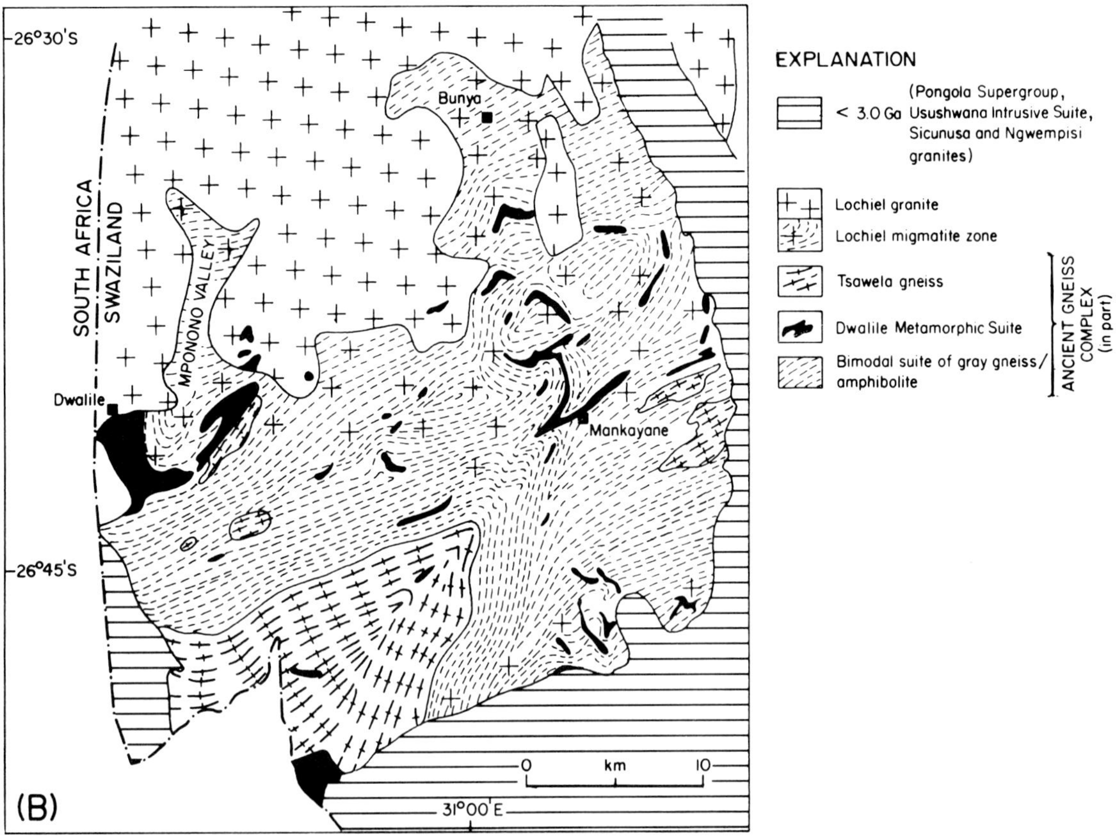

(B) Geologic map of the Mankayane inlier in southwestern Swaziland. The inlier represents the type area for the bimodal suite, Dwalile Metamorphic Suite, Tsawela tonalite gneiss, and the Mponono Intrusive Suite (outcrops too small to show). (After mapping by D.R. Hunter, A.C. Wilson, and M.P.A. Jackson.)

migmatites grade into the more regularly banded gneisses of the bimodal suite, to which they are mineralogically similar.

The main outcrop of the Mkhondo Valley Metamorphic Suite consists of a layered sequence of amphibolite, cordierite–garnet–biotite gneisses, quartz–hornblende–biotite–diopside gneisses, iron-formation, quartz–diopside and plagioclase–diopside granofelses, anthophyllite–cummingtonite gneisses, and metaquartzites. Coarse-grained quartzofeldspathic gneisses, containing thin amphibolitic and garnet-bearing layers, underlie an extensive area to the northeast of the main syncline in the Mkhondo Valley (Hunter, 1970). Their mineral assemblages reflect metamorphism at high temperature (~650–750°C) and low pressure (0.3–0.4 GPa or ~3–4 kbar). The relationships of the metamorphic suite to the other subdivisions of the Ancient Gneiss Complex are obscure because of lack of outcrops, but small lenses of quartzofeldspathic gneiss, which may be tectonically intersliced, are interlayered with gneisses of the bimodal suite at a number of localities in central Swaziland.

The Dwalile Metamorphic Suite in the Mankayane district of Swaziland consists predominantly of amphibolite with subordinate calc–silicate gneiss, serpentinite, pelitic schist, and metaquartzite. The most extensive outcrop is in the vicinity of Dwalile border beacon in western Swaziland, where they have undergone at least three periods of folding (Fig. 2-2B). The field evidence suggests that this group of mainly metavolcanic rocks was intruded by the Tsawela tonalite gneiss.

The quartz monzonites are best developed south of Mankayane where they crop out as elliptical lenses about 1 km long, their long axes

Table 2–2. Mean Analyses of Siliceous Gneisses in the Ancient Gneiss Complex[a]

	1	2	3	4	5	6
SiO_2	76.98	76.11	76.81	71.57	66.48	74.14
TiO_2	0.22	0.25	0.31	0.36	0.52	0.12
Al_2O_3	11.11	11.76	10.97	14.68	15.24	13.91
Fe_2O_3	1.86	0.88	0.61	0.69	1.01	0.23
FeO	2.21	2.22	2.44	2.01	3.24	0.97
MnO	0.04	0.04	0.04	0.04	0.08	0.04
MgO	0.39	0.73	0.69	0.87	2.34	0.23
CaO	2.23	1.44	1.00	3.20	4.29	1.00
Na_2O	3.66	4.11	2.57	4.71	4.18	3.85
K_2O	0.89	1.76	4.06	1.37	1.78	4.87
H_2O^+	0.39	0.45	0.53	0.49	0.95	0.38
H_2O^-	0.02	0.04	0.04	0.06	0.06	0.07
P_2O_5	0.03	0.03	0.04	0.08	0.14	0.03
CO_2	0.01	0.06	—	0.05	—	0.01
Cl	0.01	0.01	—	0.01	—	0.01
F	0.03	0.10	—	0.05	—	0.02
less ≡ ^{t}O	0.01	0.04	—	0.02	—	0.01
Total	100.07	99.95	100.08	100.22	100.31	99.87
Rb	42	60	58	64	67	231
Sr	113	80	85	278	339	130
K/Rb	189	247	581	210	236	177
K/Na	0.27	0.48	1.77	0.34	0.48	1.41
Rb/Sr	0.37	0.75	0.68	0.23	0.21	1.79
Na/Na+K	0.86	0.78	0.49	0.84	0.78	0.54

[a] 1, High-Si, low-K gneiss, bimodal suite, 2 analyses (type Ia, Table 2–4). 2, High-Si, inter-K gneiss, bimodal suite, 2 analyses (type Ib, Table 2–4). 3, High-Si, high-K gneiss, bimodal suite, 1 analysis (type II, Table 2–4). 4, Normal-Si gneiss, bimodal suite, 5 analyses (type III, Table 2–4). 5, Biotite-hornblende tonalite gneiss, 5 analyses (Tsawela gneiss). 6, Quartz monzonite, 1 analysis.

trending east-northeast, parallel to the foliation of the bimodal suite. Their isotopic age of approximately 3150 Ma and lack of deformational fabric suggest that the quartz monzonite lenses were emplaced late in the evolution of the Ancient Gneiss Complex. The quartz monzonite consists of plagioclase, microcline, and quartz and less than 5 percent biotite.

The limited geochronologic data suggest that units of the Ancient Gneiss Complex were formed over a period of at least 400 Ma and possibly much longer.

Early crustal processes: Geochemistry and petrogenesis. The bimodal gneiss suite in Swaziland shows the bimodality of composition that is typical of many high-grade gneiss complexes (columns 1–4 in Tables 2-2 and 2-3). Siliceous gneisses of the bimodal suite in Swaziland have SiO_2 contents ranging from 69 to 77 percent, whereas the interlayered amphibolites contain a markedly different range between 48 and 55 percent SiO_2; an exceptional dioritic gneiss in the bimodal suite has 57 percent SiO_2.

Of the 10 available analyses of siliceous gneisses from the bimodal suite, five have contents of SiO_2 > 75 percent and the remainder contain <75 percent. The gneisses containing >75 percent SiO_2 can be subdivided into: (1) those with low to intermediate contents of K_2O and K/Na ratios less than unity, and; (2) those with high K_2O content and K/Na ratios greater than unity. Only one sample of the latter type has been identified to date.

Two distinct types of rare earth element (REE) patterns that can be correlated with types I and III in Table 2-4 have been recognized (Hunter *et al.*, 1978). The highly siliceous type I has La and Ce contents 100–300 times that of chondrites, heavy REEs 20–50 times chondrites, and pronounced

Table 2–3. Mean Analyses of Basic Gneisses in the Ancient Gneiss Complex.[a]

	1	2	3	4	5	6
SiO_2	48.05	50.35	52.20	52.04	57.31	54.22
TiO_2	0.67	0.66	1.24	0.80	0.30	0.93
Al_2O_3	5.99	15.16	12.58	15.37	17.84	13.05
Fe_2O_3	5.69	1.80	5.30	0.78	0.89	1.56
FeO	6.87	8.00	9.76	9.26	4.65	9.45
MnO	0.21	0.18	0.18	0.16	0.13	0.18
MgO	17.23	7.66	5.05	9.07	4.21	6.01
CaO	10.79	11.25	8.86	7.88	8.47	8.60
Na_2O	1.34	2.83	1.95	1.94	4.11	3.34
K_2O	0.25	0.56	0.74	0.14	0.60	0.59
H_2O^+	2.32	1.49	1.65	2.37	1.16	1.18
H_2O^-	0.08	0.04	0.02	0.07	0.08	0.09
P_2O_5	0.06	0.13	0.14	0.06	0.05	0.48
Total	99.55	100.11	99.67	99.95	99.74	99.64
Rb	<5	—	16	1.2	63	2.4
Sr	94	168	119	203	276	257
Ba	—	107	114	71	—	160
K/Rb	>415	—	384	968	79	2040
Rb/Sr	<0.05	—	0.13	0.006	0.23	0.007
Sr/Ba	—	1.6	1.0	2.86	—	1.61

[a]Bimodal suite: 1, Hornblendite, 1 analysis. 2, Amphibolite (olivine normative), 2 analyses. 3, Amphibolite (quartz normative), 3 analyses. 4, Amphibolite (quartz normative), 1 analysis. 5, Diorite gneiss, 1 analysis. Mkhondo Valley Metamorphic Suite: 6, Amphibolite (quartz normative), 2 analyses.

Eu anomalies. The type III gneisses have La and Ce contents <100 times that of chondrites, heavy REEs <10 times chondrites, and small positive or negative Eu anomalies. Gneisses of both types I and II have low $\delta^{18}O$ values with an average of 6.9 permil (F. Barker *et al.*, 1976a). The siliceous gneisses of the bimodal suite have a primitive $^{87}Sr/^{86}Sr$ ratio of 0.6999, indicative of a mantle source (Barton *et al.*, 1980). Their low $\delta^{18}O$ values have been interpreted to mean that these rocks were derived from the mantle by igneous and metamorphic processes without a weathering stage (F. Barker *et al.*, 1976a; Hunter *et al.*, 1978).

The interlayered amphibolites in the bimodal gneiss suite also can be divided into olivine-normative and quartz-normative types (Table 2-3, columns 1–4). The olivine-normative amphibolites include one with high MgO and low Al_2O_3 contents and two with basaltic chemistry. Three

Table 2-4. Major- and Trace-Element Characteristics of Siliceous Gneisses in the Bimodal Gneiss Suite[a]

Type	Major-Element Characteristics	Trace-Element Characteristics
I(a)	$SiO_2 > 75\%$; $Al_2O_3 < 14\%$; $K_2O < 1\%$ K/Na ratio ~ 0.27	50–130 ppm Sr; K/Rb ratio 150–230 Rb/Sr ratio 0.2–0.6
I(b)	$SiO_2 > 75\%$, $Al_2O_3 < 14\%$; $K_2O > 1\%$ K/Na ratio ~ 0.5	50–120 ppm Sr; K/Rb ratios 200–280 Rb/Sr ratio 0.5–1.3
II	$SiO_2 > 75\%$; $Al_2O_3 < 14\%$; K_2O ~ 4% K/Na ratio greater than unity	~85 ppm Sr; K/Rb ratio > 300 Rb/Sr ratio > 0.5 < 1.0
III	$SiO_2 < 75\%$; $Al_2O_3 > 14\%$; K_2O 1–2.25%; K/Na ratio < 1	100–150 ppm Sr; K/Rb ratio variable 130–400; Rb/Sr ratio <0.3

[a]From Hunter *et al.* (1978 and unpublished data).

of the quartz-normative amphibolites have, in addition to their enrichment in iron, enhanced contents of TiO_2, P_2O_5, and Rb. The amphibolites typically have flat REE patterns at about 10 or 20 times that of chondrites for the iron-rich quartz-normative amphibolites (Table 2-3, column 3).

The low $^{87}Sr/^{86}Sr$ ratio and $\delta^{18}O$ values of the siliceous gneisses conform in these respects to compositionally equivalent gneisses in other similar terranes for which magmatic origin is proposed. Models for the generation with <75 percent SiO_2, which is the most common variety in gray gneiss complexes, include fractionation of wet basaltic magma, partial melting of quartz eclogite, and partial melting of amphibolite. Hunter *et al.* (1978) prefer the third model for the Swaziland gneisses because Archean terranes apparently contain no eclogite whereas amphibolites are common. This is not a unique solution because it is impossible to discriminate between this model and one involving quartz eclogite.

The low-K_2O, high-SiO_2 gneisses in the bimodal suite are rare in Archean gray gneiss complexes (Hunter *et al.*, 1978). Generation of such liquids enriched in large-ion-lithophile elements (LILE), which contain 15–30 times the light REEs and 4–5 times the heavy REEs of typical Archean tholeiites, is not yet understood. The absence of intermediate rocks from the bimodal suite presumably excludes models involving fractionation of basaltic liquids. Partial melting of relatively REE-enriched basalt containing neither residual hornblende nor garnet has been proposed for similar gneisses interlayered with amphibolite in the Webb Canyon gneiss of Wyoming (F. Barker *et al.*, 1976b). Alternatively, partial melting of a basaltic parent of granulite-facies mineralogy, including plagioclase, quartz, biotite, and pyroxene, may generate liquids of a suitable composition.

The Tsawela tonalite gneiss is chemically more homogeneous and is the least siliceous (62–68 percent SiO_2) sialic gneiss in the Ancient Gneiss Complex. The consistent K/Rb ratios have a mean value of 236 (Table 2-2, column 5). The geochemistry of the Tsawela tonalite gneiss is similar to that of other intrusions for which origins by partial melting of metabasalt or quartz eclogite have been proposed (Arth and Hanson, 1975; Condie and Hunter, 1976), which suggests that similar processes may have been responsible for the generation of these Swaziland rocks. Their low initial $^{87}Sr/^{86}Sr$ ratio and $\delta^{18}O$ values (F. Barker *et al.*, 1976a; Davies and Allsopp, 1976) suggest a mantle source for the parent magma.

The migmatitic gneisses from central Swaziland are tonalitic or trondhjemitic and, despite their nebulitic character, their REE patterns are like those of nonmigmatitic rocks of these compositions. The patterns are steep with La 60–100 times that of chondrites, Yb and Lu close to chondrites, and small to large positive Eu anomalies. Although these gneisses have similar REE patterns, other trace elements vary considerably: Rb ranges from 9 ppm to 70 ppm, Ba from 270 ppm to 2070 ppm, and K/Rb ratios from 300 to 1037. There is no correspondence of LILE and REE abundances (Hunter *et al.*, 1978). The nebulitic gneisses have distinctly higher values for $\delta^{18}O$ ($\bar{x}$ = 8.6 permil) than the bimodal suite gneisses ($\bar{x}$ = 6.9 permil) (F. Barker *et al.*, 1976a).

The quartzofeldspathic gneisses of the Mkhondo Valley Metamorphic Suite (Table 2-5, columns 1 and 2) plot on the Qz–Ab–Or diagram close to the H_2O-saturated minimum-melt compositions for low pressures (Hunter *et al.*, 1978). The gneisses have Rb/Sr and K/Na ratios greater than unity and Rb contents generally higher than those in the siliceous bimodal gneisses. Their REE patterns show enrichment in light REEs, prominent negative Eu anomalies, and slight depletion and gentle to flat slopes of heavy REEs. These quartz monzonitic gneisses may have been formed by partial melting of older trondhjemitic–tonalitic gneisses. However, the parental rocks must have contained little or no residual hornblende or garnet, as implied by the gentle slopes of the heavy REEs, and plagioclase must have been in the residuum. If older gneisses were partially melted, they are likely to have been the low-Al_2O_3 siliceous gneisses of the bimodal suite, consisting largely of feldspar, pyroxene, and quartz at granulite grade.

A granoblastic diopside–hornblende–biotite gneiss in the Mkhondo Valley Metamorphic Suite has calc–alkaline affinities and its REE pattern is similar to those of the Tsawela tonalite gneiss. Its origin is uncertain; it may represent an intrusion into the metamorphic suite, although its relatively high $\delta^{18}O$ value of 9.2 permil suggests that it represents a metagraywacke.

The amphibolites in the metamorphic suite are quartz-normative tholeiites; they have high K/Rb ratios ($\geq$ 1000) (Table 2-3, column 6) and anomalous REE patterns that lie between the fields

A

B

C

Figure 2-3. (A) Intrusive contact between Tsawela tonalite gneiss and amphibolite of the bimodal suite. Location: Mhlatane River, 2 km south of Mankayane village. (B) Amphibolite dike (intruded during early basic-dike emplacement, event 7 of Table 2-6) cutting layered gneisses of the bimodal suite (extreme top) and highly strained Tsawela gneiss. Folding of the dike took place during D2 (event 9) followed by strong finite flattening during D4 (event 12). Location: small tributary, 1.5 km south of Tsawela River bridge. (C) Weakly deformed anorthosite and anorthositic gabbro of the Mponono Intrusive Suite (intruded during event 8). Cumulus crystals of plagioclase are set in a dark matrix metamorphosed to horneblende and biotite under upper amphibolite-facies conditions. A transecting vein of pegmatitic Lochiel granite is visible at upper right. Small faults on the right formed during post-Lochiel brittle tectonics along the Mponono Valley. Location: Mponono River, 12 km northeast of Dwalile village.

Table 2–5. Analyses of the Mkhondo Valley Metamorphic Suite[a]

	1	2	3	4	5
SiO_2	71.60	76.85	67.35	78.15	55.21
TiO_2	0.52	0.15	0.56	0.27	0.51
Al_2O_3	12.39	12.06	14.58	10.40	12.73
Fe_2O_3	2.07	1.45	0.78	0.61	0.39
FeO	3.18	0.74	3.29	3.91	9.46
MnO	0.08	0.03	0.07	0.05	0.12
MgO	0.42	0.30	2.20	2.34	15.01
CaO	2.02	0.57	4.73	Tr	2.45
Na_2O	3.12	3.85	3.82	0.54	0.42
K_2O	3.91	3.53	1.32	2.09	0.29
H_2O^+	0.36	0.41	0.79	1.00	3.45
H_2O^-	0.05	0.03	n.d.	0.08	0.05
P_2O_5	0.12	0.03	0.14	0.03	0.11
CO_2	0.03	0.02	0.27	—	0.10
Cl	0.02	0.01	—	—	—
F	0.05	0.01	—	—	—
less ≡ 'O	0.02	0.00	—	—	—
Total	99.92	100.04	99.90	99.47	100.30
Rb	129	90	50.7	—	—
Sr	111	73	207	—	—
Ba	~500	900	347	—	—
K/Rb	258	325	216	—	—
K/Ba	~65	32.6	31.4	—	—
K/Na	1.40	1.03	0.38	4.32	0.77
Sr/Ba	~0.22	0.08	0.60	—	—
Rb/Sr	1.16	1.23	0.24	—	—

[a] 1, Quartzofeldspathic gneiss, 1 analysis. 2, Quartzofeldspathic gneiss, mean of three analyses. 3, Quartz–biotite–hornblende–diopside gneiss, 1 analysis. 4, Siliceous biotite–garnet gneiss, 1 analysis. 5, Cummingtonite–cordierite–quartz gneiss, 1 analysis. Tr, trace.

of low-K tholeiite and continental tholeiite (Hunter *et al.*, 1978). This distinguishes these amphibolites from those of the bimodal suite.

The leucocratic quartz monzonite forming lensoid bodies near Mankayane is similar in major-element chemistry to the quartz-monzonite gneisses in the Mkhondo Valley Metamorphic Suite (Table 2-2, column 6, and Table 2-5, column 2). The REE pattern is steep with strong depletion of the heavy REEs. This rock is chemically similar to quartz monzonite of Late Archean age from Minnesota that Arth and Hanson (1975) deduced to have formed by 20–25 percent melting of graywacke. In Swaziland the common siliceous gneisses in the bimodal suite have Rb, Sr, and REE abundances suitable for parent rocks, although graywacke compositions are rare to nonexistent in the older parts of the Ancient Gneiss Complex. The quartz–monzonite lenses are therefore likely to have formed by partial melting of hornblende- or garnet-bearing gneisses of the bimodal suite at the transition between upper amphibolite and granulite facies. This quartz monzonite has yielded an isotopic age of about 3150 Ma and an initial $^{87}Sr/^{86}Sr$ ratio of 0.7048 (Davies and Allsopp, 1976). The high initial ratio is compatible with crustal contamination; the age suggests that quartz–monzonite magmas may have been generated during tectonism in the Ancient Gneiss Complex contemporaneous with emplacement of tonalitic and trondhjemitic plutons farther north near Barberton between 3.3 and 3.2 Ga.

Structural evolution. Detailed structural study of the Ancient Gneiss Complex has been confined to the Mankayane area and the Mkhondo Valley.

In the former area a complex superposition of

high strains can be recognized in the bimodal suite, Tsawela tonalite gneiss, Dwalile Metamorphic Suite, and the Mponono Intrusive Suite (Jackson, 1979a) (see Fig. 2-2B and Table 2-6, summary). Isoclinally folded gneissic layering and quartz–feldspar veins with axial-planar schistosity in the bimodal gneisses are crosscut by intrusive contacts of the Tsawela tonalite gneiss, demonstrating that the bimodal gneisses in this area were deformed at least once before the intrusion of the Tsawela gneiss. The heterogeneous sequence of Dwalile supracrustal rocks accumulated prior to the intrusion of the Tsawela gneiss. Repeated high strains have obscured the relationship between the bimodal gneisses and the supracrustal sequence but their preferred occurrence in the cores of synforms suggests that the supracrustal rocks originally overlay the bimodal suite. After intrusion of the Tsawela gneiss, the Mponono anorthosite intrusion, and a suite of basic dikes, a high-strain event (D2) resulted in finite fabrics in all lithologic units (Fig. 2-3A, B, C) and folding about flat-lying axial surfaces (Fig. 2-4A). A second generation of basic dikes cut across these D2 structures but are themselves offset by D3 ductile shear belts. Even-grained gray mesocratic dikes intruded the bimodal gneiss, Tsawela gneiss, and the Mponono Intrusive Suite and were metamorphosed at amphibolite grade during the widespread D4 high-strain increment, which resulted in major folding of the D2 fabric. This folding is the chief cause of the present outcrop pattern, in which Tsawela gneiss is exposed in antiformal cores and Dwalile supracrustals are exposed in the cores of synforms (Fig. 2-2B). Folds show a change in orientation from near upright in the northeast, through inclined plunging in the center, to reclined in the northwest of the Mankayane area. This trend, together with qualitative evidence of higher finite strains in the northwest, suggests that the XY plane of the D4 bulk strain ellipsoid was rotated with progressive deformation from an initially steep orientation in the east to a gentle south-southeastward dip in the west. This variation in geometry could be explained by large-scale heterogeneous simple shear upward and toward the north and northwest, as proposed for the Laxford front in northwest Scotland (Beach *et al.*, 1974).

The highly strained gneiss terrane underwent further shape changes during D5 by means of heterogeneous shear displacements along shallow dipping planes of two main types (Fig. 2-4C): (1)

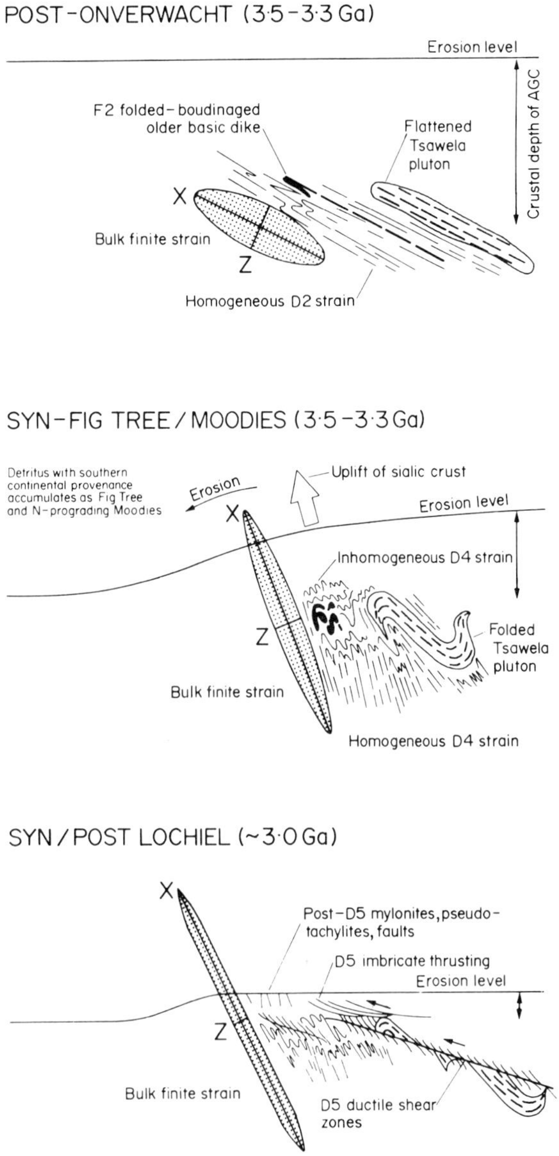

Figure 2-4. Schematic portrayal of the structural evolution of the Ancient Gneiss Complex (based on the Mankayane area of southwestern Swaziland) from 3.5 Ga to 3.0 Ga. Homogeneous ductile strains (e.g., D2) are overprinted by heterogeneous strains in the form of major folds and shear belts (e.g., D3/D4) and then by heterogeneous brittle deformation in the form of imbricate faults and pseudotachylites (e.g., D5). This trend suggests deformation during progressive uplift to the surface. Further uplift, and probably lateral spreading, is inherent in the gross shape changes of the crust represented by the bulk finite strain ellipsoid. It is probable that this massive uplift provided the provenance for the thick terrigenous clastic units of the Swaziland Supergroup to the north. [Adapted from Jackson (1979a, 1980).]

ductile shear belts with a thrust sense and (2) brittle imbricate thrust wedges tectonically interleaved along knife-sharp thrust surfaces. The nonpenetrative nature of these displacements suggests that shortening and uplift toward the northwest took place under conditions of lower ductility (Jackson, 1979a, 1980).

The supracrustal rocks in the Mkhondo Valley just west of Nkweni Hill (Fig. 2-2A) contain well-preserved primary structures such as bedding, cross-lamination, and scour channels. Metasediments and metavolcanics are exposed on the eastern flank of a doubly plunging synformal syncline (Jackson and Clarke, in preparation). The first recognizable period of deformation appears to have been most intense. The strains were of a flattening type with XY bulk strain planes flat-lying and approximately parallel to the stratification. The heterogeneity of the deformation has allowed survival of primary structures in the most competent rocks, but all primary structures are destroyed in the least competent rocks, such as biotite-cordierite-almandine gneisses, and are overprinted by strong schistosity. Two periods of upright folding followed: a synform with a north-striking axial plane and axial-planar fracture cleavage and axial lineation was cross folded by a synform with a west–northwest-striking axial surface and sporadic crenulation cleavage, which caused the older axial surface to become strongly curved.

Discussion and synthesis. Tonalitic and trondhjemitic rocks constitute most of the pre-3.0-Ga crust in Swaziland. It has been suggested that some of the liquids from which these rocks crystallized could have been derived by partial melting of amphibolite or eclogite with hornblende or garnet as residual phases. The environment in which this process may have operated is uncertain but the geochemical data apparently preclude direct analogy between Archean trondhjemite–tonalite suites and Cenozoic island arcs or Andean continental margins. As in similar gneiss terranes elsewhere, the siliceous members of the bimodal gneisses are distinguished by their low initial $^{87}Sr/^{86}Sr$ ratios which, together with their low $\delta^{18}O$ values, suggest that these rocks were derived from the mantle by igneous and metamorphic processes. It is concluded that no older sialic crust existed in Swaziland prior to the formation of the bimodal suite, which is considered on structural grounds to represent the oldest unit of the Ancient Gneiss Complex.

The preferred model for evolution before 3.0 Ga envisages an early stage involving the development of a highly metastable, hydrous lithosphere. Local zones of higher heat flow could have initiated partial melting of mantle material and generated basaltic liquids. As the stability of the early lithosphere and crust gradually increased, relatively thick piles of the crystallization products of these liquids accumulated. Metamorphism of the lower parts of the basaltic pile started at relatively shallow depths because of the steep Archean geothermal gradient and the abundance of water. Depression of amphibolites at the base of the pile resulted in partial melting and generation of tonalitic liquids, which rose and were extruded as flows or intruded into the upper basalts. In this manner a bimodal sequence was generated and was subsequently metamorphosed and deformed to become the bimodal gneiss suite.

The Tsawela tonalite gneiss is now intensely deformed, but reconstruction of its predeformational geometry suggests that it had the form of a pluton intruded into the bimodal suite. The Tsawela geochemistry implies that it was derived by partial melting of metabasalt or quartz eclogite. Metabasalts are preserved at Dwalile and similar rocks are intruded by the Tsawela tonalite gneiss near the southwestern border of Swaziland. Although they are likely parental rocks from which tonalitic magma could have been generated, it is necessary to assume that they were originally far more abundant than their present restricted extent.

The stratigraphic position of the Mkhondo Valley Metamorphic Suite is uncertain; the trace-element composition of its quartzofeldspathic component suggests that it could have formed from liquids generated by partial melting of the low-Al_2O_3 gneisses in the bimodal suite. The requirement that hornblende was absent from the residuum implies anhydrous conditions. After metamorphism to granulite facies, these gneisses would have consisted largely of feldspar, pyroxene, and quartz; formation of garnet was precluded by their low-Al_2O_3 content. The Mkhondo Valley Suite may have accumulated on the older units of the Ancient Gneiss Complex while greenstone volcanism was active in the adjacent region, but this cannot be confirmed because age data on the suite are lacking. Alter-

Table 2–6. Provisional Sequence of Events in the Evolution of the Early Granitoid Terrane and Mkhondo Valley Metamorphic Suite in Swaziland[a]

Episode	Event	Geologic Process
A. MANKAYANE AREA		
I	1	Partial melting of mantle generates basaltic liquids.
	2	Accumulation of relatively thick basaltic pile, the base of which is metamorphosed to amphibolite and partially melted to yield tonalitic liquids.
	3	Tonalitic liquids intruded or extruded to give bimodal association of basalt and tonalite.
II	4	Deformation (D1); folding (F1), and metamorphism of tonalite–basalt sequence; minor partial melting to generate older quartz–feldspar veins (?).
	5	Extrusion of Dwalile basic and ultrabasic lavas; minor clastic and chemical sedimentation.
III	6	Intrusion of biotite–hornblende tonalite pluton (Tsawela gneiss) at ~3.3 Ga.
	7	Crustal dilation; intrusion of older basic dikes.
	8	Intrusion of Mponono Intrusive Suite.
	9	Deformation (D2); intense F2 folding and boudinage of older basic dikes.
IV	10	Crustal dilation; intrusion of younger basic dikes.
	11	Deformation (D3); development of local ductile shear zones.
V	12	Deformation (D4); widespread inhomogeneous strains; tight to isoclinal folding (F4); high homogeneous strain superimposed to give finite flattening (east–central part of area) and finite constriction (northwest part of area).
	13	Development of ductile shear zones (D5) with displacement upward toward northwest; imbricate thrusting.
	14	Partial melting of deeper parts of bimodal suite generates quartz monzonitic liquids at ~3.2 Ga.
		Early Archean evolution ends with multiphase intrusion of Lochiel quartz monzonite at ~3.0 Ga; local mylonitic refoliation of Ancient Gneiss Complex accompanies intrusion.
B. MKHONDO VALLEY		
I	1	Deposition of clastic sediments and iron-formation on older crust of earlier parts of the Ancient Gneiss Complex (?) accompanied by extrusion of basalts. Quartz-monzonite gneisses derived from rhyodacitic magma, possibly partial melt of low-Al_2O_3 gneisses of bimodal suite.
	2	Deformation (D1), metamorphism and partial melting of pelitic units; XY bulk strain planes approximately parallel to stratification, resulting in heterogeneous deformation as consequence of contrasting ductilities of supracrustal rocks.
II	3	Upright folding (F2) produces synform with north-striking axial plane; metamorphism and partial melts intruded along cleavage planes.
	4	Upright folding (F3) deforms F2 synform about west-northwesterly striking axial planes.

[a]Note: No equivalence between episodes in the two areas is implied; the timing of events in Mkhondo Valley is not known but predates ~2.6 Ga.

natively the metamorphic suite may have been deposited and metamorphosed much later but prior to about 2.6 Ga. Certain members have compositional similarities to the post-3.0-Ga Pongola Supergroup, despite their much higher metamorphic grade. The distinctive deformation style of the Mkhondo Valley Metamorphic Suite shows characteristics intermediate between the Ancient Gneiss Complex in the Mankayane area and the Pongola Supergroup. For example, the high flattening strains on flat-lying XY planes and upper amphibolite metamorphic grade are characteristic of the Mankayane area. Conversely, the open upright folding resulting in large periclinal structures and the local survival of primary structure are characteristics of the Pongola Supergroup. A reliable correlation of the Mkhondo Valley Metamorphic Suite with either the Pongola Supergroup or with older supracrustal metamorphic rocks in the Mankayane area will only be possible after detailed mapping of the Mkhondo Valley and comprehensive geochronologic study.

Table 2-6 summarizes the sequence of tectonic, metamorphic, and other events during the evolution of the early sialic crust in the Mankayane area. Figure 2-4 attempts to represent this sequence in a series of schematic cross sections that relates the different deformational styles in the Ancient Gneiss Complex to progressively shallower crustal levels.

The presence of rocks with K/Na ratios close to or greater than unity in this gneiss terrane has implications for the growth of early continental crust. The evidence from Swaziland suggests that it is not necessary to infer or to require the prior existence of a sedimentary cycle of deposition to account for rocks with these K/Na ratios. Igneous and metamorphic processes could have been solely responsible for the early development of continental crust.

Table 2–7. Mean Composition of Gneisses from the Johannesburg Dome and North of Nelspruit[a]

	1	2	3
SiO_2	62.28	67.57	68.53
TiO_2	0.53	0.37	0.31
Al_2O_3	15.91	16.45	17.70
Fe_2O_3	1.44	0.79	2.92[b]
FeO	3.40	2.06	—
MnO	0.10	0.06	0.06
MgO	3.50	1.31	1.81
CaO	5.76	3.62	2.56
Na_2O	3.83	5.04	4.36
K_2O	1.57	1.44	1.54
H_2O^+	1.63	0.95	1.23
H_2O^-	0.07	0.06	—
P_2O_5	0.27	0.10	0.01
CO_2	0.14	0.07	—
Rb	136	112	—
Sr	845	686	—
K/Rb	96	114	—
K/Na	0.46	0.32	0.40
Rb/Sr	0.17	0.18	—
Na/Na+K	0.69	0.76	0.72

[a] 1, Mesocratic hornblende–biotite tonalitic gneiss, average of 2 analyses, Johannesburg Dome (Anhaeusser, 1973). 2, Leucocratic biotite tonalitic gneisses, average of 4 analyses, Johannesburg Dome (Anhaeusser, 1973). 3, Tonalitic gneisses and migmatites (Robb, 1977), average of 5 analyses, north of Nelspruit.
[b] Total iron as Fe_2O_3.

Johannesburg Dome

Tonalitic gneisses of the Johannesburg Dome (Fig. 2-1) comprise two main varieties (Anhaeusser, 1973): (1) mesocratic biotite tonalitic gneisses and (2) leucocratic biotite tonalite and trondhjemite gneisses. In places the gneisses show layering which defines tight fold closures with axial-plane foliation. The gneisses become more homogeneous locally (Anhaeusser, 1973). Layered gneisses or migmatites underlie nearly the entire northern half of the Johannesburg Dome. The gneisses display polyphase folding, although their foliation strikes systematically southeastward with steep or near-vertical dips.

The gneisses are composed of biotite, quartz, and sodic plagioclase, and some contain minor amounts of potash feldspar, hornblende, and chlorite. More homogeneous and porphyritic phases, which have higher contents of quartz and feldspar, are present.

Table 2-7 lists the average major-element chemistry of these rocks together with Rb and Sr concentrations. These gneisses differ in several respects from the bimodal suite and the gneissic tonalites in Swaziland. As might be expected in gneisses with the SiO_2 content of the Johannesburg Dome, Al_2O_3, MgO, and CaO are slightly higher than in the Swaziland bimodal gneisses. The Johannesburg gneisses have twice the Rb content and thrice the Sr content of the Swaziland rocks and a consequently lower Rb/Sr ratio. An anomalous feature of the Johannesburg gneisses is their very low K/Rb ratios. Anhaeusser (1973) attributed the formation of these layered gneisses to a complex interplay of processes involving stoping and assimilation of preexisting supracrustal greenstones by granitic liquids followed by potash metasomatism. The deformation in the gneisses was attributed to partial mobilization that resulted in "flowage folding." This interpretation therefore differs fundamentally from that proposed for the Swaziland gneisses.

Nelspruit Terrane

Tonalitic gneisses and migmatites in the area north of Nelspruit are found only in areas adjacent to supracrustal remnants (Robb, 1978).

The mean major-element composition of these gneisses (Robb, 1978) is given in Table 2-7, wherein it can be seen that they are chemically similar to the gneisses in the Johannesburg Dome. Robb (1978) also attributed the formation of the tonalitic gneisses and migmatites to "interaction between invading tonalitic material and pre-existing greenstone (mafic) minerals."

2.2. Swaziland Supergroup: A Uniquely Preserved Early Archean Supracrustal Pile

Relicts of volcanosedimentary sequences metamorphosed at greenschist facies are preserved in six principal greenstone belts in the Kaapvaal Province (Fig. 2-1). The Barberton belt is not only the largest of these but is also the best preserved and exposed in the province. Each belt contains volcanic or sedimentary units that are lithologically similar to those at Barberton and it has been general practice to correlate the volcanosedimentary rocks of each belt with the Swaziland Supergroup for which the type area is the Barberton belt. The lithologic similarities between belts has led to the suggestion that the present-day distribution of the belts represents either the preserved synformal keels of an originally more widely distributed rock sequence or the remnants of discrete depositories in which there was contemporaneous accumulation of volcanic and sedimentary rocks. However, there are insufficient geochronologic data to confirm whether these belts were contemporaneous with the Barberton succession. Isotopic ages from the Barberton belt confirm that it exceeds 3.0 Ga in age and that the earliest volcanism occurred there at approximately 3.5 Ga. A minimum age of about 3.0 Ga has been tentatively proposed for the Murchison belt (Saager and Köppel, 1976) but no geochronologic data are available from the other belts. No specific stratigraphic terminology has been proposed for the Sutherland, Amalia, and Muldersdrif belts, mainly because of their poor outcrop, but the term Pietersburg sequence has been introduced in the southwestern part of the Pietersburg belt (Grobler, 1972). These uncertainties regarding relative age and stratigraphic detail place a constraint on the precise definition of the Swaziland Supergroup, a name that should provisionally be restricted to the Barberton belt until more information becomes available.

The Swaziland Supergroup in the Barberton belt is divided as follows, showing present-day thicknesses:

(3) Moodies Group (3500 m thick)
(2) Fig Tree Group (2000 m thick)
(1) Onverwacht Group
(b) Geluk Subgroup
  (iii) Swartkoppie Formation (920 m thick)
  (ii) Kromberg Formation (1920 m thick)
  (i) Hooggenoeg Formation (4850 m thick)
(a) Tjakastad Subgroup
  (iii) Komati Formation (3500 m thick)
  (ii) Theespruit Formation (1980 m thick)
  (i) Sandspruit Formation (3200 m thick)

The Tjakastad Subgroup is preserved around the periphery of the Barberton belt with the Geluk Subgroup being confined to the southern and central parts (Fig. 2-5). The sedimentary Fig Tree and Moodies Groups are preserved in the synclinal core of the belt.

Onverwacht Group: Primordial Ocean-Type Crust

Field relations and stratigraphy. The subdivision of the Onverwacht Group was based on the distinctive assemblage and association of rock types recognized to be characteristic of each formation (Viljoen and Viljoen, 1970). The main lithologic units constituting the six formations are summarized in Table 2-8. The three lower formations are distinguished by the abundance of ultrabasic and basic rocks which, in the Komati Formation in particular, have a chemistry that distinguishes them from typical peridotites and picrites. The basaltic volcanics in the Geluk Subgroup are normal tholeiites. The persistent (at least in the south) Middle Marker, which separates these two subgroups, was interpreted to represent a major time break when fundamental changes, probably related to a thickening of the sialic crust, took place in the mantle (R.P. Viljoen

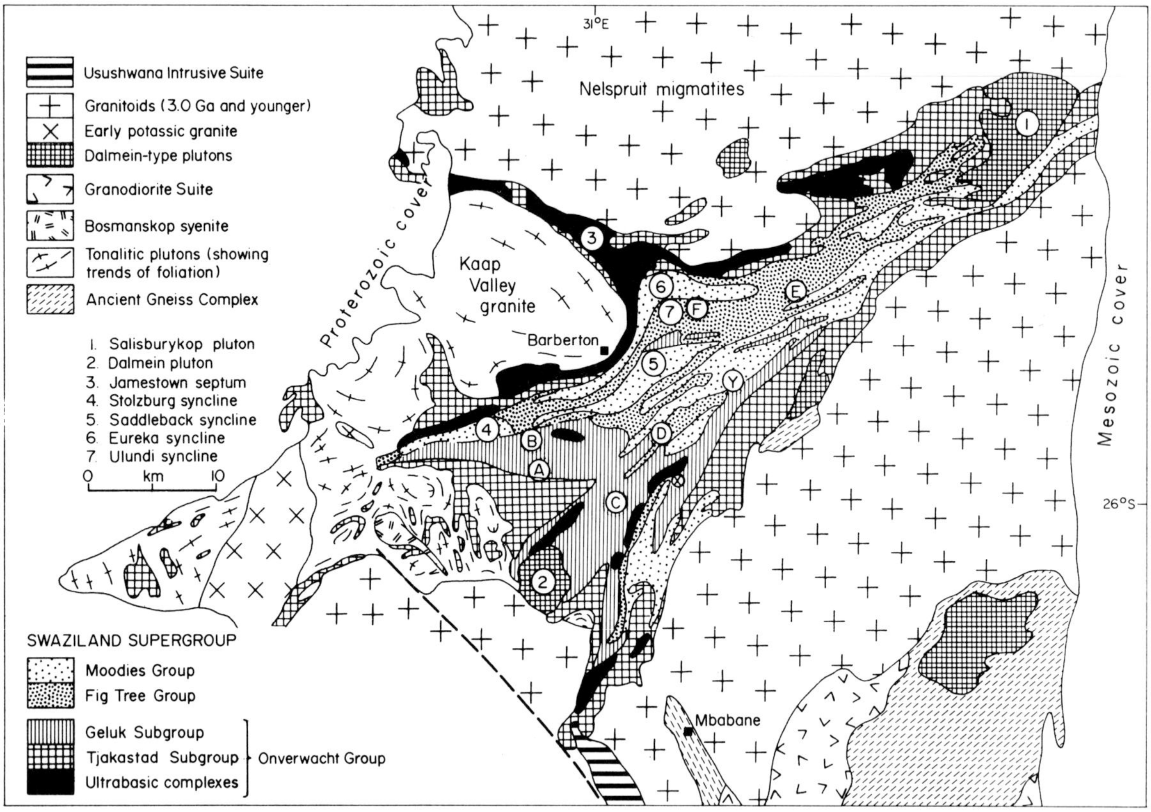

Figure 2-5. General geology of the Swaziland Supergroup. X and Y are localities of Moodies Group stratigraphic columns (see Fig. 2-8). A–F refer to localities of Onverwacht sedimentary units (see text and Fig. 2-7). [Adapted from Hunter (1968), Viljoen and Viljoen (1970), Anhaeusser and Robb (1978).]

and Viljoen, 1969). This interpretation has been questioned by Williams and Furnell (1979), who recognized in the Hooggenoeg Formation a thick (1000 m) sequence of basaltic komatiites texturally and chemically identical to those in the Komati Formation. This unit has a minimum strike length of 15 km and these authors noted that basaltic komatiites are reported by R.P. Viljoen and Viljoen (1969) in a similar stratigraphic position in the Hooggenoeg Formation some 5 km southeast of the area studied by Williams and Furnell (1979). Basaltic komatiites are also found in the Kromberg Formation (R.P. Viljoen and Viljoen, 1969). This broad-scale alternation of komatite and tholeiite is regarded as being typical of greenstone belts elsewhere and demonstrates that the Middle Marker does not represent a major crustal or mantle adjustment (Williams and Furnell, 1979).

Williams and Furnell (1979) also concluded that there was a random and complex interfingering of basaltic komatiite, dacite, rhyodacite, tholeiite, and basaltic andesite in the Hooggenoeg Formation instead of repetitions of a series of basalt–dacite–chert cycles, the intercalated cherts being more continuous along strike than the lavas. These authors therefore regarded the Hooggenoeg Formation as resulting from a complex interplay of overlapping, compositionally different, volcanic products. Waxing and waning of activity at different volcanic centers seem to them to be the cause of the apparent cyclicity.

Acid lavas, feldspar and quartz porphyries, and acid pyroclastic rocks constitute the top of the Hooggenoeg Formation and pass along strike into water-worked acid aluminous tuffs with agglomeratic interlayers (Viljoen and Viljoen, 1970). The contact of this unit with the underlying basalts is described as irregular and transgressive by Williams and Furnell (1979) who note the presence of undeformed conglomerate containing dacite pebbles and resting on sheared and carbonated basalts and dacites. This interpretation implies the existence of an unconformity or tec-

Table 2–8. Summary of Main Lithologies in the Onverwacht Group[a]

Formation	Lithology	Comments
Swartkoppie	Acid, intermediate, and basic schists; banded black and white chert; minor pyroclastic rocks: main rock types in center of belt banded, black, white, gray, and green cherts: sericitic and talc-carbonate schists.	Strongly sheared in south; conformably overlain by Fig Tree Group but relationship to Kromberg Formation not known.
Kromberg	Basalt with thin intercalations of pillowed basalt; acid lava associated with chert, calc–silicate, and carbonate layers; palagonitic tuff, cross-bedded basic tuff, agglomerate; black, carbonaceous siliceous shaly layers in some chert units.	Best exposed and developed south of Barberton belt; basaltic flows dominantly massive; some pillow breccia.
Hooggenoeg	Base marked by thin, persistent chert and carbonate sediment (the Middle Marker); cycles consisting of basalt, andesite, dacite, rhyodacite, and banded chert; minor ultrabasic units overlie some chert layers; upper part of formation composed of thick zone of acid lava, clastic sediments, and quartz and feldspar porphyries.	Metamorphosed at lower greenschist facies; pillow structures in both basalt and dacite; amygdales, spherulites, and variolites.
Komati	Alternation of amphibolitized, pillowed, and massive basalt and ultrabasic layers; ultrabasic rocks serpentinized with olivine kernels; basic rocks: amphibolite composed of variable amounts of tremolite-actinolite, cummingtonite, anthophyllite, talc, chlorite.	Ultrabasic rocks about 30% of formation; spinifex textures and less abundant pillow structures in both basic and ultrabasic rocks.
Theespruit	Ultrabasic rocks: serpentine, talc–chlorite schist; basic rocks: actinolitic, tremolitic, chloritic, carbonate-bearing schist; acid rocks: sericitic and pyrophyllitic schist, siliceous schist, intercalations of dark chert, agglomerate.	Formation widely developed throughout Barberton belt; pillow structures preserved but not abundant; formation characterized by acidic rocks interpreted to be tuff.
Sandspruit	Ultrabasic rocks: serpentinite, antigorite–chlorite schist, tremolite–chlorite schist; basic rocks: amphibolite.	Preserved as large rafts or screens between intrusive trondhjemite–tonalite plutons; ultrabasic rocks constitute ~70% of formation.

[a] Adapted from Viljoen and Viljoen (1970).

tonic slide in the upper part of the Hooggenoeg Formation. As a result Williams and Furnell (1979) questioned the "layer-cake" stratigraphic model of Viljoen and Viljoen (1970) and suggested the possibility of greater complexity in the sequence of volcanic events in the Barberton belt than had been previously recognized. Confirmation of these conclusions (Williams and Furnell, 1979) is clearly necessary if the geology of the Onverwacht Group is to be properly understood.

Where spinifex textures are present it is possible to estimate the thickness of individual flows. Peridotite units 200 m thick consist of a pile of spinifex-capped flows, which vary in thickness from 2 to >20 m. Well-defined basaltic flow units, commonly 2–10 m thick but locally up to 30 m thick, form zones up to 700 m thick. In places the spinifex-textured basaltic flows are interlayered with fine-grained pillowed basalts or with more-massive variolitic basalts.

In addition to flows, layered sills and dikes are found in the Tjakastad Subgroup and the Hooggenoeg Formation (M.J. Viljoen and Viljoen, 1969a; Williams and Furnell, 1979). The Komati Formation, for example, contains thin (<5 m thick) but persistant sills of peridotite with coarse poikilitic pyroxenes. Two-pyroxene pyroxenites and norites are found at the top of some of the peridotite sills (Williams and Furnell, 1979). Peridotitic dikes appear to be associated with these layers of poikilitic peridotite.

Layered sills up to 200 m thick intrude all units of the Hooggenoeg Formation (R.P. Viljoen and Viljoen, 1969; Williams and Furnell, 1979). The Rosentuin sill in the southern part of the Barberton belt has been traced for 11 km along strike and consists of a lower peridotitic layer, 75 m thick, composed of serpentinized dunite, wehrlite, or lherzolite. The upper zone consists of orthopyroxene crystals set in a matrix of clinopyroxene, the latter constituting about 70 percent of the rock (R.P. Viljoen and Viljoen, 1969). A 200-m-thick sill near the top of the Hooggenoeg Formation consists of the sequence: (1) olivine–clinopyroxene cumulate, (2) clinopyroxene cumulate with poikilitic plagioclase, (3) clinopyroxene–plagioclase (–orthopyroxene) cumulate with intercumulus quartz, and (4) a coarser grained variety of layer 3 with abundant iron oxides (Williams and Furnell, 1979).

In addition to these layered sills, M.J. Viljoen and Viljoen (1969a) recognized a number of larger, well-layered intrusions that are closely associated in time and space with the largely extrusive rocks of the Tjakastad Subgroup (see Fig. 2-5). These intrusions are most prominent adjacent to the northwestern margin of the Barberton greenstone belt. Three main types have been recognized (M.J. Viljoen and Viljoen, 1969a), namely: (1) Kaapmuiden type, (2) Noordkaap type, and (3) Stolzburg type.

The Kaapmuiden type is represented by three layered ultrabasic sills each over 600 m thick that crop out in the northeastern portion of the belt and are intrusive into the Komati Formation with which the presently near-vertical sills have been deformed and metamorphosed.

Each sill consists of the following sequence of rock types (Viljoen and Viljoen, 1970):

	Maximum thickness (m) in each sill
7. Late-phase pegmatoids	
6. Upper dunite–peridotite zone	365–616
5. Anorthositic gabbro-norite zone	12–36
4. Websterite zone	18–24
3. Orthopyroxenite zone	52–137
2. Main dunite–peridotite zone	166–381
1. Peridotitic chill zone	12–36

The dunite and peridotite zones are extensively altered to serpentinite and talc. The bulk chemical composition of these intrusions is almost identical to the average composition of peridotitic komatiite flows in the Komati Formation (M.J. Viljoen and Viljoen, 1969a).

The Noordkaap type is represented by two main intrusions north of Barberton and a number of smaller bodies in greenstone septa between the Kaap Valley "granite" and the Nelspruit migmatite terrane. Anhaeusser (1969) recognized a number of cyclic differentiation units that commence with peridotites, which are succeeded by layers of pyroxenite and gabbro, the latter being locally anorthositic. Not all cyclic units contain the full range of rock types. The primary mineral phases are rarely preserved because of extensive alteration to serpentine or amphibole.

Intrusions of the Stolzburg type are found at three localities (Fig. 2-5), the type area being the Stolzburg intrusion at the southwestern end of the greenstone belt. The Kalkloof body has intruded a remnant of lower Onverwacht volcanic rocks west of the main outcrop of the Barberton belt. Several bodies of this type have been recognized at the western end of the greenstone septa that contain the Noordkaap-type intrusions. The fault-bounded Stolzburg intrusion exceeds 16 km in length and has a mean outcrop width of 1 km. Anhaeusser (1978a) has subdivided the steeply dipping layered sequence into a lower division, consisting of an alternation of serpentinized dunite and orthopyroxenite and an upper division of serpentinized harzburgites and peridotites. Layers of pyroxenite (websterite), gabbro or norite, and, locally, gabbroic to noritic anorthosites are associated with the peridotite layers, which include lherzolites and wehrlites. Calcium metasomatism has occurred along the boundary between monomineralic cumulates of the lower division and layers of the upper division containing two or more cumulus minerals (Anhaeusser, 1978a). Here rodingite dikes cut serpentinized dunite or form irregular pods or replacement bodies which grade into gabbro. The rodingite dikes vary in size from veinlets a few centimeters thick to dikes up to 5 m wide. The rodingite contains a complex assemblage of garnet, diopside, epidote, zoisite, nephrite, prehnite, plagioclase, magnetite, and sphere. Anhaeusser concluded that Ca-metasomatism played a dominant role forming the rodingite.

Intrusive quartz and feldspar porphyries and keratophyric dikes and sill-like bodies are abundant in the Komati, Hooggenoeg, Kromberg, and Swartkoppie Formations. Those in the Komati Formation are roughly conformable, varying in length from a few meters to 2000 m, and are generally elongate but irregular (M.J. Viljoen and Viljoen, 1969a). Albitic feldspar porphyries are the most common variety in the Komati Formation and are usually altered, with sericitization of feldspar, corrosion of quartz grains, and abundant carbonate.

The porphyries in the upper formations reflect at least two generations. Sodic feldspar porphyries intimately associated spatially with the acid flows in the Hooggenoeg Formation are considered to be genetically related to these flows (R.P. Viljoen and Viljoen, 1969). Other microcrystalline quartz porphyries and coarser grained feldspar porphyries are prominent adjacent to a major fault near the Transvaal–Swaziland border in the southern part of the Barberton belt (R.P. Viljoen and Viljoen, 1969). These bodies range in size from a few hundred to over 1000 m in diameter. It is suggested that their emplacement is associated with the major faulting of the area (R.P. Viljoen and Viljoen, 1969).

Geochemistry and petrogenesis. Interpretation of the chemical data of the volcanic sequences is constrained by the mobility of major and trace elements during alteration, low-grade metamorphism, and reaction with sea water. The presence of high-magnesium rocks in the Onverwacht Group and the problems of their genesis have resulted in the komatiites receiving more detailed study.

Whereas the Tjakastad Subgroup consists predominantly of ultrabasic and basic lavas with minor acid tuffs and porphyries, the Geluk Subgroup is a calc–alkaline suite (Anhaeusser, 1978b). McIver and Lenthall (1973) noted the presence of nepheline-normative basalts in the Hooggenoeg Formation based on two analyses given by R.P. Viljoen and Viljoen (1969). It is not certain whether this normative mineralogy is a reflection of the pristine chemistry of these rocks or is a consequence of alteration or metasomatism.

The Tjakastad Subgroup is characterized by chemically distinct ultrabasic and basic volcanic komatiites with CaO/Al_2O_3 ratios exceeding unity, high contents of MgO (>9 percent), and a very low content of alkalis (M.J. Viljoen and Viljoen, 1969a). The high CaO/Al_2O_3 ratio (>1.0) has been shown to be peculiar to the Barberton rocks; basaltic komatiites from Mt. Monger, Western Australia and Munro Township, Canada have ratios of 0.76 and 0.78 (Williams, 1972; Arndt *et al.*, 1977).

Basaltic komatiites have lower contents of Al_2O_3, FeO (t), and TiO_2 and higher contents of SiO_2, MgO, and Cr than basalts (Table 2-9). The analyses of the basaltic andesites (Williams and Furnell, 1979) show higher contents of TiO_2 and total alkalis and lower contents of MgO and CaO than average basalts. Williams and Furnell (1979) noted that one of their samples of basaltic andesite, intercalated with Hooggenoeg basaltic komatiites, contains an unusual abundance of apatite for which no explanation is offered.

Hooggenoeg basaltic komatiites have a similar chemistry to those of the "Barberton type" in the Komati Formation (Table 2-10) except for higher FeO (t) and lower CaO and TiO_2 contents in the Hooggenoeg samples. The CaO/Al_2O_3 ratio is therefore decreased to a level comparable to those in similar rocks from Canada and Western Australia, which nevertheless have higher contents of Al_2O_3 compared to both Komati and Hooggenoeg basaltic komatiites.

The mean major-element composition of the metabasalts from the Onverwacht Group is similar to that of the Western Australian examples, but the iron-rich tholeiites from Munro Township contain appreciably higher concentrations of TiO_2 than the Barberton and Western Australian tholeiitic basalts (Table 2-10).

Table 2–9. Mean Compositions and Estimated Bulk Composition of Volcanic Rocks in the Barberton Greenstone Belt and in the Superior Province, Canada.[a]

	1	2	3	4	5	6	7
SiO_2	46.97	52.28	51.85	53.37	67.23	52.14	56.9
TiO_2	0.41	0.57	0.99	1.48	1.05	0.92	0.93
Al_2O_3	3.33	7.38	14.06	14.88	14.54	12.13	15.4
FeO (t)	11.35	10.78	12.23	12.45	5.94	9.04	9.42
MnO	0.20	0.32	0.21	0.19	0.08	0.22	0.18
MgO	31.69	15.88	6.84	4.21	2.71	10.21	5.12
CaO	5.67	10.13	9.96	5.51	2.30	8.91	7.14
Na_2O	0.31	1.43	2.91	3.62	2.50	2.48	3.09
K_2O	0.06	0.16	0.32	2.36	2.65	0.53	0.87
P_2O_5	0.02	0.05	0.06	1.10	0.15	0.14	0.20
Rb	1.7	4.5	7	34	62	10	—
Sr	18	87	166	1018	374	220	214
Rb/Sr	0.09	0.05	0.04	0.03	0.17	0.05	—
K/Rb	293	295	320	576	355	337	—
Cr	2545	1063	243	25	—	—	220
Ni	1225	405	164	85	—	—	131
Co	93	50	50	35	—	—	32
No. of analyses	12	18	15	2	7	54	1629

[a] 1, Peridotitic komatiite mean. 2, Basaltic komatiite mean. 3, Metabasalt mean. 4, Basaltic andesite mean. 5, Intermediate and acid volcanic mean. 6, Weighted-average composition of Onverwacht volcanic rocks (ultrabasic, 6.6; basic, 19.5; acid, 1; ratios from Anhaeusser, 1976a). 7, Weighted-average composition of volcanic rocks, Superior Province, Canada (Goodwin, 1977). Sources: M.J. Viljoen and Viljoen (1969a), R.P. Viljoen and Viljoen (1969), Williams and Furnell (1979), Herrman *et al.* (1976), Glikson (1976), and Hunter *et al.* (1978).

Table 2–10. Comparison of Major-Element Chemistry of Basaltic Komatiites and Basalts from Barberton, Western Australia, and Canada

	Basaltic Komatiites[a]				Tholeiitic Basalts[b]		
	1	2	3	4	5	6	7
SiO_2	52.55	53.15	52.61	51.6	51.85	51.93	51.1
TiO_2	0.45	0.79	0.64	0.65	0.99	0.97	1.70
Al_2O_3	10.27	10.14	12.67	13.3	14.06	14.98	13.3
FeO (t)	13.35	11.27	10.40	11.7	12.23	10.63	15.0
MnO	0.23	0.24	0.17	0.19	0.21	—	0.28
MgO	10.95	10.89	11.22	10.0	6.84	6.78	5.92
CaO	8.67	10.74	9.61	10.4	9.96	10.93	8.6
Na_2O	2.70	1.90	2.53	2.16	2.91	2.73	3.9
K_2O	0.17	0.30	0.16	0.11	0.32	0.18	0.25
P_2O_5	0.03	0.06	0.10	—	0.06	0.12	—
CaO/Al_2O_3	0.84	1.06	0.76	0.78	0.71	0.73	0.65

[a] 1, Hooggenoeg Formation, mean of 4 analyses (Williams and Furnell, 1979). 2, Komati Formation, mean of 6 analyses (M.J. Viljoen and Viljoen, 1969b). 3, Mt. Monger, Western Australia, mean of 5 analyses (Williams, 1972). 4, Munro Township, Canada (Arndt *et al.*, 1977).

[b] Onverwacht Group, mean of 15 analyses from Table 2–9, column 3. 6, Western Australia, mean of 337 analyses (Hallberg and Williams, 1972). 7, Munro Township, Canada (Arndt *et al.*, 1977).

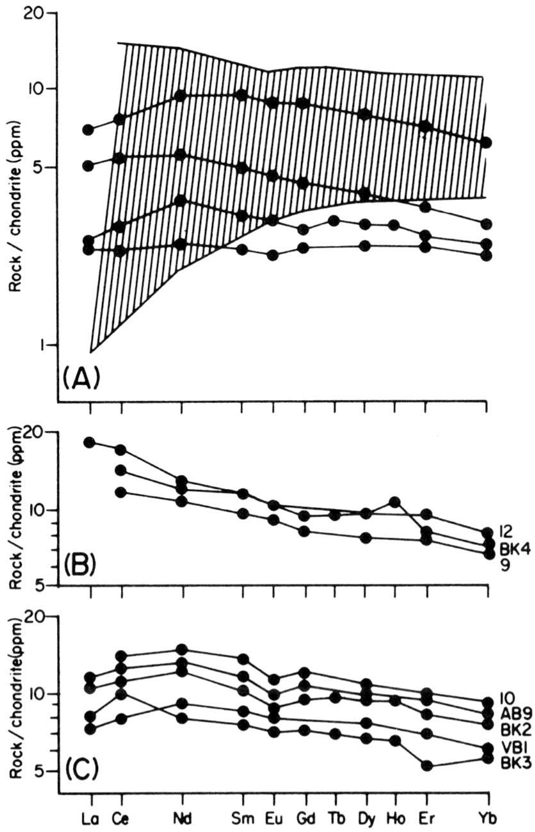

Figure 2-6. Chondrite-normalized rare-earth plots of Onverwacht Group volcanic rocks. (A) Peridotitic komatiites: shaded area is field of Munro Township komatiites. (B) Basaltic komatiites: BK4, Barberton type. (C) Basaltic komatiites: AB9, Badplaas type; VB1, BK2, BK3, Barberton type. Samples 9, 10, 12 are Komati amphibolites; chemistry suggests sample 9 may be of Badplaas type and samples 10 and 12 of Barberton type. [Sources: Hermann *et al.* (1976), Hawkesworth and O'Nions (1977), Sun and Nesbitt (1978).]

Data on the chemistry of the intermediate and acid rocks are very incomplete due to carbonation and silicification (R.P. Viljoen and Viljoen, 1969). Dacitic, rhyodacitic, and keratophyric varieties have been recognized. Unusual alkali contents are revealed by some analyses, with seemingly normal values for other major elements. Rare earth element (REE) data are sparse. One sample from the Hooggenoeg Formation has a generally uniform slope from La at 37 times that of chondrite to Lu at 2.6 times that of chondrite (Hunter *et al.*, 1978). Analyses of albite porphyries at the level of the Middle Marker have steep REE patterns, with La at about 100 times that of chondrite and Yb at about the chondrite mean (Glikson, 1976). Porphyries from the Komati and Theespruit Formations have flatter slopes for heavy REEs than those from the Middle Marker, with Yb ranging between two and five times that of chondrites (Glikson, 1976). A possible cogenetic relationship between the albite porphyries and the post-Swaziland trondhjemitic–tonalitic plutons has been suggested by Glikson (1976) but he cautions that further studies are needed.

Rare earth element data for the ultrabasic and basic volcanic rocks are not extensive but have been reported by Herrmann *et al.* (1976), Hawkesworth and O'Nions (1977), and Sun and Nesbitt (1978). The plots of peridotitic komatiites (Fig. 2-6A) show samples with strong depletion in heavy REEs and convex upward curves suggestive of a source slightly depleted in REEs, as well as flatter curves, one of which has a weak peak at Sm-Nd. The depletion in heavy REEs in the 3.5-Ga Onverwacht peridotites contrasts with the flat patterns of heavy REEs in Munro Township komatiites (Arth *et al.*, 1977) (Fig. 2-6A) and with the lack of heavy REE depletion in komatiitic peridotites from 2.7-Ga greenstone (Sun and Nesbitt, 1978). This apparent difference in REE patterns between peridotitic komatiites of different ages has not yet been shown to be consistent but may suggest an important evolutionary trend in the genesis of these magmas.

M.J. Viljoen and Viljoen (1969a) distinguished Badplaas- and Barberton-type basaltic komatiites on the following basis:

Badplaas type: relatively high silica ($\bar{x}$ = 52% SiO_2), magnesium ($\bar{x}$ = 15.25% MgO), and calcium ($\bar{x}$ = 12.8% CaO) contents; low alumina ($\bar{x}$ = 5.4% Al_2O_3) and alkali ($\bar{x}$ = 1.3% $Na_2O + K_2O$) contents.

Barberton type: approximately equal contents of calcium, aluminum, and magnesium (about 10 percent for each oxide); high silica ($\bar{x}$ = 52.7% SiO_2) and alkali ($\bar{x}$ = 3.1% $Na_2O + K_2O$) contents.

The REE patterns do not discriminate between these two types (Figs. 2-6B and 2-6C). Five samples (BK2, BK3, VB1, AB9, and 10) have patterns showing a slight to moderate convex upward curve for the light REEs with a small negative Eu anomaly and some depletion of the heavy REEs. In contrast, the three remaining samples show an enrichment in the light REEs. The contrast in REE patterns is best demonstrated by reference to the La_n/Sm_n ratio. The value for this ratio for the first group is close to unity, whereas it is about 1.5 for the second group of samples.

High CaO/Al_2O_3 ratios in the komatiites of the Barberton belt have been ascribed to garnet fractionation during diapiric uprise prior to segregation of magma (Green, 1975). Several other explanations have been offered, including: (1) polybaric olivine and orthopyroxene fractionation (McIver and Lenthall, 1974); (2) polybaric assimilation of a crystalline fraction into an initial Mg-rich liquid (Bickle *et al.*, 1975; Cox, 1978); (3) multistage melting, the first-formed liquids being tholeiitic in composition, followed by further melting of the residual mantle to yield komatiitic liquids (Arth *et al.*, 1977); (4) derivation from ultrabasic magmas by low-pressure fractional crystallization (Whitford and Arndt, 1978); and (5) partial melting of heterogeneous mantle under differing pressure and temperature (P-T) conditions (Sun and Nesbitt, 1978).

The trace-element characteristics of the komatiites from the Barberton belt show a close similarity to abyssal and island-arc basic and ultrabasic rocks (Herrmann *et al.*, 1976). These authors regard the high nickel contents in their samples as consistent with a partial melting model and rendering a fractional crystallization model unlikely. Their data do not permit exact modeling of the genetic relations between the komatiites and basalts but Herrmann *et al.* (1976) suggest that the basalts may correspond to 8–19 percent partial melting compared with 60–80 percent to produce peridotitic liquids.

Sedimentation in a volcanic environment. Cherts consisting of green, black, white, and gray layers are prominent in the Geluk Subgroup and have been interpreted as the terminal phase of the volcanic cycles recognized by Viljoen and Viljoen (1970). Subsequent sedimentological studies of cherts provide critical information on the evolution of the Early Archean crust with the recognition of three main sediment types (Lowe and Knauth, 1977):

(1) Terrigenous gravel, sand, and silt altered to a chert–sericite mosaic
(2) Coarse-grained dolomite, and possibly fine-grained carbonate, now altered to homogeneous chert beds
(3) Primary nonterrigenous silica that accounts for only a minor proportion of the sequence

One of the most persistent sedimentary units in the southern part of the Barberton belt is the Middle Marker, forming the base of the upper subgroup of the Onverwacht and averaging about 10 m thick. A typical section (Fig. 2-5 locality A, and Fig. 2-7A) described by Lowe and Knauth (1977) consists of dark gray, fine-grained chert interlaminated with dark yellow, orange-weathering chert that shows a medium-grained granular texture on weathered surfaces. The finely laminated gray chert contains a high proportion of microcrystalline chert intergrown with microquartz, whereas the orange chert contains irregular grains of hematite and ghosts of silicified primary detrital grains visible in a microquartz matrix. Orange chert layers greater than 1 cm thick are commonly lenticular and cross laminated. The upper part of the Middle Marker consists of a sequence of beds, up to 15 cm thick, of silicified coarser grained detritus (Lowe and Knauth, 1977). Grains composed of a chert–sericite mosaic and disseminated carbonate and hematite are commonly spherical to subspherical, moderately well-sorted, and between 0.5 and 5 mm in diameter. These grains are interpreted as water-worked accretionary lapilli mixed with angular fragments of chert and silicified volcaniclastic sand cemented by microquartz (Lowe and Knauth, 1977, 1978).

Sedimentary units at the top of the Hooggenoeg Formation have been studied at localities B and C (Fig. 2-5). The uppermost 100–200 m of the Hooggenoeg Formation at B consists of alternating white and medium-gray chert units 1–5 cm thick (Lowe and Knauth, 1977). The cherts consist of microquartz with finely disseminated carbonate in the gray chert bands. Impurities, primary sedimentary structures, and detrital grains are absent; but postdepositional features, such as load structures, soft-sediment folds, and breccias are abundant and indicate deposition of soft ooze (Lowe and Knauth, 1977).

At locality C along the Komati River the top of the Hooggenoeg Formation consists of up to 174 m of agglomerate, rhyodacitic tuff, and chert (R.P. Viljoen and Viljoen, 1969). The uppermost 50 m consists of fine-grained silicic rocks that overlie well-bedded tuff, locally showing large-

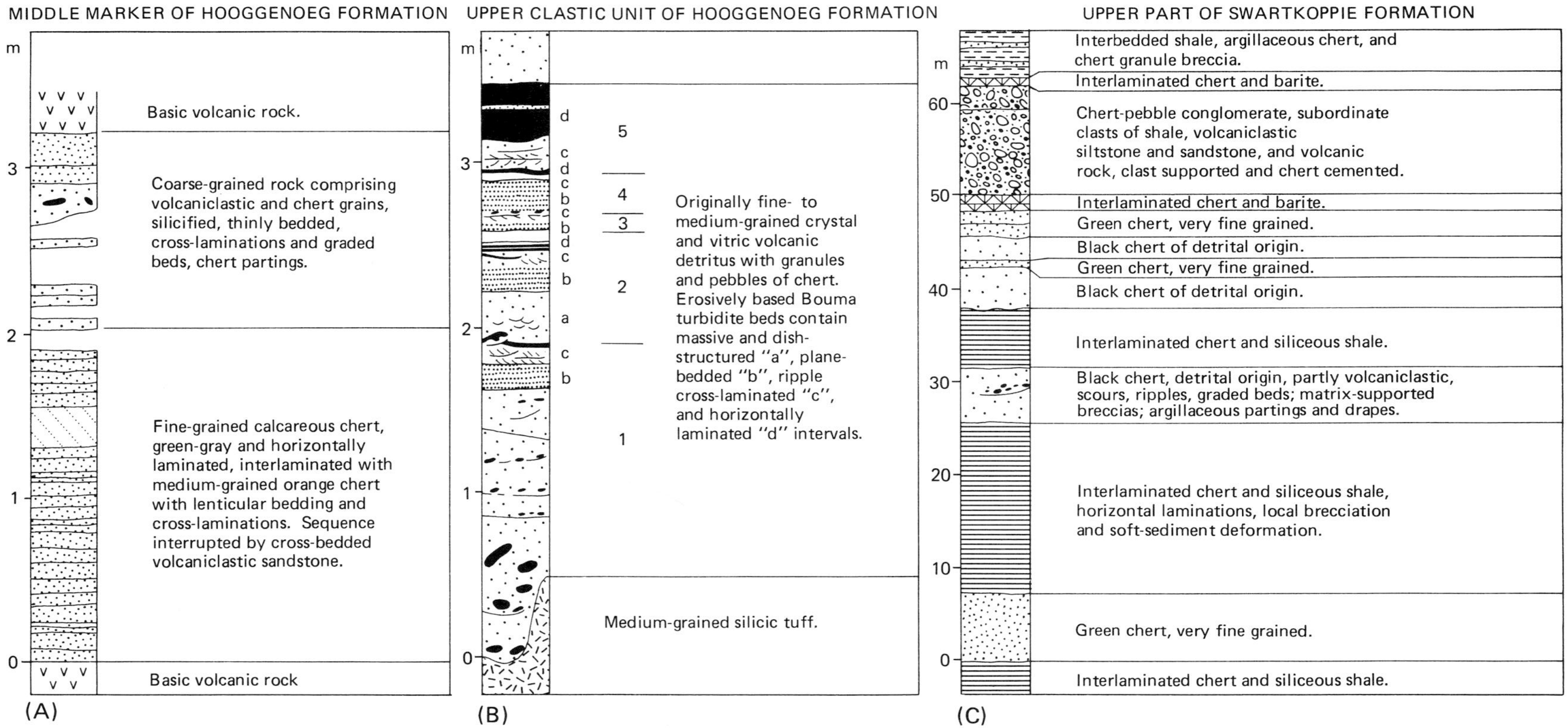

Figure 2-7. Onverwacht stratigraphic sections (localities shown on Fig. 2.5; refer to Table 2-8 for general succession). (A) Middle Marker at locality A; (B) upper part of acid sequence at the top of Hooggenoeg Formation at locality C showing stacked turbidite units (letters a–d indicate Bouma turbidite subdivisions); (C) uppermost part of the Swartkoppie Formation overlying thick cherts at locality D. [After Lowe and Knauth (1977).]

scale cross-bedding. The silicic rocks comprise five cyclic sedimentary units which resemble Bouma sequences (Lowe and Knauth, 1977). Individual sedimentation units 15 cm to 2 m thick show erosive bases, normal size grading, and sedimentary structures similar to those in turbidites (Fig. 2-7B). The cherts appear to represent the finest debris deposited at the end of each sedimentation cycle. Silicification after burial is evidenced by a chert ripup clast showing soft-sediment deformation, which is incorporated in the next overlying sedimentation unit (Lowe and Knauth, 1977). The volcaniclastic grains include alkali feldspar and quartz phenocrysts and altered pumice fragments and shards.

Sedimentary units up to 30 m thick in the Kromberg Formation consist of alternating black carbonaceous chert and brownish dolomite. Sedimentary structures including upper flow regime plane-beds and trough cross-beds are abundant in the dolomite layers and suggest a clastic origin (Lowe and Knauth, 1977). The black cherts appear to have accumulated as very fine-grained detritus (Lowe and Knauth, 1977).

The heterogeneous sedimentary intervals in the Swartkoppie Formation in the southern part of the Barberton belt contrast with mainly massive black chert overlying calcareous phyllites in the north (Lowe and Knauth, 1977). In the south the Swartkoppie Formation contains chert units up to 400 m thick (Fig. 2-5, locality D, and Fig. 2-7C), whereas in the north it is represented by black cherts only 5–50 m thick (Fig. 2-5, localities E and F). The southern outcrops include pebble conglomerates comprising rounded clasts up to 6 cm in diameter, debris-flow deposits, bedded barite, trough cross-bedded dolomite, calcareous chert, and finely interlayered siliceous shale and hematitic chert. The Swartkoppie sections described by Lowe and Knauth underlie thick shale sequences of the Fig Tree Group. The black cherts in the northern outcrops of the Swartkoppie Formation show no evidence of traction current deposition, their argillaceous character indicating the onset of deep-water sedimentation in the north (Lowe and Knauth, 1977).

Although the Hooggenoeg and Kromberg Formations include sedimentary units that resemble turbidites, Lowe and Knauth (1977) argue against a deep-water depositional environment because the detritus was probably of local origin and not transported over long distances. The sediments in the Swartkoppie Formation appear to have accumulated in shallow to deep water. Accumulation in shallow water is implied by the presence of subaerially formed accretionary lapilli, intraformational unconformities, thick deposits of dolomite showing large-scale cross-bedding, and locally derived chert-clast conglomerate in the southern part of the belt. In contrast, the massive carbonaceous chert, devoid of current structures and overlain directly by graywackes and shale of the Fig Tree Group, suggests deposition in quiet, and possibly deep, water in the northern part of the basin.

Lowe and Knauth (1977) concluded that all the detritus in the sedimentary units of the Onverwacht Group was derived from penecontemporaneously erupted pyroclastic material; the source terrane and locus of deposition was exclusively oceanic and isolated from any continental or sialic rocks that may have existed. This is in marked contrast to the situation that prevailed during deposition of the Fig Tree graywackes where, in addition to material of volcanic origin, the presence of abundant K-feldspar grains and granitoid lithic detritus indicates subaerial exposure of a silicic plutonic terrane.

Primitive life forms. The first suggestion that primitive life forms might have existed during the Archean in South Africa was the discovery of "oolitic" microstructures resembling *Calcisphaera* (Ramsay, 1963). Spheroidal and cup-shaped algal-like bodies and filamentous structures associated with amorphous carbonaceous matter were reported by Engel *et al.* (1968) from cherts interlayered with lavas in the Theespruit and Kromberg Formations. Reservations have been expressed about the biogenic origin of these microstructures. Nagy and Nagy (1969) noted that some fall in a wide size range of 6 to 196 μm but that others, lying within a narrow size range, are morphologically more suggestive of fossil organisms. However, Brooks *et al.* (1973) and Muir and Hall (1974) considered the paired spheroids, agglomerations of cells, and filaments in these rocks to be of biogenic origin. The organic material occurs mainly in the form of kerogen and contains aromatic components (Nagy and Nagy, 1969). Brooks *et al.* (1973) suggested that these components are related to sporopollenin of known biogenic origin. Sporopollenin yields aromatic compounds when heated so it is not necessary to propose derivation of these aromatic compounds from a lignin source. Furthermore, the carbon isotopes from the lower Onverwacht rocks are anomalously heavy compared with those from the upper Onverwacht, suggesting that the difference may be related to evolution from nonphotosynthetic to photosynthetic organisms (Oehler *et al.*, 1972).

The problem of establishing whether life forms were present in the Onverwacht rocks is not yet resolved. Their apparently simple morphologies make it difficult to distinguish them from nonbiogenic microstructures. Furthermore, the composition and evolution of kerogen are imperfectly known and different kerogen degradation processes yield different products. Until these uncertainties are resolved, the assertion that primitive life forms existed in these 3.5-Ga rocks remains unproved.

Fig Tree and Moodies Groups: Sedimentation Along the Oldest Continental Margin

Stratigraphic succession. A gradational relationship between the Onverwacht and Fig Tree Groups characterizes all areas except northern Swaziland, where the Onverwacht and Moodies Groups are in unconformable contact. The Fig Tree Group comprises two facies tracts. A southern facies consists of coarse terrigenous clastics, whereas along the northern flank of the mountain land, graywackes and shales with subordinate cherts and iron-formations are present. The Fig Tree Group attains its greatest thickness in the north where a threefold subdivision is recognized (FT1-FT3 in Fig. 2-8). Units FT1 and FT2 are characterized by different graywacke to shale ratios; unit FT3 is largely tuffaceous and, in the Eureka syncline, partly conglomeratic.

The Fig Tree Group generally coarsens upward into the Moodies Group. Where thickest, the Moodies Group is divided into five stratigraphic units termed MD1 through MD5 (Fig. 2-8). A prominent zone of amygdaloidal lavas at the base of unit MD4 facilitates correlation between the Eureka, Saddleback, and Stolzburg synclines. Conglomerate and arkose are present at the base of the Moodies Group and dominate the stratigraphy in the south. Subarkose and quartz arenite become more abundant northward, with shale and iron-formation being important constituents in the Eureka and Stolzburg synclines.

Sediments in the Swaziland Supergroup as a whole display an upward increase in maturity from volcaniclastics in the Onverwacht Group to quartzofeldspathic arenites in the Moodies Group. An upward increase in lithophilic elements in the Fig Tree graywackes and shales also reflects this increasing maturity, which is probably a response to progressive unroofing of a sialic provenance (Condie *et al.*, 1970).

Submarine-fan construction. Various submarine-fan facies in the Fig Tree Group have been interpreted in terms of Walker's (1978) model (Eriksson, 1980a,b). Upper-fan facies in the south comprise an upward-coarsening terrigenous clastic sequence. These sediments gradationally overlie iron-formation and pinch out laterally over 5 km, allowing the Moodies to overlap directly onto the Onverwacht Group.

Arenites at the base of the sequence coarsen upward to a conglomeratic interval at the top. The arenites are composed predominantly of rock fragments and chert; also present are quartz, a matrix of deformed rock fragments, and orthoclase. Pebbles in the conglomerates generally display moderate rounding and comprise banded and brecciated chert, jasper, hematite, acid volcanics, tuffs, and botryoidal vug-filling quartz. The compositions of the arenites and conglomerates indicate derivation mainly from the Onverwacht Group with a minor granitoid contribution.

In the arenaceous interval, stacked upward-fining cycles between 4 and 6 m thick are developed through a 45-m stratigraphic thickness. Small-pebble basal conglomerates are poorly sorted and angular; these beds thicken upward in the stratigraphy coinciding with improved rounding of the clasts. Plane-bedded sandstones and horizontally laminated siltstones dominate the cycles with rare thin intervals of ripple cross-lamination. The arenaceous interval is analogous to the upper-fan pebbly sandstone facies of Walker (1978).

Massive conglomerates with thin intercalated sandstones constitute the upper 40 m of the southern facies. Individual depositional units comprise conglomerate–sandstone couplets and extend laterally for tens of meters with sharp but nonerosive bases. Clasts are commonly supported in a sand-granule matrix and imbricated. Both normal and inverse grading are developed. A prograding upper-fan model has been suggested for the coarse clastics of the southern Fig Tree

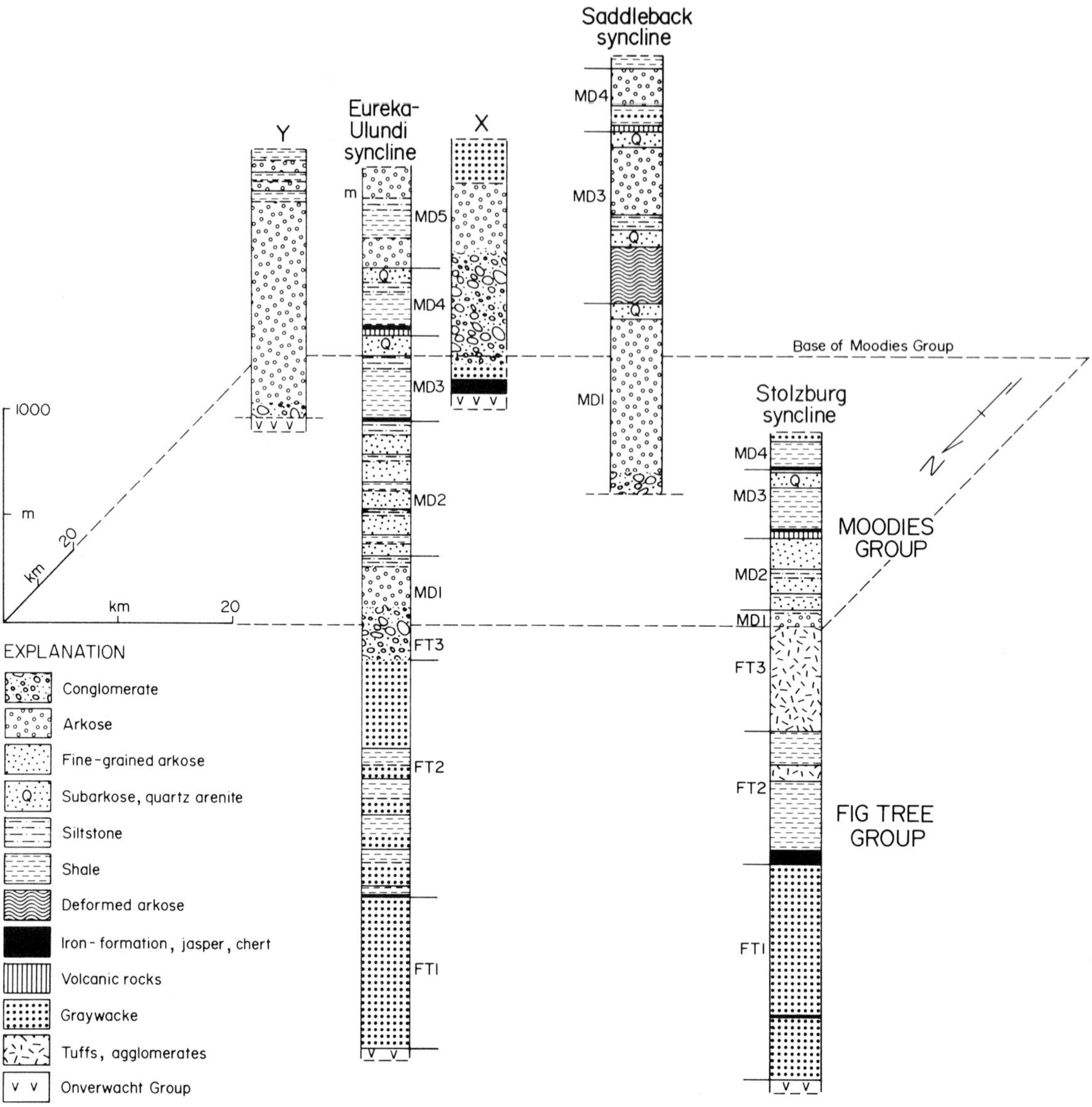

Figure 2-8. Generalized stratigraphic columns of the Fig Tree and Moodies Groups. [After Eriksson (1980b).]

facies (Eriksson, 1980b). Northward progradation occurred over iron-formations that contain abundant tuffaceous material and probably developed in response to waning volcanism. Ferruginous sediments on the floor of the Red Sea (Fryer and Hutchinson, 1976) are possibly analogous (Eriksson, 1980b).

Midfan and slope environments are represented in the northern Fig Tree facies and are dominated by quartz-rich graywackes (Eriksson, 1980a). Grains of plagioclase are abundant. Modal analyses of rock fragments show an increase in granite and decrease in volcanic constituents from unit FT1 to unit FT2. Furthermore, the graywackes in unit FT2 are enriched in Sr and depleted in Ni relative to those in unit FT1 (Condie *et al.*, 1970). These criteria indicate an upward-increasing contribution of sialic source material to the Fig Tree Group. The matrix of the wackes formed by decomposition of labile volcanic constituents, indicating that the Onverwacht was still an important source area during accumulation of the Fig Tree Group (Reimer, 1972). Reimer has shown that up to 70 percent of the Fig Tree constituents were derived from the Onverwacht Group.

The graywackes and subordinate shales in unit FT1 are arranged in stacked, partial, or complete

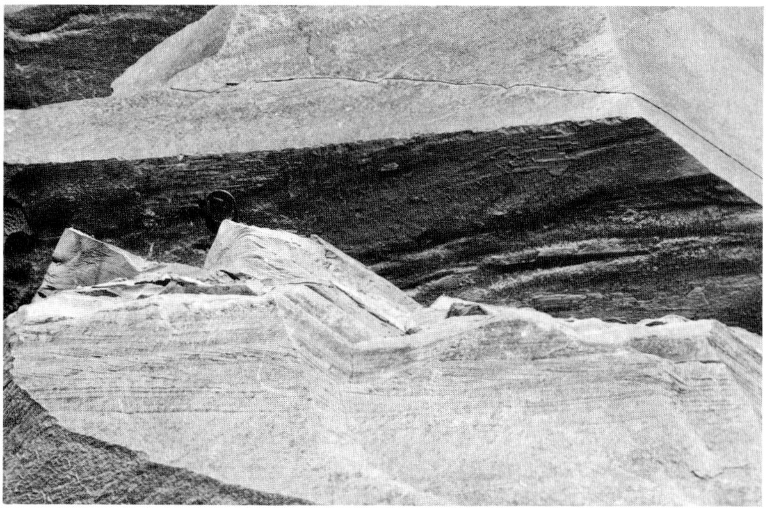

Figure 2-9. Upper half of Bouma turbidite bed consisting of massive "a", plane-bedded "b", ripple cross-laminated "c", horizontally laminated "d", and argillaceous "e" intervals, and overlain by a massive "a" interval with basal groove casts of the succeeding turbidite bed. Fig Tree Group, Ulundi syncline. Coin is 18 mm in diameter.

Bouma sequences which are locally capped with chert. These sequences were deposited from waning turbidity currents followed by background suspension and chemical sedimentation (Fig. 2-9). Sole marks indicate northward-directed paleocurrents. The graywackes of the unit FT1 have been interpreted as midfan deposits (Eriksson, 1980a). Randomly interbedded "a-e" and complete or "b-c-d-e" Bouma sequences probably represent midfan channel and interchannel deposits, respectively.

The overlying unit FT2 consists predominantly of fine-grained terrigenous clastics with intercalated iron-formation and chert (Fig. 2-8). In the Ulundi and Eureka synclines, increasing numbers of lensoid beds are associated with an overall upward coarsening into graywackes and conglomerates of the Schoongezicht Formation. Progradation can be inferred from the facies relationships in the upper Fig Tree Group. The sediments in unit FT2 display characteristics common to both lower fan–basin plain and slope environments (cf. Ingersoll, 1978). Soft-sediment folding of iron-formations reflects gravity-induced displacement in a slope environment (Eriksson, 1980a). The conglomerates in unit FT3 (Fig. 2-8) represent the most proximal sediments of the northern Fig Tree facies. They are believed to have developed through local reworking of graywackes and volcanics in channels incised into the slope sediments.

Fluvial systems. A gradational contact between the Fig Tree and Moodies Groups, best developed in the Eureka syncline, is marked by rounded pebbles in clast-supported conglomerates. The conglomerates increase in thickness upward through an interval approximately 50 m thick. Conglomerates are also present higher in the Moodies Group (Fig. 2-8). Clasts consist of black, white, and banded chert; vein quartz; and acid volcanic rocks with subordinate jasper, basic volcanic, and granitoid pebbles (Eriksson, 1978).

Although poorly sorted, the conglomerates are commonly well stratified and display internal grading and weak imbrication in places. In addition the base of the 1- to 5-m-thick conglomeratic units is abundantly channelized; intercalated sandstones are either plane-bedded or, more commonly, trough cross-bedded. These characteristics suggest that traction processes controlled deposition of the Moodies conglomerates and sandstones (mass-flow processes were more important in accumulating the Fig Tree deposits). The upper reaches of Holocene braided alluvial plains invariably contain longitudinal bars composed of pebbles, cobbles, and boulders, separated by channels (Boothroyd and Ashley, 1975) that resemble the coarse Moodies facies. Sands generally cap these modern bar surfaces and, where associated with a rapid lowering of water level, are structured by plane-beds (Rust, 1972). Adjacent channels commonly contain dunes which continue to migrate at low-flow stages after gravel movement has ceased (Boothroyd and Ashley, 1975). The intercalated trough cross-bedded sandstones in these Archean sediments are thought to represent similar low-water channel deposits over which gravel bars migrated as they shifted laterally (Eriksson, 1978).

Mid-alluvial-plain sediments consist almost entirely of sandstones with subordinate conglomerates and rare shale partings. The fluvial sandstones in the Moodies Group are arkosic with up to 25 percent potash feldspar and minimal amounts of plagioclase (Eriksson, 1980b). The matrix content is significantly lower than in the underlying graywackes which, together with fewer rock fragments and common microcline, reflects continuing emergence of a sialic provenance. Trough cross-bed azimuths are directed consistently to the north.

Examples of sandstone motifs illustrated in Figure 2-10 are similar to channel-fill sequences in Holocene rivers that form either by aggradation (Cant, 1978) or lateral point-bar accretion (Allen, 1970). The lack of evidence for point-bar accretion and overbank sedimentation in these mid-alluvial-plain deposits favors aggradation and midchannel braid-bar formation as the dominant depositional processes. The considerable vertical thicknesses (Fig. 2-10) may be a result of facies stacking.

The stacked trough cross-bed sets probably resulted from dune migration and bar formation within channels, whereas sandwedge accretion off bar margins during falling water stages could account for the rare planar cross-bed sets (Eriksson, 1978). Basal conglomerates are interpreted as channel lags that formed during highest discharge and over which dunes migrated as stream velocities decreased. Large-scale bedforms migrate primarily during high-water stages. With decreasing discharge and falling water level, a variety of hydraulic regimes are generated on braid-bar surfaces and are reflected in the diverse sedimentary structures that cap the depositional sequences (Fg. 2-10). In-phase ripple-drift cross-lamination and shale laminae are similar to overbank levee accumulations on modern braided alluvial plains (Boothroyd and Ashley, 1975). Overbank sedimentation was confined to abandoned reaches of the alluvial plain.

Although aggradation was an important process forming the mid-alluvial-plain sandstones, the lateral persistence of the trough cross-bedded

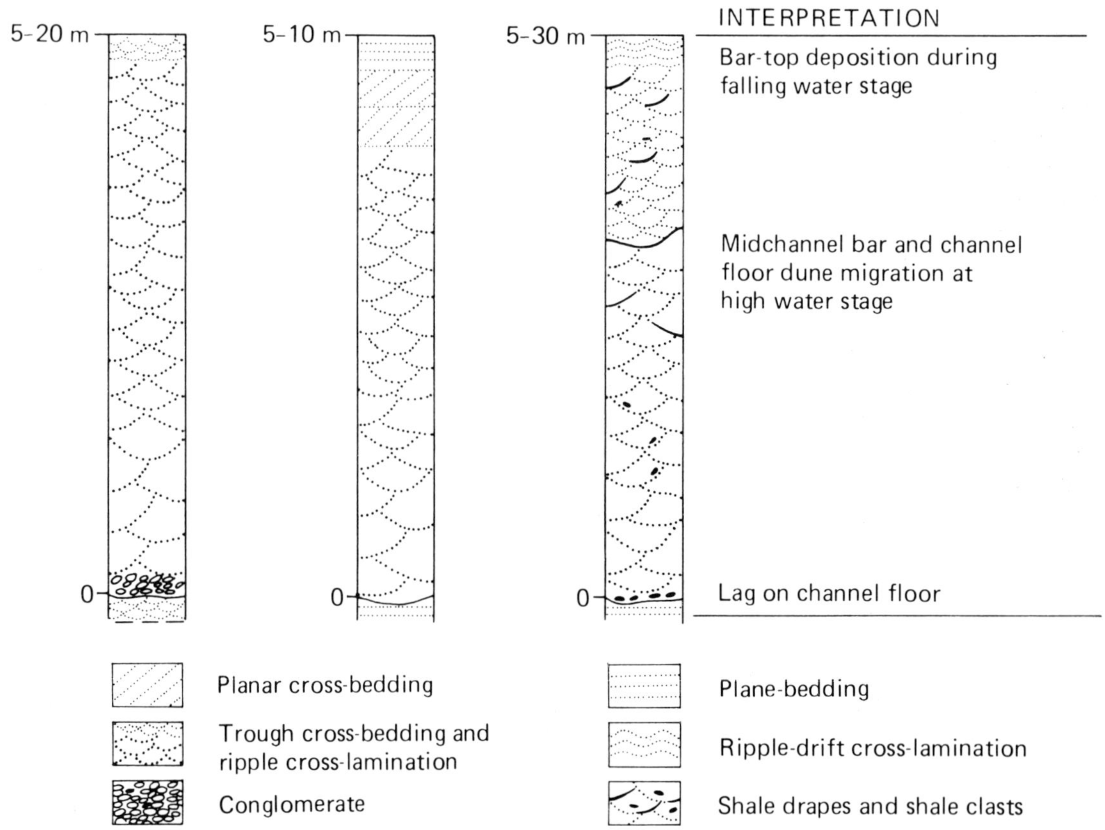

Figure 2-10. Representative vertical sequences of lithologies and sedimentary structures and inferred depositional processes for mid-alluvial-plain facies. Moodies Group. [After Eriksson (1978). © Can. Society of Petroleum Geologists, Memoir Five, "Fluvial Sedimentology," pp. 287–311.]

sandstones indicates that downstream and lateral migration were equally significant. Sand bars in the Brahmaputra River are known to migrate downstream 1700 m during a single flood, whereas channel migration rates of 800 m a year are common (Coleman, 1969). Even greater migration rates are known for the Kosi River (Gole and Chitale, 1966). Braided floodplains are thus highly active depositional regimes consisting of active and abandoned channels. Through constant shifting of stream courses, thick accumulations of sediment form.

Alternating siltstones and shales, with intercalated conglomerate-sandstone associations, are interpreted as lower alluvial floodplain and channel deposits, respectively (Eriksson, 1978). The coarse-grained sediments form 2- to 6-m-thick, upward-fining sequences which are generally continuous for tens of meters. Their lateral persistence and limited thicknesses suggest that the conglomerate–sandstone sequences formed by lateral accretion of point bars rather than by aggradation and channel bar development. Pebble conglomerates formed as channel lag, across which lower point-bar dunes migrated to produce the coarse- and medium-grained trough cross-bedded sandstones. A continuing decrease in flow velocity at shallower water depths resulted in deposition of finer grained sandstones of the upper point bar. These are structured by small-scale trough cross-beds and ripple cross-lamination. Sporadic current lineation on plane-bedded sandstones indicate short-lived upper-flow-regime conditions on the upper point bars as a result of shallowing.

The overbank siltstones and shales accumulated under conditions of suspension and waning bedload sedimentation which is reflected in graded units between 1 and 5 cm thick. There is no evidence of desiccation in these overbank sediments, suggesting that they were continually saturated by repetitive overbank flooding and that the climate may have been abnormally humid (Lowe and Knauth, 1978).

Deltaic outbuilding. A progradational assemblage of deltaic sediments in unit MD3 in the Saddleback syncline (Figs. 2-5 and 2-8) includes delta-front, tidal sand shoal, and delta-plain environments (K.A. Eriksson, 1979).

The lower part of the delta prism consists of argillaceous sandstones deposited in a delta-front milieu (Fig. 2-11). Individual depositional units vary from 1 to 2 cm in thickness and are separated by thin shale partings; primary sedimentary structures are lacking. Coarse-tail grading is common and is suggestive of liquefaction and gravity-flow processes. Deformation structures include water-escape pipes and sheets; their vertical persistence indicates simultaneous dewatering of depositional sequences over 5 m thick. The ragged contacts between individual depositional units and the small flame structures (Fig. 2-11) probably formed as a result of loading and water escape. Repetitive gravity-flow surges are attributed to oversteepening of the delta front.

Thick lenses of quartz arenite overlying the delta-front facies are interpreted as tidal sand shoals (Fig. 2-11). The lenses have a north–south orientation perpendicular to the paleoshoreline and resemble the elongate sand ridges at the mouth of the Elbe River in the German Bight of the North Sea (Reineck and Singh, 1973, p. 318). The Elbe estuary is open mouthed and is influenced by the strong tidal currents of a macrotidal regime. Dunes form on the surface of the shoals below low-water level and smaller bedforms and clays accumulate above low-water level (Dörjes *et al.*, 1970). Dominant northward-directed planar cross-beds and pebble stringers within the sandstones (Fig. 2-11) possibly reflect strong ebb currents across the surfaces of submerged sand shoals. Periodic reversals of the tidal currents are shown by herringbone cross-bedding around the margin of the shoals (Fig. 2-12A). Associated ripple-drift cross-lamination provides further evidence of strong currents capable of transporting appreciable quantities of sand in suspension (Gustavson *et al.*, 1975). The predominance of low-amplitude sedimentary structures and the eventual appearance of desiccation-cracked shale drapes in the upper 40 m of the facies sequence (Figs. 2-11 and 2-12B) indicate a shallowing and ultimate emergence of the sand shoals.

The delta-plain facies consist of upward-fining arkosic channel fill enclosed within rhythmically interlayered sandstones and shales (Fig. 2-11). The channel-fill sediments display some characteristics typical of fluvial point bars, such as immaturity and the vertical arrangement of sedimentary structures above channel lag. However, other features are more compatible with a tidal channel origin. Although individual outcrops display unimodal paleocurrent patterns, thin herringbone cross-bedded units are indicative of periodic

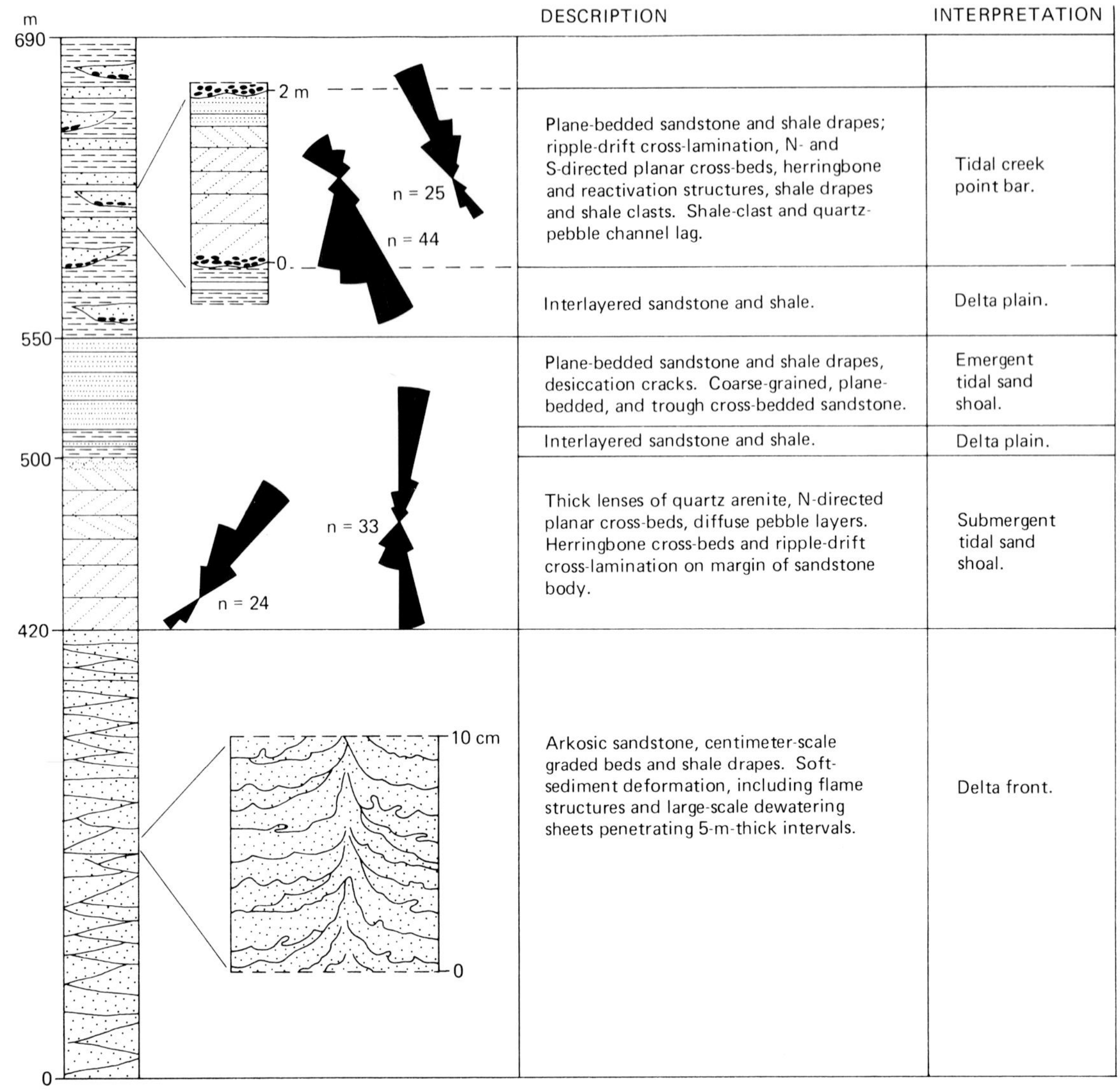

Figure 2-11. Tide-dominated deltaic sediments in lower half of unit MD3, Moodies Group, Saddleback syncline. [After K.A. Eriksson (1979).]

current reversals (Fig. 2-11). Furthermore, adjacent outcrops show a distinct bipolarity of cross-beds with southerly and northerly modes reflecting segregation of flood and ebb currents. Shale drapes are ubiquitous on foresets, between cosets, and within plane-bedded sandstones; they are diagnostic of tidal-channel processes (Barwis, 1978). Shale drapes on foresets indicate intermittent dune migration and slack-water suspension sedimentation. Late-stage emergence is believed to have formed the plane-beds near the top of the point-bar sequences.

The rhythmically interlayered sandstones and shales enclosing the tidal-channel sequences commonly display evidence of desiccation and are interpreted as overbank deposits that accumulated on estuarine tidal flats as a result of alternating bedload and suspension sedimentation. Shale flakes in the channel deposits were probably derived by reworking of the desiccated tidal-flat muds.

The deltaic assemblage constituting unit MD3 in the Saddleback syncline displays few, if any, characteristics in common with river-dominated deltas, such as the Mississippi. However, deltas that experience high tidal ranges display tidal-flat depositional features such as those represented in the delta-plain and tidal sand shoal facies. Tidal

(A)

(B)

Figure 2-12. (A) Herringbone cross-beds in tidal sand shoal facies. Unit MD3, Moodies Group, Saddleback syncline. Scale 15 cm. [From Eriksson (1977a).] (B) Desiccation cracks in emergent sand shoal sediments. Unit MD3, Moodies Group, Saddleback syncline.

processes at river mouths produced intense meandering above the zone of tidal influence, causing the tidal channels to be sand filled, and resulted in the accumulation of large, linear sand ridges directed seaward from the river mouth. The Ord River Delta in northern Australia is an example (Coleman and Wright, 1978).

Barrier-beach progradation. Thick, upward-coarsening depositional sequences constitute unit MD3 in the Eureka and Stolzburg synclines and are interpreted as progradational barrier island and shallow-shelf deposits (Figs. 2-8 and 2-13).

The iron-formations and shales at the base of the sequence are believed to have formed on the deeper part of the shelf under conditions favoring chemical and suspension sedimentation in quiet water (Fig. 2-13); iron-formations were precipitated farthest from areas of terrigenous influx. Finely alternating ferruginous and siliceous layers in iron-formations most readily accumulate in water with cyclically fluctuating Eh and pH (Walter, 1972b). The overlying shales with lenses and layers of jaspilite indicate an offshore envi-

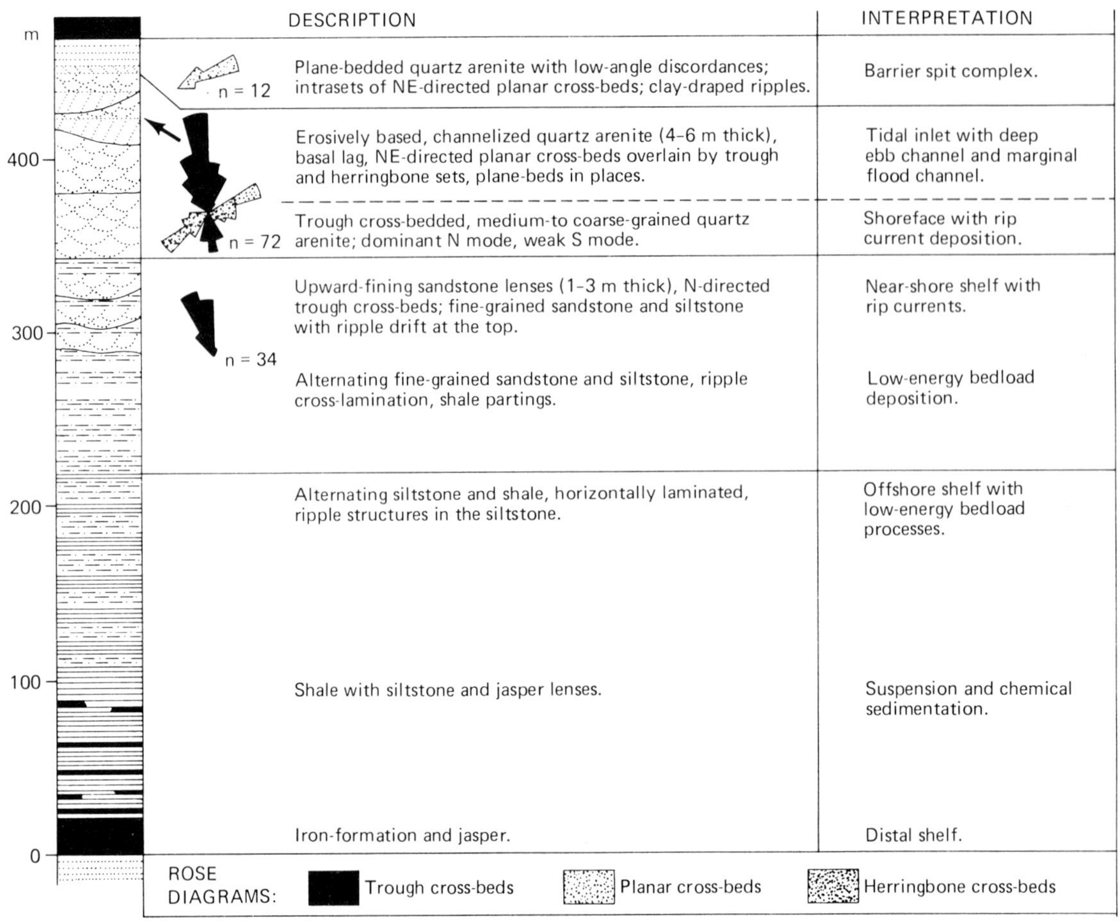

Figure 2-13. Progradational barrier beach–shelf sediments in unit MD3, Moodies Group. [After K.A. Eriksson (1979).]

ronment intermittently receiving fine-grained suspended detritus alternating with short periods of siliceous chemical sedimentation at relatively low pH and high Eh. At shallower depths a continuous supply of fine-grained detritus prevented the formation of precipitates. Traction sedimentation was limited to formation of "starved" siltstone ripples in the shales. Low-energy bedload processes became more important with increased shallowing, resulting in interlayering of rippled and horizontally laminated siltstones and shales.

The overlying fine-grained sandstones and siltstones with thin shale partings and ripple cross-lamination (Fig. 2-13) are analogous to the nearshore deposits of Sapelo Island, Georgia (Howard and Reineck, 1972); they indicate low-energy bedload deposition and infrequent suspension sedimentation. The upward-fining quartz arenite lenses, common higher up in the nearshore shelf facies, imply periodic higher energy currents. Cross-bed azimuths reflect a predominant northward or offshore flow, probably generated by waning rip currents.

In progradational barrier-island sequences, nearshore deposits are commonly overlain by shoreface sands. However, the sharp basal contact (Fig. 2-13) and the dominant offshore paleocurrent trend of the trough cross-bedded quartz arenites are atypical of shoreface sediments. Their strong offshore paleocurrent mode suggests deposition by rip currents. A weak onshore component can be related to the shoreward motion of shoaling waves (Clifton *et al.*, 1971).

Stacked upward-fining units, which overlie the shoreface sandstones (Fig. 2-13), closely resemble tidal-inlet deposits (cf. Kumar and Sanders, 1974). Tidal inlets migrate in the direction of longshore drift, with resultant spit accretion on the updrift side of the inlet. Basal conglomerates in these facies sequences are interpreted as channel-floor deposits. Overlying planar cross-beds were formed by northeastward migration of

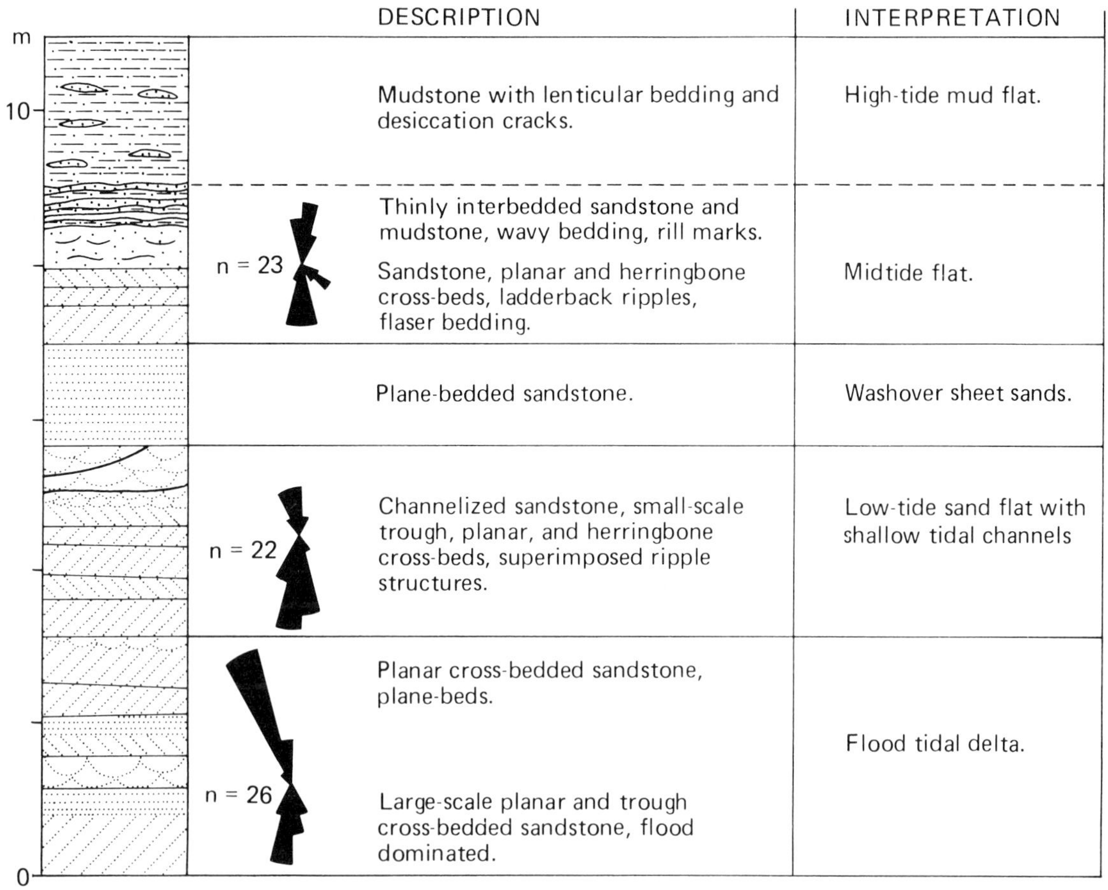

Figure 2-14. Backbarrier lithofacies in unit MD2, Moodies Group, Eureka and Saddleback synclines. [After K.A. Eriksson (1979).]

ebb-oriented sandwaves in the deeper parts of the inlets. Whereas the deeper parts of tidal inlets are generally characterized by unidirectional ebb flow, the shallower marginal flood channels contain bimodal currents related to ebb and flood tides (Hayes, 1975). The herringbone cross-bedding at the top of the units can be related to these processes; shale drapes on foresets provide further evidence of tidal influence. Paleocurrent patterns show that the inlets were oriented in a northeast-southwest direction, or oblique to the paleoshoreline, indicating strong longshore drift to the east.

The dominantly plane-bedded sandstones at the top of the sequence (Fig. 2-13) superficially resemble foreshore deposits. However, planar cross-beds with a longshore dip direction and clay-draped ripples are atypical. A depositional environment influenced by swash and longshore currents and containing isolated areas of suspension sedimentation is indicated. A barrier spit setting is envisaged. Subaqueous spit platforms contain sandwaves generated by longshore currents (Kumar and Sanders, 1974). Spit beaches are subjected to wave processes and are structured by plane-bedded sand with low-angle discordances. Swales between the spit beach ridges are subject to current ripple migration and periodic suspension sedimentation.

Backbarrier environments. Tidal processes largely controlled deposition of Moodies unit MD2 in the Eureka and Stolzburg synclines (Figs. 2-5 and 2-8). Bimodal–bipolar paleocurrent patterns are ubiquitous (Fig. 2-14), indicating reversals of flow direction. Other criteria, including thinly interbedded sandstones and shales, which point to alternating bedload and suspension sedimentation, and desiccation-cracked shale drapes, reflecting periodic exposure, are indicative of tidal-flat deposition (K.A. Eriksson, 1979).

On extensive tidal flats, such as along the north German and Netherlands coasts, argillaceous sediments accumulate near the high-water line and coarser sandy sediments nearer the low-water level (Reineck, 1967). The mixed mid-tidal-flat environment is characterized by interbedded sands and muds. Representatives of each of these environments are recognized in the Moodies Group (Fig. 2-15). Lenticular-, wavy-, and flaser-bedded sandstones and shales accumulated on upper and midtidal flats as a response to alternating bedload and suspension sedimentation. Interference and ladderback ripples and rill marks are diagnostic of ebb runoff accompanying rapid lowering of water level prior to exposure (Klein, 1971). Small-scale planar and herringbone cross-bedded sandstones probably accumulated on the lower tidal flats (Fig. 2-14) where higher current velocities precluded suspension sedimentation. Low bed-shear conditions are indicated by the low amplitude of cross-bed sets. Other back-barrier environments include cross-bedded sandstones of flood-tidal delta origin and thick plane-bedded sandstones ascribed to washover processes; both facies are intercalated with tidal-flat deposits (Fig. 2-14).

Paleogeography. The Fig Tree and Moodies sediments were derived from a mixed sialic-volcanic-chert terrane in the south and accumulated along a northward-facing continental margin (Eriksson, 1980b). Two paleogeographic models emphasize the progressive evolution of the Fig Tree and Moodies continental margins. The abrupt transition from submarine fan to braided alluvial plain (probably as a fan delta) in the Fig Tree and basal Moodies Groups, and the absence of shallow marine facies, reflect sedimentation along a steep continental margin with a narrow shelf (Fig. 2-15A). Higher units in the Moodies Group, however, display abundant evidence of shallow-marine processes, implying the evolution of a wider continental shelf. The inferred relationships of the different depositional environments in the Moodies Group are illustrated in Figure 2-15B; this is a palinspastic reconstruction based on time planes represented by laterally persistent lavas at the base of units MD3 and MD4 in the Moodies Group (Fig. 2-8).

The paleogeographic model for the Moodies Group recognizes the interactions of four major depositional environments (K.A. Eriksson, 1979). Braided alluvial-plain and deltaic sedimentation represent the constructive elements of the model. Marine reworking formed barrier complexes with wide backbarrier tidal flats. Low-energy and storm-surge sedimentation in the near shore and suspension and chemical sedimentation on the deeper shelf complete the depositional model.

The association of depositional facies in the Moodies barrier succession is most similar to modern microtidal barrier complexes; features in common include shallow tidal-inlet channel fills (4–6 m), abundant barrier washover sheet sands, arenaceous backbarrier deposits, and only rare backbarrier tidal channels (cf. Hayes, 1975). Klein (1971, 1972) suggested that the thickness of an upward-fining tidal-flat sequence was a measure of paleotidal range. On this basis the composite 6-m-thick tidal-flat sequences (Fig. 2-14) indicate macrotidal conditions. However, the association of these tidal flat deposits with inferred microtidal barrier deposits suggests that the thickness of the tidal-flat sequences is exaggerated and resulted from stacking of facies. Coastal embayments produced local amplification of tides, resulting in tidal channel and tidal-sand shoal deposits in nonbarred estuaries. The coexistence of macrotidal and microtidal ranges is thus implied (Fig. 2-15); the German Bight of the North Sea is a modern example.

Swaziland Supergroup Tectonics

Structural and metamorphic studies of the Barberton belt are generally lacking except for a few areas, mainly in the vicinity of the Jamestown greenstone septum and the Eureka syncline (Fig. 2-5). The metamorphic grade of the Onverwacht rocks is typically low grade except for a narrow aureole adjacent to intrusive granitoids, where medium grade is reached. The contact aureole of the Lochiel granite in Swaziland shows two generations of metamorphic minerals. In the older generation the minerals are aligned parallel to the foliation, whereas the second-generation minerals are randomly oriented and postkinematic. Pyroxene hornfels is found locally within the Lochiel contact aureole in Swaziland. Fig Tree and Moodies rocks are metamorphosed at similar low grades as those of the Onverwacht Group except where immediately adjacent to the Lochiel granite

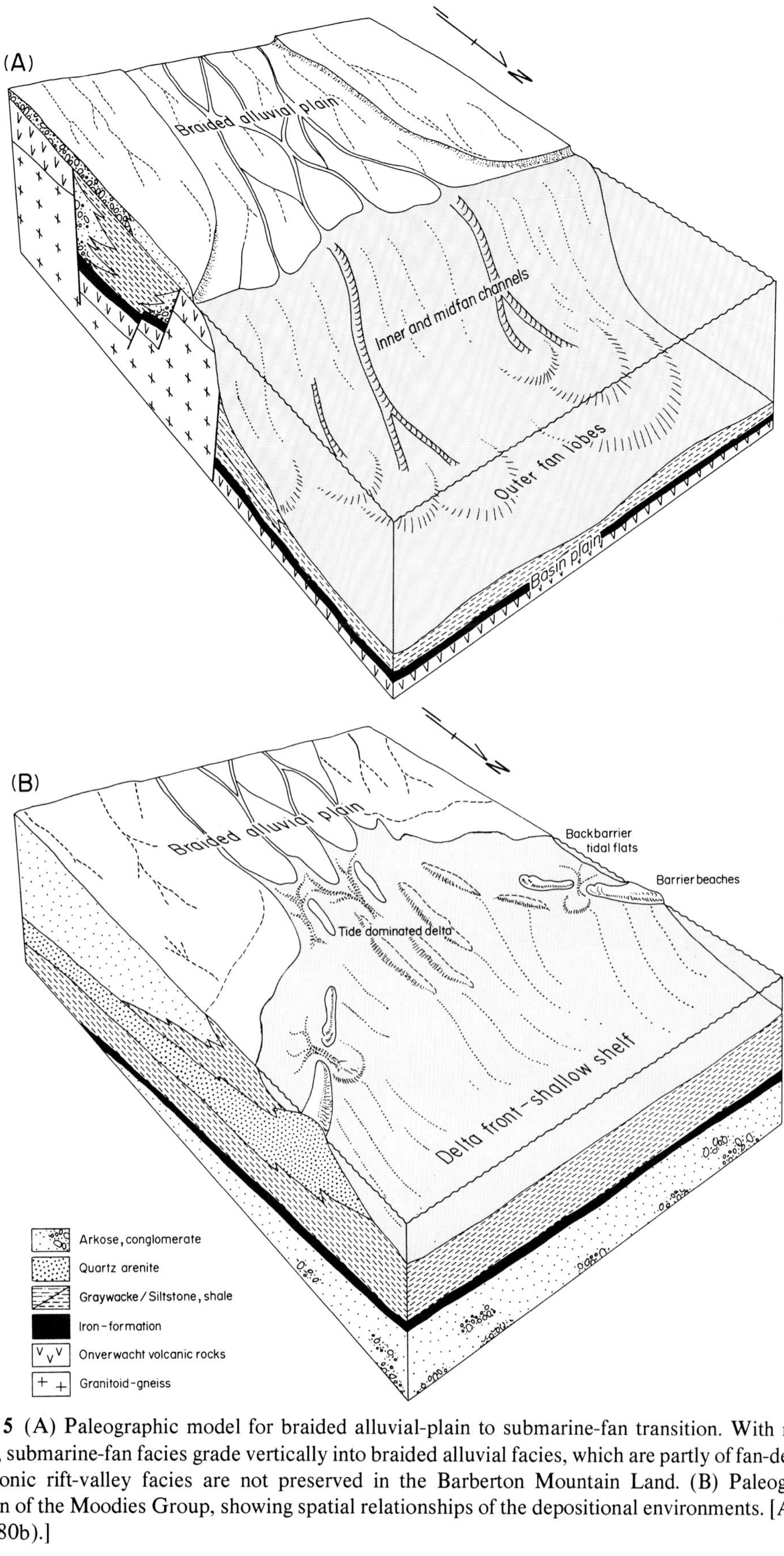

Figure 2-15 (A) Paleographic model for braided alluvial-plain to submarine-fan transition. With northward progradation, submarine-fan facies grade vertically into braided alluvial facies, which are partly of fan-delta origin. Inferred cratonic rift-valley facies are not preserved in the Barberton Mountain Land. (B) Paleogeographic reconstruction of the Moodies Group, showing spatial relationships of the depositional environments. [After K.A. Eriksson (1980b).]

in Swaziland. Strong recrystallization of the quartzose rocks and the development of hornblende and pyroxene hornfels from rocks of appropriate compositions are typical.

The regional structure of the Swaziland Supergroup is that of a broad synclinorium trending east-northeast to north-northeast and consisting of tight or isoclinal synforms, generally overturned to the northwest, that are separated by tectonic slides rather than antiforms. Many of these slides show evidence of later reactivation as transcurrent faults. Two consequences of this structure are that, first, the Swaziland Supergroup is now preserved in a series of northeast-trending segments (Anhaeusser *et al.*, 1969), and second, the lowermost rocks of the Onverwacht Group are preserved around the periphery of the belt adjacent to surrounding granitoids. Septa of strongly deformed Onverwacht rocks are found between the tonalitic plutons that intrude the greenstone belt along its southwestern and northwestern flanks (Fig. 2-5).

In the Motjane schist belt, a highly flattened synform of greenstones west of Mbabane and extending southwestward from the main Barberton synclinorium (Fig. 2-5), Jackson and Robertson (1982) identified several periods of deformation in the Onverwacht Group. Two generations (termed F1a and F1b) of reclined and recumbent mesoscopic folds formed under amphibolite–facies conditions and suggest southward translation of the volcanic pile away from the center of the Barberton greenstone belt. The Motjane greenstones were subsequently engulfed by the Lochiel granite (refer to Section 2.4) and subjected to subhorizontal compression during D2, which created the synformal structure of the Motjane schist belt. Severe D2 flattening, with up to 80 percent shortening, resulted in upright folds being imposed on the Lochiel granite and dikes and superposed on the older flat-lying folds in the Onverwacht Group; axial–planar schistosity and crenulation-cleavage fabrics were annealed by residual heat from the Lochiel granite. The D2 strain involved up to 300 percent stretching of the Motjane greenstones downward and toward the sagging mass of the main Barberton synclinorium (Jackson and Robertson, 1982). The Motjane schist belt was subsequently cross folded during D3 about upright axial surfaces striking southeast, possibly during emplacement of the Usushwana Intrusive Suite at 2.8 Ga.

In the southwestern part of the main synclinorium a series of open antiforms and synforms, plunging northeast or north-northeast, have been identified by R.P. Viljoen and Viljoen (1969). Their map (1969, their Fig. 3) also indicates the presence of dip reversals and apparent fold closures in the upper part of the Hooggenoeg Formation on the western limb of one of the open antiforms, suggestive of an older period of recumbent folding which has since been confirmed by remapping (De Wit and Stern, 1980). This is the same area where Williams and Furnell (1979) concluded that an unconformity or tectonic slide may be located.

Remapping of parts of the Onverwacht Group has led to a fundamental reinterpretation of these rocks, which, if correct, has major structural and stratigraphic significance. De Wit and Stern (1980) questioned the prevailing sixfold division of the stratigraphy on the following structural grounds: (1) The Hooggenoeg Formation (upper Onverwacht Group) has been tectonically repeated by polyphase isoclinal folding and faulting; (2) the Theespruit Formation (lower Onverwacht Group) has been similarly deformed and may be coeval in part with the Hooggenoeg Formation; (3) westward thinning of the Komati Formation (middle Onverwacht Group), formerly attributed to stratigraphic thinning (R.P. Viljoen and Viljoen, 1969), has been tectonically induced, as evidenced by strain analysis of deformed variolites. In addition De Wit and Stern (1980) drew analogies between the volcanic and hyperbyssal Onverwacht rocks and Phanerozoic ophiolite suites by regarding the moderately to steeply dipping Onverwacht sequence as a broad expanse of oceanic crust, probably originally less than 2 km thick and preserved close to its original orientation; prevailing views hold that the sequence has a present stratigraphic thickness of about 15 km. They reinterpret the undeformed type "section" of the Komati Formation as an intrusive rather than an extrusive unit because of: (1) observed crosscutting contacts; (2) recognition of intrusive pillows, common in Phanerozoic sheeted-dike complexes, separated by septa of truly extrusive pillow lavas; (3) symmetrical textural variations from the core of each "pseudoflow" outward to its margins, typical of dikes rather than sills or flows; and (4) recognition of mineral layering at right angles to intrusive contacts of ultrabasic bodies in the Komati Formation. If the 15-km thickness of the Onverwacht Group is regarded as a lateral, rather than vertical, thickness, this explains the widespread preservation of such delicate primary structures as spinifex textures and the almost ubiquitous low-grade metamorphic imprint; this would be impossible under burial of a 15-km thickness of basic volcanics capped by an additional thickness of at least 5 km of sediments in the form of the Fig Tree and Moodies Groups. The Komati Formation is therefore envisaged by De Wit and Stern (1980) as an originally vertical sheeted-dike complex that fed overlying and flanking basic volcanic sequences in the form of the Theespruit and Hooggenoeg Formations, all of which have undergone ocean-floor metamorphism rather than regional metamorphism. As might be expected, there are also significant differences between the Onverwacht igneous rocks and Phanerozoic ophiolites: more MgO-rich liquids, fewer cumulates, distinctly wider dikes, and inferred polydirectional spreading rather than bilateral extension. All these differences may eventually throw valuable light on how Archean crust formed. It is clear from the foregoing that neither the structure nor the stratigraphy of the Onverwacht Group can be fully understood independently of each other. It is vital to establish whether the layering was originally vertical, as in a sheeted-dike complex, or horizontal, as in an extrusive-sill complex because this layering serves as a primary indicator of the rotation of supracrustal rocks during folding. At the present stage it is obvious that there are sufficient anomalies to warrant further detailed structural-stratigraphic work on the Onverwacht Group, especially on the less pristine areas, which have previously been regarded as unattractive for study.

The sequence of deformation in the overlying sedimentary units forming the core of the synclinorium is better understood. Most attention has been directed toward the Eureka syncline, which is cored by the Moodies Group and represents one of several synforms referred to above (Fig. 2-5). In the first published application of geometric analysis to multiply deformed rocks in southern Africa, Ramsay (1963) recorded the imprint of three principal deformation events:

D1. Major and minor inclined folding about axial planes originally striking northeast or north–northeast and dipping southeast, forming the Eureka syncline and adjoining Ulundi syncline.

D2. Shortening along north-northwesterly axes oblique to the D1 shortening under greenschist metamorphism led to the development of a transecting slaty cleavage and schistosity (cf. Borradaile, 1978) and concomitant downward-facing structures (cf. Shackleton, 1956). Steeply plunging F2 folds formed sporadically on the flanks of the Eureka syncline.

D3. A large northwestward-trending synform buckled the major synclines, their transecting foliations, and the intervening tectonic slides to produce crescentic interference patterns. Minor conjugate folds and faults may have formed during D3 or they may constitute a separate D4 event.

Ramsay regarded the D1 and D2 events as closely related phases of the same progressive deformation. His inference that the transecting cleavage formed after the main D1 folding proved well founded, and this structural relationship has since been found to be common in other greenstone belts in the Kaapvaal Province (Graham, 1974) and the Zimbabwe Province (Coward and James, 1974; Coward *et al.*, 1976a).

Subsequent structural studies (Anhaeusser, 1966, 1976; Gay, 1969) have largely confirmed Ramsay's (1963) findings and have been directed largely toward quantifying the finite strains, using deformed Moodies conglomerates, and establishing the role of the adjacent Kaap Valley pluton in the deformation. Although generally of the flattening type, the strains vary widely from place to place, depending on the distance from the granite contact and position in major structures; they are compatible with the proposal by Anhaeusser *et al.* (1969) that intrusion of the Kaap Valley pluton was responsible for the D3 arcuation of the regional structure and for the flattening of the northwest-trending Jamestown schist belt (Anhaeusser, 1972). The intrusion may also have been responsible for conversion of the Barbrook fault east of the Ulundi syncline into a décollement zone and for the thrusting northwest of the Eureka and Ulundi synclines (Anhaeusser, 1979).

Fripp *et al.* (1980) drew attention to a late-formed major dislocation zone along much of the northern boundary of the Barberton belt and the

Jamestown septum. Mylonite-bounded lenses of gneisses are intersliced with greenstones containing a nonpenetrative fabric that is mylonitic in places. This fabric is deformed by subhorizontal northward-verging folds whose axial surfaces are parallel to the granite–greenstone contact. Fripp *et al.* linked this folding, which took place under amphibolite-facies conditions, to diapiric rise of the Kaap Valley pluton and correlated it with the D2 cleavage-forming events described above. Continued north–south shortening flattened the D2 folds and intensified fabrics to the point where mylonitic shear belts, pseudotachylites, and thrust faults formed in imbricate fashion at successively higher crustal levels. It is possible that these zones of discontinuity represent upward splays that connect at depth with a large low-angle sole thrust (Fripp *et al.*, 1980), the existence of which is suggested by gravity surveys that indicate the greenstone belt is only 4–6 km thick and that its northern margin dips gently to the south (Darracott, 1975). All these structures indicate repeated episodes of upward and northward movement of the Barberton greenstone belt relative to the Nelspruit gneiss terrane to the north. This large-scale translation of crust is closely analogous in orientation to possibly earlier crustal movements farther south in the Ancient Gneiss Complex (Fig. 2-4) and corroborates a mid-Archean episode of major crustal shortening, which Fripp *et al.* (1980) correlate with uplift of the Limpopo gneiss terrane north of the Kaapvaal Province.

More structural studies throughout the Barberton belt are clearly critical to unraveling its stratigraphy. There is also a strong economic motive in that the important gold mineralization in the Eureka syncline and Jamestown schist belt is largely structurally controlled. The source of the gold is uncertain, but it seems likely that it migrated into its present environment from the Onverwacht volcanics, although some gold mineralization may be associated with iron-formations in the Fig Tree Group. Not only is gold mineralization structurally controlled, but major deposits of chrysotile asbestos and iron ore have also been shown to be structurally controlled. The intense deformation of ultrabasic rocks has been responsible for the generation of small, but economically exploitable deposits of talc.

2.3. Other Kaapvaal Greenstone Belts

The Murchison, Pietersburg, Sutherland, Muldersdrif, and Amalia belts (Fig. 2-1) have been examined in much less detail than the Barberton area. Outcrop is poor and it is possible to delineate only the broad outlines of their geology.

Pietersburg

The Pietersburg belt extends in a northeasterly direction from beneath the cover of the Transvaal Supergroup for a distance of about 100 km and has an average width of approximately 10 km. At its northeastern end it tapers out between intrusive granitoids in a poorly known area of limited exposure. Grobler (1972) subdivided his Pietersburg sequence at the southwestern end into:

(2) Uitkyk Formation: Dominantly conglomerate and quartzite with interbedded shale
(1) Mount Robert Group: Basic lava with minor ultrabasic and acid lava
  (b) Landsberghoek Formation: Two major cycles of ultrabasic lava and minor intercalated iron-formation and chert
  (a) Eersteling Formation: Metatholeiite lava

The ultrabasic and basic lavas of the Mount Robert Group are chemically similar to the komatiitic lavas of the Barberton belt (Grobler, 1972). The Uitkyk Formation, which rests unconformably on the Mount Robert Group, resembles the Moodies Group, but equivalents of the Geluk Subgroup and the Fig Tree Group of the Barberton succession are apparently absent in the Pietersburg belt.

Three periods of deformation were recognized by Grobler (1972), the first preceding deposition of the Uitkyk Formation. Isoclinal overfolding to the northwest followed deposition of the Uitkyk Formation and is related by Grobler to the development of prominent slaty cleavage and schistosity. However, subsequent mapping by Graham (1974) has shown that the dominant cleavage is transecting and postdates the gently plunging syncline that contains the Uitkyk metasediments. This is indicated by reversals in younging direction without change in bedding-cleavage sense, steep fold hinges, and parallel intersection and mineral lineations. The foliation was crenulated during a fourth deformation (Grobler, 1972). Granitoid plutons embay the schist belt but little is known of their geometry, distribution, or petrologic characters, although they appear to predate the regional fabric (Graham, 1974).

Like other volcanosedimentary remnants in the Kaapvaal Province the Pietersburg belt has been affected by greenschist metamorphism; amphibolite-facies rocks adjacent to granitoid intrusions may reflect superimposed contact metamorphism, although there is a regional increase in metamorphic grade northeastward toward the high-grade Limpopo Province (Graham, 1973).

The Pietersburg belt was the site of the first discovery of "reef" gold in South Africa in 1871 but, although there has been sporadic production since then, the deposits have yielded only small amounts of gold. No other significant mineralization is known.

Murchison

The Murchison belt is a narrow, 140-km-long low-grade greenstone belt, which has a maximum width of about 15 km at its southwestern end and which bifurcates into two tapering septa at its northeastern end, where it is covered by Karoo strata. The South African Committee for Stratigraphy has attempted to rectify the previous lack of defined lithostratigraphic units by proposing that the complete succession be named the Gravelotte Group and divided into:

(6) Rubbervale Formation: Acid pyroclastic rock, acid lava, quartz and quartz–feldspar porphyry with opalescent blue quartz grains.
(5) Mac Kop Formation: Quartzite, feldspathic sandstone, quartz–sericite schist, minor conglomerate, and chloritic schist found in the eastern part of the belt.
(4) Weigel Formation: Chlorite schist, quartz–chlorite schist, quartzite, quartz–muscovite schist, minor sericite schist, iron-formation (both oxide and sulfide facies).
(3) La France Formation: Grit, quartzite, conglomerate, and chloritic schist.
(2) Leydsdorp Formation: Massive and schistose tholeiitic and komatiitic metalavas which are extensively carbonatized; some evidence of tholeiitic andesites.
(1) Mulati Formation: Mainly magnesium-rich basalt and ultrabasic metalava, some of which have a similar chemistry to those in the Onverwacht Group; minor quartz–sericite schist and iron-formation are interlayered.

The structure of the Murchison belt is dominated by strong east striking D2 slaty cleavage, which transects two F1 gently plunging isoclinal synclines (Graham, 1974). As with the Pietersburg greenstone belt, the younger age of the regional fabric is indicated by reversals in younging directions in cross-bedded metaquartzites without concomitant change in bedding-cleavage sense, thus resulting in downward-facing structures. A further similarity is the finite constrictional strain caused by superposed flattening, with stretching parallel to the steeply plunging bedding-cleavage intersection. Younger F3 folds deform the cleavage. The northern margin of the greenstone belt is a left-lateral shear belt (Graham, 1974), cut by younger pseudotachylite veins (Fripp *et al.*, 1980). Younging directions indicate that the Mulati Formation, which crops out only on the southern flank of the belt, is the oldest unit. The Rubbervale Formation, confined to the northern flank of the belt, is believed to be the youngest lithostratigraphic unit. The stratigraphic relationships of the Weigel and Rubbervale Formations and of the Weigel, Mac Kop, and Mulati Formations are uncertain. It is possible that the Mac Kop and La France Formations may be lateral equivalents of the Weigel Formation.

The Rooiwater Complex intrudes the Murchison greenstones along their northern flank. It consists of a lower, mainly gabbroic, layered unit and an upper hornblende granitoid. The gabbro contains two magnetite layers, each about 4 m thick. The complex has been metamorphosed, folded, and sheared with the metavolcanic rocks to the south. Various granitoids that are considered to be intrusive into the Murchison belt are described in Section 2.6.

The Murchison belt contains important deposits of antimony with associated gold mineralization. The major deposits coincide with a number of steeply plunging F2 folds and are located along the so-called Antimony Line , an apparently stratiform zone of talc, talc-carbonate, and chloritic phyllites associated with green fuchsite–chert marbles. The Antimony Line can be traced for about 30 km along strike in the Weigel Formation in the core of the Murchison belt. The presence of major antimony mineralization distinguishes the Murchison from the Barberton belt, where, except for scattered uneconomic traces of stibnite, no antimony mineralization is known. The origin of the antimony sulfide (stibnite, berthierite, tetrahedrite, and chalcostibnite) is uncertain, and both epigenetic (hydrothermal) and syngenetic (exhalative) origins have been suggested (Viljoen *et al.*, 1978). An important unanswered question is why antimony should have been mineralized in the Murchison belt and not in the Barberton belt.

Prominent pegmatites in the granitoid terrane immediately south of the Murchison belt contain deposits of emeralds, mica, beryllium, lithium, tantalum, corundum, feldspar, and silica.

Sutherland

The poorly exposed Sutherland belt lies to the northeast of the Pietersburg belt near the fringes of the Limpopo Province. It is about 60 km long and up to 15 km wide and bifurcates into two tapering arms in the southwest. Little is known of the geology of this belt, which apparently consists mainly of various schistose and massive basic and ultrabasic rocks with minor intercalations of iron-formation, chert, and quartz-sericite schist. Middle to upper greenschist metamorphism has affected the Sutherland belt. East-trending F2 upright folds deform recumbent north-trending F1 folds, forming mushroom-type interference structures. A widespread D2 crenulation cleavage has overprinted the D1 cleavage (Fripp *et al.*, 1980). As is the case with the Murchison belt, these folds are tightened and rotated into a shear belt that dips southward at shallow angles along the northern margin of the greenstone belt.

Muldersdrif and Amalia

The Muldersdrif belt underlies a small area (about 150 km^2) on the edge of the Johannesburg Dome. The dominant rock types are basic and ultrabasic massive and schistose metavolcanic rocks together with remnants of layered ultrabasic intrusions (Anhaeusser, 1977, 1978c).

The Amalia belt in the western Transvaal is preserved in an inlier surrounded by the Ventersdorp Supergroup. The belt extends for about 50 km in a northward direction across the inlier; the belt is only about 5 km wide. The main rock types are graywacke, grit, shale, iron-formation, lava, and tuff (Van Eeden *et al.*, 1963), all of which are poorly exposed.

2.4. Archean Cratonization: Granitoid Emplacement in the Eastern Kaapvaal Province

The granitoid terrane covering some 25,000 km^2 around the Barberton belt has been mapped in considerably more detail than the remainder of the Archean basement in the Kaapvaal Province. Outcrop quality is generally good and deep river valleys enable the granitoid intrusions to be examined in three dimensions. Section 2.6 summarizes what is presently known of the remaining areas where granitoids were emplaced from about 3.0 Ga to 2.6 Ga.

After the volcanism and sedimentation of the Swaziland Supergroup, various granitoids were intruded into the greenstone belt and adjacent gneisses (Fig. 2-16). On the basis of field relations and limited geochronologic data, the provisional sequence of events (oldest to youngest) is:

(1) Emplacement of the Granodiorite Suite into gneissic terrane in Swaziland at about 3.35 Ga (J.M. Barton, personal communication, 1979; Rb-Sr).
(2) Emplacement of leucotonalitic–trondhjemitic and tonalitic plutons mainly along the southwestern and northwestern peripheries of the greenstone belt between 3.3* Ga and 3.1 Ga (Oosthuyzen, 1970; U-Pb).
(3) Emplacement of the Nelspruit porphyritic granite (including the Nelspruit migmatite terrane) and Hebron granodiorite north of the Barberton belt at about 3.2 Ga (Robb, 1977; provisional Rb-Sr age).
(4) Emplacement of the Lochiel quartz monzonite along the southeastern flank of the Barberton belt at about 3.0 Ga (Davies and Allsopp, 1976; Rb-Sr).

*Since this volume went to press, one of the largest of these intrusions, the Kaap Valley "granite," has yielded a Rb-Sr whole-rock age of 3491 ± 166 Ma (J.M. Barton, personal communication, 1981).

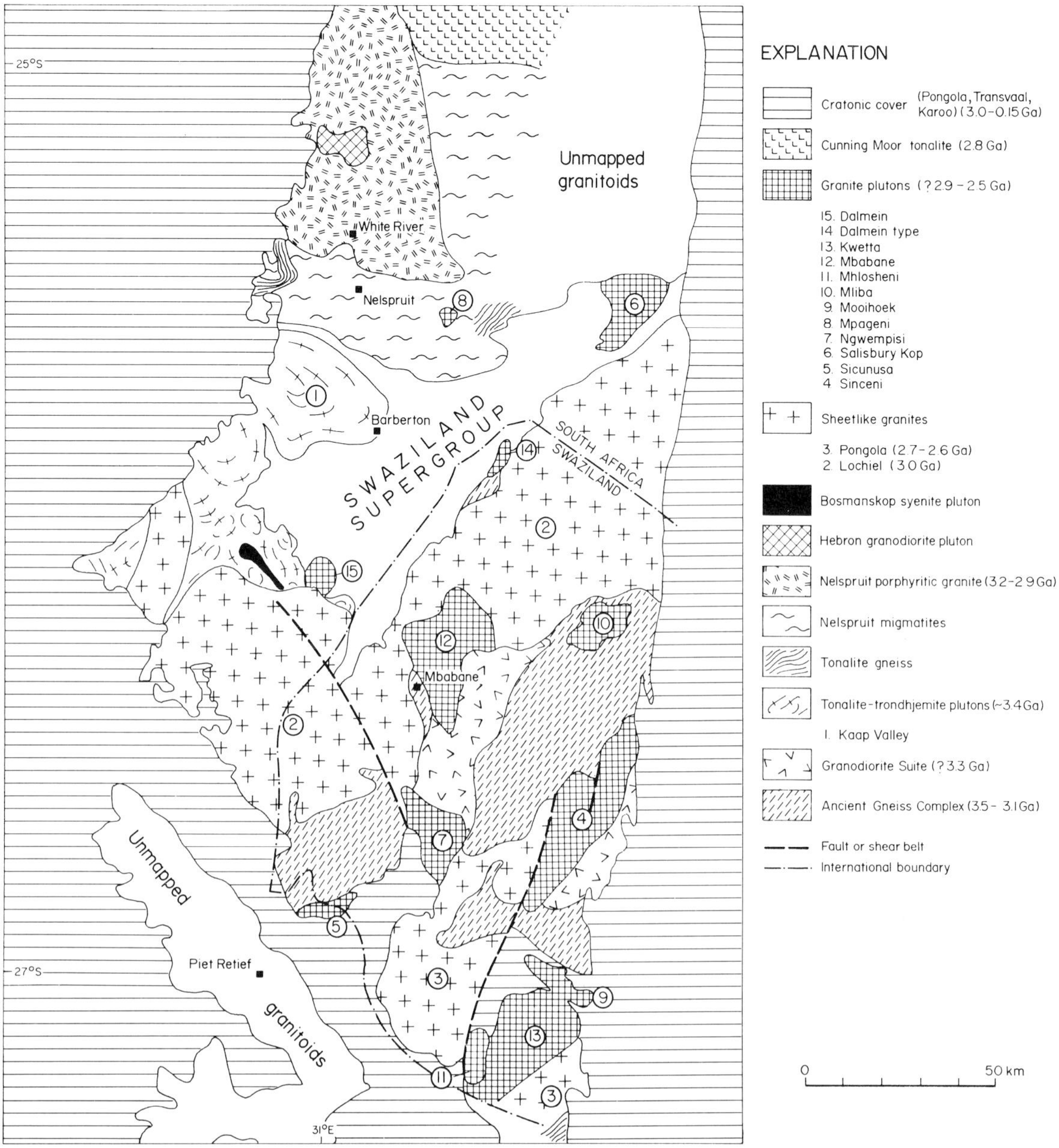

Figure 2-16. Simplified geologic map of the granitoids of Swaziland and the eastern Transvaal. [Adapted from Hunter (1968), Viljoen and Viljoen (1970), Robb (1977), Anhaeusser and Robb (1978).]

(5) Emplacement of coarse-grained granodiorite plutons of Dalmein type between 3.1 Ga and 2.9 Ga (Oosthuyzen, 1970; Davies and Allsopp, 1967; U-Pb and Rb-Sr).

Granodiorite Suite

The Granodiorite Suite, which intrudes the bimodal gneisses of the Ancient Gneiss Complex, has been recognized only in central Swaziland. It consists of coarse-grained, poorly foliated rocks that range in composition from hornblendite to granodiorite. Three discrete intrusions are elongated in a general north-northeast direction. The age of the Granodiorite Suite is uncertain, for it is nowhere found in juxtaposition to the Swaziland Supergroup, but it is intruded by plutons of Mpageni type which set a minimum age of about 2.5 Ga. The parallelism of the elongation of the three intrusions and their foliation with the main axial trace of the Swaziland Supergroup suggests possible contemporaneity of the Granodiorite Suite intrusion with folding of the Swaziland Supergroup (Hunter, 1974a).

Granodiorite and tonalite account for about 70 percent of the area underlain by the Granodiorite Suite, ultrabasic rocks amounting to 0.5 percent, and the remainder consisting of quartz diorite and minor gabbro. Hornblendite and peridotite (the latter consisting of olivine and actinolite) are found as small coarse-grained xenoliths within tonalite and granodiorite. Plagioclase, diopside, hornblende, and chloritized biotite with variable, but minor amounts of quartz are present in the dioritic and quartz dioritic rocks of the suite. The tonalitic and granodioritic varieties consist of plagioclase and quartz with minor amounts of hornblende and biotite.

Leucotonalitic, Trondhjemitic, and Tonalitic Plutons

Plutons, which range in diameter from 2 km to 40 km, intrude the basal members of the Onverwacht Group along the southwestern and northwestern margins of the Barberton greenstone belt. The largest of these plutons is a hornblende-bearing variety to which the name Kaap Valley "granite" was given (Hall, 1918), although it is now known to be tonalitic or quartz dioritic in composition; the other plutons are leucotonalitic or trondhjemitic (Table 2-11). The plutons are generally elliptical or circular in plan, commonly with deformed septa of lower Onverwacht rocks separating individual plutons, each of which is lithologically homogeneous and medium to coarse grained with a strong peripheral schistosity parallel to the margins. The schistosity is defined by biotite or, in the case of the Kaap Valley "granite," by hornblende. The foliation is accentuated near margins of the plutons by the parallel alignment of deformed basic xenoliths. A strong subvertical lineation has been recognized during recent structural studies of some of the leucotonalitic plutons at the southwestern termination of the Barberton belt. This geometry is typical of the solid-state rise of granitoid diapirs (Dixon, 1975; Schwerdtner *et al.*, 1978) and has been interpreted as indicating rise of tonalitic basement rather than intrusion (R.E.P. Fripp, personal communication, 1978). Kaap Valley "granite" consists of hornblende, oligoclase, and quartz with minor amounts of biotite and microcline, whereas the leucocratic varieties contain no hornblende but have a greater abundance of biotite.

Two ages of trondhjemite have been recognized in the Theeboom Valley southwest of the end of the Barberton belt (Anhaeusser and Robb, 1978; Anhaeusser, 1980), but the absolute age difference of the two trondhjemitic intrusions is not known. The mineralogic and chemical compositions of the two trondhjemites are similar (Table 2-12) except for the slight enrichment in Sr and impoverishment in Ba in the younger trondhjemite (Anhaeusser and Robb, 1978; Anhaeusser, 1980).

At the same locality boudinaged quartzofeldspathic layers 15–40 cm wide are present in amphibolite intruded by the older trondhjemite. The quartzofeldspathic rocks are chemically similar but not identical to oceanic plagiogranites (Anhaeusser and Robb, 1978) except that in places they are quartz monzonitic in composition. The origin and relationships of these quartzofeldspathic rocks are not known.

Nelspruit Granitoid–Gneiss Terrane

The Nelspruit porphyritic granite and the Nelspruit migmatites and gneisses underlie a large area north of the Swaziland Supergroup (Fig. 2-16). The migmatites and gneisses crop out along major river valleys, whereas the porphyritic granite is generally confined to higher ground (McCarthy and Robb, 1977). The migmatite–gneiss terrane exhibits a wide variety of layered, porphyroblastic, and nebulitic textures with mafic schlieren, together with more-homogeneous quartz–monzonitic phases. Quartz–feldspar veins and dikes are common. The layered gneisses are complexly folded. Robb (1977) reported compositions that range from leucotonalitic to granitic or quartz monzonitic and suggests that the migmatitic terrane developed as a result of the intrusion of granitic magma, possibly cogenetic with the Nelspruit porphyritic granite. This granitoid is coarse grained and is characterized by the presence of microcline megacrysts, up to 30 mm in length, plagioclase, biotite, and quartz. The granitoid ranges from leucotonalitic to granitic (*sensu stricto*) despite its uniform textural appearance.

The Hebron granodiorite is a medium-grained rock that crops out as a small pluton within the Nelspruit porphyritic granite. Veins and dikes of the granodiorite cut the Nelspruit porphyritic granite, but preliminary Rb-Sr dating suggests that the

Table 2–11. Mean Chemical Analyses of Post-Swaziland Granitoids[a]

	1	2	3	4	5	6	7	8	9	10	11	12	13	14
SiO_2	71.05	64.72	69.72	67.70	70.78	70.94	68.97	70.12	69.01	72.10	78.05	72.76	72.70	71.98
TiO_2	0.27	0.50	0.32	0.42	0.32	0.33	0.36	0.42	0.61	0.32	0.13	0.24	0.30	0.37
Al_2O_3	15.09	15.41	15.52	15.83	14.52	14.51	15.80	13.84	14.17	13.61	11.54	14.17	14.24	14.20
Fe_2O_3	0.68	1.97	2.18[b]	2.83[b]	0.91	0.62	2.52[b]	1.01	1.82	0.71	0.85	0.54	0.55	0.73
FeO	1.23	2.43	—	—	1.24	1.64	—	2.43	2.22	1.61	0.76	1.11	1.22	1.49
MnO	0.05	0.04	0.05	0.05	0.04	0.08	0.06	0.10	0.25	0.04	0.06	0.05	0.05	0.06
MgO	0.92	2.59	0.95	1.72	0.84	0.65	1.61	0.67	0.87	0.49	0.15	0.37	0.39	0.46
CaO	2.48	4.29	2.02	2.41	1.97	1.62	2.72	1.85	2.16	1.41	0.54	1.48	1.53	1.68
Na_2O	5.83	4.88	4.31	4.22	4.55	4.10	4.98	3.22	3.18	3.47	2.62	4.18	4.31	4.25
K_2O	1.73	1.51	3.65	3.18	3.60	4.43	2.13	5.30	4.86	5.29	4.91	4.14	3.99	4.08
H_2O^+	0.56	0.90	0.48[c]	0.59[c]	0.69	0.68	0.45[c]	0.77	1.06	0.43	0.37	0.70	0.63	0.60
H_2O^-	0.13	0.20	—	—	0.12	0.07	—	0.10	0.07	0.08	0.06	0.06	0.05	0.07
P_2O_5	0.06	0.18	0.04	0.11	0.14	0.31	0.01	0.09	0.20	0.11	0.02	0.07	0.10	0.12
CO_2	—	—	—	—	—	—	—	—	—	—	—	0.14	0.08	0.04
Total	99.65	99.59	99.24	99.03	99.72	99.98	99.61	99.92	100.48	99.67	100.06	100.01	100.14	100.13
Rb	51	45	—	—	107	206	—	261	216	255	267	268	340	317
Sr	517	545	—	—	482	240	—	124	328	194	17	221	226	274
Ba	346	450	—	—	682	591	—	610	1150	797	300	—	—	—
K/Rb	281	280	—	—	280	178	—	169	187	172	153	132	102	117
Sr/Ba	1.49	1.21	—	—	0.65	0.41	—	0.20	0.28	0.24	0.06	—	—	—
Rb/Sr	0.098	0.08	—	—	0.20	0.86	—	2.1	0.66	1.31	15.71	1.31	1.50	1.38
K/Na	0.33	0.34	0.95	0.84	0.90	1.21	0.48	1.84	1.71	1.71	2.09	1.11	1.04	1.07
No. of analyses	11	7	18	9	13	9	9	7	4	17	6	4	5	4

[a]1, Leucotonalitic–trondhjemitic plutons. 2, Kaap Valley "granite." 3, Nelspruit porphyritic granite. 4, Hebron granodiorite. 5, Dalmein-type granodioritic plutons. 6, Lochiel granite. 7, Cunning Moor tonalite. 8, Pongola granite. 9, Kwetta-type plutons. 10, Mpageni-type plutons. 11, Mooihoek-type plutons. 12, Homogeneous granodiorite, Johannesburg. 13, Homogeneous granodiorite–adamellite, Johannesburg. 14, "Castle Koppie" granodiorite, Johannesburg. Sources: M.J. Viljoen and Viljoen (1969b), Hunter (1968), Anhaeusser (1973), Robb (1977), Condie and Hunter (1976).

[b]Total Fe given as Fe_2O_3.

[c]Loss on ignition.

Table 2–12. Analyses of Granitoids in Theeboom River, Southern Margin of Barberton Greenstone Belt[a]

	1	2	3	4
SiO_2	74.03	72.70	77.52	73.68
TiO_2	0.19	0.25	0.06	0.12
Al_2O_3	15.08	15.42	14.19	14.92
Fe_2O_3 (t)	1.34	1.56	0.19	0.97
MnO	0.07	0.12	0.01	0.03
MgO	0.45	0.46	0.06	0.48
CaO	1.82	2.30	1.54	1.32
Na_2O	4.98	5.32	6.46	4.64
K_2O	1.83	1.45	0.53	4.17
P_2O_5	0.08	0.11	0.02	0.07
LOI[b]	0.37	1.41	0.39	0.62
Rb	61	52	14	118
Sr	608	759	81	338
Ba	424	226	5	510
K/Rb	260	231	314	301
Rb/Sr	0.10	0.07	0.17	0.35
K/Na	0.42	0.30	0.09	1.00
Sr/Ba	1.43	3.36	16.2	0.67

[a] 1, Mean of 2 analyses of older gneissic trondhjemite (Sr, 6 analyses). 2, Younger gneissic trondhjemite intrusive into 1 (Sr, 3 analyses; remainder, 1 analysis). 3, Quartzofeldspathic boudins in amphibolite (1 analysis). 4, Mean of 2 analyses of quartzofeldspathic layers in amphibolite. Source: Anhaeusser and Robb (1978).

[b] Loss on ignition.

granodiorite and granite are radiometrically contemporaneous. The chemically homogeneous Hebron granodiorite consists of plagioclase, quartz, and biotite with minor amounts of microcline and muscovite (Robb, 1977).

Late-Syntectonic Lochiel Sheeted Batholith

The Lochiel batholith typically consists of massive, medium- to coarse-grained quartz monzonite containing quartz, plagioclase, microcline, and biotite; it underlies a broad zone about 65 km wide on the southeastern side of the Barberton greenstone belt. The quartz monzonite is cut by a complex stockwork of quartz–feldspar pegmatites which rarely exceed 1 m in width. The pegmatites, located within a 5-km-wide belt extending northwestward across the outcrop of the Lochiel quartz monzonite in the vicinity of Mbabane, are mineralized with cassiterite, beryl, yttrotantalite, and monazite.

The deeply incised valleys in Swaziland provide excellent outcrops that reveal the sheetlike form of the Lochiel intrusion (Hunter, 1974a). The Ancient Gneiss Complex in the floors of valleys is cut by pegmatite dikes which increase in abundance together with dikes and sheets of quartz monzonite toward the base of the Lochiel intrusion. The dikes and sheets of quartz monzonite increase in abundance to the extent that only xenoliths of gneiss remain. At higher elevations the quartz monzonite contains progressively fewer gneiss fragments until the rock becomes homogeneous. The complex contact zone between gneisses below and quartz monzonite above occupies a vertical distance of about 200 m. In this respect the Lochiel quartz monzonite has apparent similarity to the Nelspruit porphyritic granite. Although the base of the Lochiel sheet can be readily identified, it is uncertain whether rocks of the Swaziland Supergroup formed the roof everywhere. At the southern end of the Barberton belt in the Motjane schist belt, the Swaziland Supergroup constitutes a local roof but there is no evidence to indicate the original extent of these rocks.

The Lochiel quartz monzonite was emplaced syntectonically in several stages in western Swaziland. Pegmatite dikes cutting the Ancient Gneiss Complex mark the initial stages. The discordant dikes show signs of increasing strain when traced toward the Lochiel sheet, as evidenced by their folding, lesser degree of discordance, and transection by cataclastic zones (Jackson, 1979a). At the contact the effects of cataclasis become penetrative and protomylonitized pegmatites have been rotated to complete concordance with the reworked gneisses. This fabric is truncated by younger phases that are weakly foliated and themselves intruded by massive quartz monzonite.

In northwest Swaziland west of Mbabane (Fig. 2-5) the geometry and relative timing of D2 strains in the Onverwacht Group, the Lochiel granite, and all five generations of dikes and veins injected from the granite correspond very closely (Jackson and Robertson, 1982). Consequently there can be little doubt that the Lochiel granite was emplaced syntectonically with respect to D2 deformation of the Onverwacht Group in the Motjane schist belt. Jackson and Robertson (1982) inferred that the sheetlike batholith spread outward from an entry point into the upper crust southwest of the Dalmein pluton (Fig. 2-16) and that this magmatic emplacement was extremely forceful and nondiapiric.

Posttectonic Dalmein-Type Plutons

Three plutons of the Dalmein type have been recognized, two of which are intrusive into the Swaziland Supergroup: the Salisbury Kop pluton in the northeast and the Dalmein pluton in the southwest. The Mliba pluton lies 40 km east of the Barberton belt and intrudes the Ancient Gneiss Complex. The unfoliated plutons truncate structures in the country rocks, commonly without pronounced disruption. The Salisbury Kop pluton has the largest exposed area, of some 250 km^2. The plutons consist of quartz, plagioclase, microcline, and biotite and are devoid of pegmatites.

Bosmanskop Syenite

The syenitic Bosmanskop pluton intrudes the granitoid terrane near the southwestern end of the Barberton belt and has yielded an age of 3.13 Ga (Oosthuyzen, 1970). The pluton forms an oval area of low hills measuring approximately 6 km by 4 km and is elongated in a northwest direction. Although no detailed account of the geology of the Bosmanskop pluton has yet been published, Anhaeusser (1975) has briefly summarized the salient features. The rocks vary in composition from syenite through syenodiorite to granodiorite and are very coarse grained in the north, coarse grained in the center of the pluton, and medium to fine grained in the south. Locally the tabular feldspar and elongate ferromagnesian minerals show parallel alignment, a feature interpreted as a flow structure by Anhaeusser (1975). Many of the large feldspar crystals are zoned and graphic quartz–feldspar intergrowths are common. Mafic minerals include augite, biotite, and hornblende–actinolite. Present geochemical data suggest similarity to the syenitic Linden and Icarus plutons in Minnesota which also form posttectonic intrusions (Arth and Hanson, 1975). Preliminary data from the Bosmanskop pluton show a low initial $^{87}Sr/^{86}Sr$ ratio, high contents of Ba and Sr (>2000 ppm), and low Rb contents (about 140 ppm) for the syenitic rocks (Anhaeusser, 1975). The primitive initial ratio together with the low $\delta^{18}O$ values suggest that the pluton has been derived from partial melting or differentiation of basic rock of mantle origin.

Petrogenetic Processes During Cratonization

Table 2-11 lists the mean compositions of granitoids from Swaziland, eastern Transvaal, and the Johannesburg Dome, where the most detailed geochemical studies have been undertaken.

The post-Swaziland tonalitic and trondhjemitic plutons are characterized by K/Na ratios <0.5, variable K/Rb ratios (200–350; $\bar{x}$ = 280), Sr/Ba ratios > 1.0, and a limited range of Rb/Sr ratios close to 0.1. They exhibit light REE enrichment of between 20 and 45 times the chondrite mean, but without significant Eu anomalies (Fig. 2-17A). The leucocratic tonalites and trondhjemites show extreme depletion in heavy REEs in contrast to the flatter, heavy REE pattern of the Kaap Valley "granite." The REE patterns of the leucotonalites and trondhjemites are similar to those reported for other Archean tonalites. Condie and Hunter (1976) used the geochemical data to test models of fractional crystallization and partial

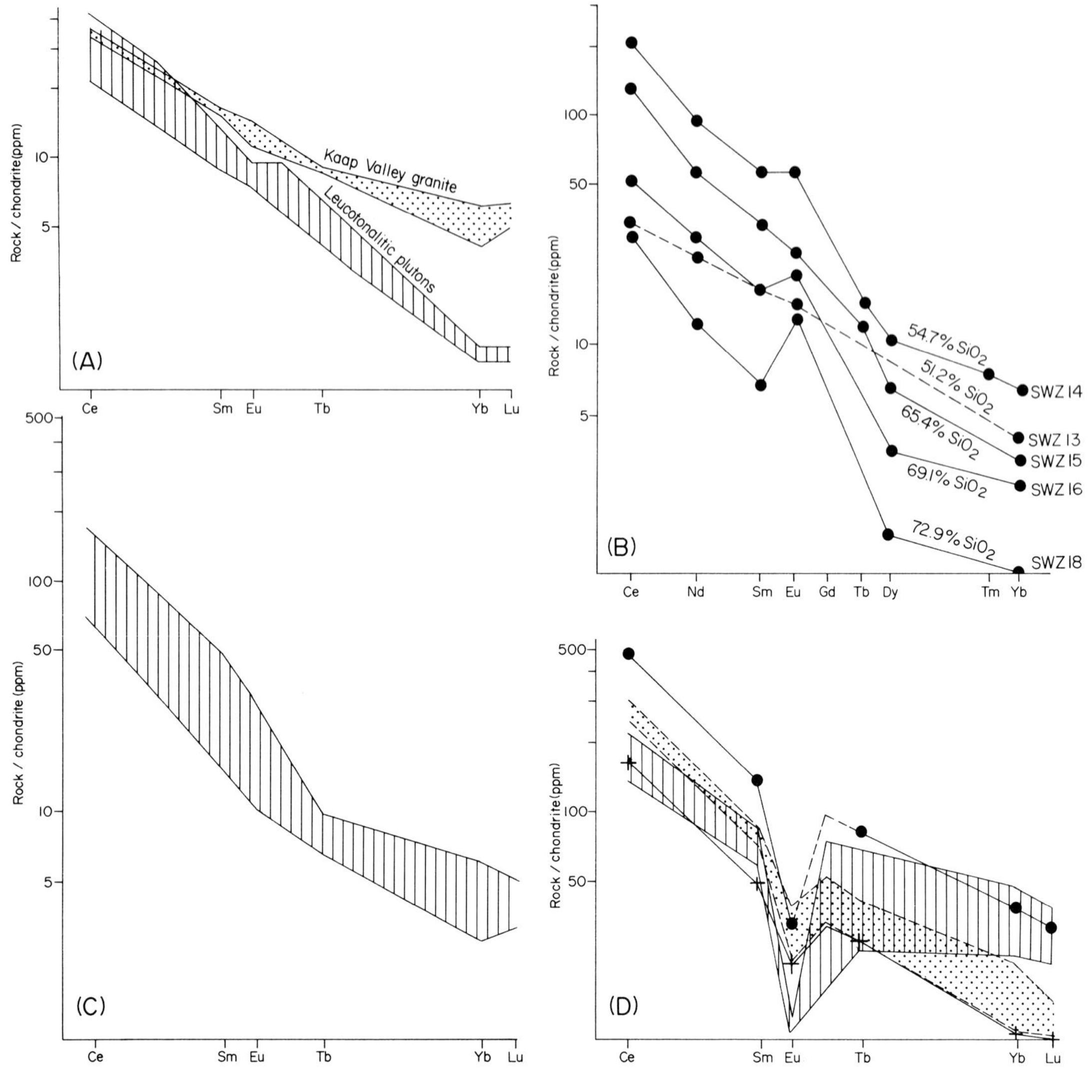

Figure 2-17. Chondrite-normalized REE plots of granitoids in eastern Transvaal and Swaziland. (A) Kaap Valley "granite" and leucotonalite plutons. (B) Granodiorite Suite. (C) Field of Dalmein-type plutons. (D) Lochiel quartz monzonite (crosses), Pongola granite (circles), field of Mpageni, Mbabane, Kwetta, and Ngwempisi plutons (stipples) and field of Mhlosheni, Mooihoek, and Sinceni plutons (vertical ruling). [Sources: Condie and Hunter (1976), Glikson (1976), Hunter *et al.* (1978).]

melting for the tonalitic magmas, the best match being obtained in a model with about 10 percent melting of eclogite of average Archean tholeiite composition. An alternative origin by partial melting of Onverwacht Group tholeiite basalt, deformed and metamorphosed to amphibolite grade, has also been suggested in view of the apparent absence of eclogite in Archean terranes (Hunter *et al.*, 1978). Both proposals imply a mantle origin for the tonalitic liquids, which is consistent with the low $\delta^{18}O$ values for these tonalites (Taylor and Mordeckai, 1975).

The wide compositional range reported for the Nelspruit porphyritic granite may result from fractional crystallization of a slowly cooling magma (McCarthy and Robb, 1977) derived from the mantle or lower crust (Robb, 1977). McCarthy and Robb (1977) based their proposal on the behavior of Ba, Sr, Rb, K_2O, and N_2O in theoretical models involving fractional crystallization of the assemblage plagioclase–quartz–biotite followed by plagioclase–quartz–biotite–K-feldspar.

Robb (1977) suggested that the Nelspruit migmatite terrane represents intrusion of a magma cogenetic with the porphyritic granite into gneisses; the origin of the latter is not specified. The possibility cannot be excluded that the Nelspruit migmatite terrane is a counterpart of the mixed zone of older tonalitic gneisses and Lochiel quartz monzonite seen in western Swaziland. It has been noted that intrusion of the Lochiel quartz monzonite was accompanied in places by strong reworking of the contiguous gneisses. Similar reworking may account for the diffuse appearance of the gneisses reported by Robb (1977).

Robb (1977) suggested that the Hebron granodiorite may have been intruded broadly synchronously with the Nelspruit porphyritic granite. The Hebron granodiorite differs in major-element chemistry from the Nelspruit porphyritic granite mainly in respect to its higher MgO and lower SiO_2 contents, but the available data are inadequate to postulate genetic models.

The Granodiorite Suite in Swaziland ranges in composition from ultrabasic to acid. Table 2-13 illustrates the compositional variations of five samples for which REE data are available. The K_2O contents reach a peak at intermediate values of SiO_2 and generally decrease with increasing SiO_2. Strontium contents are high, whereas Rb contents are low, but both trace elements show an irregular decrease with increasing SiO_2. The K/Rb ratios show a wide scatter, but Rb/Sr ratios have a narrow range (Hunter *et al.*, 1978). The five REE patterns show a distinctive trend (Fig. 2-17B). The gabbro (SWZ-13) has a rather steep pattern similar to that of alkali basalts, whereas diorite (SWZ-14) is enriched in light REEs. Patterns for rocks of increasing SiO_2 content are successively depleted in all REEs except Eu. This unusual trend distinguishes the Granodiorite Suite from other tonalitic rocks in the Barberton area but it is similar to that shown by the gabbro–trondhjemite suite of the Uusikaupunki–Kalanti area in Finland (Arth *et al.*, 1978) except for the greater abundances of light REEs in the Granodiorite Suite (Hunter *et al.*, 1978). It is presumed that the Granodiorite Suite was generated by differentiation of a mildly alkaline basaltic liquid in which hornblende and biotite were the major crystallizing phases (Hunter *et al.*, 1978).

The Dalmein-type plutons (Table 2-11) are granodioritic in composition, differing from the

Table 2-13. Representative Analyses from the Granodiorite Suite[a]

	1	2	3	4	5
SiO_2	51.25	54.73	65.44	69.15	72.85
TiO_2	0.28	0.75	0.70	0.41	0.27
Al_2O_3	17.69	20.69	16.42	15.61	14.60
Fe_2O_3	1.16	2.04	1.74	1.15	0.33
FeO	5.24	3.74	2.88	2.30	1.28
MnO	0.13	0.10	0.06	0.06	0.02
MgO	7.70	2.21	1.43	1.07	0.68
CaO	9.88	5.51	3.98	3.86	2.72
Na_2O	2.82	5.22	5.03	4.71	4.79
K_2O	0.88	2.97	1.24	0.93	1.41
H_2O^+	2.41	0.60	0.45	0.39	0.36
H_2O^-	0.12	0.06	0.04	0.02	0.03
P_2O_5	0.15	0.38	0.25	0.12	0.07
CO_2	0.04	0.36	0.05	0.07	0.24
Cl	0.02	0.02	0.01	0.01	0.01
F	0.05	0.13	0.06	0.02	0.02
Less ≡ 'O	0.02	0.05	0.03	0.01	0.01
Total	99.80	99.46	99.75	99.87	99.67
Rb	60.0	59.1	18.2	9.9	29.3
Sr	697	1249	727	414	529
K/Rb	126	406	555	786	416
Rb/Sr	0.09	0.05	0.03	0.02	0.06

[a] 1, Gabbro. 2, "Quartz diorite." 3 and 4, Tonalite. 5, Trondhjemite.

closely contemporaneous Lochiel quartz monzonite with respect to their lower contents of K_2O and Rb and higher contents of Sr. Compositional variations between individual plutons are minor. Their Rb/Sr ratios range from 0.13 to 0.36, K/Rb from 270 to 380, and Sr/Ba from 0.6 to 0.7. The REEs are enriched compared with the tonalitic plutons and they have a relative REE distribution pattern not unlike the leucotonalitic plutons (Fig. 2-17C). A model involving about 50 percent melting of a siliceous garnet-bearing granolite gives the best fit for all elements except Sr (Condie and Hunter, 1976). It was found that a wide range of mineral proportions was allowable, particularly for quartz, biotite, and K-feldspar (Condie and Hunter, 1976). However, the predicted Sr content is too low in the models tested, and three factors may be responsible (Condie and Hunter, 1976):

(1) The feldspars had lower distribution coefficients than those used in the calculations.
(2) The assumed Sr content of the parent is low by a factor of two.
(3) The magmas were selectively contaminated with Sr.

If the magmas were selectively contaminated with Sr, the initial $^{87}Sr/^{86}Sr$ ratios (0.7020 and 0.7033) reported for the Mliba and the Salisbury Kop plutons (Oosthuyzen, 1970; Davies and Allsopp, 1976) would support a crustal origin for these magmas.

The Lochiel intrusion ranges from quartz monzonitic to granitic in composition with SiO_2 contents of 64–75 percent. K_2O is present in equal or greater amounts than Na_2O except for those samples from the area that contains tin-bearing pegmatites. These samples also show significant enrichment in Sr (up to 955 ppm) compared to an average content of about 240 ppm in the remaining samples. The Rb content of all samples falls within a narrower range, from 153 to 274 ppm. The REE pattern for a composite sample of the Lochiel quartz monzonite (Fig. 2-17D) is similar to that of the Dalmein-type plutons except that it has a large negative Eu anomaly and flatter slope for the heavy REEs. Nevertheless, the Lochiel granite differs from the latter by its higher K_2O, Rb, Rb/Sr, and K/Na and, with the exception of the tin belt samples, by its lower Sr content. The initial $^{87}Sr/^{86}Sr$ ratio for the Lochiel quartz monzonite is high, 0.7054 (Davies and Allsopp, 1976). These data, together with the relatively high $\delta^{18}O$ values reported from the Lochiel quartz monzonite (Taylor and Mordeckai, 1975), suggest that it was derived from crustal sources.

2.5. Pongola Supergroup: The Oldest Cratonic Cover

The earliest widespread cratonization occurred in South Africa over the southern half of the Kaapvaal Province. The Archean Pongola Supergroup with many Proterozoic characteristics developed on this stabilized basement. This sequence is exposed in Swaziland and the southeastern Transvaal, and in northern Natal as scattered inliers surrounded by Karoo cover (Fig. 2-18). The supergroup unconformably overlies Archean granites. A twofold division of the Pongola Supergroup into the lower Nsuze Group and the upper Mozaan Group is recognized in the Vryheid–Piet Retief and White Mfolozi areas (Matthews, 1967). The contact between the two groups is gradational in the north (Watchorn, 1978) and unconformable farther south (Matthews, 1967). In the most northern and southern regions only the Nsuze Group is developed.

The age of the Pongola Supergroup is not precisely known. Nsuze lavas have yielded an age of 3.09 Ga (Burger and Coertze, 1973; U-Pb), whereas to the east of Ermelo the Mozaan Group rests nonconformably on 3.07 Ga granites, which provide a maximum age (Hunter, 1974b; Rb-Sr). The succession is intruded by the 2.87 Ga Usushwana Intrusive Suite, which provides a minimum age (Davies *et al.*, 1970; Rb-Sr). These age determinations and the stratigraphic relations with other dated intrusions are consistent with a general age of 2.9–3.1 Ga for the Pongola Supergroup.

Detailed stratigraphic and geochemical studies on the Pongola Supergroup are confined to the White Mfolozi and Vryheid–Piet Retief areas (Figs. 2-18 and 2-19). The Nsuze Group is about 7000 m thick in the type area east of Paulpietersburg and thins northward into Swaziland. In the White Mfolozi inlier terrigenous clastic, calcareous, and cherty sediments constitute a significant proportion of the attenuated volcanic sequence. A major quartzite that rests on older granitoids forms the base of the group east of Paulpietersburg, but its continuity is disrupted to the north and south by the Usushwana Intrusive

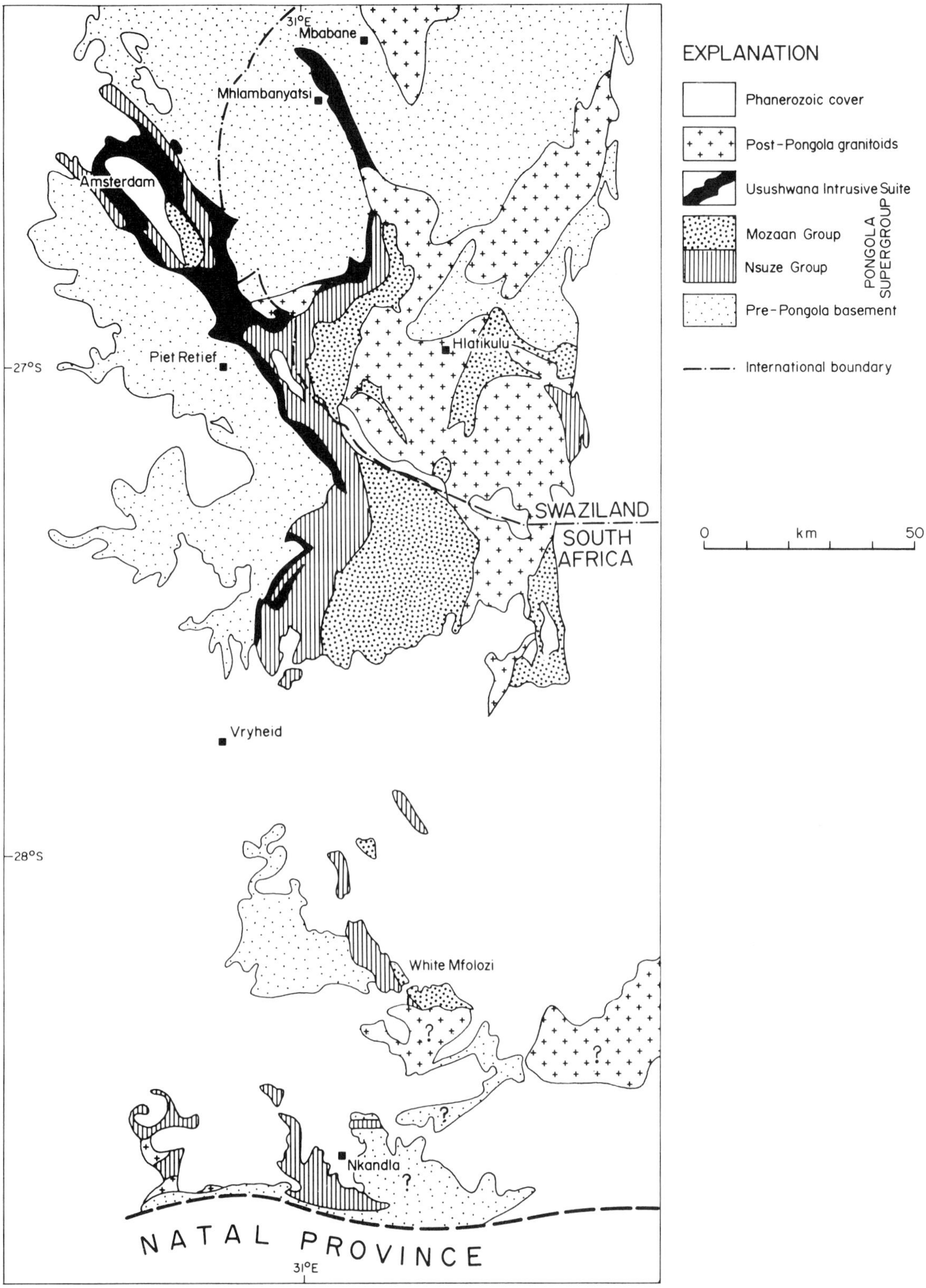

Figure 2-18. Distribution of the Pongola Supergroup, Usushwana Intrusive Suite, and post-Pongola granitoids.

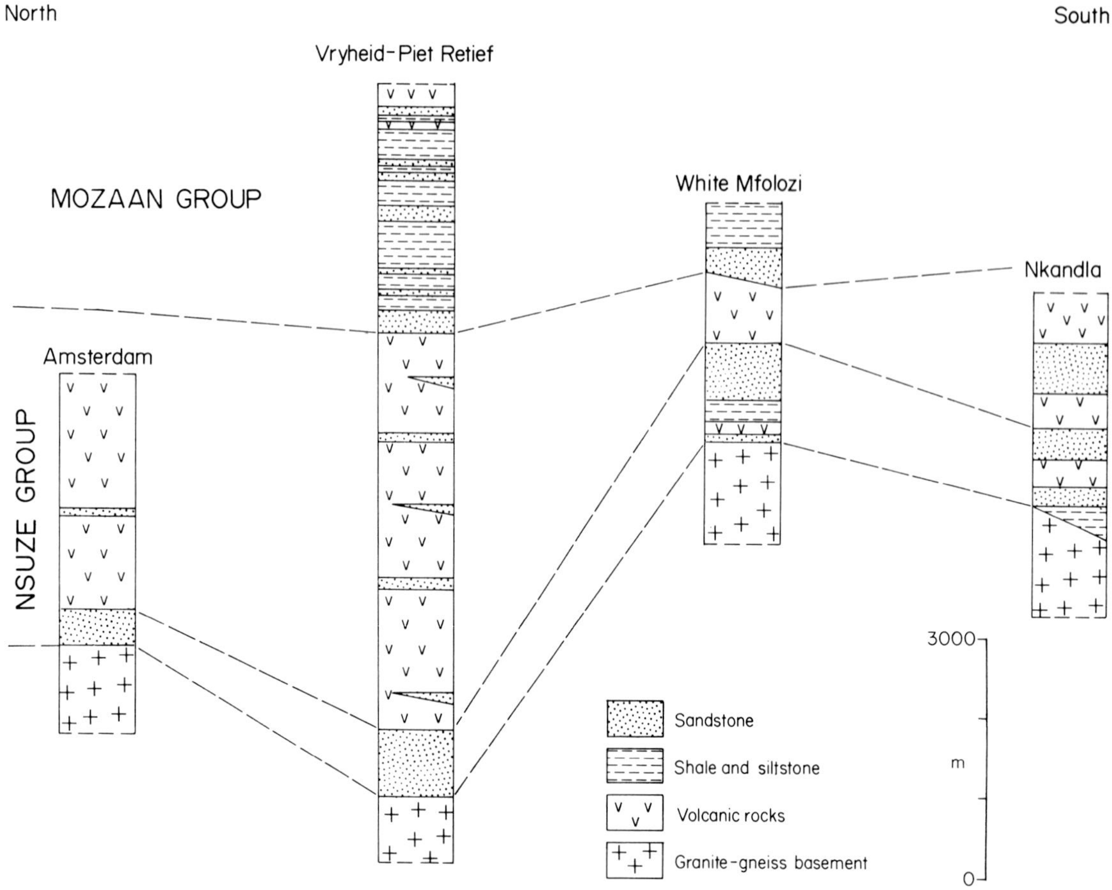

Figure 2-19. Stratigraphic sections through the Pongola Supergroup (see Fig. 2-18 for locations). [Adapted from Matthews (1967), Watchorn (1978, 1980a).]

Suite. In the type area tuffaceous lenses become more abundant toward the top of the group. In Swaziland prominent aluminous phyllitic and schistose beds are interlayered with the volcanic rocks and pinch out near the Swaziland–Transvaal border.

Sandstone and shale are the main rock types in the Mozaan Group, which reaches a maximum thickness of 1800 m in the Vryheid–Piet Retief inlier (Figs. 2-18 and 2-19). Subordinate iron-formation is commonly associated with the shale. Alumina-rich sediments are also present at the base of the Mozaan Group and at stratigraphically higher levels.

Nsuze Volcanism: Tholeiitic or Calc-Alkaline?

The Nsuze volcanic sequence consists of interfingering basalt, basaltic andesite, dacite, and rhyolite. Indisputable pillow structures have been recognized only in the White Mfolozi inlier, but amygdaloidal and phenocrystic varieties are more common. Thicknesses of individual flows range from less than 10 m in the rhyolites to between 50 m and 100 m in the basalts. Flow-top structures and features suggestive of gas streaming are present.

Chemical data from the type area (Table 2-14) show that SiO_2 ranges from 53 to 73 percent. Other major-element oxides, with the exception of Na_2O, K_2O, and Fe_2O_3 (t), also show well-defined regular trends with respect to crystallization and modified Larsen indices. In contrast to the Barberton supracrustal belt, high-MgO lavas are apparently absent and MgO ranges from about 9 percent in some basalts to less than 1 percent in the rhyolites.

The mean compositions of the Nsuze lavas (Table 2-14) are characterized by high concentrations of Fe_2O_3 (t) and relatively low contents of Al_2O_3. Although the lavas are tholeiitic on an A–F–M diagram, REE patterns show closer affinities with calc–alkaline suites. The most primitive

Table 2–14. Major and Trace-Element Chemistry of Nsuze Lavas (Calculated Anhydrous)

	Basalt	Basaltic andesite	Andesite	Dacite	Rhyolite
SiO_2	53.36	55.30	62.18	68.31	73.40
TiO_2	0.90	1.07	0.98	0.73	0.46
Al_2O_3	14.65	14.86	13.38	12.73	12.08
Fe_2O_3	1.99	3.00	3.14	2.37	1.72
FeO	9.95	10.00	7.85	5.91	3.41
MnO	0.27	0.28	0.31	0.14	0.10
MgO	6.25	4.24	2.08	0.93	0.24
CaO	9.18	6.94	5.06	3.03	1.19
Na_2O	2.48	2.87	2.66	2.30	2.23
K_2O	0.74	1.21	2.10	3.37	5.09
P_2O_5	0.18	0.22	0.26	0.18	0.10
Ba	147	247	344	391	610
Rb	19	35	78	114	149
Sr	325	255	160	145	95
Nb	6.3	8.7	14	15	26
Y	26	34	38	40	63
Zr	131	177	231	268	423
Cu	57	50	28	22	15
Ni	98	48	17	11	10
Zn	97	109	70	82	87
No. of analyses	12	12	8	7	9

Source: N.V. Armstrong (personal communication, 1980).

basalt yet analyzed has a Ce content nearly 30 times the chondrite mean, with Yb about 7.5 times that of chondrites. With increasing SiO_2 the total REE contents increase, as does the negative Eu anomaly. Rhyolites have Ce and Yb contents 150 and 250 times those of chondrites.

Attempts to describe the tectonic setting of the lavas have proved inconclusive. The scatter of Sr, Ba, and Fe_2O_3 (t) on Larsen and crystallization index diagrams presumably reflects the mobility of these elements as a result of alteration at the time of extrusion or during subsequent metamorphism. The Nsuze lavas have been metamorphosed to greenschist facies but they are rarely intensely deformed. Plots of Nsuze basalts on Ti–Y–Zr and Ti–Zr diagrams fall in the field of calc-alkaline basalts, but the reliability of these tests as a means of discrimination is suspect because Ti may be mobile during submarine alteration. The REE patterns for the various magma types suggest calc–alkaline characteristics, which apparently conflicts with the marked iron-enrichment trend on an A–F–M diagram, the high Ni and low Al_2O_3 contents, the moderate TiO_2 contents reaching a maximum at intermediate compositions, and the low rate of SiO_2 increase with increasing FeO (t)/Mg ratios. The weight of evidence thus favors a tholeiitic rather than a calc–alkaline affinity.

Fluvial Feeder Systems

Fluvial sediments in the Mozaan Group of the Vryheid–Piet Retief area comprise conglomerate, cross-bedded sandstone, and shale (Figs. 2-19 and 2-20A). Quartz arenite is the dominant lithology and is structured mainly by planar cross-beds in sets 30–100 cm thick. Tangential foresets commonly contain regressive ripples indicating flow separation (Jopling, 1965). Trough cross-beds are less common. Slipface orientations indicate deposition by southeastward flowing rivers. Plane-beds—displaying internal grading but lacking primary current lineations—are analogous to the structures produced by the migration of low-amplitude, long-wavelength sand waves (N.D. Smith, 1971b; McBride *et al.*, 1975). Intercalated lenticular mudstones, less than 30 cm thick, are horizontally laminated. Local conglomerates consist of pebbles and cobbles of chert, quartz, and quartzite forming pebble stringers or beds up to 2 m thick. The conglomerates are intercalated within cross-bedded arenites, and the thicker units are generally overlain by plane-bedded sandstones.

Watchorn (1978, 1980a) suggested that this lithofacies and the association of sedimentary structures (Fig. 2-20A) are similar to the Platte River-type model of braided-stream deposits erected by Miall (1977). The ubiquitous planar cross-beds represent superimposed linguoid bars with the foreset characteristics determined principally by stream velocity. The trough cross-beds probably represent channel-floor migration of dunes during falling discharge. A characteristic of the Platte River is the sporadic development of longitudinal gravel bars; the Mozaan conglomerates may have formed in a similar manner during intense discharge. The plane-bedded sandstones reflect bar-top accretion at falling water stage. Braided river systems experience rapid switching of the active channels. Abandoned portions of the floodplain become stagnant overbank environments in which suspension settling of clays produces laminated mudstones of the type intercalated within this fluvial sequence.

Tide-Dominated Near-Shore Sedimentation

Shallow subtidal and tidal-flat sediments (Fig. 2-20B) are common in the Mozaan Group in the Vryheid–Piet Retief and White Mfolozi areas and in the Nsuze Group of the Mpongoza and White Mfolozi inliers (Von Brunn, 1974; Von Brunn and Hobday, 1976; Mason and Von Brunn, 1977; Von Brunn and Mason, 1977; Watchorn, 1978, 1980a).

Isolated or stacked upward-fining sequences were probably formed by progradation of tidal-flat sediments across subtidal environments on a nonbarred macrotidal coast; the northwest coast of Australia is a modern example (Gellatly, 1970). The lower part of the facies sequence comprises quartz arenite structured by southeastwardly-directed planar and trough cross-beds and ripple cross-lamination (Fig. 2-20B). The cross-bed azimuths are indicative of strong ebb currents which, at shallower depths, formed scour surfaces, channels, and pebble sheets. Ripple trends (Fig. 2-20B) reflect flow modification about the large bedforms during shallowing and emergence. The ripples include linear and interference forms. Herringbone cross-bedding and reactivation surfaces are evidence of periodic current reversals. Above the arenites, thin sandstones with interbedded mudstones display an upward transition from flaser through wavy to lenticular bedding, which was produced by alternating bedload and suspension sedimentation processes on a midtidal flat. The sandstones contain flat-topped, double-crested, and interference patterns, and the mudstones show evidence of desiccation. These features indicate ebb runoff with emergence and periodic exposure of the tidal-flat surface. Thick mudstones that cap the facies sequence represent upper tidal-flat suspension sediments formed during slack-water periods. Associated wavy lamination and low-amplitude ripples in siltstones were generated by waves. Chemically precipitated jaspilites are sporadic.

In the Nsuze Group of the White Mfolozi, 30-m-thick carbonate units are developed within the volcanics. These comprise carbonate sands and stromatolites; sedimentary structures in the sands include herringbone cross-bedding. Algal-chip conglomerates are evidence of subaerial exposure, desiccation, and subsequent erosion and, together with indications of current reversals, imply deposition on a tidal flat (Von Brunn and Mason, 1977).

Shelf Environment

Shelf sediments in the Mozaan Group of the Vryheid–Piet Retief area consist of iron-formation and shale with subordinate siltstones, characteristically arranged in an upward-coarsening pattern (Figs. 2-19 and 2-20C). Iron-formation at the base directly overlies Nsuze volcanic rocks and comprises interlaminated chert–jasper and magnetite with rare hematite. These sediments have been attributed to chemical precipitation in distal shelf environments farthest from terrigenous clastic influx. Cyclic Eh and pH fluctuations are inferred for the alternating ferruginous and siliceous laminae, subtle chemical variations being related to the pulsatory flow of a reduced circulating brine (Watchorn, 1978). During periods of inactive circulation, oxygen was possibly produced on the sea floor by algal photosynthesis—or, less probably, through the dissociation of water molecules—and was responsible for oxidation of divalent iron and precipitation of magnetite and hematite. Photosynthesis

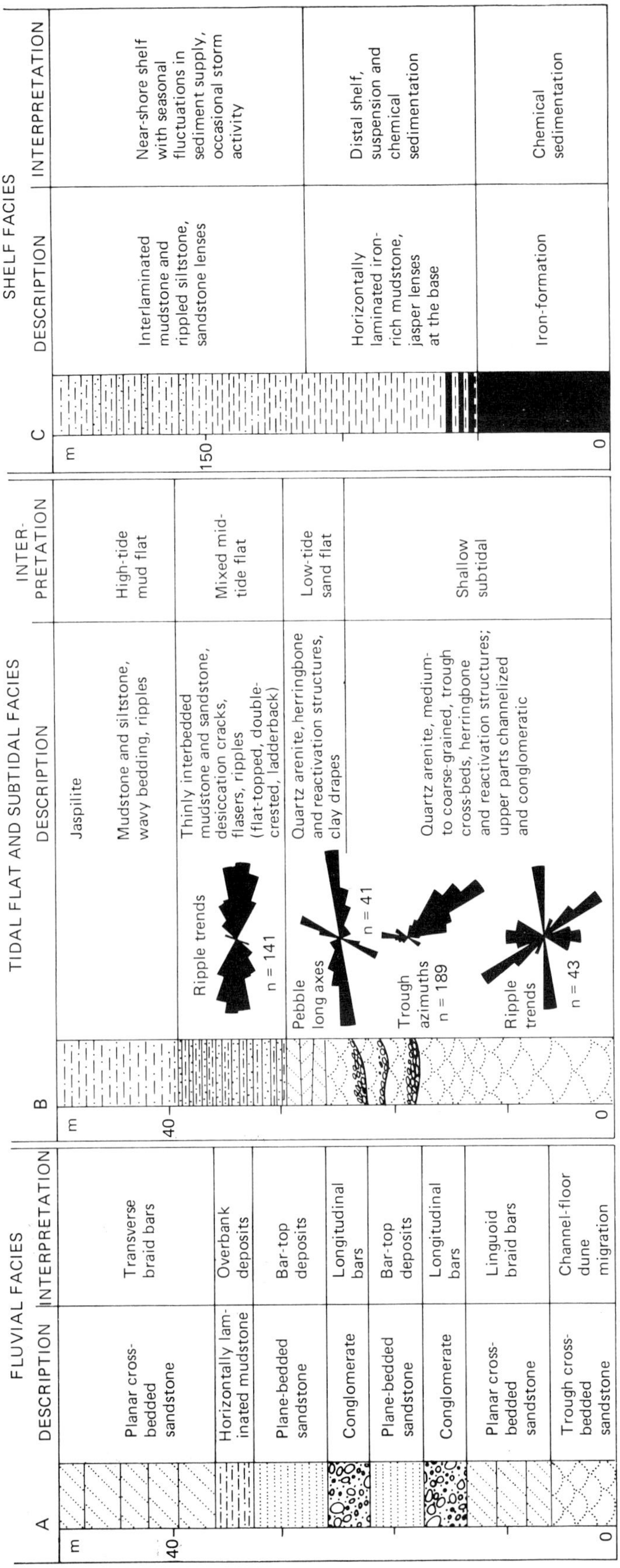

Figure 2-20. Interpretative stratigraphic sections for Pongola Supergroup. [After Von Brunn and Hobday (1976), Watchorn (1978, 1980a).]

would also have increased pH through removal of CO_2 from sea water, resulting in a greater solubility of silica. With renewed brine circulation, equilibrium between a reducing CO_2-rich atmosphere and the hydrosphere would have lowered pH and promoted silica precipitation. The iron and silica in the iron-formation were probably contributed by late-stage volcanic exhalations and were undoubtedly highly concentrated in the hydrosphere. Iron would have remained in its soluble divalent state under the reducing atmospheric conditions, and silica probably concentrated in solution in the absence of silica-secreting organisms.

The proportion of chemical sediments decreases upward in the succession as mudstones become more prominent (Fig. 2-20C). Lower in the sequence the mudstones are highly ferruginous and manganiferous but become progressively less so with upward-increasing siltstone content. Grading is common in the siltstones, which also display sinusoidal and type-B ripple-drift cross-lamination and current ripples. These may represent storm deposits.

The vertical ordering of facies (Fig. 2-20C) reflects increasing proximity to a clastic supply on a dominantly low-energy shelf (cf. Coleman and Gagliano, 1965).

Implications to Earth History

Despite the uncertainty surrounding the identity of the Nsuze magma type, there is little doubt that the environment of extrusion was a volcanic arc. Whether this was in an island arc or continental setting is less certain, although the presence of granitoid basement and thick arenite at the base of the Nsuze Group strongly favors the latter. Whatever the setting, the chemistry of the extrusives is very different from those in the Barberton supracrustal belt. Furthermore, extrusion of the Nsuze lavas was essentially contemporaneous with the more typically Archean greenstone volcanisms in Zimbabwe, indicating that two distinct environments prevailed 3.0 Ga ago.

The textural and mineralogical maturity of the subtidal and intertidal sandstones in the Pongola Supergroup is a consequence of prolonged abrasion by tidal processes. However, fluvial sediments in the Mozaan Group are also mature (Watchorn, 1978, 1980a), suggesting that provenance was equally important. The Pongola sediments were transported by southeasterly flowing rivers, probably from a Moodies source terrane containing texturally mature feldspathic sandstones.

Complete tidal-flat sequences in both the Nsuze and Mozaan Groups vary from 12 m to 25 m thick. Von Brunn and Hobday (1976) suggested that the thick tidal facies sequences in the Pongola basin were a response to abnormally high tidal ranges in the Precambrian (cf. Cloud, 1972). This conforms to the model of a closer moon in the Precambrian following lunar capture (Wise, 1969; Singer, 1970). Nevertheless it has already been shown that parts of the Moodies Group accumulated on a microtidal coast and that basin configuration probably determined tidal range. Along Holocene coasts tidal amplification is associated with embayments (Hayes, 1975; Mofjeld, 1976) or with wide continental shelves (Redfield, 1958). In contrast to the continental-margin setting of the Fig Tree and Moodies Groups, the Pongola Supergroup accumulated on a partly stabilized craton where the frictional effects of a broad cratonic shelf caused amplification of the tidal wave. The "open sea" probably lay to the south in an area now occupied by the high-grade Natal Province (Fig. 1-2).

Stromatolites in the Pongola Supergroup are among the oldest yet discovered (Mason and Von Brunn, 1977) and, like others predating the development of extensive shelf carbonates at approximately 2.3 Ga (Button, 1973a,b,c), they are associated with volcanic rocks (Macgregor, 1940; Joliffe, 1965; Schopf *et al.*, 1971; Henderson, 1975; Martin *et al.* 1980; Orpen and Wilson, 1981). This has led to speculation that liquid volcanic emissions, accompanied by exhalative gaseous emanations, may have provided sufficient nutrients for biogenic processes to have taken place in the Early Precambrian seas. In an overall reducing environment, such as has been proposed for the Early Precambrian (Cloud, 1976), photosynthesizing organisms could have produced sufficient oxygen to account for the red pigmentation in jasper associated with tidal-flat mudstones (Von Brunn and Hobday, 1976) as well as the hematite of the offshore shelf facies (Watchorn, 1978, 1980a).

2.6. Post-Pongola Magmatism

Volcanism and sedimentation in the Pongola basin were terminated at about 2.8 Ga by the intrusion of the basic–ultrabasic Usushwana Intrusive Suite, which was followed by renewed granitoid magmatism between 2.7 Ga and 2.5 Ga.

Usushwana Intrusive Suite

The Usushwana Intrusive Suite, first identified in Swaziland and named the Usushwana Complex by Hunter (1950), crops out in three discrete areas separated by two younger plutons (Fig. 2-18). The intrusion has been dated at 2874 ± 30 Ma with an initial $^{87}Sr/^{86}Sr$ ratio of 0.7029 (Davies *et al.*, 1970).

The northernmost outcrop is a northwest-striking dike 40 km long and with a maximum width of 6.5 km. An arcuate outcrop of gabbro and granophyre, apparently sill-like, extends for 25 km toward the southwest; there it is intruded by a small granitic pluton separating this outcrop from a second dike that strikes northwest along the Swaziland–Transvaal border. The Usushwana Intrusive Suite continues toward the southeast into Natal as a sill at the base of the Pongola Supergroup.

Layering in the complex has been largely destroyed by widespread intrusion of quartz gabbro and is best preserved in the northwestern part of the northernmost dike. The oldest phase is a coarse-grained olivine pyroxenite consisting of olivine, hypersthene, augite, and minor interstitial plagioclase. Rocks of gabbroic composition exhibit the most prominent layering, which results from variations in grain sizes as is shown by the following sequence from the northwestern flank of the gently synclinal northernmost dike:

(9) Coarse-grained hypersthene gabbro with layers up to 60 m thick of medium-grained hypersthene gabbro (at least 750 m thick)
(8) Leucocratic spotted, plagioclase rock (3 m)
(7) Coarse-grained hypersthene gabbro; locally with disseminated sulfides (12 m)
(6) Fine- to medium-grained gabbro (75 m)
(5) Discontinuous lenses of serpentinite (thickness not known)
(4) Medium- to coarse-grained quartz gabbro (450 m)
(3) Zone of variably grained gabbro (including pegmatoid) with disseminated sulfides (10 m)
(2) Medium- to coarse-grained quartz gabbro (105 m)
(1) Medium-grained gabbro (15 m)

The quartz gabbro in layers 2 and 4 represent intrusions that swamped and disrupted the regularity of the layering so that in most outcrops layered rock types are found as xenoliths within quartz gabbro.

The southeastern part of the northernmost dike consists mainly of quartz gabbro overlain by granophyre, with magnetite-rich, altered gabbro at the interface between these units. Remnants of the Pongola Supergroup are preserved above the granophyre and in this respect these outcrops of the suite resemble the southernmost dike in the Transvaal. The altered gabbro consists of highly saussuritized plagioclase surrounded by micropegmatitic intergrowths and of anhedral magnetite grains enclosed in laths of green-brown hornblende, in which relicts of pyroxene are locally preserved. As the abundance of intergrowths between quartz and plagioclase increases, the altered gabbro grades into granophyre. Granophyric textures are prominent except at the top of the sill at its southwestern end in Swaziland, where the rock is microgranitic in texture.

Aplogranitic dikelets from 1 m to 20 m in width fill joint planes in the gabbroic rocks. The aplogranite consists of plagioclase, quartz, and epidote, with minor biotite and rare micropegmatitic intergrowths. Less abundant dikes of coarse-grained pyroxenite, consisting mainly of hypersthene, appear to have filled joints in the layered rocks in the northwestern part of the northernmost dike.

Disseminated chalcopyrite and pyrrhotite are prominent particularly in the well-layered part of the northernmost dike, but no mineralization of economic grade is known. The sulfides are present either as coarse-grained aggregates associated with gabbro pegmatite or are more finely disseminated in a melanocratic gabbro in which hypersthene laths measure up to 3 cm long.

Gravity data from the northernmost outcrop (Burley *et al.*, 1970) indicate that the intrusive suite has the form of a steep-sided dike but the geometry of the other parts is less clear. The outcrop between the Ngwempisi River and the Sicunusa pluton (Fig. 2-18) in Swaziland is apparently a sill intruded at the base of the Pongola Supergroup. Large xenolithic blocks of the basal arenite of the Nsuze Group are found in the gabbro. Although mainly sill-like in form, there is one locality where gabbroic rocks flank a central zone of granophyre which is suggestive of a synformal arrangement. Lack of prominent layering and, hence, of indications of dip, does not permit confirmation.

Insufficient data are available to construct a model for the intrusion as a whole, which apparently consists of two dikes linked by a gently dipping sill. The dikes contain remnants of Pongola rocks and are paralleled by major faults in the adjacent granitoid terrane. The possible sequence of intrusion of the northernmost dike is shown schematically in Figure 2-21. Rocks of the Pongola Supergroup were apparently preserved in two graben-like structures extending northwestward, the presence of which provided the channel ways through which gabbroic magmas were intruded. Whether initial layering developed by gravity settling, convective overturn, or some other mechanism is not known. Subsequently intrusion of large volumes of quartz gabbro disrupted the original layering. Renewed movement along the bounding faults resulted in erosion of the uplifted flanks. Selective preservation of the Pongola rocks in the reactivated graben may have been aided by cross folding, evidence for which is found in the southernmost dike.

The significance of the Usushwana Intrusive Suite lies in its gross similarities to the Late Archean Great Dyke of Zimbabwe. Both were emplaced after a major period of granitoid emplacement that terminated the typical Archean volcanosedimentary processes. Both are faulted dikelike intrusions with, particularly in the Great Dyke, gently dipping layers. The Usushwana Intrusive Suite does not display the same abundance of ultrabasic rocks and associated chromitite layers as does the Great Dyke; but the geometry and relative time of emplacement of both intrusions suggests that they mark a stage in the evolution to more-stable and brittle crust.

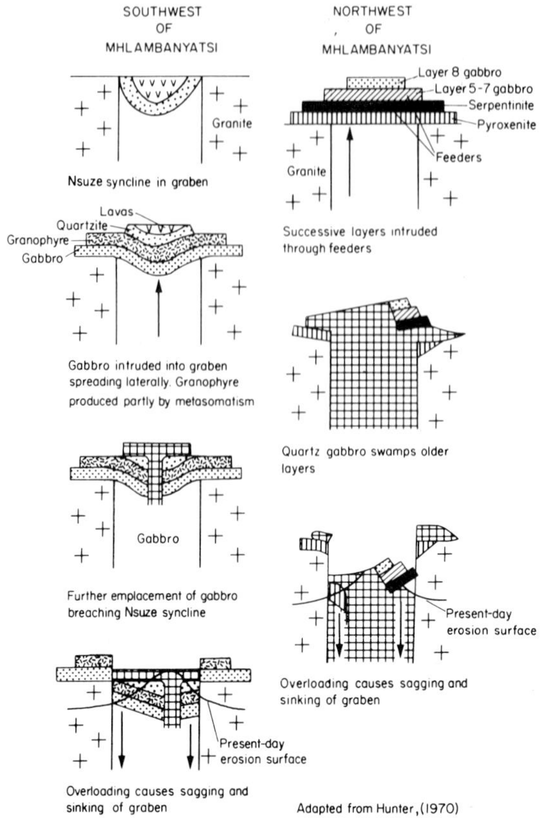

Figure 2-21. Diagrammatic sections showing the possible evolution of the Usushwana Intrusive Suite in Swaziland.

Posttectonic Granitoid Plutons

Field relations and lithology. More granitoids were emplaced at about 2.7–2.5 Ga after the Usushwana Intrusive Suite. Late Archean granitoids are more abundant south of the Barberton greenstone belt than immediately to the north. Granites of similar age have also been reported from the northwestern and southwestern edges of the Kaapvaal Province (Burger and Coertze, 1973).

In southern Swaziland and northern Natal the intrusions are the sheetlike Pongola granite and a number of small high-level plutons, which have been subdivided on the basis of their mineralogy and lithology into (1) Kwetta-type plutons, (2) Mpageni-type plutons, and (3) Mooihoek-type plutons. All but the small Mpageni stock are south of the Barberton belt. On the basis of geochemical analyses from eight of these plutons, Condie and Hunter (1976) suggested a broad subdivision into two groups: The Mpageni group consists of the Kwetta-type and three of the Mpageni-type plutons, and the remaining four plutons constitute the Sicunusa group (Table 2-15). The relative ages of the plutons cannot be fully determined from field evidence, but the Mooihoek-type plutons intrude the Kwetta-type plutons in southern Swaziland.

The Pongola granite is lithologically similar to the Lochiel quartz monzonite but is compositionally more potassic. Like the Lochiel quartz monzonite, the Pongola granite is sheetlike. Whereas the nature of the roof of the Lochiel intrusion is uncertain, the Pongola Supergroup forms the roof of the Pongola granite in southern Swaziland. It is therefore possible to estimate the thickness of the Pongola sheet in this area as between 300 m and 1000 m.

Table 2–15. Mean Compositions of the Two Geochemically Defined Groups of Post-Pongola Plutons[a]

	Mpageni Group[b]	Sicunusa Group[c]
SiO_2	70.39	75.39
TiO_2	0.45	0.19
Al_2O_3	13.93	12.69
Fe_2O_3	1.09	0.62
FeO	1.88	1.18
MnO	0.11	0.05
MgO	0.68	0.26
CaO	1.97	0.90
Na_2O	3.48	3.09
K_2O	5.19	5.16
H_2O^+	0.64	0.37
H_2O^-	0.08	0.06
P_2O_5	0.14	0.02
Total	100.03	99.98
Rb	230	280
Sr	283	45
Ba	1083	371
K/Rb	190	154
Sr/Ba	0.27	0.15
Rb/Sr	0.87	7.6
K/Na	1.69	1.88

[a] Analyses are based on lithologic classification of the plutons. After Condie and Hunter (1976).

[b] Mpageni group includes the Mpageni, Mbabane, Ngwempisi, and Kwetta plutons.

[c] Sicunusa group includes Sicunusa, Sinceni, and Mhlosheni-type plutons.

Seven coarse-grained, commonly porphyritic granitoid plutons were emplaced at about 2.5 Ga in Swaziland contemporaneously with the small Mpageni pluton in the north. The Kwetta-type pluton in the south is characterized by microcline megacrysts between 25 and 50 mm in length, some of which are mantled by narrow rims of plagioclase (An_{18}).

The Mpageni-type granites are grayish-pink rocks with microcline megacrysts set in a matrix of quartz, plagioclase, and biotite. Opalescent blue quartz grains are common. A further distinguishing feature is the occurrence of medium-grained biotite-rich enclaves with a disklike shape and containing randomly oriented microcline megacrysts.

The Mhlosheni and Mooihoek plutons in southern Swaziland are examples of the third type of pluton. They are siliceous, coarse-grained, gray-green rocks with minor biotite. These plutons intrude the Kwetta-type pluton but are not in juxtaposition with the Mpageni-type plutons.

None of the plutons markedly deforms the rocks it intrudes. The intrusions lack foliation and are surrounded by narrow contact aureoles. In the case of the Mooihoek pluton, which intrudes gently dipping Mozaan sediments, it is possible that the pluton has made room for itself by displacement along bedding planes of the Mozaan rocks. Movement along these planes can be observed with upper beds being displaced away from the granite over lower beds. No similar evidence has been recorded in country rocks adjacent to other plutons. Individual plutons range in size from 40 km^2 to 650 km^2. Most plutons are elongate but the direction of elongation appears to be near random, with a slight preference for northwest and north-northwest.

The Cunning Moor tonalite recognized by Robb (1977) crops out in the far north (Fig. 2-16) and its relationships to the other granitoids in the area are not clear. A provisional Rb-Sr age determination quoted by Robb suggests that the tonalite was emplaced at about 2.6 Ga. The tonalite is a massive, medium- to coarse-grained rock with an equigranular texture consisting of quartz and plagioclase with subordinate biotite and microcline.

Geochemistry and petrogenesis. Geochemical data for the post-Pongola granitoids are summarized in Table 2-15. Genetic models can be proposed only for the Swaziland granitoids and the Mpageni pluton north of the Barberton supracrustal belt.

The sheetlike Pongola granite has a range of SiO_2 contents from 65 to 74 percent, like the Lochiel quartz monzonite. It differs from the Lochiel intrusion in having higher concentrations of K_2O (up to 6.5 percent), TiO_2, and total Fe, and low Sr contents. In this last respect it resembles granite dated at about 2.7 Ga (and hence of similar age) from the western Transvaal.

The Mpageni group has K_2O/Na_2O ratios up to 1.5 whereas in the Sicunusa group this ratio exceeds 1.5 (Condie and Hunter, 1976). The Sicunusa group is strongly depleted in Sr ($\bar{x} = 45$ ppm) and Ba ($\bar{x} = 371$ ppm) in comparison with the Mpageni group, which has mean concentrations of 283 ppm Sr and 1083 ppm Ba (Table 2-15). The REE patterns also distinguish these two groups. The Mpageni group has a pattern similar to the Lochiel quartz monzonite, whereas the Sicunusa group has a larger negative Eu anomaly and flatter slope of heavy REEs (Fig. 2-17D). The most acceptable model for the generation of these magmas is one involving a high degree of fractional crystallization of a granodioritic parent magma similar in composition to the Dalmein type (Condie and Hunter, 1976). The major differences between the Mpageni and Sicunusa groups can be accounted for by a combination of:

(1) A slightly lower f_{O_2} during crystallization of the Sicunusa group than that associated with the Mpageni group
(2) The presence of liquidus K-feldspar in the Sicunusa group
(3) Crystallization of more plagioclase from the Sicunusa group magmas

Initial $^{87}Sr/^{86}Sr$ ratios have been reported from only the Mpageni (0.7065; De Gasparis, 1967) and the Sicunusa (0.7006; Davies and Allsopp, 1976) plutons, indicating a crustal source for some plutons but not others.

A model similar to that for the Mpageni and Sicunusa group magmas was tested for the Pongola granite and found to be broadly consistent with the observed range of compositions of the granite (Condie and Hunter, 1976). There are insufficient data to speculate on the origin of the Cunning Moor tonalite, which is reported to have a relatively high initial $^{87}Sr/^{86}Sr$ ratio (Robb, 1978).

Granitoids Elsewhere in the Kaapvaal Province

Although granitoids of various ages underlie large areas of the Kaapvaal Province, details of their geology and chemistry are not everywhere fully documented.

Minnitt (1975) reported the presence of tonalite gneisses, tonalitic-trondhjemitic plutons, and small potassium-rich granitoid plugs and plutons adjacent to the eastern portion of the Murchison greenstone belt (Fig. 2-1).

The granitoids of the Johannesburg Dome have been divided into five main units (Anhaeusser, 1973):

(1) Tonalite gneisses
(2) Homogeneous granodiorites
(3) Gray granodiorite suite
(4) Transitional zone granodiorites, gneisses, migmatites, and porphyritic granites
(5) Migmatite and gneiss terrane

The first and last units have been described briefly in Section 2.1 because they have closer similarities in some respects to the early granitoid terrane in Swaziland.

Homogeneous granodiorites around the southern and southwestern flanks of the dome appear to grade into the tonalite gneisses (Anhaeusser, 1973). The granodiorites are medium to coarse grained and consist of quartz, plagioclase, microcline, and biotite. The gray granodiorite suite includes both porphyritic and nonporphyritic varieties and is found mainly in the southern part of the dome, where it contains remnants of layered gneisses, migmatites, and basic rocks. Quartz, albite–oligoclase, orthoclase, and microcline with subordinate amounts of biotite and muscovite are present. Granodiorites, gneisses, migmatites, and porphyritic granites occupy a transitional zone between the migmatite and gneiss terrane that underlies most of the northern half of the dome (Anhaeusser, 1973). The granitic and granodioritic rocks range from fine grained to relatively homogeneous,with or without megacrysts, to coarse grained. These rocks weather to form characteristic boulder-strewn hills and tors between which are outcrops of gneiss and migmatite.

The granodiorites of the Johannesburg Dome are approximately coeval with the Lochiel quartz monzonite, with which they have gross chemical similarities except for their low K/Rb ratios. Although Anhaeusser (1973) suggested that the granodiorites represent the final stages of a complex process involving "rheomorphism, anatexis, granitization and metasomatism," the possibility cannot be excluded that they were generated by a mechanism similar to that suggested for the Lochiel quartz monzonite.

Granites dated between 3.1 Ga and 2.9 Ga crop out as inliers around the flank of the Early Proterozoic Witwatersrand depositional basin near Ventersdorp, west of Prieska at the southwestern edge of the Kaapvaal Province, and at Jacobsdal west of Bloemfontein (Fig. 2-1). Boreholes have intersected granites of similar age at Glen and Strydenburg in the Orange Free State. Although little is known of the mineralogy and chemistry of these granites, available data suggest mineralogical similarities with the Lochiel quartz monzonite.

Granites cropping out in the Schweizer Reneke and Ventersdorp inliers have yielded ages between 2.6 Ga and 2.8 Ga. Other granitoids of this age are reported from Mahalapye in Botswana, from pegmatites near the southern margin of the Limpopo Province, and from west of Prieska (Fig. 2-1). In the Zimbabwe and Limpopo Provinces there was widespread intrusion of granite and quartz monzonite about this time (Chapter 3).

From the meager data presently available it appears that both the granitic events at about 3.0 Ga and 2.6–2.8 Ga were widespread in the Kaapvaal Province. The main loci of emplacement of these rocks, which are dominantly granodioritic or quartz monzonitic in composition, may have been in linear zones. This supposition needs more rigorous testing, but it may be significant in that it implies a northwestward migration of intrusion that preceded a similar migration of the depositional axes of the Archean and Proterozoic ensialic basins across the Kaapvaal Province (Chapter 1).

2.7. Broad Implications of Archean Crustal Development in the Kaapvaal and Zimbabwe Provinces

The salient features of the Archean geology of the Kaapvaal Province in this review show that interpretation of both field relations and environments of formation are commonly equivocal. Age relations between the Early Archean bimodal gneiss suite in Swaziland and the supracrustal rocks of the Barberton belt are not yet clear. The contact between these high-grade layered gneisses and the generally low-grade supracrustal rocks is either faulted, obscured by repeated high strains, or masked by younger intrusions. Geochronologic data yield ages of 3540 ± 30 Ma (Hamilton *et al.*, 1979; Nd-Sm) and 3555 ± 111 Ma (Barton *et al.*, 1980; Rb-Sr) for the Onverwacht Group and bimodal suite, respectively, indicating their approximate contemporaneity. Amphibolitic interlayers in the bimodal suite have been variously interpreted. For example, they have been regarded as xenoliths of Onverwacht rocks in a pervasive leucotonalite intruded into the roots of the Swaziland Supergroup (Viljoen and Viljoen, 1970; Anhaeusser, 1978b). In this model the leucotonalite plutons and the layers of similar composition in the bimodal gneisses are considered to be cogenetic, differences in structural style being attributed to exposure at different crustal levels. Another interpretation visualizes the amphibolitic and siliceous layers as the products of bimodal igneous activity, either intrusive or extrusive (Hunter *et al.*, 1978). Other amphibolites are demonstrably dikes that crosscut the layering of the gneisses (Hunter, 1970; Jackson 1979a). The geochronologic data now demonstrate the greater antiquity of the tonalitic bimodal gneisses compared with the tonalitic plutons that intrude the Onverwacht rocks. The concept of one major tonalitic event entirely postdating the Swaziland Supergroup is no longer tenable. The problem of age relations between the bimodal gneisses and the Onverwacht rocks still remains. Nevertheless available structural and metamorphic evidence (Jackson, 1979a) suggests that the bimodal suite is likely to represent a basement to a greenstone-type cover at Dwalile (Fig. 2-2B). Farther south, similar supracrustals are intruded by the Tsawela tonalite gneiss, which yields an age of 3.3 Ga contemporaneous with tonalitic plutonism postdating the Onverwacht Group. It therefore seems probable that the Dwalile Metamorphic Suite is coeval with the Onverwacht Group. Deformation of the bimodal gneisses and Dwalile supracrustal rocks, together with the Tsawela tonalite gneiss, in the Dwalile area has resulted in the preservation of the supracrustals in cores of synforms. This common geometric arrangement suggests that the supracrustal rocks, which are of lower metamorphic grade than the bimodal gneisses, constituted a cover sequence. Although deformed basic dikes appear to be more common in the bimodal suite, it has not yet been possible to demonstrate unequivocally that any of the dikes predate the Dwalile supracrustals.

If parallels are drawn with other Archean terranes, such as the Zimbabwe Province, it seems probable that there has been repetitive development of siliceous gray gneisses and greenstone volcanism. The important geochronologic data reported from Zimbabwe suggest that there were at least three cycles of the sequence: layered gneiss → greenstone volcanism → intrusive tonalitic plutons (Table 2-16). The oldest of these cycles in Zimbabwe appears to have formed at about 3.5 Ga and is represented by the Tokwe and Shabani

Table 2–16. Timing of Events in the Kaapvaal, Limpopo, and Zimbabwe Provinces[a]

Time (Ma)	Limpopo Province		Zimbabwe Province		Kaapvaal Province: Eastern Transvaal		Kaapvaal Province: Swaziland	
2400								
	2460	Great Dyke	2460	Great Dyke **Cratonization**		**Cratonization**		**Cratonization**
					2500	Mpageni-type granite plutons	2500–2610	Mpageni-type, Mhlosheni-type, Kwetta-type granite plutons. Pongola "hood" granite
			2570	Chilimanzi Granite Suite				
2600	2600	Matok granite						
	2630–2710	Bulai granitoid gneiss	2630	Somabula–Sesombi tonalites				
			2660	Late Bulawayan greenstones				
			2700	Gwenoro–Rhodesdale gneisses				
2800	2790–2870	Gneisses (Bangala/ S.W. Messina)		Mashaba Intrusive Suite			2870	Usushwana Intrusive Suite
					2880	Dalmein–Salisbury		Pongola Supergroup
			2910	Mashaba tonalite Early Bulawayan greenstones Chingezi gneisses	2930	Kop granite plutons	2930	Mliba granite pluton

3000 —					— 3000	Lochiel "hood" granites	— 3000	Lochiel "hood" granite
	— 3060	Basic dikes						
					— 3130	Bosmanskop syenite[b]		
	— 3150	Messina Intrusive Suite					— 3150	Quartz monzonite lenses
3200 —								
					— 3220	Leucotonalite		
					— 3250	Plutons[b]		
					— 3310	Kaap Valley "granite"[b]		
		Beitbridge sequence (undated)	— 3350	Mont d'Or granite			— 3350	Granodiorite Suite Mponono anorthosite (undated)
						Fig Tree–Moodies Groups		Basic dikes (undated) Tsawela tonalite gneiss Dwalile Metamorphic Suite Basic dikes (undated)
3400 —								
			— 3450	Mushandike granite Sebakwian greenstones				
3500 —			— 3500	Tokwe–Shabani gneisses				
					— 3540	Onverwacht Group[c]		
							— 3555	Bimodal gneiss suite
	— 3560	Basic dikes						
3600 —								
	— 3790	Sand River gneisses						

[a]Rb-Sr whole-rock isochron ages except those indicated.
[b]U-Pb ages.
[c]Nd-Sm ages.

gneisses west of Fort Victoria and Sebakwian greenstones; the supracrustal rocks are best preserved at Selukwe and contain stromatolites near the Mushandike Dam west of Fort Victoria (Orpen and Wilson, 1981). The Mushandike stromatolites are the oldest unequivocable organic remains known. No equivalents of the intrusive tonalitic plutons have yet been recognized, but the strong deformation these ancient rocks have suffered renders recognition of this component extremely difficult. The Tsawela tonalite gneiss in Swaziland, which appears to have had the initial form of a pluton, has been so strongly deformed that in an area of poor outcrop its separate identity would be difficult to recognize. However, the early cycle in Zimbabwe is terminated by the granodioritic Mushandike "granite" and the Mont d'Or granite, which have been dated at 3.5 Ga (Hickman, 1974; Rb-Sr) and 3.35 Ga (Moorbath *et al.*, 1976; Rb-Sr), respectively.

The second cycle in Zimbabwe seemingly developed close to 2.9 Ga (Hawkesworth *et al.*, 1979). Structural evidence supports the view that the Chingezi gneisses predate the overlying Early Bulawayan greenstones of the Belingwe belt (Hawkesworth *et al.*, 1975). Tonalitic plutons intrude the Chingezi gneisses and the Early Bulawayan greenstones. The gneisses have low strontium initial ratios, implying that they are not likely to represent reworked older crust. The third cycle, between 2.7 Ga and 2.6 Ga, is the most complete (Table 2-16). The layered gneisses have primitive strontium initial ratios and are considered on structural grounds to predate the volcanosedimentary supracrustal rocks of the Late Bulawayan greenstones (Stowe, 1973). The sedimentary basal unit of the predominantly volcanic Late Bulawayan sequence rests nonconformably on 3.5-Ga basement represented by the Shabani gneiss and Mushandike "granite" in the Belingwe (Bickle *et al.*, 1975) and Fort Victoria (Orpen and Wilson, 1981) greenstone belts, respectively. Both basement and cover are intruded by tonalitic plutons. Geochronologic data cannot discriminate the ages of the three events constituting this cycle, which may have been completed within a period as brief as 200 Ma (Hawkesworth *et al.*, 1975). Data presented in Table 2-16 indicate the presence of a gneiss-forming event in the Limpopo Province still older (3.8 Ga; see Chapter 3) than those recognized in the adjacent structural provinces.

These repetitions of the cycle represented by basement layered gneisses → greenstones → tonalitic plutons in Zimbabwe contrast with the Kaapvaal Province, where present interpretations suggest a nonrepetitive evolution (Anhaeusser, 1978b). However, it is quite possible that this reflects the dearth of radiometric ages on Kaapvaal greenstone belts apart from Barberton. It is significant that in both provinces there is radiometric evidence that gneiss formation and greenstone volcanism were virtually contemporaneous. The data summarized in Table 2-16 reveal that sialic crust existed prior to greenstone volcanism and that the trinity of gneisses-greenstone-tonalitic plutons continued to form in Zimbabwe while more-stable tectonic conditions prevailed in the Kaapvaal Province. This implies that the evolution of the two provinces differed fundamentally after 3.5 Ga.

The continuum in the formation of layered gneiss suites and greenstone volcanism in the Barberton area and the three cycles in Zimbabwe suggest the following sequence of events in each cycle:

(1) Early basaltic volcanism in a highly unstable environment was accompanied by partial melting, which generated dacitic liquids that were either intruded or extruded.
(2) Ultrabasic and basic volcanism continued in a progressively more stable environment favoring its more complete preservation.
(3) Subsequent deformation of the gneisses and supracrustal belts was accompanied by either solid-state diapiric rise of the sialic basement or partial melting of the greenstone volcanic rock to yield further tonalitic liquids which were intruded as homogeneous plutons.

The environment in which these events took place is uncertain. The volcanological and sedimentological data from the Tjakastad Subgroup require a marine environment distant from any existing sialic crust. There is no reason to suppose that a similar marine environment would have been inappropriate for extrusion of basalts that ultimately formed one component of the layered gneisses. However, the subsequent spatial arrangement of the products of these two events is uncertain. At Dwalile in western Swaziland circumstantial evidence suggests that parts of the "greenstone-type" volcanic pile accumulated on a sialic crust of bimodal gneisses. Sun and Nesbitt (1978) opined that geochemical data alone cannot uniquely define the tectonic setting of either modern or Archean volcanism. Hawkesworth and O'Nions (1977) have adopted a similar stand, believing, however, that the geochemistry of Archean volcanics places important constraints on the thermal history of these belts. These authors suggest that the presently available data imply eruption of higher temperature magmas, reflecting the greater heat production in the mantle during the Archean. The association of MgO-rich liquids, such as the Tertiary basalts of Baffin Bay and the Karoo Nuanetsi basalts within areas of continental rifting or incipient rifting, together with the presence of the highest MgO-rich rocks near the base of these volcanic piles suggest to Hawkesworth and O'Nions (1977) that the Archean volcanics also have formed in a rifting environment. Rapid decompression of the underlying mantle could have resulted in high degrees of partial melting of the mantle source, the liquids produced being erupted as ultrabasic lavas. Arth *et al.* (1977) suggested that tholeiitic liquids were the first products of melting of mantle peridotite and that more MgO-rich komatiitic liquids were subsequently produced. Alternatively, advanced degrees of mantle melting may have been achieved only where the thermal gradient was adiabatic to near the surface (O'Nions *et al.*, 1978), thereby producing komatiitic liquids. There may have been less partial melting to yield basaltic liquids at an earlier stage as a consequence of a different thermal regime. These hypotheses may account for the appearance of the most magnesium-rich lavas near the top of the Tjakastad Subgroup.

On the basis of detailed study of the Tertiary rochas verdes in southern Chile, Tarney *et al.* (1976) have argued in favor of a marginal basin as the environment in which Archean greenstones extruded, provided that subduction occurred in the Archean. Other authors have proposed an island arc setting for these rocks (e.g, Jahn *et al.*, 1974). Condie and Harrison (1976) suggested a model for the Bulawayan greenstones of Zimbabwe involving formation of basic volcanic rocks above a spreading center in a marginal basin, and other tholeiitic, andesitic, and rhyolitic volcanics in an adjacent maturing island arc.

The subsequent history of the basin in which the Swaziland Supergroup accumulated is further constrained by the lack of detailed geochemical data for the Geluk Subgroup, which is reported to be calc–alkaline (Anhaeusser, 1978b). If modern environments for the generation of calc–alkaline rocks are applicable to the Early Archean, this implies convergence and collision of lithospheric plates. However, there is no certainty that Phanerozoic plate-tectonic models can be applied to the Archean.

The evolution of the Onverwacht Group has been summarized by Lowe and Knauth (1977). They envisage accumulation of the Tjakastad volcanic pile in an open ocean, far from land or terrigenous sediment; the evidence for this environment is the paucity of sedimentary or pyroclastic deposits and the presence of pillow lavas. Accumulation of the Hooggenoeg and Kromberg Formations, primarily in the southern part of the Barberton greenstone belt, was marked by the onset of acid volcanism and the formation of subaerial volcanic vents. The development of volcanic slopes led to gravity flows that redistributed pyroclastic debris. The scarcity of argillaceous sediments suggests that the volcaniclastic material was not weathered before redistribution. Localized uplift and erosion of older Onverwacht rocks during deposition of the Swartkoppie Formation was accompanied by the initial uplift of a granitoid-gneiss terrane to the south of the basin and deepening of the depository to the north. Erosion of this terrane was minimal and sediments in the Swartkoppie Formation consist largely of locally derived detritus.

By the time the Swartkoppie Formation was deposited, the Barberton basin had deepened to the north (Lowe and Knauth, 1977). Eriksson (1980b) suggested that this was the first evidence of spreading. Onset of accumulation of iron-formation in the southern Fig Tree facies was also attributed to basin deepening, the iron-formation accumulating in local grabens in response to waning volcanism. Ferruginous sediments on the floor of the Red Sea have a similar setting (cf. Fryer and Hutchinson, 1976). Concomitant uplift of older volcanic, sedimentary, and silicic plutonic rocks in the south resulted in the progradation of a submarine-fan–slope complex across the underlying chemical sediments.

The basal Moodies conglomerates conformably overlie the Fig Tree submarine-fan–slope sediments in most areas and display fluvial and fan-delta characteristics. The exclusion of shallow-marine facies in this submarine-fan–slope to fan-delta transition is interpreted as the result of sedimentation on a "steep rift-type" continental margin with a narrow shelf (cf. Burke, 1972; Brown and Fisher, 1977). Subsequent rise of sea level coupled with the building of the underlying sedimentary wedge produced a wider continental shelf. Marginal marine facies are conspicuous in this part of the Moodies Group and indicate deposition on a "shelf rise-type" continental margin.

The sediments in the Fig Tree and Moodies Groups show a vertical change from volcaniclastic to quartzofeldspathic arenites (K.A. Eriksson, 1979) and an increase in lithophilic elements in the Fig Tree graywackes (Condie *et al.*, 1970), reflecting the progressive uplift of a plutonic–metamorphic terrane to the south. The detritus could have been derived either from unroofing of leucotonalitic plutons intrusive into the Onverwacht Group or from unroofing of an older sialic terrane, such as the Ancient Gneiss Complex. The Fig Tree and Moodies sediments contain only small amounts of plagioclase. There is no detrital plagioclase in the Moodies beach sandstones and less than 10 percent plagioclase in the Moodies delta front facies and Fig Tree sediments. In addition the Moodies conglomerates contain clasts of potassic granite, and other Moodies sediments contain up to 30 percent detrital potash feldspar. The leucotonalitic plutons, impoverished as they are in potassium, could not have provided detritus of this nature. The source of the K-rich granite clasts in the Moodies conglomerates is unknown, and attempts to date these clasts have provided contradictory results that do not resolve the problem (Van Niekerk and Burger, 1978a). A possibility was the existence of a sheetlike, pre-Moodies potassic granite, similar in geometry to the Lochiel quartz monzonite, that was destroyed by erosion. Clasts of this composition imply the existence of a pre-Moodies potassic granite event and suggest that repetitive tonalitic and granitic events may have occurred in the Kaapvaal Province as they did in the Zimbabwe Province.

Condie *et al.* (1970) showed that detritus in the basal graywackes of the Fig Tree Group was derived from a provenance that contained an abundance of igneous or metamorphic rocks depleted in Sr relative to K, Ba, Ca, and Rb and included an ultrabasic component to account for their high Ni content. Table 2-17 summarizes the mean values of relevant ratios in the bimodal suite, migmatized gneisses and post-Swaziland tonalitic plutons. These data indicate that the bimodal suite has the appropriate chemistry for a source area. Amphibolites interlayered in this suite could have contributed Ni to the Fig Tree graywackes.

Structural data from the Mankayane area south of the Barberton supracrustal belt strongly suggest uplift of this terrane (Jackson, 1979a, 1980). High finite strains are the cumulative result of successive strain increments from D1 to D5 (Table 2-6) with the younger episodes being the more heterogeneous and less ductile. The *XY* flattening planes of finite bulk strain dip gently to the southeast and south (Fig. 2-4). Whether the deformation took place by pure shear or simple shear or a combination of these processes, considerable lateral extension in the XY plane took place. The updip component of this extension would have resulted in displacement of the gneisses toward the north and northwest, with the vertical component of this displacement vector being many kilometers. The inevitable result of these large shape changes was uplift and erosion of sialic crust.

The evidence is thus consistent with the view

Table 2–17. Mean Element Ratios of Siliceous Bimodal Gneisses, Migmatized Gneisses, Leucotonalitic Plutons, and Fig Tree Graywackes[a]

	1	2	3	4	5
Rb/Sr	0.46	0.08	0.10	0.64	0.27
Ca/Sr	105	37	36	151	66
Ba/Sr	—	1.3	0.67	3.2	1.7
K/Sr	134	30	30	136	55
K/Rb	255	380	281	216	204

[a] 1, Siliceous gneisses (bimodal suite), mean of 9 analyses. 2,Migmatized gneisses (Ancient Gneiss Complex), mean of 3 analyses. 3, Leucotonalitic plutons (post-Swaziland), mean of 11 analyses. 4, Graywackes (Sheba Formation, base of Fig Tree Group). 5, Graywackes (Bellevue Road Formation, upper part of Fig Tree Group). Sources: Columns 1 and 2, Hunter *et al.* (1978) and unpublished data; column 3, M.J. Viljoen and Viljoen (1969b) and Condie and Hunter (1976); columns 4 and 5, Condie *et al.* (1970).

that the Swaziland gneiss terrane was the source of detritus for the Fig Tree and Moodies Groups, which represent a single protracted depositional event. Active northward progradation accompanied by an increasing amount of granitoid detritus suggests continuing uplift in the south.

Sedimentation in the Swaziland depository was terminated by major granitoid plutonism that reached its peak at about 3.0 Ga. Tonalitic plutons intrude the lower formations of the Onverwacht Group but it has been suggested that these plutons represent gneiss diapirs. If this is correct, it is likely that they continued to rise after Fig Tree and Moodies sedimentation.

Granitoid intrusion at about 3.0 Ga took place on both flanks of the Barberton belt which contrasts with the asymmetry of the preceding episode of sedimentation and with orthodox plate tectonic models of plate convergence. The Nelspruit porphyritic granite appears to have a similar sheetlike geometry to the Lochiel quartz monzonite in the south, the Nelspruit migmatite terrane being the result of granitic intrusion into an older gneiss sequence. If some of these granitoid liquids were generated by partial melting of sialic crust, this would entail depression of sialic crust on the flanks of the Barberton supracrustal belt to depths of at least 15 km. An early model envisaged the sagging of sialic crust along the margins of a rift in response to the collapse of the thermal plume or mantle diapir that provided the source of the Onverwacht magmas (Hunter, 1974a). However, this model is seemingly incompatible with the sedimentological data. It is possible that accumulation of a thick sequence of Fig Tree and Moodies sediments may have created a thermal blanket sufficient to increase the geothermal gradient so that transformation of gabbroic oceanic crust to eclogite was possible. A dense eclogite "sinker" may have had the effect of dragging sialic crust down into a zone of partial melting. However, the asymmetry of the Fig Tree and Moodies sediments, which thicken to the north, suggests that generation of granitic liquids should be asymmetric to the supracrustal belt.

Any model must take cognizance of the wide extent of the 3.0-Ga granitoid event. Intrusions of this age are found apparently occupying a broad belt across the Kaapvaal Province (Fig. 2-1). It is certain that this magmatism promoted cratonization, resulting in sufficiently stable crustal conditions for the accumulation and subsequent preservation of the Pongola cratonic cover.

A fundamentally different style of volcanism is found in the basal Nsuze Group, which consists of lava ranging through the complete spectrum from basalt to rhyolite. Such a sequence is typical of Phanerozoic volcanic arcs, implying that plate convergence was dominant. Outcrops of the Pongola Supergroup are confined to the southeastern margin of the belt of 3.0-Ga granitoid intrusion. Moreover sediment transport was toward the south in contrast to the direction that prevailed during the accumulation of the sediments of the Swaziland Supergroup. The confinement of Pongola volcanics and sediments to the southeastern flank of the 3.0 Ga granitoid belt suggests that the buoyant rise of these granites closely controlled the development of the Pongola depositional basin. Sediment supply could have been maintained by continuing isostatic rise of the granitoid belt and older sialic crust to the north.

The main site of post-Pongola granitoid emplacement appears to lie to the north of the 3.0-Ga belt (Fig. 2-1) and is most widespread in Zimbabwe (Chapter 3). This apparent migration of granitoid emplacement is accompanied by a related migration of the sedimentary basins across the Kaapvaal Province and a progressive inundation of the craton in a northwesterly direction by epicontinental seas.

Pongola sedimentation ended with intrusion of the Usushwana Intrusive Suite. The first indication of a return to basic igneous activity is the basalt at the top of the Mozaan Group. The cycle of granitoid intrusion → sedimentation → basic plutonism or volcanism is repeated throughout the Early Proterozoic in the Kaapvaal Province (cf. Hunter, 1974b). This cycle suggests a complex interplay between processes in the mantle and responses in the crust. In its simplest form it appears to involve the periodic accumulation of subcrustal heat that induces melting in the asthenosphere, the heat source being depleted when magmas are transferred to the crust. The rate of thermal dissipation as surface heat flow and magmatic transfer to the crust exceeds the rate at which heating in the asthenosphere returns to a peak level, resulting in the development of sedimentary basins. A further magmatic transfer to the crust is preceded by inflation or thermal doming of the crust, which causes regression and ultimate closure of a preexisting sedimentary basin. Thermal doming causes fracturing of the crust, permitting egress of mantle-derived magmas as flows or intrusions, and facilitates partial melting of the lower parts of the sialic crust to generate granitoid liquids that rise buoyantly. Widespread generation of granitic liquids by partial melting of the lower crust can be expected to produce a refractory residuum, the presence of which could contribute to stabilization and cratonization. The apparent migration of the main site of granitic emplacement implies either a migration of the continental plate over a linear mantle heat source or the migration of the latter. This concept requires rigorous testing to establish its validity but, until this is done and more complete data are available, no entirely satisfactory model for Archean development can be proposed. The Kaapvaal Province provides a nearly complete spectrum of events spanning the first billion years of geologic time and is therefore a critical area in which to attempt to unravel the early history of the Earth.

Chapter 3

Granulite-Gneiss Terrane: Limpopo Province

High-grade gneisses and granolites are exposed in two contrasting settings in southern Africa: (1) as scattered areas in granite–greenstone terranes (e.g., Kaapvaal Province) intruded by more-homogeneous granitoids and associated with predominantly low-grade supracrustal sequences (Chapter 2), and (2) as extensive gneiss terranes in "mobile belts" with subordinate intrusive rocks.

The paucity of preserved stratigraphy has greatly hampered interpretation of southern Africa's high-grade terranes. Within the last decade, however, the synthesis of geochronology, geometric and strain analysis, geochemistry, and detailed mapping has resulted in an explosion of data and new ideas on the complex evolution of the high-grade gneiss terranes.

The granulite–gneiss terranes have evolved, in some cases, over billions of years. Although many of the gneisses represent highly metamorphosed supracrustal sequences, as in west Greenland (e.g., McGregor, 1973; Myers, 1978), some layered gneisses may also represent highly deformed plutonic and hypabyssal intrusions. Lithologic components include clastic and chemical sediments, volcanic and volcaniclastic rocks, hypabyssal basic and acid intrusions, and a full range of granitoids and gabbroids, many of which were layered intrusions. This wide variety of rocks has been repeatedly subjected over a long period to strain and metamorphism, the effects of which differ greatly in intensity from place to place. Central to our understanding of the gneiss terranes is the recognition of three fundamental tectonic units:

(1) A basement of crust formed and modified in a previous tectonic cycle
(2) A precratonic cover representing supracrustal sequences that accumulated during the evolution of the high-grade terrane, commonly in the early stages
(3) Precratonic intrusions emplaced prior to cratonization (attainment of stability).

Of these three tectonic units, the basement is commonly the most difficult to recognize, principally because of the blurring of the basement–cover interface by geochronologic, metamorphic, and structural homogenization. Evolution of a gneiss terrane terminates with cratonization, uplift, and erosion, but this stage can be reliably estimated only by dating the oldest nonorogenic cratonic cover sequence that has accumulated on the uplifted gneiss terrane.

Granulite–gneiss terranes can be divided into tectonic zones, each of which has a distinct character and history. Where possible, established boundaries are used here for these zones, although in some cases their geographic names have been changed in the interest of uniformity.

3.1. Extent of Limpopo Province

The tectonic evolution of the Limpopo Province covered the pre-3.8-Ga to 2.0-Ga time span. The Limpopo Province trends east-northeast and has an exposed length of about 690 km and exposed widths of about 170 km in the west and 220 km in the center (Fig. 3-1). It extends from eastern Botswana through southern Zimbabwe and northern South Africa. The northern and southern margins grade metamorphically and structurally into the Archean Zimbabwe and Kaapvaal Provinces. Both ends of the Limpopo Province are obscured by cratonic cover sequences, the ages of which are Precambrian (Waterberg, Soutpansberg, Palapye, and Umkondo Groups) and Phanerozoic (Karoo, Cretaceous, and Kalahari sequences).

The true length of the Limpopo Province is therefore not known. However, Key and Hutton (1976) estimated that west of longitude 27° 30′ E displacement on the relatively young Tuli–Sabi shear belt decreased to zero. In the east the Lim-

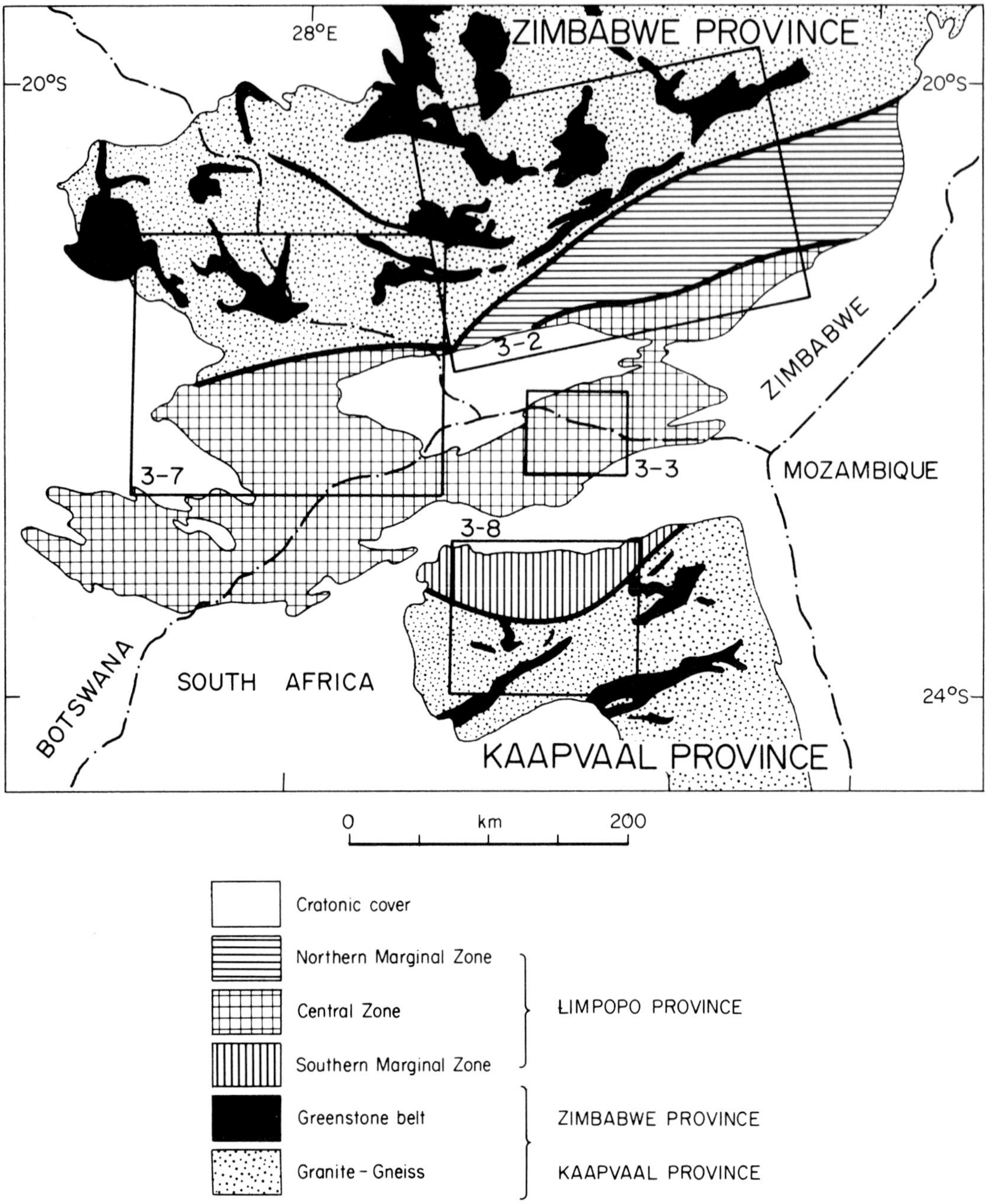

Figure 3-1. Tectonic zones within the Limpopo Province and their relation to granite–greenstone terranes of the adjacent Zimbabwe and Kaapvaal Provinces. Rectangles delimit areas shown in subsequent figures.

popo Province is probably truncated by Pan African crustal reworking in the Mozambique Province.

The tectonic zonation of Cox *et al.* (1965) which was later rationalized by Mason (1973), provided a valuable framework for later workers. The Limpopo Province can be divided into a central zone with well-developed precratonic cover sequences metamorphosed to granulite and amphibolite facies and deformed into large north-trending periclinal structures (Table 3-1); this zone is flanked by two marginal granulite terranes containing large proportions of deformed granitoids and subordinate precratonic cover sequences of greenstone affinity, deformed into strongly linear or periclinal structures parallel to the margins of the zone (Fig. 3-1). The symmetry of the belt emphasized by Mason (1973) is now generally disputed and there have been a number of proposals for minor modifications of the borders recognized for the tectonic zones. For example, the "Shashe Mobile Belt" in the Matsitama area of Botswana (Fig. 3-7) was believed by Crockett (1968) to be a small north-trending offshoot of the Limpopo Province and incorporated in Mason's zonation. The Shashe area is now regarded as a relatively high-grade part of the Zimbabwe Province (Key and Hutton, 1976; Key *et al.*, 1976; Coward *et al.*, 1976a).

This account is based on the structural history outlined by Coward *et al.* (1976a). By its nature this history is a summary and simplification. Nevertheless, because it is based principally on deformation fabrics, it provides an objective framework for the entire belt without suggesting any time equivalence of, for example, the regional D2 finite fabric in different parts of the province.

3.2. Northern Marginal Zone

This tectonic zone was delineated by Mason (1973) as a parallel-sided linear tract extending from Zimbabwe to Botswana. This disposition is upheld by geologists of the Botswana Geological Survey (Key, 1977), who show the Northern Marginal Zone lying to the north of the Tuli–Sabi shear belt. However, we have not extended the Northern Marginal Zone westward into Botswana for two main reasons: First, the characteristic pervasive granulite metamorphic imprint is absent; second, although there are "straightened" structures superficially similar to those in the Northern Marginal Zone, these are of D4 age and are not equivalent to the D2 "straightened" structures in the Northern Marginal Zone. The Northern Marginal Zone is therefore shown as a wedge tapering westward to die out south of the Great Dyke in Zimbabwe; the Central Zone of the Limpopo Province in Botswana is considered to grade northward directly into the Zimbabwe Province north of the Tuli–Sabi shear belt (Fig. 3-1). These boundaries are similar to those of Coward (1976) and Coward *et al.* (1976a). As shown in Figure 3-2, two structural boundaries ("Area of negligible D4 strain" and "Structural domain boundary defined by orientation of maximum extension") and one metamorphic boundary ("Orthopyroxene-in isograd") confirm the wedge shape of the Northern Marginal Zone. Any of these three boundaries can be arbitrarily chosen as the zone boundary, but we have selected the orthopyroxene isograd as the northern boundary (following Robertson, 1973a) and the D4 Tuli–Sabi shear belt as the southern boundary.

The structural and metamorphic effects of the Limpopo episode spread into the Zimbabwe Province (Fig. 3-2) for at least 120 km (Coward *et al.*, 1973; Robertson, 1973a; Coward, 1976). As shown below, the Northern Marginal Zone of the Limpopo Province apparently had a similar early history to the adjoining part of the Zimbabwe Province and did not form a separate entity until the Late Archean. For this reason, and because the geochronologic and structural history of the adjoining Zimbabwe Province is now well established, although not fully understood, we will draw on this history for clues to the evolution of the Northern Marginal Zone (Hawkesworth *et al.*, 1975, 1979; Hickman, 1976, 1978; Barton and Ryan, 1977; Moorbath *et al.*, 1977). The original extent of the precratonic supracrustal sequences can only be surmised, but it is possible that they once formed a nearly continuous cover (Litherland, 1973; Coward *et al.*, 1976b).

Greenstone-Type Precratonic Cover and Possible Basement

Cover rocks in the Northern Marginal Zone are represented by linear relicts of granulite-grade greenstone belts. Metasediments are represented by compositions that are ferruginous (quartz–magnetite–pyroxene rocks), calcareous (quartz–bytownite–diopside rocks) and, possibly, pelitic (cordierite–sillimanite–biotite–sapphirine rocks and sapphirine–hypersthene rocks), and by rare metaquartzites (Robertson, 1977). More common are orthopyroxene-bearing basic granolites (Winkler, 1974) which may represent basic volcanic rocks forming part of the precratonic cover sequence (Robertson, 1968). This association of basic volcanic rocks and chemical sediments is typical of the lower parts of greenstone belts and suggests a subaqueous volcanic environment. Pods and slices of metamorphosed basic and ultrabasic intrusive suites with rare anorthosite and chromitite layers probably represent layered bodies penecontemporaneously intruded into the greenstone belts (Worst, 1962; Robertson, 1973b; Odell, 1975).

Table 3-1. Characteristics of Tectonic Zones in the Limpopo Province and Contiguous Areas of Adjacent Provinces

	Zimbabwe[a] Province	Limpopo Province: Northern Marginal Zone	Limpopo Province: Central Zone	Limpopo Province: Southern Marginal Zone	Kaapvaal[a] Province
Age span (cover to cratonization)	2.9–2.6 Ga	?2.9–2.6 Ga	>3.2–2.6 in E, >2.7–2.0 in W	?–2.5 Ga	?–2.5 Ga
Basement	3.5-Ga tonalitic layered gneiss, granite, Sebakwian greenstone volcanics and stromatolitic limestone	Not recognized	3.8-Ga granodioritic and dioritic layered gneiss, 3.6-Ga basic dikes	Not recognized	Not recognized
Precratonic cover	2.9-Ga (E. Bulawayan), 2.7-Ga (L. Bulawayan and Shamvaian) greenstone volcanics, stromatolitic limestone, and clastic sediments	≥2.9-Ga scattered greenstone-type basic lavas and layered intrusions, pelitic, chemical sediments	3.6–3.2 Ga widespread clastic and chemical sediments, basic and acid lavas	Scattered greenstone-type ultrabasic and basic lavas	Greenstone volcanics, clastic and chemical sediments
Precratonic intrusions	2.9-Ga tonalites, 2.7-Ga layered basic intrusions, dike swarms	2.9-Ga charnockitic rocks, sheeted megacrystic granitoids	3.2–3.1 Ga abundant anorthositic layered intrusions and basic dikes, 2.7-Ga granitoids, in places megacrystic and charnockitic	2.6-Ga megacrystic granite with gabbroids	Granitoids
Dominant structures	Upright folds in NE, nappes in SW; pinched greenstone apophyses and transecting cleavage; shear belts along margin: main events 2.7–2.5 Ga	Upright folds cut by widespread ENE anastomosing shear belts of two ages: main events 2.7–2.6 Ga	Nappes deformed into N-trending elongated periclinal interference structures, rotated and tightened in N by dextral shear belt: mainly 3.2–2.5 Ga in E, >2.7–2.0 Ga in W	Periclinal interference structures cut by shear belts: main events 2.6–2.5 Ga	Upright folds, pinched greenstone apophyses and transecting cleavage
Dominant metamorphism and facies series	Greenschist, with marginal amphibolite; mainly low P/T with medium P/T in SW	Two granulite events, retrogressed in shear belts; low P/T	Two granulite events → upper amphibolite; medium P/T → low P/T	Two granulite events → upper amphibolite; medium P/T → low P/T	Greenschist, with marginal amphibolite

[a] Refers only to areas immediately adjacent to the Limpopo Province.

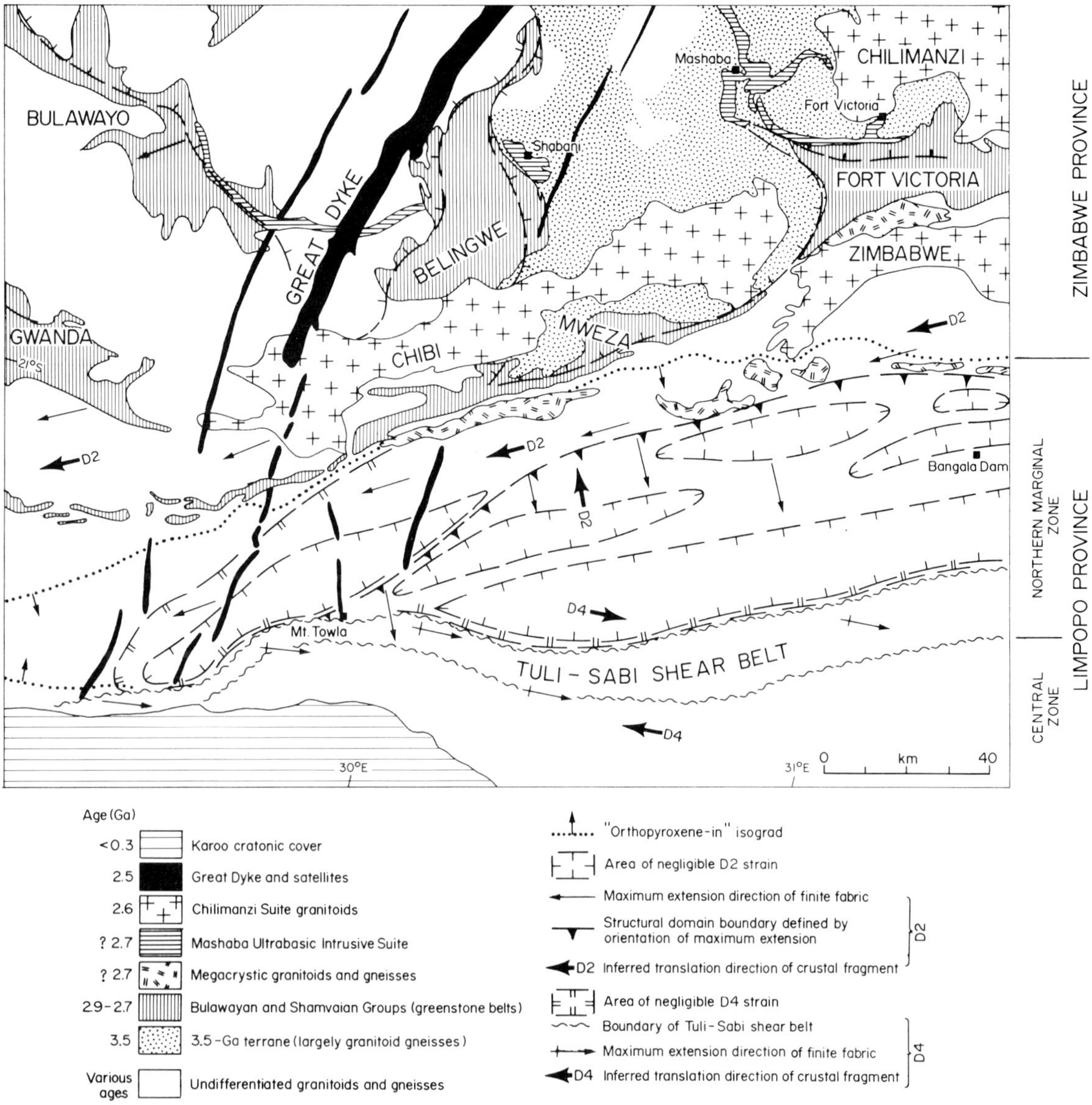

Figure 3-2. Relations between structural domains of various ages, inferred directions of crustal translation, and the orthopyroxene isograd in the Northern Marginal Zone and the contiguous Zimbabwe Province. Note that the extent of D2 deformation in the north is controlled by the distribution of greenstone belts and in the south by anastomosing shear belts between lenticular low-strain enclaves. [Compiled from Coward and Shackleton (1975), Coward (1976), Coward *et al.* (1976a), Orpen (1976), and maps of the Rhodesia Geological Survey.]

No basement to the metasupracrustal rocks has yet been positively identified. Some of the dark granulite-facies quartzofeldspathic gneisses grouped by Robertson (1973a) as "granulite gneiss" may well represent the remains of such a basement, although this group also includes precratonic intrusions younger than the cover. The possible ages for this precratonic basement–cover sequence are discussed under Age Relations but are likely to be 2.9 Ga or older.

Prefabric Early Tectonics

There is at present no evidence that D1 deformation in the Northern Marginal Zone was different in style from that in the Zimbabwe Province. In the latter province the imprint of this deformation is largely preserved in the Late Bulawayan greenstones, which yield radiometric dates of 2.8–2.5 Ga but which are likely to have a minimum age of 2630 Ma (Hawkesworth *et al.*, 1975). The D1 deformation in the Zimbabwe Province is likely to have taken place in the period 2.6–2.7 Ga. However, in the Northern Marginal Zone D1 appears to have preceded a widespread period of metamorphism and intrusion (see the following two sections) at about 2.9 Ga (Hickman, 1978).

The tectonic style of the D1 event was highly variable and the strains have not yet been quantified. In the Zimbabwe Province (Fig. 3-2), autochthonous greenstones on gneissic or granitoid basements, such as the Belingwe (Bickle *et al.* 1975) and Fort Victoria (Orpen and Wilson, 1981) belts, were tightly folded into large upright synforms (James, 1975; Coward *et al.*, 1976b). Farther southwest in this province, allochthonous basement gneisses were thrust over greenstone cover sequences in the Lower Gwanda belt (Fig. 3-7) (Coward *et al.*, 1976b). Nappe folding (e.g., Tati belt) and imbricate thrusting (Matsitama and Gwanda belts; Figs. 3-2 and 3-7) have led to large areas of inverted stratigraphy and spectacular thickening of the supracrustal successions (Wright, 1975; Coward *et al.*, 1976b).

In the Northern Marginal Zone the D1 event resulted in major upright folds, similar to those in the adjacent parts of the Zimbabwe Province (James, 1975). A striking feature of the D1 deformation in both provinces is the paucity of strain fabrics. This paradox of concomitant high strains and translation is most easily ascribed to folding and thrusting in a water-saturated pile at high crustal levels.

The 2.9-Ga Granulite Metamorphism

Two-pyroxene assemblages in basic rocks were produced by low-pressure granulite metamorphism throughout the Northern Marginal Zone (Robertson, 1973a,b) and sporadically farther southwest in Botswana (Mason, 1973). There is a rapid metamorphic transition northward into the Zimbabwe Province (Fig. 3-2), passing through a marginal zone of amphibolite-facies greenstones (Mweza belt) into zones of greenschist facies typical of the Zimbabwe Province (Saggerson and Turner, 1976). Robertson (1973a,b) explained this rapid transition in terms of *in situ* granulite metamorphism with a "telescoped" metamorphic facies series under a thin crustal cover. However, it is probable that at least part of this shortening was tectonically induced during the D2 event.

With regard to the timing of metamorphism, James (1975) suggested that it may have taken place during D1 deformation or soon after. Granulite metamorphism certainly took place before D2 and before the widespread 2.9-Ga intrusive event because basic granolite has been intruded by megacrystic granite (Robertson, 1973a) and both units have been deformed by D2 (Coward *et al.*, 1976a).

The 2.9-Ga Intrusive Event

Much of the crust constituting the Northern Marginal Zone was intruded or remobilized during a series of magmatic events that possibly began during the early granulite metamorphism and continued under conditions of amphibolite-facies metamorphism and partial melting. Granulite metamorphism of the granulite gneiss was closely followed by segregation and injection of quartzofeldspathic and granite veins, and magmas crystallizing under conditions of low water pressure to form charnockites and enderbites (Robertson, 1973a; Odell, 1975). The enderbites are dark-brown granitoid gneisses containing both hydrous and anhydrous mafic minerals and are commonly well layered (Robertson, 1973b), possibly because of high strains (cf. Myers, 1978). They intrude previously deformed basic granolites but are themselves intruded by quartz–feldspar mobilizate (Robertson, 1973a).

After intrusion of the charnockitic magmas, megacrystic granites were intruded as sheetlike bodies on both sides of the orthopyroxene isograd (Fig. 3-2). The granite sheets dip southward concordantly with the gneisses around them and have elongated outcrop shapes. Robertson (1973a) suggested that the megacrysts grew by replacement of plagioclase porphyroblasts and his name "porphyritic granite" is therefore inappropriate. The brown color of the feldspars suggests that the megacrystic granites in the granulite zone have been subjected to high-temperature metamorphism, but the scarcity of orthopyroxene indicates amphibolite-facies conditions of moderate water pressure.

The very strong D2 fabrics developed in the megacrystic granitoid gneisses, closely similar in attitude to those of the granulite gneisses (Robertson, 1973a), show that these intrusions predate the D2 deformation. Robertson (1973a) suggested that they were younger than the even-grained Chibi granite in the Zimbabwe Province (Fig. 3-2) with a radiometric age of approximately 2.6 Ga. However, this is unlikely because the Chibi granite, a composite batholith (Coward, 1976; Hawkesworth *et al.*, 1979) is posttectonic and cuts across D2 structures in the Belingwe greenstone belt.

D2 Regional Fabric Formation and Crustal Translation at 2.7 Ga

This extremely important tectonic episode can be bracketed between two ages. The first is about 2.7 Ga, the age of the Late Bulawayan greenstones and Shamvaian metasediments, both of which have been deformed during D2. The second is 2.5 Ga, the age of posttectonic basic dikes shown in Figure 3-2, which were intruded as satellites of the Great Dyke, and which transect even the D4 structures in both the Northern Marginal Zone and the Zimbabwe Province.

The regional fabric produced during D2 is of fundamental significance because it represents a datum to which all older or younger structures may be referred, even if it varies in age from place to place. It also provides a basis for a structural delimitation of the borders of the Northern Marginal Zone (Coward, 1976). Areas affected by the D2 event are shown in Figure 3-2. The effects of D2 in the southern part of the Zimbabwe Province have been well documented (Coward, 1976; Coward *et al.*, 1976a). Analysis of the magnitude and orientation of this fabric across a wide area indicates a southwestward movement of the Zimbabwe Province relative to the embryonic Limpopo Province. The relative displacement is estimated at approximately 200 km in the east and 50 km in the west. This displacement was accompanied by pure-shear flattening, increasing toward the southwest, and culminating in an estimated mean shortening of 65 percent in the Matsitama greenstone belt of Botswana (Fig. 3-7). To accommodate the sinistral displacement, simple-shear strains with near-horizontal shear directions (Fig. 3-2) were set up along the boundary of the Zimbabwe Province, the intensity of which must have increased toward the northeast. Acting as an east-northeast-trending zone of weakness, the boundary of the Northern Marginal Zone began to be defined.

Another major translation of the crust took place concomitantly along this same zone of weakness. In the northeast the Northern Marginal Zone crustal block was displaced northwestward (Fig. 3-2) with an estimated vertical component of 24 km and a horizontal component of 26 km (James, 1975). As a result the northern margin of the Northern Marginal Zone south of the Fort Victoria and Mweza belts is marked by a major zone of simple shear dipping steeply southward (James, 1975) along which the 3.5-Ga basement (Table 3-1), a precratonic cover of Bulawayan greenstones, and their probable feeders, the Mashaba-Chibi basic dikes, have all been mylonitized in an extensive area (Coward and Shackleton, 1975). The strike of the foliation in the Northern Marginal Zone is similar to that in the adjoining province (east-northeast), but fold hinges, lineations, and maximum extension directions are all perpendicular to those in the Zimbabwe Province, plunging southwestward downdip (Fig. 3-2).

Although the bulk of the Northern Marginal Zone escaped D2 deformation ("Areas of negligible D2 strain" in Fig. 3-2), this crustal block is crossed by numerous anastomosing shear belts, each with the same steep southward dip, downdip lineation, and movement sense (James, 1975; Coward, 1976; Coward *et al.*, 1976a). Within the D2 shear belts granulite parageneses were retrogressed (Coward *et al.*, 1976a). The megacrystic granites were highly strained to form augen gneisses with a strong L > S fabric (Flinn, 1965) plunging downdip (Robertson, 1973a). Between the anastomosing shear zones are lenticular prisms up to 20 km wide and 60 km long that show no D2 fabrics (Fig. 3-2). It is clear that major uplift and northwestward movement of the granolites took place along a basal shear belt at their northern margin and that this movement took place along discontinuities within the granulite crustal block as well. This uplifting explains the presence of granolites at the same present crustal level as the greenschist-facies greenstone belts in the adjoining province. Tectonics of this type probably resulted in the "telescoped" metamorphic facies series referred to above. Robertson (1973b) found no evidence for thrust faulting at the southeastern end of the Northern Marginal Zone (as proposed by Worst, 1962) and suggested that the granulite zone was formed *in situ*. However, the tectonite fabric in this end of the zone, with its characteristic downdip lineation (Robertson, 1973a), indicates that the allochthonous granulite zone proposed by James (1975) and Coward (1976) may be present in the entire Northern Marginal Zone. The northwestward and upward movement of the granolites probably took place by penetrative ductile simple shear at this crustal level and there is no necessity to invoke thrust faulting.

A Younger Granulite Event?

The possibility of a second episode of granulite metamorphism is indicated on two independent grounds. First, within the Northern Marginal Zone the megacrystic granite has dark feldspars and, in places, a granulite-facies assemblage (Robertson, 1973b; Coward *et al.*, 1976a). The megacrystic granite intrudes previously metamorphosed granulite-facies rocks (Robertson, 1973a). Second, there is textural evidence for two episodes of granulite metamorphism in the basic granolites (Cliff, 1975). A two-pyroxene granulite assemblage was deformed to produce a strained texture. The texture was partially annealed under static conditions and recrystallized to produce a symplectic rim of two pyroxenes between earlier formed hornblende and pyroxene. The partial annealing indicates that temperatures were not sufficiently high between the two granulite events for full recrystallization to take place. The strain effects could be the result of D1 or D2 or a combination.

D4 Shearing

Within the Northern Marginal Zone the effects of deformation younger than D2 are negligible (Fig. 3-2). The D3 event, forming sporadic crenulation or mylonitic fabrics elsewhere, left no trace in this zone (Coward *et al.*, 1976a). However, large-scale crustal movements along the southern edge of the zone during D4 have had a profound effect. The Tuli–Sabi shear belt (Fig. 3-2) dips gently southward and consists of flat-lying mylonitized gneisses. A D4 fabric was formed in previously unstrained rocks, whereas older fabrics were rotated and modified (Coward *et al.*, 1976a). A stretching lineation, presumably close to the shear dircction, plunges consistently eastward despite large variations in the strike of the foliation (Fig. 3-2). Movement can therefore be inferred to be dextral with a slight updip component of slip so that the upper southern block (i.e., the Central Zone) moved westward over the northern block (i.e., the Northern Marginal Zone). The minimum true thickness of the shear belt increases northeastward from about 2 km south of Mt. Towla to about 10 km southeast of Bangala Dam (Fig. 3-2). Estimates of displacement increase from at least 8 km in the west to 50 km in the east (Coward *et al.*, 1976a).

The flat-lying D4 mylonites at the southern margin of the Northern Marginal Zone grade upward and southward into less sheared, gently dipping gneisses of the Central Zone. Farther south at a higher structural level these gneisses have been folded into upright periclinal folds with hinges trending north-northeast. This geometry suggests that folding in the Central Zone was accommodated by a décollement zone between the Northern Marginal Zone, which acted as a rigid basement, and the overlying gneisses of the Central Zone undergoing D4 folding (Coward *et al.*, 1976a).

The 2.6-Ga Posttectonic Granitoid Batholiths

At the close of Limpopo tectonic activity in the Northern Marginal Zone, even-grained, elongated granitoid batholiths were emplaced just north of the Northern Marginal Zone (Fig. 3-2). These granitoids are of broadly similar composition and comprise the Chibi, Chilimanzi, Zimbabwe, and Victoria batholiths; together they constitute the Chilimanzi Intrusive Suite (Robertson, 1973a; Wilson *et al.*, 1978). Both the Chibi and the Zimbabwe batholiths have gneissic margins but are predominantly isotropic (Robertson, 1973a). The fact that undeformed Chibi granite truncates the southern margin of the strongly cleaved Belingwe greenstone belt indicates that the bulk of the batholith is posttectonic with respect to D2.

The Chibi granite has gradational contacts with tonalitic gneisses in the southwest but is clearly crosscutting in the northeast. Robertson (1973a) interpreted this to indicate a deeper level of crust exposed at the southern end of the batholith nearer the mobile belt. There appear to be analogous differences in erosion level of the narrow elongated Mweza greenstone belt, which is metamorphosed to amphibolite facies, as compared with the much broader and thicker Fort Victoria belt, which is metamorphosed to greenstone facies. Hawkesworth *et al.* (1979) have shown that Chibi is an isotopically composite batholith. This is consistent with the hypothesis of an oblique section presently exposed through the batholith.

The granitoids of the Chilimanzi Suite are pertinent because although they crop out entirely within the Zimbabwe Province, they may well have originated at crustal depths similar to those presently exposed in the Northern Marginal Zone. Robertson (1973a) intuitively suggested that they formed by partial melting of Limpopo gneisses within the Northern Marginal Zone and rose through the interface to crystallize within the Zimbabwe Province. An origin by melting of this type of parent is supported by the isotopic data of Hickman (1978), described in the following section.

Age Relations

The origin of the high-grade gneisses and granolites that constitute the basement-cover units of the Northern Marginal Zone is speculative. Robertson (1973a) suggested that the precratonic cover rocks formed at least 3.0 Ga ago. However, if we accept the evidence that the early history of the Northern Marginal Zone was similar to that of the Zimbabwe Province, three ages of adjoining crust invite analogy (Wilson *et al.*, 1978; Hawkesworth *et al.*, 1979):

(1) 3.5-Ga terrane (Fig. 3-2), including the Tokwe and Shabani gneisses, the stromatolitic Sebakwian greenstones, and the Mushandike "granite"
(2) 2.9-Ga terrane, including the Chingezi gneisses, the Early Bulawayan greenstones, and the Mashaba tonalite
(3) 2.7- to 2.6-Ga terrane, including the Gwenoro gneisses, the Late Bulawayan greenstones and Shamvaian metasediments, the Chilimanzi granitoids, and various basic intrusions, such as the Mashaba and Shabani basic layered intrusions and the Mashaba–Chibi dike swarm.

What is significant is the close lithologic similarity between all three terranes. Lithologically and structurally there is little to choose between any of these groups being the likely parent material for the Northern Marginal Zone. However, the isotopic data of Hickman (1978) indicate that the granulite gneisses of the Northern Marginal Zone are most likely to have affinities with the 2.9-Ga group. Near Bangala Dam (Fig. 3-2) granulite gneisses, which escaped penetrative deformation during D2, yield a whole-rock Rb-Sr age of 2870 ± 60 Ma and an initial $^{87}Sr/^{86}Sr$ ratio of 0.7014; their low initial ratio precludes an age of intrusion much before this date. From field relations (Robertson, 1973a,b) the charnockitic and megacrystic granitoid intrusions are likely to be of similar age or younger, whereas the supracrustal rocks are probably similar in age or older. Although greenstones of 3.5-Ga age are comparatively rare in the Zimbabwe Province (Wilson *et al.*, 1978), they might be more common at the lower crustal levels exposed in the Northern Marginal Zone.

There is good evidence that rocks similar to the 2.9-Ga granulite gneisses of the Northern Marginal Zone served as parent material for the widespread Chilimanzi granitoid intrusions along the southern border of the Zimbabwe Province (Hickman, 1978). These granitoids give ages of 2.5–2.6 Ga with high initial ratios, indicating origins by partial melting of older crustal material (Hickman, 1976, 1978; Hawkesworth *et al.*, 1979). It is possible that melting of ancient (3.5-Ga) crust with admixtures of much newer crustal material produced the initial $^{87}Sr/^{86}Sr$ ratio of 0.7040. However, it is more reasonable to assume that they formed by partial melting of 2.9-Ga material, presumably the granulite gneisses of the Northern Marginal Zone (Hickman, 1978). Hawkesworth *et al.* (1979) suggest that the 3.5-Ga crust escaped melting at 2.6 Ga because of its high crustal level after underplating by the 2.9-Ga intrusions.

Given the ages of the syntectonic 2.9-Ga intrusions in the Limpopo Province and the posttectonic 2.5-Ga satellites of the Great Dyke (Robertson and Van Breemen, 1970), the D2 event in the Northern Marginal Zone can be bracketed between 2.9 Ga and 2.5 Ga. Analogous deformation in the Zimbabwe Province took place 2.7–2.5 Ga ago.

3.3. Central Zone—Limpopo Valley

Knowledge of the Central Zone of the Limpopo Province is drawn from two main areas: the Limpopo Valley along the common border between South Africa and Zimbabwe, and eastern Botswana (Fig. 3-1). Because correlations between these two areas are still highly tentative, each is described separately here, with cross reference where appropriate. The principal rock units of the Central Zone in the Limpopo Valley are summarized in Table 3-2.

The 3.8-Ga Sand River Basement: An Ancient Sedimentary Sequence?

Some of Earth's oldest rocks are exposed in the Sand River 10 km southeast of Messina (Fig. 3-3). Beneath a complexly deformed mantle of precratonic cover, an ancient basement crops out as scattered inliers in antiformal cores. This basement, known as the Sand River gneisses, is of bimodal character (Figs. 3-4A, D), consisting of interlayered homogeneous gray gneiss of quartz dioritic composition and more-leucocratic, weakly layered, gray gneiss of granodioritic composition (Barton *et al.*, 1977, 1978; Fripp, 1980). Both gneisses yield statistically identical ages and their combined whole-rock Rb-Sr age is 3786 ± 61 Ma, initial $^{87}Sr/^{86}Sr$ ratio = 0.7012 (Barton *et al.*, 1978). This is a minimum age, thought to reflect a metamorphic event that homogenized the Sr isotopes. The subsequently robust nature of the Rb-Sr system is remarkable considering that the gneisses

Table 3–2. Principal Archean Lithostratigraphic Units and Lithologies in the Limpopo Valley[a]

Messina Area (South Africa)		Beitbridge Area (Zimbabwe)	
Bulai granitoid gneiss: Tonalitic–granodioritic gneiss interlayered with megacrystic granitic gneiss; opx-bearing felsic to mafic core and opx-bearing charnockitic margins in places.			
Messina Intrusive Suite: Metamorphosed anorthosite and leucogabbro with subordinate gabbro, pyroxenite, dunite, magnetitite, chromitite			
Beitbridge sequence		Beitbridge sequence	
Messina Metamorphic Suite	Metaquartzite (± feldspar, magnetite)	Nulli Metamorphic Suite	Metaquartzite (± sillimanite, magnetite, fuchsite, diopside, garnet)
	Aluminous gneiss		
	Magnetite iron-formation		Magnetite–grunerite iron-formation
	Calc-silicate rock		Calc–silicate rocks and marble
			Mafic granolite
	Amphibolite and hornblendite, Mafic granolite, Serpentinite } Some basic dikes		Amphibolite, Hornblende gneiss, Serpentinite } Some basic dikes
Shanzi Metamorphic Suite	Singelele gneiss: Quartz–feldspar leucogneiss (± hornblende, biotite, garnet)	Diti Metamorphic Suite	Singelele gneiss: Quartz–feldspar leucogneiss (± hornblende, biotite, garnet)
	Quartz–feldspar–biotite–garnet leucogneiss with minor aluminous gneiss, calc-silicate rock, and basic dikes		Quartz–feldspar–biotite–garnet leucogneiss with minor aluminous gneiss, metaquartzite, iron-formation, calc-silicate rocks, marble, pyroxene-garnet granolite, and basic dikes
Basement: Sand River gneisses: Nebulitic layered gray gneisses of granodioritic, tonalitic, and quartz dioritic composition with metabasite dikes			

[a] "Formations" and "Subgroups" have been more realistically renamed metamorphic suites (following Sohl, 1977).

Compiled from publications by SACS and the Rhodesia Geological Survey; Barton *et al.* (1977, 1978, 1979 a,b); Fripp *et al.* (1979); and M. K. Watkeys (personal communication, 1979).

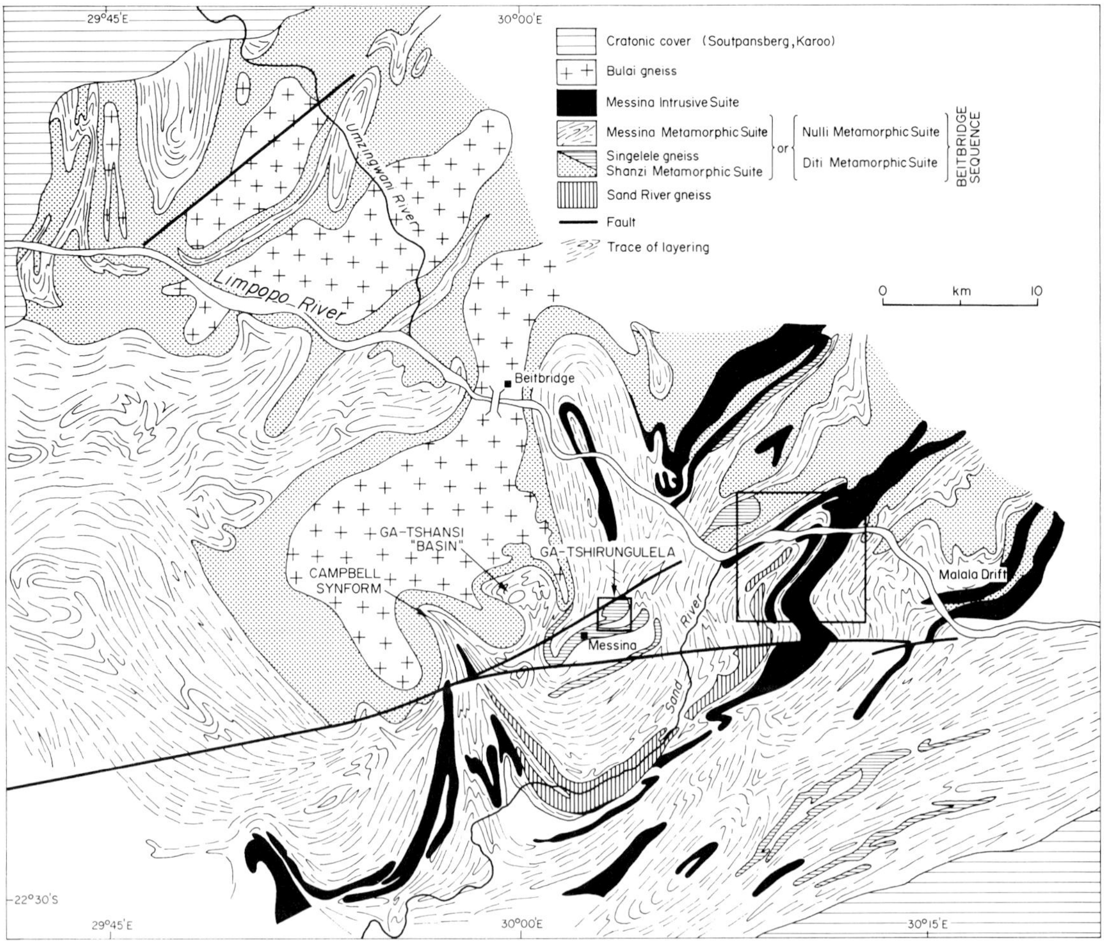

Figure 3-3. Geologic map of the Central Zone in the type area of the Limpopo Valley. Rectangles delimit areas shown in subsequent figures. [Compiled from maps of the Geological Surveys of South Africa and Rhodesia according to additional mapping by M.K. Watkeys and R.E.P. Fripp (1977–1979).]

have undergone at least six periods of deformation from 3.8 Ga to 2.6 Ga. The Sand River gneisses were intuitively believed to be of sedimentary origin (Barton *et al.*, 1977), a conclusion reinforced by their apparent inability to form a meaningful Sm-Nd isochron (Barton, 1979a; personal communication, 1979).

These gneisses, representing undisputed basement, are exposed in an arc to the south and east of Messina along the course of the Sand River. Elsewhere in the Limpopo Valley, gneisses of widely varying composition have also been designated as basement (e.g., Bahnemann, 1972; Light and Watkeys, 1978), but the structural relations of these units are obscure or controversial and include components now known to be part of the younger Bulai gneiss (M. K. Watkeys, personal communication, 1979).

Precover Basement Tectonics and Dike Intrusion

Partial melting and segregation of quartz–feldspar veinlets took place during high-grade metamorphism of the basement gneisses (Barton *et al.*, 1977; Light and Watkeys, 1978). Folding and penetrative deformation of the gneisses and their enclosed veins resulted in the development of a deformation fabric at 3.8 Ga. Subsequent cooling allowed the emplacement of abundant quartz tholeiite dikes (Fig. 3-4A), which yield a Rb-Sr whole-rock isochron of 3560 ± 100 Ma, initial $^{87}Sr/^{86}Sr$ ratio = 0.7014 (Barton *et al.*, 1977). The basic dikes and country rocks were then folded, accompanied by the development of ductile shear belts, together with anatexis and granulite-facies meta-

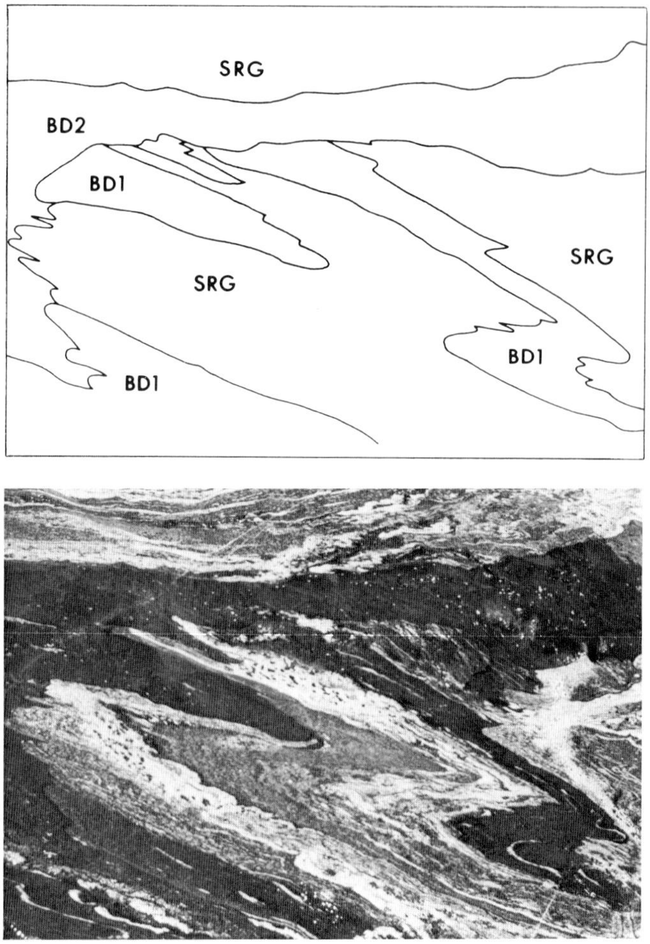

(A)

Figure 3-4A. Tholeiitic dikes, metamorphosed to amphibolite, intruded and infolded with the 3790-Ma Sand River gneisses (SRG). An older basic dike (BD1), dated at 3560 Ma, transects the layering and quartz–feldspar neosomes in the gneisses. Both units are crosscut by a younger basic dike (BD2), dated at 3060 Ma. Locality: Sand River, 2 km upstream from bridge on Tshipise road; outcrop width 3 m. (B) Metapelite gneisses near the base of the Messina Metamorphic Suite boudinaged by D1 extension and subsequently shortened by younger folding (scale is 10 cm). Location: sand River, 0.25 km upstream from bridge on Tshipise road. (C) "Football" anorthosite preserved in the layered Messina Intrusive Suite, dated at 3150 Ma. Metamorphism at upper amphibolite facies has retrogressed the igneous mineralogy to labradorite in a dark matrix of hornblende and clinopyroxene (scale is 10 cm). Location: near road to Malala Drift, 6 km east of Sand River. (D) Two boudins of metagabbro, probably representing the deformed remains of a feeder dike to the Messina Intrusive Suite, in a country rock of multiply deformed Sand River gneisses. The sampled boudin falls on the same Rb-Sr isochron (3150 Ma) as the intrusion and its composition falls within the same range, apart from enrichment in Na_2O and K_2O, possibly caused by wallrock reactions. Location: Sand River, 1.5 km upstream from bridge on Tshipise road.

morphism of the dikes (R. E. P. Fripp, personal communication, 1978).

The Beitbridge Precratonic Cover

Forming most of the Central Zone is a layered sequence of metasupracrustal gneisses known as the Beitbridge sequence (Fig. 3-3). Table 3-2 lists the main rock types in this sequence. Pseudostratigraphy has been recognized in places (e.g., Bahnemann, 1971, 1972). However, it seems unrealistic to attempt a detailed stratigraphic subdivision over a large area because of the absence of reliable younging criteria coupled with stratigraphic duplication (by polyphase isoclinal folding and imbricate thrusting) and stratigraphic excision (by partial melting, thrusting, and boudinage).

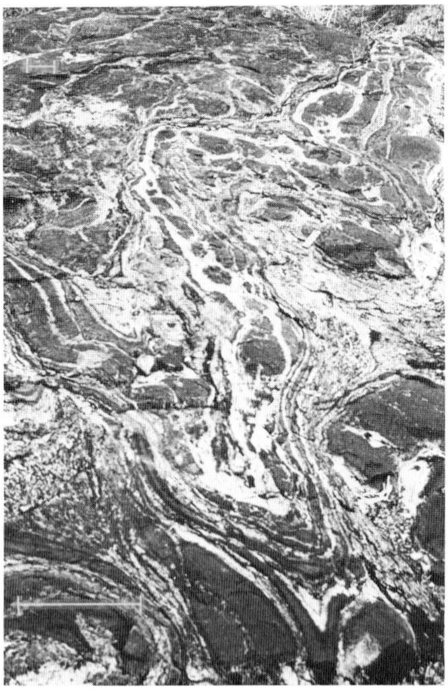

(B)

(C)

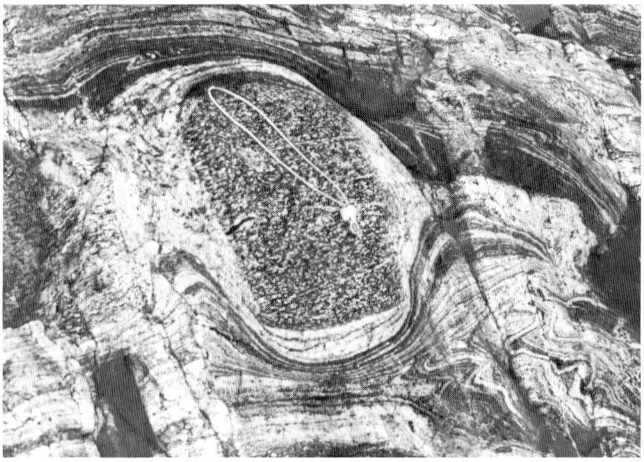

(D)

Figure 3-4. (B)–(D)

The Beitbridge precratonic cover rocks are thought to represent a relatively thin, predominantly sedimentary sequence deposited in a broad basin under shallow water (Söhnge *et al.*, 1948; Bahnemann, 1972; Mason, 1973; Coward *et al.*, 1976b; Light and Watkeys, 1978). Bahnemann (1972) asserted that the metabasites of the cover sequence (together with hornblendite and anorthositic gneiss) represented the restite residue from partial melting of less basic parent rocks, whereas most other workers regard them as forming from volcanic or hypabyssal basic igneous rocks.

Reconstruction of the sedimentary basin is hampered by the lack of preserved sedimentary structures, the only reported occurrence being some disputable graded bedding in metaquartzites of the Nulli Metamorphic Suite (M. K. Watkeys, personal communication, 1979). Nevertheless, it is clear that around Messina the basin shallowed eastward because the carbonate facies is only abundant to the east of the main mass of the Messina Intrusive Suite shown in Figure 3-3 (Söhnge *et al.*, 1948; Light and Watkeys, 1978). There is no reliable indication whether the quartzofeldspathic gneisses of the Diti–Shanzi Metamorphic Suites were formed from acid volcanic rocks or arkosic quartz arenites, but the intercalations of iron-formation, marble, and metaquartzite indicate that some sedimentation took place. Shallow-water parental sediments postulated for the Messina and Nulli Metamorphic Suites are arenaceous and argillaceous (Fig. 3-4B) in the western part of the basin (Bahnemann, 1972; Light and Watkeys, 1978). However, as shown below there may be more deep-water eugeosynclinal characteristics than generally realized. The association of basic volcanism (now represented by amphibolite, hornblende gneiss, serpentinite, and basic granolite) with chemical sediments such as Algoman-type iron-formation and possible cherts (some of the thinner metaquartzites) and graywackes (aluminous gneisses) is not unlike that of the greenstone upper Onverwacht and Fig Tree Groups of comparable age (Chapter 2).

The Singelele Controversy

The Singelele gneiss is conformably enclosed within the Beitbridge cover sequence as relatively thin, folded sheets throughout the Limpopo Valley (Fig. 3-3). The gneiss is of granitic composition, consisting mainly of quartz, microcline, and plagioclase with a generally diffuse layering defined by varying proportions of biotite, hornblende, and garnet. According to Bahnemann (1972, 1973), the gneiss is alaskitic in fold crestal zones and biotite-rich in fold limbs. A well-developed quartz-leaf fabric imparts a strong schistosity to the gneiss (Fripp *et al.*, 1979).

Following Söhnge (1946), Bahnemann (1972, 1973) concluded that the Singelele gneiss was of intrusive origin, having been derived by partial melting of an underlying tonalitic parent such as the Bulai gneiss, and had risen upward into antiformal cores of the folded metasupracrustal rocks. Evidence cited for this process includes the apparent restriction of the Singelele gneiss to antiformal cores, the prevailing high grades of metamorphism, and a vertical sequential change in bulk composition compatible with progressive anatexis toward a melt of hornblende syenitic composition. However, Bahnemann (1973) acknowledged the absence of intrusive contacts between the Singelele and Beitbridge gneisses. In addition, he argued that the presence of metabasite dikes in the Singelele gneiss, coupled with their apparent absence from the Beitbridge sequence, indicates that the former was an older basement to the latter. This reasoning casts further doubt on an intrusive origin for the Singelele gneiss.

The absence of younging criteria makes it impossible to assess the true stratigraphic position of the Singelele gneiss, but more-recent work in the type area (Fig. 3-5) has shown that the gneiss occupies the cores of early-formed antiforms and synforms, with the major late structure at Ga-Tshirungulela being synformal (Fripp *et al.*, 1979; M. K. Watkeys, personal communication, 1979). The lack of discordant contacts or xenoliths further negates an intrusive origin. Conversely, the conformable relations with the surrounding gneiss, fine layering parallel to the contacts, similar structural history, and presence of similar metabasite dikes all indicate that the Singelele gneiss is of supracrustal origin and of comparable age to the Beitbridge sequence (Fripp *et al.*, 1979). The gneiss occupies a similar structural level to, and grades laterally into, the Diti and Shanzi Metamorphic Suites (Light *et al.*, 1977; M. K. Watkeys, personal communication, 1979). It is therefore logical to include the Singelele gneiss in the Diti and Shanzi Suites (Table 3-2). It could be argued that the heterogeneous composition of the gneiss indicates a clastic or volcaniclastic parent (Fripp *et al.*, 1979).

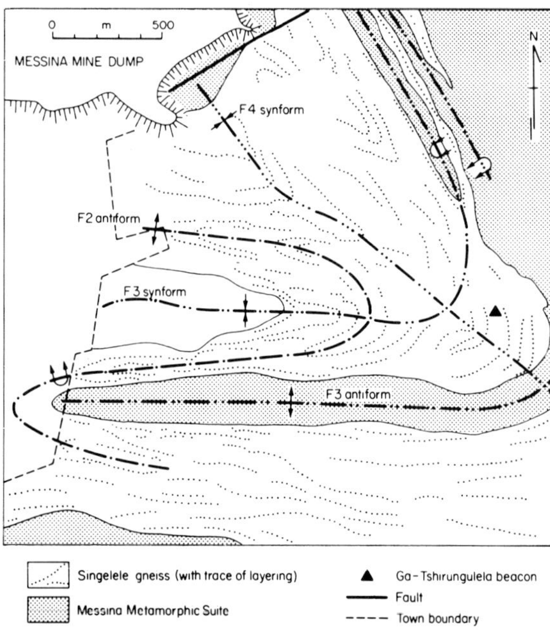

Figure 3-5. Detailed structural map of the Ga-Tshirungulela area showing structural relations between the Singelele gneiss and the Messina Metamorphic Suite in the type area. F2 and F3 show type 3 interference, whereas F3 and F4 show type 2 interference, indicating F3 folds were gently inclined before D4. [After Fripp *et al.* (1979).]

Van Breemen and Dodson (1972) obtained a whole-rock Rb-Sr errorchron of 2690 ± 60 Ma for both the Singelele and the Bulai gneisses. Subsequent dating of the Singelele gneiss by whole-rock Rb-Sr and Pb-Pb and U-Pb zircon methods by Barton *et al.*, (1979b) has produced even younger radiometric ages that bear no relation to the time of emplacement suggested by field relations because of thorough isotopic resetting. Initial $^{87}Sr/^{86}Sr$ ratios are variable and high because of postemplacement homogenization of the Sr isotopes. This loss of isotopic memory may result from the fine-grained leucocratic nature of this gneiss (Barton *et al.*, 1979b)

Early Deformation and Metamorphism of the Precratonic Cover

North of the Limpopo River, evidence for intense deformation (D1) and high-grade metamorphism is preserved in rocks of the Beitbridge sequence but is not recorded in the anorthositic Messina Intrusive Suite (Light and Watkeys, 1978). Isoclinal recumbent folding and flat-lying ductile shearing resulted in the formation of a deformation fabric and folding of quartz–feldspar anatectic veins. Medium-pressure granulite metamorphism accompanied D1 and formed the paragenesis "orthopyroxene + kyanite + quartz" at the expense of an older, low-pressure "cordierite + sillimanite" paragenesis. Metamorphism continued interkinematically, as evidenced by a rim of randomly oriented kyanite and sillimanite around garnet porphyroblasts.

This high-grade preanorthosite deformational event has not yet been recorded south of the Limpopo River. The oldest deformation (D1) commonly recognized in either the Beitbridge sequence (including the Singelele gneiss) or Messina Intrusive Suite is regarded as a low-grade event with only sporadic cleavage formation (Barton *et al.*, 1979a; Fripp *et al.*, 1979; Fripp, 1980). Nevertheless, these authors also record D1 as involving recumbent folding together with possible thrusting.

Emplacement of the Anorthositic Messina Intrusive Suite and Basic Dikes at 3.2–3.1 Ga

Exposures of highly deformed and metamorphosed precratonic layered intrusions are widespread throughout the Limpopo Valley and westward into Botswana (Fig. 3-3). Barton *et al.* (1979a) have suggested that all of these exposures may be the remains of a single extensive sheet duplicated by thrusting and several periods of folding. Although Hor *et al.* (1975) regarded the main exposure east of Messina as a tilted sheet younging consistently westward, it is clear that the sheet has undergone polyphase folding and stratigraphic duplication. Two structural interpretations are shown in Figure 3-6.

The bulk of the intrusion consists of interlayered metaanorthosite and metaleucogabbro and about 10 percent consists of more-mafic phases, such as metagabbro, metamelagabbro, metapyroxenite, hornblendite, and serpentinized dunite (Barton *et al.*, 1979a). Bahnemann (1972, 1973) regarded the metaanorthosites and hornblendites as basic residues resulting from partial melting of calcareous metasediments. However, evidence for a plutonic igneous origin is overwhelming (Söhnge *et al.*, 1948; Van Eeden *et al.*, 1955; Windley, 1973; Hor *et al.*, 1975; Coward *et al.*, 1976a; Light *et al.*, 1977; Light and Watkeys, 1978; Barton *et al.*, 1979a). The rocks contain size- and phase-graded layering that gives consistent younging directions, and laterally extensive magnetitite and chromitite layers in anorthositic and ultramafic hosts, respectively; abundant cumulus plagioclase grains are preserved in the less strained, more competent anorthositic rocks (Fig. 3-4C). In addition the rocks contain inclusions of metaquartzite and metapelite, and an inferred feeder dike of metagabbro (falling on the same isochron as the intrusion) cuts the Sand River gneisses (Fig. 3-4D) (Barton *et al.*, 1979a). Similarly a dike of metagabbro transects the foliation in anorthositic gneisses to the northeast near Malala Drift (Fig. 3-3) (R.E.P. Fripp, personal communication, 1978).

Barton *et al.* (1979a) obtained a whole-rock Rb-Sr isochron of 3153 ± 47 Ma, which they interpret as being close to the age of intrusion. However, this must be reinterpreted in view of a more recent Pb-Pb age of about 3350 Ma (J. M. Barton, personal communication, 1979). The relatively high intitial $^{87}Sr/^{86}Sr$ ratio (0.7033) indicates a magma source from a Rb-enriched mantle. Fractionation trends calculated by these authors indicate a quartz–tholeiitic parental magma. The intrusion was emplaced magmatically or tectonically as a sill near the base of the supracrustal succession. From consideration of the compositions of crystallizing phases and rock mechanics, intrusion is thought to have taken

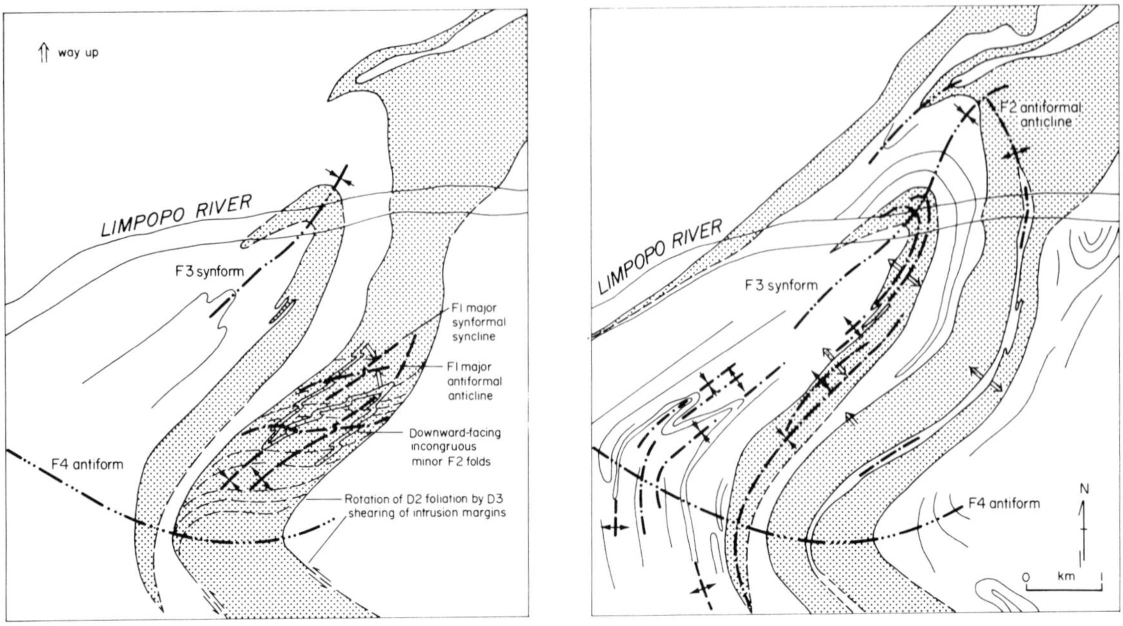

Figure 3-6. Two possible explanations of the structure of a large folded sheet of the anorthositic Messina Intrusive Suite (stippled). [Left, adapted from Graham (1974) and Coward *et al.* (1976a); right, after Barton *et al.* (1979a).]

place at relatively shallow depths of less than 12 km (Barton *et al.*, 1979a). Flat-lying shear belts parallel to the contacts (Graham, 1974; Coward *et al.*, 1976a; Light and Watkeys, 1978) may be connected with tectonic emplacement of the sheet in an environment of ductile thrusting as proposed for the similar Fiskenaesset intrusion in west Greenland (Myers, 1976).

The date of 3150 Ma for the time of intrusion of the Messina Intrusive Suite provides a minimum age for the Beitbridge sequence because inclusions of it are present within the former. The maximum age for the supracrustal rocks may be that of the 3.6-Ga basic dikes in the Sand River gneisses, which have not been found in the Beitbridge sequence. Intrusion of the anorthositic Messina Intrusive Suite was closely followed by the injection of basic dikes (Fig. 3-4A) into the basement at 3060 Ma, an age isotopically indistinguishable from that of the anorthosites (Barton *et al.*, 1977). This generation of basic dikes has been observed in the Messina Metamorphic Suite and Singelele gneiss (Fripp *et al.*, 1979). It is possible that the shallowly emplaced layered intrusion, the gabbroic feeder dikes, and younger basic dikes may constitute plutonic and hypabyssal phases of the same period of igneous activity, possibly analogous to greenstone-belt formation (Barton *et al.*, 1979a). If so, this episode of greenstone-type igneous activity would not be coeval with any of the three greenstone cycles reported from the Zimbabwe Province (Wilson *et al.*, 1978; Hawkesworth *et al.*, 1979) or with the Onverwacht volcanism in the Kaapvaal Province (Hamilton *et al.*, 1979).

The Bulai Enigma: Ancient Basement or 2.7-Ga Intrusion?

The Bulai granitoid gneiss is exposed extensively northwest of Messina (Fig. 3-3). The name was originally applied to megacrystic varieties (Söhnge, 1946) but now embraces all rock types constituting the heterogeneous migmatite complexes. Variable strains during D2 created pervasive mineral fabrics ranging from weak to strong. The core of the Bulai migmatite complexes typically contains weakly foliated acid and basic orthopyroxene–plagioclase–hornblende–biotite–garnet gneisses (M. K. Watkeys, personal communication, 1979). Surrounding these granulite-facies rocks is a broad expanse of tonalitic or granodioritic gneisses interlayered with more-potassic megacrystic granitoid gneisses which intrude the surrounding country rocks of the Diti and Shanzi Metamorphic Suites. Granites with rectangular megacrysts were converted by high strains to augen gneisses and, ultimately, to finely layered gneisses. Near its margins the Bulai granitoid gneiss is particularly heterogeneous with abundant nebulitic layering defined by biotite and hornblende. At the contact, in places infolded with the Diti Suite, the gneiss grades into an orthopyroxene-bearing charnockitic gneiss containing small xenoliths and infolded inclusions of the Diti–Shanzi Metamorphic Suites (M. K. Watkeys, personal communication, 1979).

The Bulai granitoid gneiss generally occupies the cores of domes and underlies the supracrustal Diti and Shanzi Metamorphic Suites. Large cuspate D4 structures such as the Campbell synform (Fig. 3-3), have been used as evidence for a viscosity contrast between a highly ductile, and possibly wet, supracrustal cover (Beitbridge sequence) and an older, more competent, dry crystalline basement (Bulai granitoid gneiss) by Bahnemann (1971, 1973). However, subsequent mapping by M. K. Watkeys has shown that around the Ga-Tshansi "basin" other cusps are reversed in direction to that forming the Campbell synform (Fig. 3-3). Bahnemann concluded that the Bulai gneiss was older than the Beitbridge sequence because of the Bulai's structurally lower position, the general absence of garnet within it, the nebulitic nature of the layering, and the presence of folded basic dikes that were apparently absent from the supracrustal gneisses. This evidence can be disputed on a number of grounds. First, garnet is present in the granulite-facies core, the main granodioritic phase, and the charnockitic envelope of the Bulai granitoid gneiss (M. K. Watkeys, personal communication, 1979) and, in any case, cannot be considered a valid criterion for a basement (cf. Mason, 1973). Second, similar dikes are present within the Messina Metamorphic Suite (R.E.P. Fripp, personal communication, 1978). Third, the nebulitic layering can be attributed to partial melting (Light *et al.*, 1977), which is consistent with field relations, metamorphic grade, and Sr-isotopic composition.

Age relations between the various phases of the Bulai granitoid gneiss are still uncertain and it is not clear what proportion is of intrusive origin. Neither is the time of intrusion known for certain. Van Breemen and Dodson (1972) produced an errorchron of 2630 Ma, whereas Barton *et al.* (1979b) reported whole-rock Rb-Sr and Pb-Pb isochron and errorchron ages between about 2950 Ma and 2700 Ma. These are consistent with ages estimated from field relations. The initial $^{87}Sr/^{86}Sr$ ratio (0.7030) suggests that the maximum age for Bulai granitoid gneiss may be around 3000 ± 270 Ma (Barton *et al.*, 1979b). In view of the strikingly similar appearance of the megacrystic phase of the Bulai to the megacrystic grantoids of the marginal zones of the belt, which give similar radiometric ages, most geologists regard the Bulai granitoid gneiss as having been intruded or remelted around 2.7 Ga (Van Breemen and Dodson, 1972; Mason, 1973; Light *et al.*, 1977; Light and Watkeys, 1978; Robertson and Du Toit, 1981).

Notwithstanding this consensus that the Bulai granitoid gneiss was emplaced at about 2.7 Ga during high-grade metamorphism (and is possibly of anatectic origin), some enigmas remain. The granitoid gneiss contains metabasite dikes that have been folded and boudinaged during D2 (Bahnemann, 1973). This could be explained by the intrusion of a third set (younger than the 3.6-Ga and 3.1-Ga sets) of basic dikes after 2.7 Ga. More difficult to explain is a boudin train of metaanorthosite and metagabbro in the Bulai granitoid gneiss between Beitbridge and Messina. There are three possibilities: (1) The boudins represent a deformed sheetlike xenolith of the 3.2-Ga Messina Intrusive Suite in 2.7-Ga Bulai granitoid gneiss; (2) the boudins represent deformed dikes of a hitherto unrecorded, post-3.2-Ga episode of anorthositic intrusion; (3) the boudins represent deformed dikes of the Messina Intrusive Suite intruding a pre-3.2-Ga gneiss. It is quite probable that the heterogeneity of the Bulai granitoid gneiss has disguised the presence within it of relicts of Singelele gneiss or ancient basement equivalent to the Sand River gneisses. The third explanation above would be in accord with the existence of older inliers or large xenoliths, as would the presence of deformed basic dikes.

D2 Regional Fabric Formation, Folding, and Metamorphism

Soon after 2.7 Ga a widespread and intense episode of crustal shortening followed the Bulai intrusion coevally with similar tectonic activity in the Zimbabwe and Kaapvaal Provinces. Predominantly flat-lying gneisses were folded into tight or isoclinal upright folds with generally northeast axial trends throughout the Limpopo Valley

(Graham, 1974; Coward *et al.*, 1976a; Light and Watkeys, 1978; Barton *et al.*, 1979a; Fripp *et al.*, 1979). A finite fabric resulted from interaction with D1 fabrics in older units, such as the Beitbridge sequence, and new fabrics were formed in younger units, such as the Bulai granitoid gneiss. A strong axial-plane schistosity, represented by flattened quartz leaves, formed in the Singelele gneiss (Fripp *et al.*, 1979). the Messina Intrusive Suite became further infolded with the supracrustal gneisses (Fig. 3-6). After major F1 upright folding, incongruous F2 minor folds and a transecting fabric formed on the major fold limbs, followed by intense ductile shearing and rotation of the fabric along the margins of the intrusive sheet during D3 (Graham, 1974; Coward *et al.*, 1976a). Medium-pressure metamorphism changed to a low-pressure granulite metamorphism in which kyanite was replaced by sillimanite and cordierite became stable (Light and Watkeys, 1978). Geothermometry and geobarometry suggest temperatures around 875°C and total pressures of 800–1000 MPa (8–10 kbar) for the medium-pressure metamorphism (Horrocks, 1980).

D3–D4 Periclinal Interference Folding and Metamorphism

The present outcrop patterns result largely from interference between the D3 and D4 fold sets. Tight F3 folds trending originally north-northeast were folded around axial surfaces trending west to northwest to form a periclinal interference set with gently dipping enveloping surfaces (Bahnemann, 1971, 1972; Graham, 1974; Coward *et al.*, 1976a; Light and Watkeys, 1978; Barton *et al.*, 1979a; Fripp *et al.*, 1979). At Ga-Tshirungulela a macroscopic type 2 interference pattern (Ramsay, 1967) between F3 and F4 folds suggests that the former set was uncharacteristically gently inclined here (Fig. 3-5) (Fripp *et al.*, 1979). Farther west around Ga-Tshansi major D3 folds of supracrustal gneisses were drawn into the macroscopic cusps bordering the Bulai granitoid gneiss and now have axial traces normal to their original trend (Fig. 3-3) (M.K. Watkeys, personal communication, 1979).

During this major folding, metamorphic conditions changed from low-pressure granulite facies to middle-amphibolite facies with sporadic development of D4 axial-plane cleavage (Light and Watkeys, 1978). Horrocks (1980) recorded a possibly analogous retrogression south of the Limpopo River with estimated temperatures and total pressures of 600–700°C and 500–700 MPa, respectively. A subsequent interkinematic period of greenschist-facies metamorphism was followed by the generation of flat-lying mylonite belts and the growth of postkinematic staurolite during lower amphibolite-facies metamorphism (Light and Watkeys, 1978).

Postcratonization Tectonic Activity

There is no evidence that the events described above continued after 2.5 Ga and this time probably marks the close of widespread tectonic activity in this part of the Central Zone. From about 2.6 Ga to 1.9 Ga this part of the now-cratonized crust underwent progressive uplift from catazonal depths to the surface, at which time the Soutpansberg cratonic cover sequence was deposited (Chapter 7). In addition this extensive and prolonged unroofing supplied detritus for older cratonic sequences in the Kaapvaal Province, such as the Witwatersrand and Transvaal Supergroups (Chapters 4 and 5). Uplift was initially along steeply dipping mylonite belts that evolved into faults controlling the shape of the Soutpansberg depository (Mason, 1973; Jansen, 1975, 1976, 1977; Light and Watkeys, 1978). Between 2220 Ma and 1960 Ma, posttectonic diabase dikes intruded the Central Zone, followed by extrusion of the basal lavas of the Soutpansberg Group at 1770 Ma (Barton and Ryan, 1977; Barton, 1979b). Radiometric ages of 1.9–1.8 Ga obtained from various paragneisses (Van Breemen and Dodson, 1972; Barton and Ryan, 1977) may represent a widespread static metamorphism or may mark the closing of certain isotopic systems during the later stages of uplift.

3.4. Central Zone—Botswana

As is the case farther east in Zimbabwe, the boundary between the Limpopo and Zimbabwe Provinces is transitional and different margins may be delineated according to arbitrary criteria. In the absence of well-defined metamorphic isograds, we have chosen a structural boundary, namely the northern margin of the Tuli–Sabi shear belt (Fig. 3-7).

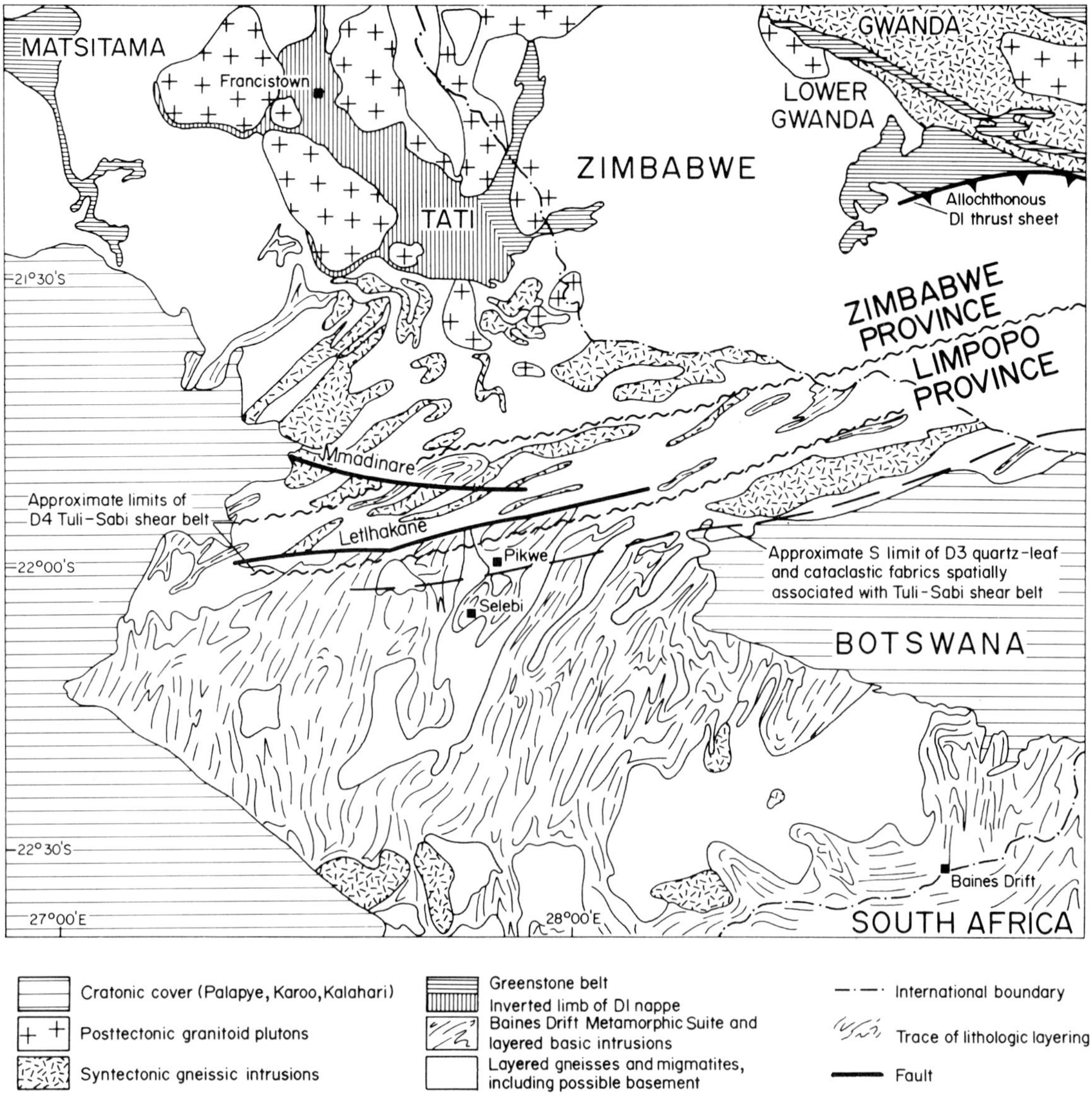

Figure 3-7. Geologic map showing the extent of the D4 Tuli–Sabi shear belt and older cataclastic rocks along the boundary between Zimbabwe and Limpopo Provinces in northeastern Botswana. [Adapted from maps of the Geological Surveys of Botswana and Rhodesia with structural amendments from Wakefield (1974, 1977a), Coward (1976), and Coward *et al.* (1976a).]

Basement

The extent and nature of the basement is unclear at present. Although Mason (1973) interpreted the bulk of the migmatitic gneiss terrane as basement, it now appears that a large proportion of these rocks consists of strongly deformed postcover intrusive granitoids (Wakefield, 1974; Wright, 1974). Foliated basic dikes that are apparently absent from the precratonic metasedimentary cover have been used as evidence for a basement (Mason, 1973; Key, 1977). However, recognition of tightly folded postcover basic dikes near Pikwe (Wakefield, 1974, 1977a; Key, 1977) indicates that this criterion should be used with caution. Wright (1974) and Coward *et al.* (1976b) reported the presence of old structures in layered gneisses northwest of Baines Drift (Fig. 3-7) that were not found in the metasedimentary cover or intruding granitoids. These gray layered gneisses, which probably represent the basement, are of tonalitic composition and contain anatectic phases of pink granite (Key, 1977).

Neither the age of this basement nor that of the metasedimentary cover is known. Clues may be present in the adjoining Zimbabwe Province where the supracrustal Matsitama sequence apparently underlies the 2.5-Ga Tati greenstone belt (Fig. 3-7) (Van Breemen and Dodson, 1972; Coomer *et al.*, 1977). A boulder bed at the base of the shallow-water Matsitama sequence contains clasts derived from an even older greenstone sequence and a previously deformed and migmatized gneiss terrane (Coomer *et al.*, 1977). This greenstone-gneiss terrane may be analogous to the 2.9-Ga terrane farther east in the Zimbabwe Province. The absense of 3.5-Ga Pb ages from the Matsitama area may indicate the absence here of the 3.5-Ga Zimbabwean basement (Fig. 3-2) (Coomer *et al.*, 1977). The Central Zone basement therefore may be approximately 2.9 Ga old, as in the Northern Marginal Zone, or approximately 3.8 Ga old, as in the Limpopo Valley.

The Baines Drift Precratonic Cover: A Greenstone-Belt Facies or Beitbridge Correlate?

Mason (1973) introduced the term "Baines Drift Formation" for metasediments in eastern Botswana resembling those of the Messina and Shanzi Metamorphic Suites. It is advisable to retain a separate name for these rocks because of the dangers in correlating rocks of unknown age with others more distant on purely lithologic grounds.

The metasediments are lithologically similar to those of the Messina Metamorphic Suite, consisting of metaquartzite, magnetite quartzite, metapelite, and metacalcareous rocks (Mason, 1973). In addition, amphibolites of tholeiitic composition grade into anorthosites, suggesting a layered intrusive parent (Key, 1977). Quartzofeldspathic gneisses similar to those of the Shanzi and Diti Suites are considered to be metaarkoses because of their association with metaquartzite (Mason, 1973; Key, 1977) and their high quartz content in places (Wakefield, 1974). These gneisses can be traced into the Zimbabwe Province as the Tutume sequence, which conformably underlies the mainly volcanic Tati greenstone belt (Key *et al.*, 1976). However, some of these gneisses are known to be intrusive granitoids (Wright, 1974).

The Baines Drift Metamorphic Suite has generally been regarded as the high-grade equivalent of shallow-water facies of the 2.7-Ga-old greenstone belts such as the Matsitama belt (Fig. 3-7), and it has been suggested that the metasediments in both provinces once formed a more or less continuous cover (Bennett, 1970; Van Breemen and Dodson, 1972; Litherland, 1973; Mason, 1973; Coward *et al.*, 1976a, b; Key *et al.*, 1976; Saggerson and Turner, 1976; Key, 1977). Clearly there is a difficulty in reconciling these correlations with recent age determinations from the Limpopo Valley that show that the lithologically similar Beitbridge sequence is 3.5–3.2 Ga in age (Barton and Ryan, 1977; Barton *et al.*, 1979a). Consequently the Baines Drift Metamorphic Suite is of comparable age either to the Beitbridge sequence or to the much younger 2.7-Ga greenstones, such as the Matsitama and Tati belts.

Precratonic Intrusions

Large sheets of layered metagabbro–anorthosite intrude the metasedimentary cover near Baines Drift and Pikwe (Mason, 1973; Wakefield, 1974; Wright, 1974; Key and Hutton, 1976). The sill-like intrusions young downward into structurally lower, but stratigraphically higher, metasediments because of subsequent D1 nappe folding (Coward *et al.*, 1976b). If these intrusions are time equivalents of the Messina Intrusive Suite, the supracrustal Baines Drift country rocks must be significantly older than the greenstone belts in the adjoining Zimbabwe Province.

The Pikwe stratabound Ni-Cu sulfide deposit (Fig. 3-7) formed during this intrusive event and owes its preservation to emplacement in an area of relatively low deformation (Wakefield, 1974, 1976). Immiscibility textures between the ore sulfides and silicates indicate that the sulfide magma was emplaced together with an ultramafic crystal mush into an older but genetically related troctolitic intrusion.

A wide range of syntectonic granitoid gneisses is present in the Central Zone (Fig. 3-7) but these rarely show crosscutting field relations and probably mostly predate the D2 regional fabric-forming event at about 2.6 Ga (Wright, 1974). Some of these are megacrystic and near the Tuli-Sabi shear belt have been strained to form augen gneisses similar to those in the Northern Marginal Zone (Mason, 1973). Many of the granitoid gneisses are associated with migmatites and appear to have formed by anatexis during the high-grade metamorphism roughly coeval with D2 (Coward *et al.*, 1976b; Key *et al.*, 1976; Key, 1977).

D1: Early Nappe Tectonics

A widespread period of horizontal tectonics, culminating in the development of fold nappes and imbricate thrust faults, took place at about 2.6 Ga in the majority of greenstone belts in the southwestern part of the Zimbabwe Province (Fig. 3-7) (Mason, 1973; Wright, 1974; Coward, 1976; Coward *et al.*, 1976a, b; Coomer *et al.*, 1977). Analogous nappe tectonics appears to have taken place in the Central Zone because the sheetlike basic intrusions south of Pikwe partly represent the inverted limbs of pre-D2 fold nappes (Coward *et al.*, 1976b). Thrusting at this time could also be responsible for the repetitious sheeted nature of these intrusions (Wright, 1974). Major and minor isoclinal recumbent folds of pre-D2 age in the paragneisses and migmatites have also been reported (Key and Hutton, 1976).

D2: Regional Fabric Formation, Folding, and Metamorphism at 2.6 Ga

A regional finite fabric formed in the Central Zone roughly at the same time as elsewhere in the Limpopo Province and the adjoining Zimbabwe Province (Hickman and Wakefield, 1975; Coward *et al.*, 1976a, b; Key *et al.*, 1976; Coomer *et al.*, 1977). The mean crustal shortening in the greenstone belts of the Zimbabwe Province has been estimated at almost 50 percent across cleavage (Coward, 1976). The strains in the Limpopo Province are likely to have been even higher because of the increased ductilities consequent on the higher temperatures of metamorphism.

Folding accompanied formation of the regional fabric. The F2 folds plunged gently toward the north-northeast (Wakefield, 1974, 1977a; Wright, 1974); enveloping surfaces of F3 folds suggest that the axial surfaces of the F2 folds dipped gently northward (Wakefield, 1977a) and therefore originally were reclined. North of the Tuli–Sabi shear belt the F2 folds were originally upright, an attitude characteristic of the Zimbabwean greenstone belts (Wright, 1974; Coward *et al.*, 1976b; Key, 1977).

The formation of the flat-lying fabric in the Central Zone was accompanied by moderate-pressure metamorphism that varied from upper amphibolite (Wright, 1973) to granulite facies (Wakefield, 1974, 1977a; Key, 1977). Water pressures were sufficiently high to enable widespread partial melting to take place.

The Majante suite of basic to intermediate dikes was emplaced prior to subsequent D3 folding during which they were strongly deformed and amphibolitized (Wakefield, 1974, 1977a; Key, 1977).

D3–D4: Interference Folding

The geologic map of eastern Botswana (Fig. 3-7) (Key, 1977) reveals a variability in structural trends caused by superposition of at least two generations of post-D2 folds and distortion as well as rotation of the resulting interference structures within the domain of the Tuli–Sabi shear belt. The interference has been studied in most detail south of the shear belt (Wakefield, 1974, 1977a; Wright, 1974; Hickman and Wakefield, 1975). Around Selebi the flat-lying D2 regional fabric was folded into steeply inclined major folds trending west-northwest. Northward through the Pikwe area the F3 major folds become overfolded and progressively more recumbent in the same direction. Cataclastic and quartz-leaf fabrics formed locally north of Pikwe as a result of these high strains (Fig. 3-7). The presence of garnet in the deformed Majante amphibolite dikes indicates that this deformation took place under medium-pressure upper amphibolite conditions. The deformation style became less ductile with time and in a northward direction. This episode of horizontal tectonics may have been responsible for the early thrust movements observed on the Tuli–Sabi shear belt (Mason, 1973; Wright, 1974).

Large upright F4 folds trending northeast were superposed on the variable F3 structures so that type 1 oval or triangular periclinal structures formed around Selebi and type 2 "mushroom" structures formed around Pikwe. Lower amphibolite metamorphism partially recrystallized and annealed the D3 cataclastic fabric to produce new strain-free textures (Wakefield, 1974, 1977a).

The Botswana Geological Survey believed that the interference structures formed by "coaxial overfolding" of upright folds plunging gently to the north (Key *et al.*, 1976; Key and Hutton, 1976; Key, 1977). However, it is difficult to see how this type 3 interference could result in either the observed periclinal structures with closed outcrop patterns or the planar dispersion of axial lineations that they report (Key *et al.*, 1976). These authors further suggested that this folding resulted from movement along the Tuli–Sabi shear belt, a view which is questioned in the following section.

The orientations of the F3 and F4 folds formed by crustal shortening both across and parallel to the trend of the Limpopo Province are similar to those in the Limpopo Valley (cf. Figs. 3-3 and 3-7) and it is tempting to equate them (cf. Coward *et al.*, 1976a). However, as described in the following section, it is possible that this deformation in Botswana occurred more than half a billion years later than in the Limpopo Valley.

The Tuli–Sabi Shear Belt

This shear belt forms the northern boundary of the Limpopo Province in Botswana. The rather simple map pattern, suggesting dextral transcurrent movement, hides a complex interplay of vertical, horizontal, rotational, and translational components of displacement and highly variable strain states (Mason, 1973; Wakefield, 1974, 1977b). The belt may have developed at the southern margin of the comparatively little-deformed granitoids south of the Tati greenstone belt, thereby maintaining compatibility between highly deformed rocks of the Central Zone and less deformed rocks of the Zimbabwe Province (Coward *et al.*, 1976a). The eastern extension of the shear belt farther east may represent a similar décollement zone between the Northern Marginal Zone and the Central Zone (Fig. 3-2).

The rocks deformed in the shear belt show progressively more brittle behavior and concomitant lowering of metamorphic grade from lower amphibolite facies to greenschist facies, probably because of uplift and erosion (Wakefield, 1977a, b). For example, a penetrative ductile mylonitic fabric with a shallow-plunging L component is successively overprinted by nonpenetrative conjugate crenulation cleavage, fracture cleavage, minor shear belts, and dextral wrench faults, such as the Letlhakane fault (Fig. 3-7). The geometrically related attitudes of these structural elements all indicate a maximum compressive stress oriented consistently northwest–southeast throughout the process of simple shear (Wakefield, 1977a, b). The gross sigmoidal trend of the mylonite fabric and the shallow-plunging extension directions are compatible with a model of transcurrent ductile simple shear (Ramsay and Graham, 1970). Rotation of the D3–D4 interference structures within the shear belt enables estimation of dextral displacement of 30 km in the west and 50 km farther east (Wakefield, 1974, 1977b; Coward *et al.*, 1976). Dextral displacement on the late-stage Letlhakane fault is estimated at about 8 km (Wakefield, 1974).

Overfolding in opposed directions on each side of the shear belt led Key and Hutton (1976) to believe that the F3–F4 interference folding took place simultaneously with shearing. They suggested that horizontal shortening by pure shear with opposed directions of translation led to the formation of the subvertical shear belt. This model raises severe space problems and fails to account for the observed rotation of the F3–F4 structures (Wakefield, 1974; Wright, 1974). The model of Key and Hutton can also be disputed on the grounds that (1) the widespread flat-lying D3 mylonites are entirely incompatible with their model and (2) ductile folding outside the shear zone could not have taken place simultaneously with mylonitization in the shear belt without anomalously low heat flow along the shear belt (Wakefield, 1977b).

The Tuli–Sabi shear belt shows a westward decrease in width and intensity of the quartz-leaf fabric, indicating a decrease in strain and displacement (Coward *et al.*, 1973; Key and Hutton, 1976). Key and Hutton (1976) found no evidence for shearing west of longitude 27° 30′ E and suggested that the Limpopo Province, as they define it, also dies out there. Geophysical interpretations of the deep structure both corroborate (Reeves and Hutchins, 1975) and negate (Pretorius, 1979) this hypothesis. Both the characteristic east-northeast-trending rotated structures and the younger transcurrent faults appear to continue farther westward toward the Phanerozoic cover (Key and Hutton, 1976; Key, 1977). These observations suggest that whereas the mylonitic core of the shear belt dies out, the shear belt as a whole, manifested by inhomogeneous rotation, may continue westward at least as far as the cratonic cover (Fig. 3-7).

The age of the Tuli–Sabi shear belt and the preceding D3–D4 deformation is of critical importance because it was at this time that the histories of the Central Zone and the adjoining Zimbabwe Province first diverged strongly. In the Northern Marginal Zone deformation of this type had ceased by the time that the satellite dikes of the Great Dyke were emplaced at 2.5 Ga. Despite the presence of much younger radiometric ages from Botswana, the tendency has been to regard the Tuli–Sabi shear belt as also forming at about 2.6 Ga (e.g., Coward *et al.*, 1976a; Key, 1977). On the other hand, Wakefield (1974, 1977a) and Hickman and Wakefield (1975) have strongly advocated a much younger age of about 2.0 Ga. Mineral ages of this date were first obtained from a wide variety of gneisses in Botswana by Van Breemen and Dodson (1972).They considered the mineral ages to represent a metamorphic imprint, possibly resulting from intrusion of the Bushveld Complex, rather than a "cooling age" coincident with uplift. Similar mineral ages have subsequently been obtained from around Pikwe (Hickman and Wakefield, 1975) and Matsitama (Coomer *et al.*, 1977) and these authors suggest that isotopic homogenization accompanied the ductile D3 and D4 deformation at 2.1–2.0 Ga. It has been argued that the survival of nonannealed D3 quartz-leaf fabrics would be impossible if they had formed at 2.6 Ga and been thermally metamorphosed at 2.0 Ga (Hickman and Wakefield, 1975).

It is therefore possible that whereas the D2 regional fabrics were formed at approximately 2.6 Ga throughout the Limpopo Province and parts of the adjoining cratons, subsequent deformation took place much later in Botswana than elsewhere.

Comparatively minor younger deformation in the Central Zone has been documented by Key (1977). The most important was the development of dip-slip faults that created depositories for the accumulation of the clastic Palapye Group at 1.8 Ga, a probable time equivalent of the Soutpansberg Group farther west (Chapter 7) (Ermanovics *et al.*, 1978).

3.5. The Southern Marginal Zone

The high-grade metamorphic imprints characteristic of the Limpopo Province continue south of the Soutpansberg Range into a poorly exposed, relatively little-understood gneiss terrane known as the Southern Marginal Zone. The boundary with the Central Zone is obscured by the Soutpansberg Group cratonic cover. The boundary with the Kaapvaal Province to the south is gradational and there are no fundamental structural or stratigraphic discontinuities between the provinces (Mason, 1973; Graham, 1974; Du Toit and Van Reenen, 1977). There is a gradational increase in the number of deformation events and in the strain magnitudes northward into the Limpopo Province (Graham, 1973, 1974; Du Toit and Van Reenen, 1977). The most significant transition is metamorphic and the boundary between the provinces may be conveniently taken at the orthoamphibole isograd, reflecting the southern limit of a late high-grade Limpopo metamorphic event (Van Reenen and Du Toit, 1977).

The few Rb-Sr whole-rock age determinations in this zone (Barton and Ryan, 1977) suggest that deformation and metamorphism took place between 2.6 Ga and 2.5 Ga. The rocks represent the high-grade equivalents of the Kaapvaal Province, as is the case for the Northern Marginal Zone and the Zimbabwe Province respectively (Mason, 1973; Graham, 1974; Du Toit and Van Reenen, 1977). However, the linear structural trends and elongated granitoid intrusions of the Northern Marginal Zone are not present in the Southern Marginal Zone.

Precratonic Greenstone-Type Cover

Because of the gradational transition between the Kaapvaal and Limpopo Provinces, the metasupracrustal sequences in the latter are best understood by initial reference to their well-preserved lithologic equivalents in the Kaapvaal Province (Fig. 3-8). In both the Pietersburg and Murchison greenstone belts a sequence of quartz arenites and conglomerates in the cores of major synclines is underlain by basic volcanic rocks and chemical sediments, which in turn are underlain by basic and ultrabasic extrusive and intrusive rocks (Grobler, 1972; Graham, 1974). The greenschist grade of these belts rises along the northeastern extension of the Pietersburg belt to amphibolite facies, characteristic of the metasupracrustal rocks around Renosterkoppies farther north (Fig. 3-8). The latter rocks consist of basic, ultrabasic, ferruginous, and quartzose compositions and appear to be the high-grade equivalents of the greenstone belts (Graham,

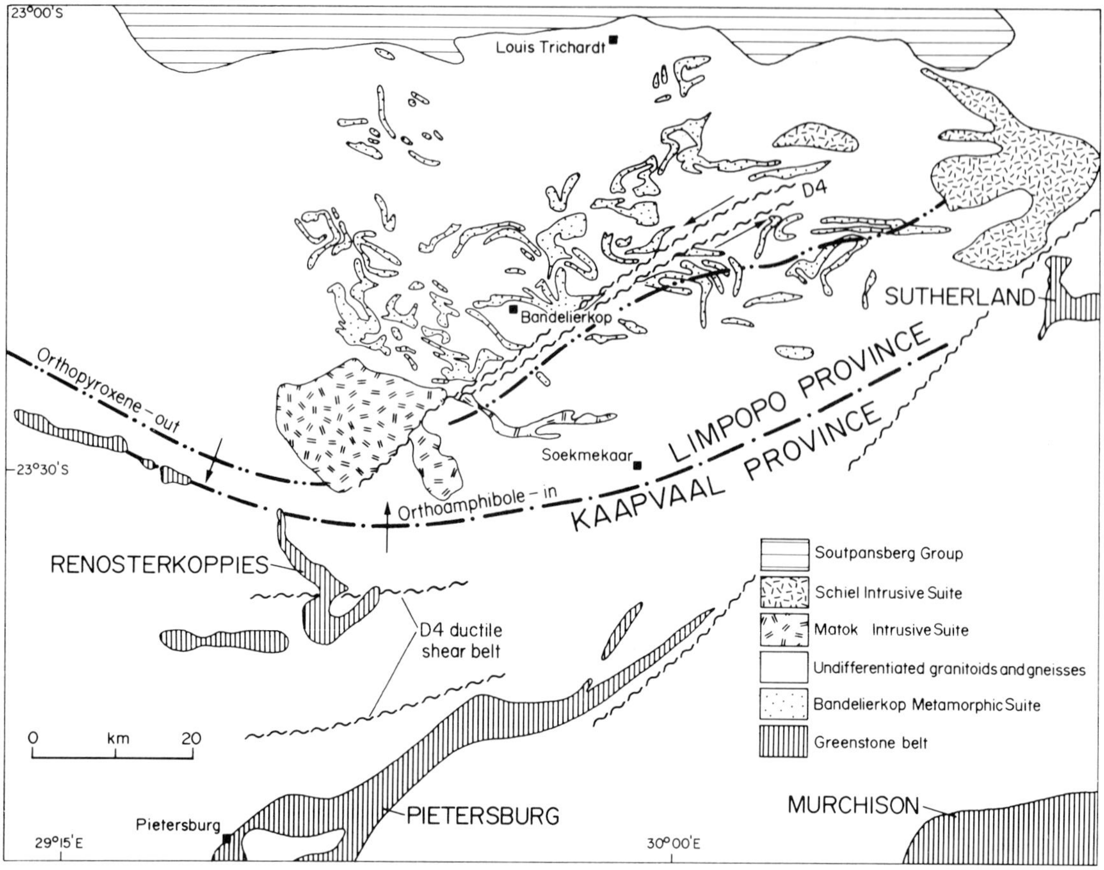

Figure 3-8. Geologic map of the Southern Marginal Zone showing M3 isograds based on metapelites. [After Du Toit and Van Reenen (1977), with structural amendments from Graham (1974).]

1974; Du Toit and Van Reenen, 1977). Farther north, lying largely within the granulite-facies Limpopo Province, are widespread synformal relicts of the greenstone-type cover known as the Bandelierkop Metamorphic Suite (Fig. 3-8). The suite is heterogeneous and consists mainly of basic and pelitic granolites with ferruginous granolites and ultrabasic rocks. Dismembered ultrabasic bodies contain pillows and lenses of iron-formation, indicating an extrusive origin (Du Toit and Van Reenen, 1977). They are apparently structurally overlain by komatiitic basic rocks and thin iron-formation (Van Reenen and Du Toit, 1977). Aluminous gneisses indicate the association of pelitic rocks, possibly graywackes, with these extrusive rocks and chemical sediments.

Surrounding the Bandelierkop Metamorphic Suite is a heterogeneous group of layered and enderbitic gneisses, grouped under the name of Baviaanskloof gneiss (Du Toit and Van Reenen, 1977). It is not known what proportions of these gneisses represent precratonic basement, cover, or intrusions.

The Matok Intrusive Suite

The Matok Suite consists mainly of megacrystic biotite granitoid gneisses and a possibly older gabbroic phase (Du Toit and Van Reenen, 1977). It constitutes an irregular batholith straddling the orthopyroxene isograd (Fig. 3-8). The relative and absolute times of intrusion are unclear. There is no contact aureole and the granitoid gneisses are foliated, suggesting that they were intruded prior to (Graham, 1974), or during (Du Toit and Van Reenen, 1977), the main D2 deformation. The Rb-Sr age of 2.6 Ga obtained from the Matok granitoid gneiss could represent either a metamorphic event or an intrusive event (Barton and Ryan, 1977).

Early North–South Shortening and Granulite Metamorphism

The deformation sequence associated with the Limpopo tectonic activity is best preserved in the Pietersburg and Murchison greenstone belts adjoining the Limpopo Province (Fig. 3-8). Regional horizontal shortening during D1 resulted in the formation of east-northeast-trending upright horizontal synclines (Graham, 1974). Continuing compressive stresses, oriented north–south and slightly oblique to those of D1, resulted in a D2 regional fabric in the form of an upright transecting cleavage, associated steeply plunging F2 folds, and mineral and intersection lineations.

This D2 regional fabric can be traced northward into the Southern Marginal Zone and is present in the majority of granitoid gneisses, indicating that these are likely to be syn- or pre-D2 in age (Graham, 1974). The deformation sequence is not well understood but has resulted in two generations of tight to isoclinal folds that are apparently coaxial (Du Toit and Van Reenen, 1977). The F2 folds are upright and fold an older D1 fabric; they have an axial-planar quartz-leaf fabric (Graham, 1973).

Evidence for two periods of granulite metamorphism is well preserved in metapelites of the Bandelierkop Suite (Van Reenen and Du Toit, 1977; Van Reenen, 1979). Medium-pressure granulite metamorphism during deformation resulted in the formation of orthopyroxene and almandine. A subsequent lowering of pressure caused disequilibrium between these minerals and the formation of cordierite and second-generation orthopyroxene in symplectic intergrowths. The radial structure of the symplectites suggests that the second granulite metamorphism (M2) took place after the D2 fabric-forming event. The survival of relict orthopyroxene beyond the present limits of the orthopyroxene zone, the boundaries of which are a function of a younger M3 metamorphic event, suggests that this granulite metamorphism extended farther southward than the present orthopyroxene isograd (Fig. 3-8). The age of 2580 Ma obtained from the Bandelierkop Metamorphic Suite may reflect the time of the M1 or M2 granulite metamorphism.

Late Folding and Retrogressive Metamorphism

Open upright folding during D3 formed periclinal interference structures throughout much of the Southern Marginal Zone (Graham, 1973). Folding was followed by the development of large D4 ductile shear zones, the most obvious of which is east of Bandelierkop with a sinistral displacement (Fig. 3-8) (Graham, 1973; Du Toit and Van Reenen, 1977). Southward through Renosterkoppies (Fig. 3-8), vertically dipping supracrustal gneisses with low strain states pass into highly strained flat-lying gneisses which revert to their steep attitudes and low strains in the extreme south. This transition can be interpreted as a traverse through a 15-km-wide D4 ductile shear zone, dipping gently to the south with normal displacement, which rotated and intensified the steeply dipping, less strained D2 foliation (Graham, 1974).

Closely linked in time to the D3 deformation was an episode of retrogressive upper amphibolite medium-pressure metamorphism. This event, designated M3, is regarded as merely a continuation of the earlier metamorphic processes (M1 and M2) without an intervening period of cooling and uplift (Van Reenen and Du Toit, 1977). The M2 symplectites were hydrated during M3 so that anthophyllite formed at the expense of hypersthene and gedrite at the expense of cordierite. The M3 event is responsible for the present metamorphic pattern and the northward retreat of the "orthopyroxene-out" isograd (Fig. 3-8). The M3 orthoamphibole laths were rotated, but not broken down, by a locally developed D3 strain fabric (Van Reenen and Du Toit, 1977), indicating that M3 conditions prevailed before and during the D3 deformation.

From the limited radiometric data available it appears that the entire tectonic history of the Southern Marginal Zone can be bracketed between 2.6 Ga and 2.5 Ga. A granite with posttectonic features near Soekmekaar yields an isochron of 2460 Ma, which is probably close to the time of emplacement (Barton and Ryan, 1977).

STAGE 2

Early Proterozoic Supracrustal Development

Fluvial and shallow-marine sedimentation was ubiquitous on the cratonized Kaapvaal Province from 2.8 Ga to 1.8 Ga. Most of the carbonates in the Transvaal epeiric sea contain stromatolites, including large subtidal mounds up to 40 m long and 20 m wide west of Nelspruit, South Africa.

Chapter 4

The Golden Proterozoic

The Dominion Group, Witwatersrand Supergroup, and Ventersdorp Supergroup constitute a volcanosedimentary sequence that is famous for its fossil gold placers. It has been estimated that since gold-bearing conglomerates were discovered near Johannesburg (Fig. 4-1) in 1886, mines in the Transvaal and Orange Free State have produced more than 55 percent of all the gold ever mined in the world (Pretorius, 1976). The level of activity in this South African industry may be gauged from the fact that 300,000 workers develop 1000 km of underground tunnels at an average depth of 1650 m every year; 10^8 metric tons of ore are mined to produce approximately 700 metric tons of gold annually (Gay and Jager, in press).

Nearly all of the other giant placer deposits in the world are of Tertiary age. Examples cited by Henley and Adams (1979) include: California (42 million ounces of gold), Klondike (9), Fairbanks (8), and British Columbia (6) in North America; Columbia (32), Chile (11), Bolivia (10), and Peru (4) in South America; Lena (40) in Siberia; Otago (8) and Westland (5) in New Zealand. Their gold was derived from Phanerozoic host rocks during erosion initiated by uplift along convergent plate boundaries in continental areas such as the circum-Pacific belt. Proterozoic placers in the Kaapvaal Province dwarf these Tertiary deposits, having produced more than 843 million ounces of gold. The reason for this prodigious enrichment is not understood, but Pretorius (1976) has drawn attention to two factors that may have had some influence: geologic age and structural setting. Of these the more important is considered to be the great age of the deposits, which is reflected in the nature of the source area.

Radiometric age determinations of Late Archean basement granitoids and of lava in the Ventersdorp near the top of the Witwatersrand triad have bracketed the age of the Witwatersrand sediments between 2.8 Ga and 2.3 Ga (Van Niekerk and Burger, 1964, 1978b). Granites underlying the Witwatersrand and Ventersdorp Supergroups in the western Transvaal have yielded ages between 3.10 Ga and 2.70 Ga (Allsopp, 1964; Rb-Sr recalculated ages). Van Niekerk and Burger (1978b) recorded even younger ages for granites of 2.66 Ga and 2.64 Ga in the same general area. Igneous rocks in the Dominion Group have yielded U-Pb mineral ages between 2.83 Ga and 2.80 Ga (Van Niekerk, 1968; Van Niekerk and Burger, 1969). There is no evidence that any of these rocks intrudes either the Witwatersrand or Ventersdorp Supergroups. Ventersdorp lavas near the top of the Witwatersrand triad have yielded U-Pb zircon ages between 2.30 Ga and 2.24 Ga (Van Niekerk and Burger, 1964; Van Niekerk, 1968; Rundle and Snelling, 1977). However, a U-Pb zircon age of 2.62 Ga on Ventersdorp rhyolite (Van Niekerk and Burger, 1978b) is at variance with the radiometric ages of pre-Witwatersrand igneous activity because it seems improbable that the thick Witwatersrand and Ventersdorp successions could have accumulated in so brief a period of time even allowing for the analytical uncertainty in the quoted ages. If this anomalous Ventersdorp age is valid, both the Witwatersrand and Ventersdorp Supergroups should be regarded as Archean as defined in this volume. However, their geological character is that of cratonic cover sequences typical of the Proterozoic (the Archean–Proterozoic boundary is diachronous). It is possible that, like the Pongola Supergroup (Chapter 2), the Witwatersrand and Ventersdorp Supergroups represent a response to the attainment of local crustal stability during the Late Archean, before cratonization became widespread at 2.5 Ga. This

Figure 4-1. Witwatersrand gold and the rich Main Reef placers were discovered on the farm Langlaagte by George Harrison in 1886. The site of this discovery and an old stampmill are shown in the foreground. Spawned from these discoveries Johannesburg is now the hub of mining and finance in South Africa. [Courtesy of Johan Ross and Anglo American Corporation of South Africa.]

cratonic stability allowed preservation of the cover rocks with minimal alteration and deformation. The sedimentary rocks of the Witwatersrand Supergroup have been metamorphosed only to lower greenschist facies. Fluid-inclusion studies indicate that temperatures probably did not exceed 250°C (Hallbauer and Kable, 1979); consequently the fabric, texture, and primary sedimentary structures in these ancient rocks are unusually well preserved (Fig. 4-2).

The Dominion, Witwatersrand, and Ventersdorp deposits are superimposed over the center of the craton, covering progressively larger areas (Fig. 4-3). The Ventersdorp Supergroup is estimated to have originally covered an area of 300,000 km^2 (Pretorius, 1976).

The broad tectonosedimentary characteristics of the intracratonic Witwatersrand basin have been documented by Pretorius (1976). The structural setting envisaged was a half-graben, with a progressively more active fault scarp along the northwestern margin and a relatively passive southeastern margin. Continued uplift in the northwest caused the sedimentary prism to offlap across the shrinking basin. An alternative to this fault-controlled model is suggested by recent studies (Hutchison, 1975; Minter, 1976, 1978, 1982) which indicate that the geometry of the sedimentary units, unconformities, and paleocurrent patterns of the entire basin can be accounted for by epeirogenic tilting and warping, followed by basin closure as a result of diapiric granite doming. Major faulting was initiated later than previously thought, possibly during Late Witwatersrand and Early Ventersdorp deposition, with fault-controlled erosion and sedimentation in subsidiary yoked basins. The Dominion, Witwatersrand, and Ventersdorp successions reflect a

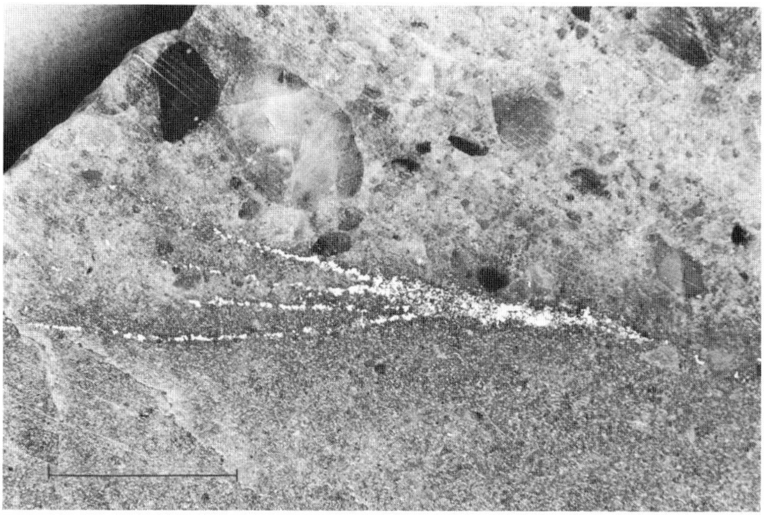

Figure 4-2. Detrital gold concentrated on the toesets of trough cross-beds. This photograph illustrates the exceptional state of preservation of these Early Proterozoic sediments. Basal placer in Free State Geduld gold mine, Welkom goldfield. Scale equal to 2 cm. [Courtesy of Johan Ross and Mike Lain.]

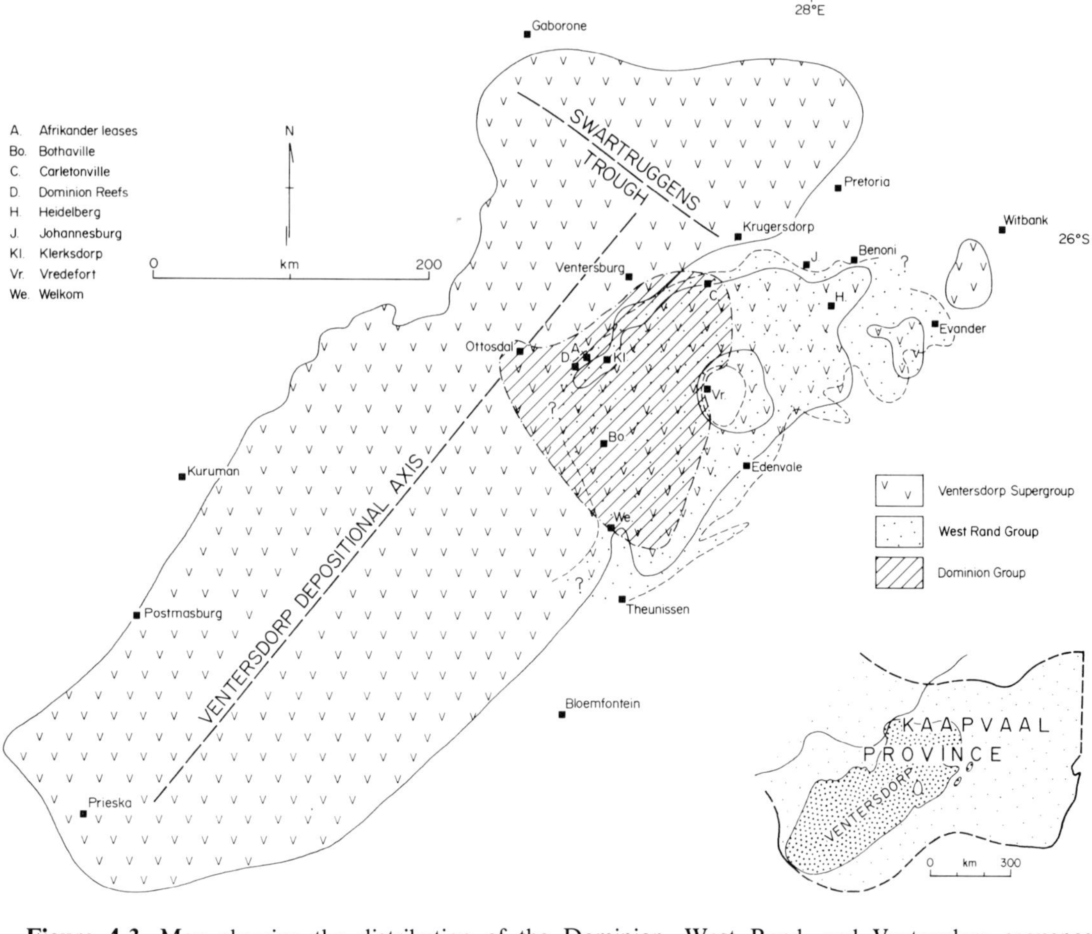

Figure 4-3. Map showing the distribution of the Dominion, West Rand, and Ventersdorp sequences. [Adapted from Hutchison (1975), Tyler (1979a), M. Strydom (personal communication, 1981).]

Table 4–1. Stratigraphy of the Early Proterozoic Groups[a]

Supergroup	Group	Thickness (m)	Locality
Ventersdorp	Pniel	1170	Bothaville
	Platberg	5000	Bothaville
	Klipriviersberg	1690	Johannesburg
Witwatersrand	Central Rand	2420	Johannesburg
	West Rand	4670	Klerksdorp
	Dominion	2710	Klerksdorp

[a] Sources: Hutchison (1975), Minter (1976), Pretorius (1976), Wyatt (1976).

progressive increase in tectonic influence from gentle warping to deep crustal fracturing and associated volcanism.

The Proterozoic succession is composed of six groups that collectively comprise 12,400 m of terrigenous clastic sediments and 5260 m of lava (Table 4-1). Economic geologists working in the Witwatersrand basin have defined formational units (or interpretative genetic packages) which are related to climatic–tectonic controls. According to H.D. Hedberg (1979) these formational units represent synthems. This genetic approach is particularly applicable to the Central Rand Group, where detailed information is available concerning the unconformities that separate the formations. These unconformity surfaces are in many cases associated with extensive placers that cover hundreds of square kilometers. Group and subgroup divisions are distinguishable throughout the depository because of major epeirogenic or climatic controls. However, individual formations that are lithologically, mineralogically, and chemically distinct are commonly related to localized events that cannot be correlated throughout the depository.

In the absence of vegetation and in a humid or semiarid climatic regime, multichannel (braided) bedload rivers dominated the landscape of alluvial fans (Fig. 4-4) and fan deltas which prograded across finer grained lacustrine deposits (Vos, 1975). The scale and gradient of the alluvial fans varied with the changing tectonic and climatic controls. The largest fans of the Witwatersrand basin were smaller but otherwise comparable to the modern Kosi River fan of India and Nepal. The Kosi fan (Fig. 7-5) is a vast, low-gradient, wet alluvial fan characterized by seasonal floods. The more local fault-controlled and volcanically active basins of the Ventersdorp Supergroup were fed by multiple smaller fans in yoked settings probably resembling the modern Basin and Range province of the western United

Figure 4-4. Braided rivers dominated the landscape of alluvial fans and alluvial plains in the Witwatersrand basin. Shown is a modern example of these bedload river systems, Markafljot, southern Iceland. [Courtesy of Brian Bluck.]

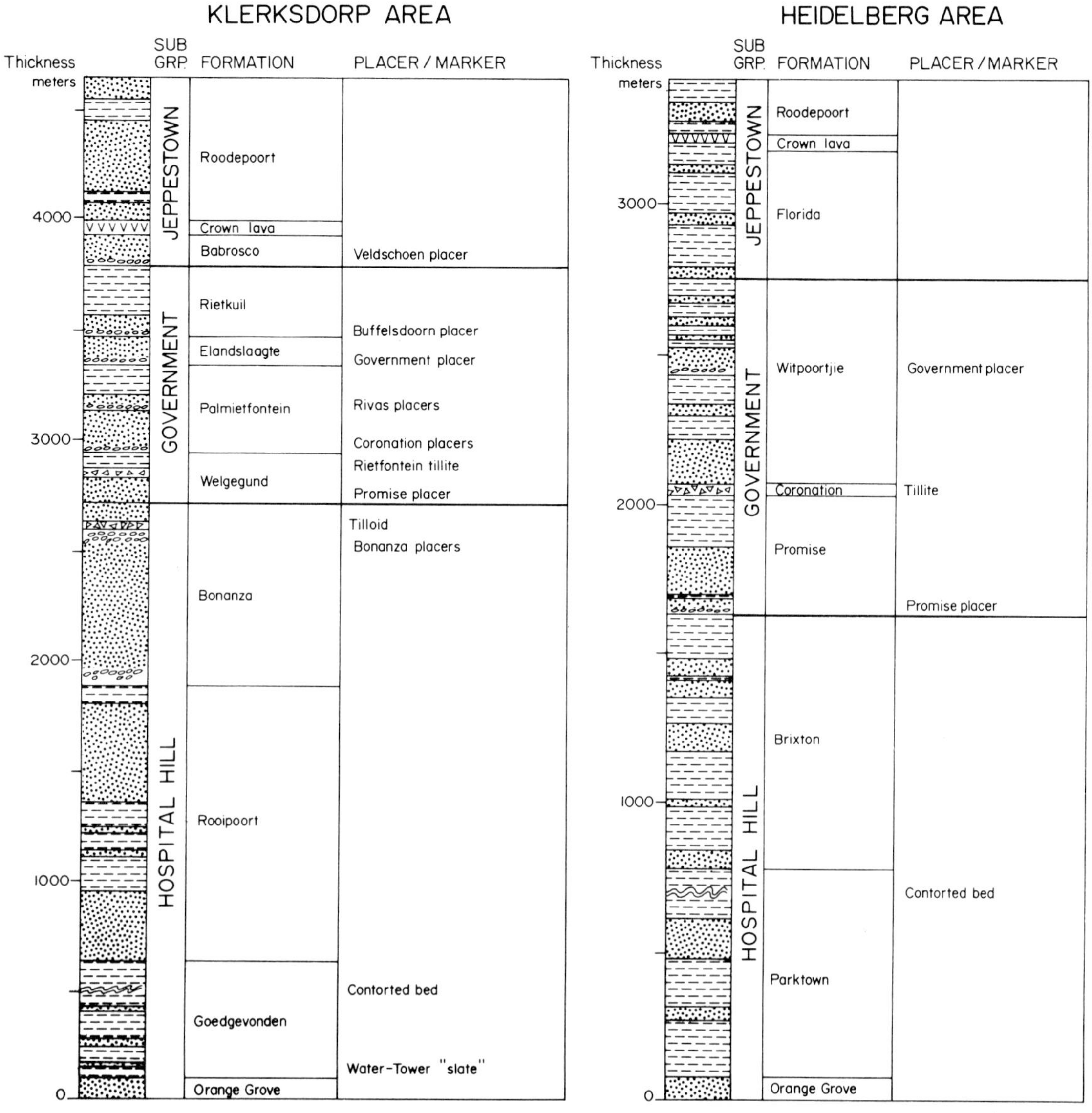

Figure 4-5. Stratigraphic columns of the West Rand Group in the Heidelberg and Klerksdorp areas. [Adapted from Camden-Smith (1980), Antrobus (1982).]

States. Tidal and wave reworking was important along the fan-delta shorelines during Witwatersrand deposition. Finer grained sediments accumulated offshore and included important biogenic carbonate facies in the Ventersdorp lacustrine and alluvial plain systems.

4.1. Dominion Group: The Witwatersrand Protobasin

The Dominion Group is believed to represent a protobasinal phase of the greater Witwatersrand basin. It is preserved over an area covering approximately 15,000 km^2 in the western Orange Free State and the southwestern Transvaal (Fig. 4-3). The group is 2710 m thick at Klerksdorp (Table 4-1). Sediments and lavas of this group rest nonconformably on Archean basement granites 3.0 Ga to 2.7 Ga in age (Hunter, 1974b). Intrusive and extrusive rocks in the Dominion Group have been dated at 2.82 Ga (Van Niekerk, 1968; Van Niekerk and Burger, 1969: U-Pb). A pink, coarse-grained granite directly underlying the Dominion Group contains 10 ppm uranium; erosion of the upper levels of this intrusion may have provided the uraninite in Dominion placer deposits.

The Renosterspruit Formation at the base of

the Dominion Group consists of up to 60 m of sandstone overlain by a mixed volcanosedimentary interval. The conformably overlying Renosterhoek Formation comprises 1100 m of basaltic andesite and tuff. A further 1550 m of acid lava with subordinate layers of tuff, andesitic lava, volcanic breccia, and quartz-feldspar porphyry lava constitute the Syferfontein Formation.

Volcanic Facies

The volcanic activity that followed initial sedimentation may reflect greater instability and basin subsidence after gentle southwestward tilting. The Renosterhoek Formation conformably buried the Renosterspruit sediments and overlapped protruding basement rocks. Boundaries between lava flows are marked by paleosols, interbedded tuffs and quartzites, and swirled lava textures (Button and Tyler, 1981). The overlying Syferfontein Formation represents additional volcanic activity of more acid composition (Table 4-2). Paleosols, tuffs, and flow breccias define separate flows of rhyolite and dacite in the 1550-m-thick succession (M. Strydom, personal communication, 1981). Source vents of the Dominion Group volcanics have not been located, but preliminary trace-element analysis (Table 4-2) indicates that the magmatic source was different from those that subsequently supplied the Crown lava in the Witwatersrand Supergroup and the

Table 4–2. Some Representative Analyses of Early Proterozoic Lavas[a]

	1	2	3	4	5	6	7	8	9
SiO_2	51.86	77.31	61.70	49.50	49.02	54.17	72.63	60.72	54.92
TiO_2	0.67	0.43	1.04	1.50	1.04	0.81	0.45	1.27	1.22
Al_2O_3	14.26	10.31	12.61	15.65	8.44	14.60	12.66	15.31	14.72
FeO	10.27	4.48	12.76	16.70	11.72	10.23	4.65	6.47	12.70
MnO	0.16	0.03	0.17	—	0.18	0.14	0.06	0.12	0.19
MgO	7.63	0.46	2.29	4.00	14.42	5.48	0.00	3.18	4.93
CaO	9.86	0.17	1.98	5.45	7.91	7.18	0.73	8.10	6.88
Na_2O	4.55	1.75	1.88	1.00	0.83	3.04	3.55	2.25	3.51
K_2O	0.61	2.83	2.62	0.30	0.18	1.47	4.90	2.03	0.68
P_2O_5	0.14	0.12	0.40	—	0.21	0.19	0.14	0.58	0.27
LOI	—	—	2.55	—	—	—	0.33	—	—
H_2O^+	—	—	—	6.00	5.37	2.79	—	—	—
H_2O^-	—	—	—	—	0.11	0.04	—	—	—
Rb	12.2	—	67	—	1	50	—	59.1	12.4
Sr	568	—	63	—	90	274	—	282	311
Y	16.3	—	34	—	15	17	—	35.4	22.8
Zr	88	—	218	—	88	97	—	336	174
Nb	1.2	—	9.5	—	4	4.7	—	13.0	4.9
Co	—	—	11.0	—	—	—	—	—	—
Cr	—	—	4.0	—	1867	1867	—	—	—
V	—	—	31.0	—	196	168	—	—	—
Cu	—	—	—	—	88	71	—	—	—
Ni	—	—	—	—	804	163	—	—	—
Zn	—	—	—	—	86	84	—	—	—
Ba	—	—	—	—	129	644	—	—	—

[a]Sources: 1, Renosterhoek lava (H. Va. Eales, personal communication, 1980; average of 2 analyses). 2, Syferfontein lava (Von Backström, 1952). 3, Crown lava (H. V. Eales, personal communication, 1980). 4, Bird amygdaloid, Sub Nigel Mine (Hall, 1938, p. 97). 5, Meredale Member, lower Klipriviersberg Group (McIver *et* 1980; average of 8 analyses). 6, Upper Klipriviersberg Group (McIver *et al.*, 1980; average of 14 analyses). 7, Rhyolite, Makwassie Formation (Tyler, 1979a, his table V; average of analyses 2, 3, and 4). 8, Rietgat lava (H. V. Eales, personal communication, 1980; average of 2 analyses). 9, Allanridge lava (H. V. Eales, personal communication, 1980; average of 2 analyses).

Klipriviersberg lava in the Ventersdorp Supergroup (H.V. Eales, personal communication, 1980).

Fluvial Environments

The Renosterspruit Formation is composed predominantly of sandstone, with minor conglomeratic and argillaceous components, deposited in shallow valleys during erosion of a subdued granitic terrane. Extensive weathering of the granitic paleosurface is evidenced by paleosols (Button and Tyler, 1981). Deposition of the coarse, immature sediment derived from these granitic paleosols took place dominantly in shallow braided streams that followed dendritic drainage systems and by ephemeral sheet wash.

The drainage pattern does not appear to have been influenced by local structural features or by greenstone remnants, as in the Elliot Lake deposits of Canada (Roscoe, 1969). Deposition of the thin clastic sediments at the base of the group was controlled principally by a gentle southwestward paleoslope. Evidence of this is particularly prominent in areas of thickest sedimentation (e.g., near Dominion Reefs Mine and Afrikander Leases Mine; Haughton, 1969, p. 70). ("Reef" is a colloquial term for a variety of mineral deposits and is synonymous with "placer" in the Witwatersrand goldfields.)

At the base of the Renosterspruit Formation, medium to large pebbles of vein quartz accumulated in channels as single-pebble layers and multiple, thin sheetlike gravel bars. These gravels were associated with trough cross-bedded quartz arenite and placer mineral concentrations. Above the basal placer is a coarse-grained quartz arenite with a yellow, slightly sericitic matrix.

Pebbly sandstones were also deposited 18 m higher in the valley fill sequence. Small-pebble conglomerates in a zone up to 4 m thick mark a widespread disconformity. Densely packed pebble beds as thin as 30 cm are particularly enriched in uraninite toward the top of the zone. The gold and silver contents are very low, but an appreciable amount of detrital pyrite and monazite is associated with the uraninite. In contrast the basal placer is more highly enriched in gold and silver. The sandstones overlying the upper placer are darker and more argillaceous than in the lower member. Argillaceous units with thin mud drapes are arranged in crude upward-fining sequences which may represent abandoned channel fill. These placers were confined by the paleorelief to areas between basement hills. The basal placer, up to 3 m thick, was originally mined for its gold content in 1888 and for its uranium content from 1953 to 1963 (M. Strydom, personal communication, 1981).

The upper part of the Renosterspruit Formation has been intruded by diabase and acid porphyry sills which show a preference for argillaceous horizons. The sediments are interbedded with tuffs and basic lava flows as the formation grades into the overlying lavas of the Renosterhoek Formation.

4.2. West Rand Group: The Witwatersrand Sea

The West Rand Group occupies a roughly oval area 42,000 km^2 in extent. The longer northeast-trending axis extends 350 km from south of Theunissen in the Orange Free State to beyond Witbank in the Transvaal, whereas the shorter axis is 200 km long (Fig. 4-3). West Rand Group sediments conformably overlie the Dominion Group from the northwest to the southwest and overlap the granitic basement elsewhere. Their extent beneath a thin cover of younger rocks has been delimited by exploration drilling and by magnetic anomalies exhibited by certain beds in the Government and Parktown shales where they are tilted along the margin of the basin (Roux, 1967) (Figs. 4-5 and 4-6). However, the northeastern and southwestern margins of the basin are obscured by a thick cover of younger rocks. Pretorius (1981) estimates that the West Rand Group may have extended originally over an area of 100,000 km^2 because Bouguer gravity anomalies associated with the Witwatersrand basin extend beyond the northeastern and southwestern limits.

Knowledge of the West Rand Group is limited to studies of geographically isolated areas that lack lateral control because of poor exposure of the faulted and folded strata. Considerable latitude in interpretation is therefore possible.

The thickness of the West Rand Group averages 4650 m but varies from 830 m in the Evander area to 7500 m northwest of Krugersdorp. The sequence contains shales and sandstones in equal proportions and 250 m of volcanic rocks (the

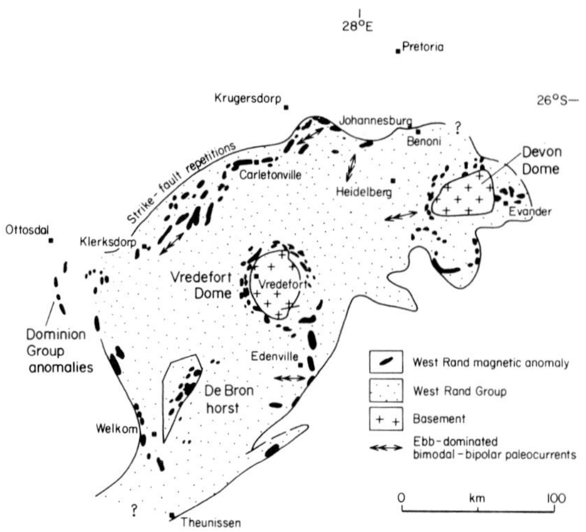

Figure 4-6. Map showing the distribution of the West Rand Group based on outcrop and magnetic anomalies and showing regional paleocurrent trends. [Source: aeromagnetic survey compilation, South African Geological Survey.]

Crown lava). Deposition took place predominantly in marine shelf and tidal environments with minor episodes of fluvial accumulation. As a consequence of the marine setting, the Hospital Hill, Government, and Jeppestown Subgroups (Fig. 4-5) are widespread and can be identified across most of the basin. In particular a contorted iron-formation in the Hospital Hill Subgroup and a lava bed in the Jeppestown Subgroup provide distinctive markers.

Volcanism

The Crown lava is found throughout the Witwatersrand basin. The 250-m-thick sequence comprises 1- to 20-m-thick amygdaloidal lava flows, many of which are capped by flow breccia. Ratios of $Al_2O_3/(FeO + MgO)$ (cf. Arndt *et al.*, 1977) indicate that they are relatively tholeiitic in composition and much more differentiated than the Dominion or Ventersdorp lavas (Fig. 4-7, Table 4-2) (H.V. Eales, personal communication, 1980).

Fluvial Environments

The Promise and Witpoortjie arenites (Fig. 4-5) are fine- to medium-grained subarkoses, which are arranged in upward-fining sequences with basal gravels, multiple sets of planar and trough cross-bedding and local herringbone structures, overlain by shales. These sequences are interpreted as fluvial lobes or fan deltas which prograded over tidal deposits. Paleocurrents in fluvial sediments of the Heidelberg area are to the southwest, coinciding with the ebb direction of the herringbone cross-bedded tidal sandstones. In the downdip Klerksdorp area fluvial transport was to the southeast and southwest.

The bases of these upward-fining sequences contain placer concentrates that were mined on a small scale at the turn of the century. Total gold production from 10 different placers exceeds 35 metric tons. (Well-known deposits are the Bonanza, Promise, Coronation, Rivas, Government, Buffelsdoorn, and Veldschoen placers.)

The lower part of the Florida Formation of the Jeppestown Subgroup in the Johannesburg area is composed of fine-grained subarkose. Trough and planar cross-bedded basal units at Heidelberg (Camden-Smith, 1980) are overlain by thick shales with interbedded sandstones and siltstones. Paleocurrents are weakly bipolar with a dominant, probably fluvial, transport direction toward the south.

Tidal Environments

In the Heidelberg area along the northeastern margin of the basin, the Orange Grove Formation rests on basement granite (Fig. 4-5). Small quartz-pebble lag is overlain by trough cross-bedded, very fine-grained quartz arenite, and by plane-bedded quartz arenite bearing wind-obstacle marks. The maturity of these Orange Grove quartz arenites is attributed to wind and tidal reworking. This 35-m-thick lithofacies sequence, characterized by bimodal–bipolar paleocurrent directions, has been interpreted as an ebb-dominated tidal-inlet deposit succeeded by beach-swash deposition (Camden-Smith, 1980).

Further trough cross-bedded units, representing the stacking of migrating tidal-channel deposits, are succeeded by 5 m of shale. At this contact, flat-topped dune formsets have runzel marks preserved on their lee slopes and ladderback and interference oscillation ripples in the scour troughs. These associations reflect modification of primary bedforms on tidal flats by emergence and runoff processes. The overlying shelf muds contain a few thin sandstone lenses with hummocky cross-stratification, possibly generated by

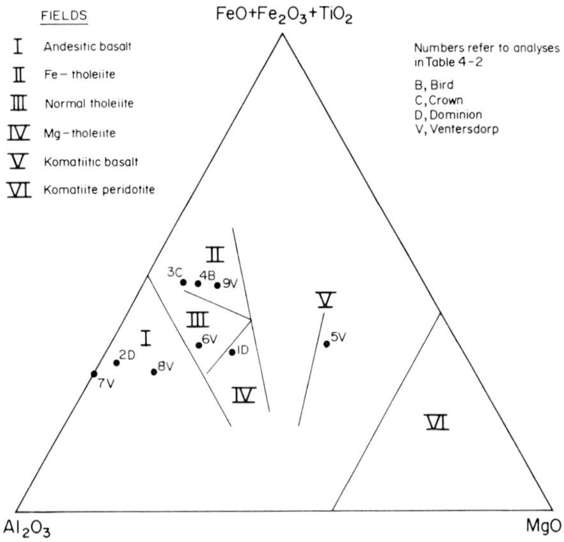

Figure 4-7. Ternary diagram illustrating compositional variations from andesitic basalt to komatiitic basalt in Dominion, Witwatersrand, and Ventersdorp lavas. [$FeO + Fe_2O_3 + TiO_2$—Al_2O_3—MgO diagram based on Viljoen *et al.* (1981). Data from McIver *et al.* (1981) and H.V. Eales (personal communication, 1980).]

storm surges below fair-weather wave base. The transport direction was generally ebb dominated to the northwest, perpendicular to the paleoshoreline around the Devon granite dome (Fig. 4-6). Farther south around the curve of the dome the transport mode was to the west and southwest, but still perpendicular to the shoreline.

Sandstones at the top of the Orange Grove Formation in this area are more feldspathic and contain larger scale trough cross-bedded sets with solitary planar cross-beds. Transport was unimodal to the southwest. Sandstone units higher in the Hospital Hill Subgroup display well-developed herringbone cross-bedding with a southwest ebb-current mode.

Tidal-inlet processes have also been inferred for the Orange Grove Formation in the Zwartkop area northwest of Johannesburg (Eriksson *et al.*, 1981). Two laterally persistent upward-fining sequences are represented above a basal conglomerate facies. The scale of trough cross-bed sets in the overlying quartz arenite facies decreases upward in 1- to 3-m-thick units. Siltstone and shale partings are believed to have been deposited during prolonged episodes between storm-augmented tidal events. Plane-bedded quartz arenite, attributed to beach-swash processes, and a second trough cross-bedded facies, complete the Orange Grove sequence. Paleocurrent directions of the quartz arenites are ebb dominated to the southwest.

In the Klerksdorp area tidal sediments of the Orange Grove Formation transgressively overlie the Dominion Group and basement granite. Eastward thinning of the Dominion Group is attributed to deposition rather than to erosional truncation. A paleocurrent analysis of the Orange Grove arenites by Watchorn (1980b) indicates that ebb-dominated vectors to the southwest prevailed, but there is weak bipolarity. Two upward-fining quartz arenite sequences are preserved, in places followed abruptly by shales. Where the quartz arenites pinch out, the shales rest directly on basement granite. This is also the case near Edenville, along the eastern margin of the basin, where the Orange Grove quartz arenite is completely absent (Hutchison, 1975).

In the upper Orange Grove and lower Brixton Formations (Fig. 4-5) in the Johannesburg area, medium-scale planar and trough cross-bedded quartz arenites overlie proximal shelf deposits. Complex flow patterns prevailed with herringbone cross-bedding near the top of the sequence indicating periodic reversals and dominance of a southerly offshore mode. Suspension settling of clay, probably during high-tide slack-water periods, produced shale partings that alternate with sandstone beds. Above these units are upward-fining sequences composed of quartz arenite and shale. Herringbone cross-bedding, shale clasts, and flaser bedding are common. Seaward progradation of tidal-flat subenvironments has been interpreted from these upward-fining packages (Eriksson *et al.*, 1981).

In the Government Subgroup (Fig. 4-5) cross-bedded quartz arenites are present at several stratigraphic levels intervening between fluvial and offshore deposits. Their transitional position and bimodal–bipolar paleocurrent patterns suggest a tidal origin. Lithic arenites of probable tidal origin also occur in the Jeppestown Subgroup.

Shelf Environments

A number of lithofacies attributed to shelf deposition have been recognized in the West Rand Group. The magnetic shale facies are perhaps the most important because they represent markers whose geophysical response provides a means of outlining the subsurface margin of the Witwatersrand basin. The most famous of these markers is probably the Water-Tower slates used by Krahmann (1936) to trace the northwestern margin of the Witwatersrand basin from Krugersdorp to Carletonville and which led to the discovery of the Carletonville goldfield.

Magnetic shale beds are present within massive shales at a number of stratigraphic levels. In the Klerksdorp area they are recognized in the Goedgevonden, Rooipoort, Welgegund, Palmietfontein, and Rietkuil Formations (Fig. 4-5); in the Johannesburg area, in the Parktown, Coronation, and Witpoortjie Formations; and in the Heidelberg area, in the Parktown and Coronation Formations. Because of their partly chemical origin and the absence of primary structures other than horizontal lamination, the magnetic shales are thought to have been deposited below wave base in a distal shelf environment. The laterally persistent and repetitious nature of this magnetic shale facies records several marine transgressions in the West Rand Group.

The contorted bed in the Parktown shale is an iron-formation and shale facies that underwent folding and décollement after lithification (Fripp and Gay, 1972). It has been interpreted as the base of a progradational upward-coarsening sequence (Eriksson *et al.*, 1981) and consists of interlaminated chert, jasper, magnetite, and shale. The proportion of shale increases upward until thin chert layers constitute the only chemical component. Horizontally laminated siltstone and shale layers are devoid of primary sedimentary structures and are attributed to deposition below wave base in a distal shelf environment where chemical and suspension sedimentation dominated. Thinner shales and siltstones represent more proximal equivalents of the rhythmically interlayered siliceous and ferruginous chemical sediments.

In the Parktown and Orange Grove Formations (Fig. 4-5) a thick sequence of horizontally laminated siltstone with subordinate shale contains ripple cross-lamination only at the top. Such low-energy bedload deposits, associated with offshore shelf sediments, suggest a proximal shelf environment. Sedimentation units 5–10 cm thick consist of horizontally laminated and rippled siltstones, many of which display an internal cyclicity from type A to type B ripple-drift cross-lamination. Sporadic hummocky cross-stratification suggests that these currents may have been storm generated.

Shale-siltstone facies in the Jeppestown shales display graded bedding and contain lenses of lithic arenite. Bipolar paleocurrent azimuths in the sandstones show ebb dominance toward the southwest. A proximal shelf setting is inferred.

The Earliest Known Glaciation

Diamictites associated with alluvial-fan deposits are not uncommon in the upper part of the Witwatersrand succession, where they have been referred to as "puddingstones" and "tilloids." However, diamictites are also present at two or three stratigraphic levels in the West Rand Group. In particular, the Coronation Formation in the middle of the Government Subgroup contains a diamictite which has equivalents in the Bonanza and Welgegund Formations in the Klerksdorp area (Fig. 4-5). Striated pebbles have apparently been recognized in the Klerksdorp diamictites (P. Strydom, personal communication, 1980). This suggests a glacial source, representing the oldest recorded glaciation (2.7 Ga to 2.3 Ga). In both areas the diamictites are overlain by distal shelf magnetic shales. The most likely mode of deposition for these West Rand Group diamictites in a distal shelf environment is by submarine debris flow, possibly from accumulations of ice-rafted moraine.

Geochemistry of Shale

Chemical analyses of the shales in the West Rand Group have provided little additional information on their depositional milieu. Their average boron content of 11 ppm, which Nel (1968) thought to be indicative of a freshwater environment, is not diagnostic because the illite content of the shales is very low. The major clay phases identified by Camden-Smith (1980) are chlorite and illite, with traces of kaolinite. The geochemistry of shales in the Government and Jeppestown Subgroups shows similarities to the Archean Fig Tree sediments (Chapter 2), particularly with respect to average MgO, total FeO, P_2O_5, Na_2O, Ni, Co, Cr, and the Ni/Co and V/Cr ratios (Danchin, 1970; Hofmeyr, 1971). The source terrane for the West Rand Group was probably similar to that of the Fig Tree Group but contained fewer ultrabasic rocks because the Ni contents and the Ni/Co ratios are significantly lower. Fuller *et al.* (1979) were also able to show provenance differences between the shales of the West Rand Group at Heidelberg and Klerksdorp, the Booysens shale at Heidelberg, and the Fig Tree Group sediments on the basis of selected groupings of major elements and the trace elements Sr, Ni, Rb, Ca, Zn, and Cr.

Depositional History

The Hospital Hill Subgroup contains beach, tidal-inlet, and shelf-mud facies which reflect a protracted transgression followed by south to southwestward progradation of a macrotidal shoreline. The lateral persistence of these macrotidal sediments (M.B. Watchorn, personal communication, 1980) indicates that the stable Kaapvaal craton provided a wide, open continental shelf across which the tidal wave was amplified (Eriksson *et al.*, 1981). A modern example is the northwest coast of Australia (Coleman and Wright, 1975).

The Government Subgroup contains fan-delta, proximal shelf, and distal shelf facies. It is more arenaceous than the successions above or below, reflecting an alternation between progradation of fan deltas and marine transgression.

Eight transgressions of magnetic-shale facies are recorded in the Klerksdorp area. In the Johannesburg area only four transgressive complexes are known, implying a more proximal situation.

The Jeppestown Subgroup contains fan-delta, beach, and proximal shelf facies in the northern part of the basin, and a distal shelf facies represented by a magnetic shale in the Klerksdorp area. Prograding fan deltas were generally less common in the Jeppestown Subgroup than in the Government Subgroup but became increasingly important during transition into the Central Rand Group.

Paleocurrents were ebb dominated with respect to the eastern and northern shorelines of the West Rand Group. However, the most western outcrops display southwestward ebb currents without evidence of a shoreline in close proximity. The basin is therefore thought to have extended much farther west than its present limits.

4.3. Central Rand Group: Alluvial-Fan Environments

Central Rand Group sediments are preserved in an area of 9750 km^2 above the West Rand Group (Fig. 4-3). The deposit measures 290 km along a northeast-trending axis and 150 km along the northwest, but much of it is buried beneath younger rocks. The Central Rand Group consists predominantly of coarse-grained subgraywacke with less than 10 percent conglomerate and quartz arenite, subordinate lacustrine or shallow marine shale and silt, and minor lava. Maximum cumulative thickness is approximately 2880 m near Vredefort. The Central Rand Group gradationally overlies the West Rand Group in the Welkom, Klerksdorp, and Krugersdorp areas, suggesting that the finer grained, marginal marine deposits of the Jeppestown Subgroup (West Rand Group) may represent distal facies equivalents of the Johannesburg Subgroup (Central Rand Group). Paleocurrent data show that sediments of the Johannesburg Subgroup prograded into the Witwatersrand basin from around its margin, possibly as fan-delta complexes. This reflects progressive shrinking of the basin and eventual closure of its margin. Unconformities merge toward the margin of the basin, substantiating the inference that the present margin probably corresponds closely to the original basin edge. The parallel strikes of unconformity surfaces beneath successively onlapping formations reveals the progressive changes in size of the Central Rand Group basin (Fig. 4-8).

Paleocurrent analyses indicate that known goldfields in the Central Rand Group are located over major entry points in the basin (Pretorius, 1976). Source areas lay northwest of Klerksdorp, Carletonville, Krugersdorp, and Benoni, and southwest of Welkom and Evander (Fig. 4-9). The southwestern source at Evander is based on information (Tweedie, 1978) that contradicts the previous assumption of a northwestern source (Pretorius, 1976); this newer concept, as well as recognition of an entry from the eastern side of the basin at Edenville (Hutchison, 1975), indicates that closure also took place along the eastern margin of the Witwatersrand basin.

The geometry and distribution of distinct sediment packages deposited from basin entry points can be defined using isopachs, pebble-size isopleths, clast lithologies and cross-bed distribution.

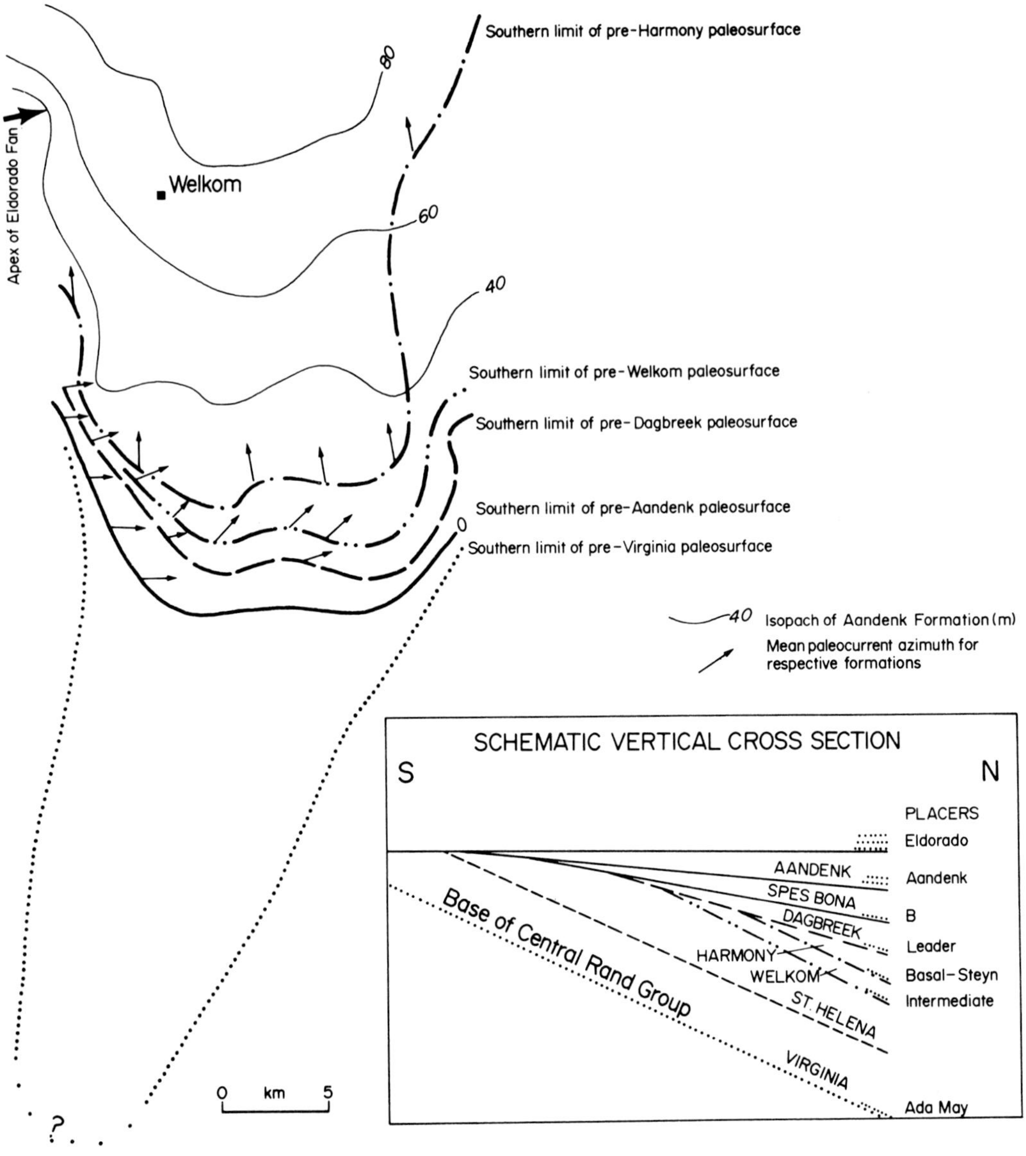

Figure 4-8. Closure of the Central Rand basin in the Welkom goldfield shown by formation subcrop limits against younger onlapping formations. Unconformities intersect at gradients of 1:100.

These studies indicate that placer deposits previously correlated for great distances across the basin are discontinuous and are restricted to fans generally less than 30 km wide. Many of the entry points were structurally influenced by the unsynchronized rise of basement granite domes (Fig. 4-9) (Pretorius, 1976). The number and stratigraphic positions of unconformities therefore vary from one lobe to another.

The effect of periodic uplift at the entry points is clearly evident from the presence of unconformities. However, the hypothesis (Pretorius, 1964) that synclinal warping superimposed across the main northeast axis of the Witwatersrand basin controlled the distribution of sediment appears to be without foundation. The placer accumulations within fans were probably controlled simply by bar and channel processes and, on a larger scale, by geomorphic processes on the developing fans.

The only synchronous response to subsidence evident in the sedimentary record of the Central Rand Group is a basinwide upward-fining sequence from arenite to argillite midway through

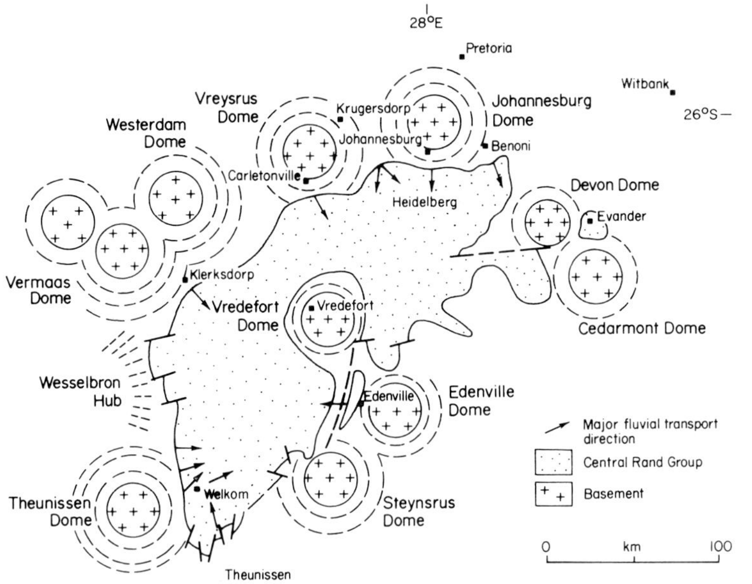

Figure 4-9. Map showing the distribution of the Central Rand Group, adjacent granite domes, and sites of major fluvial influx. [Adapted from Pretorius (1974), Hutchison (1975), and Minter (1980).]

the Central Rand Group. This persistent lithologic marker is a composite feature resulting from coalescence of the upward-fining units at each entry point; it is known as the Upper Shale Marker at Welkom, the Modderfontein Member at Klerksdorp, and the Booysens Shale Formation from Carletonville to Evander. The Booysens shale probably represents a number of coalescing units. These fine-grained sediments undoubtedly represent the distal facies equivalents of the formations underlying them and were probably a product of widespread subsidence of the basin. The position of this basinwide lithologic marker is used to divide the Central Rand Group into the Johannesburg and Turffontien Subgroups. (Fig. 4-10).

Volcanism

Two or more sequences of lava flows separated by minor arenite beds were extruded during deposition of the Johannesburg Subgroup. The lower and upper sequences of the Bird amygdaloid are up to 30 m and 170 m thick, respectively, and extend from Johannesburg to Heidelberg (Fig. 4-3). They are amygdaloidal and basic to intermediate in composition (Table 4-2; Fig. 4-7), containing sodic plagioclase in a devitrified matrix of chlorite, actinolite, brown mica, epidote, and traces of quartz. The lavas thicken in synforms and thin against antiforms, indicating that folding took place prior to or during extrusion (Pretorius, 1964). This is compatible with the tectonic activity reported from the northeastern part of the basin.

Fluvial Environments

The Central Rand Group comprises numerous tectonogenetic sedimentary packages separated by unconformities. These packages range in thickness from 30 m to 600 m, averaging about 250 m, and consist almost entirely of sandstone and minor gravel-sized units (Fig. 4-10). The arenites vary from brownish with a muddy sericitic matrix to white or light gray colored with very little sericite in the matrix. The compositional variations are thought to reflect changes from ephemeral to more perennial conditions.

These genetic packages generally consist of a scour-based pebble lag or gravel-bar deposit overlain by trough cross-bedded sandstone in which cosets are separated by scour surfaces and mud drapes up to 10 cm thick. The 50-m-thick Mapaiskraal package (formerly known as the MB 5) at Klerksdorp has an average coset thickness of 170 cm (Minter, 1972), whereas the 270-m-thick Welkom package (formerly known as the Livingstone) has an average coset thickness of 70 cm (Minter, 1978). In both cases the average thickness of trough cross-bed sets is 8 cm. Planar cross-beds form rare, solitary sets (Fig. 4-11). The trough cross-bed paleocurrent vectors in all

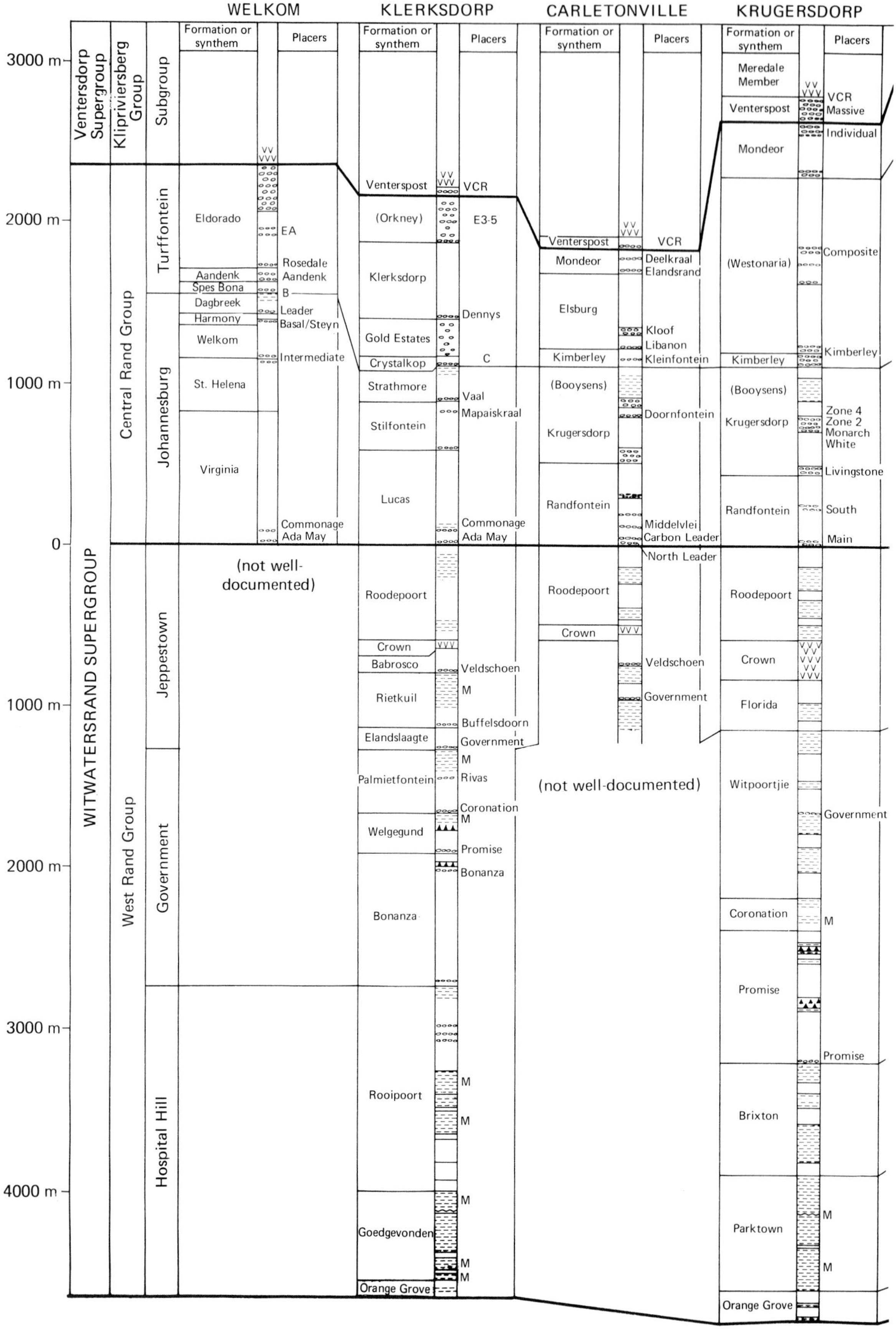

Figure 4-10. Correlation of the stratigraphic columns of the major goldfields, showing the principal placers.

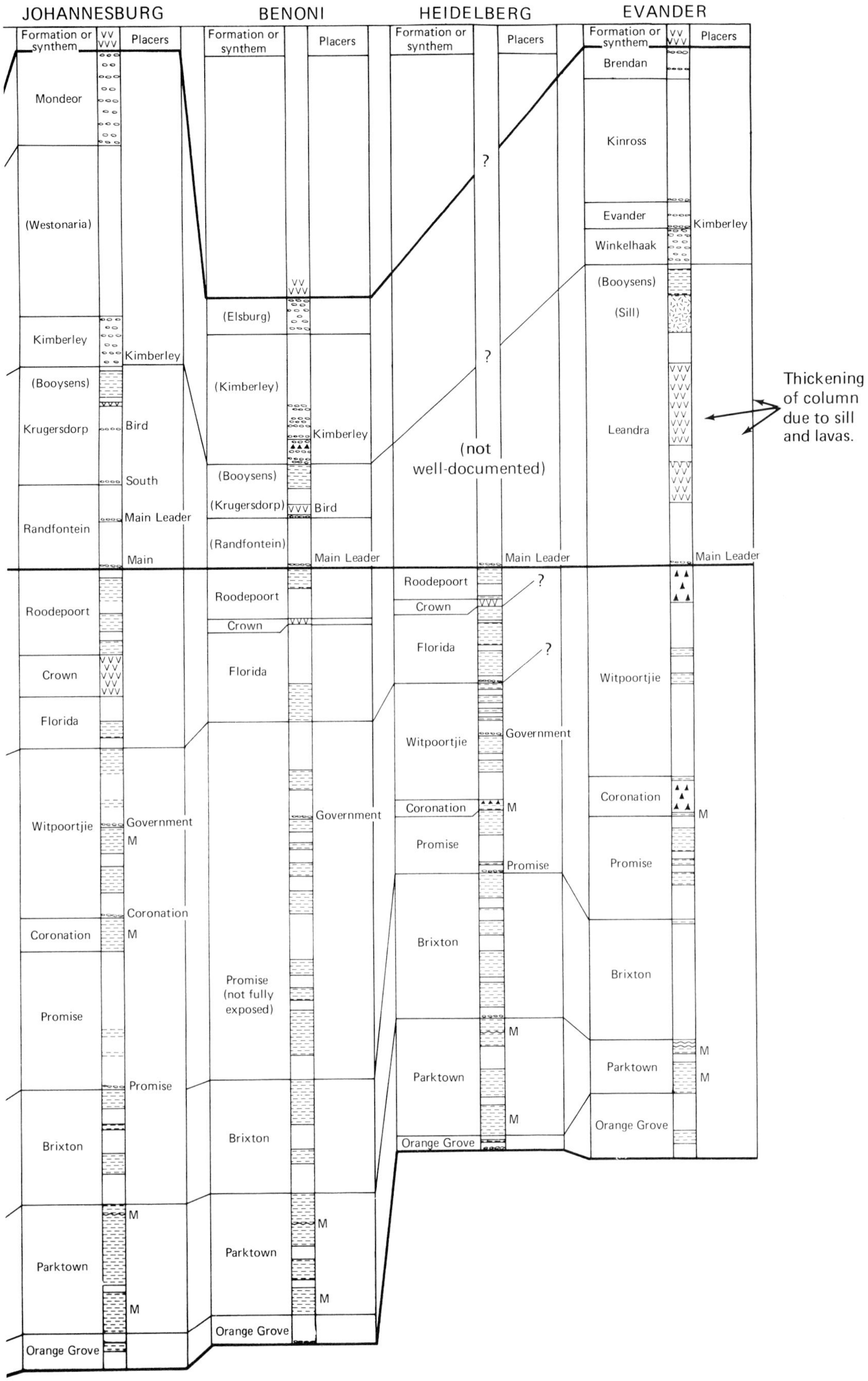
JOHANNESBURG
BENONI
HEIDELBERG
EVANDER
Formation or synthem
Placers
Mondeor
(Westonaria)
Kimberley
(Booysens)
Krugersdorp
Randfontein
Bird
South
Main Leader
Main
Roodepoort
Crown
Florida
Witpoortjie
Government
M
Coronation
Promise
Brixton
Parktown
Orange Grove
(Elsburg)
(Kimberley)
(Krugersdorp)
(Randfontein)
Promise (not fully exposed)
?
(not well-documented)
Brendan
Kinross
Evander
Winkelhaak
(Sill)
Leandra
Thickening of column due to sill and lavas.

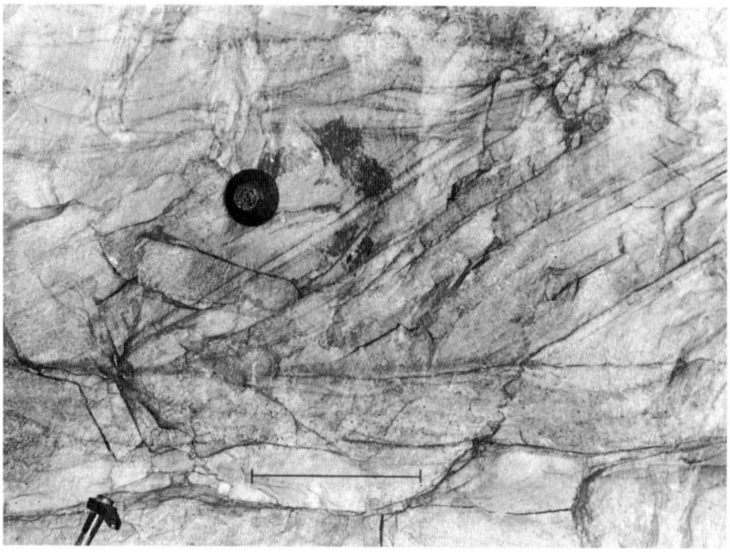

Figure 4-11. Sandstone containing a solitary planar cross-bed set between trough cross-bed sets attributed to transverse bar and channel-floor dune migration respectively. Welkom Formation, Free State Saaiplaas goldmine, Welkom goldfield. Scale bar = 50 cm.

studies are unimodal. Planar cross-bed vectors have a much greater dispersion, suggesting lateral growth, in concert with downstream accretion in shallow braided streams.

The trough cross-bedded units reflect dune migration in wide, shallow channels. Accretion of bar margins produced the sporadic planar cross-bedded sandstones. The planar cross-beds are up to 100 cm thick and display reactivation surfaces, in places draped by clay, indicating intermittent growth and shifting of the bar margins. Regressive ripples at their base are evidence of flow separation; slumped foresets are common. These sequences resemble Miall's (1977, 1978) South Saskatchewan braided-stream model.

Stacked planar cross-bed sets, resembling the distal braided Platte sequence (Miall, 1977, 1978), have also been observed in the downdip part of the Benoni goldfield (formerly known as the East Rand basin) near Heidelberg in 60 m of sandstone overlying the Main Reef Leader placer (Button, 1979b). Basinward equivalents are apparently present only at the top of the Johannesburg Subgroup. For example, the Dagbreek Formation in the Welkom goldfield fines upward for 150 m from the Leader placer into the Upper Shale Marker where very fine-grained subgraywacke and rippled siltstone are interbedded with thin, black, loadcasted shales. Generally the paleoslopes conform with the marginal dips into the basin. Erosional paleorelief is rarely more than 0.5 m in distal areas and up to 6 m in proximal areas. Depositional surfaces were therefore relatively flat and sediment dispersal was unconfined.

Deposition of recycled sediments, including clasts and placer minerals, above the unconformities suggests that the older deposits were probably not lithified but simply consolidated. This is substantiated by loadcasting of channel-fill deposits of the Leader placer indicating that sands of the underlying Harmony Formation remained uncemented. The Central Rand Group deposits are viewed as a succession of superimposed and coalescing fanlike bodies composed predominantly of sand.

Channel patterns were generally unconfined and braided. However, in the Klerksdorp goldfield the Vaal placer occupies erosional depressions scoured into the paleosurface of an angular unconformity (Minter, 1978). The depressions average 30 cm in depth but in places they are over a meter deep. The channels were shallow and wide, with width/depth ratios ranging from 50 to 1000. Between depressions the placer sediments are only a few centimeters thick. The drainage pattern (Fig. 4-12A) is almost linear, with interconnecting low-sinuosity channels oriented directly down the paleoslope. There is a slight convergence toward the northwestern edge of the deposit. In the Welkom goldfield underground mapping of the B placer also indicates a low-sinuosity pattern, despite the presence of channels up to 2 m deep entrenched into a cohesive argillite that could have stabilized banks and promoted meandering (Minter, 1978). Drainage patterns on

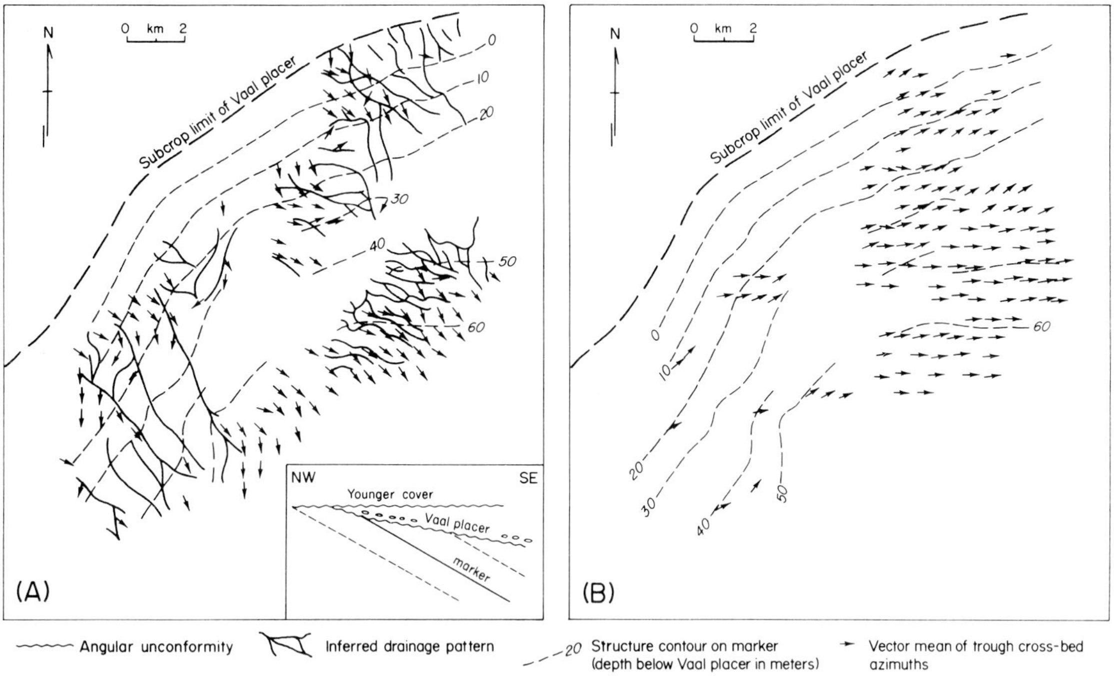

Figure 4-12. (A) Subcrop map of the Vaal placer in the Klerksdorp goldfield showing the drainage pattern above the angular unconformity. The transgression is shown by isopachs of the interval between the placer and an underlying marker. [After Minter (1978).] (B) Subcrop map showing longshore paleocurrent directions in shallow-shelf sediments overlying the Vaal placer. Flow directions subparallel the strike of isopachs between the unconformity and underlying marker. [After Minter (1978).]

other paleosurfaces occupied by placers have been interpreted from contour maps of gold or uranium mineralization or from plans of mined-out areas. The main trends of mineralization in the Steyn and Basal placers (Minter, 1978) reflect braided channels, some details of which have been confirmed by underground observation (Fig. 4-13). These trends do not represent single channels but they do show the broad drainage patterns across the fan. Discrete channel systems prevailed in the proximal area of the Steyn placer, where bed relief was high and footwall islands separated channels. With a downslope change to fan-plain environments, channels merge intricately into a tabular body. The mapping of basal scour surfaces and internal structures in the distal fan area indicates that sediment was deposited in wide, shallow channels with relatively low bed relief.

The unconfined drainage pattern described above also defines the separate Basal placer fan, whose easterly distribution pattern is overlapped by the younger north-to-northeastward trend of the Steyn placer, thus substantiating other sedimentological information supporting this relationship (Fig. 4-13).

The Klerksdorp goldfield is the only part of the Central Rand Group basin where there is evidence of longshore currents (Minter, 1976). Trough cross-bed azimuths ($n = 500$) in subgraywacke, deposited above the Vaal placer after submergence along a subsiding coast, reflect bedload transport parallel to the strandline (Fig. 4-12B). However, there is no evidence to prove that this current also redistributed the Vaal placer, which is oriented down the paleoslope. The notion that placer fans are skewed in a clockwise direction around the margin of the basin (Pretorius, 1976) cannot be supported.

The genetic packages of sediment in the Central Rand Group indicate rapid aggradation. Some, like the Leader placer in the Welkom goldfield (Minter, 1982), commence with a gravel or placer gravel and then fine upward. Repetition of gravel beds in the sequence reflects shifting of major channels across the fan. Placers are locally associated with these deposits but generally they are best developed on the unconformity surfaces

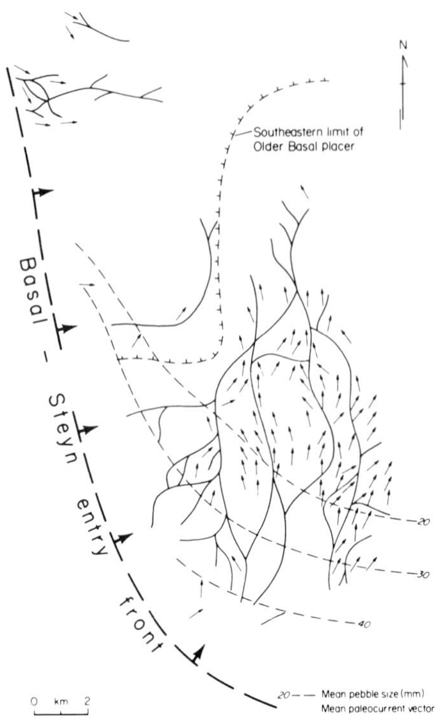

Figure 4-13. Drainage of the Steyn placer established on the basis of mineralization trends and cross-bed azimuths. Welkom goldfield. [After Minter (1978).]

where degradation prevailed initially. Adams *et al.* (1978) have shown that during aggradation of a humid fan the slope reaches a geomorphic threshold beyond which channels begin to retrench by headward erosion. This process effectively taps pebble and heavy mineral accumulations from the source area drainage as the base level is lowered. In the Aandenk fan of the Welkom goldfield a number of "A" placers accumulated in degrading channels at various stratigraphic elevations in this way. This response also explains the position of the Steyn placer as a terminal gravel on top of the Welkom Formation, which it resembles in terms of paleocurrent directions and clast lithologies. The lithology of placer sediments is conspicuously different from the bounding strata because of the placer's greater maturity, evidenced by lower sericite contents.

Individual beds within placer units tend to be irregular or lenticular over short distances, particularly in sections transverse to the paleocurrent direction. This irregularity, as well as obvious bar and channel morphology and grain-size changes that reflect vertical and lateral changes in flow regime, resembles the deposits of modern gravel-bed streams characterized by narrow channels and longitudinal braid bars (Fig. 4-14). An example of this internal morphology (Fig. 4-15) has been described from the Elsburg No. 5 placer in the Klerksdorp goldfield by Smith and Minter (1980). Highly variable flow conditions and the lack of vegetation to stabilize channel banks and regulate sediment supply promoted braiding. In coarse proximal reaches bars are sites of sediment accumulation during flood discharge. Emergence of these bars during waning flow causes braiding and dissection. Sediment is therefore stored temporarily in the bars which become local sources of sediment and placer minerals during the constant shifting of channels. Channel scour surfaces with thin gravel lags, planar cross-bedded sandstone, and, more commonly, trough cross-bedded sandstone are the other most common sediment facies observed in Witwatersrand placer deposits (Fig. 4-15).

Carbon seams associated with some of the placer deposits consist mainly of hydrocarbons with organic sulfur and oxygen compounds. Zumberge *et al.* (1978) have shown that the kerogen (a term preferable to "carbon seam") originated by polymerization of biochemical compounds from decayed, primitive microorganisms. The kerogen comprises at least two types of organisms (Hallbauer and Warmelo, 1974; Hallbauer, 1975; Hallbauer *et al.*, 1977). *Thuchomyces lichenoides* is a fibrous columnar form 0.5–5.0 mm long and 0.2 mm in diameter and has spheroidal vegetative diaspores 0.2 mm in diameter. *Witwateromyces conidiophorus*, representing hyphae of saprophytic fungi, occur in amorphous carbon that may represent organic debris.

The association between kerogen and mineralization appears to be largely circumstantial because the plants grew preferentially on channel scour surfaces where detrital heavy mineral lags also concentrated. The kerogen is virtually restricted to the deposits of perennial streams. Although most commonly preserved in distal facies, such as the richly mineralized Carbon Leader placer of the Carletonville goldfield, disaggregated kerogen in proximal placers shows that it was originally more widespread. The kerogen characteristically forms drapes and specks on channel floors and other scour surfaces, encapsulating gravel lag, on slipfaces of planar cross-beds (Minter, 1978), and draping gravel bars (Smith

Figure 4-14. Modern example of the sheetlike longitudinal bars and shallow channels in the proximal facies of Witwatersrand and Ventersdorp alluvial fans, Klifandi fan, southern Iceland. [Courtesy of Brian Bluck.]

and Minter, 1980). Modern braided rivers, such as the Sabi River of Zimbabwe, display similar patterns of algal growth and accumulation (Button, 1979a). Although the kerogen seams are paleontologically, chemically, and mineralogically interesting, the average cumulative thickness of seams in the combined placers of any goldfield probably does not exceed 1 cm. Compaction has contributed little to this preserved thickness because pebble and sand grains in the kerogen have provided a resistant framework. Of the 37 placers currently mined only 15 contain kerogen.

Placer minerals were effectively concentrated in pebble-supported gravel bars that were compound units modified over a long period. The matrix, especially at the upper surface of individual gravel beds, is well mineralized as are scour surfaces in the gravel upon which heavy minerals were concentrated between successive events of gravel deposition. Scour surfaces with gravel lags representing the armored floors of channels were the next best sites of placer concentration, particularly where confining flow created turbulent sorting conditions. The lateral migration of these conditions to produce well-mineralized bank-hugging gravels is not uncommon (Smith and Minter, 1980). Planar cross-bedded sandstone, which resulted from flow divergence and rapid dumping of sediment, contains very poor placer mineral concentrations. Trough cross-bedded sandstone represents the migration of dunes in channels and over shallow bars and is associated with somewhat higher, although still low-grade, mineral deposits. Placer minerals are concentrated on cross-bed foresets, at the crests, on the toes as a result of backflow (Fig. 4-2), and particularly on the scoured bottom surfaces of the sets. The sorting mechanisms that controlled the concentration and deposition of small, dense particles are not understood but these empirical relationships are repeatedly observed (Fig. 4-16).

In the more distal parts of the Steyn placer, 20 km downslope from the basin margin, the bedforms of the proximal reaches have flattened to thin small-pebble bars, only a single pebble thick

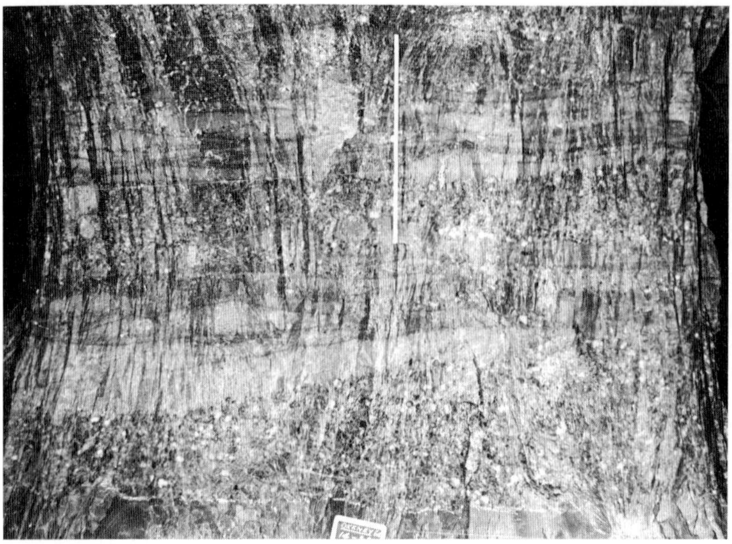

Figure 4-15. Bar and channel deposits in the proximal Elsburg No. 5 placer. The compound gravel bar at the base is overlain by planar and trough cross-bedded deposits with multiple scour surfaces. Vaal Reefs West mine in the Klerksdorp goldfield. Scale bar = 1 m. [After Smith and Minter (1980).]

in places, on the unconformity and scour surfaces. Distal placer deposits are composed predominantly of trough cross-bedded sandstone (Fig. 4-17). Economic minerals are concentrated on the scour surfaces, particularly at the base. The association of placer mineral concentrations with bedforms oriented parallel to flow makes paleocurrent analysis essential in projecting payshoots. For the same reason, more representative valuations can be obtained by sampling transverse sections because the maximum number of bed-

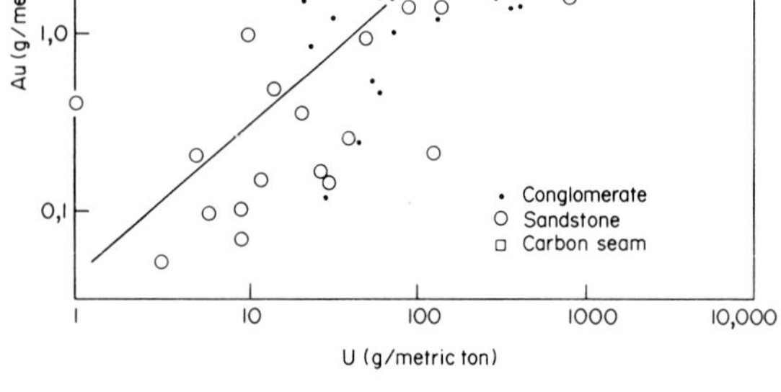

Figure 4-16. Graph showing linear correlation between gold and uranium in the Leader Placer and the relationship between sedimentary facies and mineral concentration. Welkom goldmine, Welkom goldfield. [After Smith and Minter (1980).]

relief elements are exposed. Much of the high variance in sample values can be attributed to these bed-relief elements and to the point-source control provided by temporary storage, since reworking processes may either disperse or concentrate the heavy minerals.

On a regional scale the distribution of gold and uranium in Witwatersrand placers is related to alluvial-fan development (Minter, 1978). In the Steyn placer, for example, gold is concentrated in the upper to lower fan plains, whereas uraninite is concentrated farther downslope in the distal fan (Fig. 4-18A and B). The uranium/gold ratio changes from less than 5 in proximal environments, where the mean size of the largest pebbles is larger than 20 mm, to more than 25 in the finer grained distal environments. This distribution may be related to the original size of the placer minerals supplied to the basin. Uraninite, zircon, and chromite were transported farther basinward than the gold, which was hydraulically the coarser grained.

The Vaal placer displays many characteristics different from those of the Elsburg No. 5 placer in the Klerksdorp goldfield and the Leader and Steyn placers in the Welkom goldfield.

(1) The Vaal placer rests on a paleosurface that has only slight relief and a discernible angular unconformity for more than 20 km downdip. The gradient of the unconformity changes downdip from 0.010 to less than 0.003.
(2) It displays neither a downdip decrease in pebble size, although paleocurrents are uni-

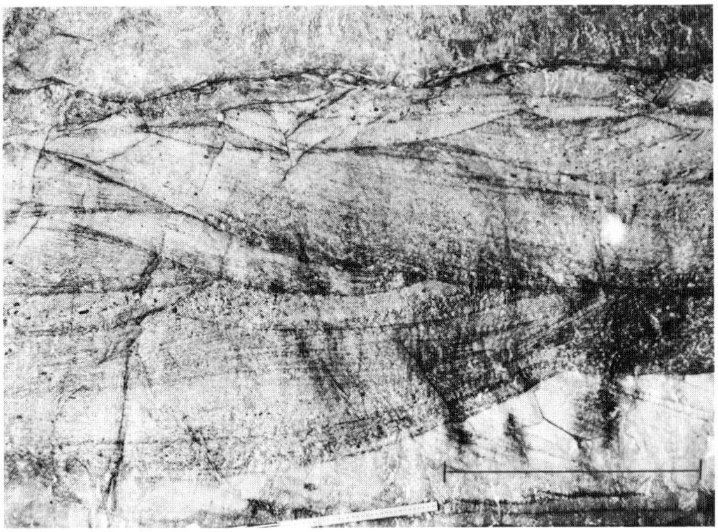

Figure 4-17. Channel containing trough cross-bedded sandstone of the Saaiplaas placer in the Harmony Formation. Note concentration of heavy minerals and small pebbles in the scour troughs. Overlying the Saaiplaas placer is the Harmony diamicite. Free State Saaiplaas mine, Welkom goldfield. Scale bar = 50 cm.

modal, nor a significant downdip decrease in pyrite or zircon size.

(3) It does not display a marked change in the pattern of gold or uranium content that can be related to downslope sorting.

(4) It contains recycled pebble ventifacts (small pebbles with multiple facets whose edges were rounded during aqueous transport) in proportions up to 5 percent.

Minter (1972, 1976) attributed these features to the upslope migration of the distal fan plain of a fan delta during transgression. These fan-delta plain deposits were gradually submerged and buried beneath sediments that were derived by transgressive shore-zone erosion and transported by a unidirectional longshore current. The upper surface of the sandy placer body is planar and commonly has placer mineral concentrations which may reflect winnowing of the surface after submergence.

This type of placer, of which the Vaal placer is a classic example, is referred to as a transgressive distal placer. It is very important economically because of the consistency of its grade and downdip persistence. The Carbon Leader placer in the Carletonville goldfield is similar, except that evidence of longshore paleocurrents has not been recognized in the overlying sediments.

Debris Flows

Debris flows, characterized by clasts supported in a high-density muddy suspension, may accumulate downslope to form diamictites. These deposits are found in alluvial-fan, glacial, and submarine canyon environments. Interpretation of their depositional setting depends on relations with surrounding facies. Diamictites are well known in arid or semiarid alluvial-fan environments, where weathered rock debris with a sparse vegetation cover is infrequently wetted and destabilized by rainstorms or melting snow; flow is restricted to ephemeral events.

Conditions that favored the development of diamictites, frequently referred to locally as puddingstones, were achieved at least twice during deposition of the Central Rand Group: after deposition of the Jeppestown shale in the Johannesburg Subgroup, and after deposition of the Booysens shale or its equivalent in the Turffontein Subgroup. Subgraywackes and shales were exposed and weathered to produce local accumulations of sericitic sands and muds, which were later moved into channels as debris flows.

The older of these diamictite-bearing paleosurfaces can be traced along the uplifted northern margin of the basin from above the Carbon Leader placer in the Carletonville goldfield, through the Krugersdorp and Johannesburg goldfields where major channels have eroded through the Main Reef placer, in places, to the Benoni goldfield where erosion completely removed the Main Reef placer (cf. Strydom, 1951; Papenfus, 1957). These sinuous channels are enormous, being more than 1 km wide and up to 100 m deep. This paleosurface may be represented by a shale unit above the Commonage placer in the Klerksdorp and Welkom goldfields.

In addition to diamictites, the channels contain

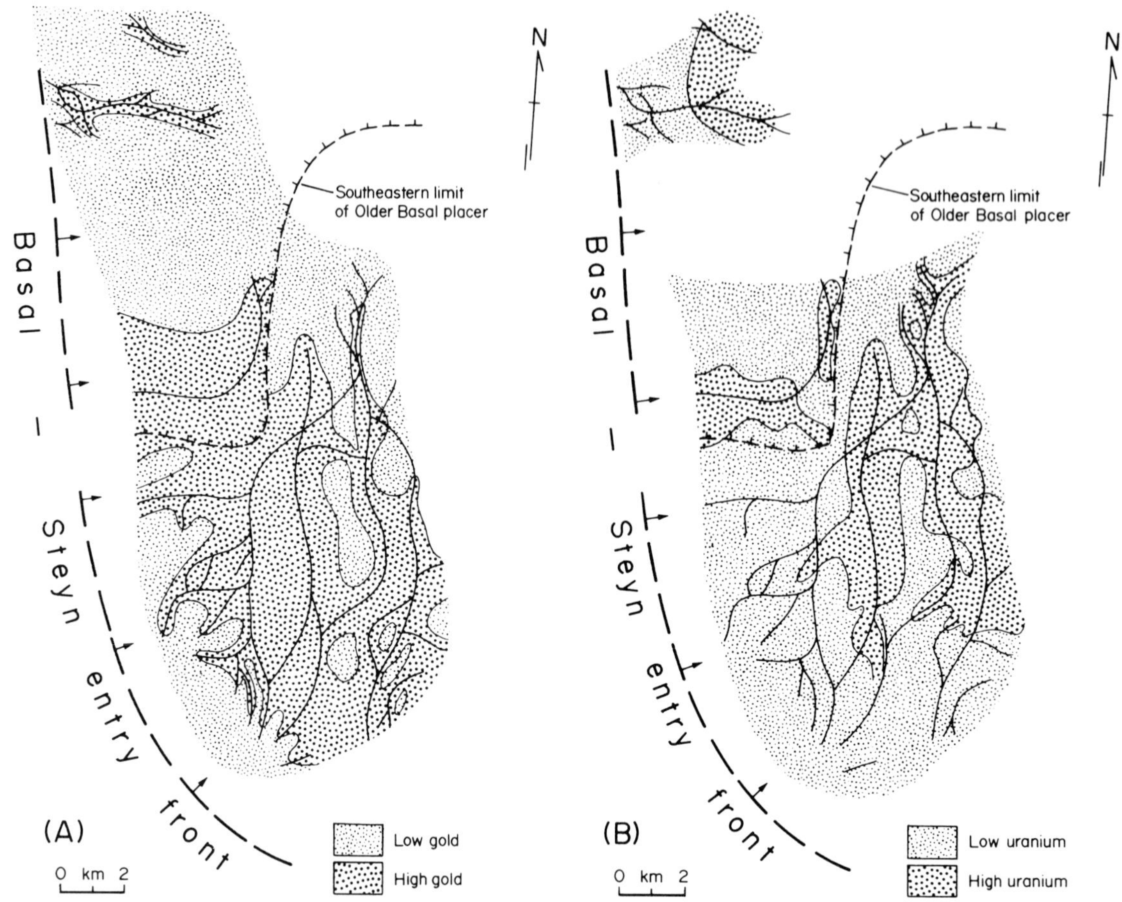

Figure 4-18. Subcrop maps of the Steyn placer paleosurface showing the distribution of detrital gold (A) and uranium (B) related to fan geometry. The preferential concentration of uranium in the finer-grained distal facies reflects the lack of coarse-grained uraninite in the source terrane. Welkom goldfield. [After Minter (1978).]

shale beds and placers of conglomerate and quartz arenite. The placer sediments which reflect episodes of increased fluvial activity, commonly foundered into the soft diamictites. Strydom (1951) described large-scale load casts in channels at Brakpan goldmine.

The younger of the two paleosurfaces mantled by diamictite underlies the Aandenk Formation in the Welkom goldfield and can also be recognized at the base of the Gold Estates in the Klerksdorp goldfield, beneath the Kimberley Reefs in the Benoni goldfield (the MK 1 Member), and in the Evander goldfield (the MK 3 Member).

The top of the Jeppestown shale is visibly discolored for more than 4 m beneath its contact with the Main Reef Leader placer in the Benoni goldfield (Antrobus and Whiteside, 1964; De Jager, 1964). Button and Tyler (1981) considered the shale to have been chemically weathered *in situ* before placer deposition; the observed alteration is associated with an enrichment in Al_2O_3, in the form of sericite, pyrophyllite, and chloritoid, and TiO_2 in residually concentrated ilmenite, partly altered to leucoxene. Button and Tyler suggested that a warm, humid climate with alkaline ground waters probably prevailed. However, because humid conditions are unlikely to have produced diamictites and because these two paleosurfaces are present around the entire Witwatersrand basin, it has been argued that a semiarid climate was more likely (Minter, 1980).

These diamictites may be the closest approximation to isochronous surfaces in the Witwatersrand basin. From their use as time data, it appears that the genetic packages of sediment with which many of the placer deposits are associated were responses to local processes rather than to basinwide epeirogeny.

Diamictites also form extensive sheetlike bodies in association with alluvial-fan deposits. In the Harmony Formation in the Welkom goldfield the debris-flow interval comprises a progradational succession of shale, sericitic sand, and pebbly sericitic sand which overlapped the Steyn placer. More mature fluvial sediments containing patchy placer concentrations, which have channeled into the diamictites and reworked them, are in many places preserved only as load casts.

Tidal Environments

The upper surfaces of many of the placer deposits in the Witwatersrand basin are planar and commonly associated with a thin layer of pebbles and concentrations of placer minerals. The clasts and heavy minerals appear to have been winnowed from the underlying placer sediment by reworking, possibly related to wave action (cf. McGowen, 1970).

Paleocurrent analysis of the quartz arenite immediately above the Kimberley placer in the Benoni goldfield (the UK 9A Marker) indicates polymodal transport directions (Armstrong, 1966). Wave ripples and ladderback ripples, such as are associated with intertidal mud flats, suggest that marginal marine conditions prevailed after placer deposition.

In the Heidelberg area a 60 to 70-m-thick subgraywacke above the Main Reef Leader placer is capped with 20 m of quartz arenite (the Houtpoort Member), which is believed to have originated by tidal working (Button, 1979b). Herringbone cross-bedding and the bimodal-bipolar paleocurrent distribution indicate southwestward, or ebb-dominated, tidal circulation.

Sedimentology of Selected Placers

A detailed account of each goldfield and placer is beyond the scope and intent of this chapter but a brief review is necessitated by their economic importance.

In the Klerksdorp goldfield, the 1155-m-thick Johannesburg Subgroup is divided into three formations (Fig. 4-10). Although the Ada May and Commonage placers have been mined to a small extent in shallow ground, the Vaal placer is the only economic deposit. It is a distal transgressive placer (Minter, 1972, 1976, 1978).

The Carletonville goldfield is similar to the Klerksdorp goldfield, and the Carbon Leader placer closely resembles the Vaal placer (De Kock, 1964). A second ore body known as the Middelvlei placer is being mined on a small scale. It is a more proximal deposit resembling the Leader placer in the Welkom goldfield.

In the Krugersdorp goldfield the Johannesburg Subgroup has a similar thickness to that in the Klerksdorp and Carletonville goldfields but seven important placers indicate more activity at its entry point. The Main Reef placer is a coarse proximal deposit at the base of a sandy sequence; the South placer is a transgressive distal facies deposit. The Livingstone placers are more confined in a thick sequence containing up to 12 proximal placers (Steyn, 1964). Four uranium-bearing placers are present in the upper part of the Johannesburg Subgroup at West Rand Consolidated. The lowest of these, known as the White placer, was probably a response to tectonic influences and appears to rest on an unconformity; proximal and distal facies are distinguished. The Monarch placer is thin and sheetlike and may be a transgressive distal placer. The Zone 2 and Zone 4 placers are distal facies within fan accumulations that could reflect local channel processes.

Little sedimentological information is available for the Johannesburg goldfield. However, it is believed that the placers were distributed in lobes that spread from the Krugersdorp and Benoni goldfields (Pretorius, 1976). This concept is based partly on the narrow width of the ore zone around the rim of the Johannesburg granite dome. It is equally likely that the Main Reef, Main Reef Leader, and South placers are distal parts of coalescing fans that prograded southward from the dome where the proximal facies equivalents have been stripped by erosion.

The Johannesburg Subgroup becomes progressively thinner from Krugersdorp to Evander. The Main Reef placer is truncated by an erosion surface so that at Benoni only the Main Reef Leader placer is preserved; the erosion surface is accentuated by major channels and diamictites. This placer fan, or series of fans, is distributed over an enormous area to the southeast. The grade of mineralization decreases to below economic levels in the distal sandy facies in the Heidelberg area.

In the Welkom goldfield the Johannesburg Subgroup is very thick and contains four important placers. The west and southwest provenance provided abundant polymictic components not found in the other goldfields. A placer stratigraphically equivalent to the Ada May placer of the Klerksdorp goldfield is present near the transitional contact with the Jeppestown shale, but it is unevenly distributed and generally deeply buried. It appears to represent the initial progradation of alluvial fans into the area. Distal placers formed by fluvial processes at the top of the St. Helena fan, and proximal placers at the base of the Welkom Formation which formed in response to tectonism, comprise a uranium-bearing zone in close stratigraphic proximity. The Basal and Steyn placers at the top of the Welkom Formation together cover 400 km^2 and represent terminal gravels in two coalescing sheets with significant concentrations of gold and uranium. They are probably the result of channel processes and display both proximal and distal characteristics. This sequence is interrupted by a 20-m-thick debris-flow complex that was derived from near the point of closure in the south; it was followed by the Leader placer, which entered the basin from the west. The Leader placer probably covers 200 km^2 and offlaps the Harmony diamictite and Steyn placer.

The period of epeirogenically induced transgression is reflected in upward-fining successions throughout the basin; a ubiquitous argillite in this interval is used as a widespread stratigraphic marker. The Turffontein Subgroup above this marker represents renewed shrinking of the basin, with coarse alluvial fans encroaching from all margins.

The Turffontein Subgroup in the Klerksdorp goldfield (Fig. 4-10) contains three placer deposits of limited economic importance. A sandy distal sequence containing a basal placer known as the Cristalkop placer is truncated by the Gold Estates angular unconformity in the southern part of the goldfield. The Gold Estates are coarse proximal deposits with patchy gold concentrations that have been surface mined around Klerksdorp.

The Orkney placers are mainly gold-bearing proximal deposits within a multistoried gravel succession. These placers are best developed in the western part of the goldfield and have been mined locally at Orkney in conjunction with Ventersdorp placer mining. The No. 5 placer has been described by Smith and Minter (1980).

In the mining area of the Carletonville goldfield much of the Turffontein Subgroup has been truncated and overlain by the Ventersdorp succession, but to the north at Western Areas and to the south at Deelkraal several placer sequences are present, some of which are being explored and mined. These were derived from the northwest.

A 20-m-thick accumulation of proximal gold placer deposits at Western Areas (known as the "Massives") has been assigned to the Turffontein Subgroup. However, it is possible that they are part of the Venterspost Formation of the Ventersdorp Supergroup. They probably represent coarse gravel sheets deposited by braided streams in a humid alluvial-fan setting strongly influenced by tectonism.

Farther north between the Carletonville and Krugersdorp goldfields a newly discovered placer deposit is being mined. It overlies an unconformity at the base of the Gemsbokfontein Formation. The gold and uranium placer was deposited in a proximal environment by braided streams flowing to the east and southeast from the Panvlakte fault (Tucker, 1980). Many Kimberley placers have been mined extensively in the Krugersdorp goldfield. Absolom (1979) has shown that proximal gravels at South Roodepoort were transported to the south from an axis near West Rand Consolidated.

Placers are poorly developed in the Turffontein Subgroup in the Johannesburg goldfield despite the thick accumulation of sediment. In the Benoni goldfield only one placer, known as UK 9A, May Reef, or Kimberley Reef, is present. It overlies diamictite deposits on a channeled paleosurface. Shallow marine interaction with fan delta deposits is inferred (Armstrong, 1966). These Kimberley sediments appear to offlap to the southeast where placer deposits are low grade.

In the Evander goldfield the only placer is in the Turffontein Subgroup. Tweedie (1978) has shown that the placer was a response to nearby uplift and was deposited by northeast-flowing streams in a sandy braid-plain environment.

In the Welkom goldfield three important placers in the Turffontein Subgroup show transport to the northeast. The B placer is a gold-bearing deposit at the base of a tectonically induced sedimentary sequence; the proximal placer gravels are confined to discrete channels eroded into a fine-grained arenite (Minter, 1980). The A placers are channel deposits in the Aandenk Formation. The Eldorado placers occur in a humid alluvial-fan sequence and comprise multiple gravel deposits. These deposits display many internal angular unconformities along a western structure, previously named the Border fault, which was subject to repeated uplift.

Placer Minerals

More than 70 ore minerals are recognized in Witwatersrand placers, of which at least 40 are detrital (Feather and Koen, 1975). The most important economic detrital minerals are, in order of decreasing abundance, pyrite, uraninite, brannerite, and gold. Other abundant detrital minerals are arsenopyrite, cobaltite, chromite, and zircon. With the exception of the platinoids, which are concentrated with free gold on Wiffley tables, the remainder are of no commercial value.

Pyrite, chromite, uraninite, gold, and zircon are the only detrital minerals that can readily be seen in hand specimen. These heavy minerals are concentrated within the matrix of conglomerates, on lag surfaces, winnowed surfaces, scour surfaces, foresets, and trapped within the kerogen of fossil algal mats. Hydrodynamic sorting processes apparently varied from one sedimentary facies to another with the result that detrital mineral ratios may vary accordingly. Although many of the detrital minerals show signs of metamorphic recrystallization, they do not appear to have moved from their sites of deposition.

Intrusive Events

The Witwatersrand succession is interlaced with numerous dikes and sills, classified into five main groups. The oldest are associated with the early Klipriviersberg (Ventersdorp Supergroup) outpourings of lava and probably represent feeder dikes. A second group is post-Ventersdorp in age and is thought to have been emplaced during the igneous activity associated with the Bushveld Complex at about 1950 Ma (Pretorius, 1976) (Chapter 6). The third group is post-Transvaal and possibly contemporaneous with the Pilanesberg period of activity (Chapter 13). Karoo diabases of Jurassic age constitute a fourth group (Chapter 11), and rare lamprophyric kimberlite dikes of probable Cretaceous age form a fifth group (Chapter 13).

An Overview

Since 1965 systematic sedimentological studies have been conducted in the newer goldfields, but unfortunately little comparative information is available from the older goldfields. What is missing in particular in the mined-out areas are paleocurrent, compositional, and fabric data with which to delimit specific fans and to determine their degree of coalescence. Although these aspects remain conjectural, the general sedimentary processes were probably similar to those deduced in the well-documented newer areas.

In each of the major goldfields the Johannesburg and Turffontein Subgroups contain numerous sedimentary packages that are separated by unconformities. Each sedimentary package represents a basin response to tectonically controlled erosion and aggradation. All Witwatersrand placers were probably deposited by shallow braided streams in proximal to distal environments. The Scott, Donjek, South Saskatchewan, and Platte models established by Miall (1978) are all represented in one or other of the genetic packages. In cases where extensive paleoslopes have been preserved, such as the Steyn placer in the Welkom goldfield (Minter, 1978), the full range of facies models are developed. Placers overlying the unconformities were generally produced by degradation, but not all are sufficiently mineralized to be economic. Placers within some of these packages reflect processes on alluvial fans that were probably less than 400 km^2 in area; consequently their extent was limited to particular goldfields.

4.4. Ventersdorp Supergroup: Crustal Fracturing

The Ventersdorp Supergroup occupies a large elliptical basin that exceeds 200,000 km^2 in area and encompasses most of the surface and subsurface distribution of the Dominion, West Rand, and Central Rand Groups (Fig. 4–3). It appears to overlie the Central Rand Group conformably in the Johannesburg, Benoni, Evander, and Welkom goldfields. However, along the northwestern margin of the Witwatersrand basin from Krugersdorp to Klerksdorp an angular unconformity with up to 5 degrees discordance has been recognized. Block

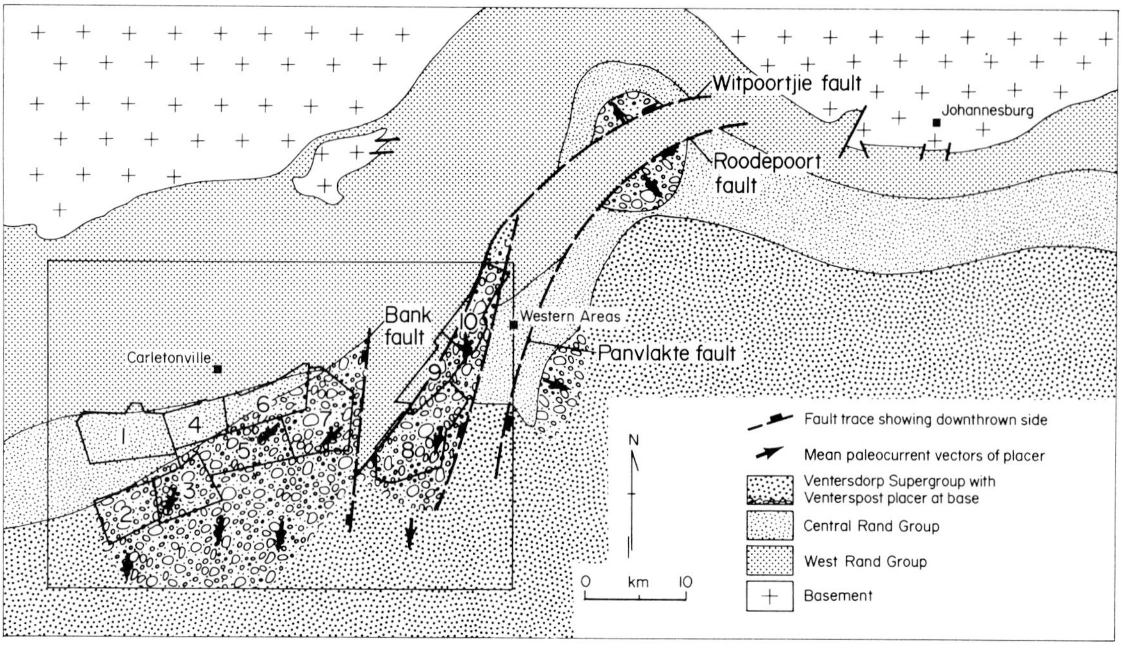

Figure 4-19. Subcrop map of the Witwatersrand and Ventersdorp Supergroups below the Transvaal Supergroup. Venterspost placers at the base of the Ventersdorp Supergroup are confined to grabens. Gold mine properties shown are: (1) Doornfontein, (2) Deelkraal, (3) Elandsrand, (4) Blyvooruitzicht, (5) Western Deep Levels, (6) West Driefontein, (7) East Driefontein, (8) Kloof, (9) Libanon, (10) Venterspost. The area within the rectangle is expanded in Fig. 4-24. [Adapted from Borchers (1961), Minter (1978), and Krapez (1980).]

Figure 4-20. Venterspost placer with mudstone partings occupying a steep-sided channel eroded into quartz arenite. Jointing and resistant bedding planes controlled channel morphology. Western Deep Levels goldmine, Carletonville goldfield. Scale bar = 50 cm. [After Minter (1978).]

faulting that produced grabens and half grabens, and unconformities that strike obliquely to those in the Central Rand Group were initiated during late Witwatersrand time (Fig. 4-19). These structures are of local extent. On a regional scale the Witwatersrand and Ventersdorp Supergroups have not been tectonically disturbed. Although the succession has been metamorphosed to lower greenschist facies, tectonite fabrics are lacking. The yoked basins contain placers derived mainly from reworking of Witwatersrand deposits which were lithified prior to this erosion. Rugged paleorelief was locally controlled by joints and bedding planes (Fig. 4-20). In the Carletonville goldfield more than 1500 m of Witwatersrand strata was truncated, and well-rounded clasts of distinctive arenite markers were incorporated in the overlying gravel accumulations. The oldest lava flows slumped into these sediments (Fig. 4-21). The remaining paleosurface irregularities above the Witwatersrand were filled with lava which subsequently overlapped older rocks beyond the limits of the Witwatersrand basin.

Although the Ventersdorp Supergroup extends for about 750 km along a northeast axis between Prieska in the northern Cape almost to Thabazimbi in the northern Transvaal and westward to Gaborone in Botswana (Fig. 4-3), only about one fourth is exposed. The Ventersdorp succession exceeds 7860 m in thickness in the Bothaville area between the Welkom and Klerksdorp goldfields. It comprises basaltic, dacitic, and rhyolitic lavas interbedded with agglomerate, subgraywacke and conglomerate, and subordinate shale and limestone. Local unconformities, lenticular and wedge-like geometries, repeated lithologies, overlaps, and contemporaneous folding and faulting have made it difficult to unravel the Ventersdorp stratigraphy using outcrop data alone. Fortunately, deep drilling for subsurface Witwatersrand strata has provided hundreds of stratigraphic sections through the Ventersdorp (Winter, 1965). A threefold subdivision is recognized comprising, in ascending order, the Klipriviersberg and Platberg Groups and the Pniel Sequence (Fig. 4-22). Age dating of lavas supports a correlation between the former Zoetlief Formation in the Cape with the Platberg Group, and the former Pniel Series with the Allanridge and Bothaville Formations of the Pniel Sequence. Radiometric ages are separated by a controversial

Figure 4-21. Spectacular deformation structures were found where the unconsolidated Venterspost placer yielded to loading of the Klipriviersberg lava. Kloof goldmine, Carletonville goldfield. Scale bar=1 m. [Courtesy of Stuart Buck.]

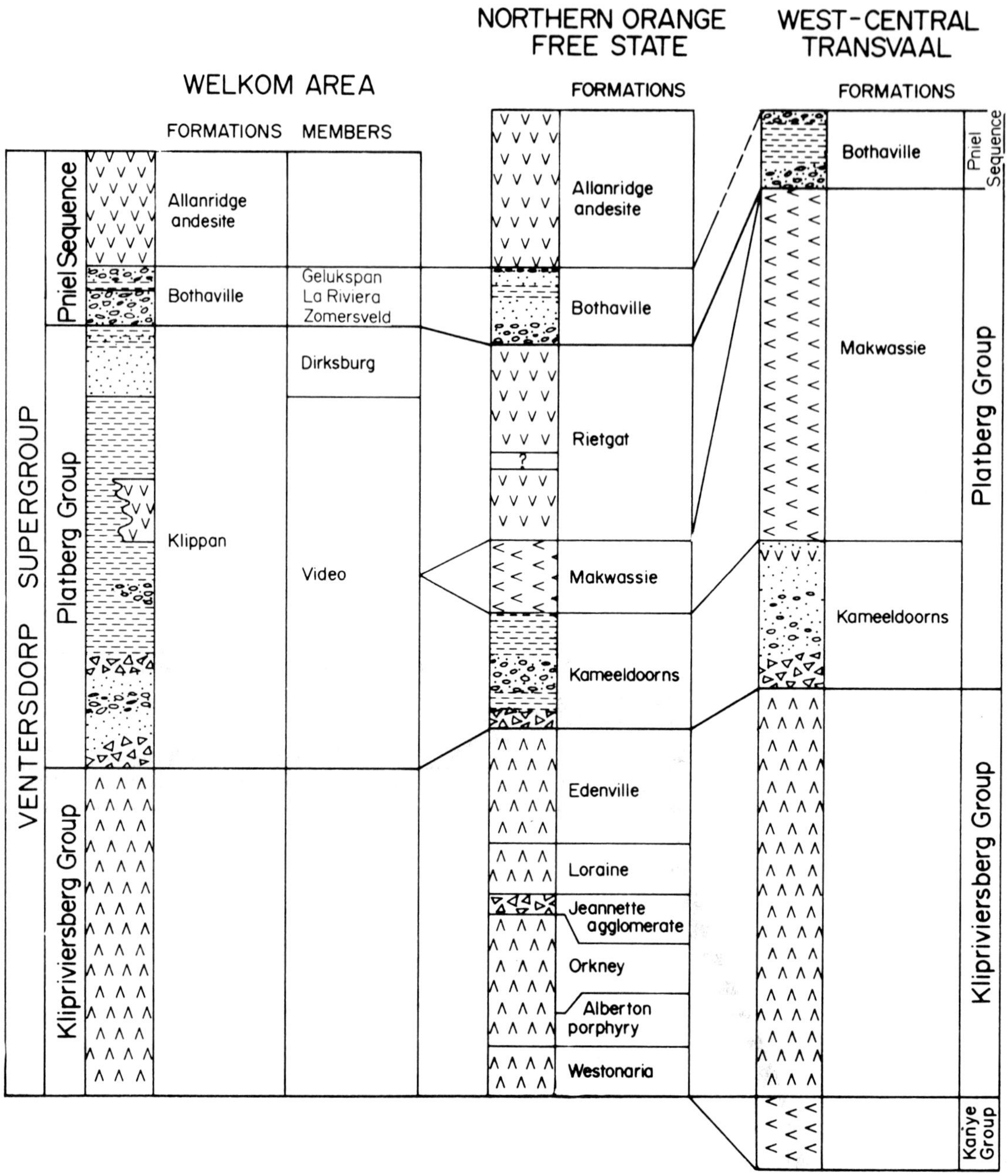

Figure 4-22. Stratigraphic correlation of Ventersdorp formations between the Welkom goldfield and the west-central Transvaal. [Adapted from Winter (1976), Tyler (1979a), and Buck (1980).]

200–400 Ma discrepancy and fall into two groups clustered around 2300 Ma and 2500–2700 Ma, respectively (Van Niekerk and Burger, 1964, 1978b; U-Pb). Allsopp and Welke (in press) maintain that the older U-Pb dates obtained from zircons in the Makwassie quartz porphyries can be accounted for by the fact that the zircons represent xenoliths. However, it is not clear whether the younger ages should be regarded as primary or reset. In view of the presence of internal unconformities and thick paleosols, it is also possible that deposition of the Ventersdorp Supergroup was very protracted. Burial metamorphism of Ventersdorp volcanic rocks beneath several kilometers of the

Griqualand West Supergroup caused isotopic resetting to a radiometric age of 1920 Ma (Cornell, 1978a; Rb-Sr isochron).

The Klipriviersberg Group, composed mainly of lavas, filled irregularities on the underlying Witwatersrand Supergroup and onlapped older rocks outside the Witwatersrand basin. The Platberg Group contains immature clastic sediments; local stromatolitic limestones; extensive, thick quartz–porphyry flows; and andesitic lavas in genetically associated rift basins that extend far to the west and north of the Witwatersrand basin. Because of its tectonic setting, the stratigraphy of this group is very irregular. A protracted period of erosion terminated accumulation of the Platberg Group and established the gently folded and faulted paleosurface on which the Pniel alluvial-plain sediments were deposited. These contain well-rounded clasts of basement rocks and Witwatersrand and earlier Ventersdorp strata. These terrigenous clastic sediments, constituting the Bothaville Formation, were buried beneath andesitic plateau lavas of the Allanridge Formation.

Rift Volcanism

During the final stages of Witwatersrand sedimentation the central Kaapvaal Province was in a state of stress as a result of subsidence under the mass of the sedimentary succession deposited in this basin. The stress in the crust was relieved by tensional faults which acted as conduits for ascending Ventersdorp magmas (Visser, 1957). On reaching the surface the volcanics were extruded over an area of at least 155,000 km^2 and attained a maximum thickness of 5775 m (Winter, 1976).

At the type locality in the northern Orange Free State, the volcanic assemblage comprises four compositionally discrete units (Fig. 4-7) interstratified with sediments (Fig. 4-22). The Klipriviersberg Group at the base of the column is composed largely of basic lavas. Acid volcanics of the Makwassie Formation grade into intermediate to basic lavas of the Rietgat Formation, and the volcanic pile is capped by basic lavas of the Allanridge Formation (Winter, 1976). In west-central Transvaal and southeastern Botswana an assemblage of acid volcanics (the Kanye Group, Fig. 4-22) overlain by the Ventersdorp Supergroup possibly represents a separate protobasinal phase of Ventersdorp volcanism (Tyler, 1978a).

Voluminous continental tholeiitic basalts constitute the 1830-m-thick Klipriviersberg Group. The greenish-gray lavas are widely distributed throughout the Cape, Orange Free State, Transvaal, and southeastern Botswana. They probably originated as fissure-fed flood basalts extruded over very large areas. Only local fluctuations in the chemical composition of the flood lavas have been recognized; for the most part they are classified as low-magnesia tholeiitic basalts (Wyatt, 1976). On the central Rand a basal unit comprising relatively high-magnesia basalt (the Meredale Member) contains textures indicative of liquid immiscibility (Wyatt, 1976). These high-magnesia basalts (Table 4-2) are enriched in Cr and Ni but are depleted in Al_2O_3 and therefore represent komatiites (Fig. 4-7). The major compositional break between these two lava types suggests that the komatiitic rocks were not parental to the upper extrusive rocks (Wyatt, 1976).

Although the chemical composition of the tholeiitic assemblage varies little with stratigraphic level or areal extent, distinctive porphyritic, amygdaloidal, tuffaceous and breccia facies are the basis of a sixfold subdivision of the group (Fig. 4-22); formation thicknesses range between 75 m and 570 m (Winter, 1976). The scarcity of sedimentary intercalations within the volcanic assemblage suggests continuous extrusion. A local basalt breccia and boulder conglomerate interbedded within the upper flows of an overthickened 2600-m-thick Klipriviersberg section in the Tshwene-Tshwene graben of west-central Transvaal suggest contemporaneous faulting with subsequent alluvial-fan sedimentation during volcanism in the marginal portions of the Klipriviersberg basin (Tyler, 1978b).

During the waning phases of Klipriviersberg volcanism lavas of intermediate composition were extruded together with the tholeiitic flood basalts. One of these volcanic units is a feldspar porphyry containing 2- to 3-mm sized ellipsoidal accretionary lapilli consisting of concentric layers of elongate microlites around a structureless core of volcanic ash (Tyler, 1979a). The genesis of accretionary lapilli has been compared to the development of hailstones in a thundercloud (Moore and Peck, 1962). During volcanism, clouds of ash and water vapor rise to great heights, where cooling causes condensation of water vapor and agglutinization of the ash, forming the nuclei of accretionary lapilli. As each embryonic lapillus falls through the eruptive cloud an outer shell is built up by ash sticking to the moistened surface of the lapillus. Turbulence within the eruption cloud may lift the lapillus into the higher reaches of the cloud, resulting in repeated accretion.

Stress in the crust was unalleviated by the voluminous outpourings of the Klipriviersberg flood tholeiites, and graben-style faulting continued with subsequent alluvial-fan sedimentation. In areas unaffected by Kameeldoorns sedimentation, the Klipriviersberg volcanic terrane was subjected to subaerial weathering and soil formation (Tyler, 1978b). One of these paleosols in the west-central Transvaal comprises a distinctive 25-m-thick lime-green marker unit composed almost exclusively of sericite with minor leucoxene. Paleosols derived from volcanic rocks of this nature contain up to 35 percent alumina (Tyler, 1979a, his table 3; Button and Tyler, 1981) and may constitute an important resource in the future.

The Makwassie Formation (Fig. 4-22) of the Orange Free State and western Transvaal overlaps the underlying Kameeldoorns Formation and rests on progressively older Klipriviersberg, Witwatersrand, and Archean terranes away from the basin axis. A maximum thickness of the Makwassie Formation of more than 2100 m has been recognized in the western Transvaal (Tyler, 1978a).

The Makwassie Formation comprises a succession of quartz porphyry flows with intervening quartz-free porphyritic and nonporphyritic lavas, ash flows and subordinate sediments. Winter (1976) recognized distinct centers of volcanism associated with cryptocrystalline porphyritic flows as much as 600 m thick. The flows thin away from the extrusive vents, with a concomitant decrease in the proportion of quartz porphyry to other lava types. Features indicative of subaerial extrusion include flow structures, agglomerates, and graded tuffs (Winter, 1976).

The Makwassie quartz porphyry is widely represented in adjoining areas. Correlative felsites in the west-central Transvaal mapped into four volcanic units up to 1200 m thick are interbedded with a 400-m-thick lensoid volcaniclastic assemblage (Tyler, 1978b). Granular, massive, porphyritic, and variolitic felsites and feldspar porphyries all approximate rhyolite in composition (Table 4-2), although alteration under conditions of greenschist metamorphism led to plagioclase breakdown with loss of sodium and calcium and enrichment of alumina (Cornell, 1977; Tyler, 1979a).

The style and composition of Ventersdorp volcanism altered dramatically at the onset of Makwassie extrusion. During Klipriviersberg time, volcanism was continuous and rapid, with tholeiitic flood basalts being extruded over vast areas from fissures. In Makwassie time, however, rhyolitic volcanics were intermittently and forcibly ejected from extrusive vents leading to local lenticularity of units. During periods of volcanic quiescence fluvial channels 8 m deep and 30 m wide were scoured into the felsites and filled with upward-fining alluvial sequences.

The explosive nature of Makwassie volcanism is indicated by the abundance of agglomerates. Two agglomerate types have been recognized in the west-central Transvaal (Tyler, 1979a). Explosive pyroclastic breccias are characterized by ill-sorted, heterogeneous angular blocks of siliceous material, with a great range in size and no recognizable stratification. The second breccia type consists of chaotic, angular to subrounded heterolithologic siliceous clasts bedded in a schistose shaley matrix; these characteristics are typical of laharic breccias.

In the center of the Ventersdorp basin, lavas that may be correlatives of lateral occurrences of the Makwassie quartz porphyry were assigned to the Garfield Member of the Rietgat Formation on the basis of their intermediate chemistry (Winter, 1976). The Rietgat Formation (Fig. 4-22) is a mixed volcanosedimentary assemblage up to 1320 m thick that rests conformably on, and may have been in part contemporaneous with, the Makwassie Formation. The basal Garfield Member (Winter, 1976), a 360-m-thick volcanic unit in the type borehole near Bothaville, consists of porphyritic and nonporphyritic lavas and sediments and is regarded as being transitional between the Makwassie and the Rietgat Formations.

The basal Rietgat lavas give way to an increasing proportion of sediments. Intercalated lavas of intermediate composition (Table 4-2) include amygdaloidal types in addition to the quartz porphyries characteristic of the lower units. Individual flows show oxidized, weathered, and eroded upper surfaces, attesting to episodic extrusion (Winter, 1976).

The Allanridge Formation, composed of fine-grained amygdaloidal and porphyritic lavas of tholeiitic composition (Table 4-2), records the final episode of Ventersdorp volcanism. The lavas rest conformably on the Bothaville Formation. Flood basalts are preserved in gently undulating basins containing Bothaville sediments in their lower parts but lap onto peripheral or interbasin arches of basement rocks (Winter, 1976). Voluminous but relatively quiescent lava effusions are indicated by the 750 m thickness, the rarity of agglomerates and tuffs, and the absence of sedimentary partings.

Alluvial Fans

Grabens produced by extensive rifting provided the setting for multiple alluvial-fan deposits and extrusive volcanics in the Platberg Group. This 5000-m-thick succession overlies the Klipriviersberg Group with a pronounced unconformity. The Kameeldoorns Formation, and to a lesser extent the Rietgat Formation, are confined to grabens so that younger sequences, such as the Makwassie quartz porphyry, may overlap directly onto pre-Kameeldoorns surfaces. Three grabens have been outlined by Buck (1980) in the Welkom goldfield. Fans within these troughs (Fig. 4-23) display an overall upward-fining pattern that is attributed to fanhead retreat. The Makwassie quartz–porphyry marker is not represented in these troughs and consequently a new formation name, the Klippan Formation, equivalent to the Kameeldoorns and Rietgat Formations, has been proposed (Fig. 4-22) (Buck, 1980).

The Klippan Formation comprises coarse scree deposits, sandy graywackes, and debris flows which were shed off horst blocks into adjacent grabens. Facies changed from boulder conglomerates near the faults to black calcareous shales inferred to have been deposited in lakes only a few kilometers distant. The lithology of the proximal boulder conglomerates, including talus breccia, reflects local provenance ranging from basement greenschist and quartzite, to pyritic Witwatersrand strata, and Klipriviersberg rocks. The thicknesses of these conglomeratic units, together with their geometries, is a reflection of contemporaneous faulting (Winter, 1976).

The Kameeldoorns landscape probably resembled the modern Basin and Range province of the western United States with small, steep alluvial fans coalescing along the peripheries of faulted basins. Flashy discharge patterns unimpeded by vegetation generated alluvial-fan types typical of modern arid or semiarid areas irrespective of the Ventersdorp climatic regime. Rapidly shed rockfall, talus-slide, and gullyhead debris was buried soon after accumulation but some was partially sorted and rounded by *in situ* abrasion in steep-headed, ephemeral stream courses. Middle and lower fan deposits show progressive downslope fining and gradation into playa lake carbonates and mudstones. Color banding may represent seasonal changes in discharge or water chemistry.

Other clastic wedges of the Kameeldoorns Formation are present in the northern Cape, northwestern Orange Free State, southwestern Transvaal, and west-central Transvaal. More than 1000 m of conglomerate and coarse sandstone crop out along the Tshwene-Tshwene range of the west-central Transvaal (Tyler, 1979a). Chaotic boulder conglomerates are overlain by conglomerate beds which grade upward into poorly sorted sandstone and, rarely, shale. Matrix-supported conglomerates containing boulders a meter or more in diameter may have been introduced by debris flow (Tyler, 1979a). Basalt flows and volcaniclastic debris are present toward the top of the formation. Locally, grabens were filled by volcanic flows which interrupt the continuity of the Kameeldoorns Formation.

Placers and Fluvial Environments

Fluvial sediments at the base of the Ventersdorp Supergroup unconformably overlie Central Rand Group strata along the northwestern margin of the basin between Krugersdorp and Klerksdorp. Uplifted margins or horsts provided source areas for renewed erosion and deposition in grabens. The proximal facies of the underlying Witwatersrand placers were reworked and redeposited in this environment and incorporated into an important new generation of placers. Around the remainder of the Central Rand Group basin the Ventersdorp is largely conformable with the Witwatersrand.

The Ventersdorp placers generated by this reworking comprise a number of economically significant deposits collectively named the Venterspost Formation (previously known as the Ventersdorp Contact Reef). In the Krugersdorp goldfield, Venterspost sediments were shed northwestward off the steep Witpoortjie fault scarp

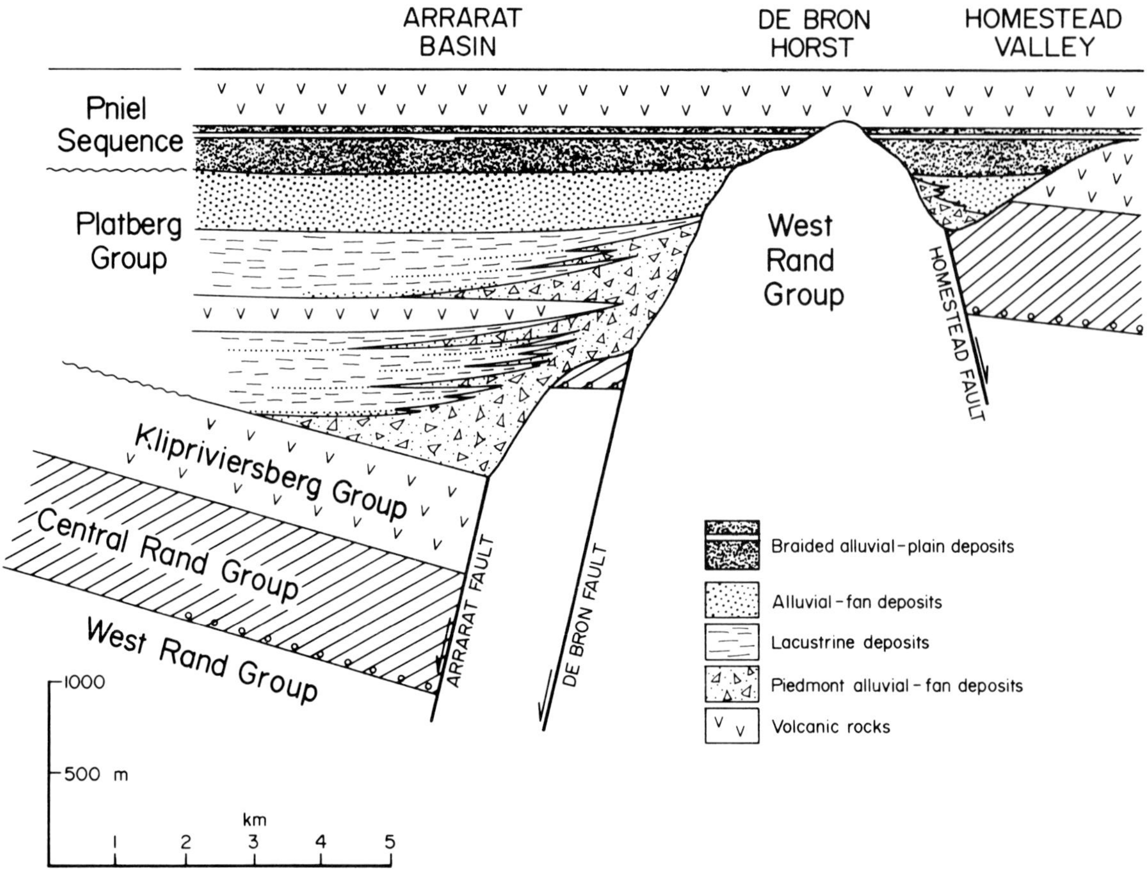

Figure 4-23. Schematic section showing alluvial fan sedimentation off the De Bron Horst, Welkom goldfield. The overall upward-fining pattern is attributed to fanhead retreat. [After Buck (1980).]

(Fig. 4-19) from a horst of Central Rand Group strata. Placers were developed in discontinuous, poorly sorted, coarse alluvial-fan accumulations close to the scarp. (They have been mined sporadically in Luipardsvlei gold mine, where they were known as the Montana reefs, and on Randfontein Estates.) A more extensive deposit was shed southeastward across the Roodepoort fault scarp on the other side of the horst (Minter, 1970). A sheet of proximal braided-stream gravels, displaying the same bar and channel facies seen in Witwatersrand placers (Fig. 4-14), was spread across an area exceeding 15 km^2. Depressions and channels in the irregular Witwatersrand surface contain placer gravels, whereas topographic highs are barren and overlain by lavas. Contact metamorphism by Ventersdorp lavas has darkened the matrix of the placer and produced a 10- to 20-cm-thick aureole on weathered Witwatersrand bedrock. Farther south, the Western Areas fan was shed off the Panvlakte fault scarp, which is a continuation of the Roodepoort fault.

A second horst occurs southwest of Western Areas east of the Bank fault (Fig. 4-19). The intervening half graben contains Venterspost placers between the Venterspost and Kloof gold mines. These conglomeratic braid bar and channel-fill deposits (Fig. 4-24) have been interpreted as an intermontane fluvial accumulation (Krapez, 1980). Similar rivers deposited a larger clastic wedge west of the Bank fault (Fig. 4-19), with gravel deposits mantling at least 30 km of the southwesterly paleoslope (from East Driefontein gold mine to the Deelkraal gold mine). Alluvial fans abutting the horsts contain cobble- and boulder-size clasts, more than half of which are locally derived from Witwatersrand sandstones. Debris-flow gravels on steep paleoslopes interfinger with crudely bedded to massive gravel sheets.

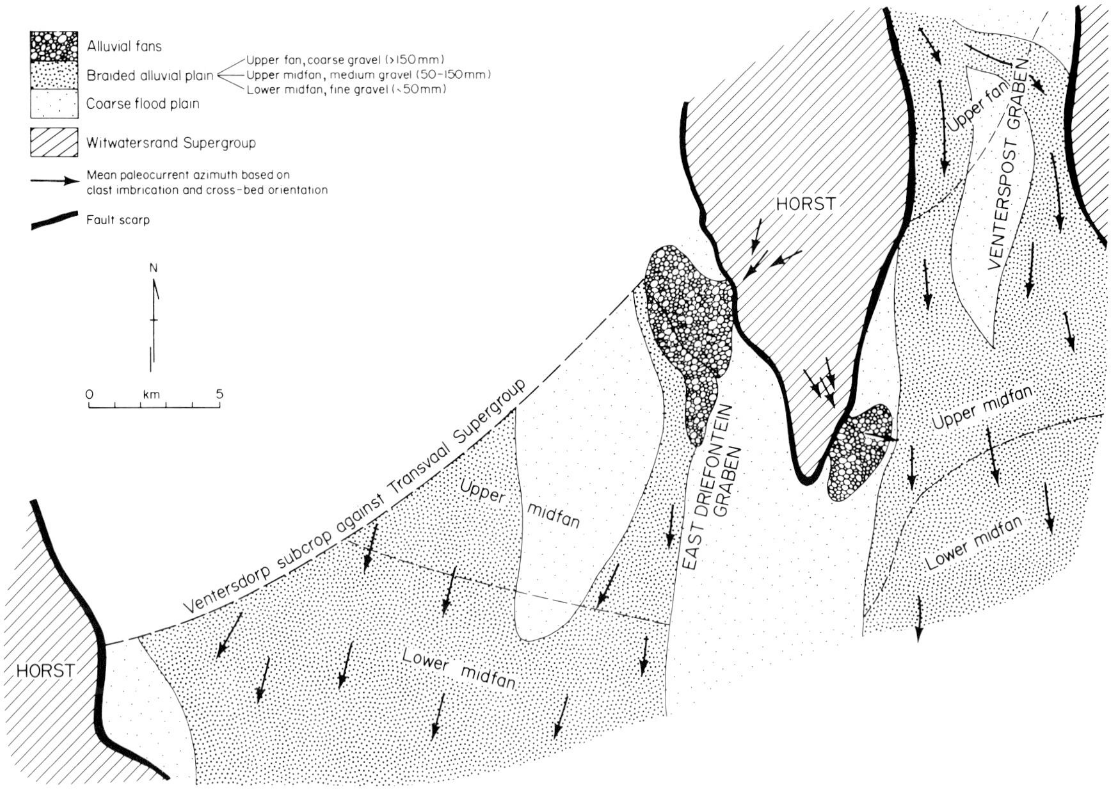

Figure 4-24. Paleogeographic reconstruction of the Venterspost Formation in the Carletonville goldfield showing the relationship of alluvial sedimentation and paleocurrents to graben formation. [After Krapez (1980).]

Between East Driefontein and Deelkraal, Venterspost sediments are rarely more than 4 m thick in the most proximal localities and average only 50–100 cm thick farther down the paleoslope. Paleorelief above Witwatersrand strata decreases from about 30 m proximally to less than 1 m distally (e.g., at Elandsrand gold mine). The slopes of the paleotopographic highs are rarely covered with sediment, the result being that they are occasionally misinterpreted in borehole intersections as either faults or widespread barren areas. Older fluvial terrace deposits commonly cap the topographic highs. A southwestward paleodip is indicated by cross-bed azimuths and clast-dimension gradients (Knowles, 1967; Minter, 1978; Krapez, 1980).

Gold content is low in the alluvial-fan deposits, averaging less than 20 ppm; it is high in the upper midfan gravel bars, averaging 30 ppm, but decreases distally to about 10 ppm. Conversely, the low uranium content increases downslope as in the Witwatersrand placers. Although the Venterspost placers largely resemble those of the Witwatersrand, they differ in two important respects: They were confined to broad valleys and they were not deposited in aggrading systems. The sediments were in transit in a degrading system when preserved by the lavas. Although 1500 m of Witwatersrand strata were removed by erosion, only a meter or so of alluvium was preserved on the paleosurface. Because these Venterspost and East Driefontein graben valleys were a considerable height above base level, elsewhere in the distal Venterspost Formation there must be a considerable volume of recycled Witwatersrand sediments.

Another important Venterspost placer deposit lies unconformably above Witwatersrand strata in the Klerksdorp goldfield (Thomas, 1977). Isopachs of the placer delineate two major braided belts with a width of 1 km and an explored length of 5 km (Fig. 4-25). Placer mineralization in these proximal braided stream deposits is related to the proportion of clast-supported gravel preserved; the gravels occur as bars, channel fill, or thin lags on the paleosurface, or on multiple scour surfaces

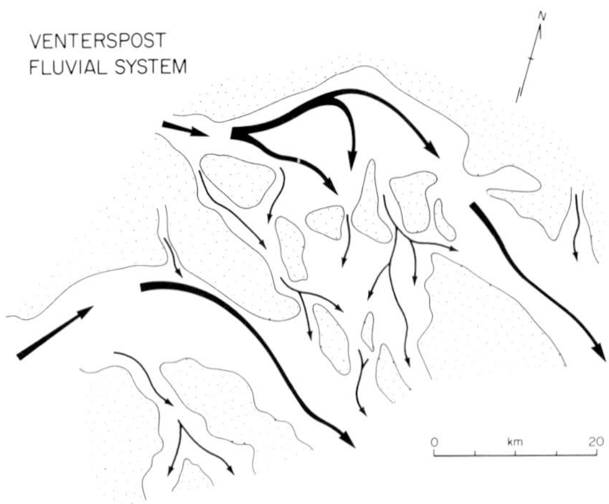

Figure 4-25. Deposition of the Venterspost placer on Witwatersrand strata by southeastward flowing braided rivers where two major channels merged. Vaal Reefs West mine, Klerksdorp goldfield. [After Thomas (1977).]

within sandy channel fills. Channel fills become more arenaceous downslope. This fluvial system was also degrading and placer sediments were evidently preserved while in transit. Channels are up to 4 m deep and minor terrace deposits overlie 2 m topographic highs.

Terrigenous clastic sediments and lavas of the Pniel Sequence (Fig. 4-23) rest unconformably on older formations of the Ventersdorp Supergroup. The two constituent formations (Bothaville and Allanridge) are lithologically distinct but are structurally concordant (Matthysen, 1953; Winter, 1976).

The Bothaville Formation, up to 430 m thick, reflects marked variations in the intensity and quality of detrital supply, probably in response to tectonic pulses and shifting fan loci. Regionally, conglomerates are best developed near the base and top of the formation (Winter, 1976). Clasts include jasper, chert, and quartz porphyry derived from the Makwassie Formation. Some of the Bothaville sandstones are compositionally and texturally more mature than those lower in the Ventersdorp succession, probably as a result of sedimentary recycling.

Lateral facies changes are less pronounced in the Bothaville Formation than in other Ventersdorp alluvial-fan and fluvial systems. This may indicate lower relief associated with reduced tectonism, more reworking, and a shift in the locus of maximum sediment accumulation away from the headward-retreating scarps to the downstream alluvial basins.

In Botswana and adjacent areas of the Transvaal a varied Bothaville rock suite includes immature conglomerates and volcaniclastic sands originating as meandering channel fill, together with flood-basin siltstones and shales (Tyler, 1979a). These sediments are referred to as the Mogobane Assemblage in Botswana (Crockett, 1971); a thick succession of shales defines the lacustrine center of the paleobasin.

Playa Lakes

This distal facies of alluvial fans in the Klippan or Kameeldoorns and Rietgat Formations comprises shale, marl, and carbonate deposited in small lakes. The carbonate contains stromatolites up to 5 cm high (Fig. 4-26A), which tend to be interlayered with chert, volcanics, and arkosic sediments deposited in the lowlying depressions by sheetflooding. The small, laterally linked domical stromatolites are elliptical in plan view, with preferred east–west elongation indicative of a general north–south paleoshoreline (Grobler and Emslie, 1976). According to these authors the stromatolites grew in isolated, shallow basins enriched in carbonate and silica from the surrounding arkoses and lavas. However, because of the andesitic nature of the upper Rietgat Formation, these lavas could only have supplied limited amounts of silica (N. Tyler, personal communication, 1981). Despite the nontidal character of these basins, the stromatolite types mimic the zonation along tidal coasts. For example, the small, domical stromatolites of low-energy, subtidal type give way elsewhere to columnar types characteristic of a higher energy intertidal situation (Grobler and Emslie, 1976). Zonation in the Ventersdorp lakes may have been caused by wind stress on a gradually shoaling bottom. Stromatolite growth was periodically inhibited by terrigenous influx. Some oolitic grains (Fig. 4-26B) are present together with detrital sediments in the upper parts of the lacustrine limestone lenses. Stromatolites are also intermittently associated with bedload fluvial deposits of the Bothaville Formation, where they grew in shallow pools behind riffles during low-flow stage (Buck, 1980). This environment did not favor ooid development.

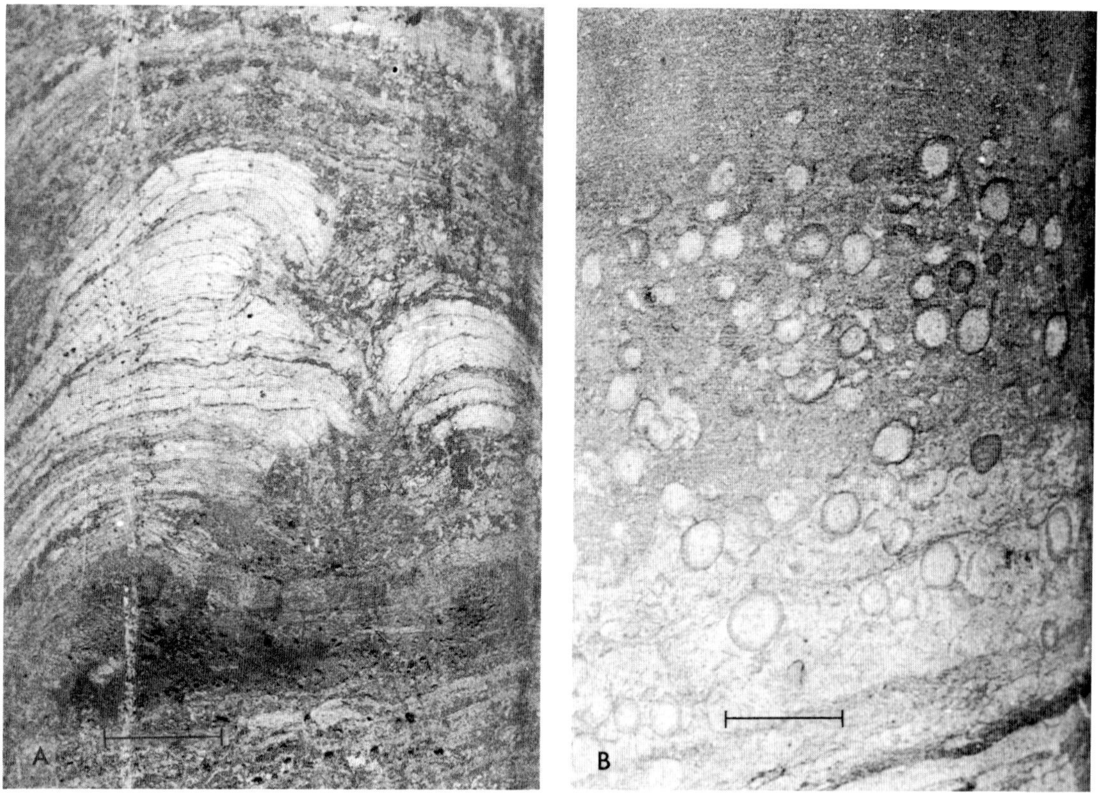

Figure 4-26. (A) Small, laterally-linked domical stromatolites which grew in playa lakes fringing some Platberg alluvial fans, Welkom goldfield. [From Buck (1980).] (B) Ooids present in the upper parts of lacustrine limestone lenses, Platberg Group, Welkom goldfield. Scale bars = 1 cm. [From Buck (1980).]

Depositional History

The three stratigraphic intervals represented in the Ventersdorp Supergroup display distinct tectonic controls. The Klipriviersberg lavas disconformably overlie the Turffontein Subgroup in many parts of the Witwatersrand basin, their age difference being concealed by structural concordance. In the Krugersdorp, Carletonville, and Klerksdorp goldfields pre-Ventersdorp tilting and erosion produced angular unconformities and subsidiary yoked basins.

The main Ventersdorp depoaxis is oriented parallel to the northeast axis of the Witwatersrand basin. However, Tyler (1978a) has identified a second Ventersdorp axis in the northern Kaapvaal Province between Kanye (Botswana) and Heidelberg (Fig. 4-3). This second axis trends at right angles to the main Ventersdorp depoaxis. Rapid subsidence about this subsidiary axis, forming the Swartruggens trough, is reflected in the preservation of two lobes of deposition. The Kanye volcanic rocks of Botswana were extruded into this trough before Ventersdorp deposition and may represent a protobasinal phase of the Ventersdorp Supergroup, in the same way as the Wolkberg and Dominion Groups represent protobasinal phases of the Transvaal and Witwatersrand Supergroups, respectively. This may be why the Gaborone granite complex, Kanye Volcanic Suite, and Ventersdorp Volcanic Suite in this area are all dated at 2750–2500 Ma (Tyler, 1979a). The Ventersdorp succession in this area contains tholeiitic lavas and alluvial-fan sediments, correlative with the Klipriviersberg and Kameeldoorns deposits, and acid lavas and alluvial-plain sediments that correspond with the Makwassie and Bothaville Formations, respectively.

Some of the late Witwatersrand structures along the western flank of the Ventersdorp basin may have been active during Klipriviersberg deposition, but there is also evidence for younger, post-Witwatersrand faults. The main strike-parallel faults in the Welkom goldfield are of late

Klipriviersberg age and they produced the scarp-bounded valleys that became Platberg Group depositories. The style of Platberg sedimentation is analogous to sedimentation in tectonically active rift valleys.

The rift valleys that formed during Platberg deposition extended far beyond the edge of the Witwatersrand basin, indicating a substantial expansion of the basin in Ventersdorp time. The wide distribution of Makwassie lavas supports this concept.

An episode of crustal stability and peneplanation was followed by deposition on the extensive Bothaville alluvial plain. Extrusion of the Allanridge flood lavas buried these Bothaville sediments. The Allanridge lavas are conformably overlain by the Transvaal Supergroup sediments without evidence of significant intervening erosion. The Pniel Sequence may represent the initial phase of deposition of the Transvaal Supergroup in the southwestern Transvaal, Orange Free State, and Northwestern Cape (Winter, 1965). It therefore may be equivalent to other pre-Black Reef deposits in the Transvaal Supergroup, such as the Wolkberg Group (Button, 1973a) and the Buffalo Springs Group (Tyler, 1979b).

Chapter 5

The Transvaal Epeiric Sea

The Early Proterozoic Transvaal and Griqualand West Supergroups (Fig. 5-1) accumulated in a vast epeiric basin, covering at least 500,000 km^2 of the Kaapvaal Province. The Transvaal Supergroup is thickest in the northeastern Transvaal, where a fourfold subdivision of the 12,000-m-thick succession is recognized (Fig. 5-2): the Wolkberg Group and the stratigraphically equivalent Buffalo Springs Group in the northwestern Transvaal, the Black Reef Formation, and the Chuniespoort and Pretoria Groups (Button, 1976a). No equivalent of the Wolkberg Group is present in the northern Cape. The Ghaap and Postmasburg Groups of the 4500-m-thick Griqualand West Supergroup correlate with the Chuniespoort and Pretoria Groups of the Transvaal Supergroup.

The Hekpoort basalts in the Pretoria Group have been dated at 2224 ± 21 Ma (D. Crampton, personal communication, 1972; Rb-Sr) and the underlying Timeball Hill shales have yielded an age of 2263 ± 85 Ma (Hamilton, 1977; Rb-Sr). Transvaal-Griqualand West sedimentation and volcanism was probably initiated after 2.3 Ga and had ceased by 2.1 Ga. The unconformably underlying Ventersdorp lavas have an age range of about 2300 Ma (Van Niekerk and Burger, 1964; U-Pb) to 2700 Ma (Van Niekerk and Burger, 1978b; U-Pb). Acid volcanics underlying Transvaal strata in the eastern Bushveld (Fig. 5-1) are dated at 2460 ± 120 Ma (Coertze *et al.*, 1978; U-Pb). Bushveld basic rocks above the Transvaal Supergroup have an age of 2095 ± 24 Ma (Hamilton, 1977; Rb-Sr); lavas in the unconformably overlying Matsap Group in the northern Cape are dated at 2070 ± 90 Ma (Crampton, 1974; Rb-Sr).

5.1. Protobasinal Phase

The terrestrial and shallow marine Wolkberg and Buffalo Springs Groups (Fig. 5-1) are developed along the northern margin of the Transvaal basin (Tyler, 1979b) where paleotopography controlled the pattern of sedimentation (Button, 1973a). The lower stratigraphic units in particular tend to pinch out against Archean granite–greenstone basement highs (Fig. 5-2A). Lithologically these groups consist of arenites, shales, and volcanics, with subordinate conglomerates and dolomitic shales.

Fluvial Sediment Supply

Sediments of probable fluvial origin constitute the Sekororo and Schelem Formations of the Wolkberg Group (Button, 1973a) and most of the Hampton Formation of the Buffalo Springs Group (Tyler, 1979b). Wolkberg paleocurrents were directed mainly to the west and Buffalo Springs paleocurrents to the southeast.

Arkoses and wackes are the dominant rock types; interbedded conglomerates consist mainly of vein-quartz pebbles, with boulders of volcanic lithologies at the base of the Schelem Formation. Siltstones and shales, commonly tuffaceous, are generally subordinate, although in the Hampton Formation they are up to 5 m thick.

Stacked upward-fining depositional sequences are characteristic and comprise basal granular to coarse-grained sandstones overlain by finer grained trough and planar cross-bedded sandstones; many of the sequences are terminated by rippled silt-

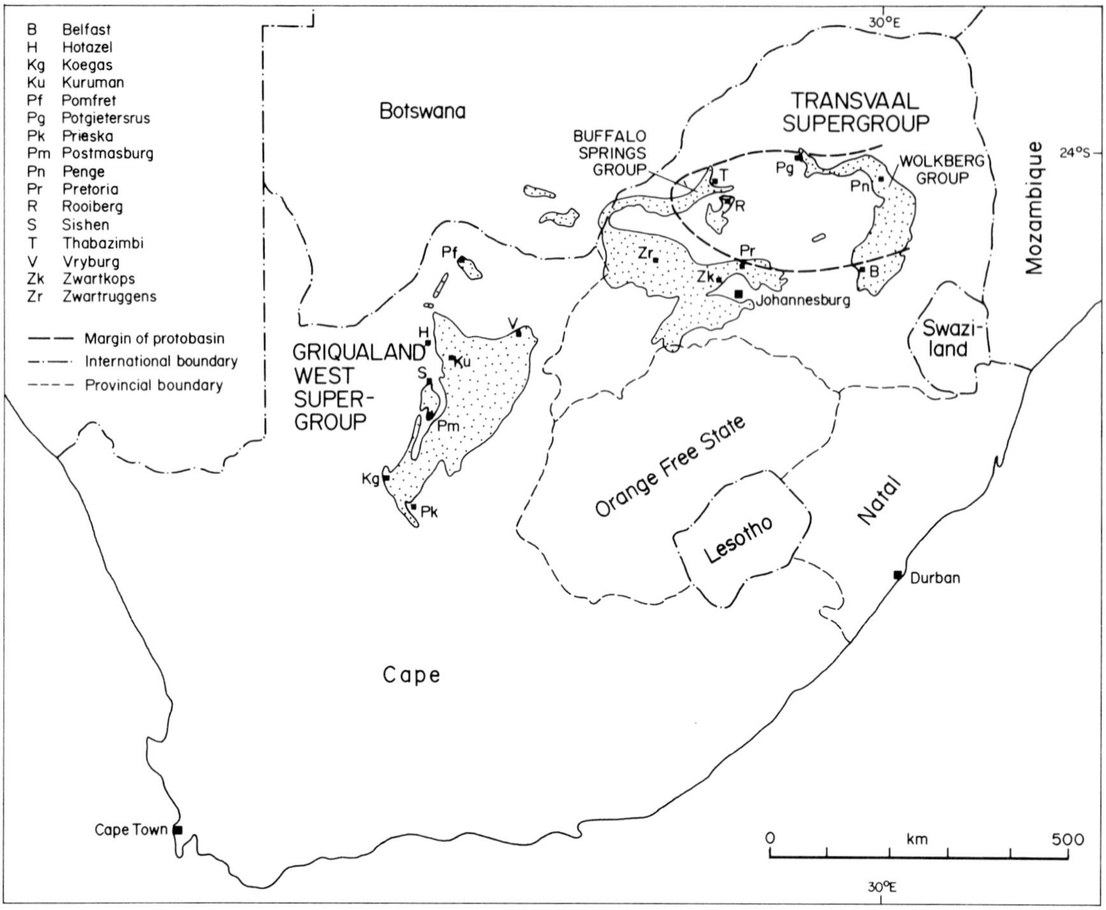

Figure 5-1. Distribution of the Transvaal and Griqualand West Supergroups. Also shown is the inferred position of the Buffalo Springs–Wolkberg protobasin. [After Tyler (1979b).]

stones and desiccation-cracked shale drapes. Sequences vary from 1 m to 8 m in thickness and lens out laterally over tens of meters. Similar facies are known from sandy reaches of many modern braided alluvial plains, where they develop as a result of aggradation or downstream migration of mid-channel bars across channel lag gravel (Miall, 1977). Thicker conglomerates in the Schelem and Hampton Formations occupy channels eroded into the underlying volcanics or form laterally extensive sheets which are interpreted as coalesced longitudinal bar deposits of the proximal braided alluvial plain.

Upward-fining sequences are particularly well developed in an arkose–shale member of the Hampton Formation (Tyler, 1979c). A typical sequence is illustrated in Figure 5-3, for which a meandering river origin has been suggested.

Volcanism

Thick volcanic sequences are present in both the Wolkberg and Buffalo Springs Groups (Button, 1973a; Tyler, 1979b). The Abel Erasmus basalts interrupt a 600-m-thick succession of fluvial sediments. The basaltic Waterval Formation and acid Witfonteinrant Formation are together 800 m thick and separate underlying fluvial from overlying shallow marine sediments.

The texture and geometry of these volcanic rocks indicate extrusion onto an irregular land surface, but rare pillows in the Abel Erasmus basalts suggest occasional subaqueous extrusion. Although generally massive, some basalts are amygdaloidal. In the Waterval Formation they contain a thin agglomerate band. The acid volcanics in the Witfonteinrant Formation comprise dark felsites and rhyolites; structures indicative of subaerial extrusion include vent agglomerates, laharic breccias and pumice with flow-top breccias. Thin basalts are interlayered with the acid volcanics.

A single analysis on a Waterval basalt reveals a continental tholeiite composition with a low Al_2O_3 content (Tyler, 1979b), a chemistry characteristic of Archean and Proterozoic basalts in southern Africa (Button, 1974). The acid volcanics of the Witfonteinrant Formation display few calc–alkaline tendencies, and instead are K_2O-enriched and Na_2O-depleted rhyolites. In the Buffalo Springs Group the volcanics are bimodal in composition. Rift-controlled emplacement is favored, with crustal contamination of basaltic magmas being responsible for the continental tholeiitic character of the basic end member (Yoder, 1976).

Thin, laterally impersistent arenite, dolomitic shale, and chert units with local domical and columnar stromatolites are intercalated with the volcanics of both groups (Button, 1973b; Tyler, 1979b). Desiccation cracks and intraformational edgewise conglomerates indicate shallow-water sedimentation. Abundant nutrients in ephemeral lakes around dormant volcanoes probably accelerated growth of blue-green algae.

Shallow-Marine Progradation and Reworking

Paleotopographic irregularities had been largely smoothed out by the time of accumulation of the Selati Formation, a succession of interbedded quartzofeldspathic arenites, siltstones, and shales (Fig. 5-2A).

The Selati arenites form lenses up to 100 m thick, which pinch out basinward into thick siltstones and shales (Fig. 5-4). The argillaceous material is mainly horizontally laminated; where associated with arenites the shales contain desiccation structures and sporadic halite hoppers (Button, 1973c). The sediments are typically upward coarsening, with giant cross-beds in sets up to 30 m thick in the upper part. These characteristics and the lobate geometry of the arenites (Fig. 5-4) suggest a deltaic origin (Button, 1973c). Paleocurrent patterns indicate deltaic progradation to the south, southwest, and west. The large-scale foresets are attributed to processes on Gilbertian deltas, which form under conditions of three-dimensional mixing of fluvial and basin waters of equal density. The fine-grained clastics are believed to be mainly prodeltaic, although sporadic shallow-water structures in the shales suggest local accumulation in interlobe bays or tidal flats.

Thin arenites in the Hampton Formation and the Kransberg quartzite were probably also deposited by shallow-marine processes. Both quartzites vary in composition from quartz wackes to quartz arenites, depending on the proportion of sericitic matrix. The Hampton arenite tends to be texturally mature; the Kransberg quartzite, in contrast, has a bimodal texture with quartz grains that are coarse and rounded or fine and angular. Plane-bedding with low-angle discordances, probably of wave-swash origin, is the dominant sedimentary structure in both suites, whereas primary current lineation and centimeter-scale graded beds are common in the Hampton arenites. Planar and trough cross-bedding are common; planar sets with abundant shale drapes in the Kransberg Formation reach 1.5 m in thickness. The characteristic bimodal–bipolar slipface orientations are mainly to the northeast and southwest, with a third mode to the southeast in the Kransberg quartzite. Shore-parallel tidal reworking on a shallow shelf is thought to have been responsible for these patterns (Tyler, 1979b). The associated shale drapes probably reflect periods of tidal deposition in slack water.

5.2. Inundation of the Kaapvaal Province

In the Wolkberg–Buffalo Springs protobasin (Fig. 5-1) the widespread Black Reef quartzite rests conformably on shales of the Selati Formation and the Kransberg quartzite, reaching more than 500 m in thickness above the depocenter (Button, 1973c). Elsewhere the Black Reef covers a regional unconformity, overstepping basement granites and greenstones, Witwatersrand strata, and Ventersdorp volcanics. Its stratigraphic equivalent in the northern Cape, the Vryburg Formation, thickens southward to more than 200 m (Beukes, 1978a).

Both fluvial and shallow-marine facies are recognized in the Black Reef–Vryburg complexes. Flu-

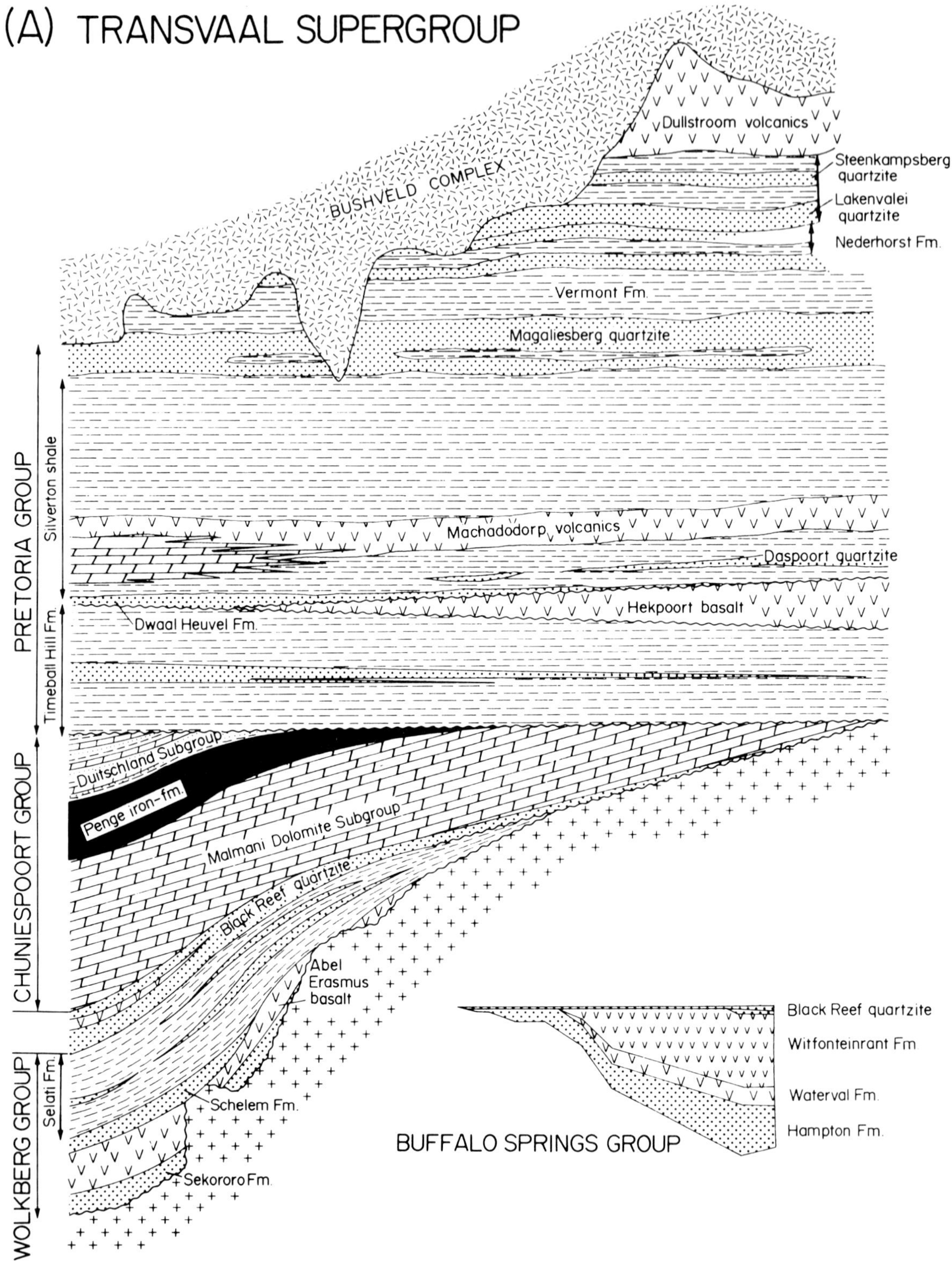

Figure 5-2. Regional stratigraphic relationships within (A) the Transvaal and (B) Griqualand West Supergroups. [After Button (1976a, b), and Tyler (1979b).]

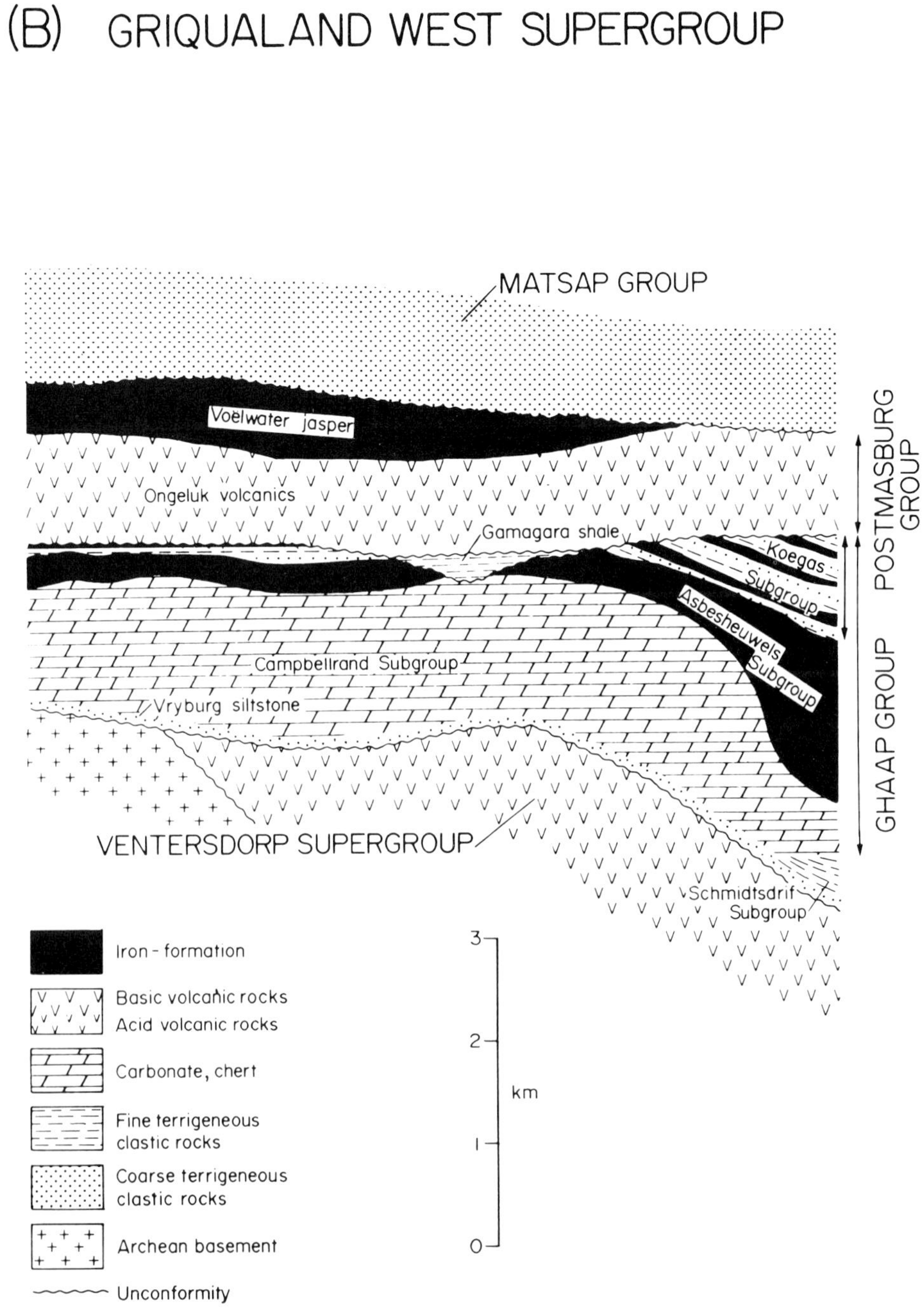

vial conglomerates and arenites are only present above the regional unconformity, where their composition and thickness reflect the lithology of the underlying formations. Shallow-marine sediments transgressively overlie the fluvial facies, overlapping older stratigraphic units over topographic highs and forming an extensive sheet sandstone (Button, 1973c). The Black Reef quartzite and Vryburg Formation fine upward into interbedded arenites and shales before grading into the overlying Chuniespoort and Ghaap Groups (Fig. 5-2A and B).

Basin-Edge Fluvial Influx

Valley-fill conglomerates up to 15 m thick overlie the basal unconformity. Polymictic varieties predominate, the clasts varying in size from peb-

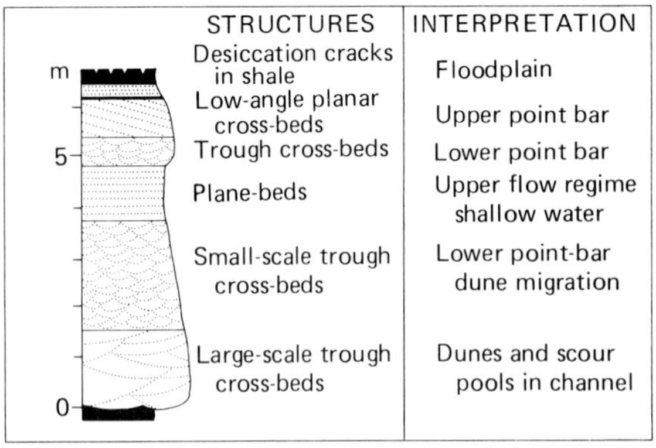

Figure 5-3. Vertical sequence of lithologies and sedimentary structures attributed to meandering river processes in the Buffalo Springs Group. [After Tyler (1979c).]

bles to boulders as large as 2.5 m in diameter. Imbrication and grading are common. Thin lenses of plane-bedded sandstone are intercalated with the conglomerates. These two rock types are attributed to accretion of longitudinal gravel bars during high discharge and deposition of bar-top sands during low discharge (cf. Miall, 1977).

Auriferous conglomerates, mapped in deep mines east of Johannesburg, occupy a north–south system of anastomosing channels (Fig. 5-6A) (Papenfus, 1964). Provenance studies and the southward-fining textural gradient suggest transportation by southward-flowing streams. The sediment, including gold, was derived from the underlying Witwatersrand Supergroup. Outcrops of similar channel-fill conglomerates along the south-

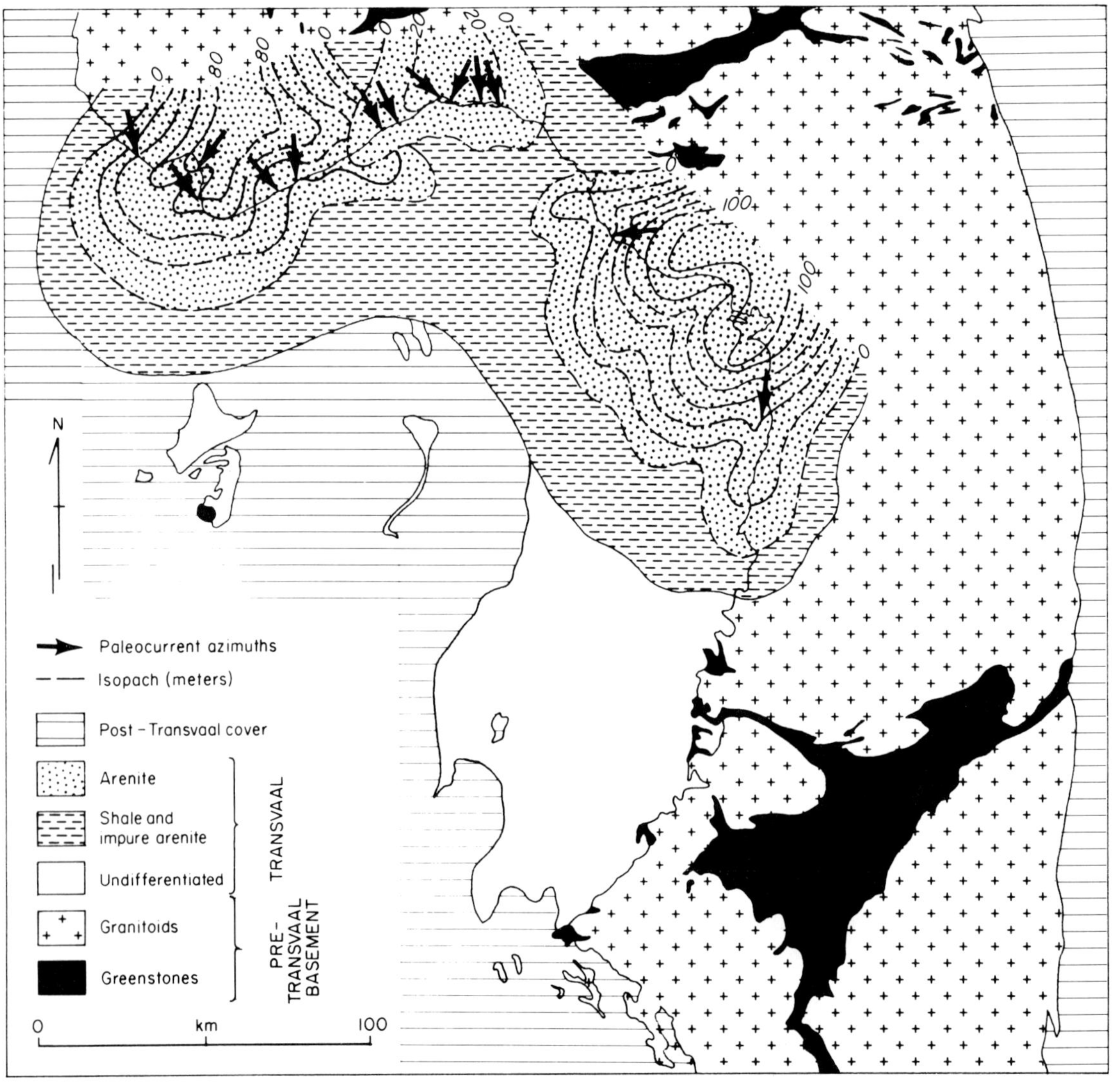

Figure 5-4. Deltaic sand lobes and paleocurrents, Selati Formation. [After Button (1973c).]

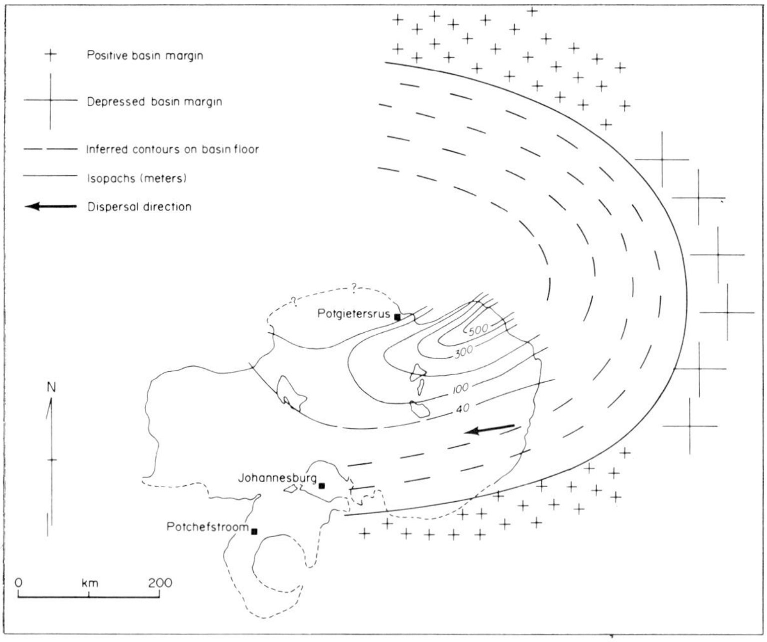

Figure 5-5. Regional and tectonic sedimentary setting of the Black Reef quartzite. Also shown are isopachs and general dispersal direction of shelf sands in the eastern Transvaal. [After Button (1973c).]

eastern Transvaal escarpment have also been prospected for gold and uranium.

Upward-fining fluvial sequences, consisting predominantly of arkosic sandstones (Fig. 5-6B), either overlie the thick basal conglomerates or rest directly on the unconformity. Channel lag conglomerates consisting of small pebbles are overlain by linguoid channel-bar deposits, dominated by planar cross-beds, and trough cross-beds formed by aggradation. Associated plane-bedded sandstones may indicate bar-top deposition in an upper flow regime. Mud drapes accumulated on abandoned reaches of the floodplain. Upward-fining arkosic sequences are also present in the southeastern Transvaal and northern Cape. Paleoflow was directed to the south in the western Transvaal and to the northwest and west in the eastern Transvaal. In the northern Cape fluvial transport was to the south (Beukes, 1977).

Shallow-Shelf Reworking

Mature quartz arenites with local thin conglomerates transgressively overlie the Black Reef braided fluvial deposits. The dominant sedimentary structure is plane-bedding with low-angle discordances, followed by planar cross-beds in sets up to 1 m thick, and less common trough cross-beds and ripple cross-lamination. Paleocurrents were directed to the west in the eastern Transvaal and to the north in the western Transvaal (Fig. 5-5).

A shallow tidal shelf environment has been inferred for the eastern Transvaal quartz arenites on the basis of the sandstone maturity, associations of sedimentary structures, and characteristic paleoflow patterns (Button, 1973c). Button attributed the unimodal paleocurrents to a clockwise-circulating tidal system in a large depository, the northern half of which has subsequently been removed by erosion (Fig. 5-5). In this context the ubiquitous plane-bedding is explained as stoss-side accretion on long-wavelength sand waves. Similar processes are believed to have prevailed in the western Transvaal where northerly tidal currents were ultimately deflected eastward. Tidal circulation patterns of this type are typical of many shallow shelves. On the northwest shelf of Australia bedforms migrate eastward in response to flood currents; ebb flow is confined to deeper waters (Jones, 1973). A rotating tidal system also dominates the North Sea, producing asymmetric bedforms (Houbolt, 1968).

The Black Reef quartz arenites grade upward into wackes and shales which, in the northwestern and central Transvaal, probably accumulated below tidal wave base following continued transgression of the shoreline (Truswell and Eriksson, 1975; Tyler, 1979d).

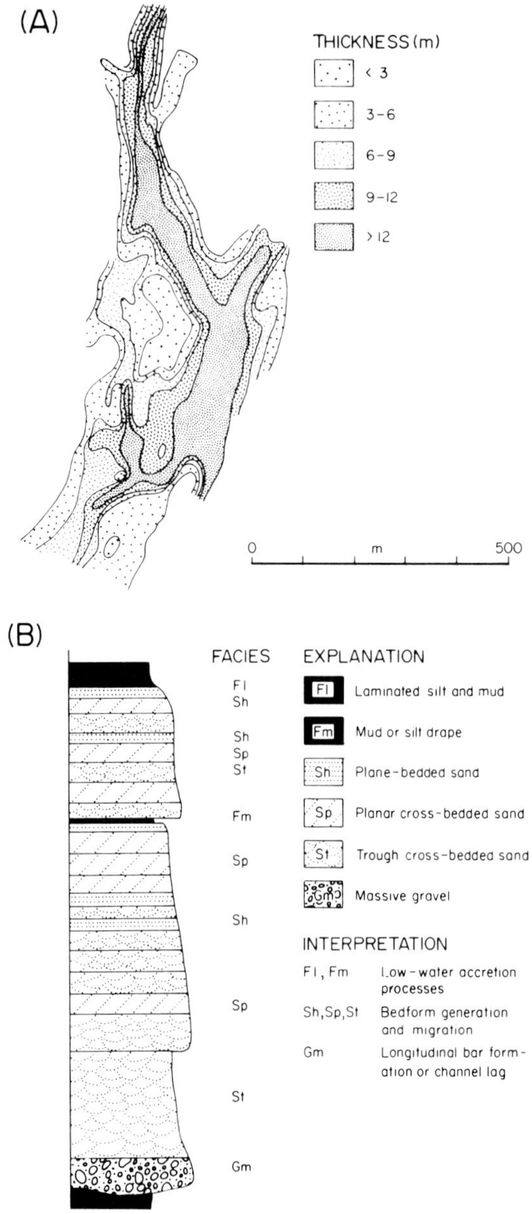

Figure 5-6. Fluvial sediments in the Black Reef Formation: (A) Isopach map based on data from underground mining at Government Gold Mining Areas to the east of Johannesburg. [After Papenfus (1964).] (B) Detailed measured section through braided river deposits in the northwestern Transvaal. [After Tyler (1979d).]

Tidal-Flat Environments

In the eastern Transvaal, tidal-flat wackes and shales overlie the Black Reef shelf quartz arenites. This facies is distinguished from the distal shelf deposits in the northwestern and central Transvaal by structures indicative of emergence, such as desiccation cracks and ripple forms modified by ebb runoff.

The fluvial arkoses of the Vryburg Formation in the northern Cape are succeeded by siliciclastic and mixed siliciclastic–carbonate facies (Fig. 5-7) (Beukes, 1977). Quartz arenites and shales in the lower siliciclastic units are arranged in upward-fining sequences 2–4 m thick, which are attributed to tidal flat progradation across shallow sub-tidal depositional environments (Beukes, 1977). Trough cross-beds in quartz arenites formed in shallow subtidal and lower tidal flats grade vertically into plane-bedding and herring-bone cross-bedding with shale flasers. Polymodal paleocurrent patterns indicate considerable dispersal of flow directions. An arenite-shale interval formed on midtidal flats contains lenticular and wavy bedding and ladderback, interference, double-crested, and flat-topped ripple varieties. The tidal sequences are usually terminated by shales of upper tidal flats.

Two subfacies are recognized in the siliciclastic–carbonate rocks. The lower of these averages 50 cm in thickness and consists of indistinct wavy- and flaser-bedded quartz arenites and wavy- and lenticular-bedded shales and arenites that display abundant evidence of exposure. The upper subfacies, which averages 2 m in thickness and consists of dolomitic arenites with shale clasts, rests with a sharp contact on the underlying quartz arenites. These herringbone cross-bedded dolarenites grade vertically into stromatolitic dolomites and dolomitic shales. A tidal-flat interpretation has been proposed for these subfacies (Beukes, 1977). The siliciclastic–carbonate cycles may have been a response to fluctuating salinities and variable rates of terrestrial sediment supply. During episodes of reduced terrestrial influx and metahaline conditions, carbonate production and algal activity were probably enhanced, resulting in progradation of the tidal flat. Blanketing by terrestrial sediment terminated the carbonate deposition.

The thicknesses of the siliciclastic and carbonate tidal flat sequences probably approximate the paleotidal range. Values close to 2 m for the siliciclastic units and of 1.0–1.5 m for the carbonate units suggest a microtidal regime. Vertical stacking of the tidal facies is a reflection of changes in relative sea level, which Beukes (1977) has attributed to faulting.

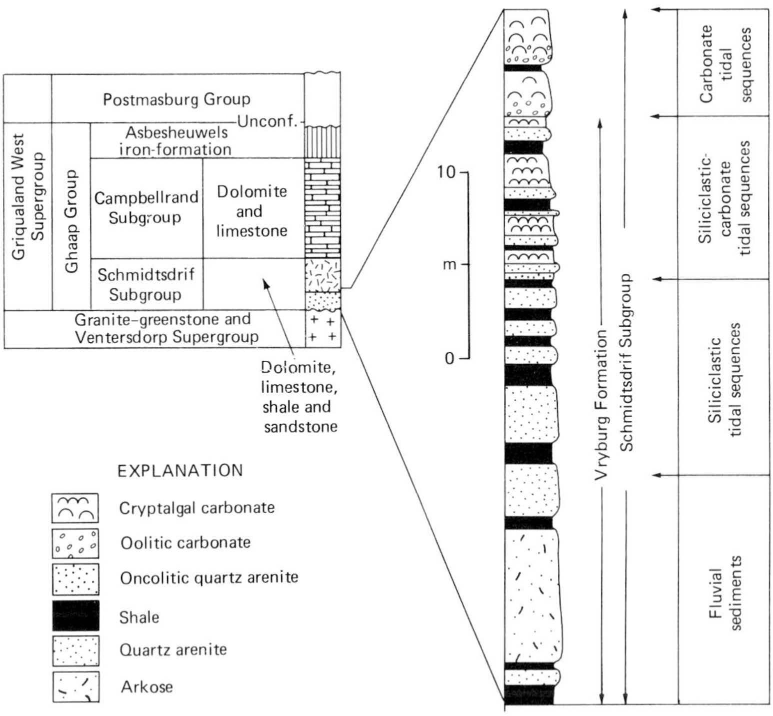

Figure 5-7. Tidal depositional sequences in the Schmidtsdrif Subgroup near Vryburg in the northern Cape. [After Beukes (1977).]

Volcanic tuff time markers show that west of Vryburg the siliciclastic sediments grade southward into mixed siliciclastic–carbonate and carbonate facies (Beukes, 1977). These transitions are analogous to lithologic changes along modern depositional coasts, such as the region between the Appalachicola Delta and the Florida Keys in the United States (Matthews, 1974). On a smaller scale, terrestrial sediment supply from the Gascoyne and Wooramel Deltas of Shark Bay, Western Australia, suppresses carbonate production (Logan *et al.*, 1970).

5.3. Sedimentation in a Clear-water Epeiric Sea

The Chuniespoort and Ghaap Groups consist predominantly of limestone, dolomite, chert, and iron-formation; terrigenous clastics comprise less than 5 percent. Stratigraphic relationships are complex (Fig. 5-2). The Schmidtsdrif Subgroup of the northern Cape, which includes the Vryburg Formation, correlates with the Black Reef quartzite in the Transvaal. Recrystallized dolomites with chert, dark iron-poor dolomites, and clastic-textured and ankeritic dolomites with limestones constitute the stratigraphically equivalent Campbellrand and Malmani Subgroups. Iron-formation constitutes the Asbesheuwels Subgroup and Penge iron-formation, whereas the Koegas and Duitschland Subgroups consist of mixed terrigenous clastic and chemical sediments (Figs. 5-2A and B).

Three intraformational disconformities interrupt the Malmani Dolomite Subgroup. They are accentuated by angular chert fragments in a shale matrix and are overlain by thin shale and occasional quartz arenite beds. A regional angular unconformity, also marked by a chert-shale breccia, separates the Chuniespoort and Pretoria Groups; it truncates the Chuniespoort succession to progressively lower stratigraphic levels from north to south (Figs. 5-2A and B, 5-8). The chert breccia is thickest above eroded chert-rich zones. These chert breccias represent insoluble residues of carbonate dissolution on subaerial erosion surfaces; a paleokarst terrain is common beneath the unconformity (Button, 1973d). A regional unconformity also separates the Ghaap and Postmasburg Groups. Mixed carbonate-chert formations are rarely truncated by this erosion surface so that chert-shale breccias are rare. Chert-shale breccias are present within the Ghaap Group, however (Fig. 5-8).

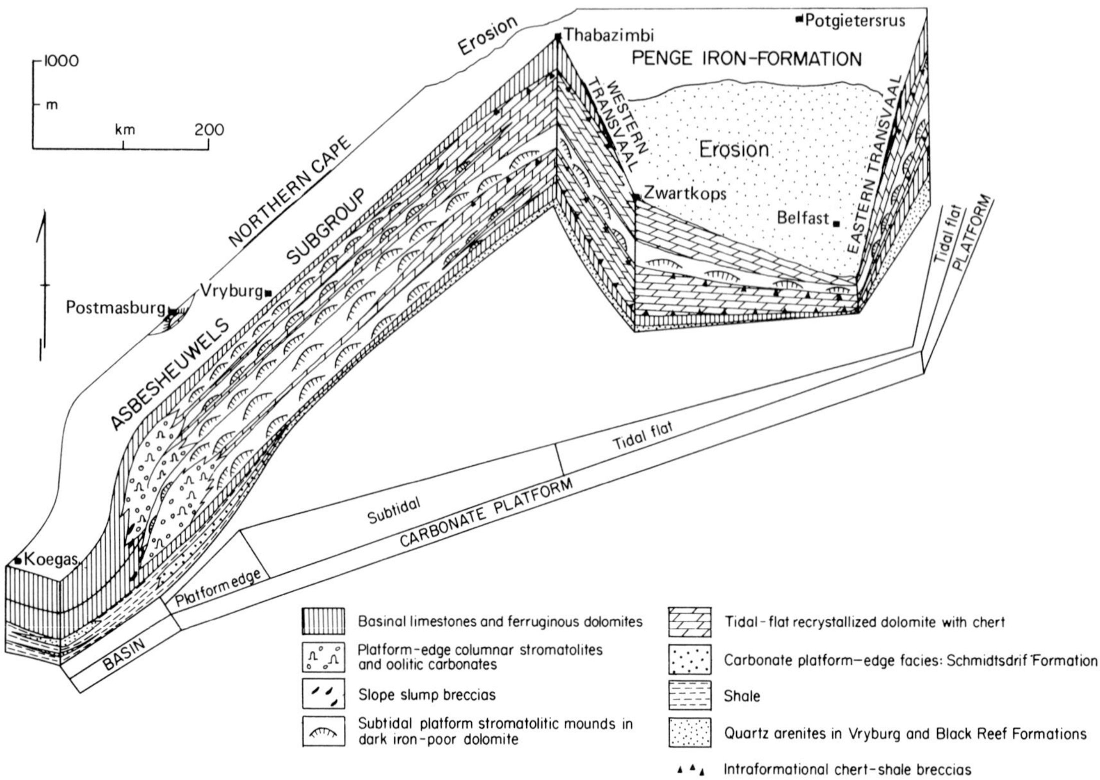

Figure 5-8. Cross section showing facies and depositional environments in the Campbellrand–Schmidtsdrif Subgroups of the norther Cape and Malmani Subgroup of the Transvaal. [After Beukes (1978a).]

Platform Setting

Tidal-flat environments. Widely distributed tidal-flat facies in the Chuniespoort and Ghaap Groups are recognized on the basis of stromatolite types, primary sedimentary structures, and paleocurrent trends.

Processes typical of shallow subtidal bars and tidal flats dominated much of lower Schmidtsdrif accumulation in the northern Cape (Fig. 5-7). The subtidal-bar facies consists of oolitic quartz arenites, quartzose oolitic dolomites, and oolitic dolomites. Cryptalgal dolomites characterize the tidal-flat sequences. In the lower parts the stromatolitic domes are markedly elongate but become more linked and decrease in relief and elongation upward. Oolites between the domes are herringbone cross-bedded. These facies sequences are invariably capped with dolomitic argillite. The Schmidtsdrif-Campbellrand transition occurs through an interval of upward-fining sequences (Beukes, 1977).

The thick recrystallized dolomite and chert intervals in the Malmani Subgroup and similar, but thinner, units in the Campbellrand Subgroup (Fig. 5-8) are attributed solely to processes on tidal flats (Eriksson and Truswell, 1974; Truswell and Eriksson, 1975; Eriksson, 1977b). Three facies are recognized. An upper intertidal and supratidal facies is characterized by the presence of flat-laminated algal stromatolites. Some of the stromatolites are entirely horizontal, whereas others contain small domes and tufts. This variation in morphology is probably a response to the growth habits of different algal species (Monty, 1972; Logan *et al.*, 1974). Associated structures include layers of edgewise conglomerate and lenses formed by reworking of desiccated algal mats. Interstices in the conglomerates are filled with coarse sparry carbonate containing swallow-tail remnants of sulfates. Fibrous, bladed dolomite crystals, believed to be evaporite pseudomorphs, are commonly developed perpendicular to the lamination. Similar structures, with gypsum still preserved, occur in the Coorong Lagoon in South Australia (C. Von der Borch, personal communication, 1979). Further evidence of evaporites is provided by fan-shaped growth forms with an internal concentric structure.

Tepee structures, typical of many evaporitic tidal flats, are also common in this facies. These form by various mechanisms, including repeated thermal expansion and contraction, desiccation, and displacive crystallization of gypsum (Horodyski and Vonder Haar, 1975). Deformed and overturned algal mats are also present; on Holocene tidal flats they are produced by storm reworking of desiccated mats.

In vertical sequence, relief of the stromatolites, and especially the domes, increases to a maximum of 1 m in response to increasing high-tide water depths on the low and midtidal flat (Eriksson, 1977b). The larger domes are markedly elongate; similar elongate structures in Shark Bay result from scouring during tidal retreat (Logan *et al.*, 1974). Ubiquitous edgewise breccias between domes provide further evidence of strong tidal currents. Oversteepening of the larger domes to the northwest is conspicuous, and, by analogy with Shark Bay (Logan *et al.*, 1974) and other occurrences in the geologic record (Hoffman, 1967), is an indication of the offshore direction.

A third tidal-flat facies in the recrystallized dolomite and chert zones comprises ripple-drift cross-laminated carbonate sands. The ripples form in 15-cm-thick sedimentation units which grade from a basal edgewise breccia through type A and type B to in-phase ripple-drift cross-lamination. These patterns suggest deposition under waning energy conditions with a high proportion of suspension to bedload transport (Jopling and Walker, 1968). Centimeter-thick intervals of markedly elongate, small domical stromatolites cap the units and probably indicate critical conditions of velocity and sediment supply which enabled algal binding to take place. Shinn (1973a) has shown that sheetlike sands, similar to the rippled units in this facies, can form by lateral migration of tidal channels.

Low relief and domical stromatolites are normally environmentally separated at Shark Bay and grow in protected embayments and semiprotected bights, respectively (Logan *et al.*, 1974). However, both stromatolite varieties are commonly present on the same tidal flat, with flat-laminated forms grading basinward into domical forms in response to increasing wave and tidal energies. This situation has significance with respect to the stacked stromatolitic tidal-flat facies in the Malmani dolomite. Above the middle intraformational unconformity (Fig. 5-8) both transgressive and progradational lower and mid-tidal-flat domical stromatolite and upper intertidal–supratidal stromatolite sequences are present. Rippled carbonate sands of probable tidal channel origin occur at various levels.

Subtidal environments. Subtidal facies constitute the dark, Fe-poor dolomites in both the Transvaal and northern Cape (Fig. 5-8). This remarkably persistent facies is characterized by large, elongate stromatolitic mounds which range in width from 5 to 20 m and are up to 40 m long (Truswell and Eriksson, 1973, 1975). The mounds have rounded crests and V-shaped troughs and are either overlapping or vertically stacked. Lamina relief is greater than 3 m and inheritances as much as 17 m. These characteristics and the absence of exposure-related features and internal erosion surfaces suggest that the mounds accumulated in subtidal environments below wave base. The elongation direction of the mounds is locally consistent, but regionally the trends vary from east–west to northwest–southeast. The nature of the transition from tidal flat to subtidal varies basinwide in response to hydrographic regime, availability of terrigenous sediment, and gradient. In the Zwartkops area Black Reef tidal-flat deposits are overlain transgressively by dark Fe-poor dolomites with stromatolitic mounds (Fig. 5-8). The Fe-poor dolomite zone separating the two recrystallized dolomite and chert zones in the Transvaal consists entirely of stromatolitic mounds. In the eastern Transvaal elongate tidal flat domes increase in size vertically and grade into the subtidal mound facies (Eriksson, 1977b). No shallow subtidal high-energy facies of the type suggested in the Irwin (1965) model is present at the tidal flat–subtidal interface. This may result from low-gradient dissipation of wave energy in the Transvaal basin. At Zwartkops a progradational tidal-flat sequence is separated from the overlying subtidal mounds by a coarse breccia consisting of dolomite and chert blocks in a carbonate matrix. The breccia has been interpreted as a talus deposit formed in front of a prograding tidal-flat slope (Truswell and Eriksson, 1975).

Platform edge. The only known occurrence in the Chuniespoort or Ghaap Groups of a high-energy facies is along the southern platform edge in the northern Cape where elongate columnar stromatolites and associated clastic-textured carbonates are developed (Fig. 5-8) (Beukes,

1978a). Structures indicative of wave activity are common along the southern margins of the Schmidtsdrif and Campbellrand Subgroups and include oolites that have been reworked into symmetrical megaripples, oncolites, edgewise ripup breccias, and current ripples. The elongate stromatolitic mounds increase in size basinward where the water is deeper. With increasing depth and decreasing bed shear, the minor structures within the domes also change from small, discrete columns to linked columns and laterally continuous algal mats (Truswell and Eriksson, 1973).

Stromatolites associated with tidal-flat facies in the northern Cape reflect more intense energy conditions than their Transvaal counterparts, probably because of steeper shore-zone gradients (Truswell and Eriksson, 1973). Columnar stromatolites adjacent to this high-energy facies are discrete and unlinked, containing oolites and intraclasts as intercolumn fill (Fig. 5-9A). Linkage by intercolumn bridging increases upward; multiple columns are elongated east–west. The tidal-flat sequences are generally terminated by low-relief symmetrical domes approximately 20 cm in diameter. These sequences and the vertical stacking of stromatolite types are attibuted to shore-zone progradation (cf. Logan, 1961). In Shark Bay columnar stromatolites (Fig. 5-9B) are most common on exposed headlands where intense tidal and wave activity on the low tidal flat precludes bridging between columns (Logan *et al.*, 1974). Shoreward-decreasing bed shear is reflected in increasing linkage and elongation perpendicular to the shoreline. Symmetrical domes on the Shark Bay headlands form on the upper intertidal and supratidal flats where bed shear is negligible.

The tidal-flat columnar stromatolites are of the Pilbara variety (Bertrand-Sarfati and Eriksson, 1977). Morphologically similar forms are present in the Duck Creek dolomite in the Mount Bruce Supergroup of Western Australia (Walter, 1972a), which is of equivalent age to the Griqualand West Supergroup (Button, 1976a).

Basinal Environments

Basinal environments, including platform slope, toe of slope, and deep basin, are well developed in the Koegas area of the northern Cape, but are also recognized above the Black Reef quartzite and in the upper parts of the Campbellrand and Malmani Subgroups (Fig. 5-8). Slope deposits south of the platform-edge facies in the northern Cape comprise slump breccias with associated U-shaped and conical stromatolites. The toe-of-slope lithologies are characterized by graded carbonate turbidites and algal debris, alternating with pelagic shales and basic tuffs. These interfinger in the deep basin with carbonaceous and pyritic shales, thick cherts, and limestones which are locally ankeritized or replaced by chert to form protobanded iron-formation. In contrast to the generally alkaline platform environment, the deep basin waters were probably mildly acidic, thus favoring precipitation of chert (Beukes, 1978a). Reducing conditions are indicated by the sulfide-facies iron-formations, namely the pyritic shales and the ankeritized limestones.

Iron-Formation Paleoenvironments

The Asbesheuwels Subgroup in the Prieska area is more than 1000 m thick and is dominated by iron-formations that thin toward the center of the Kaapvaal Province. Stratigraphic units are traceable for a distance of 1000 km between the Asbesheuwels Subgroup and its metamorphosed equivalent, the Penge iron-formation in the Transvaal. Depositional facies in the iron-formation resemble those of the Campbellrand and Malmani Subgroups. These facies comprise an upward-shoaling sequence that developed in response to westerly and northwesterly progradation across the carbonate platform following a major marine transgression (Beukes, 1978a).

Basinal iron-formations consist of stacked cycles:

Microbanded hematite and magnetite containing pods of chert, each ferruginous band being less than 1 cm thick

↑

Mesobanded siderite, magnetite, and chert

↑

Siderite with mesobands of chert greater than 1 cm thick

↑

Chert

↑

Stilpnomelane

(A)

(B)

Figure 5-9. (A), Platform-edge columnar stromatolites in the Campbellrand Subgroup. Note the partial bridging between the columns through the dark intercolumn fill. Scale in centimeters. [From Macgregor *et al.* (1974). Reprinted by permission from *Nature* **247**, No. 5442, pp. 538–540. Copyright © 1974 Macmillan Journals Ltd.] (B) Stromatolites on exposed headlands in Shark Bay, Australia.

Many of the stilpnomelane layers contain well-preserved shard structures (La Berge, 1966; Fockema, 1967), suggesting a volcanic source for the stilpnomelane. The chemistry also supports a volcanic derivation; compared with iron-formation, the stilpnomelane has high aluminum, magnesium, and potassium contents (Table 5-1). A similar interpretation for the overlying chert is unlikely because a nonterrigenous Precambrian sea would have been supersaturated with silica. The influence of volcanism decreased and chemical effects increased upward in the cycles, resulting in the precipitation of the different iron-bearing minerals; the magnetite was probably diagenetic (Beukes, 1978a). Iron-formation slope facies comprise microbanded greenalite and siderite (rhythmite) with sporadic allochemical bands, probably formed by grain flow.

Table 5–1. Chemical Composition of Iron-Formation and Associated Rock Types in the Kuruman Iron-Formation and the Koegas Formation[a]

	1	2	3	4
SiO_2	40.71	51.50	50.18	40.40
TiO_2	0.12	0.09	0.14	0.10
Al_2O_3	2.32	1.97	2.58	9.90
Fe_2O_3	21.04	1.68	17.43	3.90
FeO	20.10	17.19	15.51	26.90
MnO	0.29	0.29	0.09	0.40
MgO	3.15	3.39	3.87	7.80
CaO	2.63	5.93	0.92	0.10
Na_2O	1.05	0.45	5.25	0.10
K_2O	0.57	0.22	0.05	7.20
P_2O_5	0.25	0.14	0.14	0.03
CO_2	6.56	16.64	0.36	n.d.
H_2O^+	0.91	0.59	3.03	4.68
H_2O^-	0.07	0.04	0.43	0.46
Total	99.77	100.12	99.98	101.97

[a]Average of 9 analyses of iron-formation, Kuruman iron-formation (Hanekom, 1966). 2, Average of 2 analyses of banded ferruginous chert, Kuruman iron-formation (Hanekom, 1966). 3, Average of 9 analyses of crocidolite, Kuruman iron-formation (Hanekom, 1966). 4, Ferrostilpnomelane, Kuruman iron-formation (La Berge, 1966).

In contrast to the basinal and slope iron-formations, those formed on the platform display considerable evidence of reworking. Representatives of the Irwin (1965) high-energy zone are numerous and include massive siderite with grains (intraclasts, pellets, and oolites) and disk-shaped clasts of chert, and chert grainstones. Massive iron oxide, iron carbonate or iron silicate bands enclosing the reworked facies are attributed to accumulation under low-energy conditions. These different facies are vertically arranged in progradational platform cycles (Beukes, 1978a). Storm deposits are locally present in the overlying supratidal flat facies. The sodium silicate, riebeckite, is commonly associated with the iron–silicate zones of the basinal, slope, and particularly the platform facies, where it occurs as the asbestiform crocidolite. The sodium may be of fumarolic origin in the basinal and slope environments, but on the platform it was probably concentrated by evaporation.

The Asbesheuwels Subgroup is capped by a massive greenalite interval containing bands of chert but is devoid of riebeckite. A freshwater lacustrine environment of deposition is inferred. Finally, influx of terrigenous clastics, including thuringite clays, occurred from the southeast, forming the Koegas and Duitschland Formations (Fig. 5-2A and B).

Origin of Iron-Formations. The origin of iron-formations has been a subject of considerable debate for decades between protagonists of a direct precipitation mechanism and those favoring a replacement origin. The source of the iron has been attributed to either volcanic or sedimentary processes. It is generally believed that in the absence of secreting organisms 2300 Ma ago the hydrosphere would have been saturated with respect to silica.

The replacement origin for iron-formations, proposed by Dimroth (1977) largely on the basis of clastic textures in the Labrador trough iron-formations, requires the introduction of unrealistically large quantities of iron and silica at the expense of calcium carbonate. It is now generally accepted that iron-formations form by direct precipitation, leaving only the source of iron unresolved. Despite the presence of shardlike structures in the stilpnomelane-rich units in the Asbesheuwels Subgroup and Penge iron-formation, an exclusively volcanogenic origin for the iron is unlikely. The shardlike structures constitute an insignificant proportion of the total stratigraphic interval and are completely absent from the upper half of these subgroups. Furthermore modern volcanoes do not release such large quantities of iron and silica to the exclusion of other metals (Holland, 1973).

Alternatively, the iron could have been derived by subaerial weathering and transportation under oxygen-deficient conditions. Holland (1973) refuted this suggestion because of the inordinately large volume of Fe-bearing river water required. However, over a long period of time, say 60 million years, the oceans could have established a substantial reservoir of Fe^{2+}, so that a continental source is not impossible. Furthermore, all the metals that would have been released under conditions of deep chemical weathering from a Ventersdorp-type parent occur in the Chuniespoort-Ghaap basin, including Ca, Mg, Si, Na, Fe, and Mn. The only insoluble product of source-area weathering would have been aluminum, which is concentrated in pyrophyllite and diaspore-bearing shales in the overlying Postmasburg Group (Button, 1976a).

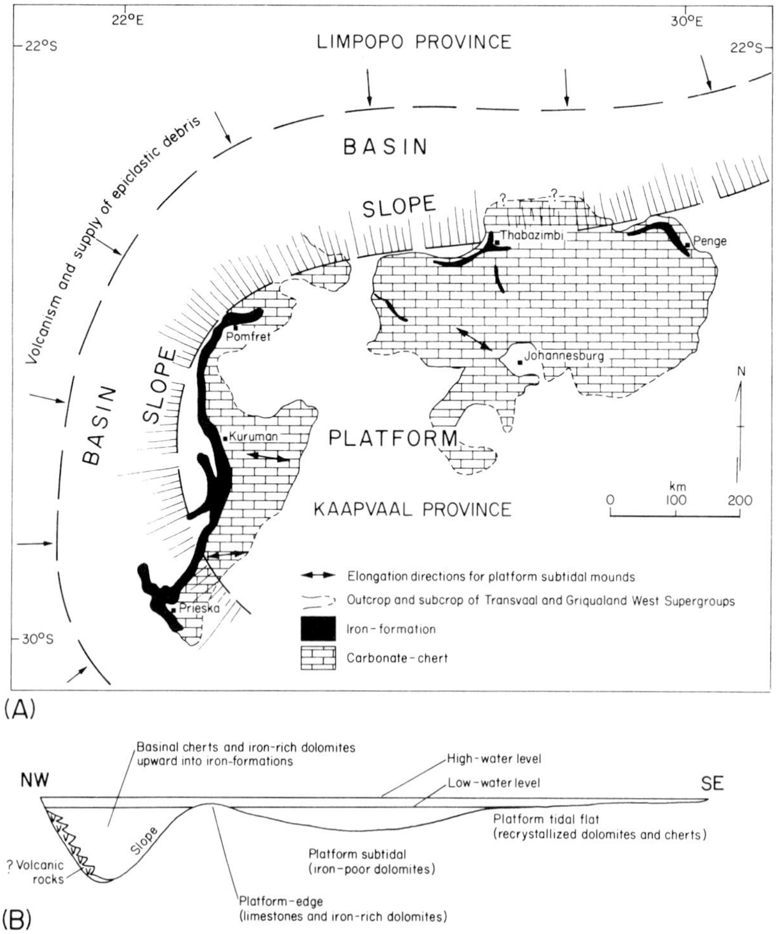

Figure 5-10. (A) Tectonic setting of the Ghaap and Chuniespoort Groups showing the inferred location of the various environments. [After Beukes (1978a).] (B) Paleoenvironmental reconstruction of the Ghaap and Chuniespoort Groups. [After Eriksson *et al.* (1976) and Beukes (1978a).]

Banding in the iron-formations probably resulted from cyclically fluctuating chemical conditions in a hydrosphere saturated with iron and silica. The necessary chemical changes for mesobanding were possibly seasonally controlled, involving arid high-pH and humid low-pH conditions (Eugster, 1969). Microbanding is less common and may have been produced by the photosynthetic activity of blue-green algae and bacteria, which are capable of locally changing diurnal pH by as much as 2 units (Walter, 1972b); microfossils have been isolated from the Kuruman iron-formation by Pflug (quoted in Fockema, 1967), Cloud and Licari (1968), and La Berge (1973). The stratigraphic changes in iron mineralogy are attributed to the variation in carbon content of the basinal water with depth. Drever (1974) maintained that if there were an excess of Fe^{3+} over available C, magnetite or hematite would result. If Fe^{3+} and C were approximately equal, Fe^{2+} silicates would form, and if C greatly predominated, siderite was produced.

Paleoenvironmental and Diagenetic Synthesis

Five lithologic associations have been discussed in the preceding sections: iron-formations, clastic-textured carbonates, limestones and ferruginous dolomites, dark Fe-poor dolomite, and recrystallized dolomite and chert. These associations can be related in a paleogeographic model that distinguishes a relatively active northwestern and western basin margin from a relatively passive southeastern one (Fig. 5-10). Basin-margin activity controlled supply of terrigenous clastic sediment as well as influx of metals in solution (Eriksson *et al.*, 1976).

Tidal-flat and subtidal environments characterized the low-gradient platform on the southeastern basin margin (Fig. 5-10). The succession consists of stacked progradational carbonate sequences with erosional contacts. Subtidal mounds and intertidal domical stromatolites are elongated radially toward the western, northwestern, and northern edges of the platform; intertidal domes are usually oversteepened in the same direction. Two generations of tidal-flat dolomitization are ascribed to schizolhaline conditions (cf. Folk and Land, 1975). Palisade pseudomorphs, probably after gypsum, indicate hypersalinity, which resulted in penecontemporaneous development of micritic dolomites. Later, coarse-grained limpid dolomites and secondary cherts formed in response to influxes of low-pH fresh water. The subtidal stromatolitic mounds comprise pure, fine-grained dolomites devoid of chert. These dolomites may be either primary or could have formed through reaction with sea water containing high concentrations of magnesium and sodium. Under these conditions sodium is preferentially hydrated, thereby releasing magnesium for dolomitization (Gaines, 1974). Increasing alkalinity, promoted by algal decay and release of urea, may also enhance dolomitization (Mansfield, 1979). In the absence of grazing metazoa in the Precambrian, subtidal stromatolites could have grown prolifically at least to the bottom of the photic zone.

Clastic-textured carbonates and associated stromatolites developed periodically on the wave-agitated platform edge. In the northern Cape the columnar stromatolites probably occupied exposed headlands. At times the platform-edge facies were poorly developed, exposing the platform to tidal currents that resulted in marked elongation of the domical stromatolites (Beukes, 1978a). Basin depths as great as 400 m were associated with a trough, possibly enhanced by faulting, which formed along the active basin margin (Fig. 5-10). Slope, toe-of-slope, and basinal limestones and ankeritic dolomites and basinal cherts and shales accumulated under both acidic and reducing conditions (Beukes, 1978a). The ankeritic dolomites are secondary and contain up to 10 percent FeO and 4 percent MnO in the carbonate lattice. Intensity of ankeritization decreases toward the platform facies, suggesting that the deep basin was the source of refluxing magnesium–iron–manganese brines.

The iron-formations may have formed contemporaneously with some of the carbonates in the deep basin (Fig. 5-10), but subsequent transgression extended the iron-formation facies across the platform as well. By this time carbonate sedimentation had been suppressed by a cooling climate (Beukes, 1978a). Progressive shallowing of the depository resulted in the sequential accumulation of toe-of-slope, slope, platform-edge, platform lagoon, supratidal, and lacustrine banded iron-formations. Late in the history of the Chuniespoort and Ghaap Groups the southeastern margin of the basin was uplifted; westward and northwestward progradation of terrigenous clastics over iron-formations resulted in deposition of the Koegas and Duitschland Subgroups (Beukes, 1978a).

5.4. Renewed Terrigenous Influx and Progradation

Renewed progradation is reflected in the 7000-m-thick Pretoria Group in the Transvaal and the 2000-m-thick Postmasburg Group in the northern Cape (Button, 1976a). Lying above a regional unconformity these lithostratigraphic units consist predominantly of shales and quartz arenites with subordinate carbonates and volcanic rocks (Fig. 5-2A and B). Substantial thicknesses of iron-formations and ferruginous and manganiferous shales are also present in the northern Cape.

Disconformities accentuated by paleosol horizons interrupt these successions. In the Transvaal the Hekpoort basalt is bounded by erosional surfaces, but the equivalent Ongeluk lava in the northern Cape has an erosional surface developed only at its base (Fig. 5-2A and B). A major unconformity separates the Griqualand West Supergroup from the Matsap Group. The Transvaal Supergroup is terminated by the Rooiberg felsites, the initial phase of the Bushveld Complex, as is impressively shown at the Rooiberg tin mine (Fig. 5-1). In the eastern Transvaal the basic phase of the Bushveld intrusion truncates progressively lower stratigraphic units of the Pretoria Group (Fig. 5-2A), eventually coming into contact with the Malmani dolomite near Potgietersrus.

Regional Facies and Stratigraphic Relationships

The Postmasburg Group is regarded as a distal-facies equivalent of the Pretoria Group (Button, 1976a). Facies patterns in these two groups were controlled by uplift of the Limpopo source terrane. Coarse terrigenous clastics are abundant along the northern margin of the Pretoria basin and interfinger southward with finer grained terrigenous material and oolitic ironstones (Fig. 5-11), eventually grading into nonclastic rocks in the northern Cape. These facies relationships are exemplified by the Timeball Hill–Gamagara shale transition below the Hekpoort–Ongeluk Volcanic Suites and by the Dwaal Heuvel–Voëlwater transition above (Figs. 5-2A and B, 5-12A and B). Stratigraphic units in the upper parts of the Pretoria Group also show a regional intertonguing of arenite and shale (Button and Vos, 1977).

The regional facies distributions shown in Figure 5-12 are repeated vertically as thick upward-coarsening sequences of shale, siltstone, and sandstone in the Timeball Hill Formation and the Lydenberg shale–Magaliesberg quartzite of the Pretoria Group (Fig. 5-2A). Similarly the Gamagara shale in the northern Cape is an upward-coarsening succession but commences with iron-formation and is terminated by a thin reworked quartz arenite. Thin upward-fining sequences, some of which are capped with stromatolitic limestones, commonly overlie the quartz arenites; examples are the Vermont Formation above the Magaliesberg quartzite, and the Houtenbek Formation above the Steenkampsberg quartzite (Fig. 5-2A and B).

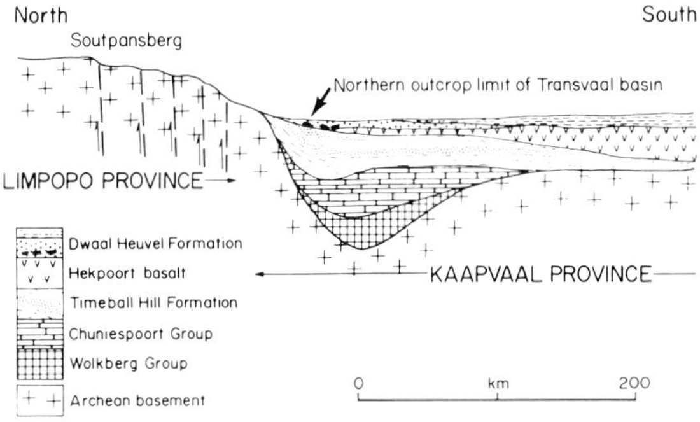

Figure 5-11. Schematic cross section showing the proposed relationship between the Transvaal basin and the Limpopo Province with the resulting facies changes away from the source area. [After Button (1975).]

Fluvial Influx

Fluvial deposits are rare in the Pretoria Group and absent from the Postmasburg Group. However, in places lenses of arkosic sandstone erosively overlie the Timeball Hill shales, immediately below the Hekpoort basalts. Channel-floor conglomerates are replaced vertically by the dominant arkoses, thin siltstones, and shales capping the sequences. The arkoses are structured by planar and trough cross-beds that display an upward decrease in set thickness. Ripple cross-lamination predominates in the siltstones, and the shales commonly display evidence of periodic exposure. Button (1973c) has attributed this pattern of sedimentation to fluvial point-bar processes.

Thick pebbly arkoses are also developed in the Nederhorst Formation in the eastern Transvaal and at a similar stratigraphic position in the Rooiberg area in the northern part of the Transvaal basin (Fig. 5-2A). Southward-directed, unimodal patterns of sediment transport and stacked, mutually truncating channels with basal pebble lags have been attributed to fluvial processes (Button and Vos, 1977; Stear, 1977). The general dearth of fluvial sediments is because the proximal facies of the Pretoria Group farther north were uplifted and reworked into younger stratigraphic units (Fig. 5-11).

Tidal-Flat Modification

Stacked, upward-fining tidal-flat sequences are common in the Pretoria Group. The lithologies include quartz arenites at the top of the Magaliesberg, Lakenvalei, and Steenkampsberg quartzites; mixed arenites and argillites in the Vermont, basal Nederhorst, and Houtenbek Formations; stromatolitic carbonates in the Vermont and Houtenbek Formations; and oolitic ironstones in the Timeball Hill and Dwaal Heuvel Formations (Fig. 5-2A) (Button, 1973c).

The quartz arenites reflect low tidal sand flat and subtidal deposition. Two subfacies are recognized in the Steenkampsberg–Houtenbek transition and in the Houtenbek Formation. In the lower subfacies, plane-bedding is the dominant sedimentary structure. Ebb-oriented cross-beds are rare, but associated bimodal–bipolar current ripples, and flat-topped, double-crested, and superimposed ripples indicate tidal processes. Desiccation-cracked shale drapes on rippled surfaces suggest periodic emergence. The abundance of plane-bedding is enigmatic, but a possible

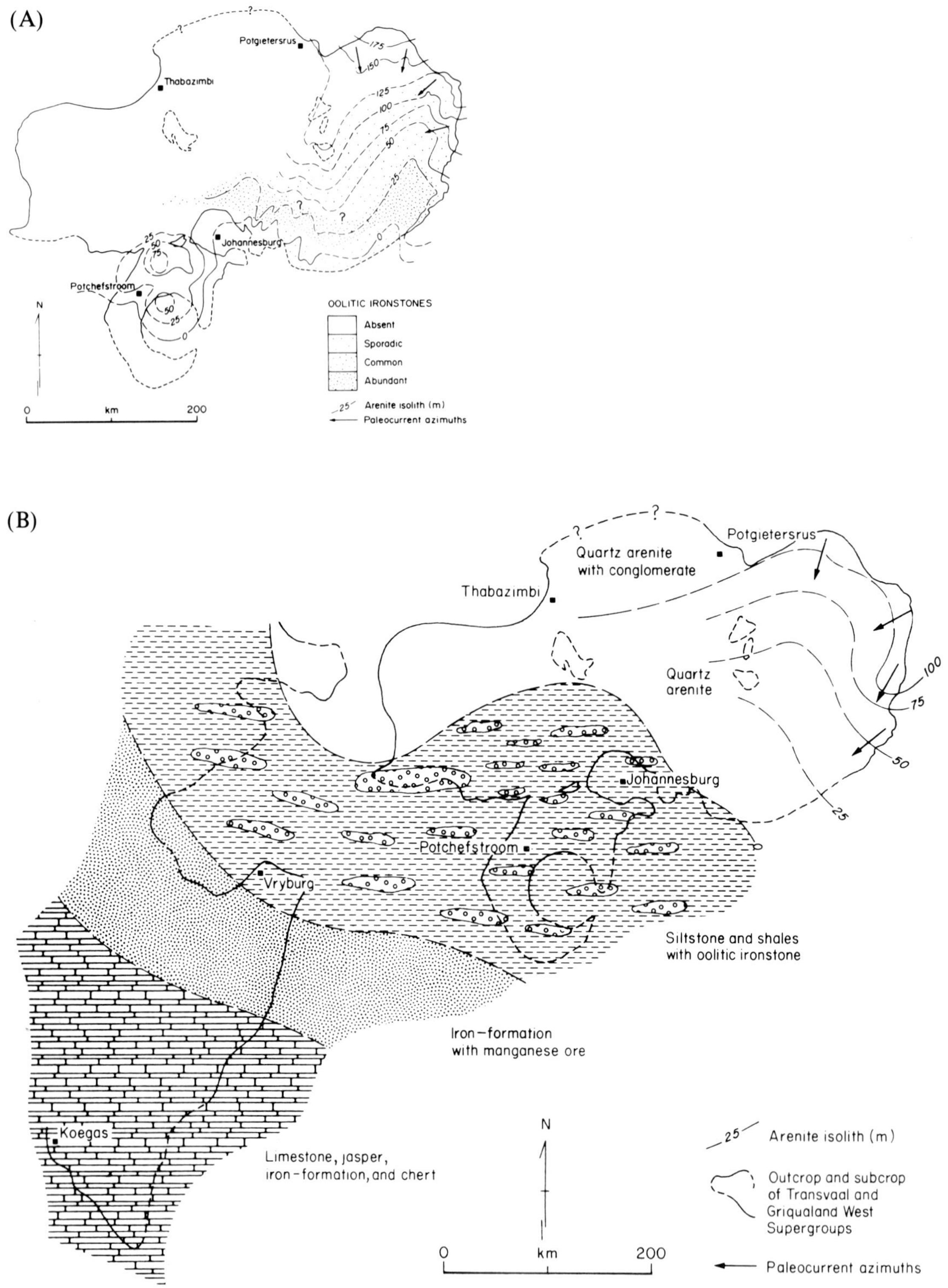

Figure 5-12. Facies patterns in the Pretoria Group: (A) Arenite isolith and oolitic ironstone distribution map for the Timeball Hill Formation. [After Button (1973c).] (B) Terrigenous clastic to nonclastic transition in the Dwaal Heuvel–Voëlwater jasper. [After Button (1975, 1976a, personal communication, 1978).]

Holocene analog may be the subtidal sand shoals of the North Sea (Button and Vos, 1977), the shallowest parts of which are exposed at low tide.

The overlying subfacies is characterized by trough, planar, and herringbone cross-bedding with paleocurrent modes to the north and south. Periodic current reversals are also suggested by reactivation surfaces and bimodal–bipolar orientations of current ripples. Flat-topped and washed-out ripples provide evidence of late-stage runoff. Slack-water suspension sedimentation and emergence are indicated by flaser and desiccation structures. These structures indicate deposition on a low tidal sand flat dominated by ebb and flood migration of dunes (Button and Vos, 1977).

The upper 20 m of the Magaliesberg quartzite consists of stacked, 1- to 2-m-thick units in which trough cross-beds are overlain by plane-beds. The sedimentation units are probably analogous to the low-tide terrace deposits of the Massachusetts coast. Dune migration during high water alternates with upper flow regime flow across the terrace at low tide (Hayes *et al.*, 1969). Ripples influenced by ebb runoff occur on the bedding planes; these include flat-topped and superimposed forms which are often draped with desiccated shale.

A mixed mid-tidal-flat facies consisting of interbedded arenites and shales overlies quartz arenites of the low tidal flats in the Vermont, lower Nederhorst, and Houtenbek Formations (Button, 1973c; Button and Vos, 1977). Reactivation structures are sporadically developed. Flaser, wavy, and lenticular bedding are ubiquitous and indicate alternating bedload transport of sand and suspension settling of mud. Sand-filled desiccation cracks are preserved in centimeter-thick shale layers. The upper tidal-flat facies comprises shales with ripple cross-laminated siltstone and rare sandstone lenses, and mud-clast breccias. Storm processes may have been responsible for transporting the sand and silt grades onto the high tidal flat (cf. Reineck, 1967).

Arkosic sandstones associated with the mid-tidal-flat arenites and shales in the Houtenbek Formation are believed to have accumulated in tidal creeks. The erosively based sedimentation units are up to 4 m thick and are structured by shallow channels up to 50 cm deep. Mud-clast channel lag is overlain by southward or ebb-oriented trough cross-beds. The surfaces of channel margin bars contain low-angle plane-beds. Fluctuating flow is suggested by reactivation surfaces and multiple scour surfaces, with occasional periods of emergence resulting in desiccated clay drapes (Button and Vos, 1977).

Stromatolitic carbonates are intimately associated with the siliciclastic tidal-flat sediments (Button, 1971; Button and Vos, 1977). Domes as high as 20 cm and crinkly algal laminations are the most common forms; solitary algal heads or bioherms reach 50 cm in height. Active reworking between these algal heads is shown by the quartz arenites which abut them. Other structures recognized are encapsulated algal biscuits, oolites, intraclastic breccias, and cross-bedded carbonate sands. Evidence of periodic emergence includes local mud-cracked shale drapes and the intraclastic breccia layers that formed by reworking of desiccated algal mats. The crinkly, low-relief algal laminations probably formed on high tidal flats. They graded seaward into domical stromatolites and algal heads, suggesting a relationship to increasing water depth and bed shear (cf. Logan *et al.*, 1974).

Oolitic ironstones occur as lenses in the Timeball Hill shales of the southeastern Transvaal and in the Dwaal Heuvel shales of the western Transvaal (Fig. 5-12). The oolites and occasional pisolites consist of quartz nucleii coated with one or more concentric ferruginous layers. The most abundant sedimentary structures are ripple cross-lamination and plane-bedding with current lineation. Cross-bedding is less common but includes trough and planar types, with some herringbone forms. Erosively based units consist of cross-lamination and cross-bedding overlain by plane-bedding with symmetrical, asymmetrical, and interference ripples. The ripples are generally draped with shale that contains desiccation and runzel structures. An environment shallow enough for agitation by waves is implied by the presence of oolites. The oolitic ironstones probably originated as shallow subtidal shoals, which were exposed at low tide in a setting starved of terrigenous material.

Shallow Subtidal Reworking

Thick quartz arenites, attributed to tidal-shelf deposition, in the Timeball Hill and Dwaal Heuvel Formations constitute the Daspoort, Magaliesberg, Lakenvalei, and Steenkampsberg quartzites (Fig. 5-2A). Thicknesses of the quartz

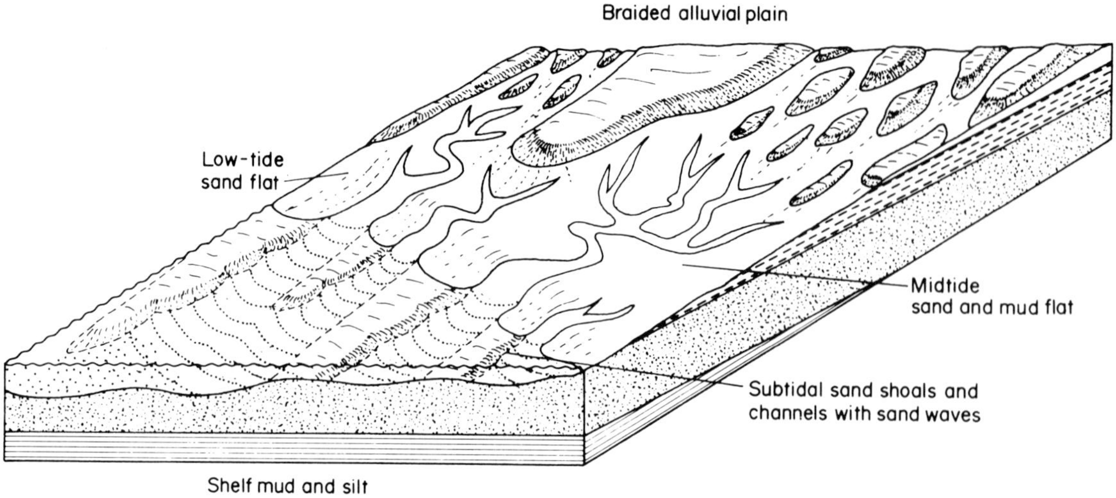

Figure 5-13. Macrotidal depositional model for the Pretoria Group showing transition from braided alluvial to tidal flat to shallow subtidal to shelf. [After Button and Vos (1977).]

arenites vary up to 600 m, thinning to the south and southwest (Fig. 5-12).

Timeball Hill and Dwaal Heuvel paleocurrents (Fig. 5-12) in the eastern Transvaal display a systematic regional variation from a southerly mode in the north to a westerly mode in the south (Button, 1973c). Northerly and southerly paleocurrent modes characterize the limited exposures of Steenkampsberg quartzite in the eastern Transvaal; large-scale planar cross-bed sets up to 3 m thick are oriented to the north, whereas trough cross-beds display a southerly preference. Planebeds and asymmetric ripples are common, the latter displaying bimodal–bipolar paleocurrent patterns. Shale partings are well developed but, unlike those in the tidal-flat facies, they are nowhere desiccated (Button and Vos, 1977).

Structures characteristic of barrier-beach progradation are conspicuously absent from the quartz arenites. Sandstones lacking evidence of exposure grade upward with increasing proportions of shale into the tidal-flat facies discussed above. A nonbarred, tide-dominated or macrotidal shoreline is indicated (Button, and Vos, 1977). Clockwise rotation of tidal currents in an epicratonic shelf sea could possibly account for the paleocurrent patterns observed in the Timeball Hill and Dwaal Heuvel Formations. The shoaling effects of the shallow shelf produced linear sand bodies oriented parallel to the tidal currents and perpendicular to the shoreline. Interbar channels contained large-scale bedforms. Current velocities were greatest in the channels; on the intervening shoals maximum velocities were ebb-directed, producing ripples, dunes, and horizontal lamination. The horizontal lamination was probably generated by settling of suspended sand and silt from wave or tidal current-agitated water.

A paleogeographic model relating the Steenkampsberg shallow subtidal facies, the Houtenbek intertidal facies, and the Nederhorst braided fluvial facies is illustrated in Figure 5-13. By analogy with the North Sea shelf (Houbolt, 1968), the large-scale planar cross-beds are attributed to onshore sand wave migration in the channels. The trough cross-bedded arenites accumulated either in the channels or on the shoals.

Stromatolitic bioherms up to 10 m high and 10 m wide are built on tidal-flat sediments in the Houtenbek Formation (Fig. 5-2A). They are markedly elongated in a north–south direction and reach 100 m in length. The bioherms are internally structured by finger-shaped columnar stromatolites. Linear channels separating the bioherms contain cross-bedded quartz arenites, siltstones, and carbonate intraclastic breccias. A shallow subtidal origin for the bioherms is suggested by their size, their degree of vertical inheritance, and the absence of erosive features on their crests (Button and Vos, 1977).

Offshore Shelf Sedimentation

The lower 3000 m of the Pretoria Group contains substantial thicknesses of shale and siltstone in the Timeball Hill and Silverton Shale Formations (Fig. 5-2A). These argillaceous rocks are an integral part of the thick upward-coarsening sequences discussed previously. A similar fine-grained terrigenous clastic facies overlies the stromatolitic bioherms in the Houtenbek Formation (Button and Vos, 1977). The shales are mainly carbonaceous and pyritic at the base, becoming chamositic upward. Monotonous horizontal lamination, sporadically interrupted by tuffaceous horizons and thin carbonate lenses, is present through hundreds of meters. Siltstones are more common upward toward the Timeball Hill and Magaliesberg quartzites. The Lydenberg shale, for example, grades into the Magaliesberg quartzite through shales with ripple cross-laminated siltstone intercalations and impure arenites.

This evidence, together with the presence of abundant pillows and water-laid tuffs in the Machadodorp volcanics, suggests deposition into a permanently subaqueous regime. The shales are ascribed to suspension sedimentation below wave base; carbonate accumulation was probably remote from terrigenous clastic influx. Low-energy bedload transport was restricted to ripple cross-laminated siltstones (Button, 1973c).

Chemical Sedimentation on the Distal Shelf

Autochthonous sedimentation characterized the distal shelf facies in the northern Cape where terrigenous clastic influx was minimal. The Voëlwater jasper is a distal facies of the Dwaal Heuvel Formation, both formations overlying a volcanic datum (Figs. 5-2A and B and 5-12B).

The Voëlwater jasper consists mainly of iron-formation with subordinate manganiferous layers (P.R. De Villiers, 1970). In the Hotazel area the iron-formations comprise alternating mesobands of hematite and microbanded jasper (N.J. Beukes, 1973). Southward, magnetite becomes the dominant iron oxide. The manganiferous sediments display an analogous facies change: manganese oxides and silicates predominate in the north and rhodochrosite increases southward. The proportion of associated limestone also increases to the south. Pseudo-oolites of calcite are locally present in a matrix of manganese oxides in the Hotazel area, but more commonly the manganiferous sediments are thinly laminated with calcium carbonate wisps. Southwest of Sishen extensive limestones are present in the Voëlwater jasper (P.R. De Villiers, 1970).

The source of the metals is uncertain. N.J. Beukes (1973) suggested a genetic relationship between the Voëlwater jasper and the underlying Ongeluk volcanics, with the metals emanating from late-stage volcanic exhalations. Alternatively the metals may have been introduced with the proximal terrigenous clastics following source area weathering. The mineralogical facies relationships indicate precipitation of the metals under highly oxidizing conditions of differing Eh and pH. Button (1976a) argued that metals held in the ancient sea water in their low valency states were precipitated with the onset of more oxidizing conditions, implying purging of Mn^{2+} and Fe^{2+} from the sea.

High Eh conditions prevailed in the more proximal Voëlwater environments, resulting in the precipitation of Fe^{3+}, $[Fe^{3+}—Mn^{4+}]$, and Mn^{4+} oxides in the Hotazel region. Oxidation of iron and manganese released H^+ (Drever, 1974), which probably inhibited limestone formation in the north. Lower Eh conditions in the south are indicated by the magnetite and rhodochrosite facies and the associated limestones.

The Timeball Hill Formation also has a distal shelf facies in the northern Cape, represented by iron-formations at the base of the Gamagara shales in the Sishen area. The overlying shales are highly ferruginous and constitute a high-grade ore deposit with greater than 68 percent FeO. Manganiferous shales rest on the Ghaap Group in the southern outcrop area of the Gamagara shale and are overlain by ferruginous shales. Although it is generally agreed that the iron-formations at Sishen are primary precipitates, the origin of the ferruginous and manganiferous shales remains contentious. Prevailing opinion favors an origin by supergene replacement involving circulating meteoric waters (Strauss, 1964). Where the Gamagara shales rest on iron-formations they are ferruginized, and where they rest on dolomites they have been enriched with manganese. Alternatively the iron and manganese ore deposits, in which the metals are present only in their higher valency states, may have formed as primary precipitates with possible later supergene enrichment. If the deposits are syngenetic, the north–south metal zonation would represent a geochemical facies transition possible only in an oxidizing environment (Krauskopf, 1967). The areal separation of the iron and manganese oxides reflects their different solubilities. That hematization preceded Paleozoic Karoo deposition is shown by glacial pavements in the hematite at Sishen (Button, 1976a).

Glacial Environments

Diamictites and associated sediments occur sporadically beneath the regional unconformity within the Pretoria and Postmasburg Groups (Fig. 5-2A and B). In the northern Cape they are overlain by the Ongeluk volcanics, but in the Transvaal they are either overlain by the Hekpoort basalt or are unconformably overlain by fluvial sandstones. The glacial diamictites in the northern Cape contain striated and rafted pebbles and locally rest on remnants of a striated pavement. Associated conglomerates, cross-bedded pebbly sandstones, siltstones, mudstones, and varved shales are interpreted as glaciofluvial and glaciomarine (Visser, 1971). Except for sporadic faceted and percussion marked pebbles, direct evidence of glaciation is lacking in the Transvaal, although Visser has inferred a glaciomarine setting. The diamictites are associated with conglomerate lenses, argillaceous sandstones, siltstones, and shales; ball-and-pillow structures are common. In the eastern Transvaal, large-scale soft-sediment folding has deformed the uppermost Timeball Hill shales (Button, 1973c). Major channels scoured into the folded shales are filled with diamictite. Button has suggested that channel scouring and diamictite deposition were the result of gravity flow generated by uplift of the basin margin.

Volcanics in the Pretoria Group

Three thick volcanic intervals are present in the Pretoria Group; these are the Hekpoort basalt and the Machadodorp and Dullstroom volcanics. The sole volcanic interval in the Postmasburg Group, the Ongeluk volcanics of the northern Cape, correlates with the Hekpoort basalt (Fig. 5-2A and B). Evidence of subaerial extrusion is plentiful in the Hekpoort basalt and Dullstroom volcanics (Button, 1973c). Distinct flows of basaltic composition have massive interiors and amygdaloidal tops, and in the Dullstroom suite they vary from 8 m to 60 m in thickness (average, 30 m). Tuffs, lapilli tuffs, agglomerates, and flow breccias are common. Laterally impersistent pyroclastic felsites in the Dullstroom volcanics were a precursor to the terminal Rooiberg felsites. The Rooiberg felsites were prized off the Dullstroom volcanics by injection of the basic phase of the Bushveld Complex (Button, 1973c).

The Machadodorp and Ongeluk volcanics were extruded subaqueously and contain abundant pillows and water-laid pyroclastics (Button, 1974; Grobler and Botha, 1976). Water-laid tuffs at the base of the Machadodorp volcanics are commonly ripple cross-laminated and pass upward into agglomerates with bombs up to 1 m in diameter, bedded lapilli tuffs, tuffs with small bombs, and a second tuffaceous interval. The basal pyroclastic phase is overlain by pillowed volcanics. Pillows vary from 0.2 m to 1.0 m in length and contain downward protuberances and tricuspate voids filled with vein quartz and opaline silica. Cooling produced concentric banding and radial contraction joints. The Ongeluk volcanics are massive at the base and pass upward into an interval containing large oval fragments of massive lava in a matrix of small glassy fragments and angular palagonitized glass shards. The fragments probably resulted from quenching and shattering of the outer parts of pillows, the matrix representing aquagene tuff. The Ongeluk volcanics are capped by nonvesicular pillow lavas; pillows from 1 m to 15 m wide and 1 m to 3 m high contain red jasper in the tricuspate spaces.

Although the Machadodorp volcanics have many characteristics of continental basalts, the few geochemical analyses (Button, 1974) demonstrate distinct oceanic affinities (cf. Pearce *et al.*, 1975). This strong oceanic character is shown on a TiO_2–K_2O–P_2O_5 discriminant diagram (Fig. 5-14). A simple ternary plot of FeO [total]–MgO–Al_2O_3 (cf. Pearce *et al.*, 1977) is consistent with this interpretation but less discriminating.

The Machadodorp lavas must have been injected through the sialic crust and the sedimentary cover with little or no contamination (Button, 1974). Pearce *et al.* (1975) have suggested that continental basalt suites showing oceanic characteristics on a TiO_2–K_2O–P_2O_5 diagram may be related to major flexure or tensional features; tectonic setting significantly influences volcanic rock chemistry. Similar basalts occur in the Ventersdorp Supergroup and Waterberg Group and, not unexpectedly, in the Archean Onverwacht Group. Other well-known examples of continental suites displaying oceanic affinities include the Deccan Traps and the Tertiary basalts of Greenland both of which are related to crustal tension (Pearce *et al.*, 1975).

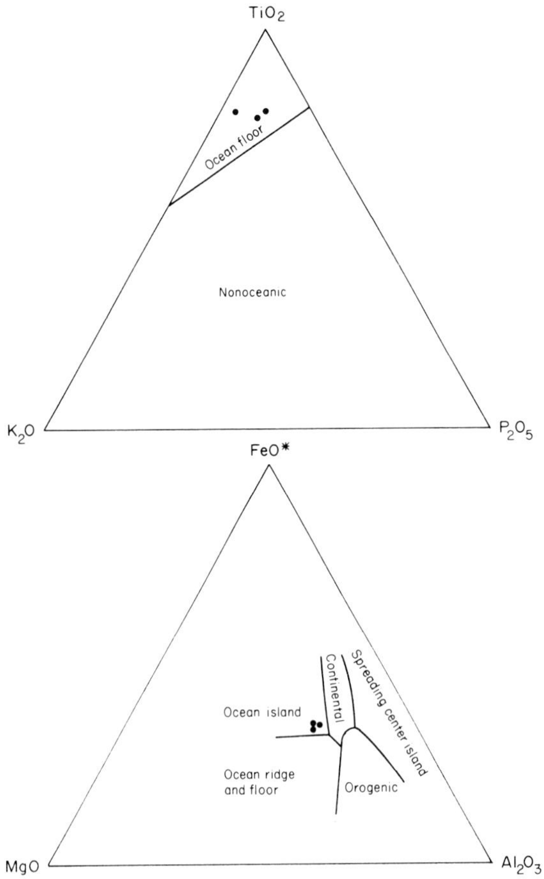

Figure 5-14. Major-element discriminant diagrams, TiO_2–K_2O–P_2O_5 and FeO [total]–MgO–Al_2O_3, for the Machadodorp volcanics from three localities showing oceanic affinities. [From Button (1974).]

5.5. Depositional History of the Epeiric Sea

Terrestrial and shallow-water sediments and subordinate volcanics dominate the Transvaal and Griqualand West successions to the exclusion of deep-water facies.

Wolkberg and Buffalo Springs sedimentation and volcanism in the protobasin occurred on a deeply incised topographic surface. However, Black Reef sedimentation produced a reduction in topographic relief and lowered gradients. Equilibrium between subsidence and sedimentation was occasionally interrupted by short-lived uplifts, which resulted in the development of intraformational unconformities in the Chuniespoort and Pretoria–Postmasburg Groups. The regional angular unconformity that separates these groups reflects a major hiatus in sedimentation and development of a karst surface with an areal extent of at least 500,000 km^2. Possibly 500 m of the Malmani dolomite were removed by solution in the southeastern Transvaal (Button, 1973c). This event is preserved in the stratigraphic record as a chert breccia, which represents the insoluble residue of the erosional episode.

Throughout the history of the Transvaal and Griqualand West Supergroups a relatively active northern basin margin can be distinguished from a more passive southern margin. Terrigenous clastic sediments were derived from the north and northeast except for a minor influx from the south during deposition of the Black Reef quartzite. The concept of active and passive basin margins is well illustrated by the Chuniespoort–Ghaap Group and Pretoria–Postmasburg Group facies patterns. Terrigenous clastic influx was considerably less substantial during deposition of the Chuniespoort and Ghaap Groups than in the earlier and later geologic history of the basin. It is also significant that the northern and northwestern margin facies in this complex of chemical sediments are markedly enriched in the relatively insoluble cations when compared with the southeastern facies.

A dramatic change in the pattern of source area tectonics is indicated by the dominantly terrigenous nature of the Pretoria Group. Renewed terrigenous influx from the north followed a major transgression across the paleoerosional surface at the top of the Chuniespoort and Ghaap Groups. Fluvial arkoses, tidal-flat quartz arenites and shales, and shallow-shelf quartz arenites accumulated in the northern, more proximal portions of the Pretoria Group depository. These graded southward into offshore shelf siltstones and shales deposited below wave base. Oolitic ironstone shoals developed along the passive southern margin remote from terrigenous clastic influx. The shoals periodically emerged so that oolites accumulated partly on tidal flats. On the most distal shelf, completely removed from influx from the north, ferruginous, manganiferous, and calcareous sediments precipitated under different chemical conditions. The inferred spatial relationships of the depositional environments above the Hekpoort basalt-Ongeluk volcanic datum is shown in Figure 5-12B.

Tide-dominated sedimentation is shown by the presence of extensive tidal-flat and shallow subtidal facies. During deposition of the Black Reef quartzite, Vryburg Formation, and Pretoria Group, fluvially introduced terrigenous clastics were reworked subaqueously by clockwise-circulating tidal currents and on tidal flats. An analogous situation exists on many shallow shelves, such as the North Sea (Houbolt, 1968). The recognition of a similar system in the Black Reef quartzite and Pretoria Group implies that the depository was open and had the form of an epicratonic sea. Facies patterns in the Pretoria–Postmasburg Group suggest that the depository opened to the south onto what is now the high-grade metamorphic Namaqua Province (Fig. 1-1). Any deep-water facies of the Transvaal and Griqualand West Supergroups that may have existed were subsequently incorporated into the complex structural and metamorphic framework.

The Transvaal and Griqualand West Supergroups may have special significance with respect to the evolution of the primitive atmosphere. The carbonates in the Chuniespoort and Ghaap Groups are, together with those in the Hamersley Group in Australia, the oldest extensive shelf sequences of such sediments in the geologic record (Button, 1973c, 1976a). These carbonates are stromatolitic virtually throughout their 1500 m thickness. In the absence of grazing metazoans during the Precambrian, stromatolites were able to populate several environments. Furthermore, well-preserved blue-green algal filaments have been recovered from these sediments (Macgregor *et al.*, 1974; Nagy, 1978), proving that large amounts of oxygen were being produced photosynthetically 2.2 Ga ago. Whereas limited evidence of oxidizing conditions exists beneath the regional angular unconformity separating the Chuniespoort-Ghaap from the Pretoria-Postmasburg Groups, there is evidence that an oxidizing hydrosphere, and possibly even an oxidizing atmosphere, developed immediately after the formation of the Chuniespoort and Ghaap Groups (Eriksson and Truswell, 1978). The chert breccia defining the unconformity contains manganese oxide in its matrix (Button, 1976a), suggesting that during subaerial weathering of the dolomites manganese held in solid solution in the carbonate lattice was not taken into solution but was precipitated upon oxidation to its higher valency state. Oolitic ironstones in the Pretoria Group and iron-formations from the upper Postmasburg Group are deficient in divalent iron. Finally, calcareous shales interlayered within the iron formations contain significant quantities of manganese in quadrivalent-state minerals.

Chapter 6

The Bushveld Complex: A Unique Layered Intrusion The Vredefort Dome: Astrobleme or Gravity-Driven Diapir?

The Bushveld Complex is unique because of its size, the persistence and regularity of the layering in the ultrabasic and basic rocks, and the importance of the economic minerals it contains. The complex underlies an area of approximately 65,000 km^2 and so constitutes by far the largest known intrusion of its kind in the world. Only the Dufek Massif in Antarctica, with an estimated minimum extent of 24,000–34,000 km^2, approaches these dimensions.

The first comprehensive account of the complex came from A.L. Hall in 1932; he visualized a major, protracted, plutonic event, the effects of which were manifest over a vast region encompassing South Africa, Botswana, and Zimbabwe. Hall suggested that this great igneous province consisted of a Main or Inner Complex represented by those plutonic rocks that form a discrete body (i.e., the Bushveld Complex). Intrusions of similar petrologic character geographically detached from the Inner Complex were regarded as constituting the Satellitic or Outer Complex, the most conspicuous of these "satellites" being the Great Dyke of Zimbabwe. The advent of radiometric age determinations led to significant modification of this concept and only those relatively small intrusions immediately adjacent to the Bushveld Complex are now provisionally regarded as being related to the Bushveld plutonic event (Fig. 6-1).

Hall (1932), and subsequently Willemse (1964, 1969), proposed that the complex embraced a number of components formed before and during its emplacement. These components can be summarized as:

(1) The volcanic rocks interstratified in the Pretoria Group of the Transvaal Supergroup, constituting the initial volcanic phase
(2) The numerous thick sheets of gabbro, diorite, and diabase that intrude the Transvaal Supergroup, forming the "floor" of the main intrusion and constituting the sill phase
(3) The Rooiberg felsites with minor interbedded sedimentary rocks, representing the epicrustal phase
(4) The main plutonic phase, consisting of layered ultrabasic and basic rocks
(5) The late plutonic phase, represented by granitic rocks.

Objections to this proposal were based on the eccentric distribution of both the volcanic rocks and the basic sheets in the Transvaal Supergroup with respect to the main complex (Truter, 1955, 1962). Frick (1973) classified the basic sills as diabases and diorites, recognizing the presence of at least two periods of diorite sill emplacement predating the intrusion of the diabase sills that are restricted in their areal distribution to the vicinity of the main complex. Frick also recognized the calc-alkaline character of the diorite sills and the uppermost volcanic unit in the Transvaal Supergroup that contrasts with the tholeiitic character of the diabases and the Bushveld Complex. Although the existence of two distinct magma types does not necessarily disprove a genetic connection, the areally widespread development of volcanic rocks and dioritic sills beyond the confines of the main complex raises doubts about the validity of events that have been regarded as heralding the intrusion of the Bushveld Complex. Consideration of these rock types is therefore excluded from this account.

The stratigraphic position of the Rooiberg felsites constituting the epicrustal phase of Willemse (1964) is still in doubt. They have been regarded either as the terminal volcanic phase of

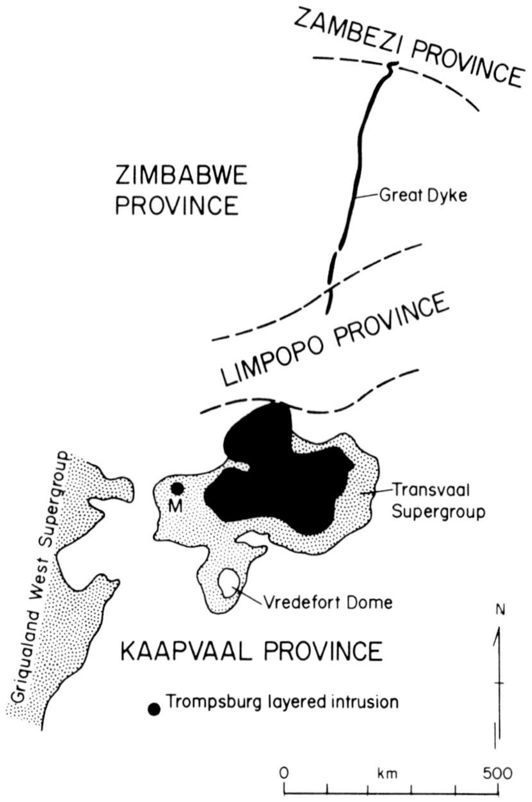

Figure 6-1. The location of the Bushveld Complex and its regional setting. M=Marico intrusion. Trompsburg layered intrusion lies under cover of the Karoo strata and is dated at 1372 ± 141 Ma (Davies *et al.*, 1970). For locations of satellite intrusions around the Vredefort Dome see Figure 6-5.

the Transvaal Supergroup, or as genetically related to the acid plutonism of the Bushveld Complex, or as unrelated to either the Transvaal Supergroup or Bushveld Complex. Geochronologic studies have failed so far to resolve this problem. The geochemical affinities of the felsites to the granitic rocks of the complex together with their confinement within the limits of the complex suggest that they should be considered as part of the Bushveld Complex.

Resolution of the problem of what actually constitutes the Bushveld event is essential to understanding the complex. Lack of general agreement imposes constraints on any interpretation of the evolution of the complex. In this account the Bushveld event is considered to consist of: (1) the layered basic–ultrabasic sequences and the granites, granophyres, and felsites forming the Bushveld Complex proper, and (2) various satellite intrusions mainly located in the vicinity of the Vredefort Dome.

6.1. Framework of the Complex

The regional gravity survey of South Africa (Smit *et al.*, 1962) demonstrated that the Bushveld Complex does not have the form of a simple lopolith. These geophysical investigations, together with those reported by Biesheuval (1970) in the western Transvaal, suggest that the complex has an approximately clover-shaped outline consisting of four lobes (see Fig. 6-2).

(1) A western lobe, extending from near Pretoria westward to Rustenburg and around the 1.4-Ga Pilanesberg alkaline complex to and along the southern flank of the Makoppa Dome of Archean granitoids and gneisses
(2) A southeastern lobe, largely covered by Mesozoic rocks
(3) An eastern lobe that continues northward from the southeastern lobe, extending to and beneath the Mesozoic cover some 50 km east of Potgietersrus
(4) A northern lobe that extends from south of Potgietersrus to Villa Nora

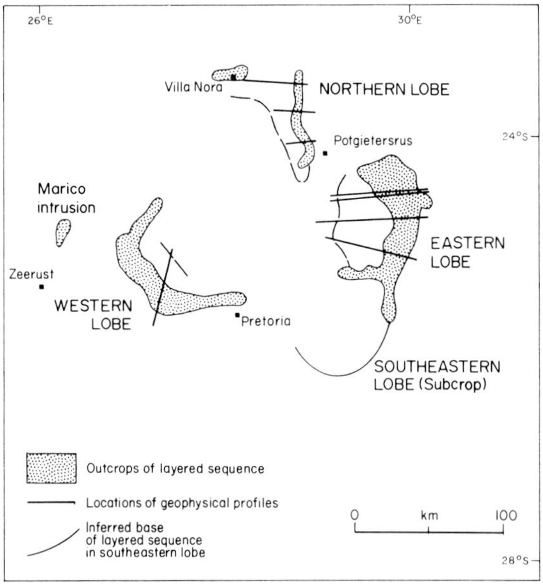

Figure 6-2. Outcrop pattern of Rustenburg Layered Suite showing positions of geophysical traverses.

These four lobes are arranged about two axes, each approximately 350 km long, trending east-northeast and north-northeast, respectively. The basic and ultrabasic rocks of the layered suite crop out on the peripheries of these lobes but do not apparently enclose the central cores of the lobes (Fig. 6-2). Geophysical data imply no continuity of the basic and ultrabasic rocks from lobe to lobe except in the east. Here the southeastern and eastern lobes can be seen in outcrop to be continuous, but the boundary between these two lobes is not yet clearly defined and may lie as far north as the Steelpoort fault (Fig. 6-3).

Beyond the limits of this main clover-like complex are the Marico intrusion (north of Zeerust; Fig. 6-l) and a number of layered intrusions around the Vredefort Dome that are regarded as formed during the Bushveld event (Abbott and Ferguson, 1965; Bisschoff, 1972a, 1973). There is no consensus as to whether the Marico intrusion is contiguous with the western lobe of the main complex. Coertze (1970) considered that a relatively thin veneer of quartz norite forming the lowermost member of the Bushveld layered sequence could be traced westward to the Marico intrusion. The absence of a substantial thickness of basic or ultrabasic rocks underlying this area is confirmed by gravity surveys (Biesheuval, 1970). However, Liebenberg (1970) and Vermaak (1970) regarded these quartz norites as sills that were deformed before the intrusion of the Bushveld Complex.

Within the cores of the western and eastern lobes domical areas of deformed rocks of the Transvaal Supergroup and older formations probably represent upfolded portions of the country rocks of the complex. However, Daly (1928) and Verwoerd (1963) considered the Crocodile River Dome to be a stoped block and a core to a Bushveld Complex cone-sheeted complex, respectively.

The dips of the layered sequence of basic and ultrabasic rocks of the complex in each of the lobes are toward the center, usually at low angles between 10° and 25°. Dips as steep as 60° are recorded in the vicinity of Villa Nora and adjacent to the Dennilton Dome (Fig. 6-3). The rocks of the Transvaal Supergroup around the peripheries of the lobes generally dip with a closely similar, gentle inclination and thus appear to form a floor to the

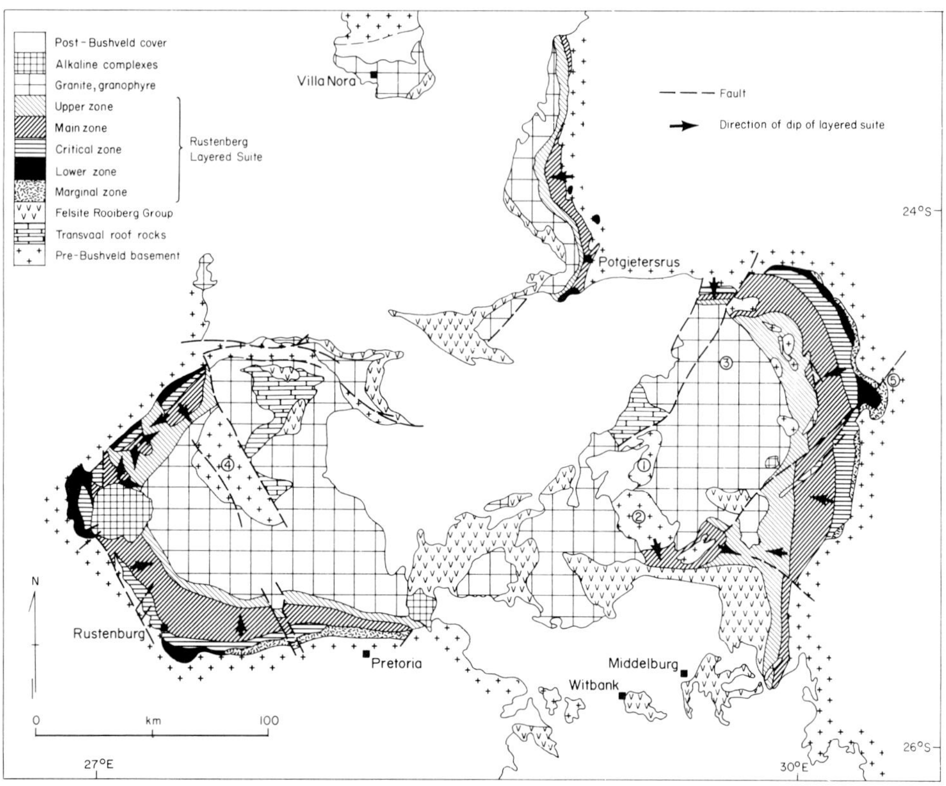

Figure 6-3. Generalized geologic map of the Bushveld Complex. 1, Marble Hall Dome; 2, Dennilton Dome; 3, Malope Dome; 4, Crocodile River Dome; 5, Steelpoort fault.

complex. At some localities in the vicinity of the eastern lobe the Transvaal rocks dip at 45°, whereas the Bushveld rocks are less steeply inclined, appearing to form a wall rather than a floor. In addition the strike of the rocks of the layered sequence is oblique to that of the Transvaal country rocks.

Recent geophysical studies (Biesheuval, 1970; Walraven and Darracott, 1976; Van der Merwe, 1976; Barrett *et al.*, 1978; Molyneux and Klinkert, 1978) have attempted to interpret the configuration of the layered sequence in the western, eastern, and northern lobes.

An interpretation of a gravity profile approximately 20 km east of the Pilanesberg alkaline suite suggests that the layered sequence has the form of a northwest-trending, elongated, deep basin, that has been gently warped about northwest axes. Such an interpretation implies that the layered sequence lies at a depth of about 500 m about 23 km south-southwest of the edge of the Crocodile River fragment. These gravity data could be interpreted to imply the presence of a feeder dike of enormous proportions, but Walraven and Darracott (1976) regard this as unlikely and prefer a model requiring an approximately constant thickness for the layered sequence (Fig. 6-4A).

Profiles derived from magnetic and gravity data generally indicate that the layered sequence in the eastern lobe is sill-like, with a maximum thickness of 5 km which thins rapidly westward under the Karoo and Nebo granite cover (Figs. 6-3 and 6-4B) (Molyneux and Klinkert, 1978). There is no convincing geophysical evidence for the presence of feeder dikes for either the layered sequence or the granitoids, but models can be constructed with a basic feeder dike located beneath the central part of the layered sequence. The southernmost profile suggests that in the eastern part of the profile the layered sequence is not as thick as to the north. A thickness of nearly 3 km of basic rocks is indicated at shallow depth beneath the Marble Hall Dome.

Gravity data from the northern lobe suggest the presence of the following features (Van der Merwe, 1976):

(1) A north-trending trough-shaped body of basic rocks, which widens northward
(2) A feeder vent west of Potgietersrus
(3) A dikelike body in the south with a sill-like body in the north
(4) A dominant northwest-trending gravity feature northwest of Potgietersrus

The preferred model profiles have an asymmetric cross section 85 km wide and 5 km deep in the north, decreasing in width to 33 km but increasing to 10 km in depth to the south (Fig. 6-4C). Dips in the western portion of the profile are gentle, whereas the eastern limb dips steeply to the west (Van der Merwe, 1976). South of Potgietersrus gravity and geologic data indicate the presence of horst and graben structures associated with strike-parallel faults (Barrett *et al.*, 1978).

The Marico intrusion is approximately kidney shaped in plan and the layered sequence has shallow centripetal dips. Biesheuval (1970) interpreted the gravity data to suggest that the intrusion is lopolithic in shape.

Although the present geophysical data do not permit a unique interpretation, it seems reasonable to conclude that the Bushveld Complex consists of at least four and possibly five intrusive centers, namely, Marico, western lobe, eastern lobe, northern lobe, and probably the southeast lobe. Evidence in support of the last named derives from the stratigraphy and petrography of the layered sequence and is discussed in Section 6.2. Precise identification of feeder dikes for the layered sequence is not yet possible but each lobe could be regarded as sheetlike in form thinning laterally from a feeder dike near the center. Maximum widths of the layered sequence sheets would appear to be of the order of 80 km in the northern and eastern lobes.

6.2. Magmatic and Volcanic Stratigraphy

Hall (1932) divided the Bushveld Complex into:

(1) First or earlier basic volcanic phase
(2) Second or later acid volcanic phase
(3) Norite lopolith or earlier plutonic phase
(4) Sill phase of the norite lopolith
(5) Granite or later plutonic phase

As discussed above, the validity of including some of these units within the complex has been questioned. The later acid volcanic phase is now equated with the Rooiberg Group, the stratigraphic position of which is still disputed. For the purposes of this account, the Rooiberg Group of dominantly felsitic volcanic rocks is included with the Bushveld Complex. The norite lopolith or norite belt (Hall,

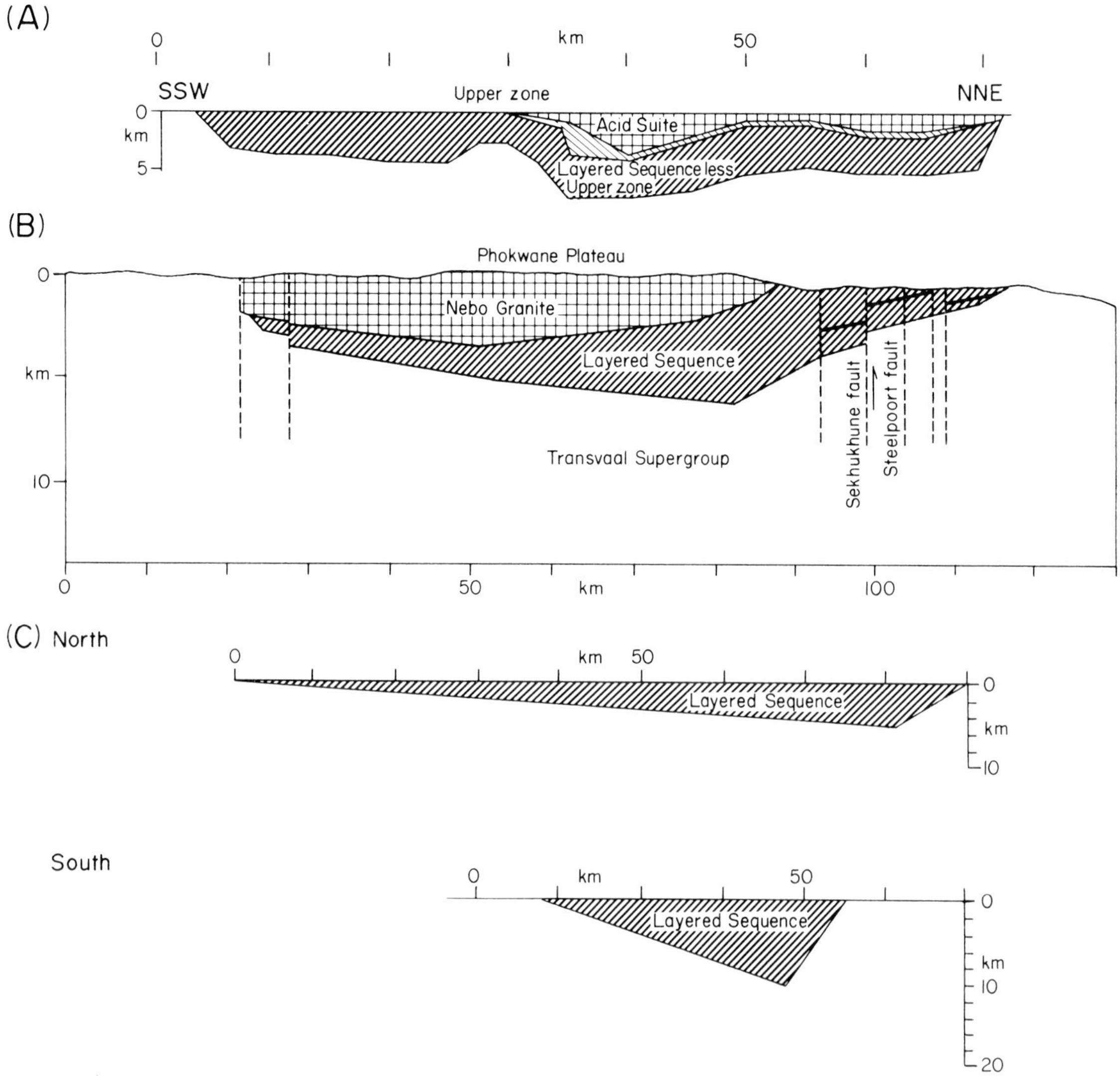

Figure 6-4. Interpretations of geophysical profiles across parts of the Bushveld Complex. For locations of profiles see Figure 6-2. (A) Western lobe: [After Walraven and Darracott (1976).] (B) Eastern lobe: [After Molyneux and Klinkert (1978).] (C) Northern lobe: [After Van der Merwe (1976).]

1932) has been called the Mafic Phase but the most recent proposal of the South African Committee for Stratigraphy is to name the plutonic basic rocks the Rustenburg Layered Suite. Granophyric rock types placed by Hall (1932) in this acid volcanic phase are now classed as members of the Rashoop Granophyre Suite and the granitic rocks of Hall's fifth phase are named the Lebowa Granite Suite. This nomenclature is an attempt to conform with recommendations laid down by the South African Stratigraphic Code and also with the general terminology applied to layered intrusions around the world. Further subdivision of these three suites is still being debated, but a recommendation has been made that rock units be given formal names and that an informal subdivision, based on a zonal nomenclature, should be retained.

Rustenburg Layered Suite

Since Hall's (1932) proposal to subdivide his norite belt into a number of zones, different workers have erected various schemes of subdivision which are summarized in Table 6-1. It can be seen that there has been no general agreement as to the boundaries between zones, a problem further com-

Table 6–1. Subdivision and Lithologies of the Rustenburg Layered Suite

Recommended formal subdivision	Important lithologic units		Approximate thickness (m) Eastern lobe	Western lobe
Upper zone				
Luipershoek olivine diorite	Diorite and olivine diorite Olivine diorite		1000	1700
Ironstone magnetite gabbro	Magnetite gabbro with layers of magnetite, anorthosite, and olivine gabbro		750	
Magnet Heights gabbronorite			250	
Main zone				
Mapoch gabbronorite	Mostly gabbro and norite	Pyroxenite marker	800	2700 to 4400
Leolo Mountain gabbronorite			1500	
Winnaarshoek norite–anorthosite		Upper mottled anorthosite Main mottled anorthosite	1200	
Critical zone				
Winterveld norite–anorthosite	Alternating layers of chromitite, pyroxenite, norite, and anorthosite	Giant mottled anorthosite Bastard "Reef" Merensky "Reef" UG1 chromitite layer	1000	450
Zwartkoppies pyroxenite	Pyroxenite interlayered with dunite and chromitite	Steelpoort chromitite Lowermost chromitite layer	500	480
Transition zone				
Serokolo bronzitite	Pyroxenite		300	50
Jagdlust harzburgite	Harzburgite interlayered with pyroxenite and dunite		500	400
Rostock bronzitite	Pyroxenite		400	140
Lower zone				
Clapham bronzitite	Norite Feldspathic pyroxenite Harzburgite layer near base		400	380
Shelter norite	Norite, feldspathic pyroxenite, quartz norite, anorthositic norite		Up to 100	Up to 240

Hall (1932)	Lombard (1934)	Schwellnus *et al.* (1962)	Willemse (1969)	Cameron (1970, 1971)	Coertze (1974)
Upper zone	Upper zone				
		Ferrogabbro	Upper zone	Upper zone	Ferrogabbro unit
Main zone	Main zone		Main magnetite layer		
		Leoloberg gabbro	Main zone	Main norite zone	Gabbro unit
Critical zone	Critical zone	Paschaskraal norite with intrusive sheets of pegmatitic feldspathic pyroxenite	Merensky "Reef"	Critical zone	Porphyritic pyroxenite unit
			Critical zone		Norite unit
					Anorthosite unit
			Main chromitite layer		
Transition zone	Basal zone	Chromiferous pyroxenite with intrusive sheets of peridotite	Basal zone	Transition zone	Pyroxenite unit
				Lower zone	
Basal or Chill zone	Norite–diabase zone	Hendriksplaas norite	Chill zone		Basal norite unit

pounded by Hall's lack of adequate definition of the upper boundary of his critical zone.

The best outcrops of the layered sequence are found in the eastern lobe where the succession from base to top is exposed. The suggested formal nomenclature in Table 6-l therefore refers only to the eastern lobe. In the other lobes, basic rocks underlie generally featureless plains or are covered by younger formations. Details of the stratigraphy of these lobes therefore are derived mainly from core drilling or mapping of underground workings. Although gross similarities in stratigraphic succession are recognized, more evidence is becoming available that indicates important differences. The simplified geologic map of the complex (Fig. 6-3) is based on the informal subdivision into zones because of the present difficulty of compiling a composite map based on formal stratigraphic terminology. In this account the informal zonal terminology will be used to facilitate correlation between lobes of the complex.

The Marginal Zone. Fine-grained noritic rocks (the Shelter norite of the eastern lobe), characterized by the presence of metasedimentary inclusions, intervene commonly between various rock units of the layered suite and the country rocks. Previously named the chill zone, Hendriksplaas norite, or Maruleng norite, these rocks have been suggested as representative of the parental Bushveld magma, but the presence of cumulus textures and their thickness render this proposal invalid. However, a thin chill facies following the shape of the floor of the complex has been recognized (Daly, 1928; Nel, 1940; Willemse, 1959; Wager and Brown, 1968; Frick, 1973). The margin of the layered suite is transgressive and rocks from this chill facies can at best represent the compositions of their contemporary magmas, which could have been modified by prior fractional crystallization.

The Lower Zone. In the eastern lobe four formal units constitute the lower zone, named from the base up:

(1) Clapham bronzitite
(2) Rostock bronzitite
(3) Jagdlust harzburgite
(4) Serokolo bronzitite

The Clapham bronzitite consists mainly of feldspathic bronzitite but includes a layer of harzburgite at the base and two thin noritic layers, one about 50 m above the basal harzburgite and the other interlayered with anorthositic norite at the top of the Clapham bronzitite (Cameron, 1978).

The overlying Rostock and Serokolo bronzitites are nearly monomineralic rocks containing 95–99.8 volume percent bronzite with minor amounts of postcumulus plagioclase and clinopyroxene. Chromite is usually lacking but constitutes a minor component of the intervening Jagdlust harzburgite (Cameron, 1978). This unit consists of harzburgite and pyroxenite, showing repetitive cyclicity of the crystallization sequence: olivine → olivine + orthopyroxene → orthopyroxene.

In the western lobe the basal rock units of the lower zone consist of norite and pyroxenite, which are overlain by cyclic units of harzburgite and pyroxenite. Plagioclase appears as a cumulus mineral in the pyroxenites of the upper cyclic units and is a common interstitial mineral in lower pyroxenitic layers. This portion of the lower zone in the western lobe appears to be a counterpart of the Clapham bronzitite. Nearly monomineralic pyroxenites with an intervening sequence of harzburgite and pyroxenite in the western lobe are lithologically similar to the upper three formal units in the eastern lobe, although their thickness is substantially less in the west (Vermaak, 1976a).

The lower zone in the southeastern lobe is known only from one borehole intersection. It is about 45 m thick and consists predominantly of orthopyroxene peridotites with thin layers of harzburgite (Buchanan, 1975). Plagioclase is intercumulus and the mafic minerals are iron poor (Buchanan, 1975).

The lower zone in the northern lobe consists of bronzitites and harzburgites, the constituent olivines and pyroxenes displaying a marked trend from magnesium rich at the base to iron rich at the top. The stratigraphically higher zones in this lobe overlap the lower zone so that ultimately the main zone is juxtaposed with metasediments (Van der Merwe, 1976; Barrett *et al.*, 1978). The presence of 40-cm-thick chromitite layers distinguishes the lower zone in this lobe from its counterparts in the other lobes. However, chromitite layers are present in the satellitic intrusion north of Zeerust in similar lower zone rocks. Van der Merwe (1976) estimated the thickness of the lower zone to be 4-6 km, which greatly exceeds the thickness of this zone in both the eastern and western lobes (see Table 6-1).

J.S. De Villiers (1970) recognized chilled main zone rocks in contact with lower zone rocks in the northern lobe, which has been taken as evidence supporting the view that the lower zone represents a discrete magmatic pulse. However, the presence of extensive strike faults has been postulated from geophysical data (Barrett *et al.*, 1978). It has been suggested that the chilled zone may be a consequence of the upfaulting of older lower-zone rocks into still liquid main-zone magma, instability of the floor during crystallization being the cause of the faulting (Barrett *et al.*, 1978).

The Critical Zone. The critical zone is subdivided into the Zwartkoppies pyroxenite and Winterveld norite–anorthosite in the eastern lobe on the basis of the appearance of cumulus plagioclase. Both can be further divided into major units that can be traced for about 45 km northward from the Steelpoort fault.

The base of the Zwartkoppies pyroxenite consists of a unit about 40 m thick in which nearly monomineralic bronzitite is interlayered with feldspathic bronzitite typical of the Zwartkoppies pyroxenite. Chromite appears sporadically as an important cumulus mineral above this unit. At least 13 chromitite layers have been recorded in the Zwartkoppies pyroxenite, the most important of which (the Steelpoort chromitite) is about 1 m thick and can be traced continuously along strike for nearly 80 km.

Igneous layering is very well developed in two harzburgite units in the Zwartkoppies pyroxenite. Cameron and Desborough (1969) noted the presence of 160 individual layers in the lower harzburgite layer which is about 20 m thick.

Whereas intercumulus plagioclase in amounts exceeding 5 volume percent is characteristic of the Zwartkoppies pyroxenite, its appearance as a cumulus phase marks the base of the Winterveld norite–anorthosite. This contact is discordant in that it is not parallel to the layering and is marked by scours and other evidence of erosion (Cameron and Desborough 1969). This contrasts with the boundary between the Serokolo bronzitite and Zwartkoppies pyroxenite, where there is no structural break and the layering in both is conformable (Cameron, 1978).

The Winterveld norite–anorthosite displays the most spectacular layering in the Rustenburg Layered Suite. Individual layers vary in thickness from several centimeters to more than 100 m. Contacts may be gradational or abrupt. The cumulus minerals chromite, orthopyroxene, and plagioclase are present in all proportions, giving rise to layers ranging in composition from norite to monomineralic layers of chromitite, pyroxenite, and anorthosite. Cumulus clinopyroxene is also present in two layers in a 200-m-thick zone (i.e., M unit of Cameron, 1970a), the top of which is about 500 m below the top of the critical zone. The cumulus minerals appear in cyclic repetition in parts of the Winterveld norite–anorthosite ideally in the order chromite → pyroxene → plagioclase, the two most complete macrocyclic units being at the top of the critical zone where the repetition of the sequence chromitite–pyroxenite–norite–anorthosite constitutes the Merensky "Reef" and Bastard Reef cyclic units, the former containing economically viable amounts of platinum metals, copper, and nickel (Vermaak, 1976a).

The critical zone in the western lobe consists of an essentially similar but thinner, lithologic sequence. This zone is apparently not present in the southernmost part of the southeastern lobe (Buchanan, 1975). The critical zone was estimated to be 400 m thick in the northern lobe, but Van der Merwe (1976) followed Willemse's (1969) subdivision and placed the top of the zone at the base of the platiniferous layer, which is 200 m thick in this lobe and which is overlain by 600 m of norites and leuconorites with four pyroxenite layers. It has been suggested (C.F. Vermaak, personal communication, 1979) that this sequence should be included in the critical zone. If this is valid, the critical zone would then be at least 1200 m thick and comparable in thickness to the critical zone in the eastern lobe.

The platiniferous layer, known as the Platreef, differs both in thickness and lithology from the Merensky "Reef," which averages about 15 m in thickness in the eastern and western lobes. At the base of the Platreef is a harzburgite or lherzolite that is rarely pegmatitic. Feldspathic pyroxenite overlies this basal unit and grades upward into a porphyritic pyroxenite. Chromitite veinlets or blebs are present at all stratigraphic levels of the Platreef.

The Main Zone. A succession of relatively homogeneous rocks overlying the critical zone has been subdivided into three subzones on the basis of the compositions of cumulus minerals (Von Gruenewaldt, 1973; Molyneux, 1974). The main zone is characterized by the appearance of abundant cumulus clinopyroxene and an absence of chromite.

The Winnaarshoek norite–anorthosite constituting the lowest subzone in the eastern lobe is similar in some respects to the underlying critical zone. Pronounced layering is primarily the result of the varying proportions of pyroxene and plagioclase, giving rise to a variety of norites and anorthosites (Von Gruenewaldt, 1973). Mottled anorthosites are prominent as a result of the presence of large intercumulus pyroxene grains. A 200-m-thick porphyritic norite constitutes an important marker layer which has been traced for 180 km and 120 km in the western and eastern lobes, respectively. The porphyritic norites contain large crystals of orthopyroxene about 5 mm in diameter, which form about 20 volume percent of the rock and are characterized by small inclusions of plagioclase (Von Gruenewaldt, 1973). The top of the subzone is taken at the uppermost mottled anorthosite layer. The position of this layer coincides with a minor reversal in the trend of the composition of cumulus orthopyroxene and plagioclase, and with the position where inverted pigeonite first appears (Von Gruenewaldt, 1973).

The next subzone (Leolo Mountain gabbronorite in the east) consists of a thick unit of gabbroic rocks that display few variations either in texture or in the proportions of cumulus minerals (Von Gruenewaldt, 1973).

The topmost subzone (Mapoch gabbronorite) starts with a layer of pyroxenite, the thickness of which cannot be determined in the field. An abrupt reversal in the composition of the cumulus minerals takes place at the base of this pyroxenite (Von Gruenewaldt, 1973; Molyneux, 1974). The remainder of this subzone consists of rock types similar to the Winnaarshoek norite–anorthosite and the Leolo Mountain gabbronorite.

The main zone in the western lobe is poorly exposed and details of its stratigraphic sequence are largely derived from borehole core, but there is an overall similarity to the sequence of rock types in the eastern lobe. In the northern lobe the main zone is about 1500 m thick and consists of gabbroic rocks with a prominent layer of noritic troctolite about 1100 m above the Platreef, a lithology that has not been recognized in other parts of the complex (Van der Merwe, 1976). A pyroxenite layer tentatively correlated with the pyroxenite at the base of the Mapoch gabbronorite is developed near the top of the main zone. The main zone has not been identified in boreholes drilled in the southeastern lobe.

The Upper Zone. The appearance of cumulus magnetite marks the base of the upper zone. Layering is prominent and results from varying proportions of cumulus magnetite, olivine, pyroxene, and plagioclase (Von Gruenewaldt, 1973; Molyneux, 1974). The upper zone in the eastern lobe is subdivided into three subzones on the basis of the appearance of significant volumes of cumulus magnetite (Magnet Heights gabbronorite subzone), cumulus iron-rich olivine (Ironstone magnetite gabbro subzone), and cumulus olivine + cumulus apatite (Luipershoek olivine diorite subzone).

The Magnet Heights gabbronorite contains the main magnetitite layer, which is about 1.5 m thick. Nine other magnetitite layers are present in the lower part of this unit but none exceeds about 30 cm in thickness. The layers generally have sharp basal contacts and grade upward through feldspathic magnetitite to magnetite gabbro. The main magnetitite layer crops out prominently and can be traced nearly 60 km along strike in the eastern lobe. A troctolite layer approximately 5 m thick below the main magnetitite layer marks the first appearance of olivine in major amounts in the upper zone and is stratigraphically the lowest of several troctolite layers that are characteristic of the upper zone (Molyneux, 1974).

Two to three hundred meters of homogeneous magnetite gabbro form the upper part of the subzone diversified only by the appearance of a layer of mottled anorthosite, the lateral continuity of which appears to be limited (Molyneux, 1974).

The base of the Ironstone magnetite gabbro is marked by the appearance of a 5-m-thick troctolite, which has been referred to in earlier literature as the "sisal marker" (Molyneux, 1974). The remainder of the subzone consists of magnetite gabbro with seven thin magnetitite layers. Cumulus olivine is found in minor amounts at various levels. The appearance of cumulus apatite in troctolite marks the base of the upper subzone. Iron-rich olivine is an important cumulus mineral through most of the subzone and the composition of the feldspar is andesine (Von Gruenewaldt, 1973; Molyneux, 1974). Seven magnetitite layers are developed in the lower half of the subzone, the topmost magnetitite layer attaining a thickness of up to 10 m (Von Gruenewaldt, 1973; Molyneux,

1974). Above this layer 200 m of olivine-bearing diorite are overlain by hornblende-bearing diorite that contains intercumulus K-feldspar and quartz with scattered fayalite. The top of the Luipershoek olivine diorite is drawn where this subzone is overlain by one of the following rock types: metasediments of the Transvaal Supergroup, Rashoop Granophyre Suite, Lebowa Granite Suite, or Rooiberg Felsite Group.

Details of the stratigraphy of the upper zone in the other lobes are less complete because of lack of outcrop. A thickness of 1100 m is given for this zone in the northern lobe, where it consists of intercalations of gabbro, magnetite gabbro, anorthosite, magnetitite, and olivine diorite (Van der Merwe, 1976). Twenty magnetitite layers have been identified (Van der Merwe, 1976). The upper zone in the western lobe also contains magnetitite, troctolite, and olivine diorite layers in addition to magnetite gabbro (Coertze, 1970).

Borehole data from the southeastern lobe enabled Buchanan (1975) to recognize four subzones in the upper zone, which overlies orthopyroxene peridotite believed to be stratigraphically equivalent to the lower zone. The stratigraphically lower subzone that immediately overlies the peridotite has been fully intersected in only one borehole and consists of gabbro and norite characterized by an abundance of anorthosite and pegmatoid, consisting of variable proportions of ortho- and clinopyroxenes, plagioclase, and magnetite. Sulfide mineralization is a feature of this subzone, pyrrhotite, chalcopyrite, and pyrite being the common species (Buchanan, 1975). Gabbro and norite constitute the bulk of the second subzone, whereas magnetite-rich gabbro and magnetite-rich diorites are characteristic of subzones three and four, respectively. Fourteen magnetitite layers ranging from 0.1 m to 1.6 m in thickness were intersected in one borehole (Buchanan, 1975). The upper zone in the southeastern lobe has a minimum thickness of about 1900 m (Buchanan, 1975).

Basic and ultrabasic pegmatoids in the form of pipelike bodies are prominent throughout the Rustenburg Layered Suite. The following types are recognized:

(1) Bronzitite pegmatoids are particularly abundant in the lower zone west of the Pilanesberg alkaline intrusion in the western lobe. The pipes in this area are marked at surface by gossans, and below the level of weathering they consist of bronzite, phlogopite, and nickeliferous sulfides. The distribution of the pipes correlates with faulting and fracturing on both regional and local scales and with the thickest accumulations in this area of Bushveld rocks (Vermaak, 1976a). Underground mapping indicates maximum development of pipes in pyroxenite layers with horizontal extension immediately below harzburgite layers (Vermaak, 1976a). Pyroxenite is intensely fractured, whereas norite and anorthosite are virtually unfractured (Vermaak, 1976b). Vermaak (1976a, b) proposed injection of sulfide fluids from below with the more compact norites and anorthositic layers constituting barriers to upward migration. He further proposed that the pipes formed as a consequence of "tectonic overpressure" along major faults which provide channel ways for circulating sulfur- or sulfate-charged fluids. Because the faulting can be related to the intrusion of the Pilanesberg alkaline suite, which postdates the Bushveld Complex by some 500 Ma, Vermaak (1976b) considers these pipes to be post-Bushveld in age, although the complex supplied the metals that make up the ore bodies.

(2) Dunite pipes are found especially in the lower zone, but platiniferous pipes with a core of hortonolite dunite and a rim of dunite crop out within the critical zone.

(3) Diallagite pegmatoids forming irregular masses, anastomosing veins, and pipelike bodies transgressive to the layering are common in the lower part of the critical zone but also are found in the main and upper zones (Willemse, 1969). Olivine, hornblende, phlogopite, plagioclase, and magnetite are also present in variable proportions.

(4) Pegmatoids with an assortment of rounded inclusions are found in the critical zone (Ferguson and McCarthy, 1970). The pegmatoid matrix consists of clinopyroxene, with lesser amounts of orthopyroxene and plagioclase. The most abundant inclusions are of leucoamphibolite and amphibolite, and other inclusions consist of chromitite and mottled anorthosite.

(5) Magnetitite pegmatoids are particularly abundant in the main and upper zones as

approximately circular bodies that taper out at depth.

(6) A vermiculite pegmatoid pipe in the eastern lobe in the upper zone contains coarse books of vermiculite and minor crystals of coarse diallage.

Different processes have been proposed for the genesis of the various pegmatoids, including replacement (Cameron and Desborough, 1964), filling of dilation fractures (Cameron, 1970b), and forceful emplacement related to aggregation of volatile fluids (Ferguson and McCarthy, 1970). Many pegmatoids have not been studied in detail, but it is evident that they have not all been formed in the same fashion.

Rashoop Granophyre Suite

Acid rocks comprising granophyre, micrographophyre, and porphyritic granophyre have been assigned to the Rashoop Granophyre Suite. Considerable confusion exists regarding the granophyric rocks of the complex, but there is no doubt that granophyric textures in acid rocks were formed by different processes. The granophyres are composed of quartz and feldspar with minor amounts of hornblende or biotite. These rocks range in grain size from fine to coarse or porphyritic. Granophyres have the following field relations.

(1) At the contact of the Rustenburg Layered Suite with overlying Rooiberg felsites. Origin: partial to complete melting of felsite.
(2) At the contact of the Lebowa Granite Suite and the overlying Rooiberg felsites or Transvaal metasediments. Origin: direct crystallization from magma genetically related to that from which the Lebowa Granite Suite crystallized (Steyn, 1962; Rhodes, 1975); or crystallization from a magma genetically unrelated to the Lebowa Granite Suite (Walraven, 1976); or metasomatic reaction (Strauss, 1947, 1954; Ianello, 1971); or metamorphic effect of recrystallization of felsite (De Waal, 1972).
(3) Associated with discontinuous lenses of metasediments between Lebowa granites above and the Rustenburg Layered Suite below (i.e., the pseudogranophyre of Strauss and Truter, 1944). Origin: metamorphism and partial melting of metasediments (Strauss and Truter, 1944; Kuschke, 1950).
(4) Within the Lebowa Granite Suite adjacent to either crosscutting plugs of younger granite or intrusive basic sheets. Origin: metamorphic effects of intrusions (Lenthall, 1973; Lenthall and Hunter, 1977).
(5) Intrusions with sharp contacts in the Rooiberg felsites. No origin specifically given but magmatic origin implied (Clubley-Armstrong, 1977).

Lebowa Granite Suite

The coarse-grained granitic rocks that clearly intrude the Rooiberg felsites and the Rustenburg Layered Suite have been previously called the Bushveld granites. Field relations have demonstrated that a number of granites of different type were included in this sack name. These granitoids are now called the Lebowa Granite Suite. Two distinct types of granite are currently regarded as constituting this suite.

The most widespread is the Nebo granite, a term which has been introduced to replace such names as Bushveld, Main, or Sekhukhune granite. The Nebo granite is largely confined within the four main lobes of the Rustenburg Layered Suite, but minor occurrences separated from the main mass are found in the eastern and southeastern lobes (Fig. 6-3).

The Nebo granite exhibits a crude stratiform layering that typically consists of coarse-grained gray granite at the base, grading up through medium-grained gray and red granite, then through red granophyric granite, to granophyre; but a full sequence of layers is not always present. The Nebo granite has a relatively simple mineralogy consisting of quartz, feldspar, and minor but variable amounts of hornblende or biotite. These mafic minerals are more abundant in the coarser grained varieties at the base of the stratiform granitoid sheet (Lenthall and Hunter, 1977). Accessory minerals are zircon, fluorite, apatite, sphene, epidote, and magnetite. Alteration of primary minerals results in the formation of biotite, chlorite, saussurite, epidote, hematite, and fine aggregates of clay minerals. Aplitic phases within the stratiform granite mass, which has a thickness of about 3000 m, belong to two types. Irregularly shaped aplitic lenses, commonly with diffuse margins, contrast with the aplitic granite that forms sharply defined sill- or dike-like bodies, the former generally dipping conformably with the layering of the stratiform granite.

North of Potgietersrus three discrete masses of granite transect the layering of the Nebo granite. The former consist of a coarse-grained red rock containing quartz and feldspars with variable amounts of chlorite and, less commonly, biotite. This granite is called the Bobbejaankop granite in this area and represents a variety of the Nebo granite. It can be distinguished from the red Nebo granite by its miarolitic texture, the presence of clusters of tourmaline crystals, and the dull, earthy-red appearance of its feldspars (Lenthall and Hunter, 1977). In the Nebo granite, quartz and feldspar form a complexly interlocking mosaic whereas the quartz grains in the Bobbejaankop granite have tended to coalesce to form discrete strings of grains. The Bobbejaankop granite has a vertical or steeply inclined contact with the stratiform granite along its eastern margin but the western margin has a shallow dip nearly conformable to that of the stratiform granite. Coarse pegmatite has intruded sporadically along this contact.

The second distinct granite type assigned to the Lebowa Suite has been named the Makhutso granite (De Bruiyn and Rhodes, 1975). It is a biotite-rich granite intrusive into the Nebo granite in the form of small dikes, sheets, and stocks. The granite is fine grained near its margin but grades into coarser grained and slightly porphyritic varieties toward the center of intrusions (De Bruiyn and Rhodes, 1975). The granite consists of quartz, orthoclase, sodic oligoclase, biotite, and hornblende.

Rooiberg Group

The problem of the stratigraphic position of the Rooiberg Group has already been mentioned. Whatever the genetic relationship of these felsites to the Bushveld Complex may be, there is no question that they predate both the Rustenburg Layered Suite and the Nebo granite. The stratigraphic sequence of the Rooiberg Group is not well understood. Rhodes and Du Plessis (1976) proposed a twofold division, consisting of a lower felsite sequence distinguished by its homogeneous character and an upper felsite sequence characterized by fluidal textures and the presence of intercalated sedimentary and pyroclastic beds. The top of the lower felsite sequence is marked by a zone of prominent quartzite xenoliths up to several decimeters in size. The xenoliths are set in vesicular or flow-banded felsite which also contains lenses of quartzite breccia. The quartzite xenoliths are regarded as being derived from the Transvaal metasediments, but it is difficult to account for these blocks by normal processes of volcanic eruption (Rhodes and Du Plessis, 1976). The base of the lower felsite sequence is commonly recrystallized and partially mobilized (Von Gruenewaldt, 1968, 1972) as a consequence of the intrusion of the Rustenburg Layered Suite. Micrographic felsite with microgranophyric sheetlike intrusions are a feature of this basal zone.

The upper felsite sequence consists of felsite flow locally interbedded with minor volcanic breccias, sandstones, and ash-fall tuffs (Rhodes and Du Plessis, 1976). The felsite flows are massive, generally porphyritic, vesicular, and commonly flow banded. The sandstone beds rarely exceed 10 m in thickness and appear to fill local channels on the surface of earlier flows. A persistent agglomerate at the base of the upper sequence is locally overlain by tuffaceous beds, frequently referred to as the "Union Tin shale" because of the presence of disseminated cassiterite.

The southern outcrops of the Rooiberg Group north of Witbank have also been subdivided into two sequences (Clubley-Armstrong, 1977). The lower unit consists of alternating layers of black felsite which is mainly pseudospherulitic, amygdaloidal, and in places glassy, overlying basal micrographic felsite and granophyre. Discontinuous sandstone beds and subordinate lenses of pyroclastic rocks are interbedded, especially in the upper part. At the top of the sequence dark brown to black felsite contains lithophysae ranging in size from 2 cm to 25 cm and filled with concentrically arranged crystals of secondary quartz, K-feldspar, calcite, or epidote (Clubley-Armstrong, 1977).

The base of the upper sequence is taken at the appearance of a prominent and continuous quartzite bed. Overlying the quartzite, red porphyritic felsite is mainly massive with sporadic quartzite and tuffaceous lenses. Xenoliths of white, massive recrystallized quartzite are scattered through the felsite and are concluded to be not of local provenance (Clubley-Armstrong, 1977). Impersistent lenses of volcanic breccia overlain by black or amygdaloidal felsite are overlain by a dark laminated mudstone with interbedded quartzite and felsite. This mudstone may be a correlative of the "Union Tin shale." The remainder of the sequence is made up predominantly of flow-banded, red porphyritic felsites. The maximum thickness is estimated to be nearly 4000 m in this area (Clubley-Armstrong, 1977).

Satellites

A number of small basic intrusions, mainly near the Vredefort Dome (Fig. 6-5), are possibly of Bushveld age. The Losberg intrusion has yielded an imprecise Rb-Sr age of 1881 ± 282 Ma (Davies *et al.*, 1970), and a bronzite granophyre dike and alkali granite in the collar of the Vredefort structure are dated 1970 Ma and 2040 Ma, respectively.

Losberg. This intrusion into sediments of the Transvaal Supergroup has an area of about 20 km^2; it is underlain by shales and capped by quartzites at its western end. Three units have been identified from the base up (Abbott and Ferguson, 1965):

(1) 15m of orthopyroxene–olivine cumulate (harzburgite)
(2) 10 m of plagioclase–orthopyroxene–clinopyroxene cumulate (quartz norite)
(3) 95 m of plagioclase–clinopyroxene cumulate (quartz gabbro)

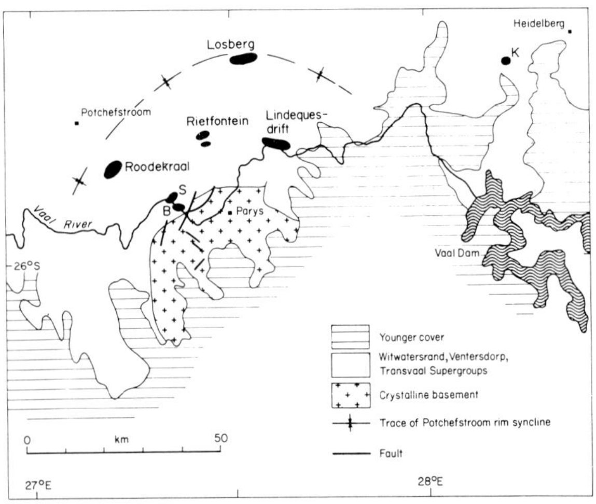

Figure 6-5. Location of presumed satellites (black) of the Bushveld Complex around the Vredefort Dome. B, Baviaanskranz; K, Kaffirskraal; S, Schurwedraai; +, Archean rocks of Vredefort Dome.

The quartz gabbro contains postcumulus quartz and micropegmatite and grades into augite granophyre with increasing quartz content. Rhythmic layering in the basal harzburgite is defined by olivine-rich and relatively olivine-poor layers. The subhorizontal layering in the harzburgitic cumulate and the disposition of the three units indicate that the intrusion has a sheet-like form (Abbott and Ferguson, 1965). It is probable that the sheet originally had a wider distribution because outcrops of quartz norite are found 3–4 km south of the main intrusion (Abbott and Ferguson, 1965).

Roodekraal. The Roodekraal Igneous Suite covers an area of 12–15 km^2. Andesitic lavas representing the volcanic phase flowed over an eroded surface of deformed sediments of the Transvaal Supergroup. Dioritic magma intruded the lavas and sediments. Small dikes of albitite represent the youngest intrusive phase (Bisschoff, 1972a). The different rock types have an approximately concentric distribution. A zone of andesite surrounds a core of coarse- to medium-grained diorite consisting of plagioclase phenocrysts in a matrix of plagioclase and augite; nonporphyritic varieties are also present. A sill of diorite intrusive into the lava near the southeastern edge of the complex has a coarse-grained light-colored upper selvage which is composed of hornblende in a graphic intergrowth with plagioclase.

The albitite dikes are porphyritic and consist of plagioclase phenocrysts set in a groundmass of plagioclase with accessory chlorite and hornblende. Other dikes contain up to 10 volume percent quartz or small phenocrysts of barkevikite (Bisschoff, 1972a).

The andesitic lavas are predominantly porphyritic but are amygdaloidal in places. Most phenocrysts are composed of pure albite but others contain remnants of oligoclase. Hornblende and chlorite are the principal mafic minerals, with accessory magnetite and apatite (Bisschoff, 1972a).

Lindequesdrift. The elongated Lindequesdrift intrusion, emplaced into dolomites of the Transvaal Supergroup, is about 11 km long and is composed of (Bisschoff, 1972a):

(1) Lamprophyre (spessartite) with phenocrysts of brown hornblende in a fine- to medium-grained groundmass of augite, plagioclase, and hornblende
(2) Medium-grained syenodiorite, composed of hornblende, plagioclase, and potassium feldspar
(3) Fine-grained albite syenite dikes, composed of albite, green hornblende, and accessory augite and brown hornblende
(4) Pegmatitic schlieren in the lamprophyre, composed of albite with accessory hornblende

The outcrop in this area is poor, but lamprophyre appears to form the major part of the intrusion with syenodiorite in the core. Contacts between lamprophyre and syenodiorite are gradational in places (Bisschoff, 1972a).

Rietfontein. The Rietfontein intrusion, about 18 km^2 in area, has intruded Ventersdorp lavas and sediments of the Transvaal Supergroup along strike faults (Bisschoff, 1973). The southern half of the complex consists of basic and ultrabasic rocks with alkali granite in the north. Layers of wehrlite, olivine gabbro, troctolite, and picrogabbro dip moderately toward the center of the intrusion. The alkali granite is intrusive into the layered sequence and is composed of feldspar (antiperthite), interstitial quartz, and arfvedsonite or aegirine, the mafic minerals constituting about 8 volume percent. Aegirine is present only in the quartz-poor phase of the alkali granite at its contact with dolomite. The syenitic phase has a patchy distribution in the fine-grained alkali granite into which it grades (Bisschoff, 1973).

Schurwedraai and Baviaanskranz. The two small Schurwedraai and Baviannskranz plutons were emplaced into the Witwatersrand Supergroup along strike faults. They consist of medium- to fine-grained alkali granite and syenite, the latter usually being coarser and more even grained than the commonly porphyritic granites. The rocks are composed of albite, microcline, quartz, aegirine, riebeckite, and arfvedsonite (Bisschoff, 1973).

Dikes and Sills. Many of the sills are differentiated and are compositionally similar to rock types in the main zone of the Bushveld Complex. They crop out in a belt about 3 km wide around the northwestern quadrant of the Vredefort Dome. The sills consist of diabase, hyperite (gabbronorite), norite, and pyroxenite with color indices ranging from 40 to 92 (Bisschoff, 1972b). Some of the gabbronorites have pronounced poikilitic texture.

Mariupolite (nepheline syenite) dikes are associated with alkali granite intrusions around the Vredefort Dome. They are intrusive mainly into the roof zones of the plutons and their sedimentary cover. The dikes consist of a fine- to medium-grained portion and a coarse pegmatitic portion which forms narrow bands or irregular elongated patches (Bisschoff, 1973). The rocks are composed of albite, nepheline, aegirine, and microcline. Texturally they range from equigranular to porphyritic, but trachytic texture is prominent in some of the finer grained varieties. The pegmatitic bands vary considerably in composition. Some are composed almost entirely of microcline or nepheline, whereas others are composed of varying amounts of microcline, nepheline, and aegirine (Bisschoff, 1973).

Dikes of bronzite granophyre represent the last intrusive event of this age around the Vredefort Dome. Whereas all the preceding intrusions are cut by pseudotachylite, the granophyre dikes postdate this brecciation (Bisshoff, 1972a). The largest dike can be traced for 16 km along strike with a maximum width of 50 m. The dikes consist of bronzite, plagioclase, and micropegmatite and contain numerous xenoliths of quartzite and subordinate granite (Bisschoff, 1972a).

These intrusions in the vicinity of the Vredefort Dome are regarded as coeval with the Bushveld Complex (Bisschoff, 1972a,b, 1973). A small layered intrusion 16 km southwest of Heidelberg may also be of Bushveld age but, except for its post-Ventersdorp age, no other indication of its stratigraphic position is available. The Kaffirskraal intrusion, 1.5 km^2 in extent, consists of a marginal zone of porphyritic norite overlain unconformably by a layered zone of magnetite pyroxenite and magnetitite (Frick, 1975a). The porphyritic norite, about 1 m thick, contains large phenocrysts of orthopyroxene and, at one locality, dips at 70° toward the center of the intrusion. The magnetite pyroxenite displays prominent igneous layering, defined by preferred orientation of clinopyroxene, which dips gently toward the center of the intrusion. An 8-m-thick magnetitite layer forms the top of the intrusion.

The Marico intrusion (Fig. 6-1) is of Bushveld

age and consists of rocks representative of the lower zone. The lowermost rocks in contact with the sediments of the Transvaal Supergroup are fine-grained bronzite norites that grade upward into pyroxenites alternating with harzburgite. Chromitite layers at the top of the sequence generally dip centripetally at shallow angles and distinguish this layered suite from the western lobe, suggesting similarity with the northern lobe.

6.3. Age of the Bushveld Event

The Rb-Sr systematics of the Rustenburg Layered Suite have been discussed by Hamilton (1977). His results from the basal upper zone, Merensky "Reef," and lower zone yield a weighted mean age of 2095 ± 24 Ma. This age is consistent with the argon age of 2098 ± 12 Ma obtained from the Layered Suite (Coertze *et al.* 1978) and with the Rb-Sr age of 2050 ± 50 Ma for biotite from the Rustenburg platinum mine (Nicolaysen *et al.*, 1958).

Initial $^{87}Sr/^{86}Sr$ ratios increase abruptly in a stepwise manner at horizons characterized by cryptic reversals in the layered suite from 0.7056 to 0.7077 (Hamilton, 1977). Ferrodiorites from the upper zone have poor-quality Rb-Sr systematics, indicating partial rehomogenization of strontium isotopes at about 1940 Ma (Hamilton, 1977). This could have resulted from contact metamorphism by the Nebo granite, which yields similar zircon ages of 1920 ± 40 Ma (Coertze *et al.*, 1978) and Rb–Sr mica and whole-rock ages of about 1950 Ma (Nicolaysen *et al.* 1958; Davies *et al.*, 1970).

The Makhutso granite has yielded a U-Pb age of 1670 ± 70 Ma (Coertze *et al.*, 1978). This age is difficult to interpret. If it represents an age of intrusion, it implies that the Makhutso granite is a shallow intrusion related to acid igneous activity during Early Waterberg sedimentation. This would imply that the Makhutso granite is unrelated to the Bushveld event. Alternatively the age could reflect resetting of the Nebo granite isotopic systematics. Seven samples of Nebo granite adjacent to the main outcrops of Makhutso granite yield a Rb-Sr age of 1780 ± 166 Ma and an initial ratio of 0.724 ± 0.016 (Hunter and Hamilton,1978). Large errors limit the interpretation of these results but they may imply isotopic resetting caused perhaps by Early Waterberg igneous activity. Published geochemical data (De Bruiyn and Rhodes, 1975), however, are not consistent with the Makhutso granite being a partial melt of the Nebo granite.

The Rooiberg Group felsites have yielded scattered Rb-Sr isochron plots indicating one or more periods of open-system behavior (Hamilton, 1975). The Rb-Sr systematics may have been perturbed during intrusion of the Rustenburg Layered Suite, or the Nebo granite, or during some later event. Open-system Rb-Sr behavior is common in acid volcanic rock and there is also evidence of at least partial melting of the felsites (Von Gruenewaldt, 1972). However, a minimum age for the felsites can be deduced from the U-Pb age of 2090 ± 35 Ma obtained on a porphyritic granophyre intrusive into sedimentary rocks overlying the Rooiberg Group north of Witbank (quoted from Clubley-Armstrong, 1977).

Zircon from granophyres of magmatic origin, mainly from the area west and north of Pretoria, yield an age of 2050 ± 30 Ma (Coertze *et al.*, 1978).

In summary it seems that the following sequence of events can be provisionally established for the Bushveld Complex:

(6) Makhutso granite	1670 ± 70 Ma?
(5) Nebo granite	1920 ± 40 Ma
(4) Granophyre	2050 ± 30 Ma
(3) Porphyritic granophyre	2090 ± 30 Ma
(2) Rustenburg Layered Suite	2095 ± 24 Ma
(1) Rooiberg Group	?

An upper age limit for the Rooiberg Group is provided by the ages of 2224 ± 50 Ma and 2263 ± 85 Ma obtained for the Hekpoort volcanic rocks and Pretoria Group shales (D. Crampton, unpublished data; Hunter and Hamilton, 1978). It is therefore possible that the Rooiberg Group is not significantly older than the Rustenburg Layered Suite and that intrusions of the Bushveld Complex occupied at most about 150 Ma.

6.4. Geochemistry

The composition of the parental Bushveld magma is not known; tholeiitic basalt has been widely accepted despite the tenuous evidence. Attempts have been made to estimate the composition of the parental magma indirectly. Hamilton (1977) based his calculations on the modal abundances and mineral compositions of some lower zone rocks and the relative volumes of rock types in the lower zone to derive a bulk composition for this zone. He concluded that the initial magma was Mg rich (MgO about 22 percent) and SiO_2 saturated (SiO_2 probably 50–55 percent). Hamilton (1977) further deduced from consideration of the petrography and mineralogy of the critical zone rocks that critical zone magmas remained relatively MgO rich.

On the basis of modal and mineral analyses Cameron (1978) estimated the MgO content of the critical zone to be about 18.5 percent. Similar calculations for the main and upper zones indicate contents of 5.8–7.3 percent MgO depending on the total thickness used in the estimation. Combination of these results suggests that the weighted-average MgO content for the layered sequence is between 13.0 and 13.4 percent. Cameron (1978) further examined the Cr content and estimated that this would be in excess of 1000 ppm without making allowance for chromium in the main and upper zones. These results strongly suggest that the parent magma must have been considerably more basic than average tholeiitic basalt. Cameron (1978) identified olivine–tholeiitic basalt for which Nockolds (1954) gave an MgO content of 14.07 percent as a more realistic parental type. He noted, however, that this assumes that the layered suite is essentially a continuous differentiation series.

Early olivines and orthopyroxenes in the lowest zone of the layered suite are highly Mg rich (Fo_{95} and En_{90} respectively; Willemse, 1969). These compositions are compatible with crystallization from a highly basic magma. Continued precipitation of orthopyroxene of relatively high Mg content (En_{84-78}) in the critical zone indicates that the magma remained Mg rich. The major- and trace-element chemistry changes markedly above the critical zone. Magnesia decreases to a mean of about 6 percent in the main and upper zones, sympathetically with decreases in mean Cr and Ni contents from about 1500 ppm and 200 ppm, respectively, to 200 ppm and 70 ppm, respectively. There is, however, local enrichment in both Cr and Ni at the level of the main magnetite layer in the upper zone. The impoverishment in Mg is reflected in the increase in Fe content of both olivine and orthopyroxene. The Fo content of the former mineral decreases from 75 to 18 percent from the base to the top of the upper zone, where the En content of orthopyroxene decreases to 29 percent. The magnetitite layers in the upper zone show a steady decrease in V_2O_5 content from approximately 2 percent in the main magnetite layer to less than 0.3 percent in the topmost layer in the eastern lobe (Molyneux, 1970).

The lower zone in the northern lobe contains orthopyroxenes with En contents as high as 95 percent and chromitite layers (Van der Merwe, 1976), unlike equivalent zones in the other lobes of the complex. Chromite is a minor to trace element in the lower zone of the eastern lobe, where its maximum concentration is about 3 volume percent in one layer of the Jagdlust harzburgite (Cameron, 1978).

The association of chromitite layers with olivine cumulates in the lower zone of the northern lobe is comparable to the occurrence of chromitite in the Stillwater, Great Dyke, and Muskox intrusions and contrasts with their association with bronzitites and anorthosites, which is more typical of the Bushveld Complex. These differences in association will be discussed in the following section.

Hamilton (1977) determined some trace-element abundances in cumulus orthopyroxene and plagioclase from the western lobe and cumulus orthopyroxene set in poikilitic plates of clinopyroxene from the eastern lobe. Relevant phenocryst–matrix partition coefficients were used to estimate elemental abundances and their ratios in coexisting magmas. These estimates are subject to some error because partition coefficients are dependent on pressure, temperature, oxygen fugacity, and magma composition. Despite these uncertainties, the predicted abundance ratios and rare earth element (REE) patterns are in good agreement for all three mineral phases from the complex (Hamilton, 1977). Orthopyroxene, the most abundant phase in the lower zone, strongly discriminates against Rb, Sr, K, and Ba, and hence abundance ratios of these elements should approximate those of the initial magma. Hamilton (1977) deduced the following abundance ratios for the initial magma of the layered suite: Rb/Sr $\sim$ 0.25; K/Rb $\sim$ 90; K/Sr $\sim$ 23; K/Ba $\sim$ 18, and Ba/Sr $\sim$ 1.4; $Ce_n/Yb_n \sim 11$. These ratios are unlike those for the majority of magma types except for kimberlites and nephelinites and those estimated for the initial Great Dyke magma (Hamilton, 1977).

The Nebo granite and those granophyres derived from magmatic crystallization have remarkably consistent major-element chemistry (Table 6-2). Although the mean values of these two groups are similar, the Nebo granite displays subtle chemical variations. The gray basal granites commonly have slightly lower SiO_2 contents and K_2O/Na_2O ratios

Table 6–2. Representative Analyses of Rocks from the Bushveld Complex[a]

	1	2	3	4	5	6	7
	Harzburgite	Dunite	Pyroxenite	Pyroxenite	Norite	Anorthosite	Gabbro
SiO_2	41.40	43.70	54.70	52.00	51.70	45.72	53.40
TiO_2	0.07	0.12	0.26	0.18	0.20	0.04	0.18
Al_2O_3	3.16	1.04	5.41	10.02	13.19	34.51	17.51
F_2O_3	0.60	1.74	1.88	0.93	1.48	0.70	1.17
FeO	6.63	11.60	6.94	7.50	7.00	0.08	6.12
MnO	0.13	0.22	0.17	0.19	0.19	Tr	0.21
MgO	42.50	38.50	24.20	20.60	17.30	0.79	7.10
CaO	0.88	1.66	3.22	6.05	6.76	15.40	11.60
Na_2O	0.22	0.06	0.87	0.56	1.20	1.64	3.79
K_2O	0.01	0.01	0.07	0.08	0.14	0.12	0.21
H_2O^+	2.86[b]	1.07[b]	1.92[b]	1.21[b]	1.04[b]	0.55[b]	0.60[b]
H_2O^-							
P_2O_5	0.03	0.03	0.04	0.02	0.04	0.04	0.02
CO_2	—	—	—	—	—	—	—
F	—	—	—	—	—	—	—
Ba	26	12	67	35	55	150	70
Co	125	116	62	64	57	—	60
Cr	6550	4080	2890	2100	1650	1200	95
Ca	190	17	32	142	10	2	12
Ni	1400	750	440	320	220	14	120
Rb	—	—	3	—	2	1.5	3
Sr	6	5	35	100	140	360	150
Y	—	—	—	—	—	—	—
Zr	Tr	45	60	30	37	12	45

[a] Columns 1, 2, and 3, lower zone; columns 4, 5, 6, critical zone; columns 7 and 8, main zone; columns 9, 10, 11, upper zone; column 12, mean of 3 analyses; column 13, mean of 5 analyses; column 14, Nebo granite, mean of 37 analyses; column 15, Bobbejaankop granite, mean of 6 analyses; column 16, Makhutso granite, mean of 20 analyses. Data sources: Columns 1–11, Liebenberg (1960); columns 12–15, unpublished data of D. R. Hunter; column 16, De Bruiyn and Rhodes (1975).

[b] Lost on ignition.

and higher contents of CaO than the more medium-grained, stratigraphically higher granites.

Trace-element contents show greater variation. The basal granites have Rb and Sr contents of about 100 and 120 ppm, respectively. Granites at stratigraphically higher levels have increased contents of Rb (up to 350 ppm) and decreased Sr contents (about 30 ppm). Barium shows a sympathetic decrease in the same sense. Some of the magmatic granophyres have contents of Rb, Sr, and Ba similar to the lower granites. Rare earth element patterns of the Nebo granite are distinguished by steep light REE and flat heavy REE slopes with an increasing negative Eu anomaly at higher stratigraphic levels.

The stanniferous Bobbejaankop granite has relatively high contents of CaO and Rb, as compared with the stratiform Nebo granite. The common occurrence of calcite and fluorite in vugs and miarolitic cavities in this granite accounts for the increased Ca content. Strontium, barium, and titanium are markedly impoverished in the Bobbejaankop granite and serve to distinguish it from the Nebo granite (Lenthall and Hunter, 1977).

Although the Rooiberg felsites have a more variable chemical composition with a greater range of SiO_2, CaO, Na_2O, and K_2O contents, their bulk chemistry is broadly similar to that of the Nebo granite and the magmatic granophyres. Barium, strontium, and titanium contents are generally high. Plots of Ti:Fe, Ti:Mg, Fe:Sc, etc, of felsites and granites reflect good linear correlations, with the felsitic rocks generally forming a cluster at the less differentiated end of the trends. Contents of Rb, Cs,

8 Norite	9 Anor- thosite	10 Trocto- lite	11 Diorite	12 Felsite	13 Grano- phyre	14 Granite	15 Granite	16 Granite
52.10	52.40	41.50	47.12	72.59	74.47	74.43	75.08	74.76
0.14	0.19	2.52	2.36	0.31	0.27	0.19	0.06	0.18
18.05	26.85	17.52	19.17	11.79	11.95	12.09	12.02	12.99
3.19	1.67	4.93	4.08	2.51	1.76	0.67	1.00	0.61
4.02	1.37	16.25	9.83	2.30	1.58	2.15	0.81	0.96
0.11	0.05	0.21	0.15	0.13	0.07	0.04	0.05	0.02
8.75	0.77	5.86	3.41	0.19	0.13	0.12	0.13	0.27
11.10	10.48	7.10	9.40	1.06	0.39	0.84	1.46	0.78
2.19	0.98	2.80	3.60	2.76	3.06	3.25	2.26	3.40
0.21	0.70	0.33	0.36	5.02	5.18	5.03	4.65	5.07
0.55[b]	1.58[b]	0.44[b]	0.61[b]	0.69	0.53	0.53	0.91	0.46
				0.09	0.12	0.08	0.16	0.14
0.04	0.11	0.08	0.12	0.04	0.03	0.04	0.02	0.04
—	—	—	—	—	—	0.04	0.65	—
—	—	—	—	0.10	0.21	0.21	0.34	0.13
120	330	120	126	1223	1213	745	166	535
30	5	100	35	5.5	—	2	1.5	—
500	47	112	104	—	—	—	—	—
10	60	5	37	—	—	—	—	—
125	20	70	55	—	—	—	—	—
3	27	4	4	181	198	205	418	420
185	340	200	205	65	55	35	8	71
—	—	—	—	67	71	91	112	—
30	25	51	45	397	437	336	307	269

Ba, and Sr plotted against the magnitude of Eu depletion yield smooth curves from felsites to the Bobbejaankop granite. Europium depletion increases as the concentrations of Ba and Sr decrease, antipathetically to the increase in Rb and Cs (Hunter, 1973). Some representative analyses of Bushveld rocks are given in Table 6-2.

6.5. Petrogenesis: Origin of Parent Magmas and Igneous Layering

A mantle-derived magmatic origin for the Rustenburg Layered Suite is now well established, but the mechanisms whereby the igneous layering evolved are more equivocal. The persistence along strike for distances of up to 100 km of individual layers and groups of layers is a remarkable feature of the Bushveld Complex. Any mechanism to account for the repetitive layering must also be able to account for this lateral persistence. Theories to account for the layered stratigraphy can be broadly divided into two groups: (1) differentiation of a magma at depth with subsequent intrusion of the separate fractions with or without further differentiation *in situ*, or (2) differentiation of magma *in situ* with or without additions of magma to the differentiating body.

Other factors that are considered either jointly or separately to have been responsible for the evolution of repetitive cyclic units are convective overturn and changes in pressure, temperature, or oxygen fugacity.

The presence of transgressive contacts between zones, inclusions of rocks of lower stratigraphic layers in upper layers, sudden reversals in mineral compositions, and geochemical and isotopic variations have all been cited as evidence favoring additions of new magma or of discrete intrusions of magma differentiated at depth.

Penecontemporaneous deformation of the crystallizing Bushveld magma is well established on both local and regional scales. Vermaak (1976a) has demonstrated that the base of the platiniferous Merensky "Reef" rests on a gently arched sequence on a regional scale in both the eastern and western lobes. Rocks of the critical and lower parts of the main zones together with underlying floor rocks are updomed at Malope (Fig. 6-3) within the eastern lobe, where upper zone rocks crosscut underlying zones, and anorthositic rocks display a conspicuous parallel elongation of their more mafic blebs and mottles (Marlow and Van der Merwe, 1977). Penecontemporaneous faults with displacements of 25 cm and a vertical extension of 3 m are reported from the critical zone in the eastern lobe (Ferguson and Botha, 1963).

Saucer-like depressions with a maximum diameter and depth of about 600 m and 26 m, respectively, are found in the anorthositic footwall of the Merensky "Reef." The depressions are filled by pegmatoid, porphyritic pyroxenite, and other layers of the Merensky cyclic unit. The walls of the depressions grade from gently inclined to overhanging and truncate the layering of the footwall anorthosite (Ferguson and Botha, 1963). Petrofabric study revealed a strong mineral lineation within the layering in the depressions, which contrasts with the isotropic fabric in the Merensky "Reef" adjacent to these depressions (Schmidt, 1952). Strong eddying currents are believed to have formed these depressions. In other parts of the layered sequence some plagioclase and orthopyroxene grains show a linear fabric, probably caused by gentle current action (Van den Bergh, 1946; Ferguson and Botha, 1963).

Current action is also held responsible for the wide range of structures such as cross-bedding, slumping, troughs, or cut-and-fill in mixed chromitite–anorthosite layers. These mixed layers in the critical zone are succeeded by a thick chromitite layer that marks the beginning of a new cycle of crystallization. The confinement of the mixed layers to the boundary between cyclic units suggests that a new phase of crystallization was initiated by currents caused by epeirogeny, instability of the floor, or some convective overturn. Winnowing of the relatively light plagioclase of the preceding cyclic unit and settling of dense chromite could have been responsible for the magmatic sedimentation structures (see Ferguson and Botha, 1963; Vermaak, 1976a).

Many of these features attributed to current action and penecontemporaneous deformation have been interpreted by other workers as intrusive phenomena in support of injections of discrete magma influxes.

Reversals in mineral compositions do occur (e.g., at the Pyroxenite Marker, Von Gruenewaldt, 1973) (see Table 6-1), but Ferguson (1969) has found that many of the postulated reversals result from comparing cumulus and intercumulus phases. He found the compositions of cumulus phases display a progressive change upward but only slight changes within sequences of at least 1100 m, which is taken as evidence for differentiation *in situ* of a body of magma of great thickness (Ferguson, 1969). Trends in the major-element chemistry of orthopyroxenes, plagioclases, and chromites across the boundary of the lower and critical zones in the eastern lobe are also consistent with differentiation *in situ* (Cameron and Desborough, 1969; Cameron, 1970a).

The progressive increase in the En content of orthopyroxene upward in a unit of the Winterveld norite–anorthosite of the eastern lobe would be expected if there has been a slow influx of undifferentiated magma mixing with partially differentiated magma in the chamber (Cameron, 1971).

The distribution of platinum-group metals (osmium, ruthenium, and iridium) in orthopyroxenes, plagioclases, and chromites above and below the boundary between the lower and critical zones was used to postulate the existence of a discrete lower zone magma (Gijbels *et al.*, 1974). This proposal has been rejected by Cameron (1978) on the grounds that:

(1) There is no evidence of a structural or compositional break.
(2) The samples analyzed by Gijbels and his co-workers are too few and too widely spaced.
(3) The ubiquitous presence of tiny inclusions of sulfides, chromite, rutile, or silicates in the main mineral phases militates against the preparation of mineral separates of sufficient purity for analysis at parts per billion level.

Hamilton (1977) identified abrupt changes in initial $^{87}Sr/^{86}Sr$ ratios that coincide with irregularities in cryptic variation in the layered sequence and interpreted the data as evidence for fresh magma influxes. Lack of adequate sampling throughout the complete thickness of the layered sequence precluded identification of the number of such influxes (Hamilton, 1977). However, Rb-Sr data for critical zone samples were found to define two different sets of systematics. One is defined by samples collected from between the UG2 chromitite layer and the top of the Merensky "Reef." The second set represents the remainder of the critical zone. Although the Merensky "Reef" has been suggested as an example of a discrete magmatic influx on other grounds, mainly because of its content of platinum metals and copper–nickel sulfides, Vermaak (1976a) regarded it as the culmination of differentiation of a substantial body of magma.

It has been shown experimentally that contamination of parental magma by siliceous rocks or melts therefrom can cause precipitation of chromite (Irvine, 1975). An alternative to this model envisages the repeated introduction of primitive magma mixing in the magma chamber with differentiated liquid (Irvine, 1977). Although there is some evidence of contamination by wall rocks (e.g., Marlow and Van der Merwe, 1977) and addition of magma, both mechanisms require rapid and uniform mixing repetitively throughout the crystallization of the critical zone.

Changes in oxygen fugacity have been proposed involving periodic assimilation of CO_2 and H_2O from the wall rocks, and periodic release of pressure through a vent. Cameron (1977) has shown that Bushveld chromites in chromite-poor silicate rocks are, in general, lower in Cr/Fe, Mg/R^{2+}, and Al/R^{3+} and higher in Fe^{3+}/R^{3+} and Cr/R^{3+} ratios than chromites in stratigraphically associated chromitites. From the experimental findings of Hill and Roeder (1974), chromite in chromitite layers should have lower contents of Al_2O_3 and lower Cr/Fe and Mg/R^{2+} ratios than chromite in stratigraphically associated chromite-poor silicate rocks if chromitite formation were caused by increase in fO_2. However, this is the reverse of what is found in the critical zone in the eastern Bushveld (Cameron, 1977). Cameron (1977) therefore proposed that change in total pressure, caused by epeirogeny affecting the magma chamber during crystallization, was a prime factor in the repetitive formation of the chromitite layers. Nevertheless oxygen fugacity is considered to be a major, if not a controlling factor in the formation of chromitite layers at the Zwartkop chrome mine in the western Bushveld (Snethlage and Von Gruenewaldt, 1977). The two layers investigated are located in the lower part of the critical zone, the stratigraphically lower layer being associated with olivine cumulates containing usually <5 volume percent chromite. Although MgO and Fe contents increase and decrease respectively in the chromitite layer, there is a decrease in Al_2O_3 content in the lower chromitite layer compared with the associated olivine and olivine–chromite cumulates. The upper chromitite layer showed a marked increase in Al_2O_3 content compared with stratigraphically associated bronzitites (Snethlage and Von Gruenewaldt, 1977), a trend which is similar to that reported by Cameron (1977) from the eastern Bushveld.

A few chromitites of the Bushveld Complex are interlayered with olivine cumulates (McDonald, 1967; J.S. De Villiers, 1970; Van der Merwe, 1976; Cameron, 1977; Snethlage and Von Gruenewaldt, 1977) but are most commonly associated with bronzitites and anorthosites. In contrast, chromitites in other layered intrusions (e.g., Stillwater and Great Dyke) are typical associates of olivine-bearing cumulates.

The lower zone in the northern lobe therefore is similar to this latter association and the question arises as to the reasons for these two different associations. Such evidence as is available is tenuous and equivocal, but Cameron (1978) suggested that the different associations are probably *not* related to:

(1) Differences in Cr_2O_3 contents of the various magmas. Cr_2O_3 contents of orthopyroxenes from the Stillwater Suite, Great Dyke, and Bushveld Complex are similar.
(2) Differences in Mg/Fe ratios. These ratios in olivines and orthopyroxenes from the Stillwater Suite and the lower zone of the Bushveld are similar, implying that difference in temperature was not a factor.
(3) Differences in total pressure. The high pressure postulated during crystallization of Rostock bronzitite should have favored chromite precipitation.

Two possibilities suggest themselves. First, the Fe_2O_3/FeO ratio in an orthopyroxene from the lower zone is given by Hess (1960) as 0.135. In contrast, this ratio is 0.057 and 0.054 in orthopyroxenes from the Stillwater Suite and the Great Dyke, respectively. If further work establishes the validity of the data, it may be that oxygen fugacity was relatively high during crystallization of the lower zone in the eastern lobe, thus suppressing crystallization of chromite in significant amounts (Cameron, 1978). It may be speculated that subtle differences in fO_2 from lobe to lobe account for the presence of chromitite layers in the lower zone of the northern lobe.

A second factor may be that because Al_2O_3 is an important component of some Bushveld chromites, the magma in the eastern lobe may have been progressively enriched in Al_2O_3 as the lower zone accumulated. Chromite precipitation may not have been possible until the Al_2O_3 content of the magma reached a sufficient concentration (Cameron, 1978).

Emphasis is placed on the lateral persistence of layers within lobes and from lobe to lobe. Although this is true on a gross scale, there is increasing evidence of important lateral variations. Von Gruenewaldt (1973) demonstrated an example in the upper zone of the eastern lobe. Although olivine and cumulus apatite appear in the layered sequence at approximately the same elevations above the main magnetitite layer in measured sections 10 km apart, thickness decreases significantly from 1870 m to 900 m between the main magnetitite and the topmost magnetitite layers. This decrease takes place in a southward direction but the total thickness of the upper zone as a whole decreases by only about 15 percent over the same distance. Von Gruenewaldt (1973) concluded that the appearance of cumulus apatite at the same stratigraphic height militates against influx of new magma. He preferred lateral variation in temperature to account for the association of cumulus apatite with lower temperature phases in the south as compared with the north.

A further example of lateral variation is provided by the fluctuations in the relative abundance of platinum group minerals in the Merensky "Reef" both within lobes and from lobe to lobe (Table 6-3).

It is clear from the foregoing that much of the evidence is apparently contradictory: some favors discrete magma additions, some, mixing of small amounts of new magma with differentiated magma, and still other lines of evidence suggest differentiation in a large magma chamber. The layering and variations in mineral composition therefore appear to be the result of a complex interplay of a number of mechanisms. It would also be unreasonable to suppose that, in a magma chamber of the size of even an individual lobe, there would necessarily be constant homogeneity in magma composition, pressure, temperature, or oxygen fugacity.

Hamilton (1977) postulated that the first Bushveld basic magma was derived by extensive partial melting of a shallow upper-mantle source as a consequence of pressure release in a tensional environment. The estimated abundance ratios for the first magma suggest enrichment of the source region by fluids with chemical characteristics akin to those of some potassic magmas such as kimberlites (Hamilton, 1977).

The high initial $^{87}Sr/^{86}Sr$ ratio (0.715 ± 0.009) reported for the Nebo granite (Davies *et al.*, 1970) suggests the liquids from which these granites crystallized were formed by melting of crust. Rb-Sr studies of the Rooiberg felsites have yielded equivocal plots indicative of open-system behavior, so that no conclusions can be made concerning a mantle or crustal origin. Lack of Rb-Sr data for the granophyres also constrains speculation as to the source of their liquids. The granophyres appear to have formed in response to a number of different processes, but those that are considered on field evidence to be magmatic have similar REE patterns and trace-element characteristics to the felsites. This might be taken to imply that they had a common origin. Certain granites near the base of the stratiform granitoid sheets have similar geochemical characteristics to the magmatic granophyres and hence, by circuitous reasoning, one could conclude that all the acid rocks were of similar derivation. This must remain speculative until more data are available. However, it seems probable that the stratiform granitoids crystallized first at top and bottom, thereby trapping volatile-rich residual liquids. Absence of oxidation and alteration phenomena in the lower parts of the stratiform Nebo granite confirms initial crystallization from a less hydrous magma than higher in the stratigraphy. Where the roof was fractured the volatiles could escape, giving rise to exogranitic veining and pipe formation as in the granophyre at Stavoren north of Groblersdal and elsewhere. If the volatiles were unable to escape, their presence would initiate oxidation of exsolved iron in feldspars and chloritization of mafic minerals. Where larger volumes of volatile-rich liquid accumulated, the melting point of the granite mush would be depressed to below the ambient temperature, leading to mobilization of this material. The amount by which the melting temperature was depressed would determine the extent of mobilization. In extreme cases

Table 6–3. Average Relative Abundances in the Percentages of Platinum-Group Minerals from the Merensky "Reef" in the Bushveld Complex[a]

	Rustenburg	Western Platinum	Atok	Northam
Braggite	35.9	6.8	60	Minor
Cooperite	13.36	10.3	25	Minor
Laurite	9.77	0.2	5	Common
Sperrylite	0.29	50.3	2	Minor
Pt–Fe alloy	0.51	6.4	2	Common
Moncheite	0.44	13.0	5	Rare
Electrum	0.40	4.2	Not observed	Minor
Merenskyite	0.07	0.2	Not observed	Rare
Kotulskite	0.02	8.1	Not observed	Rare
Michenerite	0.01	—	—	—
Laurite–sperrylite int.	0.15	—	—	—
Cooperite–Pt–Fe alloy int.	3.46	—	—	—
Pt–Fe alloy-BMS int.	35.39	—	—	—

[a]The Rustenburg, Western Platinum, and Northam mines are located in the western lobe, whereas the Atok mine is in the eastern lobe. Sources: Rustenburg Mine, Vermaak and Hendriks (1976); remainder Brynard *et al.* (1976). int, intergrowth; BMS, base-metal sulfides.

intrusive contacts would result, such as those seen along the upper contact of the Bobbejaankop stock. Bobbejaankop-type granites therefore are seen as the consequence of the accumulation of the volatile-rich liquids (Lenthall and Hunter, 1977).

In summary the Bushveld Complex represents emplacement of magma, at least as basic as olivine tholeiite, through five main vents (i.e., western, eastern, northern, and southeastern lobes and Marico). This rising magma generated by relatively shallow melting of a mantle source induced partial melting of the sialic crust to give rise to liquids from which the Nebo granite and its variants crystallized. There is less certainty regarding the origin of the felsite liquids but these too may have been derived from a crustal source. The status and origin of the Makhutso granite is uncertain.

Some of the Bushveld magmas were emplaced into and around the Vredefort Dome. Bisschoff (1973) envisaged limited reaction between this magma and the wall rocks as the mechanism whereby the associated alkali granites were generated. More-detailed geochemical studies are required, but it may be that the alkali granites and syenites reflect interaction between sialic rocks of granulite facies, in contrast to involvement of amphibolite-facies rocks in the main area of the Bushveld Complex.

6.6. Contact Metamorphism

In rocks of appropriate composition, such minerals as cordierite, chiastolite, sillimanite, staurolite, garnet, olivine, antigorite, diopside, and wollastonite appear within the contact aureole of the complex. Advanced feldspathization of orthoquartzite has been described from the eastern lobe (Schwellnus, 1956) and, north of Potgietersrus, pseudogranophyre is associated with feldspathic quartzite (Van Rooyen, 1950).

Metamorphism of the sedimentary and volcanic roof rocks frequently involved more-intense deformation in addition to thermal effects. Where the original nature of the rock has been wholly or partially destroyed the term "leptite" is frequently applied to these rocks. The basal felsites were partially melted as a result of the intrusion of the layered suite, giving rise to sheeted bodies of granophyre and granodiorite within the felsite pile (Von Gruenewaldt, 1968, 1972).

The metamorphic aureole in the floor rocks shows considerable variation in vertical thickness which increases from 360 m near Pretoria to 3100 m at the western limit of the western lobe (Hall, 1932). In the eastern lobe the vertical thickness increases from 600 m in the south to 4000 m in the north. The reasons for these variations are not clearly understood, but two periods of metamorphism have been recognized (Button, 1973c). The earlier metamorphic event is related to sill emplacement and a second metamorphism is superimposed because of the intrusion of the complex itself. Where it is possible to distinguish the two metamorphic episodes, the thermal effects of the complex alone are more uniform.

6.7. Sulfide Mineralization

The Bushveld Complex has been described as the largest repository of magmatic ore deposits in the world (Willemse, 1969). Table 6-4 confirms the validity of this statement. For details of the economic mineralization the reader is referred to Vermaak (1976a) and Crocker (1979).

The platinum–group metals, nickel, and copper, are won primarily from the Merensky "Reef" situated near the top of the critical zone. Vanadium and magnetite ore deposits are located near the base of the upper zone. Tin and fluorspar are found in the Lebowa Granite Suite.

Disseminated sulfides rarely exceed 2 volume percent in the Rustenburg Layered Suite, and the Merensky "Reef" layer presently constitutes the only economically viable deposit. In general, pyrrhotite is the most abundant sulfide mineral, followed by pentlandite, chalcopyrite, pyrite, cubanite, and mackinawite, but relative order of abundance varies with stratigraphic position. In the lower zone the ratio of pyrrhotite to pentlandite and chalcopyrite varies considerably even in rocks of the same composition. Anorthosites and norites in the critical zone contain mainly chalcopyrite, whereas harzburgites, dunites, and pyroxenites carry pentlandite and pyrrhotite.

The content of chalcopyrite increases to a maximum below the main magnetitite layer and thereafter decreases. Pentlandite decreases above the Merensky "Reef" but both pyrite and cubanite maintain a nearly constant content through the layered suite. Higher concentrations of disseminated sulfides are frequently found where a transgressive relationship is observed between the layered suite and the country rocks and is attributed to assimilation of S and As from these sediments (Liebenberg, 1970).

The sulfur content of the layered suite is generally low and decreases gradually to the upper part of the critical zone, where there is a sudden and dramatic increase at the level of the Merensky "Reef," followed by a gradual decrease to the base of the upper zone. In comparison to the lower zones the upper zone is moderately rich in sulfur (mean about 350 ppm). Two abrupt increases have been noted in this zone in anorthosite below the main magnetitite layer and below the uppermost magnetite layer (Liebenberg, 1970; Von Gruenewaldt, 1976).

The ratio of Pt/(Pt+Pd) in the Merensky "Reef" at the Rustenburg and Western Platinum mines is nearly constant and is higher than the UG2 chromitite layer in the critical zone and the mineralized anorthosite in the upper zone (Table 6-5). This ratio is significantly higher in the Bushveld Complex than in other layered complexes, although these latter complexes commonly contain

Table 6–4. Mineral Production and Resources of the Bushveld Complex[a]

	1975 Production	Estimated total resource
Platinoid metals and gold	~85,000 kg	62.89 kg × 10^6
Nickel	~18,250 tons	22.80 tons × 10^6
Copper	~10,800 tons	9.95 tons × 10^6
Chromite	2,075,378 tons	156.0 tons × 10^6
Vanadium pentoxide	19,002 tons	16.8 tons × 10^6
Magnetite iron ore	1,561,670 tons	1,030.5 tons × 10^6
Tin	5,232 tons	~0.05 ton × 10^6
Fluorspar	202,583 tons	~40.0 tons × 10^6

[a] Source: Von Gruenewaldt (1977).

higher concentrations of total platinum–group metals (Table 6-5).

6.8. Vredefort Dome

The almost circular Vredefort Dome is one of a swarm of domes that extend in a broad belt approximately 500 km wide northwestward from beneath Mesozoic cover. The dome is distinguished by a collar of steeply dipping and overturned sedimentary and volcanic rocks of the Witwatersrand, Ventersdorp, and Transvaal Supergroups surrounding a core of Archean rocks. An outer rim synclinorium of the Pretoria Group (Transvaal Supergroup) completes the structure (Simpson, 1978). The basement core of the dome has a radius of 18 km, and the widths of the collar and rim synclinorium are 17 km and 20–30 km, respectively. The rim synclinorium is only weakly deformed, as evidenced by the absence of penetrative cleavage, even in pelitic rocks, and by the lack of distortion of quartz-fiber infillings or undulose extinction away from faults (Simpson, 1978). Overturning in the collar by flexural slip about a vertical central axis took place under brittle conditions during a single progressive event. Most of the strike-slip and normal faults, which are concentric about the collar, were synchronous with folding. Radiating normal faults formed after folding and overturning.

The origin of this structure has been much debated. Tangential pressure was proposed to account for the doming of the Vredefort structure and the basining of the Bushveld (Daly, 1928). The approximately contemporaneous development of the Vredefort Dome with the intrusion of the Bushveld Complex found support on the basis of paleomagnetic data (Hargraves, 1970). The orientations of the mean remanent vectors in samples from the Losberg intrusion and a basic granophyre ring dike in the periphery of the central core are nearly identical to that of the Bushveld Complex. The orientations of the paleomagnetic vectors from the smaller intrusions situated in the collar of the structure are markedly divergent from the Bushveld direction, despite the fact that these rocks have yielded Bushveld ages (Nicolaysen *et al.*, 1963). However, if the vectors are corrected to allow for the overturning of the strata, their direction agrees within 30° to that of the basic granophyre (Hargraves, 1970). Hargraves concluded that the Vredefort structure developed in the brief interval of time between the emplacement of the layered complexes and the intrusion of the granophyre.

On the basis of sedimentological data, Brock and Pretorius (1964) deduced the presence of an an-

Table 6–5. Some Precious and Base Metal Concentrations in the Bushveld and Other Layered Complexes[a]

	Pd (ppb)	Pt (ppb)	Au (ppb)	Ni (%)	Cu (%)	Pt/ Pt + Pd	Overall grade of PGM[d] (ppm)
Bushveld Complex							
UG2 chromitite	—	—	—	—	—	0.55	—
Merensky Reef[b]	1,600	3,700	460	0.16	0.079	0.70	—
Merensky Reef[b]	1,770	4,500	370	0.19	0.11	0.72	7.4
Merensky Reef[b]	1,640	3,730	455	—	—	0.69	—
Merensky Reef[c]	1,548	3,660	198	—	—	0.70	5.4
Anorthosite	900	730	470	0.15	0.96	0.45	—
Stillwater Complex							
Banded zone	11,438	3,268	—	0.12	0.04	0.22	14.7
Sudbury Irruptive							
	364	346	123	1.5	1.3	0.49	0.79
Noril'sk							
Picritic gabbro	6,900	2,600	230	1.3	1.6	0.27	9.6
Taxitic gabbro	7,200	2,300	290	1.3	1.9	0.24	9.6

[a] Sources: From compilation by Naldrett and Cabri (1976). Merensky "Reef"[b] from Brynard *et al.* (1976).
[b] Merensky "Reef" at Western Platinum Mine.
[c] Merensky "Reef" at Rustenburg Platinum Mine.
[d] PGM, platinum-group metals.

cestral Vredefort Dome during deposition of Witwatersrand strata, thus implying that the structure had had a long history of repeated uplift.

The presence of shatter cones in rocks of the collar of the Vredefort Dome prompted the suggestion that meteorite impact cratering was responsible for the formation of the dome (Dietz, 1961). The concept was extended by Hamilton (1970) to account also for the location of the Bushveld Complex. However, shatter cones occur on fault breccia produced by downfaulting of already overturned rocks, which is not compatible with a meteorite impact origin for the dome. No shock-metamorphic effects have so far been found associated with the Bushveld Complex (French and Hargraves, 1971), but Rhodes (1974a) considered that the textural features of the Rooiberg felsites were in keeping with an origin by shock melting.

Veinlets of pseudotachylite are common in the rocks of the collar and core of the dome, following a random pattern. The borders of the pseudotachylite veinlets are sharp, individual veinlets being discordant to the fabric of the rocks in which they are found. The pseudotachylite veinlets are generally considered to represent fracture fillings by mobile, probably molten, silicate material produced by faulting of dry rocks with extremely high strain rates (Bisschoff, 1969; Sibson, 1975). A close chemical similarity between pseudotachylite and the enclosing rock has been reported (e.g., Shand, 1916), but recently Schreyer and Abraham (1978) have shown that the pseudotachylite vein studied by them is less aluminuous than the hornfels enclosing the pseudotachylite veinlet. Foreign xenoliths are numerous in the pseudotachylite, some having been transported for distances of more than 50 m (Schreyer and Abraham, 1978).

Pseudotachylite veinlets cut and therefore postdate hornfelses resulting from contact metamorphism by magmatic intrusions into both the core and collar (Bisschoff, 1969). The fact that the pseudotachylite, indicative of high-strain-rate deformation, postdates the contact metamorphism has been taken as evidence of meteorite impact, as has the presence of high-pressure SiO_2 polymorphs (coesite and stishovite) in pseudotachylite veins cutting quartzites in the collar of the dome (Martini, 1978).

The Archean granulite terrane in the core of the Vredefort structure has suffered static metamorphism and is cut by pseudotachylite. Formerly idioblastic garnet crystals in the granolites (Winkler, 1974) have been converted to symplectic assemblages of cordierite, orthopyroxene, and relict garnet. A temperature of 700°C and pressure of 500 MPa for the static metamorphism have been deduced from the garnet-cordierite pair (Schreyer and Abraham, 1978). This estimate is consistent with the interpretation of increasing temperature of metamorphism from the collar hornfelses (500–600°C) to the center of the core (>800–850°C) (Schreyer and Abraham, 1978). The apparently shallow depth of metamorphism (i.e., 500 MPa or about 18 km) is also consistent with this event having occurred during or after uplift of the dome. The hornfelsed granolite is located about 12 km stratigraphically below the sedimentary unconformity at the core–collar interface. A thickness of between 13 km and 15 km of Witwatersrand and Transvaal strata overlie this unconformity. If static metamorphism had occurred before uplift, a metamorphic depth of 25–27 km would be expected (Schreyer and Abraham, 1978).

Structural data indicate that folding, including overturning of collar rocks, occurred during a single event, simultaneously with the upward movement of the core of the Vredefort structure (Simpson, 1978). Most of the faulting in the collar occurred simultaneously with the folding, but some normal faults are late or postoverturning in the northern area of the rim syncline.

The problem of the relationship between the static metamorphism reflected by the hornfelsed granolite and the pseudotachylite has been examined (Schreyer and Abraham, 1978). Small pyroxene grains surrounding larger quartz fragments in the pseudotachylite have similar fabric to that seen in the symplectite pseudomorphs after garnet in the hornfels. These textural similarities and the identity of the pyroxene phase with those in the pseudomorphs in the hornfels suggest that pseudotachylite generation overlapped in time with the breakdown of the original garnet (Schreyer and Abraham, 1978). If this is valid, it suggests pseudotachylite emplacement during or after metamorphism (Schreyer and Abraham, 1978). The presence of stishovite in the pseudotachylite cutting collar rocks (Martini, 1978) implies doming prior to meteorite impact as stishovite can only be preserved at low pressure if temperature is also low (i.e. <250°C).

The paleomagnetic data (Hargraves, 1970) indicate that uplift of the Vredefort Dome was approximately contemporaneous with the emplacement of the Bushveld Complex (i.e., at about 2.0 Ga). If the assumption that the static metamorphism occurred during or after uplift is valid, the metamorphic event has a maximum age of about 2.0 Ga. The metamorphic conditions deduced by Schreyer and Abraham (1978) imply a high geothermal gradient of nearly 40°C/km, which is appropriate to either low P/T regional metamorphism or contact metamorphism. There is no evidence in the extensive Archean and Proterozoic terrane surrounding the Vredefort structure of a regional metamorphic event at about 2.0 Ga. A localized heat source to account for this metamorphism seems more plausible (Schreyer and Abraham, 1978). The presence of numerous small intrusions of basic and alkaline rocks, about 2.0 Ga old, lends support to the view that heat was derived from magma as it rose into the Vredefort structure.

The available evidence implies that metamorphism and uplift of the Vredefort Dome have been related to an internal process, to which it is reasonable to relate the high-strain-rate deformation. The fact that coesite can be synthesized from quartz at relatively low pressures (about 1000 MPa) at high strain rates (Green, 1972) implies that the presence of coesite need not necessarily indicate meteorite impact. An impact origin for the pseudotachylite veinlets and the Vredefort structures requires a remarkable coincidence in time and space between impact and a localized area of metamorphism and uplift. Shock metamorphism may have resulted from the explosive detonation of rapidly expanding carbonated fluids that had risen from lower crustal levels (Nicolaysen and Ferguson, 1981).

6.9. Structural Setting and Mechanics of Intrusion

Hall (1932) was the first person to draw attention to the fact that a southward extrapolation of the trend of the Great Dyke of Zimbabwe bisects the Bushveld Complex and passes through the centers of the Johannesburg and Vredefort Domes, the former dome being traversed by a major fracture zone coincident with the trend of the Great Dyke.

The alignment of a number of layered complexes along a north–northeast-trending line over a distance of at least 1500 km suggested that a major lineament controlled the emplacement of these complexes. In contrast, Truter (1955) emphasized the east–west extent of the Bushveld Complex and postulated that the complex had been intruded through a series of vents aligned in that direction.

The Vredefort Dome forms part of the broad belt of domical structures exposing Archean basement rocks known as the Vryburg arch (Hunter and Hamilton, 1978). On either side of the arch are preserved substantial thicknesses of Proterozoic and Mesozoic cratonic cover. Evidence that this arch has been present over a long period of time is listed below.

(1) The western limit of intraformational erosion during deposition of the Pretoria Group is located along the southwestern flank of the arch.
(2) The Transvaal Supergroup attenuates over the arch in Botswana.
(3) The depositional axes of the Transvaal and Waterberg basins are deflected over the arch.
(4) The main concentration of the Pilanesberg Intrusive Suite, intruded at about 1300 Ma, lies along the northeastern flank of the arch and parallels the trend of the arch.
(5) Major faults near Rustenburg, bounding the Crocodile River fragment and at Brits, are parallel to the trend of the arch.
(6) Subhorizontal allochthonous slices of Transvaal rocks from the arch are found in Botswana.

The Bushveld Complex is located adjacent to this broad arch in which the northwest–northeast fold interference pattern is most marked. Eastward from the arch east-northeast trending structures dominate over northwest trending structural directions. Uplifted areas must have existed to the north and east of the present outcrop of the Bushveld Complex because Button (1973c) has demonstrated that transport of sediment into the Transvaal basin was from the north and east. The Bushveld Complex is therefore situated in a major structural trending depositional axis of the Transvaal basin.

The northern lobe of the complex intrudes granitoid–gneiss basement north of Potgietersrus. The intrusion of the complex therefore does not depend solely on differences in physical properties between basement and cratonic cover. The gravity data do not reflect a reticulate pattern of feeder dikes that would be expected if Anderson's

(1951) hypothesis of stress reorientation had influenced the emplacement of the complex. Intrusion of the complex requires the stress distribution to be altered in such a way that the trajectories of maximum principal stress steepen toward the magma chamber(s) from which the layered sequence was derived. Roberts (1970) has drawn attention to the coincidence of major monoclinal flexuring, dike swarms, and major sill intrusion. The fact that successive layers are stacked above each other suggests that the complex was emplaced into a subsiding structural basin, bounded by major crustal flexures. The overlap by the main and upper zones, the evidence for periodic tectonic disturbances, and the paleomagnetic data indicating crystallization of the main zone in a horizontal attitude support the concept of continuing subsidence of the basin during crystallization of the complex.

The coincidence of major depositional and structural basins would contribute to the conditions favoring optimum sill intrusion. That the Vryburg arch remained an active structural feature after emplacement of the complex is indicated by the subsequent intrusion at about 1300 Ma of alkaline plugs and dike swarms along the flank of the arch. This may account for the relatively small size of the Marico intrusion, which is located within the uplifted Vryburg arch and which may therefore have suffered erosion and removal of the layers above the lower zone presently exposed. The presence of alkali granites and syenites adjacent to the Vredefort Dome may be a consequence of this structural pattern. If it is assumed that the sialic crust grades downward through amphibolite to granulite facies, it could be expected that in structural depressions the amphibolite facies (and therefore more potassic zone) would be involved in either reaction melting with the invading ultrabasic and basic magmas or partial melting as a result of depression into regions of higher temperature. Any syenitic magma produced in the underlying granulite zone would continue to react with melts produced in the amphibolite zone, with the result that predominantly potassic granite melts would be generated within the confines of the main suite. However, in the structural arch only the granulite zone would be likely to be involved in any melting or reaction process, so that syenitic magmas derived from potassium-depleted granolites would be produced. Limited reaction of this magma with the wall rocks during intrusion could lead to the generation of alkali granites, a mechanism in keeping with that described by Bisschoff (1973), but more detailed geochemical data are required to test this hypothesis.

The model suggested (Hunter, 1975) appears to satisfy much of the available, but often contradictory, evidence. It accommodates opinions that the layered sequence was formed as a result of single or multiple intrusion. The uncertainty of the stratigraphic position of the Rooiberg felsites does not affect the model. Should the extensive sills in the Transvaal Supergroup be shown to be part of the complex, the fact that they were deformed prior to intrusion of the main mass of the complex would also be consistent with the model.

Sharpe and Snyman (1980) have further developed the above model with respect to the eastern lobe of the complex. They suggested that depression of the floor of the Transvaal depositional basin into a regime of increasing compression induced conditions favorable for intrusion into the Transvaal sediments of diabase sills totaling 2.5 km in thickness, the presence of which assisted subsidence. The Bushveld basic magmas were intruded into the base of this sill complex in a regime characterized by shear failure. Initial influxes of magma formed a shallow cone and continued influxes filled the conical cell, transgressed the floor and arched up the roof rocks.

The Bushveld Complex therefore is regarded as the response to a unique combination of events coincident in time and space, the siting of the complex being the culmination of a sequence of events that preceded its emplacement. However, much additional data are required before an internally consistent hypothesis can be presented to account for all the many petrologic, geochemical, and structural features.

Chapter 7

The Earliest Red Beds

The Waterberg, Soutpansberg, Umkondo, and Matsap Groups (Fig. 7-1) are of roughly equivalent age and are characterized by the ubiquitous presence of the earliest red beds, which provide significant information concerning the character of the Earth's early atmosphere. Irrespective of whether the red pigmentation originated in the source area or developed at the site of deposition (Van Houten, 1968), oxidizing conditions must have existed at the time the sediments were deposited or shortly thereafter.

The sediment/volcanic ratios of these four groups differ according to the tectonic settings in which they were deposited. The predominantly sedimentary Waterberg Group is confined to the Kaapvaal Province. It postdates the 1913 ± 30 Ma Bushveld granites (Davies *et al.*, 1969; Rb-Sr) and is intruded by a 1420 ± 70 Ma alkaline complex (Oosthuyzen and Burger, 1964; U-Pb). A contemporaneous lava flow in the Middelburg basin with an age of 1790 ± 70 Ma (Oosthuyzen and Burger, 1964; Hunter, 1974b; U-Pb) dates Waterberg sedimentation at about 1800 Ma.

Volcanic and sedimentary rocks of the Soutpansberg Group occupy a graben within the previously uplifted and eroded high-grade gneiss terrane of the Limpopo Province. The development of the Soutpansberg Group postdated a sporadic thermal event in the gneiss terrane at 1800 ± 50 Ma (Barton, 1979b; Rb-Sr). On the western edge of the graben, upper stratigraphic units in the Soutpansberg Group overlie Waterberg sediments (Jansen, 1976). The Soutpansberg Group has probable correlatives in Botswana in the form of the Palapye Group (Ermanovics *et al.*, 1978).

The Umkondo Group is exclusively sedimentary and occurs on the southeastern edge of the Zimbabwe Province. Diabase sills give a minimum age of about 1100 Ma for this unit (Allsopp *et al.*, 1973; Rb-Sr). A maximum age is more difficult to determine, but a thermal resetting event at 1785 ± 80 Ma (Vail and Dodson, 1969; K-Ar) indicates that the Umkondo Group is probably about 1800 Ma old (Button, 1977).

Crampton (1974) suggested that the Matsap Group is older than the 1800 Ma age of the other three units as evidenced by an age of 2026 ± 90 Ma (Rb-Sr) on intercalated volcanics, but Armstrong and Cornell (personal communication, 1979) have recently dated the same volcanics at circa 1800 Ma. The predominantly sedimentary Matsap Group crops out on the western edge of the Kaapvaal Province, where it unconformably overlies the Griqualand West Supergroup.

7.1. The Intracratonic Waterberg Group

Stratigraphy and Paleocurrents

The Waterberg Group crops out in the Warmbaths and Middelburg basins, where thicknesses up to 5000 m are preserved (Figs. 7-1 and 7-2). Rooiberg felsites most commonly constitute the floor, but the Waterberg Group also rests unconformably on basement rocks of the Kaapvaal Province, Transvaal Supergroup strata, or Bushveld granites. The Waterberg Supergroup is the youngest Precambrian stratigraphic unit developed in the Kaapvaal Province and as a result its upper contact is erosional. At the base of the Waterberg Group in the southern portion of the Warmbaths basin, trachyte and quartz porphyry lavas are associated with volcaniclastic sediments (Jansen, 1970). Elsewhere, conglomerates and sandstones, characteristically displaying a red

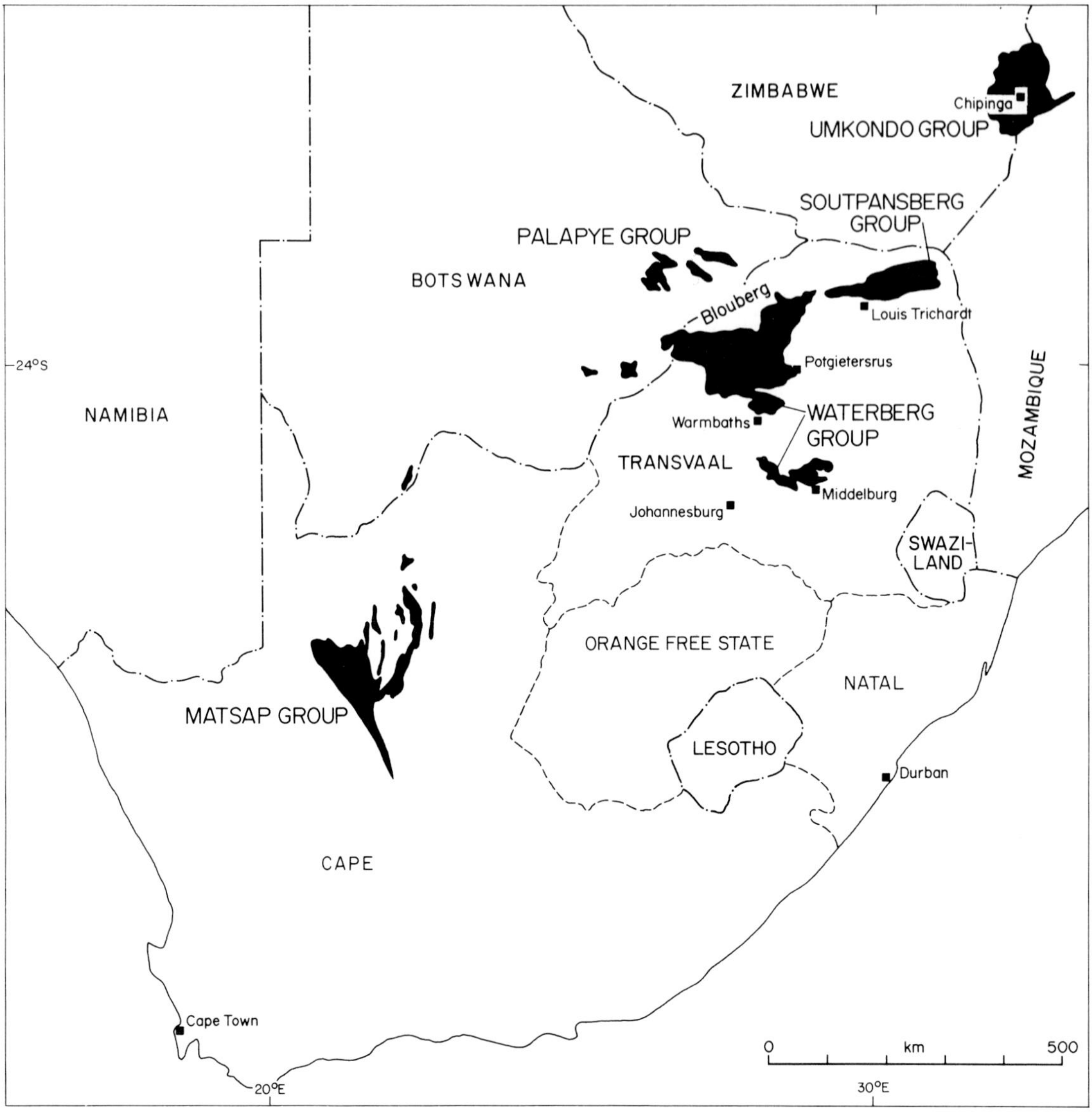

Figure 7-1. Distribution of the Waterberg, Soutpansberg, Umkondo, Matsap, and Palapye Groups in southern Africa.

pigmentation, are the predominant lithologies. Finer grained terrigenous clastics constitute less than 10 percent of the stratigraphic sequences (De Vries, 1968; Vos and Eriksson, 1977).

Conglomerates range from 3 m to 6 m in thickness and have strike continuities of hundreds of meters. The conglomerates are mainly clast supported, with pebbles up to 35 cm in diameter, and contain thin intercalated sandstone layers. Pebbles are rounded to subangular and include felsite, white quartzite and jasper, with subordinate chert, iron-formation, vein quartz, and granophyre. No Bushveld gabbroid or granitoid clasts are present. Sorting in the conglomerates is generally poor and imbrication is weakly developed in certain outcrops in the Middelburg basin.

Texturally and mineralogically immature and mature sandstones constitute more than 80 percent of the stratigraphic sequences in the two basins (Fig. 7-2). Immature red sandstones predominate and comprise arkosic and lithic arenites and feldspathic and lithic graywackes. The sandstones are medium grained to granular and poorly sorted, with monocrystalline quartz as the dominant constituent. Rock fragments include felsite, white quartzite, chert, and siltstone. Feldspars are

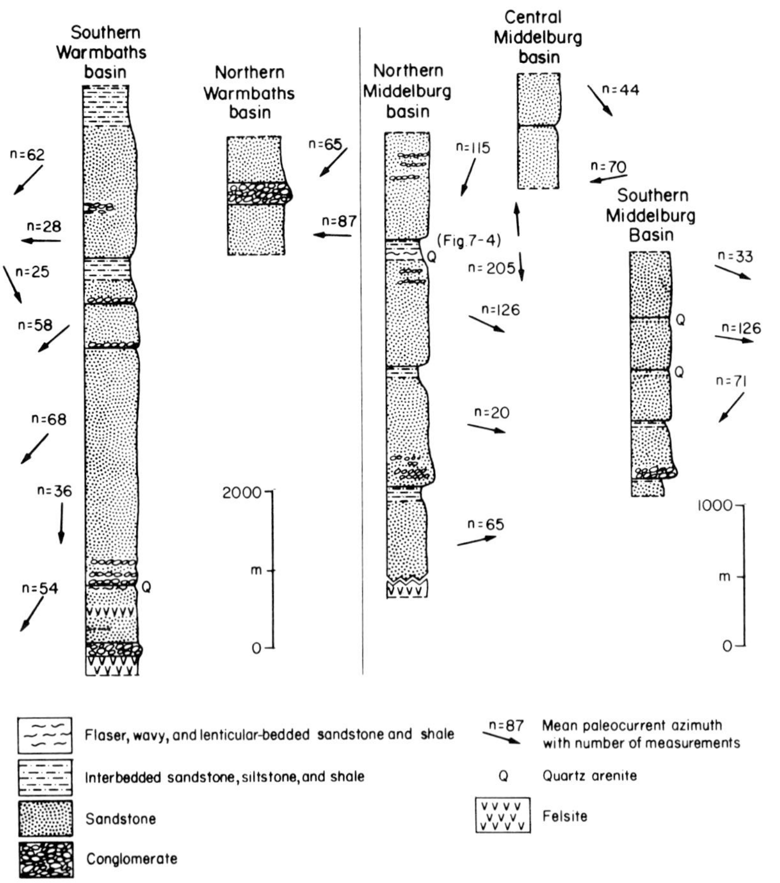

Figure 7-2. Stratigraphic columns of the Waterberg Group in the Warmbaths and Middelburg basins. Paleocurrent means determined from cross-beds and ripples are shown for various stratigraphic units. [After Eriksson and Vos (1979).]

mainly orthoclase with less abundant microcline and albite. The felsite fragments and orthoclase grains have characteristically been hydrated to clay minerals. This weathering liberated iron as hematite and limonite specks and veinlets within the grains. Nonopaque heavy minerals include pyroxene and sphene, which is in variable stages of breakdown to leucoxene. Magnetite is concentrated on and at the base of many foresets, especially within small-pebble conglomerates. The immature sandstones are cemented by hematite, typically in the form of specular laths and needles; limonite with authigenic quartz overgrowths is also common.

The second variety of sandstone comprises texturally and mineralogically mature quartz arenites, which are entirely devoid of red pigmentation. These form a single thin sandstone unit low in the Warmbaths sequence and three 10- to 15-m-thick lensoid bodies in the Middelburg basin (Fig. 7-2).

The Waterberg sandstones in both areas contain planar and trough cross-bedding, asymmetric or current ripples, and primary current lineation. Within the immature sandstones, azimuth measurements on trough cross-beds give the most consistent results at a particular locality. Less abundant planar cross-beds commonly give paleocurrent modes that are at right angles to the mode obtained from trough cross-beds. Mean paleocurrent azimuths for trough cross-bedding in the Warmbaths and Middelburg basins are illustrated on Figure 7-2. These results indicate that most of the Waterberg sediments were introduced from a northern source terrain. Southerly to westerly paleocurrents dominated Warmbaths deposition, although along the western margin of the basin transport of sediment was from the northwest. Measurements in the Middelburg basin come from a 20-km-wide belt and indicate a continual easterly to westerly alternation of transport directions. The only sedimentary structures suitable for paleocurrent determinations in the clean quartz arenites are current ripples. Measurements for one of these units in the Middelburg basin give a bimodal–bipolar pattern (Fig. 7-2).

The clast population and composition of the immature red sandstones in the Waterberg Group indicate a mixed provenance. White quartzite, chert, and iron-formation pebbles were derived from the Transvaal Supergroup; granophyre, felsite clasts, and sand-sized particles come from the Bushveld acid phase; the only evidence for contemporaneous erosion of the stratigraphically intervening Bushveld mafic phase is the pyroxene and opaque magnetite grains. The unimodal paleocurrent patterns indicate source areas to the northeast and, less significantly, to the northwest of the Warmbaths basin and to the north of the Middelburg basin, areas now underlain by the extensive granite–gneiss terranes northeast of Potgietersrus (Fig. 7-1). On the basis of facies and paleocurrent patterns Button (1973c) suggested that the Transvaal Supergroup originally extended considerably farther north of the present outcrop limit, and it is probable that the Bushveld Complex likewise had a greater areal extent than at present.

Braided Alluvial Processes

The textural and mineralogical immaturity of the red sandstones and associated conglomerates indicate a depositional environment of limited reworking. Coupled with the consistent paleocurrent patterns and distinctive assemblages of sedimentary structures, a fluvial depositional environment is suggested.

Upper alluvial plain. Clast-supported conglomerates and associated plane-bedded sandstones in the Waterberg Group developed on proximal braided alluvial plains and can be equated with the Scott model of Miall (1977). During the early stages of waning flood discharge, the load transported by the streams exceeded their carrying capacity and pebble lags accumulated on the channel floors and commonly evolved into longitudinal bars. At falling-water stage when pebbles were no longer being moved, sand-sized particles infiltrated the gravel framework and accumulated on bar tops, where upper flow regime conditions produced plane-bedding. Migration of channels is a characteristic of braided alluvial plains and results in the coalescing of longitudinal bars to form laterally persistent sheet gravels separated by low-water stage, cross-bedded channel sandstones, a common association within conglomeratic units of the Waterberg Group.

Middle alluvial plain. Vertical sequences of sedimentary structures in sandstones provide evidence of two contributing depositional processes operating in a mid-alluvial-plain setting. Thick trough cross-bedded and highly argillaceous sandstones probably formed as a result of rapid aggradation within active channels during prolonged high-water periods. Associated planar cross-beds, which give the polymodal paleocurrent directions, are thought to have formed by lateral accretion of linguoid bars in a fashion analogous to that documented by Cant (1978) from the South Saskatchewan River.

The well-defined depositional sequences are erosively based and lensoid. Aggradation, as well as downstream migration, were probably responsible for these sequences. Variations in discharge generated successive upward-fining cycles similar to the Donjek model of Miall (1977). Pebble lags accumulated during high discharge, followed by channel aggradation and migration of low-amplitude transverse or linguoid bars to produce planar and trough cross-bedded sequences respectively. Depending on the prevailing flow regime, a variety of small bedforms were generated on bar surfaces; at low-water stage, plane-beds probably formed under upper flow regime conditions associated with a rapid lowering of water level, and ripples formed in response to gradually waning current velocities. Convolute lamination resulted during subsequent compaction of the sediments. Basal mud-flake conglomerates are overlain by plane-bedded sandstones which are probably upper flow regime phenomena that formed on the channel floor during high discharge. After the formation of trough cross-beds as a result of linguoid bar migration, the active channel was abandoned and fine-grained sediment settled in stagnant overbank environments. The resulting muds were desiccated before being partially reworked into overlying conglomerates. Similar vertical sequences involving analogous depositional processes are described by Williams and Rust (1969).

Lower alluvial plain. Within modern fluvial settings, especially in the absence of vegetation, fine-grained clastics (silts and clays) are flushed out onto the lower alluvial plain to form relatively thin layers of mud. Some vertical sequences in the Waterberg Group indicate distal alluvial-plain sedimentation. Higher energy channel-fill deposits show large-scale low-angle foresets. The overlying trough cross-beds are directed to the southwest and are followed by horizontally laminated siltstones and shales. The suggested depositional processes for these sequences are lateral point-bar accretion within sinuous channels, followed by downstream dune migration and suspension sedimentation either during low-water stage or following abandonment of the active channel. Lateral accretion of the type proposed does not necessarily imply a meandering river system because point bars also form on the inside of meander curves in more sinuous braided rivers (Bluck, 1976). The intercalated sandstones, siltstones, and shales are analogous to the contemporary Bijou Creek ephemeral stream deposits of Colorado (Miall, 1977). During flash floods, upper flow regime plane-bedded sands (commonly showing primary current lineation) and lower flow regime planar cross-bedded sands are deposited in alternation with silts and clays during long periods of low-energy ripple migration and suspension settling.

Marginal Reworking of Alluvial Lobes

The thin intercalated layers of mineralogically and texturally mature sandstones (Fig. 7-2) probably developed through marginal marine reworking of the fluvial sediments (Frick, 1970; Vos and Eriksson, 1977). Bimodal–bipolar paleocurrent patterns (Fig. 7-2); flaser, wavy, and lenticular bedding (Fig. 7-3); and desiccation cracks and runzel marks are cited as evidence of low-energy tidal-flat settings (Reineck, 1972). Wave-swash deposits, recognized only in the Middelburg basin, are characterized by plane-bedding with low-angle discordances and parting lineation. Sporadic desiccated shale partings are atypical of beach foreshores and can be attributed to weak swash processes (Vos and Eriksson, 1977).

Paleogeographic Model

A paleogeographic model accounting for the relatively thin tidal and wave-swash deposits within thick alluvial-plain sequences was developed for the Middelburg basin, but it can also be applied to sediments in the Warmbaths basin. A backbarrier tidal-flat model adjacent to an alluvial plain was discounted on the basis of the absence of any vertical or lateral association between the wave-swash and tidal deposits. Furthermore, the poorly developed littoral sequence is not comparable with modern barrier deposits. Instead, the direct association of both the tidal and wave-swash deposits with the alluvial sequences appears to have been a response to marine processes operating on the shoreline of a fluvially deposited lobe (Fig. 7-4).

Shallow, gently shelving embayments separated alluvial fans (fan deltas); tidal flats probably developed in response to weak tidal action. Similar open, nonbarred tidal flats are developed in shallow embayments in Shark Bay, Australia, where the tidal range is less than 1 m (Logan *et al.*, 1970). Wave swash developed local destructive bars on the margins of alluvial lobes during episodes of low detrital influx or when the center of deposition shifted elsewhere (Fig. 7-4), as has been documented from present-day fan deltas by McGowen (1970). Similar interbedded fluvial and tidal sediments were described by Miall (1976) from the Proterozoic of Canada, and interdistributary tidal deposits have also been recognized in Pennsylvanian strata from the Appalachians (Visher, 1976).

Development of the Stratigraphic Sequence

Holocene fluvial depositional models (Coleman, 1969; Boothroyd and Ashley, 1975; Miall, 1977) adequately explain the varied sequences of lithologies and sedimentary structures observed in the geologic record but fail to account for the considerable lateral and vertical extent of many ancient alluvial deposits. The dominantly fluvial Waterberg Group in the Middelburg and Warmbaths areas, for instance, has a preserved extent of over 20,000 km^2 (Fig. 7-1) and a thickness of up to 7000 m (Fig. 7-2). Lateral migration of an active river system across an approximately 400-km-wide floodplain, coupled with prolonged aggradation, were required to form this stratigraphic sequence.

The Kosi River system in Nepal and India (Gole and Chitale, 1966) illustrates extensive

Figure 7-3. Flaser and wavy bedding in tidal-flat deposits of the Waterberg Group, northern Middelburg basin. [After Vos and Eriksson (1977).]

migration. Over the past 200 years the channel of the Kosi River has migrated more than 125 km from east to west (Fig. 7-5). Away from the Mahabharat Range (associated with a gradient decrease from 90 cm/km to 2.2 cm/km over a distance of 250 km), proximal boulders and cobbles give way to sands, mixed sands, silts, and clays, and finally to distal clays. These downstream grain-size changes, especially the transition from boulder to mixed sand–silt–clay, are analogous to those documented from the various Waterberg facies. Although the thickness of the Kosi River deposits is not known, continued channel switching over many millennia in a slowly subsiding basin would conceivably develop a considerable accumulation of sediment.

Origin of the Red Pigmentation: Paleoclimatic Significance

The red pigmentation in the Waterberg Group is confined to the immature fluvial sediments, whereas the mature marginal marine facies have a pale-buff to white coloration. These associations suggest that the source of the ferruginous coloration is unstable framework and/or matrix constituents in the fluvial sediments.

Numerous hypotheses for the origin of red beds have been proposed, of which three are considered to be most feasible. The red pigment may be (1) detrital hematite derived from a lateritic source (Krynine, 1950); (2) hematite formed *in situ* by conversion of limonite derived from a

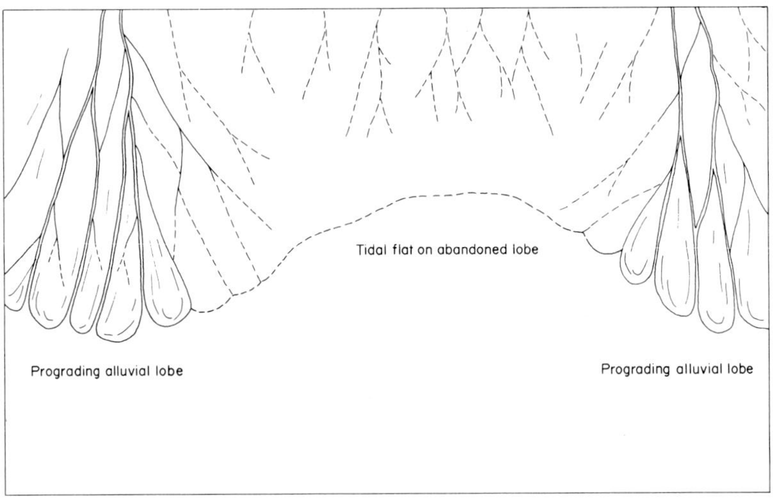

Figure 7-4. Depositional model for the Waterberg Group showing tidal flats developed on the inner margins of embayments between actively prograding alluvial lobes (fan deltas). [Adapted from Vos and Eriksson (1977).]

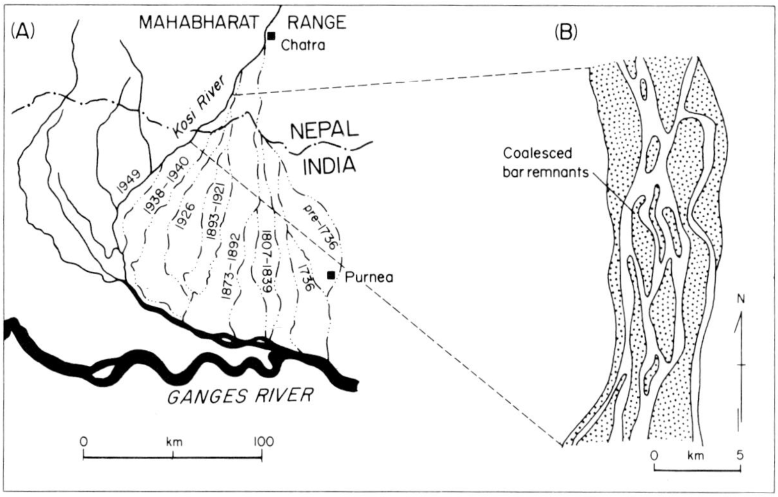

Figure 7-5. The Kosi River alluvial fan of India and Nepal. (A) Plan view illustrating the lateral migration of the active channel over the past 200 years: (B) Morphology of the proximal fan represents coalesced bar remnants bounded by active erosive channels. [After Gole and Chitale (1966).]

lateritic source (Van Houten, 1968); or (3) authigenic hematite formed through the diagenetic alteration of iron-bearing detrital silicate and oxide minerals (Walker, 1976; Hubert and Reed, 1978).

Based on petrographic examination of the Waterberg sandstones, the red pigmentation appears to be of diagenetic origin. Iron specks and veins in orthoclase grains, pyroxene grains, and felsite rock fragments are interpreted as having formed by release of iron upon hydrolysis of these clasts. Orthoclase in Bushveld granites contains up to 0.5 percent FeO (R.G. Cawthorn, personal communication, 1978), felsite contains FeO both in feldspar and ferromagnesian minerals, and the mafic phase of the Bushveld Complex contains a

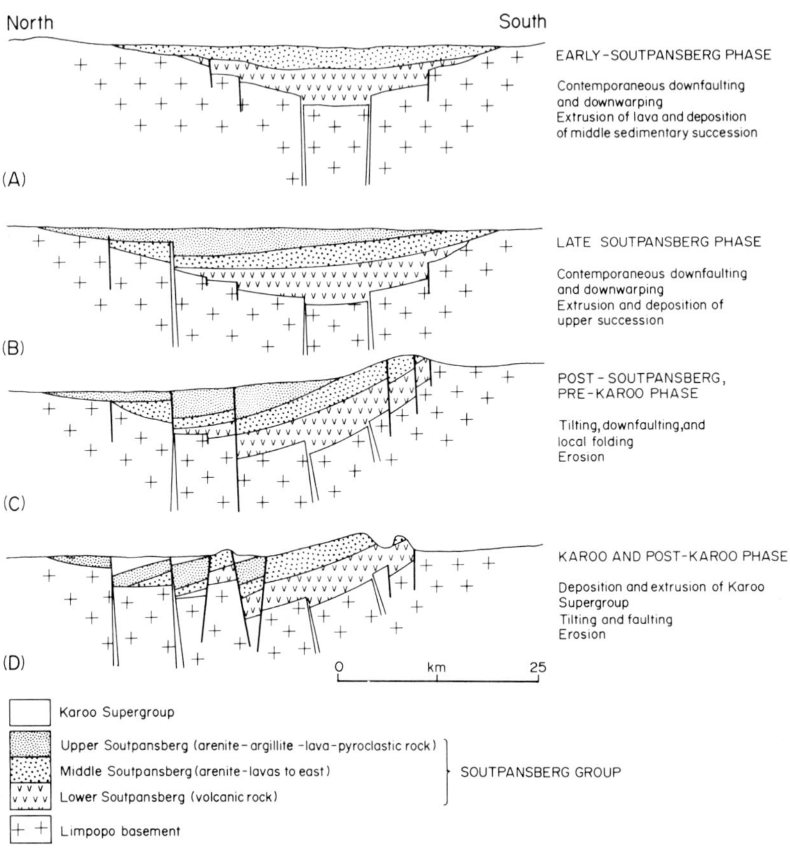

Figure 7-6. Inferred evolution of the Soutpansberg trough. [After Jansen (1975).]

vast reservoir of FeO. The suggestion has also been made that magnetite may be a possible source of hematite in red beds (Hubert and Reed, 1978) but this mineral appears to be stable in the Waterberg sediments.

Iron released from silicate minerals by hydrolysis was probably transported in mildly acidic ground waters prior to its precipitation on grain surfaces and along grain fractures at elevated Eh and pH, or in response to increased concentrations of iron in solution (Walker, 1976). Experimental studies by Paraguassu (1972) showed this to be a likely mechanism. The precipitate was probably in the form of limonite, which altered to hematite with time. The red coloration probably developed at an early stage of diagenesis, as is evidenced by authigenic quartz overgrowths that postdate the pigmentation.

The ubiquitous red pigmentation indicates that an oxidizing atmosphere was in existence 1800 Ma ago. Older immature sandstones of similar fluvial origin, such as those in the Witwatersrand Supergroup (Chapter 4) and the Moodies Group (Chaper 2), are devoid of red pigmentation.

7.2. Soutpansberg Trough

Stratigraphy and Paleocurrents

The Soutpansberg Group contains roughly equal proportions of sedimentary and volcanic rocks (Fig. 7-6) (Jansen, 1975; Barker, 1979). Lavas with subordinate sedimentary intercalations constitute the lower 3600 m of this group. The overlying subarkoses contain numerous small-pebble conglomerates and are intercalated with basaltic volcanics through a thickness of about 1100 m. The upper 800 m of the Soutpansberg Group comprises litharenites and lithic wackes, siltstones, mudstones, and pyroclastics consisting of tuff- and lapilli-sized particles. The sandstones in this interval become more mature upward as the proportion of clay and rock fragments decreases. Paleocurrent determinations from the Soutpansberg sediments indicate derivation exclusively from the north (De Villiers, 1967; Barker, 1979).

Lithology and Geochemistry of the Volcanics

The volcanics in the Soutpansberg Group are considered to have formed by cyclic subaerial extrusion along a linear vent system (Barker, 1979). Individual flows range from 15 cm to 200 cm in thickness. The flows are porphyritic at their base, containing abundant pyroxene phenocrysts, and have massive central portions and amygdaloidal tops. Amygdales consist mainly of quartz and calcite and are up to 10 cm in diameter. The flows are intercalated with basic and acid pyroclastics, which include agglomeratic and tuffaceous tephra deposits as well as lapilli and tuffaceous ash flows. Basalts are volumetrically most important, but acid phases are locally present as immiscible ocelli. Geochemically the basalts are tholeiitic, but with high potassium and rubidium contents (Barton, 1979c). Local differentiates of the basalts include andesites and dacites (Barker, 1979); minor trachytic lavas and ignimbrites are also present (Jansen, 1975).

Depositional Environments

Although detailed paleoenvironmental studies have not been undertaken, sediments in the Soutpansberg Group are believed to have accumulated exclusively in continental environments as braided-stream deposits (Jansen, 1975; Barker, 1976). The predominantly arenaceous sediments of the lower and middle Soutpansberg Group contain abundant evidence of channeling. These arenites are structured with planar cross-beds up to 2 m in amplitude, and less common trough varieties. Thin basal conglomerates are common. No internal cyclicity is recognized in these sediments and the only evidence of fluctuating discharge is the common occurrence of foreset grading. A sandy, braided, midalluvial plain similar to that of the Platte River of Nebraska (Smith, 1971a) was the probable depositional setting of these sediments.

The upper Soutpansberg sediments are arranged in upward-fining cycles between 1 m and 2 m thick. Mud-clast conglomerates are overlain by small- and medium-scale trough and planar cross-beds and ripple cross-lamination and are capped by shales with desiccation cracks and raindrop impressions. The depositional environment was probably a distal braided alluvial plain. Frequent switching of channels could have created stagnant ponds on the abandoned floodplain in which suspension settling produced argillaceous lenses.

Evolution of the Soutpansberg Trough

The stratigraphic units in the Soutpansberg Group are truncated by numerous faults, many of which represent reactivated older structures (Chapter 3). These had a profound effect on Soutpansberg volcanism and sedimentation (Jansen, 1975). Recognition of the role of faulting in the development of the stratigraphy has led to proposals that the Soutpansberg trough represents a yoked basin (Barker, 1976; 1979) or an aulacogen (Jansen, 1975). The yoked basin hypothesis invokes asymmetric faulting of the stabilized craton; in the aulacogen model (Fig. 7-6) symmetric grabens form the abandoned arm of a triple junction. Aulacogens, by definition (Tuyezov, 1967; Hoffman, 1973; Hoffman *et al.*, 1974), have a radial disposition in relation to the craton; they die out in the interior of the craton and pass into a geosyncline on the edge of the craton. In contrast, continental rift zones, including yoked basins, form in the interior of cratons and are not associated with geosynclines.

Attributes that have been cited by Jansen (1975) to support an aulacogen model for the Soutpansberg Group include the following:

(1) The Soutpansberg trough resembles an aulacogen in shape, size, and thickness of the stratigraphic succession.
(2) The trough is thought to die out westward into Botswana.
(3) Oblique-slip, as opposed to dip-slip faults are ubiquitous in the Soutpansberg Group. This fault pattern is characteristic of the Siberian and Canadian aulacogens.
(4) Basalts in Canadian aulacogens are characteristically tholeiitic in the lower parts of the sequence and alkaline higher up. Although alkaline varieties are rare, thick tholeiitic basalt flows characterize the lower Soutpansberg Group.
(5) Regional metamorphism and batholithic intrusions are absent in the Soutpansberg.
(6) The geosynclinal facies of the Soutpansberg Group is probably represented by the Umkondo Group (see Section 7.3), the eastern extent of which was involved in the high-grade Mozambique metamorphic event.

Although the structural style of the Soutpansberg Group may support the aulacogen model, other evidence does not (Barker, 1976). The asymmetric, north to south paleocurrent patterns are more indicative of a yoked basin (sediment transport in an aulacogen is parallel to the graben axis). In addition, the absence of marine facies in the Soutpansberg Group may support an intracratonic depository.

7.3. The Miogeoclinal Umkondo Group

Stratigraphy and Paleoenvironments

The Umkondo Group is composed predominantly of terrigenous clastic sediments with thin carbonates near the base. A generalized stratigraphic column for the Chipinga area is illustrated on Figure 7-7. Also shown are composite paleocurrent patterns indicating transport of sediment from the Zimbabwe Province.

Tidal-flat–lacustrine environments. The lower 60 m of the calcareous unit above the basement comprises a basal conglomerate and arenite, representing reworked detritus, overlain by dolomite and red siltstone of tidal-flat–lacustrine origin (Button, 1977). The dolomites normally have a fine-grained crystalline texture and commonly contain a siliciclastic component in the form of rounded quartz grains. Also present are domical stromatolites, delicate crinkly algal laminations, and tepee-like structures outlined by chert. Physical structures include intraclastic breccia, ripples, and poorly defined cross-bedding. The former presence of bird's-eye vugs is suggested by elongate blebs and spheres of white, crystalline carbonate. Disk-shaped blebs up to 1 cm long and 2–3 mm thick occur with their major axes in the plane of the bedding. Also

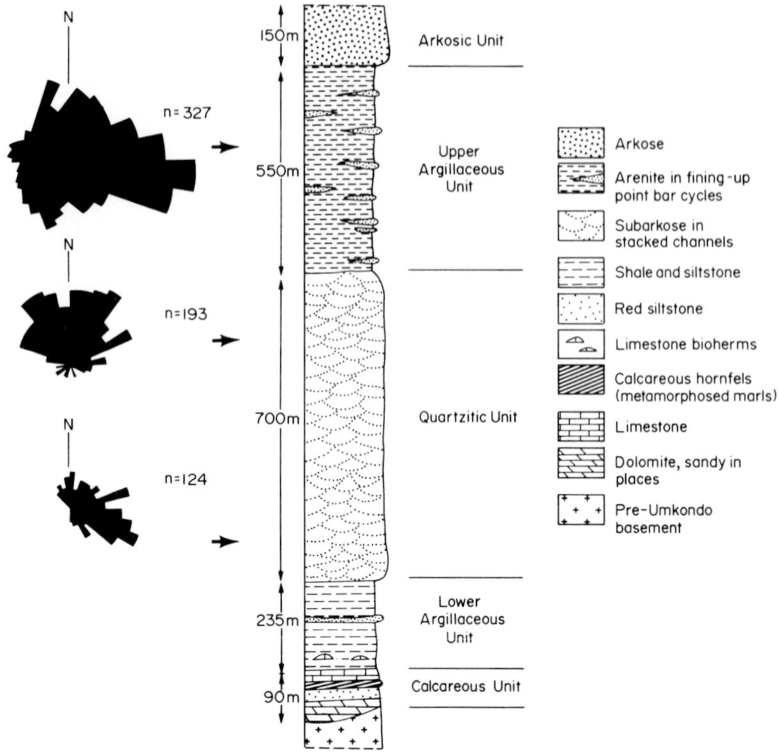

Figure 7-7. Generalized stratigraphic column and paleocurrent distributions of the Umkondo Group in the Chipinga area of Zimbabwe. [After Button (1977).]

present are spherical structures (1–2 cm in diameter) with siliceous rims and cores of sparry calcite. According to Shinn (1968) disk-shaped vugs are formed by alternate wetting and drying of carbonate mud, usually in the supratidal setting, whereas spherical vugs (1–3 mm in diameter) are produced by escaping bubbles in unconsolidated supratidal and intertidal carbonate sediments. The large size of the spherical structures in the Umkondo dolomites suggests replacement of evaporite nodules (Button, 1977).

The association of structures in the dolomite unit is consistent with a supratidal flat or sabkha depositional environment (Button, 1977). A sabkha is usually dominated by dolomitization, algal laminations, and evaporites; there is evidence of each in this unit. In addition to the spheroidal nodules of probable replacement origin, the tepee structures suggest growth of evaporite crystals beneath the depositional surface. Desiccation is suggested by the sparry calcite structures; the intraclastic breccias probably formed by drying of algal mats followed by reworking during inundation. Quartz grains commonly occur in modern sabkhas and represent sediment transported onto the tidal flat by eolian processes (Shinn, 1973b).

The overlying siltstones crop out poorly and are normally pink or red with sparse intercalations of green-gray shales and feldspathic arenites. Also present are chert with rare stromatolites, dolomite, and a clean quartzite lens at the top of the unit. The dominant sedimentary structures in the clastic sediments are horizontal lamination and less abundant ripple cross-lamination. Minor erosional surfaces are present but lack desiccation structures. The red siltstones are considered to have accumulated in lagoons or lakes associated with the carbonate supratidal flats (Button, 1977). The immature and mature arenites are interpreted as fluvial and eolian deposits, respectively.

Shallow-shelf processes. Calcareous sediments metamorphosed to low grade constitute the upper third (60–90 m) of the calcareous unit (Fig. 7-7). Both the calcareous phyllites and overlying limestone are argillaceous, with the amount of clay increasing upward until distinct shale beds appear at the top of the unit. Primary structures are rare and confined to delicate crinkly laminations, which may be of algal origin (Button, 1977). Regular, continuous bedding is commonly disturbed by faults and folds that do not affect the overlying and underlying strata. The bedding is commonly inclined in sets 0.1–2 m thick. Closely analogous features described from the Upper Cretaceous of France (Kennedy and Juignet, 1974) represent eroded crests and flanks of carbonate banks.

The overlying argillaceous unit (Fig. 7-7) consists predominantly of horizontally laminated shales which become more arenaceous toward the top. Stromatolitic limestones are commonly developed in the basal 40 m of this unit as bioherms, 2–3 m thick and up to 100 m in strike length, which are completely enclosed in shale. Most individual stromatolites are columnar, but laterally linked domes are common. In plan view the stromatolites are elongated in a northeasterly direction. Thin arenite lenses that occur near the top of the argillaceous unit are usually plane-bedded, with primary current lineation and rare cross-bedding.

The upward-coarsening sequence of calcareous and clastic sediments is believed to have developed on a shallow-marine shelf, probably below wave base but within the photic zone (Button, 1977). As increasing quantities of fine-grained terrigenous clastics were added to the shelf, carbonate sedimentation was replaced by suspension settling of clays, and bedload transport of sand, probably during storm periods. The bioherms developed during times of reduced clastic input when clear waters prevailed. These were smothered by an increased supply of mud or died when subsidence took them below the photic zone.

Progradational fan deltas. The quartzite unit dislays a transitional upward-coarsening relationship with the underlying argillaceous unit (Fig. 7-7) and is interpreted as a fan-delta complex in which a braided alluvial plain prograded directly onto the shallow-marine shelf. Apart from the upper 50 m where quartz arenites are developed, the sandstones in the quartzite unit are arkosic subgraywackes and wackes. A characteristic pattern of overlapping, channel-like lenses 0.3–2 m thick, on an erosive surface up to 1 m in relief (Fig. 7-8A) typifies all but the upper part of this unit. The lenses contain smaller depositional sequences (Fig. 7-8B) analogous to those developed on sandy reaches of braided alluvial plains (Miall, 1977). Basal mud-clast conglomerates are normally overlain by planar cross-beds, which develop in modern braided rivers as a result of transverse or linguoid bar migration at flood stage. Occasional intercalated trough cross-beds may have resulted from dune migration in channel scours developed by bar modification during falling-water stage (Boothroyd and Nummedal, 1978). Overlying plane-bedded sandstones (Fig. 7-8B) probably represent bar-top deposits and may be upper or lower flow regime structures (Smith, 1971b; McBride *et al.*, 1975). Ripple cross-laminated intervals (Fig. 7-8B) that are erosive into the plane-beds may represent a low-water stage response to bar modification processes. As a result of lateral switching of active braided channels, stagnant overbank ponds were

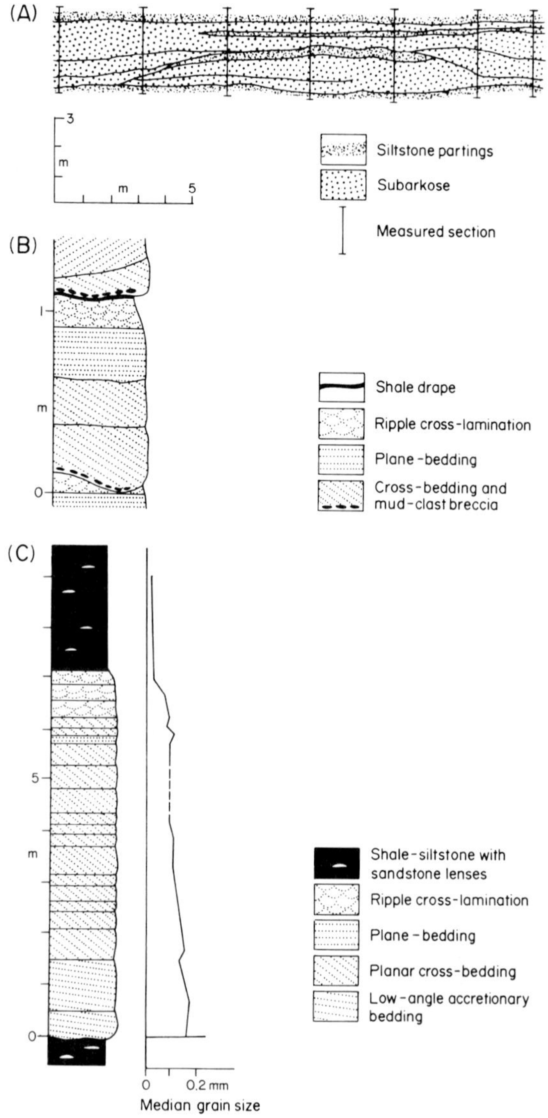

Figure 7-8. Fluvial sediments in the Umkondo Group showing: (A) channelized sedimentation units in quartzite unit; (B) idealized braided fluvial depositional cycle in arenites from the quarzite unit; (C) measured profile through point-bar deposits of the upper argillaceous unit showing upward decrease in grain-size. [After Button (1977).]

probably formed on the abandoned floodplain. Suspension settling of clays in such ponds could have produced the ubiquitous shale drapes in these fluvial sediments (Fig. 7-8A and B). Cross-bed azimuths for the lower 650 m of the quartzite unit display general easterly unimodal paleocurrent patterns at individual stations, but for the interval as a whole a considerable spread of foreset azimuths occurs (Fig. 7-7).

The uppermost part of the quartzite unit is composed of quartz arenites lacking fluvial depositional sequences and showing northerly paleocurrent patterns (Fig. 7-7). These criteria led Button (1977) to propose a shallow-marine environment in which longshore reworking of the underlying immature arenites occurred. A decrease of sediment supply to the braided fan delta, combined with continued subsidence, could have resulted in transgression of the shoreline across the delta. This is analogous to the situation in Alaska where the Copper River fan delta is periodically subjected to marine reworking (Galloway, 1976).

Meandering rivers. Meandering river environments are recognized in the upper argillaceous unit (Fig. 7-7). The depositional settings include upward-fining point-bar sequences and overbank flood basin, levee, and crevasse splay deposits. The ratio of overbank to channel sediments is about 19:1.

Floodbasin sediments comprise thick sequences of red shales and silty shales with intercalated beds of siltstone that were introduced during flooding. The siltstones are rippled, show some primary current lineation, and contain mud clasts derived from desiccated shales. Associated halite casts and syneresis cracks indicate periodic saline, evaporitic conditions in the floodplain (Button, 1976c). Interbedded fine-grained argillaceous sandstones and siltstones, mostly arranged in 1–5 cm couplets, are interpreted as levee deposits. These depositional sequences are up to 1 m thick, have erosional bases, and lens out rapidly along strike. Sedimentary structures include ripple cross-lamination, in-phase ripple-drift cross-lamination, and plane-bedding with primary current lineation. Desiccation structures and reworked mud clasts are also present. The alternation of coarser and finer grained sediment is believed to typify levee sedimentation during waning flood cycles (Button, 1977). Sandstone lenses up to 70 cm in thickness are enclosed by the fine-grained floodplain deposits. Cross-beds overlying an erosive base pass upward into ripple cross-lamination followed by siltstone drapes. Waning bedload sedimentation on the floodplain is indicated, such as is characteristic of crevasse splays associated with levee breaching during flood discharge.

Point-bar sequences comprise 3- to 7-m-thick upward-fining cycles (Fig. 7-8C) resting on a basal erosive surface. Low-angle accretionary (epsilon) cross-beds at the base of thicker cycles dip approximately at right angles to the paleotransport direction obtained from overlying cross-beds. The epsilon cross-beds are internally plane-bedded and show primary current lineation parallel to the strike of the epsilon surfaces; they are interpreted as upper flow regime deposits formed on the deepest part of the gently inclined point-bar surface (cf. Allen, 1965a). Trough and planar cross-beds either overlie the epsilon cross-bedding or, in the thinner cycles, rest directly on the erosion surface. These are believed to have resulted from dune migration on the lower to intermediate point-bar surfaces. Plane-bedded sands either succeed or occur with the cross-beds (Fig. 7-8C). Despite the presence of current lineation, Button (1977) considers these structures to represent a lower flow regime response to migration of low-amplitude, long-wavelength bedforms of the type documented by Smith (1971b). Alternatively, these plane-beds could represent upper flow regime chute deposits formed during flood events. The upper parts of cycles are ripple cross-laminated and are inferred to have been deposited on the emergent upper point bar where it merged with the natural levee. Many of the cycles are multistory and contain mud drapes. These features are taken to indicate repeated flood events followed by low-stage flow.

Paleogeography

The inferred depositional history of the Umkondo Group is shown in Figure 7-9. Initially sabkhas and lakes developed under conditions of restricted circulation (Fig. 7-9A; calcareous unit). The carbonates and siltstones are cut by channel-like bodies of immature sandstone, which may represent ephemeral stream deposits. The sabkha was then drowned under a marine shelf during eustatic rise of sea level (Fig. 7-9B; lower argillaceous unit). Carbonates and marls accumu-

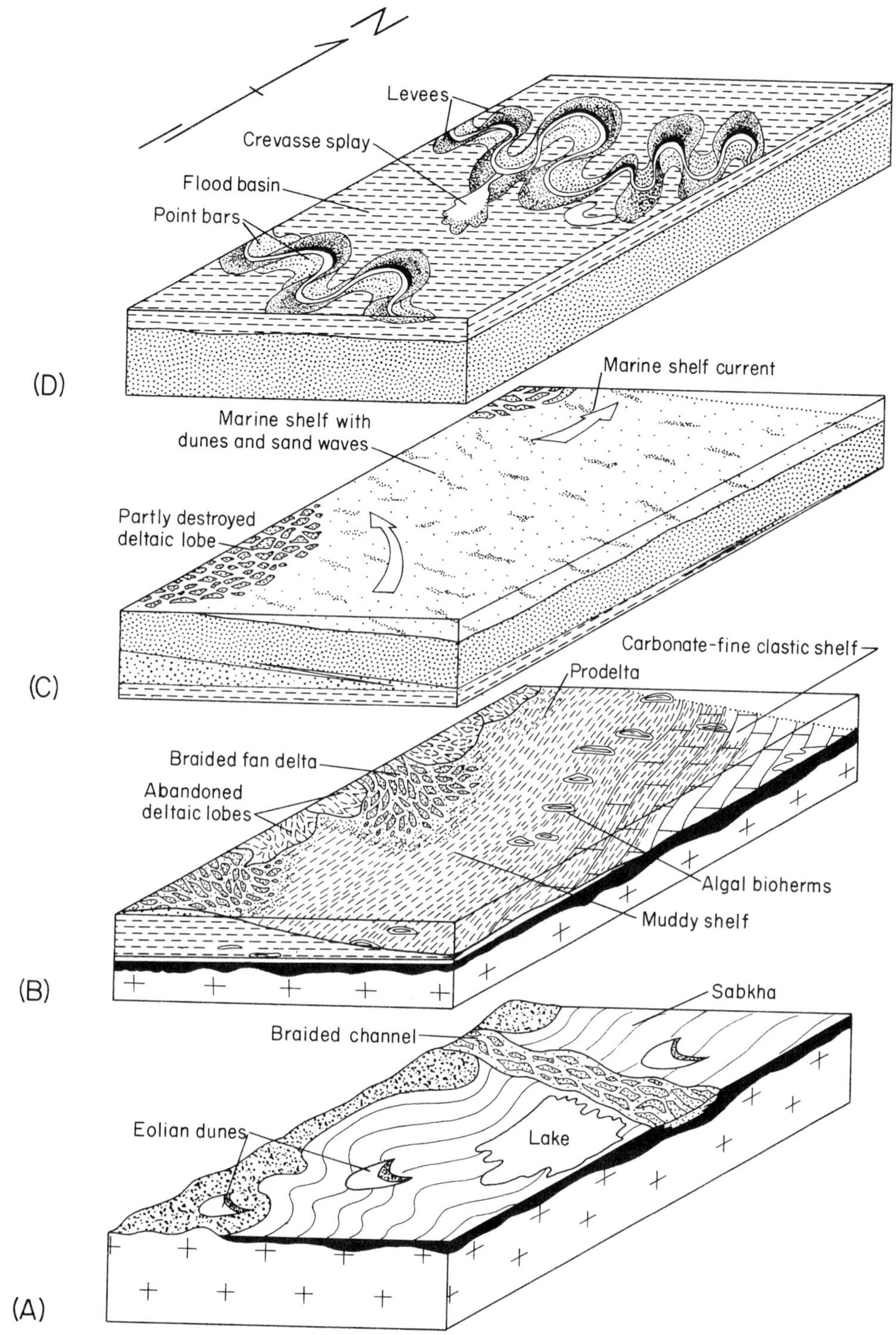

Figure 7-9. Depositional models inferred for the Umkondo Group in the Chipinga area. Individual paleogeographic reconstructions refer to: (A) calcareous unit; (B) lower argillaceous and quartzite units; (C) upper part of quartzite unit; (D) upper argillaceous unit. [After Button (1977).]

lated on the outer shelf; these developed into banks off which soft-sediment slumping occurred. Prodeltaic muds gradually prograded across the carbonate shelf. Algal bioherms developed in the transitional carbonate–clastic zone but were eventually eliminated as further progradation brought muds, delta-front silts, and braided fluvial sands across the shelf, the sands in the form of fan deltas (quartzite unit). Upon abandonment of individual deltaic lobes (Fig. 7-9C, upper quartzite unit), the immature fluvial sands were reworked by longshore marine currents into quartz arenites. Renewed clastic input down the preexisting easterly inclined paleoslope produced a meandering fluvial system on a low-gradient alluvial plain (Fig. 7-9D, upper argillaceous unit). Data from the upper arkosic unit shown on Figure 7-7 are sparse, but it is probable that a higher gradient braided river system eventually prograded eastward across the older meandering river deposits.

The transition from a braided to a meandering pattern of sedimentation is unusual in a progradational system such as existed during deposition of the Umkondo Group. However, a similar transition occurs in the Mississippi River valley upstream of the progradational birdfoot delta, where meandering river deposits represent only a thin veneer over Early Holocene sands and gravels deposited by braided rivers (Fisk, 1944).

The presence of meandering river deposits in Precambrian sediments is unusual because stabilization of banks by vegetation is thought to be an important prerequisite for meandering (Schumm, 1968a). In the absence of vegetation during deposition of the Umkondo Group, bank stabilization was possibly achieved through the supply of high proportions of silt and mud.

7.4. The Craton-Edge Matsap Group

Stratigraphy

Process-related studies have not been undertaken on the Matsap Group but the stratigraphy in the type area of the Langeberg Range is well-established (Fig. 7-1) (Botha *et al.*, 1976). The basal Lucknow Formation reaches 1000 m in thickness and is dominantly sedimentary, comprising sandstones and shales with subordinate conglomerates, dolomitic limestones, and volcanics. A basal conglomerate contains pebbles of quartz, quartzite, jasper, diabase, granite, and iron-formation. The overlying Hartley Formation averages 1300 m in thickness and consists of amygdaloidal lavas and intercalated tuffs, quartzites, and brecciated quartzites. Arkosic sandstones predominate in the uppermost 1500 m thick Volop Formation. Scattered jasper and chert pebbles are common and the sandstones are structured mainly by medium-scale trough cross-beds. Cross-beds indicate general southerly transport. The stratigraphy of the Matsap to the southwest of the Langeberg Range, together with its correlation with metasediments in the Namaqua Province, are discussed in Chapter 8.

7.5. Synthesis

The Waterberg, Soutpansberg, and Umkondo Groups accumulated contemporaneously in different tectonic settings. Waterberg sedimentation occurred in gently subsiding intracratonic basins which, unlike the Pongola, Witwatersrand, and upper Transvaal Supergroup depositories, were probably not connected to major water bodies. Reworking of the dominantly fluvial sediments was achieved by weak waves and tides, probably generated by local winds. Rifting played a major role in the development of the Soutpansberg Group. The Soutpansberg trough formed rapidly about 1770 Ma ago (Barton, 1979c). Although no marine influence is recognizable, the proposed intracratonic rift-type depository may have been connected to a distant open sea. Thus the Umkondo Group probably accumulated along a miogeoclinal continental margin on the active spreading arm of a triple junction. Both early and postrift stages of passive continental margin development are identified in the Umkondo Group. These are represented by the lacustrine and fluvial–shallow-marine intervals respectively. The Matsap Group accumulated along the western margin of the Kaapvaal Province, possibly as long ago as 2100 Ma, but is more likely coeval with the other three stratigraphic units. Certain stratigraphic units in the Matsap Group may possibly be traced into the Namaqua Province (Chapter 8). This is a unique phenomenon in the Proterozoic geology of the Kaapvaal Province because older stratigraphic sequences, such as the Griqualand West and Transvaal Supergroups, are abruptly terminated on the craton edge.

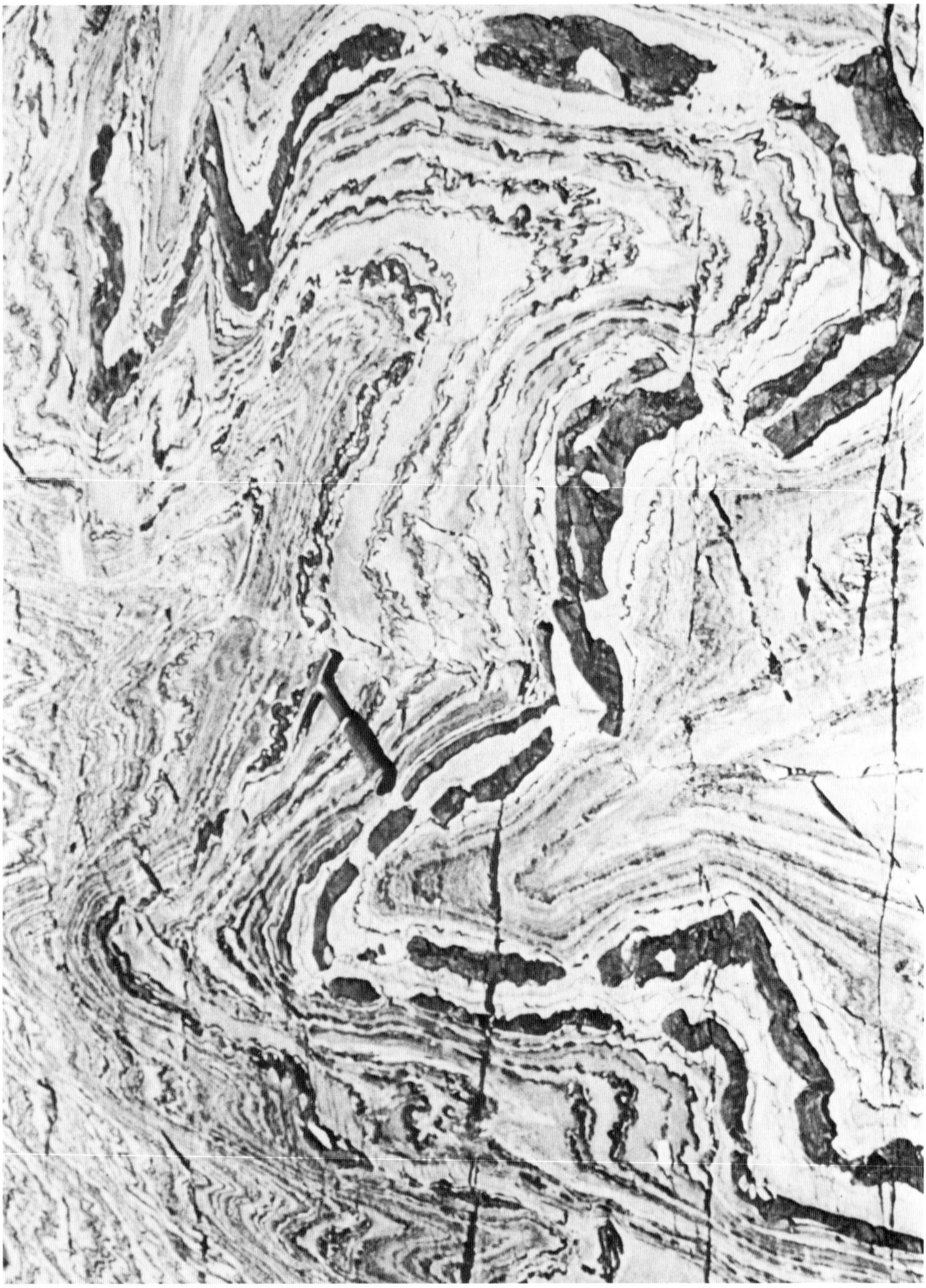

STAGE 3

Proterozoic Orogenic Activity

Forming the longest structure in the subcontinent, the Late Precambrian Pan African orogen is intermittently exposed from Angola to the eastern Cape. Folded boudins of calc-granofels in Karibib marble of the eugeosynclinal Swakop Group in the Damara Province (Klein Aukas farm, Namibia). [Courtesy of A. R. N. Jackson.]

Chapter 8

Namaqua–Natal Granulite–Gneiss Terranes

8.1. The Natal Province

Inliers of a Proterozoic gneiss terrane, known as the Natal Province, are exposed in eastern Natal and constitute an elongated cross section some 280 km long subparallel to the present coastline (Fig. 8-1). Phanerozoic cratonic sediments obscure the western, southern, and most of the eastern margins of the Natal Province. The gneiss terrane may extend westward beneath the cratonic cover skirting the southern margin of the Archean Kaapvaal Province to link up with the high-grade Namaqua Province (Fig. 1-4), which has yielded comparable Mid Proterozoic radiometric ages (Nicolaysen and Burger, 1965; Burger and Coertze, 1973). Based on reconstructions of pre-Mesozoic Gondwana (Smith and Hallam, 1970; Piper, 1974; Norton and Sclater, 1979), the eastern extension of the Natal Province is most likely to be the high-grade terrane of Dronning Maud Land, Antarctica.

Only the northern margin of the Natal Province is exposed. Here, allochthonous rocks of the Natal Province have been thrust onto a foreland represented by the Kaapvaal Province in a spectacular Precambrian nappe zone. The few radiometric age determinations are close to 1.0 Ga. However, some of the precratonic supracrustal suites resemble greenstone belts typical of the Archean (Chapter 2); gneisses apparently forming a basement to these suites are likely to predate them.

Because of a dearth of recently published work on the Natal Province, the present account draws heavily on an excellent summary by Matthews (1981) that is based largely on unpublished work. The tectonic zonation proposed by Cain (1975) and Matthews (1981) is adopted here, but the names of some of the zones have been modified to conform with those in other southern African gneiss terranes. The characteristics of the tectonic zones are presented in Table 8-1 and their extent is shown in Figure 8-1.

Northern Marginal Zone

This tectonic zone consists of large thrust nappes that have been transported northward onto the stable Kaapvaal foreland (Table 8-1). The narrow, southward-dipping Natal thrust belt marks the northern limit of this tectonic zone (Fig. 8-2A). The thrust belt consists of imbricated slices of two unconformable supracrustal sequences metamorphosed to greenschist facies. Para-autochthonous slices of the younger Ntingwe Formation represent a transgressive sequence deposited unconformably on a stable shelf represented by the crystalline Archean foreland. The Ntingwe succession has been overridden by allochthonous thrust nappes of the older Mfongosi Metamorphic Suite (Matthews, 1959). The latter comprises argillite and basic lava, with relict pillows and amygdales, and thin intercalations of chemical and arenaceous sediments.

South of the Natal thrust belt is a broad, westward-plunging nappe complex comprising four extensive thrust nappes. The lower units of the structural sequence are exposed in the east and the higher units in the west (Fig. 8-2B). The thrust sheets consist largely of the Tugela Metamorphic Suite, a supracrustal sequence metamorphosed to amphibolite facies. The suite comprises metabasite lavas represented by amphibolites and hornblende gneisses, and subordinate clastic and chemical sediments represented by metapelite, magnetite metaquartzite, and marble. The lower nappes also contain tectonic slices and infolded

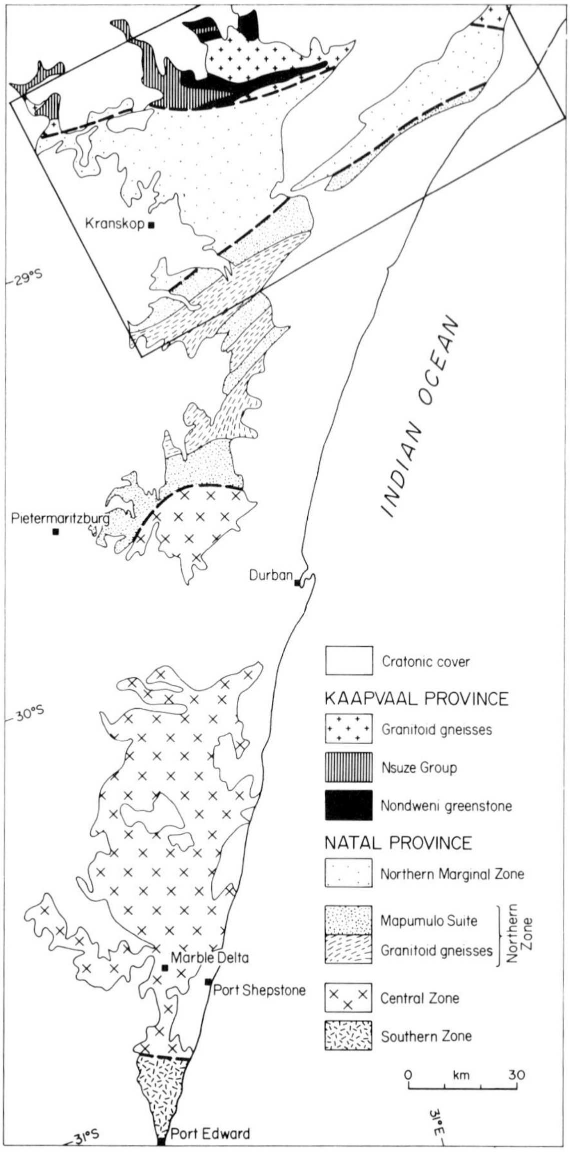

Figure 8-1. Geologic map of the Natal Province and adjoining areas of the Kaapvaal Province. Rectangle in north shows location of Figure 8-2A. [Based on Matthews (1981), Cain (1975), and McIver (1966).]

elements of a possible basement represented by granitoid gneisses. The proportion of metavolcanic (basalt) to metasedimentary (largely graywacke) rocks increases upward through the structural sequence.

Matthews (1972) suggested that the Tugela and Mfongosi Suites represented the lower and upper parts of an ophiolite suite obducted northward along a basal décollement known as the Makasana thrust fault. This contention is supported by the association of pillowed metalavas, with subordinate metachert and carbonate rocks, with layered basic–ultrabasic intrusions of the Tugela Rand Intrusive Suite (Dix, 1981), and plagiogranite (Harmer, 1979). If so, the two supracrustal suites represent transformed oceanic crust. Consequently the relicts of granitoid "basement" in the Northern Marginal Zone may represent part of the Kaapvaal Province overridden by the allochthonous ophiolite suite.

Submarine extrusion of basic lavas was possibly accompanied by intrusion of layered basic bodies and plagiogranite (and sheeted dikes?) and followed by accumulation of pelagic sediments. Deformation commenced with tight to isoclinal folding and fabric formation under upper amphibolite conditions accompanied by partial melting of the granitoid infrastructure and injection of neosomes into the more basic rocks. Sporadic "cordierite + garnet" in the rare pelitic rocks indicates low to intermediate pressure/temperature (P/T) metamorphism in a regime of high heat flow not compatible with present-day plate collision zones. This inferred high heat flow and granitoid infrastructure suggest that the volcanic sequence might have accumulated in a backarc ensialic basin rather than a marginal oceanic basin as suggested by Matthews (1972). Obduction had not started at this stage.

Further penetrative ductile deformation (D2) took place by regional simple shear and overfolded the F1 folds to give a strong northward vergence. Medium P/T metamorphism, caused by either decreased heat flow or increased confining pressure, is indicated by the formation of kyanite and garnet in metapelites and garnet in metabasites. Ultrabasic sheets, now represented by serpentinite and magnesian schists, were subsequently intruded subconcordantly.

The northward translation that began on a penetrative scale during D2 continued during D3 on a less penetrative scale as thrust nappes and northward verging folds (Fig. 8-2B). Thrust surfaces were apparently lubricated by the highly ductile ultrabasic intrusions. A steeply dipping zone of migmatites at the southern boundary of the predominantly shallow-dipping Northern Marginal Zone may represent part of the root zone of the nappe complex. The total displacement is estimated at more than 100 km by unstraining the nappe complex, assuming all thrust nappes to have been originally adjacent parts of the same crustal segment. If so, the root

Table 8–1. Characteristics of Tectonic Zones in the Natal Province and Contiguous Kaapvaal Province[a]

	Kaapvaal Province (foreland)	Natal Province			
		Northern Marginal Zone	Northern Zone	Central Zone	Southern Zone
Width (km)		15–50	60	150	20
Basement	Not recognized	Northern Marginal Zone allochthonous but granitoid gneisses tectonically interleaved may be basement	Possibly augen and charnockitic gneisses and basic dikes	Not recognized	Not recognized
Precratonic cover protolith	>3.1-Ga Nondweni greenstones; 3.0-Ga Nsuze Group volcanics and clastic sediments; Ntingwe marine shelf sediments are cratonic but overthrust by NMZ	Mfongosi lavas, argillites; Tugela basic–ultrabasic extrusives, subordinate sediments	Mapumulo arenites, graywackes, basalts	Marble Delta shelf sediments	Leisure Bay basic and acid lavas
Precratonic intrusions	Pre- and post-Nsuze granitoids	Layered intrusions, plagiogranite, granitoids	Granitoids	Granitoids, commonly megacrystic, locally charnockitic	Dioritic to granitic charnockites, granitoids
Dominant structures	E–W trending folds steeply inclined to N, gently plunging to W.	N-verging thrust and (?) fold nappes overriding thrust sheets on Kaapvaal foreland	Upright major synforms of cover separated by antiforms of granitoids and gneisses	Multiple batholiths, narrow septa of gneisses	Tight folds overfolded to N
Dominant metamorphism and facies series	Greenschist, medium P/T	Amphibolite, low P/T → medium P/T in Tugela; greenschist medium P/T in Mfongosi	Granulite low P/T → amphibolite; lower in N, upper in S	Locally granulite → mainly amphibolite; low P/T	Granulite, medium P/T → upper amphibolite

[a]Compiled from Cain (1975), Matthews (1981).

zone would embrace the entire Northern Zone to the south of the Northern Marginal Zone. Translations of this magnitude do not necessarily imply overall crustal shortening and may be the expression of gravitational spreading from a rising axial zone farther south (cf. Bucher, 1956; Carey, 1962; Elliott, 1976).

Downward younging structures in magmatic sediments of the Tugela Rand Intrusive Suite in the Tugela nappe indicate that this large layered intrusion possibly forms part of the inverted limb of a major recumbent fold (Dix, 1981). This fold could be either an F3 fold nappe or an F1 fold overturned during D2 simple shear.

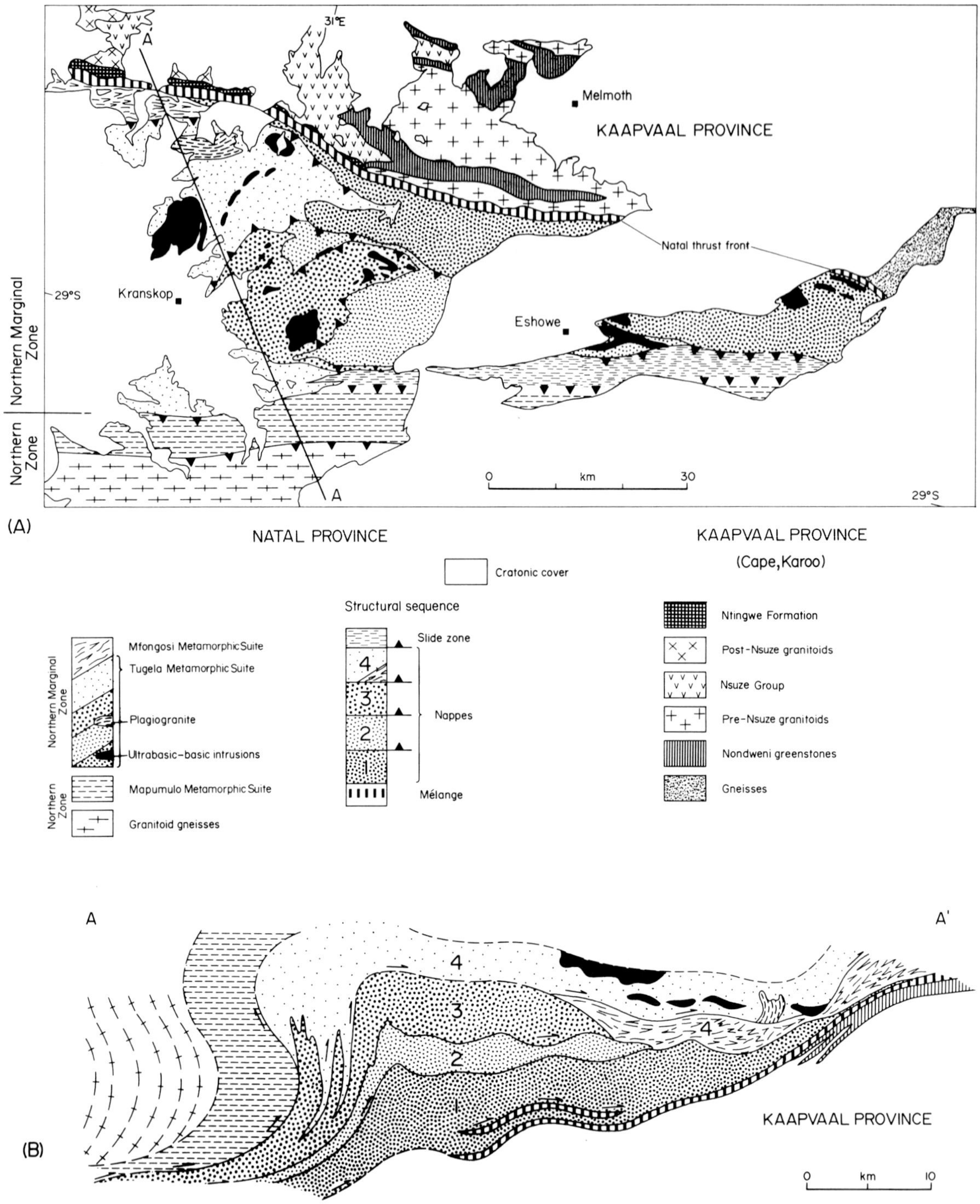

Figure 8-2. (A) Geologic map of the Northern Marginal Zone of the Natal Province and contiguous Kaapvaal Province. (B) Tectonic profile of the Northern Marginal Zone to show structural sequence of nappes and overthrusting onto the Kaapvaal foreland. Owing to the down-structure method of construction, this profile has no true line of section but the approximate equivalent is shown as AA′ on the map. [A and B after Matthews (1980).]

Northern Zone

This zone is characterized by four steeply dipping belts of precratonic cover rocks, known as the Mapumulo Metamorphic Suite (Fig. 8-1). This heterogeneous suite consists of high-grade migmatized psammitic gneiss, metagraywacke, and metabasalt. These rocks may represent clastic wedges of continental provenance deposited in fault-bounded grabens (Cain, 1975). The four supracrustal belts represent synformal relicts separated by anticlinoria of strongly foliated augen gneisses. These partially remobilized gneisses are charnockitic in places and, together with intermediate-pressure granolite inclusions and two generations of dismembered amphibolite dikes, may represent a granulite-grade basement that was deformed and metamorphosed before deposition of the Mapumulo precratonic cover (Rhodes and Leith, 1971; Cain, 1975). However, the structural and metamorphic relations between the "basement" and the Mapumulo cover strikingly resemble those between the 2.7-Ga Bulai granitoid gneiss and the >3.2-Ga cover of the Beitbridge sequence in the Limpopo Province (Chapter 3). It is therefore possible that most of the "basement" to the Mapumulo Suite consists of granitoids intruded syntectonically after the cover accumulated.

Low-pressure high-temperature metamorphism resulted in partial melting in both "basement" and cover to form migmatites and elongated granite plutons of anatectic origin (Rhodes and Leith, 1971). Folding was followed by the development of shear belts and pseudotachylites (Cain, 1975).

Central Zone

This large zone is characterized by extensive bodies of massive or foliated biotite or hornblende granitoids which are commonly megacrystic. Batholithic masses are separated by septa of augen gneisses, charnockitic intrusions, and metamorphic suites formed under high-temperature low-pressure metamorphism, locally at granulite facies. The granolites and charnockitic rocks may represent relicts of a formerly much more extensive granulite terrane that has largely been retrogressed to upper amphibolite facies (Cain, 1975). However, some of the charnockitic rocks crosscut foliated granites and are therefore younger (Du Toit, 1946). Large xenoliths of strongly flattened metaquartzite in the Bomela charnockite, described in the following section, confirm the post-cover age of some of the charnockites. Thick bodies of dolomitic marble, quartz-graphite schist, and metaquartzite, together with amphibolite dikes and lavas at Marble Delta (Fig. 8-1) may represent the remains of a precratonic stable shelf sequence (Simpson and Tregidga, 1956; Otto, 1973).

Southern Zone

This narrow zone is characterized by a more extensive development of the charnockite–granolite association than is present farther north, but the boundary is entirely arbitrary. Acid and basic granolites, known as the Leisure Bay Metamorphic Suite, constitute a precratonic cover sequence (McIver, 1966). This has been intruded by the even-grained Nicholson's Point granite. High flattening strains under granulite facies produced isoclinal, originally flat-lying folds in the Leisure Bay granolites and a quartz-leaf fabric in the granite.

An extensive suite of charnockitic (orthopyroxene-bearing) rocks, known as the Port Edward Intrusive Suite, intruded the previously deformed granolites and granite. The charnockitic rocks are partly megacrystic and range in composition from dioritic to granitic, with the former being more common along the coast (Gevers and Dunne, 1942; McIver, 1966). The intrusions contain numerous xenoliths of Leisure Bay granolites and have a minimum age of 990 Ma (Nicolaysen and Burger, 1965; Rb-Sr).

Farther north in the Central Zone is the isolated Bomela charnockite southwest of Port Shepstone (Fig. 8-1). This has the unusual assemblage (in a granitoid) of ferrohypersthene, ferroaugite, fayalite, garnet, and hornblende. Low oxygen fugacity may be the cause of the iron enrichment of the mafic minerals, which allowed the rare coexistence of olivine and quartz (McIver, 1966). Possibly related to the same phase of igneous activity is the Glenmore granite, which is megacrystic except in its marginal phases. Subsequent deformation has resulted in relatively weak fabrics in the charnockitic Port Edward Suite and the Glenmore granite and in the open-to-close overfolding toward the north.

8.2. The Namaqua Province

The Namaqua Province forms most of the crystalline basement of southwestern Africa. It covers the northern Cape and southern Namibia and is exposed in a triangular area, the longer sides of which are approximately 850 km and 700 km in length (Fig. 8-3). The full extent of the province is unknown because its margins are largely obscured by younger cover (Koras, Sinclair, Gariep, Nama, and Karoo). The Province may extend eastward beneath the Karoo cover around the southern margin of the Kaapvaal Province to link up with the Natal Province (Fig. 1-4). The only exposed boundary with an older tectonic province is in the east. At this boundary a relatively narrow (15–30 km) zone of low-grade but complexly deformed precratonic cover rocks, referred to here as the Eastern Marginal Zone of the Namaqua Province, is in faulted or gradational contact with the Kaapvaal Province (Fig. 8-3). The latter is represented by crystalline Archean basement overlain by generally flat-lying Early Proterozoic cratonic cover.

Most of the Namaqua Province is represented by the Central Zone, a complexly deformed heterogeneous group of gneisses and intrusions metamorphosed to medium and high grade and collectively known as the Namaqua Metamorphic Complex. In the west center of the Central Zone is a small wedge-shaped area of low-grade supracrustal rocks and high-level intrusions referred to here as the Western Zone (Fig. 8-3). In the extreme west, crustal reworking during the Late Precambrian truncated the Namaqua Province (Chapter 9); the zone of reworking delineates much of the western margin, the remainder being represented by the Atlantic coastline.

8.3. Eastern Marginal Zone of the Namaqua Province

This narrow tectonic zone is extremely significant in that it provides an opportunity to examine the gradational boundary between a gneiss terrane and an older craton. The zone is metamorphically transitional between the Namaqua gneisses of the Central Zone in the west and the basement and cratonic cover sequences of the Kaapvaal Province in the east (Fig. 8-4). The structural history and Mid-Proterozoic radiometric ages of this zone are typically Namaquan, which justifies its inclusion in the Namaqua Province (Vajner, 1974a; Cornell, 1975; Botha and Grobler, 1979). The zone is approximately equivalent to the area referred to as the Kheis domain by Vajner (1974b) and the combined Kheis and Matsap domains of Botha and Grobler (1979). However, the name Eastern Marginal Zone is preferred for three reasons: first, the term "Kheis" no longer has any formal or widely accepted stratigraphic standing; second, our proposal accommodates both geographic (Eastern) and geotectonic (Marginal) labels for this part of the Namaqua Province; third, this nomenclature is consistent with that applied to the Limpopo, Natal, and Damara Provinces.

The boundaries of the Eastern Marginal Zone have never been clearly defined. The western boundary of this zone is generally accepted as being formed by an array of *en echelon* faults called the Doringberg lineament (Pretorius, 1974; Vajner, 1974b; Botha *et al.*, 1977; Stowe, 1979), although there are disagreements of minor significance on the relative importance of the individual faults making up this array. We believe that the Landgoed and Brakbos faults show the largest discontinuities and therefore follow Pretorius (1974) and Stowe (1979) in defining the western margin of the zone along their course (Fig. 8-4). The tectonic features defining the boundary between the Eastern Marginal Zone and the Kaapvaal Province are more varied. In the south the Doornberg (or Doringberg) fault forms a natural boundary between the Marydale Formation (part of the Eastern Marginal Zone) and the Kaapvaal cratonic cover (Fig. 8-4). The Doornberg fault is not to be confused with the Doringberg lineament, an array of faults that does not include the Doornberg fault. Vajner (1974a, c) and Botha and Grobler (1979) trace the boundary around the western margin of the Archean Kaapvaal granites and this delineation is followed here. The northern part of the boundary is best placed along the Orange River synform, which marks the easternmost expression of Namaquan orogeny (Fig. 8-4) (Stowe, 1979); east of here F1 folds extend more than 100 km into the Kaapvaal Province but do not appear related specifically to Namaquan tectonism.

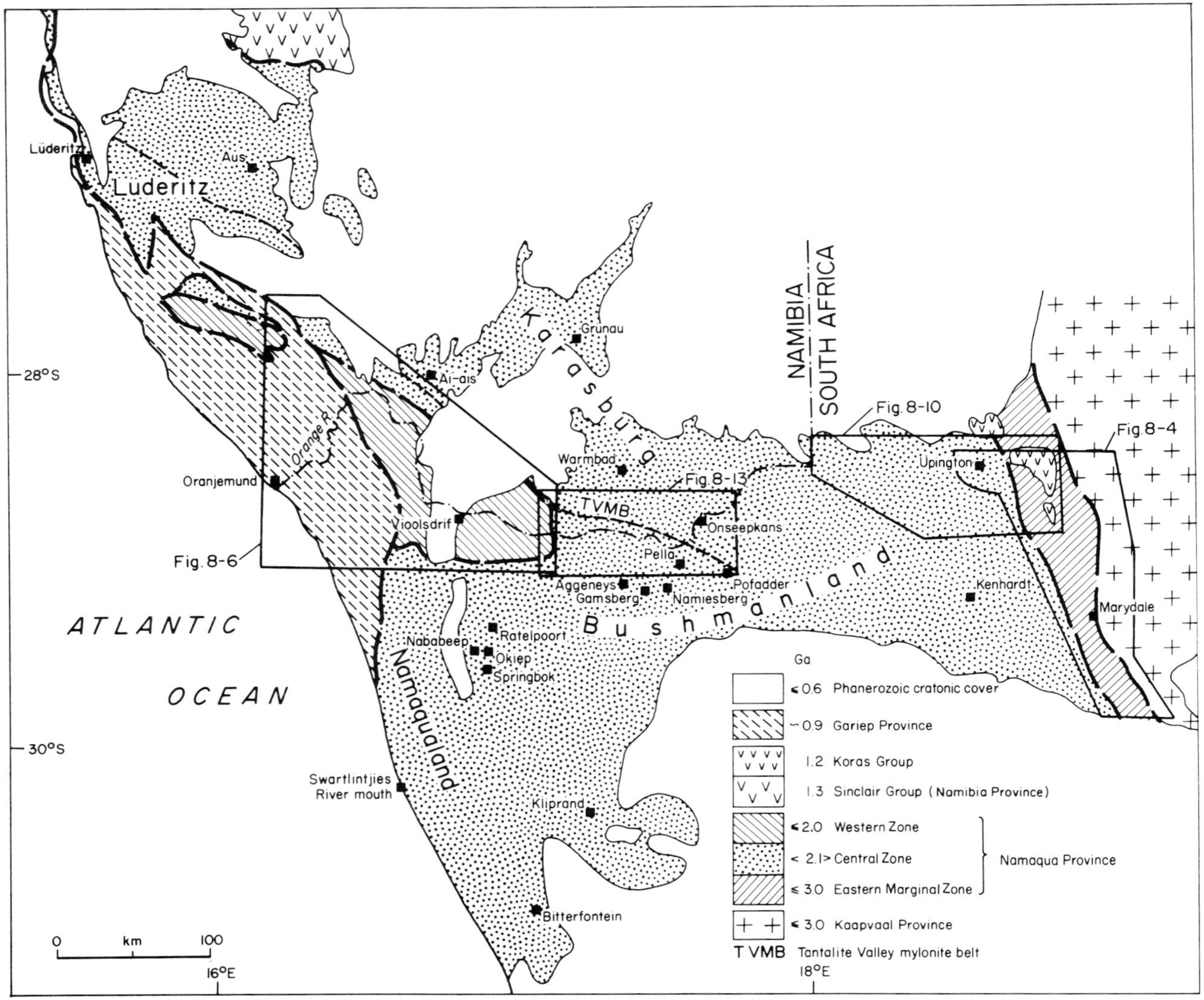

Figure 8-3. Tectonic zones within the Namaqua Province and geographic subdivisions of the Central Zone. [After Jackson (1976, 1979b) and Figures 8-4 and 8-6.]

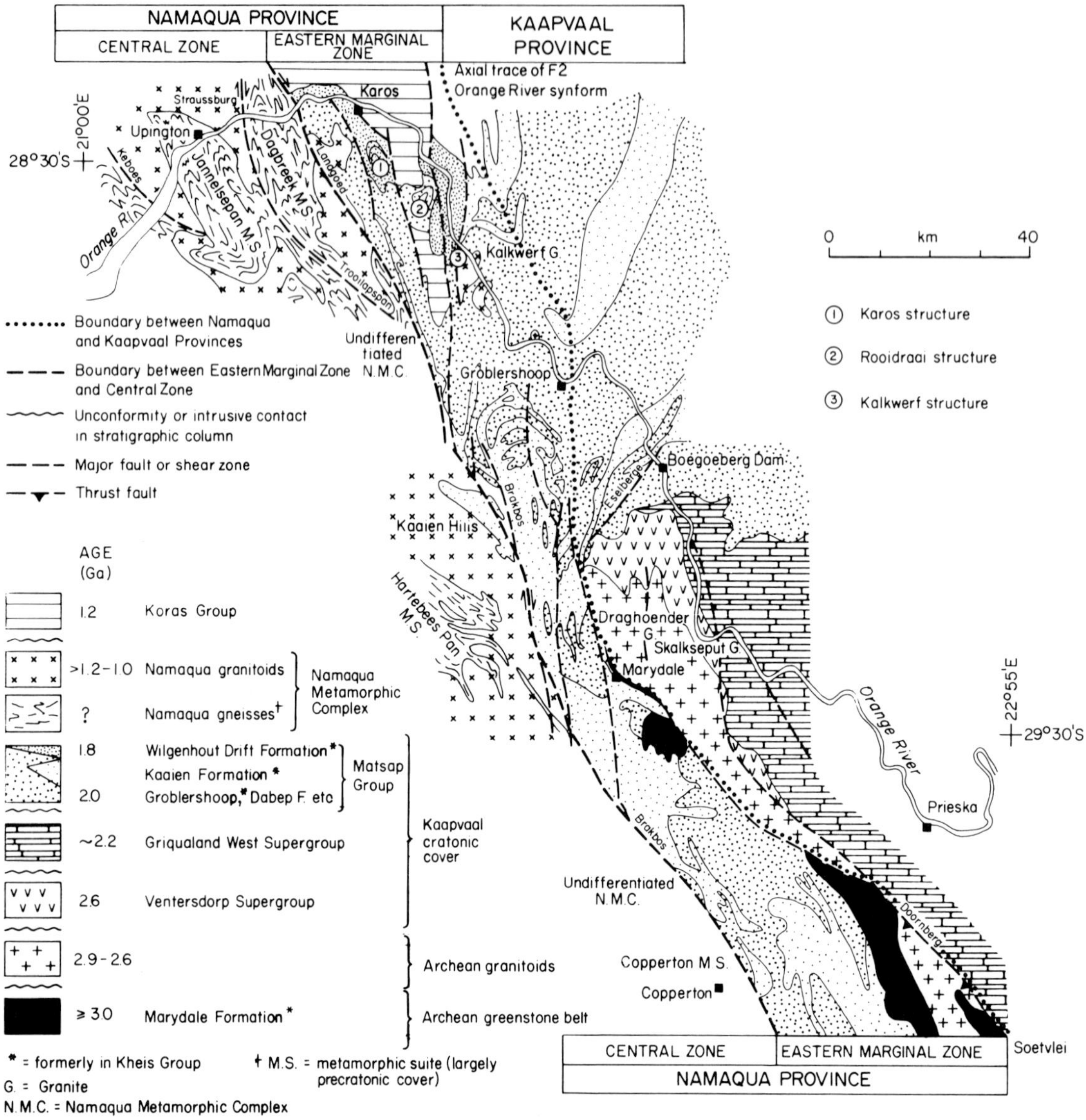

Figure 8-4. Geologic map of the Eastern Marginal Zone and adjoining areas of the Central Zone and Kaapvaal Province. [Compiled from Cornell (1975) south, Vajner (1974a) center, Stowe (1979) northwest, Botha *et al.* (1976, 1977) northeast, and Botha and Grobler (1979) northeast.]

Description of Units

The Marydale Formation, an Archean greenstone belt, represents the oldest unit in the Eastern Marginal Zone (Table 8-2, columns 5 and 6). The formation is exposed in two main areas (Fig. 8-4): an extensive southern area south of Prieska of some 8 km present thickness, and a smaller northern area around Marydale village. In the southern exposure the rocks have a structure interpreted as a monocline (Cornell, 1975, personal communication, 1980) or a synformal syncline (M. P. Coward, personal communication, 1978). The eastern margin of this structure is truncated by the Doornberg fault. Outcrops of granite, dated at 2.6 Ga (Burger and Coertze, 1973), divide the southern area into two parts (Fig. 8-4). Cornell (1975) regarded the eastern part of the Marydale outcrop (known as the Soetvlei area) as a klippe, thrust northeastward over a window of Archean granite. The western boundary of the inferred klippe is the Steenkoppie

Table 8–2. Summary[a] of Previous Correlations of Lithostratigraphic Units in the Eastern Marginal Zone with Those in the Adjoining Kaapvaal Province

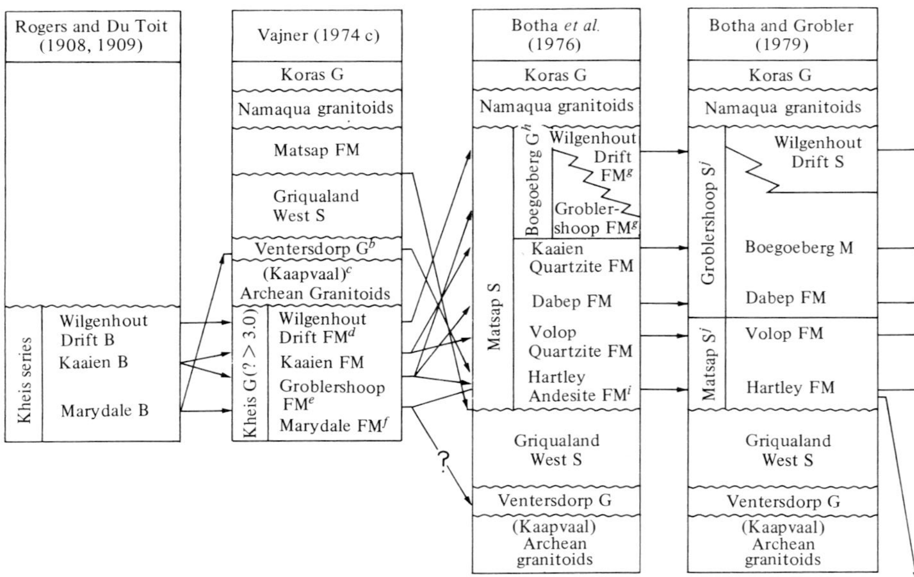

This work		Age (Ga)	Lithology
Koras G		1.2–1.1	Unmetamorphosed clastic sediments, acid to intermediate lavas and pyroclastics
Namaqua granitoids		> 1.2–1.0	Weakly foliated biotite granitoids, locally with muscovite or cordierite or feldspar megacrysts
Matsap G[k]	Wilgenhout Drift FM	?[p]	Schistose basic lava, tuff, agglomerate overlain by phyllite, polymict metaconglomerate, hematitic metaquartzite, and marble
	Groblershoop FM[m]	?1.8	Quartz–sericite schist, minor metaquartzite
	Kaaien FM[l]		"Clean" metaquartzite
	Dabep FM[m]		Quartz–sericite schist, minor chlorite schist, and metaquartzite
	Volop FM		Quartz arenite, conglomerate, and andesite overlain by hematitic lithic arenite and argillite
	Hartley FM	2.0	Quartz arenite, phyllite, metaandesitic tuff and lava, conglomerate
Griqualand West SG		~ 2.2	Basal quartz arenite grading into dolomite, banded iron-formation with crocidolite asbestos, riebeckite slate, laminated mudstone, capped by andesitic tuff and lava
Ventersdorp G[n]		2.6	Arkosic conglomerate overlain by acid lava unconformably overlain by basic to intermediate volcanics and arenites
(Kaapvaal) Archean granitoids		2.9 & 2.6	Largely even-grained cataclastic granitoids with xenoliths of metabasite and metaquartzite
Marydale FM[o]		≥ 3.0[q]	Metaarkose overlain by two metabasite members separated by banded iron-formation, and overlain by metagraywacke

Abbreviations: B, beds; FM, formation; M, member; S, sequence; G, group; SG, supergroup.

[a] Based on authors shown, with additional data from Cornell (1975, 1977) and Vajner (1978).
[b] Named by Vajner (1974 c) the Seekoebaard FM.
[c] Included Draghoender and Skalkseput granites, dated by Burger and Coertze (1973) at 2.9 Ga and 2.6 Ga, respectively, and similar granites to south of Prieska dated at 2.6 Ga.
[d] Regarded stratigraphic position as tentative.
[e] Included augen gneisses and migmatites of uncertain correlation.
[f] Included Swartkop S.
[g] Regarded as facies equivalent by Vajner (1974 c, 1978) and Botha *et al.* (1976).
[h] Because the Groblershoop FM is said to conformably overlie the Kaaien FM and is part of the same sedimentary sequence, its separation into a "Boegoeberg Group" appears unjustified.
[i] Part of Vajner's (1974 c) Ventersdorp was correctly correlated here with Matsap because the conglomerates contain clasts of distinctive Griqualand West material; correlation of Marydale FM with Hartley Andesite FM is erroneous on geochemical (Cornell, personal communication, 1979) and geochronological grounds (Crampton, 1974; Cornell and Barton, 1979).
[j] No reasons given for separation of Groblershoop S from Matsap S (see note[h])
[k] Inclusion of Kaaien, Groblershoop, and Wilgenhout Drift FM into Matsap G is convenient but tentative until substantiated by reliable geochronology and intensive remapping of the Boegoeberg area.
[l] This classic name is retained instead of Botha and Grobler's (1979) Boegoeberg Member, which is easily confused with the apparently unrelated Boegoeberg Group of Botha *et al.* (1976).
[m] Because of the early recumbent folding (Vajner, 1974 c) the Groblershoop and Dabep FM may represent the same stratigraphic unit duplicated by folding.
[n] Includes Zoetlief FM.
[o] Includes Swartkop S.
[p] A 4-point Rb-Sr isochron of 1124 ± 37 Ma with a low $R_0 = 0.7016$ obtained from highly sheared phyllitic greenstones is interpreted by Cornell (1975) as being close to a primary age (disputed by Botha *et al.*, 1976, on the grounds that the 1.2-Ga Kalkwerf [Namaqua] granite intrudes the Groblershoop FM, which they believe to be a time equivalent of the Wilgenhout Drift FM).
[q] Cornell and Barton (1979).

granite gneiss, which has yielded a minimum age of 2.6 Ga (Burger and Potgieter, 1979; U-Pb zircon).

The basal metaarkoses of the Marydale Formation in the Soetvlei area represent the oldest sedimentary rocks in the southwestern part of the Kaapvaal Province (Table 8-2). The Soetvlei metabasites, which dominate the Marydale succession, are the only rocks of unequivocal extrusive origin, as evidenced by their well-preserved volcanic textures (Cornell, 1975). The metabasites show a tholeiitic differentiation trend from low-K basalts with komatiitic affinities to dacites. Comparisons of the Marydale lavas with those of the nearby Ventersdorp southeast of Prieska (Fig. 8-4) show these rocks to be highly dissimilar in both age and magma type (D. H. Cornell, personal communication, 1980). The intercalation of clastic sediments and a persistent marker bed of iron-formation show that intrusion of the lavas was at least partly subaqueous.

Cornell (1975, 1977) interpreted a Rb-Sr isochron of 1860 ± 57 Ma from Marydale metalavas from the Soetvlei area as primary for a number of reasons. Subsequent Pb-Pb dating of the Marydale iron-formation shows that the Marydale Formation is 3.0 Ga old and indicates that the Rb-Sr date of 1.9 Ga for these rocks should be reinterpreted as a metamorphic event (Cornell and Barton, 1979).

Late Archean granitoids of the Kaapvaal Province intrude the Marydale greenstones near Marydale village. The oldest of these, the Draghoender granite (Fig. 8-4), has been dated at 2.9 Ga (Burger and Coertze, 1973). In at least three localities it contains large (up to 100 m long) xenoliths of amphibolite, hornblende schist, and metaquartzite correlated with the Marydale Formation (Rogers and Du Toit, 1908; A. Leube, in Martin, 1965; Vajner, 1974c). The adjacent Skalkseput granite (Fig. 8-4), dated at 2.5 Ga (Burger and Coertze, 1973), contains xenoliths resembling the Draghoender granite, sericite schist, and amphibolite (Vajner, 1974c). Other Archean granitoids are exposed to the southeast (Fig. 8-4).

The next youngest stratigraphic units, the Ventersdorp and Griqualand West Supergroups (Table 8-2), are part of the Kaapvaal Province and have been fully described (Chapters 4 and 5). With respect to their field relations, the Ventersdorp Supergroup nonconformably overlies the Archean granitoids north of Marydale village (Rogers and Du Toit, 1908; Vajner, 1974c), and is in turn overlain unconformably by the Griqualand West Supergroup (Fig. 8-4).

Both the Ventersdorp and Griqualand West Supergroups are overlain by the Matsap Group, a predominantly clastic sequence cropping out in both the Eastern Marginal Zone and the Kaapvaal Province (Chapter 7). According to Botha *et al.* (1976), both the Matsap and Griqualand West sequences are conformable and pinch out laterally against a basement "high" of Archean Draghoender granite and Early Proterozoic Ventersdorp strata. However, Vajner (1974c) interpreted this relationship as an overstepping of older units beneath an angular unconformity at the base of his Matsap "Formation." The rock types and lithostratigraphy of the Matsap Group are summarized in Table 8-2. The following two sections focus on two important and controversial parts of the Matsap stratigraphy, namely the uncertain relationship between the Kaaien and Groblershoop Formations, and the status of the enigmatic Wilgenhout Drift Formation.

The Kaaien–Groblershoop Confusion

The Kaaien and Groblershoop Formations are widespread throughout the Eastern Marginal Zone (Fig. 8-4). Vajner (1974a, c) differentiated the semipelitic members (quartz–sericite schists and minor metaquartzites) of the "Kaaien Beds" (Rogers and Du Toit, 1908) from the main metaquartzite unit and designated these the Groblershoop and Kaaien Formations, respectively. The Groblershoop Formation reaches maximum thickness, estimated at more than 1000 m, around Groblershoop village (Fig. 8-4) and thins southward toward Marydale village (Vajner, 1974c). Vajner tentatively included a number of gneisses (e.g., augen gneisses, muscovite–biotite gneiss, and stromatic migmatites) in the Groblershoop Formation, but their higher metamorphic grade and uncertain structural position is also compatible with their representing basement in mantled gneiss domes.

Considerable confusion exists as to the structural relation between the Kaaien and Groblershoop Formations. Vajner (1974a, c) defined the Groblershoop as underlying the Kaaien quartzites because the former occupied the cores of F3 domes formed by the latter in the Kaaien Hills. However, Vajner's (1974c) map indicates the reverse relationship just north of the Eselberge fault (Fig. 8-4). This may be explained by differing erosion levels of the recumbent F2 folds deformed into the F3 domes; sequences in the domes are arranged in structural order and are not always in stratigraphic sequence. The Groblershoop Formation is more likely to be the younger because Vajner's (1974c) mapping shows a large expanse of this unit south of Marydale village separated from the undoubtedly older Marydale Formation by the Kaaien metaquartzites, assuming that all three units are autochthonous. Botha *et al.* (1976) arrived at a similar conclusion on the grounds that the Groblershoop Formation formed a regional synclinorium enveloped by the Kaaien Formation south of the Orange River. However, their maps (Botha *et al.*, 1976, their Figs. 2, 6, and 7; Botha and Grobler, 1979, their Fig. 3) show that the core of the synclinorium north of the Orange River is occupied by Kaaien quartzite; in this volume, this is interpreted as a large F1 syncline of Kaaien quartzite underlain by Groblershoop Formation and refolded by the F2 Orange River synform (Fig. 8-4). This anomaly is not easily explained; it is possible that these quartzites are not stratigraphically equivalent to the Kaaien Formation farther south in the Kaaien Hills type area.

The lithostratigraphy proposed for the Kaaien and Groblershoop Formations is similarly confusing (Table 8-2). Botha *et al.* (1976) recorded that the schistose Groblershoop Formation conformably overlies the quartzose Kaaien Formation which rests conformably on the schistose Dabep Formation. Despite the conformable and transitional contacts, the Groblershoop Formation was separated by them from the related units below it and incorporated in a "Boegoeberg Group," part of their "Matsap sequence" (Table 8-2, note *h*). Subsequently, Botha and Grobler (1979) downgraded the Boegoeberg Group to a member, included it in an expanded Groblershoop sequence, and separated the latter from the remainder of the Matsap sequence (Table 8-2, notes *h* and *j*). The last three columns of Table 8-2 represent an attempt to rationalize the stratigraphy (see note *k*).

The Wilgenhout Drift Formation

This formation is exposed in two areas along the Orange River east of Upington (Fig. 8-4). The stratigraphic position of the Wilgenhout Drift Formation is far from clear, but it has been suggested that it may be a lateral equivalent of the Groblershoop Formation, itself of uncertain stratigraphic position (Vajner, 1974b, 1975, 1976; Botha *et al.*, 1976).

The Wilgenhout Drift Formation consists of a thick unit of metavolcanic rocks overlain by a mixed unit of metavolcaniclastic rocks and metasediments capped by a metasedimentary unit. In the lower unit a felsitic porphyry is overlain by greenschist-facies phyllitic metabasites with rare deformed pillow structures and amygdales (Vajner, 1978). The middle unit contains thin polymict metaconglomerate with flattened and folded clasts, schists derived from volcanic breccias and agglomerates, and hematite metaquartzite. The upper unit consists of metapelite phyllite and chlorite schist with minor metaquartzite and marble (Vajner, 1978).

The stratigraphic and structural relations of the Wilgenhout Drift Formation to other formations of the Matsap Group are best displayed in the Karos antiform south of Karos village (Fig. 8-4). Rogers and Du Toit (1909) interpreted this as a simple anticline plunging northwest and thereby suggested that the Wilgenhout Drift "Beds" represented the youngest unit in the Kheis sequence. Vajner (1974c) recognized that the Karos antiform was a complex interference structure and inferred that the Kaaien metaquartzites were younger than the Wilgenhout Drift Formation. Botha *et al.* (1977, their Fig. 3a) interpreted the structure as interference between two steeply inclined fold sets. However, Stowe (1979) has shown that *two* generations of flat-lying folds (F1a and F1b) have been refolded by a third set of steeply inclined F2 folds (Fig. 8-5). The implication of these findings is that the Wilgenhout Drift Formation can now be clearly shown to overlie the Kaaien Formation.

It is still unclear whether the Wilgenhout Drift Formation is coeval with the Groblershoop Formation as previously suggested (Vajner, 1974b, 1975, 1976; Botha *et al.*, 1976). An intriguing possibility is that the former formation may be distinctly younger. The Groblershoop Formation is probably of Matsap age (2.2–1.8 Ga) and is certainly older than the intruding Kalkwerf granite, which has a minimum age of 1.2 Ga (Botha *et al.*, 1976; single U-Pb zircon). In contrast, Cornell (1975) obtained a Rb-Sr isochron of 1124 Ma from four phyllitic metavolcanic samples from the Wilgenhout Drift Formation. Because of the low initial ratio of 0.7016, he interpreted this age as close to primary and accordingly regarded the Wilgenhout Drift Formation as distinctly younger than the Matsap Group.

The Kheis Controversy

Supracrustal rocks within the Eastern Marginal Zone have been referred to as the Kheis sequence since Stow (1874) proposed this name to distinguish them from the Griqualand West and Matsap sequences. However, in spite of its historical precedence, we do not favor retention of the term "Kheis" because its lowermost formation (Marydale) is a greenstone belt of Archean age, whereas its upper formations (Groblershoop, Kaaien, and Wilgenhout Drift) are of uncertain correlation but are probably Mid-Proterozoic in age.

Stow's (1974) Kheis sequence consisted of schist and metaquartzite in what is now known to be the upper part of the succession. The Kheis sequence was later extended to include underlying basic lava and divided by Rogers and Du Toit (1908, 1909) into the three well-known stratigraphic units (Table 8-2) of Marydale "Beds" (metabasic lava) overlain by Kaaien "Beds" (metaquartzite and felsic schist) and Wilgenhout Drift "Beds" (heterogeneous metavolcanic rock and metasediments). Vajner (1974c) designated the schistose members of the Kaaien "Beds" as the Groblershoop Formation to distinguish them from the metaquartzites of the underlying Kaaien Formation (Table 8-2). He had strong reservations on the stratigraphic position of the Wilgenhout Drift Formation because of structural complexities in the type area south of Karos (Figs. 8-4 and 8-5) apparently not recognized by Rogers and Du Toit (1909).

Vajner (1974c) supported previous inferences (e.g., Rogers, 1910; Du Toit, 1939; Martin, 1965) that the Kheis Group was of great antiquity and surmised that its age was about 3.0 Ga. This deduction was largely based on the occurrence of amphibolite xenoliths, lithologically similar to the Marydale Formation, in the 2.9-Ga Draghoender granite a few tens of meters from the Marydale–granite contact. Because the overlying Kheis units were apparently conformable, he deduced that the entire Kheis Group was older than the granite (Table 8-2). Vajner separated his Archean Kheis Group from the structurally *underlying* 2.0-Ga Matsap "Formation" along the Eselberge fault near Boegoeberg Dam on the Orange River (Fig. 8-4). North of this fault he recorded the presence of greenschist-facies mature Kaaien metaquartzites with the imprint of four phases of deformation. South of the fault he recorded the presence of the Matsap immature red bed sequence, which shows only two phases of deformation and is virtually unmetamorphosed. Vajner (1974b, c) proposed that the first two tectonometamorphic events took place during a pre-Matsap Kheis orogeny and the latter two events during a post-Matsap Namaqua episode, which affected both the Namaqua and Kaapvaal Provinces. He conceded that some of the formations north of the Orange River might be of Matsap age rather than of Kheis age, but he maintained that the extent of the Matsap was limited south of the river.

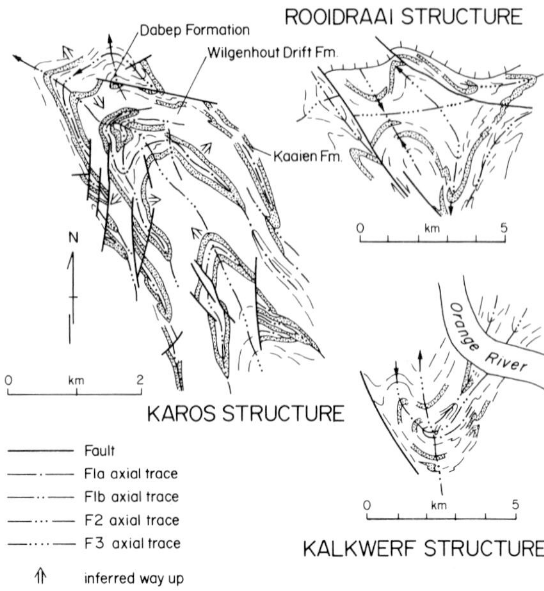

Figure 8-5. Superposed major structures between Karos and Kalkwerf villages. See Figure 8-4 for locations. Karos: Type 2 interference of upright F2 antiform with two gently inclined older fold sets (Dabep Formation structurally lowest). Rooidraai: Type 1 interference of doubly plunging upright F2 folds caused by cross folding by F3 upright antiform.. Kalkwerf: Type 3 interference between approximately coaxial plunging F1a, F1b, and F2 folds. [After Stowe (1979).]

An opposing viewpoint is that the Kheis strata are part of the upper Matsap Group (Smit, 1973; Botha *et al.*, 1976; Botha and Grobler, 1979). Vajner regarded the Eselberge fault (Fig. 8-4) as a tectonic contact between these sequences, but Botha and others asserted that this fault transected a conformable sedimentary contact although they conceded that this area was highly faulted (see column of Botha *et al.*, 1976, in Table 8-2).

Cornell (1975, pp. 172–173) suggested that the different degree of deformation in the Matsap and Kheis was caused by a greater depth of burial of the latter. According to the stratigraphic proposals of Botha *et al.* (1976), this is unlikely because the Kheis was claimed to be stratigraphically and structurally higher than the Matsap *sensu stricto*. However, Cornell (1975) envisaged the Kheis sequence accumulating on a continental margin, possibly directly on oceanic crust.

It is clear, therefore, that correlation of the units in the classical Kheis trilogy remains controversial. The Marydale Formation apparently represents an Archean greenstone belt, similar in its lithology to other greenstone belts of the Kaapvaal Province, but perhaps even more similar to the volcanic Nsuze Group which is also about 3.0 Ga in age (Chapter 2). Thick basal metaarkose of the Marydale Formation suggests that, like the Nsuze Group, it accumulated on continental crust. With respect to the remainder of the Kheis sequence, its stratigraphic position remains unclear (Table 8-2, note *k*).

Structural Evolution

Structural correlation across the Eastern Marginal Zone is vital for linking the tectonic histories of the Namaqua and Kaapvaal Provinces. Stowe (1979) attempted to integrate the diverse structural sequences from the literature; the present account is based largely on his framework with the addition of some proposed names (Kaapvaalian and Namaquan) for the deformation episodes (Table 8-3).

The oldest deformation imprint is preserved in the Archean Marydale Formation. In the northern area of outcrop, metabasites of the Marydale Formation with tight folds and axial-planar schistosity are cut by the 2.5-Ga Skalkseput granite (Vajner, 1974c). Vajner proposed two periods of Archean folding, which he referred to as F1 and F2, for the entire "Kheis" sequence. His mapping did not extend southward to the area around Soetvlei and his scheme was based mainly on the Kaaien Formation. Purely from the attitude of the folds, it is likely that the Soetvlei synform is of D2 age in Vajner's classification. There is some confusion as to the age of the main foliation in the Marydale Formation, which is recorded first as Dl (Vajner, 1974a, c) and second as D2 (Vajner, 1974b, c). The true age of this folding is difficult to define (Vajner, 1974c), but metamorphic studies throw some light on this problem (see "Metamorphism," next section).

After cratonization of the Kaapvaal Province, Early to Mid-Proterozoic cover sequences accumulated and underwent mild deformation, referred to here as Early Kaapvaalian (Table 8-3). Because the Ventersdorp and Griqualand West Supergroups are not recognized in the Eastern Marginal Zone, the Early Kaapvaalian deformation is preserved only in the Kaapvaal Province. Owing to the low intensity of deformation during the Early Kaapvaalian, these structures were strongly overprinted by the more intense Late Kaapvaalian event.

Accumulation of the Griqualand West and Ventersdorp sequences was followed by mild folding with broad wavelengths and predominantly northeastern axial trends (Vajner, 1974c; Botha *et al.*, 1977). Epeirogenic folding of this type appears to have taken place before and after deposition of the Griqualand West Supergroup, as evidenced by angular unconformities at its base (Vajner, 1974c) and top (Vajner, 1974c; Botha *et al.*, 1977). Metamorphic reconstitution of these strata appears to be limited to burial metamorphism.

Late Kaapvaalian deformation (D1) affected both the newly deposited Matsap sequence and overprinted the Early Kaapvaalian folds in the older cratonic cover (Table 8-3). The D1 event produced the oldest strains in both the Matsap sequence and the Namaqua gneisses. It is clearly a tectonic event of considerable importance in both the Namaqua and Kaapvaal Provinces but does not appear to have resulted specifically from the development of the Namaqua gneiss terrane, unlike later deformation events. Therefore, as in Stowe (1979), this event is classified as Kaapvaalian rather than Namaquan (cf. Botha *et al.*, 1977; Botha and Grobler, 1979).

Stowe (1979) has shown that the D1 event consisted of two fold generations, F1a followed by F1b, which are coaxial in places and divergent in others (see Kalkwerf and Karos structures in Fig. 8-5). The F1 folds extend from the Langeberg Range well within the Kaapvaal Province, where they trend north or northeast (Botha *et al.*, 1977), into the Central Zone in the west, where unfolding of younger folds indicates that the F1

Table 8–3. Provisional Correlation of Deformation Sequences in the Eastern Marginal Zone and Adjoining Namaqua Province and Kaapvaal Province

Proposed[a]	Vajner (1974a, 1975) EMZ, KP	Botha *et al.* (1977) CZ, EMZ, KP	Geringer and Botha (1977) CZ	Van Bever Donker (1980) CZ	Van Zyl (1979) (CZ)
Archean post-Marydale					
1. Upright folding and schistosity	F1 Archean folds F2 Archean folds (regional foliation)				
2. Kaapvaalian (early Mid-Proterozoic) Early Kaapvaalian (pre-Matsap) folding					
3. Late Kaapvaalian (post-Matsap) folding					
D1a: Isoclinal folds, trend NNE to ENE, SE vergence; schistosity					
D1b: Isoclinal folds, trend N to NE in Kaapvaal, NW in Namaqua, variable attitude; type 3 interference with F1a	F3 post-Matsap	F1 folds and regional foliation	F1	Dn+1	F1
4. Namaquan (Mid-Proterozoic)					
D2: Isoclinal folds, trend NNW, NE vergence in E, SW vergence in W; schistosity	F4 post-Matsap	F2	F2	Dn+2	F2
D3: Open NE-trending folds producing type 1 and type 2 interference with F2		F3	F3	Dn+3	F3
D4: E-trending chevron and conjugate folds					F4
D5: Faults and shear belts; NNW dextral and WNW sinistral	NNW shears	Megashears and wrench faults	Shears	Shears	Shears
5. Koras (Mid-Proterozoic) Graben faulting	Faults	Faults	Faults	Faults	Faults

[a]After Stowe (1979). CZ, Central Zone, Namaqua Province; EMZ, Eastern Marginal Zone, Namaqua Province; KP, Kaapvaal Province.

folds originally also trended northeast. The axial traces converge southward and wrap around the western flank of the resistant buttress of the Archean granites near Marydale village in an area of high strain. The folds are generally steeply inclined in the east along the Langeberg Range but become tighter and more flat lying toward the Central Zone (Stowe, 1979). In the less deformed eastern areas the folds plunge toward the north and northeast and face east and southeast (Vajner, 1974c), suggestive of translation from the west and northwest. Botha *et al.* (1977) proposed that the D1 folding was directed from the west, but this is contradicted by their map (their Fig. 2) which indicates that the F1 folds originally trended eastward or northeastward in the Eastern Marginal Zone rather than northward. Botha and Grobler (1979) similarly asserted that folding was directed toward the east but one of the two major nappes proposed (over their F2 Achterkop antiform) faces westward.

The first deformation event associated specifically with the Namaqua Province is the D2 event, which overprinted the earlier Kaapvaalian strains and initiated the Namaquan episode (Table 8-3). Hence the eastern boundary of the Namaqua Province and the Eastern Marginal Zone is marked by the limit of large-scale Namaquan deformation represented by the Orange River synform (Fig. 8-4) (Botha *et al.*, 1977; Stowe, 1979). This major fold shows up prominently on Landsat imagery and is closely mirrored by the residual gravity field which shows "folded" axes of negative and positive anomalies (Pretorius, 1974, his Figs. 7 and 10).

The steeply inclined F2 folds trend north-northwest and face northeast in the east and southwest in the west (Vajner, 1974c; Botha *et al.*, 1977; Stowe, 1979). The folds generally plunge northward but cross folding during D3 has resulted in reversed plunge in places (e.g., Rooidraai structure in Fig. 8-5). The folds increase in tightness and amplitude away from the foreland toward the Central Zone (Vajner, 1974c; Botha *et al.*, 1977; Stowe, 1979). In the high-strain area around the hub of Archean granites near Marydale village, older folds and fabrics were tightened and rotated into the north-northwesterly D2 trends (Vajner, 1974c). Because the attitude of the F1 folds varied from steeply inclined to reclined, superposition of the F2 folds has produced variable interference structures. Botha *et al.* (1977) proposed type 1 interference in the Kaaien Hills, which is substantiated by periclinal outcrop patterns (their Fig. 1), but their structural map (their Fig. 2) indicates type 2 interference on the basis of the strongly horseshoe-shaped F1 axial traces. In the Wilgenhout Drift Formation the F2 folds are generally of chevron style, whereas farther west in the Central Zone folds of the same age show evidence of greater ductility (Stowe, 1979), probably reflecting a deeper crustal level of formation. It has been proposed that the D2 deformation reflects heterogeneous simple shear in a northwest direction, concentrated in places as shear belts representing tectonic slides (Fleuty, 1964; Hutton, 1979) along the limbs of the F2 folds (Vajner, 1974c). Strike-slip and oblique-slip faults with the same trend in this area may also have formed at this time, together with fracture cleavage and minor kink folds in the Ventersdorp sequence (Vajner, 1974c).

Shortening along northwest axes during D3 interfered with the F2 folds and produced broad domes and narrow elliptical basins. The F3 folds are upright and trend northeastward, decreasing in intensity across the Eastern Marginal Zone to die out in the vicinity of the Kalkwerf granite stock (Fig. 8-4) (Botha *et al.*, 1977; Stowe, 1979). Because many of the F3 domes are cored by granite gneiss, their uprise may have been diapirically enhanced (Stowe, 1979) in the manner proposed for granitoid–gneiss upwellings in the Zambezi and East Greenland gneiss terranes (Talbot, 1973, 1979).

A period of minor folding (D4 in Table 8-3) was followed by movement (D5) along major structural discontinuities, shown as north-northeast-trending faults in Figure 8-4. The overall pattern is one of tectonic prisms separated by steep shear belts with general dextral displacement and upward movement of the southwest prisms toward the Kaapvaal block (Stowe, 1979). This process represents a continuation of the lateral shortening and northward translation of the Namaqua Province (in relation to the Kaapvaal Province) that may have begun in D2. The shear belts are high-strain zones subparallel to the lithologic layering marked by an intensification of fabrics (Stowe, 1979). A subordinate set of faults trending west-northwest with sinistral displacement appears to have formed complimentarily (Stowe, 1979).

The western boundary of the Eastern Marginal Zone is formed by an array of *en echelon* D5 faults called the Doringberg lineament (Pretorius, 1974; Vajner, 1974 a, b, c; Botha *et al.*, 1977; Stowe, 1979). In contrast to other authors, Pretorius (1974) believed that the right-lateral separation on this system of faults was caused not primarily by dextral strike-slip but by dip-slip displacement of inclined strata. This view may be simplistic in view of the complexities caused by polyphase folding, but there is strong evidence to indicate much differential uplift across this group of faults. First, considerable uplift is indicated by the absence of syntectonic Namaquan deposits at the present erosion level (Stowe, 1979) and by the isotopically indistinguishable ages of 1.2 Ga given for the syntectonic Kalkwerf granite and the postcratonic Koras Group (Botha *et al.*, 1977, 1979). Second, the increase in metamorphic grade inward from the margins of the Namaqua Province must involve considerable differential uplift in the west to expose deeper levels; higher heat flows may also be responsible, but there is no evidence of different metamorphic facies series to support this.

The D5 shear belts remained as strain-softened planar zones of weakness and were reactivated as faults that controlled the form of the depositories for the Koras Group at about 1.2 Ga (Botha *et al.*, 1977; Grobler *et al.*, 1977; Botha and Grobler, 1979; Stowe, 1979). Deposition of the Koras succession marks the final stage of cratonization in this part of the Namaqua Province. Six formations of alternating mainly acid volcanic rocks and arenites with internal disconformities were deposited in grabens (Du Toit, 1965; Grobler *et al.*, 1977).

Metamorphism

The only detailed metamorphic study carried out in the Eastern Marginal Zone is that by Cornell (1975), who concentrated mainly on metabasites of the Marydale Formation. Available metamorphic ages are post-Matsap and appear to reflect the Namaquan episode of tectonism.

The oldest recorded metamorphic date is 1860 Ma, originally regarded by Cornell (1975, 1977) as primary and now reinterpreted as metamorphic. Little is known of this event apart from the fact that it resulted in greenschist-facies parageneses. It has not been established whether it was a discrete event or merely the termination of a protracted episode of isotopic homogenization. The next metamorphic event, termed the Sonqua by Cornell (1975), resulted in a metamorphic zonation from greenschist to granulite under a low to medium P/T facies series (geotherm estimated at 40°C/km). The granulite-facies metabasites south of Marydale village are granoblastic hypersthene–diopside granolites, referred to by Vajner (1974c) as "intrusive gabbros," and dated by Cornell (1975; Rb-Sr) at 1350 Ma. Widespread retrogression some 200 Ma later (referred to by Cornell as Namaqua) resulted in amphibolite-facies parageneses and fabrics.

Farther north in the remainder of the Eastern Marginal Zone the metamorphic pattern is broadly established on a regional scale but little is known of the temporal and structural relations. The bulk of the zone shows greenschist or amphibolite-facies parageneses with an increase of metamorphic grade from the Kaapvaal Province southwestward toward the Central Zone (Vajner, 1974c; Botha and Grobler, 1979; Botha *et al.*, 1979). Only one metamorphic imprint has been reported and this has been correlated with D1 (Botha *et al.*, 1979). Nevertheless, D. H. Cornell (personal communication, 1980) reports cordierite–garnet assemblages partly retrogressed to chlorite schists in the Groblershoop area. There do not appear to be metamorphic discontinuities along either boundary of the Eastern Marginal Zone.

Precratonic Intrusions

The only significant intrusions within the Eastern Marginal Zone are the scattered bodies of Mid-Proterozoic Namaqua granitoids that become extremely abundant west of the Landgoed–Brakbos lineament in the Central Zone (Fig. 8-4). In the latter zone the granitoid intrusions are exposed along antiformal F3 arches and appear to have been emplaced during D3 (Stowe, 1979). In the Eastern Marginal Zone the granitoid intrusions have the form of small stocks such as those near Kalkwerf village (Fig. 8-4). The Kalkwerf granite has differently oriented fabrics to those of the other Namaqua granites, which may reflect differences in age or tectonic setting (Vajner, 1976). Botha *et al.* (1977) maintained that this granite was syntectonic with D1, but their cross section (their Fig. 3b) shows the granite crosscutting their F2 Kalkwerf synform.

8.4. Western Zone of the Namaqua Province

This tectonic zone is a triangular wedge projecting into the west-central part of the Central Zone (Fig. 8-3). It consists predominantly of Early Proterozoic low-grade supracrustal rocks of largely extrusive origin, known as the Orange River Group, and a composite granitoid batholith of slightly younger age, known as the Vioolsdrif Intrusive Suite (Fig. 8-6). The metamorphic and structural imprints of this zone are closely related to those of the Central Zone. As is the case for the Eastern Marginal Zone, the Western Zone is an integral part of the Namaqua Province and should be regarded as a tectonic zone within this province (Bertrand, 1976; Jackson, 1976; Joubert,

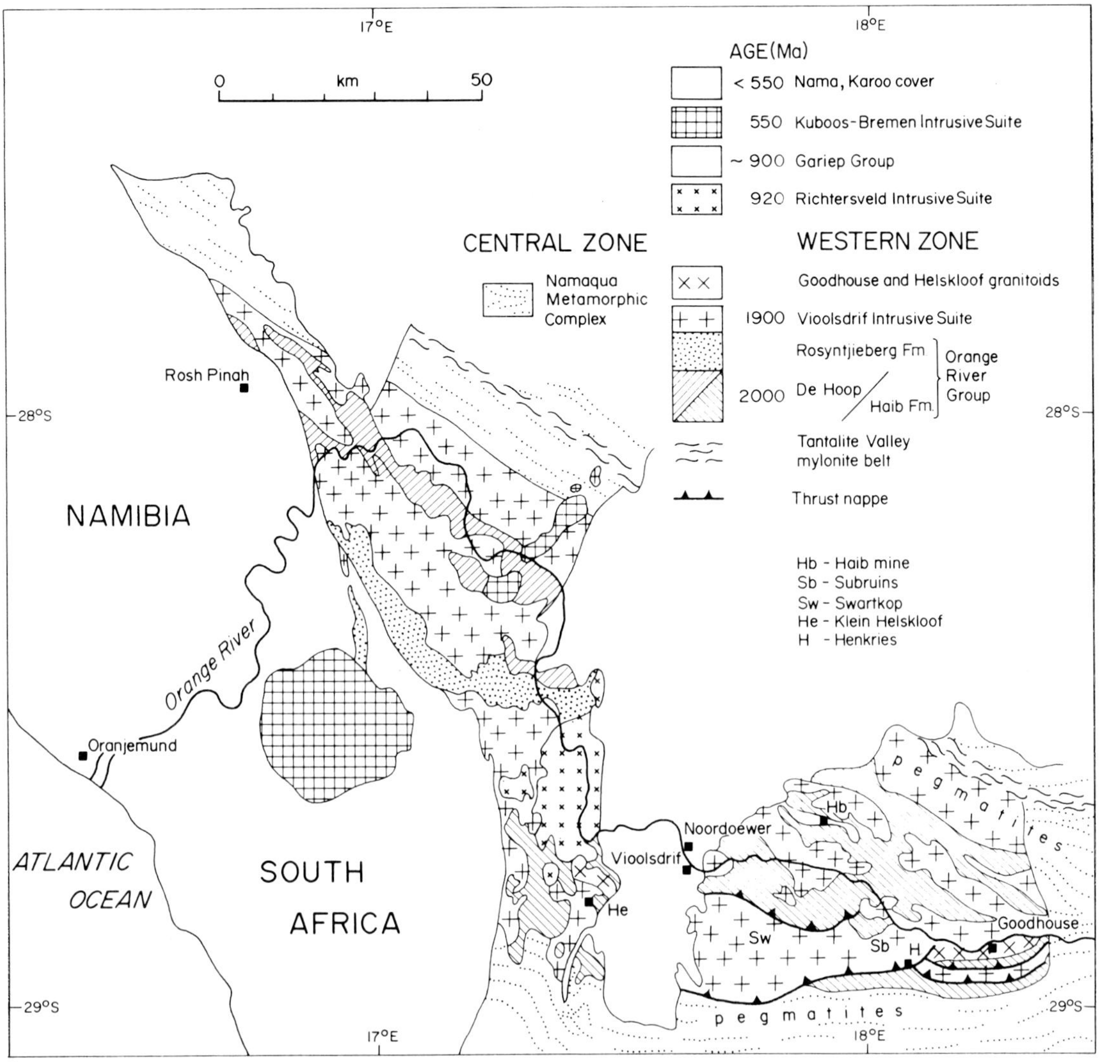

Figure 8-6. Geologic map of the Western Zone of the Namaqua Province. [Compiled from Bertrand (1976), Toogood (1976), Blignault (1977), Duplan (1978), Ritter (1978).]

1976; Toogood, 1976; Ritter, 1978) rather than being designated an entirely separate province (Blignault *et al.*, 1974; Kröner and Blignault, 1976; Blignault, 1977).

The absolute and relative timing of the igneous activity that formed these rocks is now well established. Nevertheless, two main problems of critical importance remain to be solved: first, the extent or existence of a preserved basement to the Orange River Group; and second, the age relations between the igneous complex of the Western Zone, whose geochronology is well established, and the gneisses of the Central Zone, whose isotopic systems largely reflect Mid-Proterozoic overprinting. These two outstanding problems are intertwined because alternative proposals have been made that either the Central Zone gneisses formed a basement to the Western Zone supracrustal rocks or that they were derived from them by metamorphic reconstitution.

Evidence for Possible Basement

Preserved remnants of basement are rare despite the relatively weak metamorphic and structural imprint. Principal workers in this area (Blignault, 1977; Reid, 1977, 1979a) have denied the existence of such a basement.

Nevertheless, others have reported rocks older

than the volcanic succession or the batholithic complex intruding it. Bertrand (1976) described a migmatitic layered biotite gneiss with intrafolial folds near Henkries (Fig. 8-6) cut by Vioolsdrif granodiorite, which is part of the composite batholith. Because of the evidence for partial melting prior to granodiorite intrusion, Bertrand interpreted the layered gneiss as a basement that had been deformed and metamorphosed at least once before the granodiorite was intruded at 1.9 Ga. However, because this layered gneiss is not in contact with older volcanic succession, it cannot be proved by field relations to be basement. Radiometric dating of this gneiss has yielded a similar age to that of the Vioolsdrif granitoids (Kröner, 1978a). In the same area Joubert (1976) reported that Namaqua paragneisses are intruded by Vioolsdrif granitoids and infolded with Orange River metalavas of the Western Zone, a relation consistent with—but not proof of—the paragneisses being an older basement. Similarly, Ritter (1978, 1980) described pink Namaqua paragneisses underlying and passing stratigraphically upward through a tectonically intercalated contact zone into acid metalavas of the Orange River Group in the southwest of the Western Zone with no evidence for overturning; this indicates that at least some of the Namaqua gneisses are older than the Orange River supracrustal sequence. However, there is no evidence of pre-Orange River deformation in the basal gneisses and a basement–cover relationship has not been proved

The Orange River Group: Precratonic Volcanic Cover

The Orange River Group is a lithostratigraphic unit proposed by J. De Villiers in 1972 (personal communication) to replace the older term "Kheis System" that necessitated stratigraphic equivalence to the supracrustal sequences in the Eastern Marginal Zone (De Villiers and Söhnge, 1959). The Orange River Group is a pretectonic cover sequence in the Western Zone and underlies about one third of the area (Fig. 8-6). The group was divided by De Villiers into three subunits, listed in ascending order: De Hoop Subgroup (volcanic), Rosyntjieberg Formation (sedimentary), Haib Subgroup (volcanic). The Haib Subgroup is not in contact with the other units of the Orange River Group and it has been suggested that it is equivalent to the De Hoop Subgroup and, hence, that both form the lower part of the Orange River succession (Blignault, 1977; Reid, 1977, 1979a).

In the west the De Hoop Subgroup consists mainly of intermediate and acid lavas and pyroclastic deposits underlain by quartz–feldspar porphyry (Ritter, 1976; Blignault, 1977). Conformably overlying the De Hoop Subgroup are metasediments of the Rosyntjieberg Formation, which comprises metaquartzite, with preserved ripple marks and cross-bedding and intercalations of magnetite iron-formation, chlorite schist, and metapelite (De Villiers and Söhnge, 1959; Ritter, 1976, 1980; Blignault, 1977).

In the southeast and east all the supracrustal rocks, which are predominantly metavolcanic, are included in the Haib Subgroup. The base of this unit is crosscut by the Vioolsdrif Intrusive Suite and the top is erosionally truncated by the angular unconformity beneath Phanerozoic cratonic cover. Metasediments are rare in the Haib Subgroup, but in the extreme east, metaquartzites with minor sillimanite schists and calc–silicate rocks are present (Bertrand, 1976).

Most of the Haib Subgroup consists of lavas and related pyroclastic deposits extruded close to 2000 Ma (Reid, 1979b; Rb-Sr and U-Th-Pb). Although the volcanic pile is characterized by great heterogeneity and lack of systematic compositional variations, it can be divided into a predominantly acid lower sequence (40 percent by area) and a largely basic upper sequence (60 percent by area) (Blignault, 1977; Reid, 1979a). The mainly acid sequence is interpreted as a complex mixture of ignimbrites, pumice sheets, and bedded tuffs. The largely basic sequence is a differentiated suite of calc–alkaline lava, ranging in composition from basaltic andesite to rhyolite, but is predominantly andesitic. The acid sequence has not been dated, but its field relations suggest that it formed contemporaneously with the more basic sequence (Reid, 1979b). Differentiation of the basic lavas appears to have followed a high-K calc–alkaline trend by progressive removal of phenocryst phases commencing with a more basic parent magma than the basaltic andesite (Reid, 1979a). The acid volcanic rocks are not related through fractional crystallization and may be derived from a separate magma source.

The Vioolsdrif Intrusive Suite: Precratonic Composite Batholith

Structurally below but stratigraphically younger than the supracrustal envelope of the Orange River Group is a large, composite, but largely granitoid batholith known as the Vioolsdrif Intrusive Suite (Fig. 8-6). The present erosion level is thought to represent the root zone of the Haib volcanic pile because both volcanic rocks and subvolcanic intrusions are present in abundance (Blignault, 1977). Because of the gross conformity of the granitoids and lavas, Blignault (1977) believed that intrusion was by magmatic stoping rather than forceful emplacement, but he nevertheless attributed early deformation in the lavas to intrusion of the Vioolsdrif Suite.

Members of the Vioolsdrif batholith have field relations suggesting an order of intrusion from basic to acid (Blignault, 1977; Reid, 1977, 1979c), a sequence consistent with the geochronology. The oldest but areally insignificant intrusions are the basic-ultrabasic layered suites, which are inferred by field relations to have been emplaced as sill-like bodies between 2.0 and 1.9 Ga. An Rb-Sr errorchron of 1860 ± 220 Ma (Reid, 1979c) probably reflects isotopic resetting during low-grade metamorphism accompanying the main Vioolsdrif intrusion (Reid, 1979d) in which the basic–ultrabasic bodies are present as xenolithic masses. Two of these bodies have been studied (Subruins and Swartkop; Fig. 8-6). A rare example of a contact with the original country rock (acid lava) is preserved at Subruins, where marginal finer grained melagabbros grade inward into massive coarser grained rocks and, ultimately, to a layered peridotite–pyroxenite sequence with a basal chromite-bearing peridotite layer dipping at moderate angles (Reid, 1979d).

Intermediate members, comprising rare diorites and abundant tonalites and granodiorites, were intruded from 1930 to 1870 Ma (Reid, 1979c; Rb-Sr, Pb-Pb). Intrusion of crosscutting plutons of granite and alkali-feldspar granite followed at about 1730 ± 20 Ma (Rb-Sr, Pb-Pb). These intermediate and acid magmas followed a calc–alkaline differentiation trend similar to the comagmatic Haib porphyritic lavas (Reid, 1977; 1979a). Porphyry-type copper and molybdenum sulfides are present in a porphyritic variety of the granite and provide the ore bodies of the Haib mine (Minnitt, 1979). In contrast, the older basic–ultrabasic bodies follow the typically tholeiitic differentiation trend of early Fe enrichment and are not considered to be derived from the same magma source (Reid, 1979d).

The generally low to moderate initial $^{87}Sr/^{86}Sr$ ratios (averaging 0.7031) of the abundant intermediate members indicate major additions to the crust of mantle-derived magmas in the Early Proterozoic (Reid, 1979c). Ward (1974), Reid (1977), and Blignault (1977) postulated that the Vioolsdrif Igneous Complex, comprising the Haib Subgroup and the Vioolsdrif Intrusive Suite, represented an Andean-type magmatic arc along the active margin of a continental plate (cf. Jakeš and White, 1972; Taylor and Hallberg, 1977; Pitcher, 1978). However, this is difficult to substantiate because neither the linear magmatic chain nor the underlying continental crust has been recognized (Reid, 1979a).

Structural and Metamorphic Evolution

The batholithic complex has undergone only minor deformation and contains a ubiquitous low-grade metamorphic imprint, increasing to medium grade and, probably, high grade near the margins (Blignault, 1977; Reid, 1979a; Ritter, 1980). Consequently the general sequence of the structural and metamorphic imprints in the Western Zone has been established in the east (Blignault, 1977), west (Ritter, 1978, 1980), and south (Bertrand, 1976; Duplan, 1978). However, the absolute timing of these events and their relation to events in the Central Zone are still not clear. For example, Blignault (1977) recorded the presence of only one fabric-forming event, called D1, and maintained that this is equivalent to the earliest deformation event in the Central Zone. Conversely it has been argued (Toogood, 1975, 1976) that the cataclastic regional foliation in the Western Zone was caused by late-stage (with respect to the Central Zone) shearing concentrated along the northern and southern margins but pervading the entire zone. This controversy is not easily resolved, because both older and younger fabrics are parallel (Fig. 8-6), but is clearly of great significance in determining the age of deformation in the Central Zone.

The structural and metamorphic histories proposed by various workers are summarized in Table 8-4 and certain aspects are discussed in

Table 8–4. Provisional Correlation of Deformation Sequences in the Western Zone of the Namaqua Province

Age (Ma)	Center-east (Blignault, 1974, 1977)	Southeast (Bertrand, 1976)	Southeast (Duplan, 1978)	Southwest (Ritter, 1978)
>2000		Pre-D1: Fabric formation, upper-amphibolite-facies regional metamorphism		
2000 (Reid, 1977, (1979b)	Haib Subgroup	Haib Subgroup	Haib Subgroup	De Hoop Subgroup
	Pre-D1: ? folding of Haib volcanics			D1: Schistosity and folds dipping to NNE
1930–1870 (Reid, 1977, 1979c)	Vioolsdrif intrusions	Vioolsdrif intrusions	Vioolsdrif intrusions	Vioolsdrif intrusions
	D1: Steeply dipping regional fabric, downdip linear component; greenschist-facies low P/T regional metamorphism; major upright folding on WNW trends	D1: Steeply dipping regional fabric, downdip linear component, greenschist- to lower-amphibolite-facies regional metamorphism		D2: Open upright major folds plunging steeply to N; weak axial-planar fabric; M1, possibly upper amphibolite facies, low P/T regional metamorphism
? 1730 (Reid, 1979c)		Goodhouse granite		Helskloof granite
		D2: Open horizontal major folds	Open horizontal major folds trending WNW and verging S; thrust nappes in S verging S, bounded by tectonic slides and strike-slip faults	D3: Vertical N-striking sinistral shear belts; M2 greenschist facies medium P/T regional metamorphism
	D2: Large pegmatite-filled dilation fractures along zone margin, indicating sinistral displacement in N and dextral in S	D3: Major shear belts and faults with vertical NW trend and moderately dipping W trend; dextral displacement; retrogressive cataclastic metamorphism		D4: Crenulation cleavage and intersection lineation
? 900	Gannakouriep dikes			Gannakouriep dikes
	D3: Subvertical major dextral shear belts trending NW and NNW: minor crenulation cleavage and folds trending NW			

more detail in the remainder of this section. Pre-Vioolsdrif deformation has been documented by Bertrand (1976) and Ritter (1978). Blignault (1977) speculated on the existence of such deformation because strains in the Haib metavolcanic rocks are higher than those in the Vioolsdrif granitoids. However, lack of overprinting suggests that the lower strains in the latter are a result of their lower ductility.

Post-Vioolsdrif high-grade metamorphism

culminating in partial melting has been recorded in separate areas (Bertrand, 1976; Ritter, 1978, 1980). In both cases the inferred melting gave rise to migmatitic neosomes and granitoids (the Goodhouse and Helskloof quartz monzonite–granodiorites) that do not have the imprint of early deformation (Fig. 8-6). Incorporation of Haib metavolcanic rocks and Vioolsdrif granitoids as xenoliths in the younger granitoids reset their isotopic systems around 1708 Ma (Reid, 1979c).

To the south of Vioolsdrif the Western Zone is bounded by a 5-km-thick shear belt that dips shallowly to the north–northeast; mineral lineations plunge moderately northeast, suggestive of oblique-slip displacement. Theart (1980) interpreted this belt as a broad, ductile thrust zone. There are some problems in reconciling this interpretation with the lower metamorphic grade of the crustal block (Western Zone) that has supposedly been upthrown relative to the Central Zone. Nevertheless this structure is broadly on strike from a large south-verging thrust nappe mapped by Duplan (1978) (Fig. 8-6) and provides further indication that much of the southern boundary of the Western Zone is allochthonous.

Young shear belts trending north to northwest have been reported throughout the Western Zone (Bertrand, 1976; Blignault, 1977; Ritter, 1978, 1980). There is some controversy as to whether they are a manifestation of the Late Precambrian Gariepian episode (Chapter 9) as has been speculated on account of their orientation (Kröner, 1974a; Kröner and Blignault, 1976; Blignault, 1977). However, Ritter (1978, 1980) demonstrated that the shear belts were of pre-Gariep age because, although they are roughly parallel with the angular unconformity at the base of the Gariep Group, they are overstepped by the unconformity in the south and are nowhere found to extend into the Gariep rocks. In addition the protracted shearing is likely to have taken place prior to basin formation because the associated metamorphism indicates depths of 10–15 km, after which this crust was uplifted to the surface to act as a basement for the Gariep accumulation. Ritter's (1978) M2 regional metamorphism (Table 8-4) is believed by him to be pre-Gariepian because although it is the same grade as the Gariepian metamorphism (low temperature, medium pressure), a significantly higher celadonite content in white micas in the Gariep suggest higher P/T ratios during their metamorphism (Sassi and Scolari, 1974).

Relation to the Central Zone

The Western Zone, with its well-defined geochronologic framework and low degree of structural and metamorphic reworking, represents a pristine key area that potentially could supply valuable clues to the origin of the more thoroughly reconstituted crust in the Central Zone of the Namaqua Province. Unfortunately two outstanding and related questions remain to be answered before such extrapolation can be done: first, what is the stratigraphic relation between precratonic supracrustal sequences in both zones; and second, what is the temporal relation between the early deformation events in both zones?

In the southwest, pink gneiss, a Namaqua gneiss common in the Central Zone, passes stratigraphically upward into De Hoop (renamed the Windvlakte Formation by Ritter, 1980) supracrustal rocks of the Western Zone (Ritter, 1978, 1980). Similarly in the southeast Bertrand (1976) and Joubert (1976) described pre-Vioolsdrif paragneisses of anomalously high metamorphic grade and concluded that these were older than the Orange River Group. However, both authors also postulated that the main group of paragneisses in the Bushmanland area of the Central Zone to the east were equivalent in age to the Orange River Group and represented a predominantly sedimentary facies of the same sequence (Fig. 8-3). In contradistinction Blignault (1977) hypothesized that the main paragneiss sequence in the Grünau area of the Central Zone to the north rested on an older basement of gray gneisses; the latter was thought to represent high-grade equivalents of the Orange River Group and Vioolsdrif Intrusive Suite as proposed by Martin (1965). Clearly then, if the field interpretations of these workers are accepted, paragneisses in the Central Zone formed from parent rocks that are older, younger, and contemporaneous with the Orange River Group in the Western Zone. It is therefore simplistic to generalize on the age of the precratonic cover sequences in the Central Zone.

A further complication is the probability of diachronous formation of structures of the same generation in different places at different times (Hobbs *et al.*, 1976; Blignault, 1977). In this regard, such D1 structures in an upper-crustal regime like the Western Zone might form contemporaneously with D2 or D3 structures in a lower crustal regime, such as the Central Zone, which is characterized by higher temperatures and ductilities. Consequently, the regional fabric in the Western Zone might have a maximum age of, say, 1.9 Ga, but the equivalent structure in the Central Zone could be considerably older. The possibility of structural diachronism, together with the variable stratigraphic ages of the Namaqua paragneisses, necessitates that correlations of structure or stratigraphy with the Western Zone be approached with caution.

Post-Vioolsdrif Igneous Activity

The Vioolsdrif igneous event and subsequent deformation and metamorphism were followed by limited but variable igneous activity.

Belts of *en echelon* pegmatites were emplaced late syntectonically along the northern and southern boundary of the Western Zone. Their geometry is compatible with limited eastward translation of this zone (Blignault, 1977) although this displacement is opposite to the much larger translation in the same area along major shear belts (Toogood, 1976). The pegmatites have been dated at approximately 1000 Ma (Nicolaysen and Burger, 1965; Schutte, 1972). They have provided most of the beryl produced in South Africa and also contain significant amounts of lithium and tungsten minerals (Von Backström and De Villiers, 1972; Von Backström, 1976).

Further igneous activity, in the form of the Richtersveld and Kuboos-Bremen Intrusive Suites and the Gannakouriep basic dike swarm, is closely related to the Gariepian orogeny and therefore is described in Chapter 9; however, older Richtersveld phases are cut by the pre-Gariep north-trending shear belts in the north (Ritter, 1980).

8.5. Central Zone of the Namaqua Province

The pre-Gariep gneisses, granitoids, and gabbroids of the Central Zone, collectively termed the Namaqua Metamorphic Complex, represent by far the greatest part of the Namaqua Province (Fig. 8-3). The Central Zone is so large and heterogeneous that it is impossible to generalize on its features apart from the medium to high metamorphic grade and Early to Mid Proterozoic radiometric age. Although there are no fundamental boundaries within the zone to justify a formal subdivision, the geologic evolution of three subareas will, for convenience, be described separately (Fig. 8-3).

(1) *Namaqualand* comprises the Namaqualand highlands and coastal plain. This has been regarded as the type area of the Central Zone for decades but particularly since the survey of Joubert (1971). Of particular interest are the well-documented cupriferous basic intrusions around Springbok, the presence of early-syntectonic deformed basic dikes, and a spectacular zone of regional crustal reworking along the coast.
(2) *Bushmanland* forms the largest continually exposed part of the Central Zone. The Zn-Pb stratabound mineralization in precratonic supracrustal rocks has focused much recent exploration activity on this area.
(3) *Namibia* contains major inliers in the Lüderitz and Karasburg Districts. A diverse precratonic cover sequence, containing an unusually thick sequence of carbonate rocks, may be significantly older than elsewhere in the Central Zone. The area also contains a unique syntectonic cover sequence and is transected by one of the largest shear belts in the African continent.

The evolution of each of these subareas will be described in terms of the concept of tectonic category, which has been successfully applied in different parts of the Central Zone (Jackson, 1974a, 1975, 1976; Blignault *et al.*, 1974; Blignault, 1975, 1977; Toogood, 1976; Kröner and Blignault, 1976). The merits and disadvantages of this system were discussed by Jackson (1976). Under this scheme the precratonic category can be subdivided into (1) pretectonic rocks, which contain the structural imprint of the earliest recognizable period of deformation, and (2) syntectonic rock units showing only the effects of later deformation. Posttectonic (postcratonic) units are unfoliated and undeformed.

The Central Zone: Namaqualand

A missing basement. The existence of precratonic basement has not yet been proved in Namaqualand. Joubert (1971) regarded augen gneisses such as the Nababeep granitoid gneiss as a basement on which the pretectonic cover was deposited. However, most subsequent workers (including Joubert, 1974a) regarded these granitoid gneisses as early-syntectonic intrusions for reasons given in the section "Nababeep-type early syntectonic granitoids." None of the established criteria for basement recognition, such as precover structures or dikes, derived fragments in cover, or angular unconformities, has been recognized. As in other parts of the Central Zone, the rarity or absence of a basement unquestionably results from the abundant syntectonic intrusions rather than from deep burial beneath a hypothetical thick precratonic cover, which is absent from the Namaqua Province.

The Namaqualand Metamorphic Suite: Pretectonic cover. This suite includes metasupracrustal sequences variously referred to as the "Okiep or Nababeep sequences" (Benedict *et al.*, 1964; Clifford *et al.*, 1975a), the "Namaqualand paragneiss sequence" (Joubert, 1971), and the "Kookfontein subgroup" (Jack, 1980). A consistent structural sequence is preserved over much of Namaqualand and this has since been extended eastward to Bushmanland (Joubert, 1971, 1974a). The base of the sequence, which is truncated by intrusion of younger granitoid gneisses, is a thick westward tapering wedge of pink-weathering biotite gneiss, referred to loosely as "pink gneiss" since it was first described by Lamont (1947). Whether this is of igneous or sedimentary origin is debatable, but its highly quartzofeldspathic nature suggests formation in a continental environment. Kröner (1968, 1973a) suggested that pink gneiss in southern Namaqualand may be of arkosic origin on geochemical grounds. However, this has been disputed by Lipson and McCarthy (1977) because the apparently limited compositional range of pink gneisses in northern Namaqualand is not compatible with a sedimentary origin. There is no reason to believe that all pink gneisses must have a common origin and it is significant that Joubert (1971) recorded xenoliths of basic granolite in pink gneiss, thus indicating the intrusive nature of the latter in places.

Above the pink gneisses are metasediments such as metapelite, gneiss, metaquartzite, and minor metaconglomerate, calc–silicate rocks and marble (Gevers *et al.*, 1937; Benedict *et al.*, 1964; Joubert, 1971). Commonly near the top of this structural sequence are metabasites in the form of basic granolite and amphibolite (Joubert, 1971), the chemistry and field relations of which indicate they are derived from tholeiitic lavas (Zelt, 1980a). There is no consistent variation in bulk major-element compositions across central Namaqualand toward the coast, but trace-element chemistry (notably Cr, Ti, Y and Zr) indicates that several populations exist (Zelt, 1980a, b).

The most intensively studied part of the Namaqualand Metamorphic Suite is in the copper province north of Springbok. Much of the structural sequence (Fig. 8-7) comprises syntectonic granitoid gneisses and granitoids and the original cover is largely restricted to the Ratelpoort and Springbok units, which consist of "granulite" (garnetiferous quartzofeldspathic gneiss) and metaquartzite, which is locally conglomeratic (Clifford *et al.*, 1975a; Marais *et al.*, 1975). Clifford *et al.* (1975a) simplified the succession even further by postulating tectonic repetition of both cover and intrusions in the form of fold nappes, a hypothesis discussed in the section "D1–D2 early deformation" (Fig. 8-7).

The age of formation of this cover sequence is not presently known. Clifford *et al.* (1975a) recorded that both the Namaqualand Metamorphic Suite and the younger intrusions fall on the same Rb-Sr isochron of 1190 Ma. The field relations and the high $^{87}Sr/^{86}Sr$ initial ratio (0.7191) indicate this is a metamorphic age, possibly associated with the D2 event. Contact relations with the Western Zone previously described suggest that parts of the precratonic cover may be coeval with or older than the 2.0-Ga Orange River Group (Bertrand, 1976; Joubert, 1976; Ritter, 1978, 1980).

Nababeep-type early-syntectonic granitoids. An important period of intrusion took place prior to 1190 Ma, resulting in the emplacement of extensive bodies of granitoids throughout Namaqualand. These leucocratic or mesocratic granitoids are commonly megacrystic and biotite bearing and contain D2 fabrics of varying intensity (Joubert, 1971; Clifford *et al.*, 1975a; Marais *et al.*, 1975; Jack, 1980; Jackson and Zelt, in preparation). They are variously known as Nababeep, Brandberg, porphyroblastic, and augen granitoid gneisses. Their intrusive origin is evident from the regionally transgressive nature of the Nababeep granitoid gneiss (Hälbich, 1978) and locally discordant contacts and xenoliths

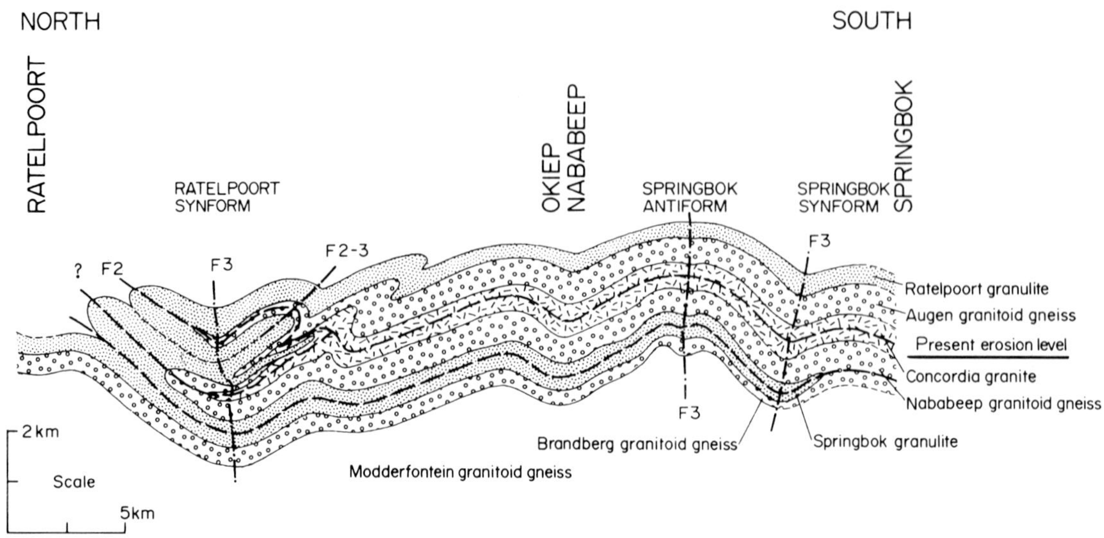

Figure 8-7. Schematic geologic cross section through part of the Namaqualand Central Zone to explain repetition of stratigraphic units by a hypothetical north-facing recumbent fold pair comprising an anticline, the Nababeep nappe, overlying a syncline. [After Clifford *et al.* (1975a), by permission of Oxford University Press.]

of the Namaqualand Metamorphic Suite. The D2 strains within these intrusions are markedly heterogeneous, with the result that they vary in appearance from megacrystic granitoids, through augen granitoid gneisses, to finely layered gneisses with increasing strain (Fig. 8-8A).

Although geologists of the Okiep Copper Company have mapped the megacrystic granitoid gneisses and granitoids as separate units, their chemistry suggests that they are broadly similar, which is compatible with—though not proof of—their being tectonically duplicated by major recumbent folding (McCarthy, 1978) as are their similar Rb-Sr ages (Clifford *et al.*, 1975a).

Buffels River dikes and related early-syntectonic basic intrusions. In his description of the Namaqualand Metamorphic Suite, Joubert (1971) referred to pre-D2 granulite-facies metabasites forming lensoid or sheetlike bodies concordant with the gneissic layering. He was not certain whether they represented intrusive or extrusive bodies but suggested that most of them represented metavolcanic rocks (Joubert, 1971, p. 39), although he considered that at least some of them were of hypabyssal origin (Joubert, 1971, p. 65).

Recent work in central and western Namaqualand has divided the metabasites into four groups that are lithologically similar but of different ages or modes of formation (Jackson and Zelt, in preparation). The oldest of these are pre-D1 tholeiitic metalavas interlayered with paragneisses of the Namaqualand Metamorphic Suite. A second group, termed the Buffels River dikes, is represented by abundant highly deformed basic dikes and sheets intruded between D1 and D2 (Fig. 8-9). These dikes closely resemble Early Archean examples in the Messina area of the Limpopo Province (Chapter 3) and in the Ancient Gneiss Complex of Swaziland (Chapter 2) and the Early Proterozoic Scourie dikes deformed by Laxfordian reworking in northwest Scotland (e.g., Park and Cresswell, 1973; Tarney, 1973; Lisle, 1977). A rare third group comprises post-D1, pre-D2 basic layered intrusions which, despite metamorphism and two periods of deformation, have retained igneous palimpsests such as layering, grading, and cumulus crystals. Both the second and third groups underwent intense distortion and rotation to positions of subconcordance during D2 so that transgressive intrusive contacts are only locally recognizable as original apophyses projecting from the intrusion. A rare fourth group of metabasites, in the form of high grade dikes, cut the D3 shear belts and hence must be post-D3 in age. As is the case for all pretectonic or syntectonic basic dikes, the Buffels River dikes provide valuable time markers for older and younger rocks as well as evidence for an extended period of granulite metamorphism.

D1–D2 early deformation. The regional survey of Joubert (1971) laid the framework for documenting the structural evolution of Namaqualand and most subsequent workers have merely extended his basic scheme, with little modification, apart from a greater emphasis on younger deformation in the coastal zone described in the section "D4 regional crustal reworking in western Namaqualand." Namaqualand is characterized by a regional finite fabric and associated D2 folds which have folded a gneissic layering of D1 age. The D2 fabric generally dips shallowly northward in the north, but farther south the dips are subvertical because of rotation by upright D3 folding.

The D1 deformation event is preserved in the metamorphic suite. It was first documented by Joubert (1971) and his conclusions have since been supported by others (e.g., Hälbich, 1978; Jack, 1980; Albat, 1979; Theart, 1980; Jackson and Zelt, in preparation). Although no major structures of this event have been reported, its imprint is widespread in the form of a gneissic layering that was subsequently folded during D2. The tectonic origin of this layering is indicated by the presence of pre-D1 intrafolially folded quartz–feldspar veins and by a D1 schistosity that is locally recognizable in D2 minor folds, but which has largely been incorporated to form a finite fabric of D2 age (Jackson and Zelt, in preparation). There is no definite evidence of the P-T conditions under which the D1 layering formed, but it is reasonable to assume medium- or high-grade conditions (Joubert, 1971; Albat, 1979).

The regional fabric throughout eastern Namaqualand has been correlated by most geologists with the D2 event, which also formed extensive major and minor tight to isoclinal folds. These fabrics were imprinted on newly intruded granitoids such as the Nababeep-type augen granitoid gneisses and even-grained biotite granitoid gneisses, and on the metabasite dikes and layered intrusions described in the previous section. Joubert (1971) recorded that, allowing for the effects of later deformation, the F2 folds were originally steeply or moderately reclined toward the north over most of Namaqualand, but in central Namaqualand the F2 folds were originally recumbent or gently reclined and the D2 bulk flattening plane was subhorizontal (Jackson and Zelt, in preparation).

The existence of major flat-lying folds is the principal evidence for tectonic duplication of the metamorphic suite and the interlayered granitoid gneisses both on a regional scale and, as proposed by Clifford *et al.* (1975a), for the area just north of Springbok. The existence of pre-D3 recumbent folds has been known for some time in this area (Benedict *et al.*, 1964) and, as further support, the macroscopic map pattern (e.g., Lombaard and Shreuder, 1978) can be interpreted as a type 2 "mushroom" interference between upright D3 folds and tight recumbent D2 folds. The chemistry (McCarthy, 1978) and lithology (Clifford *et al.*, 1975a) of different components of the paragneiss–granitoid gneiss sequence are sufficiently similar to suggest that the multiple appearance of these units at different levels of the structural sequence has been caused by duplication by large-scale recumbent isoclinal folding (Fig. 8-7). With remarkable acuity, Martin (1965, p. 63) speculated that "the generally gentle dips of the gneisses of the Namaqualand Massif may well hide more complicated structures," by analogy with Alpine nappes.

Early metamorphism and regional metamorphic zonation. Polymetamorphism has been proposed by many geologists since it was recorded by De Villiers and Söhnge (1959) in northern Namaqualand. As with other gneiss terranes, there have been difficulties in establishing to what extent the present regional metamorphic pattern is the result of any one particular metamorphic event, a problem discussed in this section.

Joubert (1971) recognized multiple metamorphic events and integrated these into his framework of structural development. His studies led him to believe that the first metamorphic event, termed M1 and coincident with D1, was ill defined and largely overprinted. By contrast the M2 event, which accompanied and outlasted D2, left an indelible stamp over Namaqualand in the form of a metamorphic zonation from greenschist facies at the coast through an intermediate P/T facies series to granulite facies in eastern Namaqualand. This zonation, which culminated in "garnet granulites" regarded by Joubert as representing high-pressure granolites (Green and Ringwood, 1967a), was accepted by Clifford *et al.* (1975a) and Lipson and McCarthy (1977). However, Jackson (1976, 1979b) and Zelt (1980a,b) preferred to regard the "garnet granulites" as calc–silicate rocks associated with intermediate-pressure granolites, which are extensively developed in the same area. More recent

(A)

(B)

Figure 8-8. (A) Early-syntectonic Nababeep-type augen granitoid gneiss cut by northtrending dextral mesoscopic shear belts spaced approximately 1 m apart. Augen granitoid gneiss has been transformed by high strains to finely layered gneiss in foreground with gradational tectonic contact at right. Location: Namaqua Province 35 km southwest of Springbok, Namaqualand, South Africa. [From Jackson and Zelt (in preparation).] (B) Pretectonic Garub metasediments of the Early Proterozoic Aus–Lüderitz carbonate shelf. Layers of diopside–plagioclase–calcite–granofels have acted as competent units surrounded by ductile forsterite marble. Isoclinal overfolds and imbricate overthrusts directed toward the left are visible. Location: Namaqua Province 30 km west of Aus village, Lüderitz District, Namibia. (C) Xenolith of early-syntectonic amphibolitized metagabbro (dark) in younger foliated late-syntectonic granite. A tongue of granite has intruded the metagabbro at upper right. Misidentification of the tongue as a xenolith in the case of incomplete exposure would lead to erroneous reversed age relations. Location: Namaqua Province at Hottentot Bay north of Lüderitz, Namibia. [From Kröner and Jackson (1974).] (D) Four generations of migmatite neosomes in Early Proterozoic Garub biotite schist (dark): N1, folded metatectic concordant quartz–feldspar veins (white); N2, discordant injected quartz–feldspar dikes (white) cutting N1; N3, biotite granite dike (dark gray); N4, veins and irregular masses of biotite leucogranite (pale gray). Location Namaqua Province 20 km north of Aus village, Lüderitz District, Namibia. [From Jackson (1976).]

Fig. 8-8 (C)

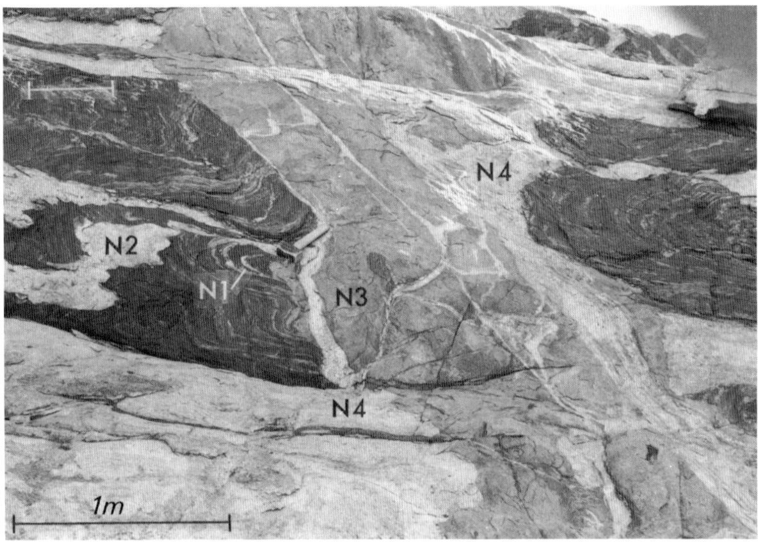

Fig. 8-8 (D)

work to the southwest of the area studied by Joubert (1971) has shown that M2 upper amphibolite-facies metamorphic imprint extends to the coastline at the mouth of the Swartlintjies River (Fig. 8-3) (Zelt, 1980a,b) and that granulite-facies rocks are preserved within 7 km of the coastline (Jack, 1980). Furthermore Zelt (1980a) has proposed that most of the regional metamorphic imprint along the coast is the result of D4 crustal reworking. Joubert (1971) also believed that many of the amphibolites in Namaqualand formed by retrogression of basic granolites.

By comparison with experimentally derived reaction curves, Clifford *et al.* (1975a) conservatively concluded that the M2 metamorphism reached its peak at temperatures of 800–1000°C and total pressures of 600–800 MPa (6–8 kbar), implying a moderate geotherm of approximately 35°C/km. Robey (1979) reported the presence of high-pressure granolites brought from depths of up to 20 km as xenoliths in kimberlites from the northern Cape; his P-T estimates imply a flattening of the geothermal gradient with depth to a value of only 10–12°C/km. Geothermometry on coexisting pyroxenes indicates that granulite metamorphic temperatures in metabasites varied

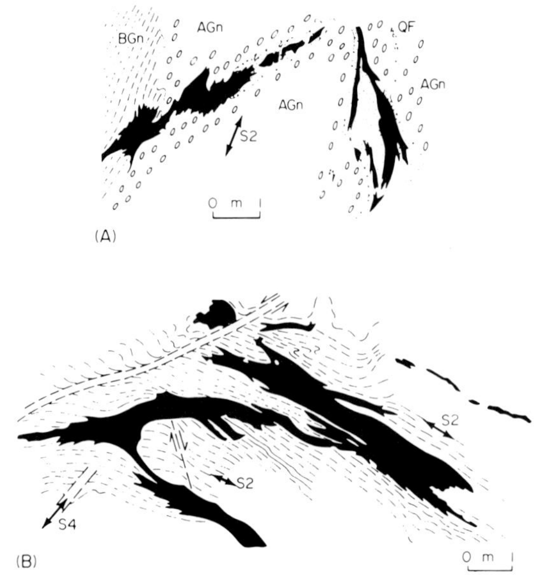

Figure 8-9. (A) Pre-D2 granulite facies Buffels River basic dikes (black) in the Namaqualand Central Zone. Metabasite dike has intruded across contact between early syntectonic biotite granitoid gneiss (BGn) and augen granitoid gneiss (AGn) with pegmatite (QF) developed along dike contact. S_2 refers to schistosity axial planar to dike cusps. Location: cutting on National Road 40 km south of Springbok. [From Jackson and Zelt (in preparation).] (B) Buffels River metabasite dike with asymmetric cusps formed during D2 shortening and boudins in uppermost limb formed by simultaneous extension. S_2 refers to schistosity axial planar to dike cusps and folds; S_4 refers to shear belts. In both figures the cusp geometry indicates that the metabasite dikes are less competent than their country rocks. Location: Swartlintjes River, 20 km northeast from mouth (Fig. 8-3). [From Jackson and Zelt (in preparation).]

between 800°C and 860°C, with no consistent temperature decrease toward the zone of D4 amphibolite-facies reworking along the coastal plain (Zelt, 1980a,b).

The P-T estimates quoted above suggest that partial melting of pelitic or micaceous quartzofeldspathic rocks in the presence of water must have been common and examples of anatexis in the form of migmatite neosomes are widespread (Joubert, 1971; Clifford *et al.*, 1975a; Jack, 1980; Albat, 1979; Theart, 1980). The composition of sapphirine-bearing aluminous rocks in the Namaqualand Metamorphic Suite around Springbok is compatible with their representing a restite formed by extraction of 75–80 percent anatectic granitic liquid from a parental pelitic sediment (Clifford *et al.*, 1975b). McCarthy (1976) argued on geochemical grounds that pink gneisses in the same area were derived by partial melting of Nababeep-type augen granitoid gneisses rather than forming part of the original cover sequence.

The M2 event left an Rb-Sr radiometric imprint of 1190 Ma, which Clifford *et al.* (1975a) interpreted as reflecting M2 isotopic homogenization. However, it is possible that this merely reflects closing of isotopic systems during later uplift and that the true age of M2 is older. In this respect it is significant that the orthopyroxene-bearing granolites define an isochron some 50 Ma older than the 1190 Ma age (Clifford *et al.*, 1975a).

Post-D2 Concordia–Rietberg granitoids. Late-syntectonic granitoids were emplaced throughout Namaqualand in the form of stocks and sheets of widely varying size. Joubert (1971) extended the name Concordia granite from the type locality north of Springbok to cover all similar gneissic granites of post-D2, late-syntectonic age. The generally weak fabrics appear to be of D3 age, but there is evidence that granitic emplacement continued up to and beyond D4 (Joubert, 1971; Jack, 1980; Theart, 1980; Jackson and Zelt, in preparation).

The granitoids are leucocratic and are biotite bearing and megacrystic in places. Nonfoliated varieties have generally been called Rietberg granite around Springbok. The Concordia and Rietberg types show a continuity of composition indicative of fractional crystallization and are of closely similar appearance (McCarthy, 1978). The Rietberg granite yields a combined Rb-Sr isochron of 1140 Ma with a moderate initial ratio of 0.708 ± 0.001, which has been interpreted as the age of intrusion and indicative of a deep crustal source (Clifford *et al.*, 1975a).

The cupriferous Okiep basic intrusions. Emplacement of late-syntectonic granitoids was followed by intrusion of more than 700 basic bodies north of Springbok around Okiep. Copper mined from 27 of these basic bodies since 1852 has made the Okiep area the center of South Africa's copper mining industry with a cumulative production of approximately 890,000 tons of blister copper from 1940 through 1972 (Hammerbeck *et al.*, 1976). Because of their historic economic importance, the petrology of the basic bodies has been intensively studied for many years and their subsurface configuration explored by diamond drilling and geophysical techniques (Strauss, 1941; Benedict *et al.*, 1964; Martin, 1965; Van Zyl, 1967, 1978).

The basic bodies are generally small, averaging approximately 1 km in length, with an irregular shape in horizontal section and an anastomosing sheetlike or pipelike vertical section (Lombaard and Shreuder, 1978). The plunge of their major axes varies from vertical to near horizontal and most bodies are known to pinch out at depth. Their distributions appear to be both lithologically controlled (they are concentrated in, but not confined to, the Concordia granite) and structurally controlled (they are generally situated in the F3 "steep structures" described in the next section). The basic bodies consist predominantly of gabbronorite, norite, and diorite with lesser amounts of hypersthenite, andesinite, and rare components containing orbicular plagioclase (Van Zyl, 1978). The bodies differ slightly in age but are clearly late syntectonic in that they are virtually undeformed but appear to have been structurally controlled in their emplacment. They are at least 1070 Ma old, this being a zircon age of sheared Brandberg augen granitoid gneiss along the contact with a basic body (Clifford *et al.*, 1975a). This age is in broad agreement with a Pb-Pb age of 1085 Ma obtained by Koeppel (1978) on copper ore from one of the Okiep bodies because, as described below in this section, copper mineralization is likely to have postdated crystallization of these bodies.

Hypotheses on the origin of the basic bodies and their mineralization were reviewed by Benedict *et al.* (1964) and Martin (1965, pp. 64–69). These hypotheses can be briefly summarized as follows: (1) purely magmatic origin (Rogers, 1916; Latsky, 1942); (2) derivation from cupriferous basic lavas that survived as pods of basic "resistors" during so-called regional granitization (Read, 1951); (3) trapping of so-called basic granitization fluids in structural traps in the form of "steep structures" (Benedict *et al.*, 1964); (4) megaboudinage of basic bodies with "steep structures" representing large folds separating the boudins (Wegmann, 1963, in Martin, 1965).

Data accumulated since 1965 have gone a long way toward providing an understanding of the basic bodies. The lack of chilled contacts noted by Rogers (1916) and used as evidence for a nonigneous origin by other geologists is most likely to result from emplacement into hot crust with concomitant slow cooling because the bodies are now known to be of intrusive origin. Not only do their contacts transect the D3 "steep structures" and younger megabreccia bodies (Lombaard and Shreuder, 1978), but Prins (1972) reported a narrow (<10 m) contact aureole around a basic body 25 km northeast of Springbok. There has been partial transformation of microcline to orthoclase in the country rock, represented by Nababeep granitoid gneiss, adjacent to the most basic, hottest magma. Despite this evidence for magma, there appears to have been little or no *in situ* differentiation, and less basic magmas preceded the more basic types (Stumpfl, 1978; Van Zyl, 1978). Although copper sulfides are present in virtually all Okiep basic bodies, economic amounts exist only in younger intrusions devoid of clinopyroxene and consisting of norite, diorite, and hypersthenite. Van Zyl (1978) ascribed this concentration to a process of unimpeded gravitational differentiation of sulfides.

Notwithstanding the evidence for the existence of basic magmas, some geologists believe that the basic bodies are not purely of magmatic origin. For example, it has been argued that the age sequence of decreasing whole-rock acidity and decreasing iron content of hypersthene, together with high Zn in spinels, and low Ni and high Co in iron sulfides, negates a process of magmatic differentiation at depth (Stumpfl, 1978). The sulfide textures are not magmatic and the ore sulfides appear to have been emplaced after crystallization and partly after hydration of the silicate and oxide phases. The lack of geophysical or geologic evidence for the presence of a basic reservoir at depth also poses problems for a purely magmatic origin. Clifford *et al.* (1975a) postulated that they were derived by partial melting of mafic granolites of the Namaqualand Suite, possibly during M2 granulite metamorphism, and perhaps emplaced during the D3 heterogeneous shearing that formed the "steep structures." In support of this, they cite the presence of strata-bound copper sulfides in the granolites, high initial $^{87}Sr/^{86}Sr$ ratios (0.71–0.81) in both the granolites and basic bodies, and similar K/Rb, Rb/Sr, and K/Sr ratios in both groups.

If modern data and ideas are compared with previous hypotheses, it appears that they are not mutually exclusive. The igneous hypothesis of Rogers (1916) can be shown to be partly correct with regard to the existence of basic magmas. Read's (1951) granitization theory also contains elements of truth in its proposed derivation of the basic bodies from parental cupriferous lavas. Wegmann's (1963, in Martin, 1965) idea that the basic bodies controlled the formation of the "steep structures" is incorrect, because the latter are now known to predate the former, but his proposal of megaboudinage remains the most reasonable hypothesis for the formation of the steep structures.

D3–D4 periclinal folding and the Okiep "steep structures". Two periods of steeply inclined folding, with axial trends of east-northeast (D3) and north-northwest (D4) formed major periclinal folds throughout eastern Namaqualand and refolded the older flat-lying F2 folds into widely varying attitudes. The F3 folds are south facing and have been described as asymmetric (Joubert, 1971; Jack, 1980, Albat, 1979), but it is also possible to account for the predominance of northerly dips by overfolding of symmetrical folds toward the south. Although Joubert (1971) reported penetrative axial-planar fabrics of D3 age, these are generally rare (Jack, 1980; Albat, 1979; Jackson and Zelt, in preparation) and refoliation is restricted to narrow east-trending shear belts, some of which have the character of tectonic slides on the limbs of major folds (Joubert, 1971; Jack, 1980; Albat, 1979). It is also possible that some of these zones represent high-strain areas where D2 fabrics were rotated on the limbs of F3 folds to positions where their shape fabrics were intensified by D3 flattening (Jackson and Zelt, in preparation). The shear belts tend to be steeply dipping and the subhorizontal attitude of the mineral lineation in their mylonitic cores (Albat, 1979) suggests a dominant strike-slip displacement.

The F4 folds are regionally insignificant in eastern Namaqualand and their crossfolding of the F3 major folds has done little more than cause the latter to be doubly plunging (Joubert, 1971; Jackson and Zelt, in preparation). As an exception in the southeast near Kliprand village west-northwest-trending F4 folds are the dominant major structures (Albat, 1979). Nevertheless, as described in the following section, D4 deformation in the coastal plain has been intense and widespread and is responsible for the regional fabric in that area.

The Okiep "steep structures" are of unknown age but may well be related to the D3 event on account of their orientation, which is similar to that of the F3 folds (Clifford *et al.*, 1975a; Hälbich, 1978; Lombaard and Shreuder, 1978). The "steep structures" are typically narrow (3–500 m) upright cusps within which the normally flat-lying gneisses have been rotated to very steep attitudes. The cusps are generally antiformal but synforms are also known. Like other examples of heterogeneous strain such as shear belts, these structures show an increase in fabric intensity and in rotation of the country rock fabric toward the core of the structure. In contrast, however, the "steep structures" have orthorhombic, rather than monoclinic, structural symmetry so that the foliation on each side curves inward in mirror symmetry, rather than sigmoidally. This suggests that they formed by inhomogeneous pure shear rather than by simple shear.

The distribution of minor structures such as conjugate shear belts and boudins suggests that the gneisses containing the "steep structures" have undergone oblate flattening, with maximum extension in a north-northeast direction (Hälbich, 1978). Evidence for metatexis, in the form of quartz–feldspar melts in areas of low stress, indicates that this extension took place under high-grade conditions. Hälbich (1978), therefore, supported Wegmann's proposal that the "steep structures" represent scar folds (Whitten, 1966) between megaboudins of gneiss undergoing layer-parallel extension. Synformal cusps would occur on the upper side of the boudinaged layer and antiformal cusps on the lower side. This is a reasonable working hypothesis, although there is apparently no consistent facing of cusps into relatively competent units like the granitoid gneisses (Lombaard and Schreuder, 1978) as would be expected in such ductility-controlled processes as boudinage. Despite their common orientation, it may be unwise to assume that the F3 folds and the "steep structures" formed at the same time because they would require very different regional stresses if the F3 folds formed by buckling. However, it is likely that, for example, the orientation of the "steep structures" has been partly controlled by the attitude of earlier formed major folds. Hälbich (1978) suggested that they were localized as crestal ruptures in an elongated dome formed by epeirogenic uplift and extending from Springbok to eastern Bushmanland. Whatever their origin, these structures are at least 1070 Ma old because they were subsequently brecciated (Lombaard and Shreuder, 1978), intruded by the Okiep basic bodies, and then sheared by this time (Clifford *et al.*, 1975a).

D4 regional crustal reworking in western Namaqualand. Several generations of shear belts in a wide variety of orientations have been documented, the youngest of these being north trending (Fig. 8-8A) (Joubert, 1971; Albat, 1979; Jack, 1980; Theart, 1980). The shear belts were regarded as essentially local, planar zones of high strain. Working in the poorly exposed northern coastal plain, Joubert (1971) was able to record that the westward trend of the F3 folds gradually changed toward a southerly direction as the coast was approached. He tentatively ascribed this rotation to left-lateral movement along the coast. These findings were echoed by Jack (1980) for the southern coastal plain.

However, mapping around the Swartlintjies River in the central coastal plain has shown that the D2 fabric, the dominant regional fabric in the east, is transposed or overprinted by a younger finite fabric, which becomes the dominant regional fabric at the coast (Jackson and Zelt, in preparation). The D4 event, which formed north-trending open folds and widely spaced shear belts in the east, was characterized by comparatively low strains in this area. By contrast in the west the D4 event appears to have constituted a major period of crustal reworking. Here D4 strains were markedly heterogeneous so that areas of low D4 strain, in which F3 folds and the D2 fabric are well preserved, alternate with areas of moderate strain, characterized by abundant north-northeast-trending upright F4 folds and nonpenetrative D4 fabrics, and extensive areas of high strain, where F4 folds attain advanced transposition, the D2 fabrics are unrecognizable, and the D4 fabrics are penetrative (Jackson and Zelt, in preparation). These authors concluded that the bulk strains comprised a concertina-type shortening of the coast along east–west axes as well as large shear translation with vertical and left-lateral horizontal components. Left-lateral movements have been tentatively postulated in the coastal zone by Joubert (1971) and Jack (1980). Theart (1980) is of the opinion that the fabric in rocks of the coastal zone due west of Springbok (Fig. 8-3) resulted from "thrusting" after deposition of the basal Gariep Group (Chapter 9). The large-scale translations and shape changes of the crust may have operated over a considerable time and space because metamorphic fabrics in the D4 structures range from upper amphibolite to greenschist (Joubert, 1971; Jackson and Zelt, in preparation). There is also evidence that the lowering of metamorphic grade in the coastal area has been at least partly caused by regional retrogression during D4 reworking (Zelt, 1980a).

This regional reworking, which appears to have taken place in the interval 1100–900 Ma, may well be a deep-seated precursor or time equivalent of basement reworking under greenschist facies that took place just prior to and just after deposition of the Gariep Group (Joubert, 1971; Joubert and Kröner, 1972) at approximately 900 Ma (Chapter 9).

The Central Zone: Bushmanland

Evidence for precratonic basement. The existence of basement has been proved in one area and inferred elsewhere. Moore (1977) described a basement unit below the Bushmanland cover sequence around Namiesberg in western Bushmanland (Fig. 8-3). Gneisses underlying the cover consist of precover layered biotite gneisses and postcover augen granitoid gneisses. The basement character of the layered gneisses is evident from the fact they contain two structural elements (a folded schistosity) that have not been recognized in either the cover rocks or the augen granitoid gneisses that intrude them. Because this occurrence represents the only example of definite basement reported from the entire Central Zone, it assumes great significance. Van Bever Donker (1980) proposed that the Kakamas Suid layered gneiss (Table 8-5) at the base of the supracrustal sequence represented a basement in eastern Bushmanland, but because no unconformity has been demonstrated this must remain speculative.

Does the existence of basement gneisses around Namiesberg imply that these gneisses represent the oldest stratigraphic unit in the Central Zone? Not necessarily, for this would assume a common age for all the precratonic cover sequences. As shown in the following section, there is evidence for very large age differences (at least 800 Ma) between the various cover sequences, and the sequence in western Bushmanland overlying basement is likely to be relatively youthful. Accordingly the preservation of basement–cover relations in this area is likely to result

Table 8–5. Precratonic Cover Sequences in Eastern Bushmanland[a]

Van Bever Donker (1980)		Geringer and Botha (1977)	Vajner (1978, unpublished)	
Kakamas	Keimoes	Gordonia	Upington West	Upington East
	(Zwart Boois Berg) Qtz-feld (incl. congl. X beds)	(? Goede Hoop)		
(Hartbees) Calc/basic	(Neuspoort) Semipelitic	(Rooidam) Semipelitic, pelitic		
(Maraisrivier) Basic		Basic		
(Boesmansrivier) Qtz-feld		(Toeslaan) Pelitic		
(Regt Kyk) Calc	(Baviaans Krantz) Quartzose/calc	(Biesiepoort) Calc/basic		(Vaalkoppies) Quartzose (± Semipelitic, basic, calc)
(Wolfskop) Qtz-feld "pink gneiss" (± semipelitic) (local augen)	(Wolfskop) Qtz-feld "pink gneiss" (± semipelitic) (local augen)	(Riemvasmaak) Qtz-feld "pink gneiss"		
(Venterskop) Pelitic		(Goede Hoop) Semipelitic (± pelitic) (incl. congl., X beds)	(Jannelsepan) Semipelitic/pelitic	
(Kakamas Suid) Acid granitoid			(Olive) Granitoid (± basic, pelitic)	(Matjiesrivier) Granitoid (± quartzose, basic)
Layered biotite gneiss			Layered bio-hbl gneiss	Layered gneiss

[a] Sequences listed in terms of their bulk compositions in structural not stratigraphic order. Lateral stratigraphic correlations between areas are extremely tenuous because of large structural discontinuities. (? Goede Hoop) indicates possible alternative position for this unit in the sequence (see text). Geographic component of lithostratigraphic name shown in parentheses. Compositional groups: basic (e.g., amphibolite, actinolite schist), calcareous (e.g., marble, calc–silicate rock), ferruginous (e.g., iron-formation), pelitic (e.g., mica schist, sillimanite gneiss), semipelitic (e.g., quartz–feldspar–biotite schist), quartzose (e.g., metaquartzite), quartzofeldspathic (e.g., "pink gneiss" low in mica), acid granitoid (e.g., layered biotite gneiss, augen granitoid gneiss), intermediate granitoid (e.g., hornblende gneiss); X bed, cross-bedding or cross-lamination; congl., metaconglomerate.

both from the relative youth of the cover and from the relative paucity of syntectonic intrusions. It is possible that the "basement" gneisses in western Bushmanland are equivalent in age to precratonic cover sequences elsewhere in the Central Zone.

Precratonic cover sequences. Thin but extensive metamorphosed cover sequences are present throughout Bushmanland and they merge with other supracrustal sequences to the west in Namaqualand and the Western Zone, to the north in the Namibian part of the Central Zone, and to the east in the Eastern Marginal Zone. The supracrustal rocks have been metamorphosed at mainly lower amphibolite facies to a variety of gneisses, schists, and metaquartzites. In order to emphasize the character of their parent rocks at the expense of superimposed features such as metamorphic grade, these sequences generally are referred to here in terms of their bulk composition, rather than in terms of their present metamorphic lithology.

Tables 8-5 and 8-6 list the bulk compositions of precratonic cover sequences in eastern and western Bushmanland. The rock types are listed in structural sequence rather than inferred stratigraphic sequence for the sake of objectivity. Some of the basal gneisses (e.g., augen granitoid gneiss) are known to be younger than the overlying cover rocks and are therefore not true basement. Where possible, each column in the tables has been "hung" on a common datum level represented by the "pink gneiss," a widespread pink-weathering biotite gneiss.

Table 8-6 shows that similar sequences have been reported in western Bushmanland (Joubert, 1973, 1974a; Moore, 1977; Rozendaal, 1978; Lipson, 1979). This is to be expected because the data are from the relatively small area of the western Bushmanland mineral province. Joubert (1974a) proposed the collective term "Bushmanland sequence" to cover the supracrustal rocks, but there has been little agreement since then as to the limits of the term, although, apart from Moore (1977), other authors have excluded the basal pink gneiss from the group (Table 8-6). Colliston (1979) proposed an elaborate stratigraphy in which he extended the Bushmanland sequence to include other sequences that he regarded as being younger (Table 8-6). In order to put all the stratigraphic proposals in perspective, it must be stressed that only one definitive younging direction has been reported (Rozendaal, 1978) over this complexly deformed and, in places, poorly exposed area.

Lithostratigraphic correlations are particularly difficult in eastern Bushmanland because of the large structural discontinuities that separate the subareas of this region (Fig. 8-10). Apart from the presence of pink gneisses in some of these subareas, they have little in common (Table 8-5). Botha *et al.* (1976) evaded this difficulty by classifying all Bushmanland cover rocks apart from the pink gneisses into two large, heterogeneous groups called the N'Rougas and Kenhardt "Formations."

Some of the sequences in eastern Bushmanland are underlain by layered gneisses, which may represent true basement equivalent to that farther west at Namiesberg. Pink gneiss is overlain by two contrasting sequences, which are of either different age or different tectonic setting. Around Kakamas village the sequence is predominantly basic and calcareous, whereas farther east around Keimoes village the cover is a metamorphosed clastic sequence with well-preserved sedimentary structures such as trough cross-bedding and pebble sheets (Van Bever Donker, 1980). By contrast, a unit lithologically similar to the latter is reported to *underlie* the "pink gneiss" northwest of Keimoes village in the Gordonia District (Geringer and Botha, 1977).

The sequences shown in Tables 8-5 and 8-6 represent volcanic and sedimentary rocks apparantly deposited on continental crust. Whether or not the pink gneiss is regarded as forming part of the sequence, its lithology suggests that its protolith was either a rhyolitic lava or tuff or an arkose derived from a sialic provenance; all suggest continental crust below the cover sequence. As in Namaqualand the origin of the pink gneisses is controversial: suggested parent rocks range from arkose (Poldervaart and Von Backström, 1949; Geringer, 1973; Moore, 1977), to rhyolite (Joubert, 1974a; Botha *et al.*, 1976), and granitoid (Coetzee, 1941; Lipson and McCarthy, 1977; Colliston, 1979). Most of these speculations are based on field relations, others on their geochemistry. However, owing to the immaturity of clastic sediments such as arkoses and graywackes, the difference in chemical composition between these rocks and igneous rocks is not pronounced (Schultz, 1978). Rare earth element and stable-isotope analysis might throw light on, for instance, whether or not the pink gneisses have been through a melting or weathering phase. Analysis of field relations is frustrated by melting and mobilization of the pink gneiss after its deposition or emplacement (Joubert, 1974a; Rozendaal, 1978).

The Bushmanland cover sequence above the pink gneiss appears to have accumulated in shallow water. Different directions of basin thickening have been reported by Joubert (1974a; north, west, southwest), Moore, (1977; southeast), and Lipson (1979; north). This might be because their observations have been drawn from different basins or different parts of the same basin. Initial sedimentation appears to have been mainly pelitic, suggesting a terrigenous provenance. An arkosic origin for the pink gneisses would suggest a fluvial origin for the pelitic rocks, whereas a rhyolitic origin for the pink gneisses would suggest that the pelitic rocks represent clastic sediments accumulated in forearc or backarc basins between a volcanic arc and its trench or on the continental side of the arc, respectively (cf. Dickinson and Seely, 1979). The association of the pelitic rocks

Table 8–6. Precratonic Cover Sequences in Western Bushmanland[a]

Rozendaal (1978) Gamsberg	Moore (1977) Namiesberg	Lipson (1979) Aggeneys	Joubert (1973, 1974a) Regional	Colliston (1979) Regional
				(Grunau) Pelitic (Pella) Quartzose/semipelitic/pelitic (incl. congl.) (Gaudam [Sill] Intermed Granitoid (± magnes) (± pelitic) (Hom) Acid granitoid (± basic/calc/magnes)
(Nousees) Basic		Granitoid "gray gneiss"	Granitoid (± basic) "gray gneiss"	
(Nousees) Qtz-feld (incl. congl. X beds) (± semipelitic)		(Spring) Semipelitic/ quartzose	Semipelitic (Pella) Ferrug	
(Gams)[Zn-Pb-Cu-Fe] Ferrug (± pelitic/ calc)		Ferrug/calc [Zn-Pb-Cu-Fe] (Fly) Pelitic (incl. ? congl.)		(Aggeneys)
(Pella) Quartzose (incl. congl.) (± pelitic/ semipelitic)	Quartzose[Fe] (± ferrug)	Quartzose	(Pella) Quartzose (± semipelitic)	
(Namies) [Sill] Pelitic (± quartzose)	Pelitic[Sill] (± calc/basic)	Pelitic[Sill]	(Namies) Pelitic (± semi-pelitic, quartzose, basic)	
(Haramoep) Qtz-feld "pink gneiss" (incl. augen and layered) (partly intrusive)	Qtz-feld "pink gneiss" (± basic/calc at base)	Qtz-feld "pink gneiss"	(Haramoep) Qtz-feld "pink gneiss" (partly intrusive)	(Hoogoor) Qtz-feld "pink gneiss" (intrudes above)
	Augen granitoid gneiss (intrudes below)	Augen granitoid gneiss	(Aroams) Augen granitoid gneiss (intrusive)	
	Acid granitoid Layered bio gneiss			

[a] Sequences listed in terms of their bulk compositions in structural not stratigraphic order. Stratigraphic extent of Bushmanland Metamorphic Suite shown between double lines. Younger intrusions and possible older basement underlie suite. Wavy lines indicate proposed unconformity or intrusive contact. Geographic component of lithostratigraphic name shown in parentheses, mineralization in brackets. Compositional groups: basic (e.g.,

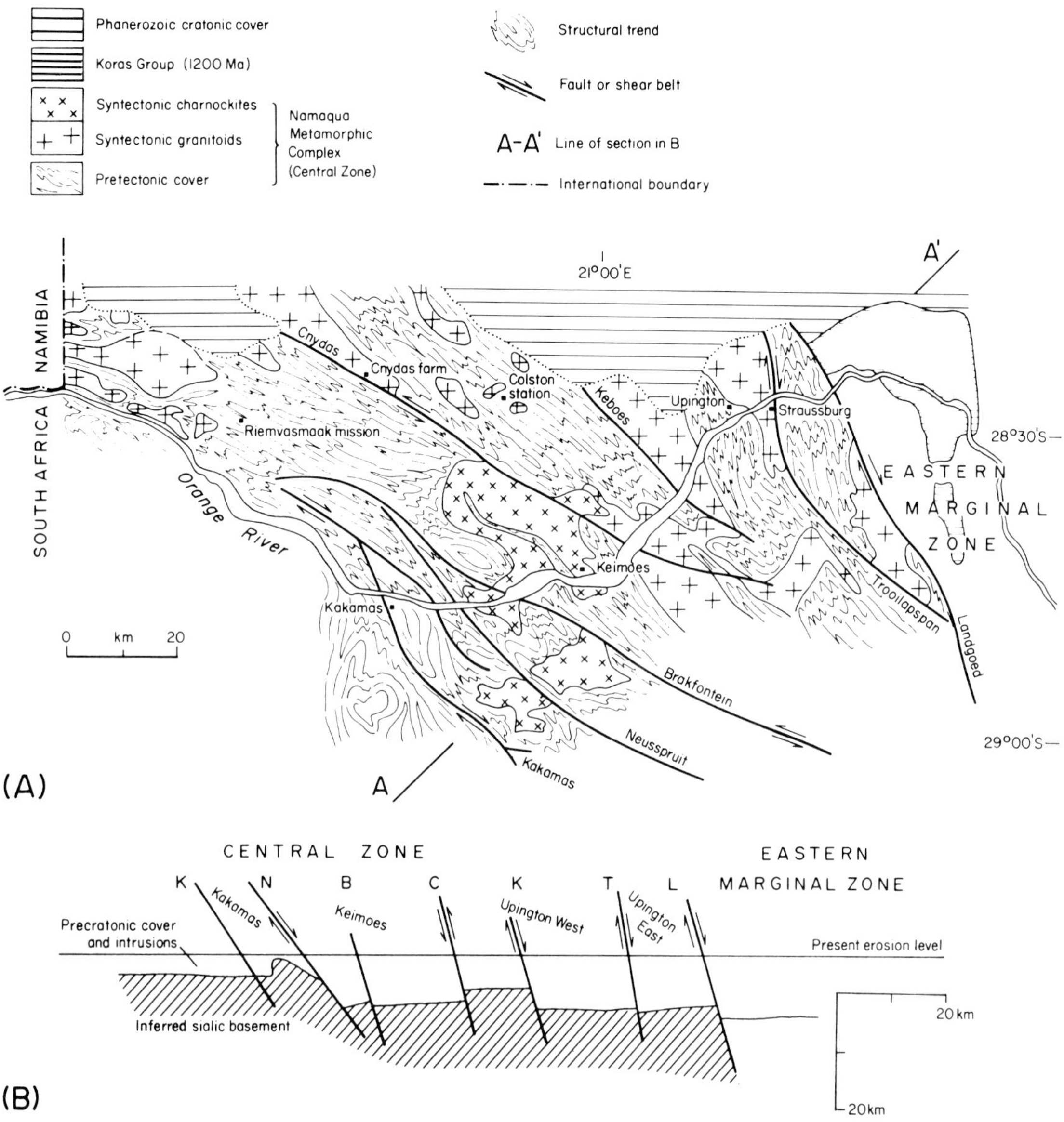

Figure 8-10. (A) Geologic map of the Central Zone in eastern Bushmanland showing distribution of granitoids and charnockites, structural trends, and principal faults and shear belts. [Compiled from maps by Vajner (1975), Geringer and Botha (1977), Schultz (1978), Stowe (1979), Van Bever Donker (1980).]
(B) Schematic cross section along line AA' to show displacements inferred by metamorphic grades and structural divisions defined by these discontinuities. [After Van Bever Donker (1980).]

---

amphibolite, actinolite schist), calcareous (e.g., marble, calc–silicate rock), ferruginous (e.g., iron-formation), magnesian (e.g., cordierite-anthophyllite rock), pelitic (e.g., mica schist, sillimanite gneiss), semipelitic (e.g., quartz–feldspar–biotite schist), quartzose (e.g. metaquartzite), quartzofeldspathic (e.g., "pink gneiss" low in mica), acid granitoid (e.g., layered biotite gneiss, augen granitoid gneiss), intermediate granitoid (e.g., hornblende gneiss). [Sill], sillimanite deposits; [Fe], magnetite, hematite, pyrite, pyrrhotite; [Cu], chalcopyrite; [Pb], galena; [Zn], sphalerite; X bed, cross-bedding or cross-lamination; congl., metaconglomerate.

with concentrations of Al, Ca, Na, and B suggests that their parent rocks consisted of clays and evaporites formed in a sabkha or playa-lake environment (Moore, 1977). An arid climate of this type would preclude the concentration of alumina (now forming economic deposits of massive sillimanite; Table 8-6) by means of tropical weathering to form bauxites as proposed by Coetzee (1942) and Frick and Coetzee (1974). Martin (1965) and Joubert (1974a) suggested that the "pelitic" rocks may have formed from acid volcanic rock, alumina being concentrated by fumarolic exhalation or leaching by thermal waters. The strata-bound nature of the sillimanite deposits within the pelitic host argues against—but does not preclude—an origin by leaching (Moore, 1977). Although the origin of these rocks is in dispute, evidence cited above suggests that the sedimentary basin was initially infilled under terrestrial or marginal marine conditions.

Prolonged reworking and increase in maturity are reflected in the accumulation of quartz arenites overlying the pelitic rocks (Table 8-6). In restricted areas this was followed by accumulation of low-energy, predominantly siliceous, clastic sediments and Kuroko-type mineralization of Fe, Mn, Zn, Pb, Cu, and S. These deposits are capped by predominantly calcareous metasediments with precipitates of Fe, Mn, and Ba (Rozendaal, 1978, 1979; Lipson, 1979). With reserves of more than 150 million metric tons of ore, the deposits around Aggeneys and Gamsberg represent South Africa's largest supply of zinc (Fig. 8-3). The iron oxide facies can be interpreted as the terminal phase of mineralization during marine volcanism when the volume of brine discharged through vents had decreased: similarly, the $BaSO_4$ is likely to have been precipitated from sea water above the vents. The curious absence of evidence for extensive volcanism up to this point in the sequence suggests a distal volcanosedimentary environment. Erosion and a renewed burst of clastic deposition following this chemogenic phase is indicated by overlying metamorphosed arenite and conglomerate. The metaconglomerate contains clasts derived from the underlying iron-formation and, together with the preservation of cross-bedding, provides evidence of younging (Rozendaal, 1978).

The top of the Bushmanland structural sequence is represented by layered gneisses of granitoid or basic composition, which may represent a suite of intermediate to basic metalavas. Colliston (1979) postulated that to the north these rocks are overlain by suites of acid and intermediate lavas, termed the Hom and Guadam gneisses, respectively. Because these gneisses are not in contact with the precratonic cover rocks to the south, their true stratigraphic position is in doubt.

In eastern Bushmanland both sedimentary and volcanic parent rocks have been proposed for the precratonic cover sequence. At Copperton mine in southeastern Bushmanland, strata-bound Cu-Zn mineralization is present between quartzofeldspathic gneisses below and siliceous to pelitic gneisses above (Middleton, 1976; Cornell, 1978b). Large ore reserves constitute the second largest supply of zinc in South Africa after those in western Bushmanland. In northeastern Bushmanland Van Bever Donker (1980) interpreted the cover sequence as representing pelitic accumulations on granitoid basement grading upward into graywackes containing basic sills or flows. The sequence is capped by shelf deposits of carbonate and quartz arenite and cross-bedded fluvial deposits (cf. Table 8-5).

The timing of this important period of sedimentation, volcanism, and mineralization is unfortunately not clear at present and there are major conflicts of interpretation. What seems increasingly likely is that a number of sequences are involved, varying in age from 2000 Ma to 1300 Ma. The older age limit is provided by the Orange River Group in the Western Zone. This 2000-Ma cover sequence can be traced eastward into Bushmanland in the form of higher-grade but coeval paragneisses (Bertrand, 1976). It has been speculated that this belt extends along the Orange River all the way into the Eastern Marginal Zone (Joubert, 1976). This idea was supported by Colliston (1979), who proposed that the belt represented a volcanic arc with proximal facies in the Western Zone and distal facies in the Central Zone in northern Bushmanland around Pofadder village (Fig. 8-3). This implies that the southern cover sequence around Aggeneys mine, which is believed to be older than the proposed northern arc by Joubert (1976) and Colliston (1979), must be more than 2000 Ma old.

This belief is at variance with the lead isotope data from ores contained in the paragneisses at Aggeneys and Gamsberg mines. Koeppel (1978) obtained scattered Pb-Pb ages of around 1300 Ma from these ores and an age of 1305 ± 100 Ma for ores at Copperton mine, which agrees well with the age of 1245 Ma obtained by Cornell (1978b; Rb-Sr) on the same ore body. The relatively low initial ratio of 0.7024 ± 0.0027 indicates that these ores are probably less than 1500 Ma old (Cornell, 1978b). Unless anomalously radiogenic lead in the ores at the time of deposition is invoked, which is not reported from elsewhere, one must conclude that the ore deposits were formed some 1300 Ma ago (Koeppel, 1978). Because the ores at Copperton, Aggeneys, and Gamsberg are held to be strata bound and syngenetic, the associated sedimentary and volcanic rocks must be of similar age and hence considerably younger than 2000 Ma rather than older as has been claimed.

Supracrustal sequences in the Eastern Marginal Zone (exemplified by the classical "Kheis" trilogy) have long been regarded as parents of the precratonic cover rocks in the Central Zone (e.g.., Gevers *et al.*, 1937; De Jager and Von Backström, 1961; Von Backström, 1964, 1967; Martin, 1965; Joubert, 1971; Clifford *et al.*, 1975a). It has also been postulated that the gneisses of the Central Zone represent eugeosynclinal equivalents of the Griqualand West Supergroup (G.J. Beukes, 1973; Geringer, 1973).

Recent workers have tended to maintain temporal links between the "Kheis"–Matsap sequence and the Central Zone gneisses, while concomitantly regarding them as forming different sedimentary and volcanic facies in contrasting tectonic settings. Thus Vajner (1974a) proposed that the Central Zone precratonic cover (referred to by him as "Hartebeest Pan Formation") formed in a backarc basin adjacent to a "Kheis" miogeosyncline. It has also been suggested that the "Kheis" sequence represents the miogeosynclinal equivalent of the eugeosynclinal Central Zone gneisses (Botha and Grobler, 1979). However, this seems incompatible with the abundant basic volcanism in the "miogeosyncline" (e.g., Wilgenhout Drift Formation) and with the presence of terrigenous shallow-water clastic sediments and carbonates in the "eugeosyncline" (their N'Rougas Formation). Lithologic differences between supracrustal rocks of the Central and Eastern Marginal Zones have been described as superficial and attributable to differences in sedimentary facies and metamorphic grade (Botha *et al.*, 1976, 1977; Botha and Grobler, 1979). These authors believed that supracrustal sequences in the Central Zone as far away as Namibia are correlates of the Matsap Group. The dangers of such long-range correlation on purely lithologic grounds have been amply demonstrated for the Beitbridge sequence (Limpopo Province), which was widely regarded as a facies equivalent of the Bulawayan or Shamvaian sequences but is now known to be 500–800 Ma older (Chapter 3). The presence of metabasites, for example, in both the Central and Eastern Marginal Zones should not be regarded as evidence for stratigraphic unity (Cornell, 1975). For example, metavolcanic gneisses in the Central Zone at Copperton (Fig. 8-4) have yielded ages of little over 1.3 Ga old (Cornell, 1978b), the Matsap volcanic rocks are 1.8–2.1 Ga old (Crampton, 1974), and the Marydale volcanic rocks are 3.0 Ga in age (Cornell and Barton, 1979). All three units occur in a relatively small area and all have been chronologically equated by geologists in the past. Although the Matsap Group contains sufficiently diverse rock types to accommodate extensive and elaborate correlations with gneisses in the Central Zone, these correlations should be based on more than similar lithology. If the precratonic cover in northeastern Bushmanland is of Matsap age (approximately 1800 Ma), these rocks would be significantly older than the cover sequences in western or southeastern Bushmanland.

D1–D2 deformation. Strains associated with the earliest periods of deformation have been recognized throughout Bushmanland and have been correlated with those in Namaqualand (Joubert, 1974a) and the Eastern Marginal Zone (Botha *et al.*, 1977; Stowe, 1979). The maximum age of the D1 and D2 deformational events is that of the pretectonic cover sequences deformed by them and hence may be 1800 Ma in the northeast, 1300 Ma in the southeast, and possibly as young as 1200 Ma in the west. There is no evidence that, for example, the D2 event was synchronous in all areas, but structures of this relative age have a consistent relation to the younger periclinal folding and so can be grouped in terms of relative age.

Joubert (1974a) extended his survey of Namaqualand eastward into Bushmanland and reported similar D1 structures in the form of transposition layering and intrafolial folds. Subsequently both major and minor F1 folds have been identified or inferred over most of Bushmanland (Geringer and Botha, 1977; Moore, 1977; Lipson, 1979; Stowe, 1979; Van Zyl, 1979; Van Bever Donker, 1980). These folds are generally inferred to have been originally recumbent or reclined with axial surfaces dipping gently northward (Moore, 1977; Lipson, 1979; Stowe, 1979; Van Bever Donker, 1980) or southeastward (Van Zyl, 1979). Most geologists have concluded that the regional schistosity is of D1 age (Geringer and Botha, 1977; Moore, 1977; Stowe, 1979; Van Zyl, 1979; Van Bever Donker, 1980), but Joubert (1974a) correlated it with D2.

The effects of the D2 event, now known in some detail, were strongly heterogeneous in both orientation and magnitude. In northeastern Bushmanland the F2 folds are commonly doubly plunging (because of D3 cross folding) to the northwest and southeast and are largely responsible for the regional "grain" in this area (Vajner, 1976; Geringer and Botha, 1977; Stowe, 1979; Van Zyl, 1979; Van Bever Donker, 1980). These folds face northeast to the north and east of Keimoes village and face southwest to the southwest of Keimoes (Geringer and Botha, 1977; Stowe, 1979; Van Bever Donker, 1980), which could imply a regional anticlinorium. Throughout most of eastern Bushmanland the F2 folds appear to have been upright on the basis of their interference patterns with older and younger folds. However, northwest of Kakamas village they form macroscopic type 2 "mushroom" interference patterns with upright F3 folds (Geringer and Botha, 1977), which suggests that in this area the F2 folds were formerly recumbent.

Despite variations in D2 strain patterns, some general trends have been observed. There is a fourfold increase in the tightness of F2 minor folds from the margins of the Central Zone toward Keimoes village (Fig. 8-10), where the folds are strongly elongated and tightly appressed (Van Zyl, 1979). Farther west, across a major structural discontinuity called the Neusspruit lineament, this trend is reversed: the F2 folds are broad and open and the bulk strains decrease (Van Bever Donker, 1980). Fold-shape analysis indicates that finite shortening for the F2 folds increased from 20 percent in the extreme east to 70 percent southwest of Upington (Van Zyl, 1979). The finite strains recorded in such deformed objects as amygdales and clasts are those of flattening. The orientations of the strains recorded are closely similar to those of F2 fold elements, which indicates the dominant influence of D2 on the regional strain pattern (Van Zyl, 1979). The finite strains were strongly influenced by the numerous shear belts described in the section "Shear belts and differential uplift" which extend through eastern Bushmanland. In the vicinity of the shear belts the strains increase in magnitude and become more oblate; the F1 and F2 folds become tighter and their axial planes converge (Stowe, 1979; Van Zyl, 1979; Van Bever Donker, 1980).

In western Bushmanland major and minor F2 folds have been recognized by Joubert (1974a), Moore (1977), and Lipson (1979) and illustrated in section by Rozendaal (1979). Unlike most of eastern Bushmanland, these folds were largely recumbent originally with shallow northward dips and east-northeastward (Moore, 1977; Lipson, 1979) or west-northwestward plunges (Joubert, 1974a). The P-T conditions under which D1 and D2 deformation took place will be discussed after the syntectonic intrusions are described.

Syntectonic granitoid intrusions. Augen granitoid gneisses are widespread throughout Bushmanland below the pretectonic cover sequences (Tables 8-5 and 8-6). As with similar rocks in Namaqualand and Namibia, their structure suggests that most of them intruded early syntectonically after deposition of the cover sequences and before D2. These strongly foliated augen granitoid gneisses may well be related to those of Nababeep type in Namaqualand, but poorer outcrop and more-extensive cover sequences in Bushmanland have meant that exposures of this rock are widely scattered but far less abundant than in Namaqualand.

In western Bushmanland younger granitoids are comparatively rare (Joubert, 1974a). Moore (1977) reported a leucogranite intruded after D2 but folded by D3. Eastward toward the margins of the Central Zone there is an increase in abundance of syntectonic granites such as the Louisvale, Straussburg, and Colston intrusions (Fig. 8-10). These are biotite granites; their textures range from even grained to megacrystic and their field relations suggest emplacement at an intermediate crustal level (Vajner, 1975; Geringer and Botha, 1977; Stowe, 1979; Van Bever Donker, 1980). The strain pattern of the relatively undeformed Straussburg granite is compatible with emplacement at the close of D2 (Van Zyl, 1979). These granites therefore contrast with those in the adjacent Eastern Marginal Zone, which are likely to be of similar age (1200–1100 Ma) but which have been emplaced as isolated plutons at a high crustal level (Fig. 8-4) (Vajner, 1975; Botha *et al.*, 1977).

Syntectonic basic intrusions. Several basic intrusions are present around Pofadder village in western Bushmanland (Joubert, 1974a). They are fairly small bodies (maximum 4 km^2 in outcrop area) found only in gray layered gneisses forming the top of the Bushmanland cover sequence. One of these bodies, the Nouzees intrusion, has been studied in more detail (Reid *et al.*, 1979). The intrusion consists predominantly of leucogabbronorite and subordinate olivine gabbronorite; both rocks have cumulus plagioclase and phase layering. Major- and trace-element compositions of rocks and minerals are compatible with derivation by plagioclase fractionation from a parental magma similar to high-alumina basalt. The western part of the Nouzees intrusion is extensively amphibolitized, as are the narrow schistose zones, which probably represent shear belts. The structural relations are poorly understood and contacts are obscured by overburden. Joubert (1974a) regarded these intrusions as relatively old and concluded that they predated the granitoids because of their absence from the latter outcrop areas.

Van Bever Donker (1980) recorded the presence of small bodies of norite in eastern Bushmanland near the Kakamas shear belt and concluded that these were intruded during the D3 event.

Syntectonic charnockites. The charnockites around Kakamas and Keimoes in eastern Bushmanland are probably the most famous in southern Africa (Poldervaart and Von Backström, 1949; Von Backström, 1964; Poldervaart, 1966). A radiometric age of 1082 Ma (Nicolaysen and Burger, 1965; Rb-Sr) on biotite from the adjacent country rock is open to several interpretations but may be roughly equivalent to the time of charnockitic intrusion.

At least five large charnockite intrusions are present in this area, the largest of which is at least 15 km long (Fig. 8-10) (Schultz, 1978; Van Bever Donker, 1980). The intrusions are of granitic aspect apart from their content of orthopyroxene, brown feldspar, and blue quartz crystals, which impart a dark color to these leucocratic rocks. They are compositionally equivalent to quartz monzonite (Von Backstöm, 1964; Schultz, 1978). Textures indicate that the orthopyroxene and feldspars formed under conditions of low water pressure and that retrogression formed clinopyroxene, biotite, and hornblende and reequilibrated plagioclase into large porphyroblasts of lower An content than previously; limited partial melting took place at this time (Schultz, 1978).

On the basis of intrusive contacts with the supracrustal rocks, Von Backström, Poldervaart, and Van Bever Donker concluded that these rocks were of magmatic origin; Van Bever Donker (1980) regarded them as post-D2. They are spatially closely related to noncharnockitic granitoids, and gradational contacts with the Straussburg granitoids are common (Van Bever Donker, 1980). Similarly, Geringer (1973) reported that parts of the Cnydas and Kourop River granitoids to the north of Kakamas are charnockitic.

This proposed igneous origin for the Kakamas charnockites was challenged in a detailed study by Schultz (1978). His remapping indicated that the distribution of orthopyroxene-bearing (i.e., truly charnockitic) components is more limited than previously supposed and that over wide areas they are interlayered with biotite–hornblende gneisses in gradational contact. Quartz segregations commonly define a weak layering that is tightly folded in places. According to Schultz (1978), crosscutting contacts are confined to dikes. The dikes were interpreted as being of the same origin as the main intrusion (Von Backström, 1964), but they do not contain pyroxenes; their place is taken by relict garnet, which is absent from the main body.

With regard to the chemistry of the charnockites, there is a low probability that they represent members of a differentiation series from a parental magma or that they were derived by anatexis. Out of 29 specimens analyzed (Schultz, 1978) only six plot in the igneous field of an Ab–An–Or ternary diagram; this minority is not representative because the six specimens were collected from the more leucocratic gradational contact zone with biotite–hornblende gneiss. None of the specimens defines trends or shows significant correlation with commonly applied indices of differentiation. Schultz therefore concluded that these rocks represented ultrametamorphosed sediments, a conclusion not at variance with the regional metamorphic grade. If an isochemical system is assumed, the parental sediments would be shales for the bulk of the charnockites and carbonate-bearing shales (marls) in the leucocratic contact zones. Van Bever Donker (1980) does not agree with these conclusions and has suggested that the anomalous chemistry of the charnockites can be explained by assimilation of country rocks in the magma as originally proposed by Von Backström. One problem with this explanation is the fact that the charnockites become more "igneous" in composition toward the supposedly contaminating country rocks. Clearly the origin of these rocks is far from resolved.

Metamorphism. Studies of metamorphism in eastern Bushmanland reveal two important features in common with other metamorphic terranes. These are the overprinting of an early high-grade event by later retrogressive metamorphic events, and a trend of decreasing metamorphic grade toward the margins of the Central Zone. Both these characteristics are compatible with progressive uplift through time and increasing amounts of uplift toward the core of the Central Zone.

The early high-grade event of regional metamorphism, referred to here as M1, has been equated with the D2 deformation event (Cornell, 1978b; Van Zyl, 1979; Van Bever Donker, 1980). Another view is held by Botha *et al.* (1979), who equated this with D1 and did not recognize more than one significant metamorphic event. A probable age for M1, and hence D2, in the Copperton area is 1285 ± 100 Ma (Cornell, 1978b; Rb-Sr isochron). In the deeper crustal levels exposed around Kakamas village this high-grade event culminated in granulite metamorphism, and orthopyroxene formed in basic and pelitic rocks (Schultz, 1978). The coexistence of cordierite and garnet in the pelitic rocks indicates pressures of 300–500 MPa (3–5kbar) and temperatures are estimated at 690–740°C. Metamorphism continued interkinematically (i.e., between deformation events) beyond D2 so that, for example, wollastonite in the calcareous rocks defines an isotropic fabric. Evidence of *in situ* partial melting from the area northwest of Upington (Geringer and Botha, 1977) may possibly be related to the M1 event. Mineral parageneses reported for M1 suggest that this formed under a low P/T metamorphic facies series. East of the Neusspruit shear belt, lesser amounts of uplift have exposed medium-grade crustal levels (Van Bever Donker, 1980).

Van Bever Donker (1980) has proposed a second metamorphic event, termed M2, responsible for the formation of garnet, sillimanite, and staurolite in pelitic rocks, and wollastonite in calcareous rocks. Both he and Von Backström (1964) attributed this event to contact metamorphism by the intruding charnockites and granitoids because of their isotropic metamorphic fabrics; this could equally well be explained by interkinematic crystallization, as proposed by Schultz (1978). Van Bever Donker estimated pressures of 550 MPa (5.5 kbar) for this event. Because this is substantially deeper than the limit of 100–300 MPa normally accepted for contact-metamorphic facies, it may be misleading to refer to this as contact metamorphism *sensu stricto*, even if the intrusions have contributed some heat to the regional geotherm or effectively altered the composition of the metamorphic fluid. Evidence of M2 preserved at distances of 10 km from the nearest known intrusion suggests that regional temperatures were high enough for this to be regarded as an interkinematic extension of the M1 regional metamorphic event, with some perturbation of the regional P-T-X (where X refers to the mole fraction of a mobile component such as CO_2 in the fluid phase) field by the granitoid and charnockitic intrusions.

Retrogression, termed M2 or M3 depending on whether the preceding "contact" metamorphism is regarded as a separate event, is responsible for the present widespread lower amphibolite (Cornell, 1978b; Schultz, 1978) and greenschist (Van Bever Donker, 1980) facies metamorphic overprint in eastern Bushmanland. These metamorphic parageneses may be ascribed either to post-D3 uplift and cooling (Van Bever Donker, 1980) or to the more local effects of water influx (Cornell, 1978b).

Differential uplift along major structural discontinuities (Fig. 8-10) has been cited as the cause of a metamorphic zonation in eastern Bushmanland (Pretorius, 1974) wherein upper amphibolite-facies rocks are present west of the Neusspruit shear belt, lower amphibolite-facies rocks between here and the Landgoed shear belt, and greenschist-facies rocks east of this shear belt in the domain of the Eastern Marginal Zone (Van Zyl, 1979; Van Bever Donker, 1980).

Metamorphic studies in western Bushmanland

have not been extensive, but they indicate that regional metamorphism of lower amphibolite facies took place during D3, followed by a greenschist to lower amphibolite overprint (Moore, 1977; Rozendaal, 1978). At this stage, correlation with the M1 or M3 metamorphic events in eastern Bushmanland is not possible.

D3–D4 periclinal folding. The large dome-and-basin interference structures that dominate maps of Bushmanland are largely the result of broad folding during D3 and D4 superimposed on tighter, older interference patterns. Because of their late formation, the D3 major folds can be traced with a high degree of confidence from similar structures in Namaqualand across Bushmanland. In eastern Bushmanland these folds have an eastward trend, forming the long axes of the canoe-shaped synforms separated by narrow antiforms of large amplitude, which are characteristic of this area (Joubert, 1974a, b; Moore, 1977; Lipson, 1979). The F3 synforms fold flat-lying F2 major folds in the supracrustal rocks and are generally upright, with local vergence to the south; axial-plane cleavage is sporadic (Moore, 1977). The F3 synforms are arranged in a right-hand *en echelon* pattern; this arrangement has been ascribed to D2 (Moore, 1977), D3 (Joubert, 1974a,b), and a younger D3b event (Lipson, 1979). Joubert (1974b) explained the pattern by dextral wrench tectonics (Wilcox *et al*., 1973), for which there is evidence in the form of a major dextral shear belt to the north. However, the sigmoidal nature of the F3 axial traces evident on Joubert's (1973) map suggests another explanation, that the *en echelon* pattern results from oblique cross folding of the east-trending F3 synforms by broad low-amplitude F4 folds trending roughly northwest; no simple shear need be involved (cf. O'Driscoll, 1964). It is therefore significant that F4 folding (mainly as minor kink folds) with this orientation has been reported by Joubert (1974a) and Lipson (1979) in this area, as well as farther east by Stowe (1979).

As the F3 folds are traced into eastern Bushmanland, their axial trends swing from eastward to northeastward (Botha *et al*., 1977; Geringer and Botha, 1977; Stowe, 1979; Van Bever Donker, 1980). The intensity of this folding also decreases eastward. Well-developed periclinal D2–D3 interference structures are present around Kenhardt village (Fig. 8-3) (Joubert, 1974b; Botha *et al*., 1977) and just east of the Neusspruit shear belt (Fig. 8-10) (Van Bever Donker, 1980) and the Cnydas shear belt (Geringer and Botha, 1977). However, near the margins of the Central Zone F3 folds are weakly developed and recognized mainly by the reversed plunges of F2 folds caused by cross folding (Stowe, 1979). That this period of folding may be diapirically enhanced is suggested by the broad open form of granitoid-cored antiforms with steeply dipping flanks passing into tight cuspate synforms (Stowe, 1979).

Younger deformation of minor extent is recorded in the presence of sporadic east-trending conjugate and kink folds (Table 8-3) (Stowe, 1979).

Shear belts and differential uplift. Planar zones of high strain and displacement developed at various stages of the structural evolution of eastern Bushmanland and divided the crust into five major tectonic prisms (Fig. 8-10). Some of the northwest- and north-northwest-striking shear belts appear to have acted as décollement surfaces because they separate zones of open D3 folding from zones of tight D3 folding (Van Bever Donker, 1980). However, most of the shear belts have accommodated strains of two main types: (1) inhomogeneous shortening along northeast axes and (2) northwestward translation of the Central Zone relative to the Eastern Marginal Zone combined with progressively greater uplift toward the southwest (Stowe, 1979; Van Zyl, 1979; Van Bever Donker, 1980). Higher grade parageneses are commonly found on the southwestern side of each shear belt and toward the southwest in general, which suggests that each tectonic prism is tilted to the southwest. Differential uplift along the Neusspruit and Cnydas shear belts (Fig. 8-10) has been estimated at 5–10 km on metamorphic grounds (Van Bever Donker, 1980). Horizontal displacements of up to 14 km for some shear belts have been estimated by analysis of shear strain (Van Zyl, 1979).

The shear belts have clearly had a long and complex history. Van Bever Donker (1980) deduced that the Neusspruit shear belt was mainly active during D2 but that motion continued during D3. The abrupt termination of the distinctive precratonic clastic cover against the shear belt and its noncontinuation on the western side suggest that the Neusspruit shear belt may mark the western edge of an ancient fracture-bounded basin (Vajner and Jackson, 1974; Van Bever Donker, 1980). Displacement in the form of faulting continued along most of the shear belts. Some of the faults controlled the formation and subsidence of the Koras basins 1200 Ma ago (Fig. 8-10). This date provides an upper age limit for folding, shearing, and granitoid intrusion in this part of the Central Zone and adjoining Eastern Marginal Zone (Vajner, 1975; Botha *et al.*, 1977; Stowe, 1979). Clastic sediments and volcanic rocks of the Koras Group accumulated in grabens with contemporaneous gentle synclinal folding (Du Toit, 1965; Grobler *et al.*, 1977).

The Central Zone: Namibia

The Central Zone comprises three large "inliers" and numerous smaller exposures extending some 580 km through southern Namibia and separated by Nama and Karoo cratonic cover. Because of the outcrop pattern, it is convenient to divide this vast terrane into four areas (Fig. 8-3): (1) western Lüderitz District, between Lüderitz port and 16° 00′ E; (2) eastern Lüderitz District, around Aus village; (3) western Karasburg District, around Grünau and Ai-ais villages in the Karasburg and Bethanien Districts; and (4) eastern Karasburg District, around Warmbad and Onseepkans villages.

The northwestern extension of the Central Zone is terminated by tectonic boundaries of three different kinds in Lüderitz District.

(1) In the north volcaniclastic and volcanic strata and granitoids of the 1300–1000 Ma Sinclair Group rest nonconformably on or intrude late-syntectonic igneous rocks of the Central Zone known as the Naisib River Igneous Suite (Von Brunn, 1969; Jackson, 1975, 1976; Watters, 1977). This most southerly extent of abundant Sinclair igneous rocks serves as an obvious boundary between the Namaqua Province and the Namibia Province to the north (Blignault *et al.*, 1974). However, inliers of pre-Sinclair igneous and metamorphic rocks, collectively termed the Kumbis Complex (Watters, 1974; Blignault *et al.*, 1974), are present at least 120 km farther north and it is quite possible that they represent remnants of the former extension of Namaqua rocks in pre-Sinclair times (Jackson, 1975). The Namibia Province comprises crystalline basement varying in age from the 950–1350-Ma Rehoboth Arc (including the Sinclair Group, Fig. 1-4) to presently known age limits of about 2000 Ma (Chapter 9, Section 9.4). This basement is overlain by the Late Precambrian Nama and Phanerozoic Karoo cratonic cover. The exposed boundaries of the province are defined in Table 1-1. Because this structural province forms the core of Namibia, the name Namibia Province is considered to be more appropriate than the name Maltahöhe Province proposed by Blignault *et al.* (1974) for a broadly similar area named after an administrative district centered on the small town of Maltahöhe. Moreover Blignault *et al.* (1974) restricted their definition to include only pre-Sinclair rocks, a definition that excluded most of the crust exposed in this structural province.

(2) The western end of the Central Zone is marked by a zone of structural and metamorphic reworking associated with Late Precambrian orogeny (Chapter 9). The extent of this zone has been delineated by structural mapping (Kröner and Jackson, 1974) and its sinuous eastern limit is shown along the coast in Figures 8-3 and 8-11. Clastic sediments correlated with the Late Precambrian Gariep Group overlie the reworked basement. Discontinuous linear zones of north–south reworking extend at least 50 km into the Central Zone but become narrower and less common eastward (Jackson, 1974b).

(3) The southern margin of the Central Zone in Lüderitz District is marked by a northwest-trending shear belt separating rocks of the Western Zone to the southwest from a granitoid–gneiss complex called the Gaidab massif (Fig. 8-12) near Witputs (McMillan, 1968; Blignault *et al.*, 1974). McMillan (1968) suggested that the Gaidab massif formed from parent rocks similar to the Orange River Group pretectonic cover and Vioolsdrif granitoids in the Western Zone. Acid to intermediate volcanic rocks intruded by granitoids in the Western Zone around the Aurus Mountains were transformed by strain and metamorphism into gneisses of the Central Zone resembling those farther north around Aus (Fig. 8-11) (Kröner, 1978b).

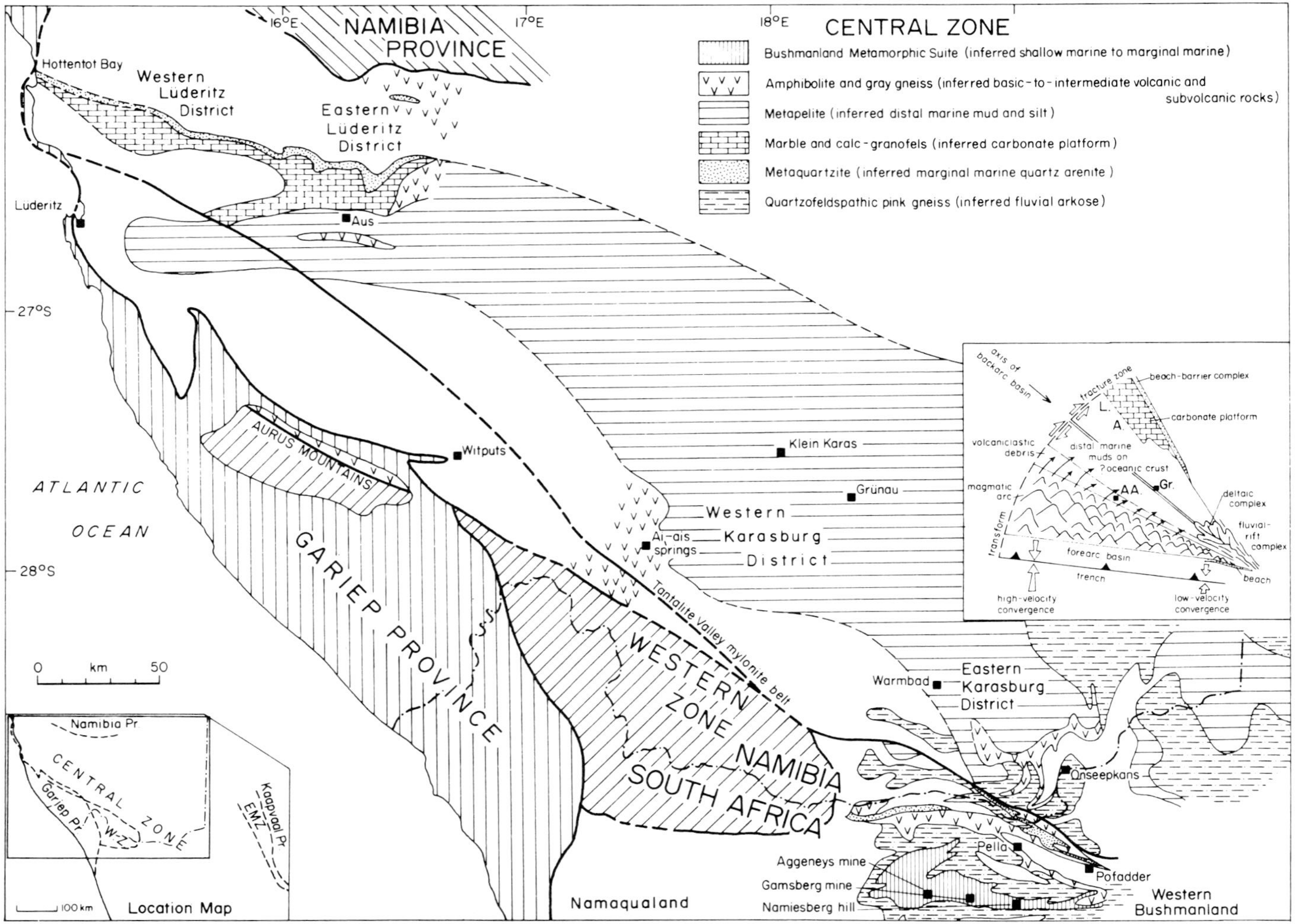

Figure 8-11. Nonpalinspastic lithofacies map of pretectonic cover sequences in the Namibian Central Zone (Namaqua Province). The map does not imply time equivalence for these rocks. Inset shows one possible geotectonic interpretation for the environments of formation for the Namibian cover sequences in a backarc basin and intracontinental rift. [Compiled from (from north to south): Kröner and Jackson (1974); McDaid (1978); Jackson (1976); Blignault (1977); G.J. Beukes (1973); Geringer and Botha (1977); Toogood (1976); and Joubert (1976).]

Table 8–7. Precambrian Geologic History for the Central Zone (Namaqua Province) in Namibia[a]

Age (Ma)	West Lüderitz (McDaid, 1978)	Age (Ma)	East Lüderitz (Jackson, 1976)	Age (Ma)	West Karasburg (Blignault, 1977)	Age (Ma)	East Karasburg (Toogood, 1976)
>2100	Deposition of pretectonic cover (Garub)	>2100	Deposition of pretectonic cover (Garub)	?2000–1900	Deposition of pretectonic cover (? Orange River) and intrusion by granitoid batholith (? Vioolsdrif)		Deposition of pretectonic cover (Onseepkans)
			Intrusion of sheetlike augen granitoid gneiss (Tsirub), metatexis				
	D1: Regional fabric in cover, metatexis		D1: Regional fabric in cover, isoclinal to tight folding, amphibolite metamorphism, metatexis; intrusion of ultrabasic plugs (Magnettafelberg)	? 1800	D1: Regional fabric in cover, isoclinal folding, amphibolite metamorphism		D1: Fabric, granulite metamorphism and intrusion of gabbroid stocks and dikes (Stolzenfels, Kum Kum, Jericho)
	? Intrusion of sheetlike augen granitoid gneiss (Tsirub)				Intrusion of sheetlike augen granitoid gneiss		Intrusion of sheetlike augen granitoid gneiss (Beenbreek)
>2000	D2: Isoclinal to tight folding		D2: Isoclinal to close folding, local refoliation, amphibolite metamorphism, metatexis	?<1800	D2: Isoclinal to close folding, local refoliation, amphibolite metamorphism		D2: Regional fabric in cover and granitoid gneiss, amphibolite metamorphism
2040[b]	Major intrusion of gabbroid (Konip) and granitoid batholiths (Kunguib, Tschaukaib), amphibolite metamorphism, major metatexis		Major intrusion of charnockitic (Jakkalskop) and alaskitic batholiths (Aus, Kubub, Anib), granulite and amphibolite metamorphism, major metatexis and diatexis		Major intrusion of charnockitic (Signalberg) and alaskitic batholiths (Grabwasser) ? intrusion of gabbroid batholith (Ai-ais), granulite and amphibolite metamorphism		Major intrusion of granitoid batholiths (Naros, Nautsis), amphibolite metamorphism
			D3: Isoclinal to open folding, regional fabric		D3: Close upright N–S folding in N,		D3: NW–SE isoclinal to close folding, local

Table 8–7. Precambrian Geologic History for the Central Zone (Namaqua Province) in Namibia[a]

Age (Ma)	West Lüderitz (McDaid, 1978)	Age (Ma)	East Lüderitz (Jackson, 1976)	Age (Ma)	West Karasburg (Blignault, 1977)	Age (Ma)	East Karasburg (Toogood, 1976)
			in granitoid gneisses, local refoliation of cover, amphibolite metamorphism		SW-verging folding in S		refoliation, amphibolite metamorphism, metatexis
		>1300	Intrusion of intermediate stocks into intermediate volcanic rocks (Naisib R.)				
>1130[b]	Deposition of late-syntectonic volcano-sedimentary cover (Konipberg)	1300[c] to 1200	Intrusion of granitoid batholiths into acid–basic volcaniclastic cover (Sinclair)				
	Intrusion of granite batholith (Glockenberg) D3: Tight E–W folding, fabric in cover						
	D4: Open NW–SE and N–S folding		D4: NE–SW open folding		D4: E–W folding, local refoliation		D4: NE–SW to NW–SE upright folding
	D5: Gentle E–W folding		D5: NW–SE open folding, retrogressive metamorphism		D5: NE–SW and NW–SE folding, retrogressive metamorphism		D5: Crescentic NNE–SSW antiformal folding strongly overturned to NW
	D4/5: NE–SW dextral shearing (Tantalite Valley)		D6: NE–SW dextral shearing (Tantalite Valley, Excelsior)		D6: NE–SW dextral shearing (Tantalite Valley, Lord Hill)		D5/6: WNW–ESE dextral shearing (Tantalite Valley) D6: WNW–ESE upright folding
900	Miogeosynclinal sedimentation (Gariep)	900	NNE postcratonic basic dike swarm (? Gannakouriep)	900	NNE postcratonic basic dike swarm (Gannakouriep)		
600	Shelf sedimentation (Nama)	600	Shelf sedimentation (Nama)	600	Shelf sedimentation (Nama)	600	Shelf sedimentation (Nama)

[a]Structural and lithologic correlations between areas are tentative. On the basis of evidence presented, deposition and early deformation of pretectonic cover took place at least 200 Ma earlier in Lüderitz District than in Karasburg District.

[b]Burger (1978).

[c]Burger (in Watters, 1977), Kröner (1976, 1978b).

In both Lüderitz and Karasburg Districts, field relations and limited geochronologic data suggest that the Central Zone evolved from ≥2000 Ma to roughly 1000 Ma ago (Table 8-7).

Pretectonic cover. In all areas of the Central Zone in Namibia the oldest rocks recognized form a heterogeneous sequence of layered gneisses representing pretectonic cover. Despite intensive search, no basement to this cover has been found. Extensive tracts of layered gray biotite gneisses are intercalated with the cover rocks and, although it has been suggested that they may have formed a crystalline basement to the cover (e.g., Jackson, 1976), no evidence of unconformity has been recognized and they must therefore be provisionally regarded as forming part of the same supracrustal pile as the cover (G.J. Beukes, 1973; Jackson, 1976; Toogood, 1976; Blignault, 1977).

There is little consistency in stratigraphic or structural sequences of the cover rocks even within small areas. It has been evident for some time that large-scale lateral variations exist and attempts have been made to explain these with paleofacies models (Jackson 1974a, 1976; Blignault, 1977; McDaid, 1978). Figure 8-11 is a nonpalinspastic map of distinctive pretectonic cover units in southern Namibia. The paleogeography is dominated by an elongated area consisting of metapelite gneisses, which trends northwest, parallel to the regional trend of the Namaqua Province as a whole. Over much of its length the northern boundary is not exposed, but it is reasonable to regard the metapelites as originating in the deep-water part of a basin supplied with terrigenous clastic detritus. The sedimentary nature of the metapelite gneisses around Aus village is confirmed by intercalations of metaquartzite and graphite schist, forming part of the Garub Metamorphic Suite (Jackson, 1974a, 1976). The margin of the basin north of Aus village is marked by coarsely crystalline Garub dolomite marbles and metaquartzites. This sequence contains the most extensive carbonate rocks in the Namaqua Province and can be interpreted as a shelf association (Fig. 8-8B) (Jackson, 1976). The hypothetical Aus-Lüderitz shelf broadens westward toward Hottentot Bay (Jackson, 1976; McDaid, 1978) and emerges from beneath the Namib Sand Sea at the coast as the Aurus Metamorphic Suite (Kröner and Jackson, 1974). To the north and east of the Aus-Lüderitz shelf, amphibolite and basic granolite may constitute metabasic volcanic rocks; fragmental textures in the granolites suggest that these rocks include metaagglomerates (Jackson, 1976). However, the close association of some of the Garub metabasites with marble and calc–silicate rocks suggests that at least some of these are paraamphibolites derived from marl.

In the Ai-ais area of western Karasburg District a 30- to 50-km-wide tract of gray biotite–hornblende gneisses separates the igneous rocks of the Western Zone from the metapelites to the north (Fig. 8-11). The gray gneisses contain intercalations of metaquartzite and more-basic hornblende gneiss with layers of deformed volcanic breccia containing clasts of porphyry. Blignault (1977) inferred that these metamorphites represent volcanic and plutonic rocks of intermediate to basic composition and that their parent rocks were correlatives of those in the Western Zone, a view echoed by Kröner (1978b) with respect to the northwest extension of these rocks north of the Aurus Mountains (Fig. 8-11). Although there seems little doubt as to the nature of the parent rocks—a magmatic arc—their age and correlation unfortunately remain a matter of speculation at this point. The gray gneisses pass conformably upward into metapelites northeast of Ai-ais springs (Blignault, 1977).

The metapelite zone terminates to the southeast against a narrow zone of sheetlike pretectonic hornblende gneiss in eastern Karasburg District (Fig. 8-11). The tonalitic composition of the hornblende gneiss and intercalated basic and calcareous layers suggest that they represent intermediate igneous rocks infolded with other members of the pretectonic cover suite, called the Onseepkans Metamorphic Suite by Toogood (1976). Similar rocks have been reported south of the Namibian border (Fig. 8-11) (Colliston, 1979). East of the hornblende gneiss is a broad zone of pink quartzofeldspathic gneiss and further metapelite. Pink gneiss, which is so common in eastern Namaqualand and Bushmanland, is confined to this southeastern end of the Namibian Central Zone. Blignault *et al*. (1974) regarded the Namibian pink gneisses as metapsammites and this conclusion is supported by Toogood's (1976) discovery of cross-bedding and intercalated metaquartzite in the pink gneiss north of Onseepkans village. No intrusive contact has been reported, but the rock is easily confused with deformed pink aplite associated with an early-syntectonic granitoid (Toogood, 1976). Some of the pink gneiss in western Bushmanland is known to be intrusive (Table 8-6), so the actual extent of this rock type is probably less than is shown in Figure 8-11. The high content of feldspar in the metapsammitic pink gneiss, some of which can be explained by high-grade metamorphism of potassic clays and micas, suggests that the parent rocks were coarse-bedload stream deposits of continental provenance. These deposits are concentrated at the southeastern end of the basin and are symmetrically disposed around the basin axis (Fig. 8-11). This distribution suggests that the pink gneisses mark the site of an ancient continental rift trending northwest via a delta complex (around Onseepkans) into a paleogulf filled with terrigenous mud and silt (Fig. 8-11, inset). As is the case in the present Gulf of Mexico, several kilometers thickness of deep-water clastics may blanket oceanic crust formed between the rifting plates. The Aus-Lüderitz carbonate platform on the northwest margin of the paleogulf may be analogous to the modern Yucatan shelf on the southern margin of the Gulf of Mexico (Fig. 8-11, inset). The southern margin of the paleogulf in Namibia appears to have been dominated by the calc–alkaline magmatic chain of the Western Zone (Figs. 8-6 and 8-11). Shedding of volcaniclastic debris from the rising mountain chain of the Western Zone directly into the paleogulf to the north would effectively prohibit the growth of a carbonate platform equivalent to that on the opposite margin of the paleogulf (Fig. 8-11, inset). The easterly extension of this magmatic arc is indicated by hornblende gneiss and amphibolite just south of the Namibian border. The massive range of metaquartzites just north of Pella village indicates reworking under marginal marine conditions. This suggests that a separate basin containing the Bushmanland Metamorphic Suite, itself of shallow-marine origin, lies to the south of the inferred fluvial-rift and deltaic complex in southeastern Namibia.

The tectonic environments proposed (Fig. 8-11 inset) to account for lithofacies variations in the pretectonic cover rocks should not be interpreted as coexisting contemporaneously because these environments may have been temporally separated by as much as 800 Ma. The oldest cover rocks known in the Namaqua Province are those of the Garub Metamorphic Suite (Jackson, 1976) partly comprising the carbonate–quartzite shelf association and pelitic rocks in Lüderitz District. Rocks of the Garub Metamorphic Suite in western Lüderitz are intruded by the Kunguib hornblende granodiorite, which cuts F2 folds in the country rock and contains a fabric correlated with D3 (McDaid, 1978). This granodiorite has a minimum radiometric age of 2040 ± 40 Ma (Burger, 1978; U-Pb zircon two-point discordiagram). The Garub cover sequence had therefore accumulated and undergone two deformation events with attendant metamorphism and partial melting by 2040 Ma ago. Considering the time interval necessary for slow accumulation of the carbonate platform and burial to sufficient depths to initiate high-grade metamorphism, an estimated primary age of at least 2100–2200 is likely. This time span is similar to that estimated for the Griqualand West Supergroup and it is not difficult to imagine the Garub dolomites accumulating on the Aus-Lüderitz shelf coevally with those of the epeiric Griqualand West Supergroup on the Kaapvaal shelf (Chapter 5). Small amounts of Garub metamorphosed iron-formation (Jackson, 1976; McDaid, 1978) further support this correlation.

If the metapelites in Karasburg District are regarded as coeval with those in Lüderitz District, the metaigneous gray gneisses that underlie them in western Karasburg District are likely to be of similar age or older than 2100–2200 Ma. However, with some justification Blignault (1977) regarded the gray gneisses as having formed from volcanic rocks and granitoid intrusions dated at 2000 Ma and 1900 Ma, respectively (Reid, 1979b, c). Accepting all the geochronologic results, it can therefore be concluded that either sedimentation and tectonic activity started at least 150 Ma earlier in Lüderitz District, or that the igneous precursors of the gray gneisses in Karasburg District are older than those of the adjacent Western Zone (Fig. 8-11).

The subsequent tectonic history of the cover sequences in different parts of the Namibian Central Zone is synthesized in Table 8-7.

Early-syntectonic intrusions: augen granitoid gneisses, charnockites, and gabbroids. The oldest granitoids in all areas studied are augen granitoid gneisses variously referred to (from west to east) as augen gneiss (Greenman, 1966), Tsirub gneiss (Jackson, 1976; McDaid, 1978), granodioritic augen gneiss (Blignault, 1977), Eendoorn granite (G.J. Beukes, 1973), and Beenbreek megacrystic granite and Grass River augen gneiss (Toogood, 1976). These biotite and garnet-bearing rocks are generally described as being of granodioritic or tonalitic composition, although G.J. Beukes (1973) reported compositions of quartz monzonite. Their most significant features are the sheetlike form of the intrusions (evident from their outcrop patterns in Fig. 8-12) and the megacrysts of plagioclase or, less commonly, K-feldspar. These granitoids were locally converted by high strains into layered or rodded gneisses. The stratiform intrusions are largely concordant with the surrounding infolded paragneisses but contain xenoliths of the latter, as is the case with the closely similar Nababeep granitoid gneiss in Namaqualand. Pregranitoid fabrics in these xenoliths indicate that in most areas (apart from eastern Lüderitz District) the augen granitoid gneisses intruded after D1 (Toogood, 1976; Blignault, 1977; McDaid, 1978). In all cases the ubiquitous fabric within the granitoid gneisses has been correlated with D2. Augen granitoid gneiss in western Lüderitz District has a minimum age of 1780 ± 35 Ma (Burger, 1978; U-Pb zircon discordiagram). Even-grained strongly foliated tonalitic to granodioritic gneisses with similar structural relations are abundant at Hottentot Bay and Lüderitz (Kröner and Jackson, 1974). These granitoid gneisses intrude the pretectonic Aurus cover sequence and an early-syntectonic metagabbro and contain amphibolite dikes. The gneisses have a minimum age of 1755 Ma (Pb-Pb zircon discordant age by A.J. Burger, in Kröner, 1975a).

Figure 8-12 shows the distribution and large size of charnockitic intrusions in eastern Lüderitz District and Karasburg District. All these intrusions contain orthopyroxene and their compositions range from charnockite *sensu stricto* to enderbite (granitoids of equivalent composition being granite to tonalite) (Jackson, 1976, 1979b; Toogood, 1976; Blignault, 1977). The rocks are weakly but pervasively foliated with a fabric ascribed to D2 (Blignault, 1977) and D3 (Jackson, 1976). Their textures have a mixture of igneous and metamorphic characteristics but their magmatic origin is confirmed by the presence of granulite-facies xenoliths of pretectonic cover. The charnockites are intruded by mid-syntectonic alaskitic granite gneisses in western Karasburg District, but there is no evidence for a significant difference in age between the charnockites and alaskites (Jackson, 1976; Blignault, 1977).

The distribution and wide range in size of gabbroid intrusive suites is shown in Figure 8-12. The intrusions have the form of stocks or batholiths and are almost entirely restricted to areas north of the Tantalite Valley shear belt, a distribution possibly resulting from uplift of deeper crustal levels north of the shear belt (Jackson, 1976, 1979b; Toogood, 1976); this hypothesis is based on a dramatic increase in regional metamorphic grade northward across the mylonitic core. Another explanation is that at least two of the gabbroid intrusions (Tantalite Valley and Kum Kum) were localized by magma tapping along the shear belt (G.J. Beukes, 1973; Kartun and Moore, 1979). However, this hypothesis has difficulty explaining the existence of a spectrum of lithologically and structurally similar intrusions up to 90 km away from the mylonite core and their striking absence immediately south of this core.

The compositions of these intrusions are generally noritic and, although igneous layering is common, they have all undergone partial or complete amphibolitization during subsequent regional metamorphism (Fig. 8-8C) (Jackson, 1979b). Only the Tantalite Valley and Kum Kum Intrusive Suites (Fig. 8-12) have been described in detail (Moore, 1975; Kartun and Moore, 1979). These intrusions show a history of episodic magma injection interspersed with deformational events. There is no compelling evidence that the separate intrusive suites tapped common magmas or were necessarily coeval and this diversity is reflected in different structural relations (Table 8-7).

Midsyntectonic granitoid intrusions. All these intrusions are foliated to varying degrees and lack chilled margins. They can be classified into three groups, listed in order of decreasing age.

(1) Alaskitic, commonly megacrystic, quartz monzonites and granites, containing only a few modal percent of biotite and garnet, form enormous batholiths in the core of the Central Zone (e.g., Aus, Kubub, and Anib granite gneisses of Jackson, 1976; Grabwasser megacrystic granite of Blignault, 1977). There is abundant field evidence that these granitoids owe their origin to partial melting of pelitic and semipelitic pretectonic cover rocks (Fig. 8-8D) (Jackson, 1976).
(2) Even-grained, gray granitic to granodioritic foliated granitoids form transgressive bodies of various size, the oldest known being the 2040-Ma Kunguib hornblende granodiorite in western Lüderitz District (McDaid, 1978).

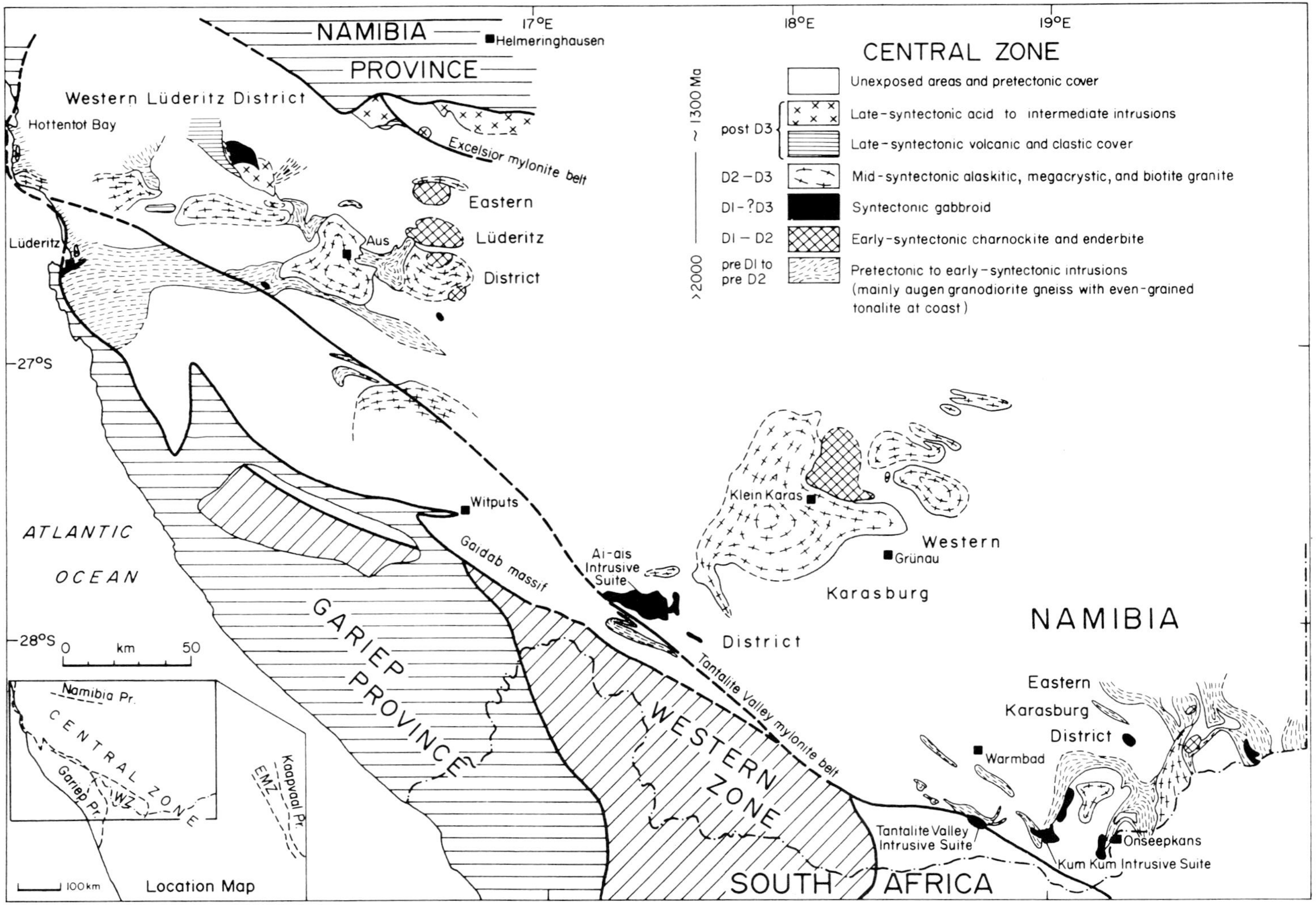

Figure 8-12. Syntectonic intrusions and a cover sequence formed during the life span of the Central Zone (Namaqua Province) in Namibia. Tectonic categories of intrusions are based on their time relations with structural fabrics of known relative age. Absolute ages are likely to show considerable variation and overlap. [Compiled from (from west to east): Kröner and Jackson (1974); Greenman (1966); McDaid (1978); Jackson (1976); Blignault (1977); G.J. Beukes (1973); Toogood (1976).]

Stocks and dikes of biotite granite gneiss intrude the alaskitic granite gneisses in eastern Lüderitz District (Jackson, 1976). The large Naros batholith of gray granite gneiss represents the youngest granitoid intrusion in eastern Karasburg District (Toogood, 1976). The large body of pink Nautsis "granite" nearby may be of equivalent age although its intrusive nature is in doubt because no intrusive contacts have been found.

(3) Dark-gray, porphyroblastic granitoid gneisses have intruded deformed and migmatized early-syntectonic tonalitic gneisses at Hottentot Bay. The former rocks range in composition from granite to diorite and are foliated to varying degrees (Kröner and Jackson, 1974.

Early deformation and metamorphism. Relations between the first three deformation events (referred to as early deformation), metamorphism, and intrusive events are shown in Table 8-7; a summary of deformation styles and tentative correlation of different schemes is also given. It should be noted that where syntectonic granitoid gneisses are abundant, the D2 or D3 fabrics imprinted on them can be described as the regional fabric, whereas where pretectonic gneisses predominate, the D1 fabric preserved in them forms the regional fabric. Both old and young major structures show considerable diversity in orientation and degree of fabric development, and variations in minor fold style are controlled by lithology rather than fold generation (Toogood, 1976).

The gross structure of the Central Zone is dominated by the major periclinal structures that formed during the late episode of deformation with the result that there is no well-defined regional "grain" to the zone (Fig. 8-12). However, along the southern margin, between the Western Zone and the Tantalite Valley shear belt in Karasburg District, a zone of uniformly northeast-dipping gray gneisses contains D1 to D3 structures, which are reported to be roughly coplanar and coaxial (Blignault, 1977). It has been proposed that this zone represents a broad regime of ductile thrusting toward the southwest in which strong flattening during D1 to D3 rotated and tightened early structures and that this thrust regime may have extended much farther northward (Blignault, 1977). Alternatively, the straightening of this marginal zone, as well as the cataclastic imprint in the Western Zone, may be caused by predominantly dextral movement along the late-formed Tantalite Valley shear belt (Toogood, 1975, 1976).

Finite-strain determinations of the early deformation events are restricted to two-dimensional estimates by Toogood (1976), who calculated D3 shortening from a wide variety of minor folds; he estimated a mean value of 67 percent in both areas east and west of Onseepkans village.

Retrogression of granulite-facies cover rocks to upper amphibolite facies led Jackson (1976) and Toogood (1976) to propose two high-grade metamorphic imprints separated in time by intrusion of the midsyntectonic granitoid gneisses described in the preceding section. The maximum P-T conditions for the granulite metamorphism have been estimated at P_t = 800–900 MPa and T = 800–860°C (Toogood, 1976) and P_t = 500–700 MPa, P_{H_2O} = 200 MPa, and T = 800–860° (Jackson, 1976). Two zones of granulite metamorphism with a combined width of 60 km are preserved in eastern Lüderitz District. The charnockitic intrusions are entirely restricted to these zones. The presence of carbonic fluid inclusions in them and the granolite country rocks suggests that dilution of the hydrous fluid phase by juvenile CO_2 may have been responsible for raising the metamorphic grade from upper amphibolite to granulite facies (Jackson, 1976).

The role of the gabbroid intrusions in the early granulite metamorphism is debatable. The two processes appear to have been intimately related in time and space because of the structural evidence for early intrusion of the charnockites and gabbroids contemporaneously with formation of basic granolites (Jackson, 1976, 1979b; Toogood, 1976). Tectonite fabrics in the granolites, together with their development tens of kilometers away from the intrusions, indicate that the granolites are the product of regional metamorphism. Another point of view is that the gabbroids are relatively young intrusions intruded during the amphibolite metamorphism that followed granulite metamorphism (Kartun and Moore, 1979; Moore *et al.*, 1979). The latter authors believed that Toogood's foliated granolites on the margins of the Kum Kum Intrusive Suite are a mix of intrusive gabbroic magma and partially melted granitoid country rock; they interpreted the strong fabric as a typical primary cumulus texture (Kartun and Moore, 1979). Nevertheless their geothermometry indicates that the pyroxenes equilibrated at approximately 800°C, a temperature indicative of granulite metamorphism rather than crystallizing basic magma (Hewins, 1975).

Moore *et al.* (1979) proposed the existence of a contact aureole around the Tantalite Valley Intrusive Suite with an estimated equilibration temperature of 800°C, in foliated orthopyroxene "hornfels" at the contact, to an estimated 650°C at 500 MPa a few meters away. The status of this aureole is questionable on two grounds: first, there is no consistent temperature gradient across the proposed aureole; second, the aureole is anomalously narrow, generally less than 20 m wide in an intrusion some 400 times as long. This has been explained by proposing extremely slow heat transfer because of dryness of the country rocks (Moore *et al.*, 1979). However, this is unconvincing because both the basic and pelitic "hornfels" contain abundant hydrous minerals, indicating that these rocks were far from anhydrous during metamorphism. It is clear from the above discussion that relations between the gabbroids and the regional metamorphism are far from resolved. The Tantalite Valley intrusion poses particularly daunting problems because it is situated in the middle of the Tantalite Valley mylonite belt. 8 km wide at this point (Figs. 8-12 and 8-13) (Moore, 1975; Toogood, 1976). The discontinuous narrow skin of metamorphic rocks around the intrusive suite may well be the remains of country rocks that have undergone regional polymetamorphism or intrusive rocks that became mylonitized, rather than constituting a contact aureole.

A later period of protracted and possibly continuous amphibolite-facies metamorphism took place from D2 through D3 in both Lüderitz and Karasburg Districts and overprinted the granulite zones. In eastern Lüderitz District the site of the metamorphic peak shifted some 30 km westward so that a thermal dome centered on the locality of Aus village, where temperatures attained 780–800°C around Aus and declined to 600–650°C north and south of the Excelsior and Tantalite Valley shear belts (Jackson, 1976). Metamorphism was roughly isobaric between estimated limits of 400–600 MPa, corresponding to a high geothermal gradient of 30–50°C/km. These estimates have been quantitatively refined to $T = 670° \pm 70°C$ and $P_t = 500$ MPa for Garub metapelites in western Lüderitz District (Waters and McDaid, 1979). Maximum P-T conditions for amphibolite metamorphism have been estimated at 740°C and 600 MPa in western Karasburg District (Blignault, 1977), and 650–700°C and 500–700 MPa north of the Tantalite Valley shear belt in the eastern Karasburg District, and 650°C and 350–500 MPa south of the shear belt (Toogood, 1976). A low-pressure facies series has been suggested throughout the Namibian Central Zone (G.J. Beukes, 1973; Jackson, 1976; Toogood, 1976; Blignault, 1977).

During amphibolite metamorphism, regional migmatization took place over most of the Central Zone. Where upper amphibolite grades were attained, *in situ* partial melting formed layered migmatites and granitoid bodies on all scales. In areas that only attained lower amphibolite conditions, as in the coastal zone of western Lüderitz District, regional migmatization took the form of injections of melts, presumably from hotter levels below (Kröner and Jackson, 1974). In eastern Lüderitz District the thermal dome is defined both by metamorphic grades (isograds) and by migmatitic grades (Jackson, 1976). The core of the dome is occupied by homogeneous biotite-bearing granitoids, whose mineralogy together with the regional metamorphic grade of the country rocks suggests an origin by diatexis (high-grade melting that includes mafic minerals; Mehnert, 1968). Surrounding these rocks is a zone characterized by less homogenized schlieric diatexites; this zone is in turn surrounded by metatexites (migmatites with *in situ* partial melting of only quartz and feldspar).

Late-syntectonic cover and related intrusions. Cover sequences that accumulated during the Namaquan tectonic episode ranging from >2000 Ma to about 1000 Ma in the northwest to <1800 Ma to about 1200 Ma in the east, are, with one exception, unknown in the entire Namaqua Province. This absence is most likely the result of massive uplift during the close of orogenic activity.

A thick sequence of quartz arenites, with well-preserved cross-bedding, and subordinate basic volcanic rocks lying unconformably on the pretectonic cover in western Lüderitz District was discovered in 1975 by M.P.A. Jackson and J.M. McDaid (Fig. 8-12). Because of lithologic similarities, they tentatively correlated this succession with the Late Proterozoic Spencer Bay Formation 75 km to the northwest at the coast in the Damara Province (Kröner and Jackson, 1974). Subsequent work by McDaid (1976, 1978) has shown that this cover, known as the Konipberg sequence, is preserved as two outliers (Fig. 8-12)

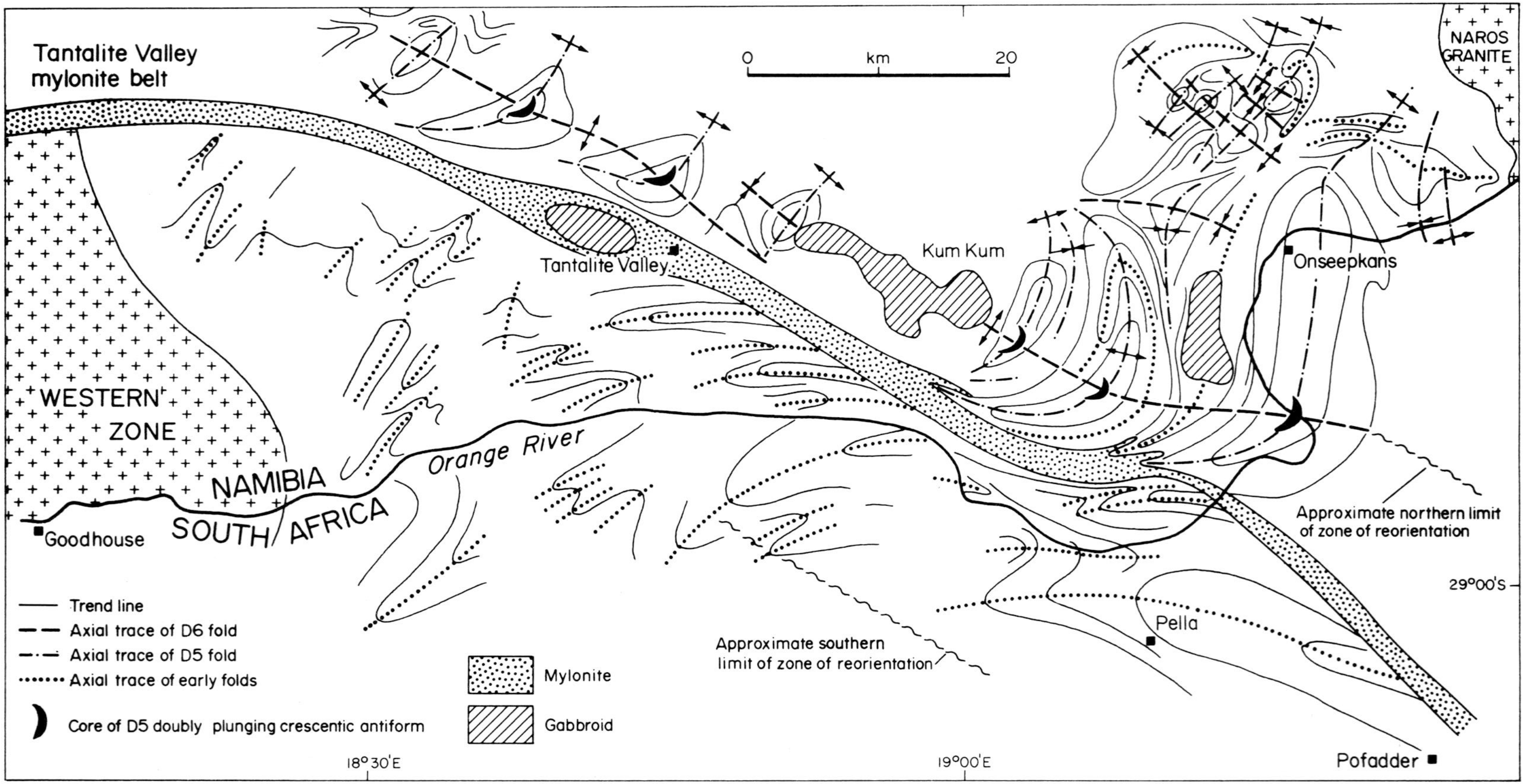

Figure 8-13. The Tantalite Valley shear belt in eastern Karasburg District and contiguous Bushmanland showing the mylonite core and zone of reorientation on either side. The doubly plunging crescentic antiforms are restricted to the northern side of the shear belt. [After Toogood (1976), based on maps by Toogood (1976), G.J. Beukes (1973), Joubert (1973), and the South African Geological Survey.]

and consists of basal conglomerate overlain by feldspathic quartzite and a large proportion of acid to basic volcanic and volcaniclastic rocks metamorphosed to lower amphibolite grade. The abundance of extrusive rocks, together with a minimum radiometric age on the acid volcanic and hypabyssal rocks of 1130–1180 Ma (Burger, 1978; U-Pb zircon) suggest that this sequence most likely can be correlated with the Mid-Proterozoic Barby Formation of the Sinclair Group (Watters, 1977) some 50 km to the northeast in the Namibia Province (McDaid, 1978). The late-syntectonic age of the Konipberg sequence is shown by the fact that McDaid's later deformation events (D3–D5 in Table 8-7) have affected this cover.

Evidence for pre-Sinclair (>1250 Ma) volcanism within the Namaqua Province is preserved near the Excelsior shear belt in northeastern Lüderitz District near the boundary with the Namibia Province (Fig. 8-12) (Jackson, 1975, 1976). Stocks ranging in composition from leucocratic granodiorite to mesocratic diorite contain inclusions of basic lava, dacite porphyry, and rhyodacite lava and jointly constitute the Naisib River Igneous Suite. The oldest unit of the Sinclair Group rests nonconformably (Von Brunn, 1979) on granodiorite containing xenoliths of greenschist-facies basic lava of the Naisib River Igneous Suite (Jackson, 1976). This suite has been affected only by late-stage mylonitization along the Excelsior shear belt and by shearing at some contacts. Late-syntectonic intrusions with weak or nonexistent fabrics in western Lüderitz District are represented by the Glockenberg granite, which intrudes the Konipberg cover sequence (Fig. 8-12) (McDaid, 1978) and stocks of three different leucocratic red or gray granite at Hottentot Bay (Fig. 8-8C) (Kröner and Jackson, 1974).

The Tantalite Valley shear belt. This structure represents the largest and most intensively studied shear belt in southern Africa. Various names have been assigned to different parts of its 510 km length (e.g., Pofadder, Pilgrim, Kanabeam, Kuckaus, and Haalenberg). We use the name Tantalite Valley shear belt after the type locality in eastern Karasburg District (Figs. 8-12 and 8-13) (G.J. Beukes, 1973; Blignault *et al.*, 1974). The shear belt extends from the vicinity of Pofadder village, where it terminates in a series of splays (Joubert, 1974b), to a point 30 km northeast of Lüderitz, where its further extension is obscured by the Namib Sand Sea (Jackson, 1974a, 1976; McDaid, 1978). Veins resembling pseudotachylite at the coast between Hottentot Bay and Lüderitz (Fig. 8-12) may mark the area where the shear belt crosses the coastline (Kröner and Jackson, 1974). The trend of the shear belt can be interpolated across areas of Nama cratonic cover by the trace of post-Nama reactivated faults visible on Landsat imagery.

The shear belt trends roughly northwest and contains a core known as the Tantalite Valley mylonite belt 0.5–8 km wide and dipping 50–80° northeast; the mylonitic core is flanked by a zone of reorientation in which older structures are tightened and drawn toward parallelism with the mylonitic core (Fig. 8-13). The width of the zone of reorientation varies from 20 to 40 km at its peak in eastern Karasburg District (Toogood, 1976) to an ill-defined but much narrower width at its northwestern end (Jackson, 1976; McDaid, 1978). Discontinuous mylonite belts persist within the zone of reorientation. The shear belt marks a zone of predominantly dextral displacement under metamorphic conditions ranging from upper amphibolite (Toogood, 1976) to greenschist facies (Jackson, 1976; Blignault, 1977). The shear belt is likely to have evolved during the final two or three deformational events (Joubert, 1974b; Toogood, 1976; McDaid, 1978), as proposed for similar shear belts in eastern Bushmanland. The mylonites are cut by posttectonic basic dike swarms correlated with the Gannakouriep Suite (Jackson, 1976; Blignault, 1977), which is thought to be 900 Ma old (Chapter 9). Reactivation of movement along this strain-softened zone is indicated by dextrally offset basic dikes, post-Nama down-to-north normal faults, and Recent fault scarps (Jackson, 1976; Blignault, 1977; McDaid, 1978).

In the zone of reorientation north of the mylonite core is a series of crescentic doubly plunging antiforms (Fig. 8-13). These D5 elongated domes are only found north of the mylonite core and they decrease in tightness away from the mylonites. The antiforms are doubly plunging because of cross folding by a D6 antiform parallel to the mylonite belt. The D5 and D6 folds appear to have formed during shearing as follows (Toogood, 1976). Initial simple shearing of layered Onseepkans gneisses formed a series of D5 folds oriented at 45° to the direction of shearing in a manner typical of wrench-fault tectonics (Moody and Hill, 1956; Wilcox *et al.*, 1973) and these folds propagated outward along their axial trends as the shear zone progressively grew in width and displacement (Ramsay and Graham, 1970). Increasing strains spreading outward from the center of the shear zone rotated and tightened the D5 folds by increasing amounts toward the center. These folds must have grown serially from southeast to northwest because they increase in amplitude toward the southeast; here they abruptly terminate against the Naros intrusive massif, which acted as a rigid buttress against which the D5 folds piled up. Pure-shear shortening, acting like a vise across the shear zone, is postulated to have arched these folds across a D6 antiform, thus producing their type 2 crescentic interference patterns. On the basis of shear–strain curves derived from the amount of rotation of older structures by simple shear, the dextral displacement is estimated to be at least 100 km (Toogood, 1976). If a 40 percent component of pure shear is added, as suggested by the D6 buckling, this estimate is reduced to a minimum of 85 km.

Absence of the D5 and D6 structures to the south of the Tantalite Valley mylonite belt in eastern Karasburg District is puzzling (Fig. 8-13). One explanation is that this area represents a large strain shadow east of the rigid mass of igneous rocks in the Western Zone, which lacked a resistant buttress like the Naros massif in the north (Toogood, 1976).

Chapter 9

The Pan African Geosynclines

Late Precambrian tectonism, commonly termed the Pan African event, affected vast parts of the African continent and adjoining areas of Brazil, Arabia, and Madagascar (Kröner, 1977, 1979; Martin and Porada, 1977a). This activity marked the final episode in a long history of spasmodic Precambrian instability, after which Africa largely retired from widespread orogeny. The term "Pan African" unfortunately is commonly used to cover events as far apart as 500 Ma. Jackson and Ramsay (1980) recommended that the term be restricted to a thermotectonic episode around 450–550 Ma ago, as originally defined radiometrically by Kennedy (1964), and that older orogenic phases be given local names where applicable. This unambiguous usage has much to commend it, and, where practicable, the term "Pan African event" is restricted to this approximate period, and its general usage is retained for geosynclines affected both by this event and by earlier tectonism.

In southern Africa, Late Precambrian orogeny resulted in the formation of a 3000-km chain of geosynclines following the western and southern coasts of the subcontinent. The chain is preserved as a series of reentrants known as the coastal branch of the Damara Province, the Gariep Province, and the Saldanian Province (Fig. 9-1). The Damara Province also extends intercratonically through Namibia toward south-central Africa as a separate but closely related branch of the geosynclinal system. These geosynclines are filled with thick, largely sedimentary, precratonic deposits. On the east coast of southern Africa the Pan African is marked by tectonothermal reworking of basement rock and widespread isotopic resetting in the Mozambique Province.

The Gariep Province, a small geosyncline with well-delineated miogeosynclinal and eugeosynclinal components, and the closely related Nama platform sequence are examined in detail, followed by the Saldanian Province, which is exposed as scattered inliers in the Phanerozoic Cape Fold Belt. In conclusion we examine tectonic models recently proposed for the Damara Province. Although this province has a close affinity to central African structural provinces such as the Zambezi Province, a tectonic overview of the Damara Province is appropriate in this volume because it plays such a pivotal role in understanding the framework of Late Precambrian sedimentation and orogeny.

9.1. The Gariep Geosyncline

The Gariep Group represents a volcanosedimentary pile that accumulated in a geosynclinal trough over at least 300 Ma. Figure 9-1 shows that the Gariep trough represents a reentrant of the much longer coastal geosyncline that includes the West Congo, coastal Damara, Gariep, and Saldanian Provinces. The Gariep Group partially overlies and is tectonically intersliced with basement rocks that have been refoliated and retrogressed during deepening of the geosynclinal trough; these rocks are referred to here as "reworked basement." The area underlain by the Gariep Group and its adjoining zone of reworked basement is referred to here as the Gariep Province. Kröner and Blignault (1976) defined this province as consisting only of the Gariep Group, but their map also includes the adjoining area of reworked basement within the domain of the Gariep Province.

The Gariep Province straddles the Orange River, which forms the international boundary between South Africa and Namibia (Fig. 9-2). For convenience we refer to the South African

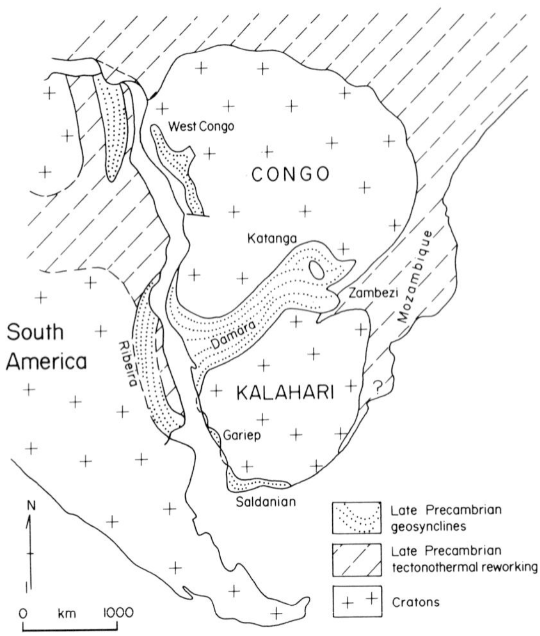

Figure 9-1. Map showing Late Precambrian structural framework of central and southern Africa and eastern South America (continental fit based on Smith and Hallam, 1970). [Adapted from Kröner (1977, 1979), Martin and Porada (1977a), Porada (1979).]

part as the Richtersveld and the Namibian part as the Sperrgebiet. Strict usage would confine the term "Richtersveld" to the mountainland within the broad arc of the Orange River, but we also include the coastal plain to the west. The Sperrgebiet (Forbidden Area), officially known as Diamond Area 1, represents a security zone restricting access to the diamondiferous terraces and beaches along the coast. A small part of the Gariep Province lies east of this zone but this need not deter use of "Sperrgebiet" as a convenient general term to include all the Gariep Province north of the Orange River.

The remote and inhospitable terrain of this arid area, together with the complex and rapid facies changes in space and time, have combined to ensure a history of stratigraphic controversy since the first detailed exploration by De Villiers and Söhnge (1959) in the 1940s. Space does not permit an examination of this history in detail (Kröner, 1975b) and emphasis here therefore is on the current consensus of stratigraphic relationships.

The Gariep Group has so far not yielded any radiometric ages that can reasonably be interpreted as primary and it has therefore not been possible to date these rocks directly. Fortunately, a number of syntectonic and posttectonic intrusions have yielded primary radiometric ages that enable the timing of at least the later Gariep history to be bracketed. A detailed discussion of these age relations is deferred until these intrusions have been described. At this stage it is sufficient to place the end of Gariep deposition at >550 Ma and its start at >780 Ma and possibly >920 Ma.

The Gariep Foundation

Along its eastern margin the Gariep Group rests unconformably on rocks of the Namaqua Province. Tectonism in this part of the Namaqua Province probably ceased at around 950 Ma, as evidenced by posttectonic pegmatites and plutons of this age (Chapter 8). There is no evidence for a substantial interval of time between the relatively rapid uplift of the lower crustal Namaqua Metamorphic Complex and deposition of the basal Gariep sediments on its eroded surface. The Gariep orogeny can therefore be regarded as a chronologic extension of Namaqua orogenic activity. However, the new tectonic cycle was initiated with structural trends oriented roughly north-northwest rather than the westward trend characteristic of most of the Namaqua Province. It is therefore significant that during the close of Namaqua orogeny south of the Richtersveld, deep crustal reworking of regional extent took place with trends parallel to the future Gariep trend (Chapter 8, Section 8.5, Central Zone of the Namaqua Province). This heralded the start of a new tectonic regime and provided a structural as well as temporal link between Namaqua orogenic activity and the Gariep orogeny that closely followed.

Where the Namaqua basement has remained unaffected after uplift, the Gariep rests on it with gross angular unconformity or nonconformity. A metamorphic discontinuity is also present between greenschist-facies Gariep and amphibolite-facies or igneous Namaqua basement. However, in many areas (Fig. 9-2) structural and metamorphic reworking of the basement during or after deposition of the Gariep Group resulted in paraconformable contacts between sheared or truncated basal Gariep and greenschist-facies reworked basement in the form of schists and phyllonites. In this high-strain contact zone, basement

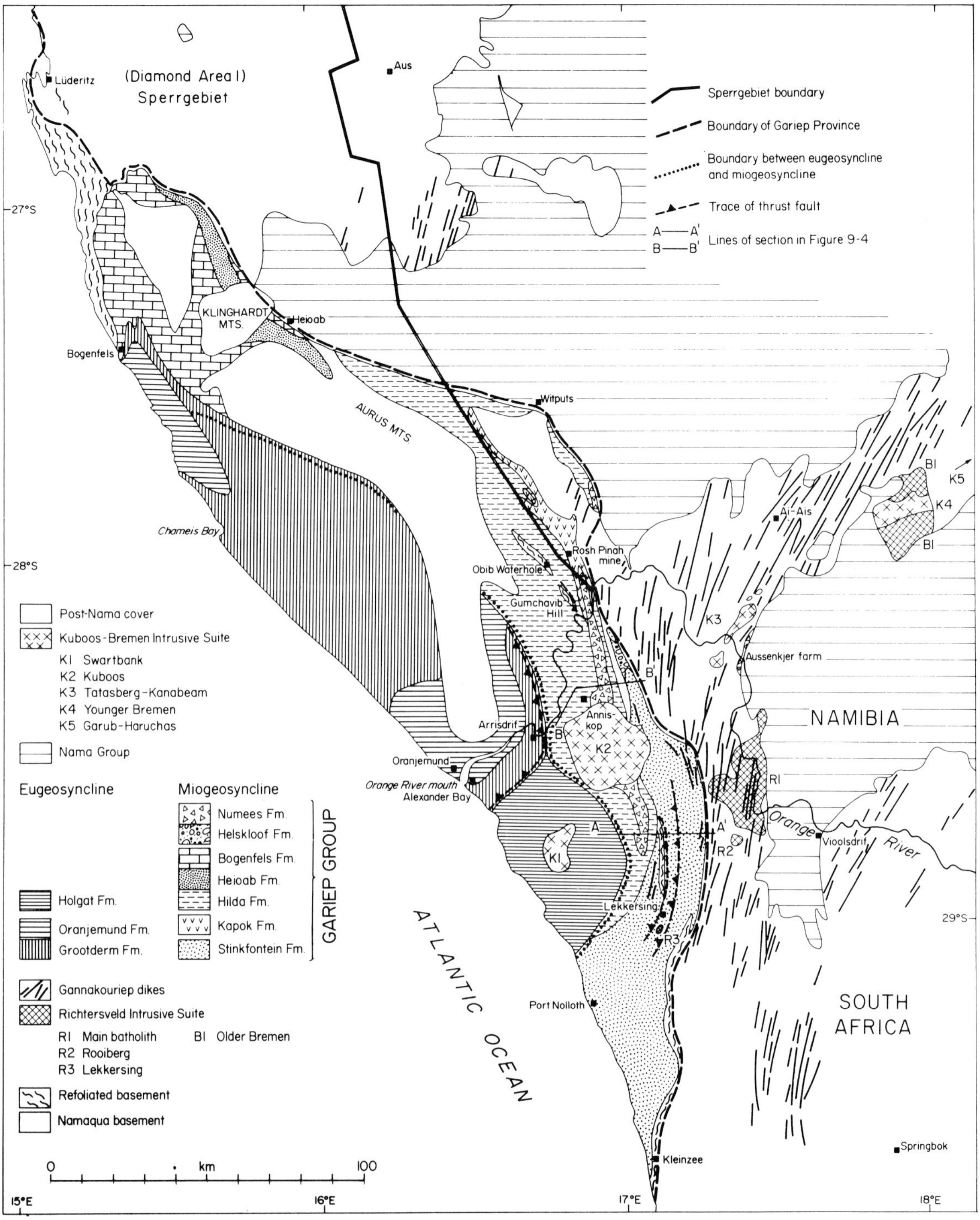

Figure 9-2. Geologic map of the Gariep Province showing extent of basement reworked during the Gariep orogeny, the Gariep Group, and early-syntectonic and posttectonic intrusions. [Compiled from De Villiers and Söhnge (1959), Martin (1965), McMillan (1968), Joubert and Kröner (1972), Kröner (1972, 1974a, b), Kröner and Jackson (1974), Jackson (1976), Kröner and Blignault (1976), Blignault (1977), McDaid (1978), Reid (1979e), Welke *et al.*, (1979).]

structures have been completely transposed to give rise to a tectonic mélange of porphyroclasts set in a schistose and commonly mylonitic matrix (Joubert and Kröner, 1972; Kröner, 1975b; Kröner and Jackson, 1974). The porphyroclasts represent the remains of disrupted competent layers, commonly quartzite and pegmatite, and attain boulder size (Fig. 9–5A). Their size, heterogeneity, and matrix-supported nature prompted earlier geologists to classify these rocks erroneously as deformed tillites in the sedimentary succession (e.g., the "Black Hills Series" of De Villiers and Söhnge, 1959, or the "Diaz Point Formation" of Greenman, 1966). Similarly the gradational and apparently conformable contacts between basement granite, reworked schistose basement, and Gariep sediments have been attributed to processes of "granitization" then in vogue. For instance granitoids and granitoid gneisses in basement slivers east of Rosh Pinah mine at Obib waterhole and near Gumchavib hill (Fig. 9-2) were thought to represent "granitized" Gariep sediments (Martin, 1965; McMillan, 1968; Kröner, 1972). Subsequent structural (Kröner, 1974b) and geochronologic work (Welke *et al.*, 1979, U-Pb zircon) indicates that these are basement rocks forming part of the Western Zone of the Namaqua Province and are approximately 1870 Ma old.

Interpretation of reworked basement as zones of "granitization" led De Villiers and Söhnge (1959) to the far-reaching but erroneous conclusion that parts of the presently defined Gariep Group were older than the granitoids of the Namaquan Western Zone, whereas the Gariep rocks are now known to be some 1000 Ma younger.

Basement reworking took place after deposition of the basal Gariep. This deformational imprint extends up sporadically through the eastern Gariep succession. The Rb-Sr systematics in the unfoliated Obib leucogranite described above were reset at 950 $\pm$ 30 Ma (Welke *et al.*, 1979). This date, corresponding to the inferred date for the start of Gariep sedimentation, may represent the approximate time of basement working and the onset of Gariepian orogeny.

Thrust-bounded slivers of basement are exposed for several kilometers west of the Gariep basal unconformity (Fig. 9-2), but the extent of basement farther west is unknown in the absence of subsurface data. Neither is the nature of the pre-Gariep crust in the west known, but the Gariep lithologic assemblages there suggest that they may have accumulated partly on oceanic crust.

The Gariep Miogeosyncline–Eugeosyncline Couple

The dual nature of the Gariep geosyncline has been recognized for some time (Martin, 1965; McMillan, 1968; Kröner, 1972, 1975b). An eastern miogeosynclinal accumulation of carbonates; terrigenous clastic shelf sediments, diamictites, and subordinate volcanic rocks is flanked by a western eugeosyncline filled with volcanic rocks overlain by fine terrigenous clastics. The eugeosynclinal trough is further characterized by higher degrees of deformation and more-intense metamorphism. The boundary between each zone is well defined in the Richtersveld (Fig. 9-2) but in the Sperrgebiet is less obvious because of poor exposure and the lack of systematic mapping.

Major Stratigraphic and Structural Relations

Lenticular units and telescoped facies combined with intense deformation in the west have resulted in a complex, linear, volcanosedimentary package. The principal stratigraphic and structural relations are shown schematically in Figure 9-3. The miogeosynclinal units thicken basinward toward a depocenter west of the Orange River mouth. Contacts between formations in the Gariep Group are generally conformable (Fig. 9-3).

In the miogeosyncline the predominantly coarse terrigenous clastic Stinkfontein Formation forms the base of the sequence in the Richtersveld, resting with marked angular unconformity on the Namaqua basement (Fig. 9-3, columns A, C, D, E). This formation pinches out northward (column E) and in the Sperrgebiet is replaced by the Kapok Formation, which comprises a lower sedimentary-volcanic and upper volcaniclastic member (column F). The latter contains stratabound and stratiform Pb-Zn sulfides of probable sedimentary exhalative origin hosted in carbonaceous chert and dolomite, with clastic intercalations at Rosh Pinah mine (Page and Watson, 1976). These formations are conformably overlain by the Hilda Formation, consisting of a heterogeneous assemblage of shelf sediments, in-

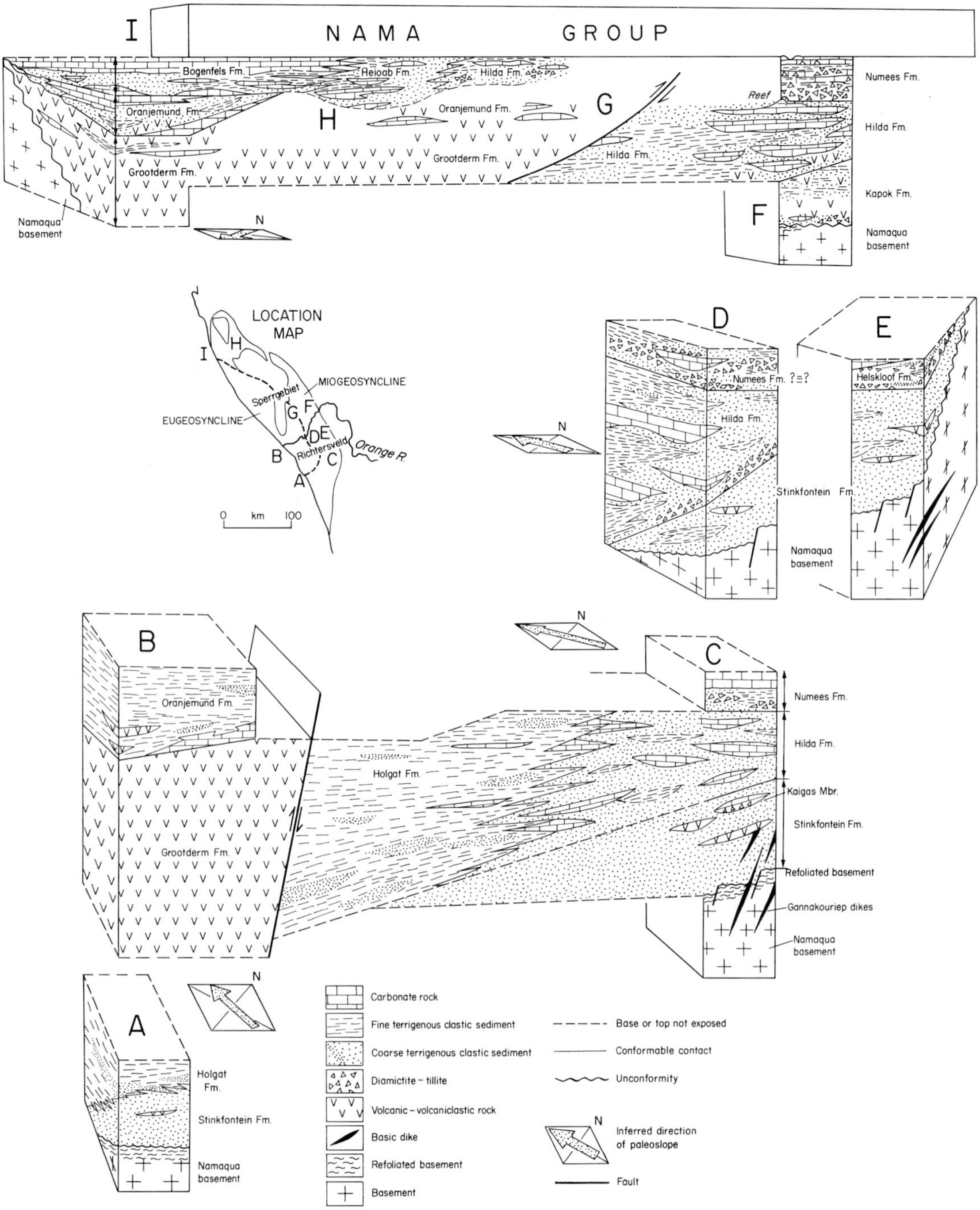

Figure 9-3. Block diagrams showing facies and structural relations in selected areas of the Gariep Geosyncline (not to scale). [Compiled from: A, Joubert and Kröner (1972), B–E, Kröner (1974a); F, McMillan (1968), Kröner (1972), Allsopp *et al.* (1979); G–I, Martin (1965), Kröner (1972).]

cluding diamictites. In the Richtersveld south of the Kuboos pluton the Hilda Formation occupies a northward-plunging syncline, whereas to the north it occupies a southward-plunging fold pair (Fig. 9-4). The Hilda Formation varies greatly in thickness but shows a general basinward thickening where it grades laterally and downward into the eugeosynclinal Holgat Formation by means of an increasing ratio of graywacke to dolomite (Fig. 9-3, column C, and Fig. 9-4). In the extreme

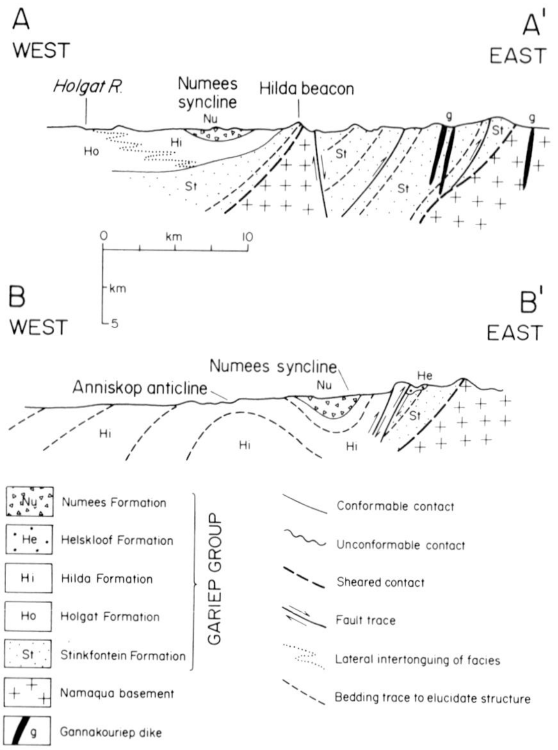

Figure 9-4. Cross sections through the Gariep Miogeosyncline. Refer to Figure 9-2 for lines of section. [After Kröner (1974a).]

southwest the Stinkfontein Formation is conformably overlain by the Holgat Formation (Fig. 9–3, column A). Near its northern termination (Figs. 9-2 and 9-4) the Stinkfontein Formation is overlain with local unconformity by the diamictite-bearing Helskloof Formation (Fig. 9-3, column E) which is probably an outlier of the Numees Formation described below. At the northern end of the Gariep Province (Fig. 9-3, column I) the carbonate-rich Bogenfels Formation (Fig. 9-5B) overlaps the eugeosynclinal Oranjemund and Grootderm Formations and the northward-rising Namaqua basement. Eastward the Bogenfels Formation grades laterally into the more-clastic Heioab Formation (Fig. 9-3, column H). The latter may be either a stratigraphic equivalent of the Hilda Formation (Martin, 1965) or a nonvolcanic facies equivalent of the Kapok Formation (Kröner, 1972).

Throughout most of the miogeosyncline the terminal unit of the Gariep Group is the glaciomarine Numees Formation (Fig. 9-3, columns C, D, F). De Villiers and Söhnge (1959) suggested an unconformable relationship between the Numees and Hilda Formations. However, this interpretation was refuted by Kröner (1975b) on the basis of a Hilda stromatolitic reef north of Anniskop (Fig. 9-2) which apparently continued to grow during deposition of the basal Numees beds. Distribution of the Numees Formation is structurally controlled, and it occupies the core of a long syncline extending from the southern Richtersveld to the eastern Sperrgebiet (Figs. 9-2 and 9-4). The Numees Formation is overlain by the Nama Group, in most places paraconformably (Martin, 1965; McMillan, 1968), although Kröner and Germs (1971) have documented a local angular unconformity on Aussenkjer farm west of the Sperrgebiet (Fig. 9-2). Local discordances of this type are a normal consequence of toplap during progradation (cf. Mitchum *et al.*, 1977) and it is debatable whether this discordance is more significant than other local unconformities within the Gariep (e.g., the Kaigas–Stinkfontein, Helskloof–Stinkfontein, and Oranjemund–Grootderm unconformities). There is no compelling evidence for separating the Numees Formation from the remainder of the Gariep Group or for significant diastrophism between deposition of the Numees Formation and the Nama Group. Germs (1972a) regarded both the Gariep and Nama Groups as different sedimentary responses to the same widespread tectonic event.

In the eugeosyncline poor exposure and a high degree of deformation have obscured stratigraphic subtleties so that only three formations have been recognized (Kröner, 1975b). Most of the Richtersveld coastal plain is underlain by the largely argillaceous Holgat Formation, a facies equivalent of the Hilda Formation. An inferred fault separates the Holgat from the basic to intermediate volcanic Grootderm Formation, which is overlain by fine terrigenous clastic and volcaniclastic deposits of the Oranjemund Formation (Fig. 9-3, column B). In the Richtersveld the base of the Grootderm Formation is not exposed but in the Sperrgebiet it lies unconformably on the Namaqua basement (Fig. 9-3, columns B and I). West of Rosh Pinah mine (Fig. 9-2) Grootderm metavolcanic schists have been thrust eastward over the miogeosynclinal Hilda Formation in a manner similar to that inferred in the Richtersveld (Fig. 9-3, column B).

The lithology of the Gariep Group has been

documented on a regional scale by De Villiers and Söhnge (1959), Martin (1965), McMillan (1968), Haughton (1969), and Kröner (1975b); lithologic descriptions for more restricted areas in the Sperrgebiet can also be found in Kaiser (1926), Knetsch (1937), and Kröner (1972, 1974b), and for the Richtersveld in Rogers (1915), Van Biljon (1939), Haughton (1961), Middlemost (1966), Joubert (1971), and Joubert and Kröner (1972). The regional lithologic variations are shown in Figure 9–3. Table 9–1 sets out the detailed lithology of the Gariep Group in the well-documented Richtersveld. Thickness estimates in the eugeosyncline are highly speculative and are not primary because of widespread and advanced transposition.

Depositional environments within the Gariep Group are interpreted in the following sections on the basis of data drawn largely from the Richtersveld.

Miogeosynclinal Alluvial Facies

During the early rifting that formed the Gariep trough, coarse alluvial-fan sediments of the Stinkfontein Formation were deposited as westward tapering clastic wedges that vary considerably in thickness along strike (Table 9-1). These upward-fining sequences contain lenticular conglomerates with clasts, up to 2 m in diameter, consisting of mixed basement and cover lithologies, the latter including shelf dolomite with algal textures. Clast-supported conglomerates consist principally of durable quartzite, whereas matrix-

Table 9–1. Stratigraphy of the Gariep Group in the Richtersveld[a]

Eugeosyncline	Miogeosyncline
Oranjemund Formation: 1000–1500 m Phyllite, quartz–chlorite schist, quartz–sericite schist, quartzite, dolomite, acid volcanics; carbonate olistostrome at base	Numees Formation: 500 m Massive dolomite up to 30 m thick Dropstone varvite Poorly stratified matrix-supported conglomerate (clasts include faceted and striated types) with intercalations of cross-bedded sandstone, boulder schist, phyllite, and lenses of massive dolomite Matrix-supported conglomerate
Holgat Formation: 5000–6000 m East: Phyllite, cross-bedded and graded arkose, thin dolomite West: Feldspathic quartzite, quartz–sericite schist, phyllite	Hilda Formation: 700–3000 m Massive dolomite up to 300 m thick Very coarse-grained quartzite and arkose, conglomerate (clasts include dolomite), phyllite, schist; intercalated carbonate rock Lenses of massive dolomite up to 6 m thick with algal textures
Grootderm Formation: ? 4000–5000 m Basaltic and andesitic lava, volcanic breccia, agglomerate, tuff	Stinkfontein Formation: 150–3000 m Well-bedded dolomite and quartz arenite Coarse-grained sandstone, conglomerate, and thinly bedded quartzite Arkose and subarkose, subordinate conglomerate and phyllite; structures include cross-bedding and desiccation cracks; interbedded volcanic rocks include lavas and pyroclastics Plane-bedded quartz arenite Conglomerate (clasts include quartzite, dolomite, lava) and very coarse-grained sandstone interbedded with finer grained sandstone

[a]Compiled from De Villiers and Söhnge (1959); Martin (1965); Middlemost (1966); Kröner (1975b).

supported conglomerates are less mature and more argillaceous, forming sequences up to 100 m thick that are channeled near the top (De Villiers and Söhnge, 1959; Kröner, 1975b).

The upper parts of the alluvial sequences consist of thickly bedded arkosic to subarkosic sandstones with subordinate conglomerate and argillite. The sandstones contain large-scale planar cross-bedding and ripple cross-lamination with westward-directed foresets (Middlemost, 1966). Sporadic desiccation cracks (Kröner, 1975b) indicate intermittent subaerial exposure. Interbedded volcanic material, in the form of lava, ash flows, tuff, and agglomerate, are mainly rhyolite or quartz latite with subordinate trachyte and andesite (Middlemost, 1966; Kröner, 1975b).

The paleogeography is envisaged as coalescing alluvial fans on the basis of lithology, consistent westward paleoflow patterns, and stacking of poorly stratified clastic wedges against the eastern margin of the basin. Syndepositional instability is evidenced by slump structures. The conglomerates represent mass-flow and fluvial deposits and there is no compelling evidence that they were deposited by glaciogenic processes (Kröner and Rankama, 1972; Kröner, 1975b) as inferred by De Villiers and Söhnge (1959). The algal structures in dolomite clasts can be interpreted as stromatolitic, which suggests sedimentary reworking of bioherms (Kröner, 1975b). Lenses of stromatolitic dolomite interbedded with these immature terrigenous clastic deposits indicate that the alluvial fans may have debouched directly into shallow-marine carbonate-forming environments as a fan-delta complex.

Miogeosynclinal Shore-Zone and Shelf Facies

Marine processes are suggested by the presence of thick units of quartz arenite and dolomite in the Stinkfontein and Hilda Formations. Well-sorted plane-bedded quartz arenites erosively overlie less-mature sediments along the eastern part of the basin (Middlemost, 1966; Kröner, 1975b). The more-distal parts of the basin (Fig. 9-3, column D) contain thick metaargillites and carbonates. Dolomites, with persistent columnar stromatolites, attain 300 m in thickness and finger out eastward into arkoses and quartz arenites. The columnar stromatolites are likely to have been generated by the upward growth of the algal mats (Schopf, 1976) and these structures, together with lensoid stromatolitic dolomite bodies some 1–6 m thick, have been interpreted as bioherms on reef platforms (Kröner, 1975b). Intermittent sedimentary reworking in the shore zone periodically broke up the reefs and gave rise to the more mature arenites (Kröner, 1975b). The stromatolitic dolomites are concentrated in the mixed volcanosedimentary sequences close to the alluvial fans. In this respect they are analogous to Mid Proterozoic stromatolites of the Ventersdorp Supergroup and Wolkberg Group in the Kaapvaal Province, which also fronted alluvial fans during periods of volcanic quiescence (Eriksson and MacGregor, 1981).

Miogeosynclinal Glaciomarine Facies

Numees diamictites gradationally overlie Hilda reef carbonates or quartz arenites and contain lenses of dolomite, cross-bedded and rippled sandstone, and argillite in the form of pristine varved shales containing dropstones. The rhythmically laminated varved shales and preservation of cross-bedding and graded bedding suggest that the argillaceous components formed under glaciomarine or glaciolacustrine conditions. The diamictites thin northward and contain matrix-supported faceted and striated clasts, consisting mainly of basement granitoids and gneisses. The largest clasts, up to 11 m long, are composed of massive dolomite (Martin, 1965) and have been widely interpreted as glacial erratics in tillite (Rogers, 1915; Van Biljon, 1939; De Villiers and Söhnge, 1959; Martin, 1965; McMillan, 1968). Nevertheless Kröner and Rankama (1972) refuted this tillitic interpretation despite the fact that the Numees Formation is accepted as a classic example of Late Precambrian glaciation. Because of the absence of striated pavements, gradational contacts with arenites, and the rounded nature of the clasts, these authors concluded that the diamictite represents redeposited glaciomarine sediments containing olistostromes of dolomite formed by fragmentation and redeposition of carbonate reefs. However, in the case of glacial deposition on a continental shelf such as the North Sea, which was exposed by lowering of sea level, it is not crystalline basement that was eroded by ice, but recently deposited sediments; therefore, striated pavements seldom form. Furthermore, well-rounded clasts are common in

Quaternary tills because of abrasion during transport and deposition of ice; the angularity characteristic of clasts in mountain glaciers is frequently the result of incorporation of frost-shattered scree (Drake, 1972). Likewise, many examples of gradational or interlaminated contacts can be seen on close inspection to have resulted from subglacial flattening and disaggregation of poorly consolidated clasts and matrix by high shear strains (Edwards, 1978; Berthelsen, 1979).

Another interpretation is that the diamictites represent aquatills deposited from suspension and ice rafting in quiet water. Such deposits form most of the Ross Sea glaciomarine sequence and can be distinguished from basal tills by their gradational contacts with normal stratified sediments, intercalations of turbidites or other beds of laterally continuous and contrasting sediments, and presence of unbroken fossils, in places in living positions (Edwards, 1978). All these features are present in the Numees Formation.

Eugeosynclinal Facies

Graywackes are abundant and carbonates are extremely rare in the distal volcanosedimentary part of the eugeosyncline (Kröner, 1975b). These are bounded by proximal deposits of phyllitic shale and cross-bedded conglomerate and arkose with thin intercalations of dolomite in the Holgat Formation. Owing to penetrative schistosity and advanced transposition of the bedding, the primary nature of the Grootderm and Oranjemund Formations is difficult to interpret. The metavolcanic rocks are almost invariably in the form of chloritic schists with sporadic preservation of primary textures, indicating that the parent rocks were intermediate to basic lava, tuff, breccia, and agglomerate. The lavas are sporadically pillowed (Porada, 1979).

The overlying Oranjemund Formation is a volcanosedimentary sequence that is typically eugeosynclinal but which has miogeosynclinal features at its northern end near Bogenfels (Fig. 9-3, column I) (Kröner, 1975b). Sedimentation was initiated with the formation of algal dolomite reefs, which were brecciated in the manner of olistostromes; the clasts are supported by a matrix of sheared volcaniclastic rocks or graywacke (Kröner, 1975b). The close association of volcanic rocks and biogenic dolomite suggests that the eugeosynclinal basin must have been locally very shallow by the end of Grootderm deposition, possibly indicating growth of carbonate reefs on guyots, or submarine quiescent volcanoes (Kröner, 1975b). Above the basal carbonates the sedimentary components are similar to those in the distal Holgat, with a preponderance of phyllitic graywacke. Metamorphosed acid volcanic rocks, in the form of quartz–sericite schists, are much more abundant than basic rocks. The northward continuations of the Grootderm and Oranjemund Formations were mapped under the names of "Chloritschiefer Formation" and "?Untere Konkip Formation" (Kaiser, 1926) and "Lower and Upper Marmora beds" (Martin, 1965). In the Bogenfels area the Oranjemund sediments are of shelf character, with dolomites and arenites being common (Martin, 1965). These sediments thin northward and pass upward into the onlapping Bogenfels Formation.

Syndepositional Deformation and Closure of the Gariep Geosyncline

Signs of strain increase westward across the Gariep Province from the mildly deformed miogeosyncline into the intensely deformed eugeosyncline. Overprinting relations are present, but all deformational events recognized can be equated either with syndepositional gravity-driven processes or progressive closure of the Gariep trough during and following its infilling.

The inference of syndepositional deformation within the miogeosyncline (Kröner, 1975b) is based mainly on slump-folded layers and slump sheets (including the inferred olistostromes) alternating with undeformed layers, soft-sediment boudinage and microfaulting, and west-facing shallowly inclined "cascade folds." In addition, lineations on bedding surfaces may have resulted from scouring by coarse slurries. The orientation of all these structures is consistent with the westward paleoslope inferred from stratigraphic relations.

Syndepositional deformation is almost inevitable during rapid clastic sedimentation, especially at the continental shelf break. However, some of the examples cited by Kröner (1974a) are not convincing. His Figure 20 illustrates large-scale trough cross-bedding in Stinkfontein quartzite distorted so that the dihedral angle is close to 90°. Although he ascribes this to westward "slumping," the strain geometry indicates movement eastward up the paleoslope. It is more likely that the distortion has been caused by simple shear during regional flexural-slip folding (Ramsay, 1967, p. 498). Similarly, folding of a "tectonic clast" in an intensely cleaved tectonic mélange (Kröner, 1974a, his Fig. 26) cannot be taken as proof of "synsedimentary slumping."

Despite being the oldest unit in the miogeosyncline, the Stinkfontein Formation is generally little deformed. This is probably a result as much of the competence of the constituent quartzites as of their marginal position. Deformation is confined to widely spaced zones of intense shearing and faulting, particularly at the basement–cover interface (see section on "The Gariep Foundation"). Within the overlying Hilda Formation, bedding is preserved in the competent rocks but has been obliterated by a strong cleavage in phyllites and schists. This cleavage is ascribed to D2, D1 being represented by the syndepositional deformation structures including a bedding-parallel foliation (Kröner, 1975b). The S2 cleavage is axial planar to major D2 folds plunging northward south of the Kuboos pluton and southward north of the pluton (Fig. 9-2). In the core of the major D2 synclinorium, Numees diamictites have been intensely flattened to give rise to a strong cleavage and distorted clasts (Martin, 1965). According to Kröner (1975b), major D2 folds in the Richtersveld show a strong westward vergence in the east, whereas farther west they are upright, a transition he attributes to a change from gravity-dominant folding to folding produced by lateral compression. Many of the north-trending faults mapped by De Villiers and Söhnge (1959) are reverse and thrust faults produced by this lateral shortening across the Gariep trough. During this process, slivers of basement were intersliced with the basal Gariep by eastward directed imbricate thrusting.

Westward into the eugeosyncline the increase in strain unfortunately coincides with a decrease in exposure. This strain trend can be deduced from an increase in fold tightness giving rise to recumbent or shallowly inclined isoclinal folds, advanced transposition and obliteration of bedding, penetrative development of shallowly inclined foliation, and axial trends toward the northeast in the Holgat Formation and northwest in the Oranjemund Formation (Kröner, 1975b). These D2 structures are deformed by open upright F3 folds in the coastal Holgat.

In the Sperrgebiet miogeosyncline mild deformation on northern trends resembles that described above for the Richtersveld (McMillan, 1968). However, near the boundary of this zone where the eugeosynclinal Grootderm volcanic rocks have been thrust over the miogeosynclinal Hilda sediments (Martin, 1965; Kröner, 1972), a spectacular increase in deformation is present (Fig. 9-2). Reconnaissance by M.P. Coward (personal communication, 1978) west of Rosh Pinah mine has revealed a sequence of progressive deformation of nappelike F1 folds. In the north of this area gently inclined major F1 folds face west and have a strong eastward dipping axial-plane cleavage. Southward these folds are progressively overprinted by a west-dipping transecting D2 cleavage, and the F1 axial surfaces become curved. This process culminates in the south with complete rotation of the F1 folds to become coplanar with a penetrative westward dipping D2 cleavage and minor folds. This rotational backfolding is analogous to flipping a horizontal pillow onto its back. These "backfolded" F1 folds were also thrust eastward over undeformed F1 folds, which retain their original west-facing orientation.

At the northern end of the miogeosyncline near Bogenfels (Fig. 9-2) the gently folded nature of the Bogenfels Formation led to its original designation as "Folded Nama" (Fig. 9-5B) (Kaiser, 1926). Folds and thrusts verge eastward and trend parallel to the coastline; anticlines are cored by Namaqua basement (Martin, 1965).

Metamorphic data from the Gariep Province are scarce, but it appears that metamorphic conditions during deformation of the Gariep trough were almost invariably low grade at the present level of exposure. In the miogeosyncline, data given by Kröner (1975b) suggest maximum metamorphic grades of lower greenschist facies in the Stinkfontein Formation and upper greenschist facies in the Hilda Formation; the same grades apparently characterize equivalent stratigraphic units in the Sperrgebiet (McMillan, 1968). Allsopp *et al.* (1979) inferred a maximum grade of lower amphibolite facies in Stinkfontein metalavas but provided no supporting data. The barometric type is obscure. De Villiers and Söhnge (1959) reported highly sericitized cordierite in Holgat schists, which would indicate amphibolite facies of low P/T barometric type in this part of the eugeosyncline. Kröner (1975b) speculated that this cordierite could be detrital rather than part of a metamorphic paragenesis. Cordierite has also been formed by contact metamorphism in the contact aureole of the Kuboos pluton (De Villiers and Söhnge, 1959).

In the eugeosyncline of the Richtersveld even less is known about the grade or barometric type of regional metamorphism, but there is no evidence of a substantially different metamorphic regime there. However, near Bogenfels in the Sperrgebiet the presence of glaucophane and other sodic amphiboles in Grootderm metalavas (Kröner, 1975b) is compatible with high P/T blueschist metamorphism, suggestive of plate convergence in this vicinity (Kröner, 1974a; Burke *et al*., 1977). Martin and Porada (1977b) argued that until other mineral parageneses characteristic of the high P/T facies series are found, there is little basis for the proposed plate convergence. Nevertheless geophysical modeling (Sydora *et al*., 1980) indicates that isotherms may actually be raised on the continental flank of a trench; it is therefore possible to have medium or low P/T metamorphism on one margin of a trench and high P/T blueschist metamorphism on the other margin. Speculations on the geotectonic regime of the Gariep and its place within the Pan-African framework will be presented in the section, "Summary of the Gariep Sedimentary Tectonic Cycle."

Syntectonic Intrusions

Immediately before or after the onset of Gariep sedimentation in the Richtersveld, syenitic to granitic intrusions were emplaced into the Namaqua basement. A large batholith and two small plutons of granite and leucogranite constitute most of the Richtersveld Intrusive Suite (R1, R2, R3 in Fig. 9-2). In addition granitic to syenitic intrusions, termed the Older Bremen Intrusive Suite (B1 in Fig. 9-2), were emplaced during the same period (Middlemost, 1964; Kröner and Blignault, 1976). The intrusions R1, R2, and B1 yield a concordia intercept age of 920 ± 10 Ma (U-Pb zircon) and a whole-rock isochron of 911 ± 39 Ma (Rb-Sr) which were interpreted as close to the age of intrusion by Allsopp *et al*. (1979). The Lekkersing granite (R3), which intrudes the Stinkfontein Formation, has yielded a U-Pb zircon age of 780 ± 10 Ma which serves as a minimum age for this formation. However, Allsopp *et al*. (1979) believed that the Stinkfontein Formation may be even older than the Richtersveld Intrusive Suite (920 Ma) because the Rooiberg granite (R2) contains xenoliths (up to 100 m long) of quartzite and lava resembling Stinkfontein lithologies.

The Gannakouriep dikes, a major northeast-trending basic dike swarm some 100 km wide to the east of the Gariep (Fig. 9-2), cut the Richtersveld batholith (R1) and, hence, postdate it (De Villiers and Söhnge, 1959). The dikes may be part of the same igneous cycle (Middlemost, 1964, 1966), but no reliable ages are presently available. The dikes average 50 m in width and are up to 100 km long (Reid, 1979e). They increase in density westward toward the Gariep Province and although they intrude the basal Gariep as a dike–sill complex up to 16 km west of the Stinkfontein base, they are generally rare in this unit (De Villiers and Söhnge, 1959; McMillan, 1968; Kröner, 1975b; Ritter, 1978; Reid, 1979e). This distribution implies a close relationship between dike injection and the formation of the Gariep trough (Reid, 1979e). Their localization along the hinge of a downwarp is compatible with their mildly alkaline to tholeiitic compositions, typical of rifted continental environments (Martin and Porada, 1977b). Some dikes in the swarm have compositions closely similar to Mesozoic Karoo basic sills farther south, and it is therefore possible that this minority reflects a much younger episode of continental rifting coincident with the fragmentation of Gondwana (Reid, 1979e). The Gannakouriep dikes are visibly deformed and recrystallized adjacent to the Gariep Group and may have been rotated from a primary northeastward trend in the east to a northward trend in the more deformed area within the miogeosynclinal downwarp (Reid, 1979e).

Syntectonic basic intrusions are apparently also present in the eugeosyncline, but intense deformation has obscured their field relations. Basic sills and plugs were reported in the vicinity of Alexander Bay (Fig. 9-2) by De Villiers and Söhnge (1959) and lamprophyric dikes and horn-

(A)

(B)

(C)

Figure 9-5. (A) Namaqua basement reworked during Gariep orogeny. The tectonic mélange comprises dismembered pegmatite veins and granite in a matrix of muscovite–chlorite schist. Diaz Point, 7 km west of Lüderitz, Namibia. [From Kröner and Jackson (1974).] (B) The arched rock of Bogenfels, consisting of landward-dipping miogeosynclinal sediments of the Bogenfels Formation, Gariep Group, Namibia (Fig. 9-2). Marine erosion formed the 60-m high sea arch by amalgamation of adjoining sea caves. (C) The Lower Nama Group and basement exposed in the 900-m-deep Fish River Canyon, Namibia (Fig. 9-6). Basement gneisses of the Namaqua Metamorphic Complex are unconformably overlain by the clastic Kanies Member (K1), the carbonate Mara Member (K2), the clastic Kliphoek Member (K3), and the carbonate Zaris Formation (K4) of the Kuibis Subgroup. (D) A migmatitic contact zone between Tygerberg slate of the Malmesbury Group and the Cape Peninsula batholith at Sea Point, Cape Town (Fig. 9-12). Xenoliths of dark hornfelsed slate with porphyroblasts of feldspar are surrounded by schlieric megacrystic granite. [Courtesy of V. Von Brunn.]

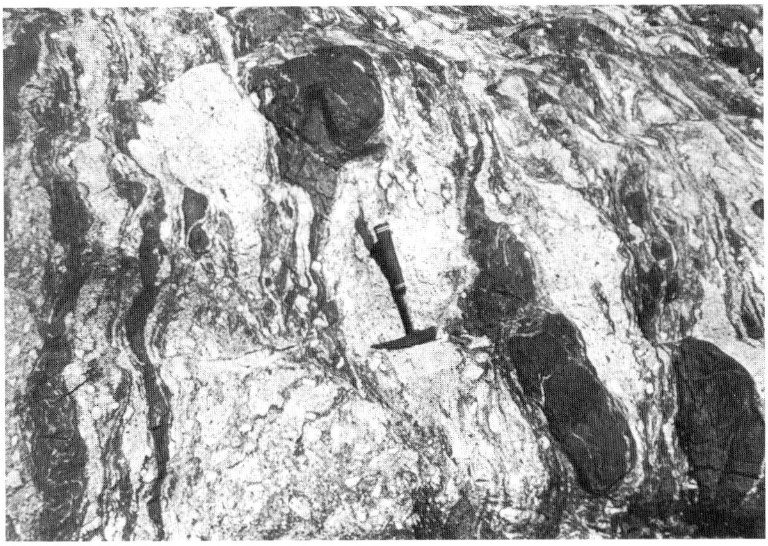

Fig 9-5D

blendite plugs were reported by Kaiser (1926) in the Sperrgebiet to the north. It is possible that these intrusions represent the closing stages of igneous activity in the Gariep eugeosyncline.

Posttectonic Intrusions

After deformation and metamorphism of the Gariep geosyncline, a chain of plutons, termed the Kuboos–Bremen line, was emplaced in the cratonized crust in five linear clusters. Two of the southernmost intrusions (K1 and K2 in Fig. 9-2) intrude the Gariep Group and the two northern clusters (K4 and K5) intrude the lower Nama Group (Kröner and Blignault, 1976). They vary in composition from granitoid to syenitoid and include carbonatite diatremes and sheets in the extreme north (Söhnge and De Villiers, 1948; De Villiers and Söhnge, 1959; Martin, 1965; Middlemost, 1967; Verwoerd, 1967; Kröner and Blignault, 1976). The Kuboos and Bremen Suites have yielded ages between 500 Ma and 550 Ma (Allsopp *et al.*, 1979; Rb-Sr whole-rock isochron) and indicate multiphase intrusion over this period. The age of the post-Nama Younger Bremen Suite is 518 ± 15 Ma (Allsopp *et al.*, 1979; mean of Rb-Sr errochron and zircon concordia intercept) and provides a minimum age for the Nama Group.

The Kuboos–Bremen line is of interest on account of its remarkably high degree of linearity (correlation coefficient based on 13 points = 0.987); the probability of this alignment being solely chance is less than one in a million. Although the northeast trend of the line is parallel to that of the Gannakouriep basic dike swarm in many places, Martin and Porada (1977b) have pointed out that this trend is not parallel either to the Gariep or to the Namaqua structural grain. For geometric and geochemical reasons, these authors rejected Kröner's (1975b) suggestion that the plutons marked the trace of a subduction zone at depth. They proposed either a deep-seated linear mantle upwelling or drift of lithosphere over a mantle hot spot. A related and attractive explanation is that this chain of high-level acid plutons and basic dikes marks the site of a failed Pan African rift (Kröner and Blignault, 1976; Reid, 1979e). The geochronology of Allsopp *et al.* (1979) suggests that around 500 Ma a major thermal event (the Pan African) resulted not only in intrusion and contact metamorphism, but in much isotopic resetting on a regional scale as well.

Summary of the Gariep Sedimentary–Tectonic Cycle

Rifting and downwarping along north–south trends at about 950 Ma appear to have been a continuation of Late Namaqua crustal reworking south of the Richtersveld. Namaqua basement to the east was progressively uplifted and the Gannakouriep basic dike swarm was injected into basement and cover, partly following the shoulder of the uplift. Simultaneously a passive continental margin developed in the west, possibly because of divergence of a Brazilian plate, onto which debris from the rising craton was shed in a westward direction (Stinkfontein). Initial rifting and proximal sedimentation was accompanied by extrusion of lavas and deposition of pyroclastic material. Shallow-shelf environments are interpreted from the occurrence of quartz arenites and stromatolitic reefs fronting alluvial fans. Debris flows suggest that gradients were tectonically maintained, probably concomitant with deformation of the base of the sequence and structural reworking of the basement. Coarse and fine terrigenous clastics and abundant carbonates accumulated on a miogeosynclinal shelf (Hilda), while basic volcanic rocks and submarine-fan turbidites accumulated in the deepening eugeosynclinal trough (Grootderm, Oranjemund, Holgat). It is unclear at present whether the volcanic rocks represent obducted or autochthonous oceanic crust, but the latter interpretation is possibly more compatible with their position at the base of the eugeosynclinal pile. Thick glaciomarine deposits (Numees) accumulated in the miogeosyncline when frigid conditions prevailed at the end of the Gariep. At this stage the extensive Nama Sea transgressed over at least the miogeosyncline. Subvolcanic emplacement of a linear chain of acid plutons from 550 to 500 Ma at right angles to the quiescent geosyncline may mark a failed rift along fractures formed earlier at the start of orogeny during emplacement of the Gannakouriep basic dike swarm.

At some undetermined stage of basin infilling, closure and plate convergence succeeded the widening and deepening of the trough in the classic geosynclinal manner. The eugeosynclinal sediments and volcanics took up most of this shortening, but folding and thrusting also extended into the miogeosyncline and the basement below it. Kröner (1972, 1975b) postulated east-facing subduction of a Brazilian plate beneath the Gariep geosyncline and a Kuboos-Bremen magmatic arc on the margins of a converging Kalahari plate. The unsuitability of the Kuboos–Bremen line as a magmatic arc has already been noted. Porada (1979) further dismissed this hypothesis by pointing to the great abundance of granitoids in the Brazilian Ribeira belt as compared with their paucity in the corresponding and once adjacent Gariep Province. Although this marked asymmetry in the distribution of granitoids may well reflect differences in exposure level, it also suggests that the Kalahari plate was subducted beneath the leading Brazilian plate during their convergence, rather than the reverse. The strong eastern vergence of folds and thrusts in the Gariep Province and farther north in the coastal Damara Province is further evidence for this hypothesis (cf. Dewey and Bird, 1970; Dewey and Burke, 1973).

9.2. The Intracratonic Nama Platform Succession

The Nama Group is a platform succession extensively preserved in southern Namibia and intermittently exposed south of the Orange River in South Africa (Fig. 9-6). Up to, and in places more than, 2500 m of sandstone, mudstone, and carbonate rocks overlie the Kalahari Province and the edge of the Gariep Province. Folding and thrusting during the Damara and Gariep orogenies affected only the northwestern and southwestern fringes of the Nama depository. Although the Nama rocks have not yet been dated directly, a minimum age of 520 Ma for the middle Nama is given by intrusion of Schwarzrand strata by the Younger Bremen pluton (Fig. 9-2) (Martin, 1965; Blignault, 1977; Allsopp *et al.*, 1979; Rb-Sr whole-rock, U-Pb zircon). The lower Nama contains representatives of the worldwide Ediacara-type fauna of soft-bodied coelenterates (Germs, 1972a, 1972b, 1974) whose age is controversial but may be Early Cambrian (Stanley, 1976) or Late Precambrian (Germs, 1972a, 1974; Glaessner, 1979), and therefore about 650–550 Ma old. Upper Nama strata contain an Early Cambrian trace fossil, *Phycodes pedum*. Paleomagnetic data indicate an age of about 650 Ma for the lower Nama Group and an Early Cambrian age for the Fish River Subgroup (Kröner *et al.*, 1981). These authors infer a considerable depositional hiatus between the upper and lower Nama, a time interval embracing the Cambrian–Precambrian boundary. The Nama Group was probably deposited during the later stages of Damara tectonism (Germs, 1972a, 1974; Hartnady, 1976, 1978a).

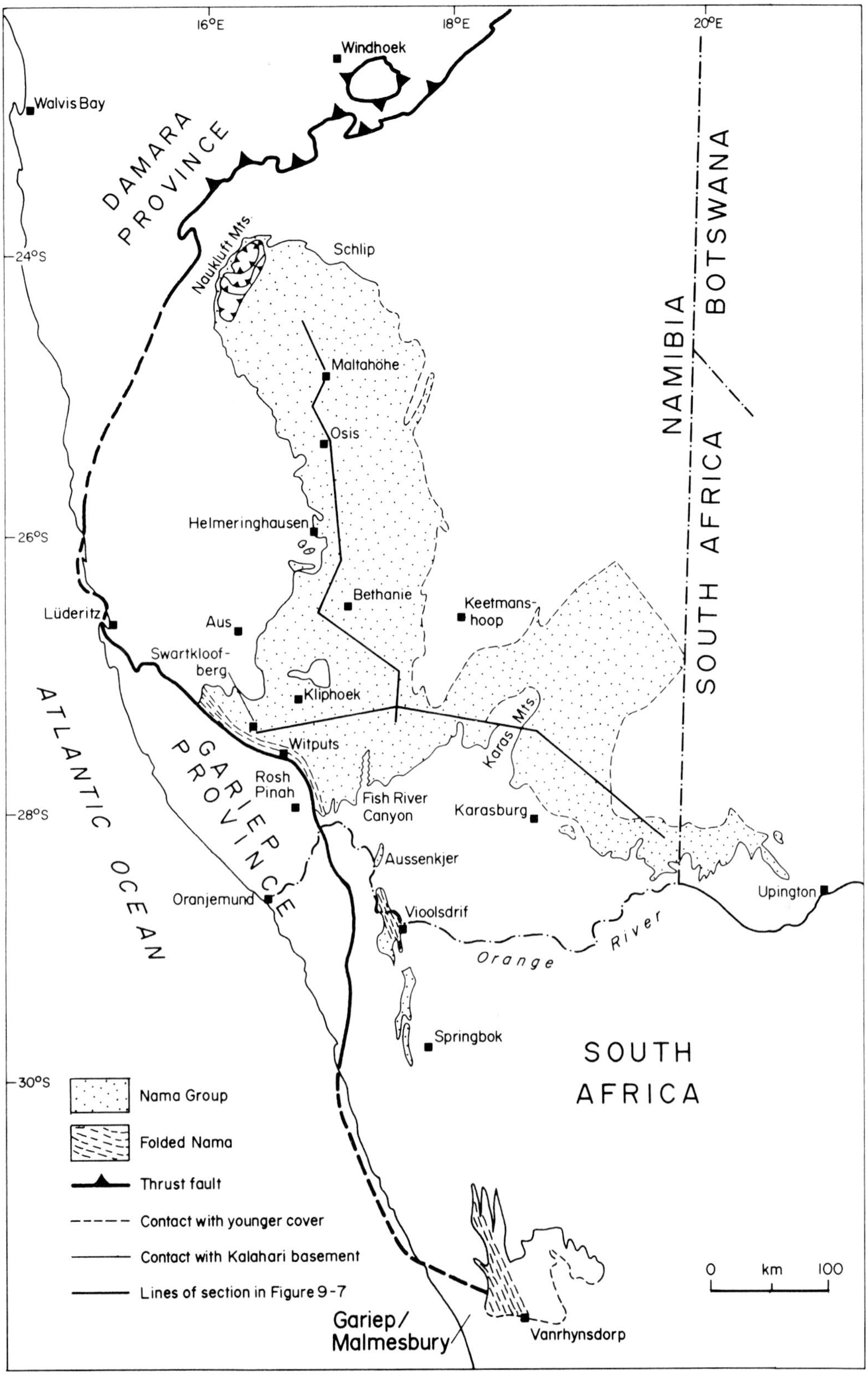

Figure 9-6. Map showing distribution of the Nama Group in Namibia and South Africa, possible southern correlates near Vanrhynsdorp, and areas affected by thrust nappes and folding. [After Martin (1965); additional data on Nama folding from Kröner (1968), McMillan (1968), and Jackson (1976); additional data on thrusting from Martin (1974), Bickle and Coward (1977), Hartnady (1978a).]

Stratigraphy

The Nama Group comprises three subgroups: Kuibis, Schwarzrand, and Fish River (Table 9-2). Progressive subsidence of the basin is reflected in rapid westward thickening of the sedimentary prism (Fig. 9-7). The basal Kuibis Subgroup overlies crystalline basement of the Kalahari Province and paraconformably overlies the Numees Formation of the Gariep Group (Martin, 1965; McMillan, 1968) with local angular unconformity (Kröner and Germs, 1971). Along depositional strike the lithosome geometries reflect the control of a west-projecting basement high known as the Osis Ridge (Germs, 1972a). The Osis Ridge basement salient divided the Early Nama Sea into two basins (Fig. 9-7). The Nama Group displays an overall upward decreasing maturity and contains no evidence of deep-water sedimentation (Martin, 1965; Germs, 1972a, 1974). A few unconformities interrupt the succession, particularly along the margin of the basin (Table 9-2).

The Kuibis Subgroup comprises a varied assemblage of conglomerate, coarse-grained arkosic sandstone, quartz arenite, green and white shale,

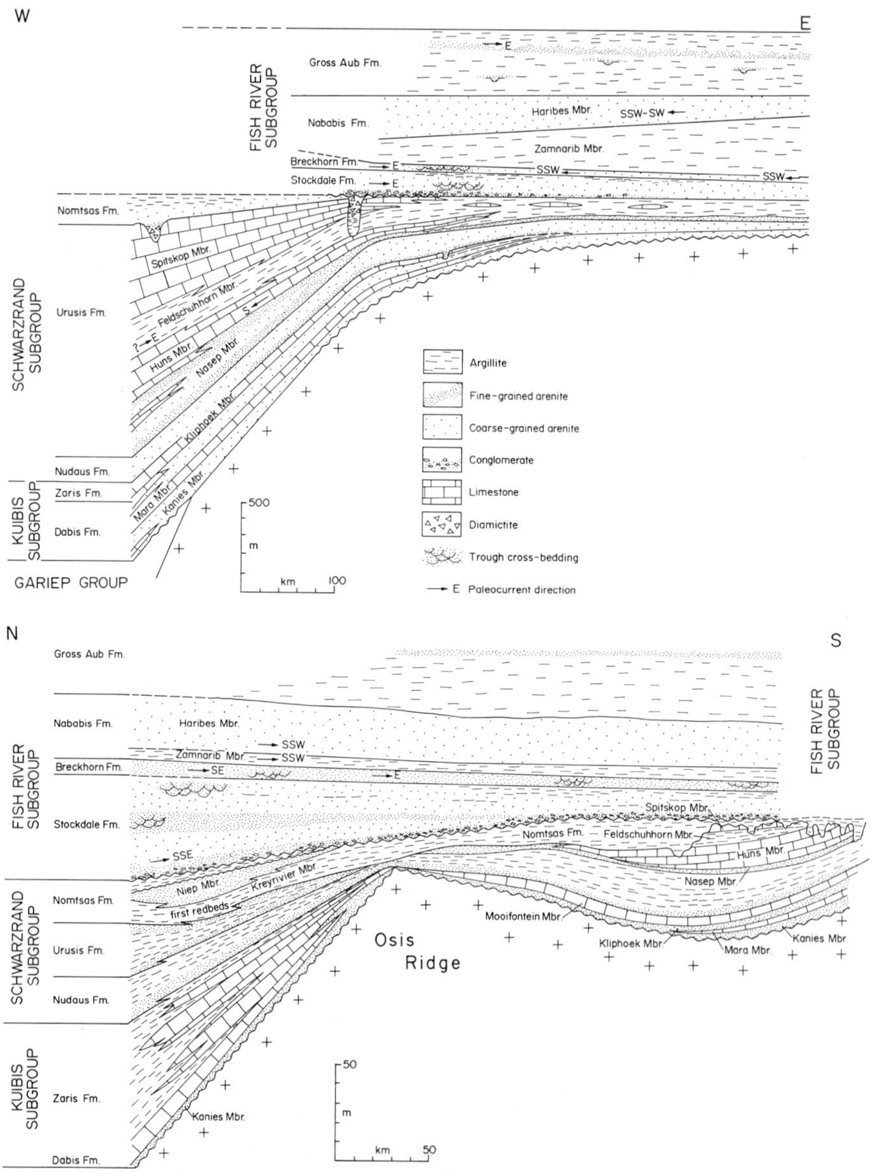

Figure 9-7. Regional geologic sections through the Nama basin. Refer to Figure 9-6 for lines of section. [After Germs (1972a, personal communication, 1980).]

Table 9–2. Stratigraphy of the Nama Group[a]

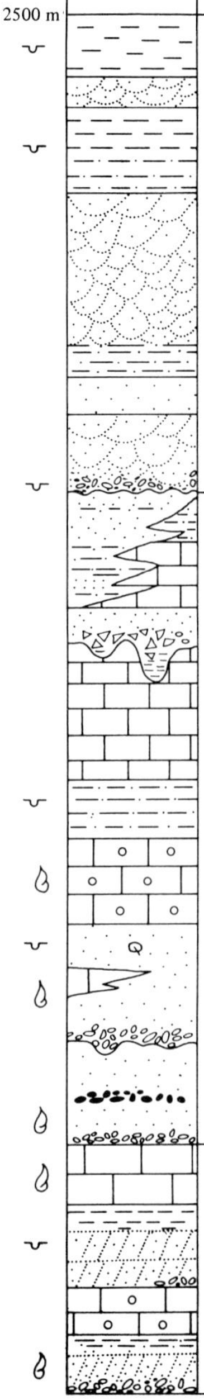

Subgroup	Formation Member	Description	Interpretation
FISH RIVER SUBGROUP	GROSS AUB	Shale and mudstone. Fossils: *Phycodes pedum*	Tidal flat and fan delta
		Sandstone (very coarse grained, cross-bedding)	Braided alluvial plain
		Mudstone with thin sandstone interbeds. Fossils: *Phycodes pedum, Skolithos, Enigmatichnus africani*	Tidal flat and fan delta
	NABABIS Haribes	Red arkosic sandstone (medium to coarse grained, channels, trough cross-bedding, asymmetric ripples); mudstone partings (desiccation cracks)	Braided alluvial plain
	Zamnarib	Red mudstone with quartz arenite interbeds (channels, ripple cross-lamination, desiccation cracks, runzel marks)	Tidal flat and fan delta
	BRECKHORN	Sandstone (trough cross-bedding, mudstone partings)	Braided alluvial plain and distal fan delta
	STOCKDALE	Red sandstone (conglomeratic and coarse grained basally, trough cross-beds directed to SE and E) and mudstone at top	Braided alluvial plain and fan delta
SCHWARZRAND SUBGROUP	VERGESIG Niep	Red mudstone and sandstone overlain by red sandstone and mudstone in N; blue-green mudstone and limestone overlain by green sandstone and shale in S. Fossils: *Phycodes* and *Planolites*	Tidal flat and fan delta followed by braided alluvial plain in N; subtidal followed by fan delta in S
	NOMTSAS Kreyrivier	U-shaped valleys, striated floors, shale, diamictite	Glacial?
	URUSIS Spitskop	Blue limestone (oolites, cross-bedded grainstones, columnar stromatolites, hardgrounds and desiccation cracks along E margin); interbedded sandstone, siltstone, shale. Fossils: *Cyclomedusa*	Clear-water shoaling platform, stromatolitic biostromes, tidal flat
	Feldschuhhorn	Green and red shale and siltstone with interbedded sandstone. Trace fossils	Tidal flat
	Huns	Blue and yellow limestone (oolites, first stromatolites) interlayered mudstones thicken eastward. Fossils: *Cloudina, Cyclomedusa*	Carbonate platform, stromatolitic biostromes, tidal flat
	Nasep	Arkosic sandstone (cross-beds directed to W overlain by mudstone-siltstone; interbedded limestone, quartz arenite and diamictite. Fossils: *Nasepia, Pteridinium, Paramedusium*	Fan delta in N; intertidal and subtidal with carbonate shoals in S
	NUDAUS	Quartz arenite and arkose (polymodal cross-bed distribution), mud-pebble conglomerate, green shale (tidal bedding structures, desiccation features). Fossils: *Rangea, Pteridinium, Planolites, Skolithos*	Distal fluvial and intertidal in N; barrier islands and carbonate shoals in S
KUIBIS SUBGROUP	ZARIS	Blue limestone (current-directed stromatolites), shale interbeds thicken to E. Fossils: *Cloudina*	Reef–lagoon complex, carbonate platform
	DABIS Kliphoek	Arkosic sandstone (planar cross-beds directed to W), flaser-bedded shale, quartz arenite, evaporite. Fossils: *Rangea, Pteridinium, Namalia, Ernietta, Orthogonium, Skolithos*	Alluvial fan, tidal flat, lagoon in W
	Mara/Kanies	Conglomerate, arkose (mudstone interbeds, cross-beds directed to W), shale, quartz arenite, organic-rich dolomitic limestone. Fossils: *Cloudina*	Braided alluvial plain; debris flows, bioherms and tidal flats in W

[a]Compiled from Martin (1965), Germs (1972a, b, 1974, personal communication, 1980).

and stromatolitic limestone (Table 9-2, Fig. 9-5C). In the southern part of the basin the suite is more mature and is dominated by a limestone–quartz arenite association (Germs, 1974). No true dolomites have been found in the Nama Group (G.J.B. Germs, personal communication, 1980). The most common sedimentary structure is planar cross-bedding followed by trough cross-bedding directed toward the west (Germs, 1972a). Westward thickening of carbonate intervals is accompanied by an increase in Ca/Mg ratios. Taxa of the Ediacara-type fauna are conspicuous (Germs, 1972a,b, 1974).

The Schwarzrand Subgroup is a conformably based succession dominated by thick limestone and terrigenous clastic sediments which become increasingly less mature upward (Table 9-2; Fig. 9-7). The carbonates form eastward-thinning wedges which become more magnesian and finger out into thick shales along the basin margin. Ediacara-type fauna is common only in the lower and middle parts of the Schwarzrand suite (Table 9-2). Two alleged glaciogenic intervals are present (Germs, 1974): the first erosively overlies the upper Dabis Formation and basal Nudaus Formation (Schwellnus, 1941; Germs, 1974; Schermerhorn, 1974); the second fills deep channels incised into the uppermost Huns limestone (Urusis Formation) (Fig. 9-7) (Germs, 1974). Cherts of possible volcanic origin are associated with the limestones. The first red beds in the Nama succession occur in the clastic Nomtsas Formation at the top of the Schwarzrand Subgroup (Table 9-2).

Fish River sedimentation is correlated with late-stage Damaran tectonism and is inferred to have had a molasse-type origin. Resting unconformably or paraconformably on older Nama rocks, the Fish River red-bed sequence consists largely of trough cross-bedded arkose and subarkose deposited by fluvial processes (Table 9-2; Figs. 9-7 and 9-8). Cross-bed azimuths reflect a diverse flow pattern in which fluvial currents flowed south–southeast toward Keetmanshoop, where highly variable paleoflow directions coincide with the presence of inferred tidal sequences (Germs, 1972a).

Westward-Prograding Alluvial Systems (Kuibis Subgroup)

During the Damara orogeny coarse-grained immature sediments were deposited on the Kalahari craton (Martin, 1965; Germs, 1972a, 1974; Hartnady, 1976, 1978a, b). These were initially conglomeratic, the largest clasts being deposited in the vicinity of the Osis Ridge, an uplift of crystalline basement (Fig. 9-8A). Within a short time a clastic wedge began to prograde westward, with the margin of the Nama Sea separating an eastern braided fluvial environment from a western shallow-marine environment (Fig. 9-8B). Alluvial deposits are preserved in the Kanies and Kliphoek Members of the Kuibis Subgroup (Table 9-2).

Thin basal conglomerates are present in alluvial sequences that show an irregular upward reduction in grain size. The clasts are angular to subrounded and include vein quartz, sandstone, and shale. Very poorly sorted, matrix-supported conglomerates with angular clasts are more typical of the lower Nama succession. These unstratified units suggest rapid dumping of sediment, possibly as debris flows. Clast-supported conglomerates are more mature, with better rounded clasts; these are braided-river deposits (G.J.B. Germs, personal communication, 1980).

The conglomerates are succeeded by reddish, medium- to very coarse-grained arkose, subarkose, graywacke, and lithic arenite. The lenticular sandstone units contain basal pebble lag and sets of westward-directed planar cross-beds and subordinate trough cross-beds, which decrease in thickness basinward and upward. Desiccation cracks are common shale partings.

These oxidized, generally mature sediments resemble the deposits of braided alluvial plains. The clast-supported conglomerates are believed to have accumulated as coarse channel fill and lag, whereas the unstratified, matrix-supported types reflect debris flows on steep gradients. Lower gradients characterized by sandy transverse braid bars are interpreted from the characteristic planar cross-bedding. Occasional overbank flooding is inferred from the shale partings with desiccation structures.

Distal Alluvial Plain–Fan Delta Complexes (Kuibis, Schwarzrand, and Fish River Subgroups)

Distally and vertically the coarser alluvial sediments grade into fine- to medium-grained feldspathic sandstone and graywacke. Coarser grain sizes are confined to thin beds and lenses.

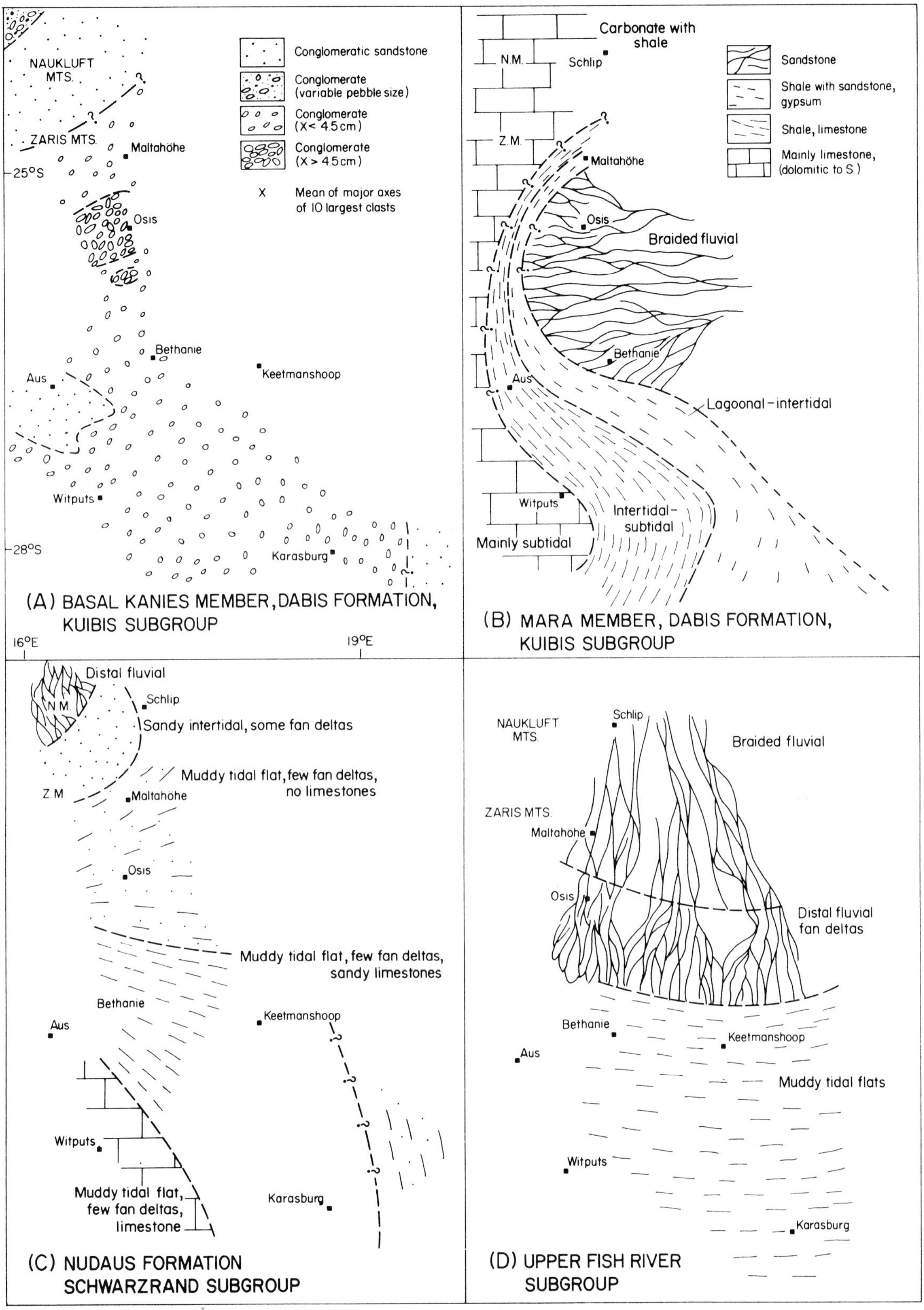

Figure 9-8. Paleofacies maps of the Nama Group. (A) Basal Kanies Member, Dabis Formation, Kuibis Subgroup. (B) Mara Member, Dabis Formation, Kuibis Subgroup. (C) Vingerbreek Member, Nudaus Formation, Schwarzrand Subgroup. (D) Upper Fish River Subgroup. [After Germs (1972a, personal communication, 1980).]

The sandstones are present either as thin beds interfingering and alternating with muddy tidal-flat facies or as thick units isolated from the tidal flats. Intercalations of limestone and mudstone are present. Sedimentary structures include trough cross-bedding, plane-bedding, ripple and ripple-drift cross-lamination, and convolute bedding.

The association of sedimentary structures, including desiccation cracks, gypsum, and fossils of marine affinity, strongly suggests that these argillaceous and carbonate rocks accumulated on tidal flats. Sandstone units in contact with tidal deposits in places show a characteristic vertical ordering; a flat erosive base with flute casts is overlain by matrix-supported mudclast conglomerates and large-scale (8–10 m) sigmoidal foresets. Other large-scale foresets overlie finer grained sediments conformably.

Distal alluvial-plain deposition was characterized by flood-stage scouring and rapid dumping of sediment load. The common intertonguing with argillaceous tidal deposits and the polymodal distribution of distal cross-bed azimuths (Germs, 1972a, 1974) suggest that the sandstones originated as tidal-channel fill and fan deltas along a low-energy shoreline. The large gradationally based, sigmoidal foresets are thought to represent fan deltas; these form an upward-coarsening association with tidal facies.

Shoaling-Platform and Tidal-Flat Environments (Kuibis, Schwarzrand, and Fish River Subgroups)

Thick sequences of platform limestones and terrigenous clastic tidal-flat sediments occur on the periphery of the Kalahari Province (Fig. 9-8C). Limestones are thickest in the southwest and thin northeastward toward the shoreline (Figs. 9-7 and 9-9). Three types of tidal flat may be differentiated: (1) sandy without carbonates; (2) muddy without carbonates; and (3) carbonate rich. Landward the carbonates grade into green and reddish shales (Germs, 1972a, 1974).

In the west the limestones are typically thickly bedded, micritic, and, in the northern Kuibis basin, oolitic. Locally bedload transport has remolded the oolitic and bioclastic material into bimodal–bipolar cross-bedding, ripple cross-lamination, and wavy bedding. Interaction of current processes with the biota generated huge, domical stromatolitic reefs subparallel to the basin margin. The stromatolites attain heights of 25 m near Witputs (G.J.B. Germs, personal communication, 1980) and occupy reef cores, whereas bioclastic material overlaps reef margins. Closely associated with many stromatolites is the serpulid-like *Cloudina*. Landward shoaling was accompanied by an overall decrease in the average stromatolite size, decreasing Ca/Mg ratios, and yellow-brown colors. Near the margin of the basin dolomitic limestones are capped by hardground exposure surfaces with abundant desiccation cracks, suggestive of periodic exposure (Germs, 1972a, 1974).

Quartz arenites also accumulated along the edge of the basin and, like the carbonate rocks, they interfinger landward with thick shale sequences (Germs, 1972a, 1974), suggesting that they may have formed local beach barriers. This inference is supported by the presence of herring-

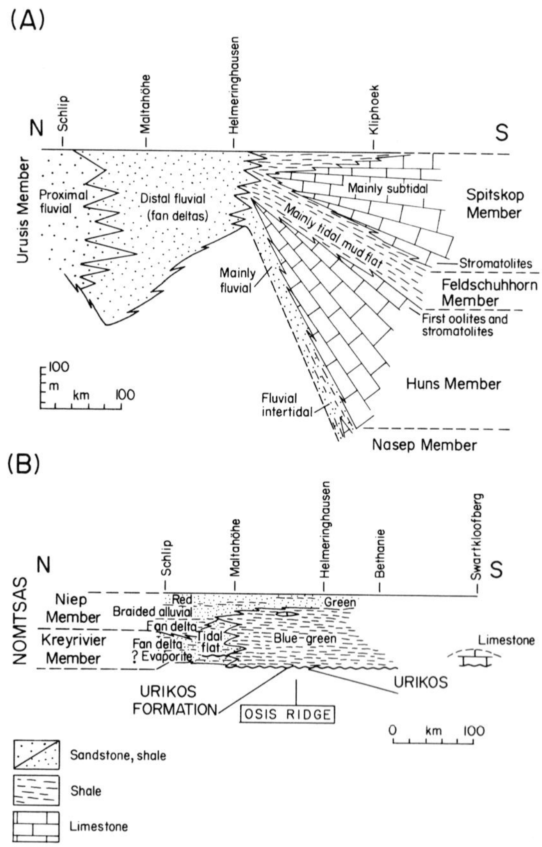

Figure 9-9. North-south geologic sections through the Schwarzrand Subgroup, Nama Group. (A) Urusis Formation; (B) Nomtsas Formation. [After Germs (1972a, personal communication, 1980).]

bone cross-bedding, polymodal ebb-dominated foreset distribution typical of barrier inlet deposits, *Skolithos*, and plane-bedding reminiscent of beach foreshore deposition. However, on the northern basin margin the flat-bedded sandstones are immature and largely unimodal in their paleocurrent distribution, suggesting fluvial fan-delta rather than shore-zone processes (G.J.B. Germs, personal communication, 1980).

Both the limestones and the quartz arenites grade eastward into thick sequences of green and slightly reddish Nudaus shales and mudstones characterized by trace-fossil assemblages (Table 9-2). Flaser, wavy, and lenticular bedding are common. Evaporitic conditions are inferred from the association of abundant desiccation cracks and rare white bird's-eye calcite and gypsum (Germs, 1972b). Thin beds of quartz arenite contain small-scale cross-bedding. The bedding surfaces preserve several structures diagnostic of shallowing and emergence: flat-topped, double-crested, and ladderback ripples, rill marks, falling-water marks, raindrop impressions, and desiccation structures.

Bahamian-type platform deposition characterized the Nama limestone shelf where traction currents redistributed oolitic and bioclastic material. Eastward shoaling is reflected in the decreasing height of stromatolitic bioherms from 20 m to 1.5 m and in crusts of dolomitic limestone along the margin of the basin. Desiccation cracks and landward interfingering with evaporite-bearing argillite indicate deposition on intertidal or supratidal flats (cf. Shinn *et al.*, 1969). The presence of marine fossils, tidal-bedding structures, emergence features, and evaporitic residues in the shale and mudstone confirm the presence of extensive tidal flats west of the braided alluvial plains (Germs, 1974).

Possible Glacial Environments (Schwarzrand)

Two glacial events may have interrupted the prevailing pattern of terrigenous clastic deposition and accumulation of dolomitic limestones in the Schwarzrand (Germs, 1972a, 1974). Glaciogenic processes are indicated by the presence of diamictites containing faceted and striated clasts deposited on grooved surfaces.

Diamictite, up to 3 m thick (average 1 m), is present in the lower Schwarzrand Subgroup in the Klein Karas Mountains (Fig. 9-6). It contains angular to rounded clasts up to 1 m in diameter, comprising quartzite, shale, limestone, and rare basement-derived granitoids and schists (Schwellnus, 1941; Martin, 1965; Schermerhorn, 1974). Martin (1965) suggested that the grooved channels containing the diamictite were formed by ice floes with embedded clasts grounding on soft sediments. This interpretation was disputed by Schermerhorn (1974) because of the absence of dropstone strata; but dropstone varvites are also uncommon in Quaternary glaciomarine deposits and glaciolacustrine deposits. The presence of syndepositional folds (Germs, 1972a) confirms some soft-sediment deformation.

The second possible glacial episode during Nomtsas times (Table 9-2 and Fig. 9-7) postdates carbonate deposition in the Nama basin. U-shaped valleys up to 200 m deep were gouged into the upper Schwarzrand carbonates, their surfaces bearing northeast-trending striations (Germs, 1972a). Green shale or conglomerate of the Nomtsas Formation mantles the channel floors; overlying matrix-supported diamictites contain clasts up to 20 m in diameter. Martin (1965) inferred an origin by debris flow for the diamictites, but G.J.B. Germs (personal communication, 1980) believes this is unlikely because columnar stromatolites are known to have grown within these deposits, which he believes to be alluvial in origin (Fig. 9-9B). Shallowing of the channels to the north-northeast has been attributed to tectonic control over fluvial style (Germs, 1972a). The apparent contradiction between sedimentological and paleomagnetic indicators of climatic seasonality in equatorial latitudes and the presence of glacial deposits is best resolved by invoking a severe world climate and drift of ice to low latitudes (Kröner *et al.*, 1981).

Fish River Molasse

A series of unconformities sets the Fish River Subgroup apart from the lower Nama. A largely alluvial genesis is reflected in the thick succession (about 1000 m) of red, medium- to coarse-grained arkosic sandstones, which display a strongly unimodal cross-bed distribution. Tidal-flat and fan-delta environments are inferred in the south (Fig. 9-8D).

Limestone and thick shale, which characterize

older Nama rocks (Martin, 1965; Germs, 1972a, 1974), are absent. Basal clastic sediments, including conglomerates, were deposited as a clastic wedge (Kabib Member) on an incised topography by currents flowing to the south-southeast. The angular to subrounded conglomerate clasts include chert, agate, jasper, silicified oolitic limestone, and mudstone. Most of the alluvial facies comprise lenticular units of feldspathic sandstone separated by desiccation-cracked mudstone. Channels are up to 4 m deep and 30 m wide. Trough cross-bedding is the dominant structure.

Channel scouring, channel-floor dune migration, and occasional overbank deposition resulted in a succession closely resembling the deposits of the lower reaches of braided alluvial plains. Although there are elements of southward flow in early Schwarzrand times (Fig. 9-8C), by the beginning of Fish River times the paleoflow had a strong southern component (Fig. 9-8D), which indicates a provenance in the Damara Province (Germs, 1972a). This suggests that the Damaran orogeny had reached a terminal stage in which the rising mountain chain was shedding molasse-type debris off its flanks to accumulate as the Fish River Subgroup (Germs, 1974). However, Germs' observation of significant amounts of detrital plagioclase in the Fish River arenites militates against their provenance being the top of a young nonvolcanic mountain chain; the Rehoboth magmatic arc, between the Damara Province and the Nama basin, is more likely to have supplied the plagioclase (H. Martin, personal communication, 1980). The Fish River Subgroup has affinities with Lower Paleozoic alluvial sediments of the Cape Supergroup (Chapter 10).

In the distal areas to the south, thin argillites are attributed to tidal-flat deposition, whereas thick quartz arenites are ascribed to shore-zone reworking of alluvial facies. The quartz arenites and related lithic sandstones are structured mainly by cross-bedding and plane-bedding. Planes of low-angle truncation, current lineation, and heavy-mineral segregations in the plane-bedded units are typical of the beach foreshore. Tidal processes resulted in a polymodal cross-bed distribution and, in the finer grained components, produced flaser, wavy, and lenticular bedding. Ripple wavelenths as small as 1 or 2 cm indicate very short-period waves. Periodic exposure is shown by ubiquitous desiccation cracks, runzel marks, and raindrop imprints. Marine trace fossils, particularly *Phycodes*, are most abundant in the uppermost strata. There is little vertical zonation of the tidal-flat sediments, suggesting that tidal ranges were very small.

Nama-Related Rocks in the Southwestern Cape

Pre-Cape rocks in the Vanrhynsdorp area of the Cape (Fig. 9-6) comprise limestone, quartzite, and shale deposited on deeply dissected basement of the Kalahari Province. Traditionally the lower facies, dominated by limestone and phyllite, have been correlated with the Malmesbury Group of the southwestern Cape (Rogers, 1904; Jansen, 1960), and the quartzose upper parts with the Kuibis and Schwarzrand suites of the Nama Group (Lamont, 1947; Brink, 1950; Jansen, 1960). Kröner (1968) recognized a conformable succession that he correlated entirely with the Kuibis Subgroup. Compressive deformation generated northwest-trending folds, thrust faults, shearing, and greenschist-facies metamorphism.

Kröner (1968) divided the stratigraphy into three components. The lowest occupies the deepest part of the basin and consists of shale overlain by great thicknesses of thinly bedded blue-black limestone and dolomite with phyllite intercalations. Northward and westward thinning of the carbonate rocks is accompanied by thickening of the phyllite. The second component comprises phyllite and slate with interbeds of quartzite and quartz-sericite schist. The uppermost component is dominated by quartzite with subordinate phyllite and shale. Arkose and conglomerate, in which gneiss and quartzite clasts are abundant, mark the base of the quartzites. The quartzites typically thin eastward and southward, where they interfinger with metaargillites. They overstep the lower two units northward to rest directly on basement gneisses.

Kröner (1968) ascribed the association of thick quartzite with coarse-grained, immature terrigenous clastics to deposition in coastal and alluvial-plain environments. Shallow-shelf paleoenvironments can also be inferred for the carbonate rocks; they grade landward into thick sequences of argillaceous rocks, which probably accumulated on tidal flats.

Nama Metazoan Fauna

Although fossils were rarely preserved in the Precambrian, certain ancient polychaete worms, such as *Cloudina* (Fig. 9-10A), developed mineralized shells or skeletons (Glaessner, 1979). The oldest megascopic multicellular animals (Metazoa) constitute the Ediacaran fauna of uppermost Precambrian or Vendian age, first discovered in the Pound Quartzite of South Australia (Glaessner, 1959). They are a distinctive marine assemblage, characterized by nonmineralized tissue, and have been reported from five of the seven continents (Lowenstam, 1980). The Ediacaran fauna usually postdates the last Precambrian glaciation and, in southern Africa, the inferred glacial episode in Numees times (Gariep Group). The fauna appears to be broadly contemporaneous with the Nama glacial episode (Germs, 1972a, b, 1974; Glaessner, 1979).

The Kuibis and Schwarzrand Subgroups contain a diverse Ediacara-type fauna (Table 9-2) which inhabited tidal flats, shallow seas, and stromatolitic biostromes (Germs, 1972a, 1974). The coelenterates *Rangea* (Fig. 9-10B), *Pteridinium*, *Namalia*, and *Ernietta*, and a possible sprigginid worm flourished on tidal flats fronting braided alluvial plains during Kuibis times. *Rangea*, *Pteridinium*, and the medusoid *Cyclomedusa* (Fig. 9-10C) extend into the lower Schwarzrand Subgroup and, with *Nasepia*, are diversified in quartz arenite–carbonate suites. Algal mats and columnar stromatolites were particularly abundant on the shallow carbonate platform, scavenged only by the polychaete *Cloudina* (Germs, 1972a). Its cylindrical wall structure, the main layer of which consists of stacked inverted cones, has been attributed to the inhibiting structure of the stromatolites. Except for *Cyclomedusa* and *Cloudina*, the body fossils in the Nama Group are Coelenterata referred to the "Petalonamae" (Glaessner, 1979). They are leaflike structures ("petaloids"), commonly with a median line or groove and lateral grooves on each "leaf." The medusoid *Cyclomedusa* is disk-shaped with concentric grooves.

Body fossils are still unknown in the Fish River Subgroup although several trace fossils are recorded. *Phycodes pedum* is most abundant in the uppermost strata (Fig. 9-10D).

The Ediacaran faunas represent an important transitional stage in evolution, which straddles the boundary between the Phanerozoic and the Precambrian. The wide range of organisms and evidence for their mobility, in the form of tracks and burrows, suggest some degree of Ca^{2+} regulation (Lowenstam, 1980). The faunas are almost invariably preserved as soft-part impressions rather than mineralized skeletons, but it is probable that any incipient mineralization would be dismissed as diagenetic precipitation. Post-Ediacaran faunas just before the Cambrian, are characterized by an absence of preserved soft parts, which suggests that scavenger and deposit-feeder-induced bioturbation had become effective by then (Lowenstam, 1980).

Summary of Stratigraphy and Paleontology

Nama sediments were deposited in a broad basin fringed in the north and west by the Late Precambrian Damara and Gariep Geosynclines during the last stages of orogeny. The succession rests conformably or paraconformably on uppermost miogeosynclinal Gariep strata (Numees and Bogenfels Formations) or, more generally, unconformably overlies crystalline Kalahari basement. The lower Nama succession comprises a seaward- or westward-thickening prism of terrigenous clastic and carbonate rocks deposited in alluvial, tidal-flat, and shallow-shelf environments. Shelf platforms were dominated by carbonate facies. A complex history of transgressions and regressions (Table 9-2) is related to tectonic control of basin geometry and glacioeustatic fluctuations of sea level (Germs, 1972a, 1974). Axial rise of the Damara Geosyncline at the close of orogenesis resulted in deposition of upper Nama alluvial red beds by southward flowing braided streams. Rapid dumping of sediment prohibited development of carbonate facies.

Paleontologically the Nama Group is unique in the subcontinent. Together with other uppermost Precambrian (Vendian) fossil localities in South Australia, the U.S.S.R., Europe, and North America, it marks the first appearance of definite metazoan body fossils (Glaessner, 1979). These Ediacaran assemblages of marine animals were dominated by coelenterates and polychaete annelids lacking mineralized shells or skeletons. Glaessner (1979) attributed the rapid evolution of metazoan animals into the Cambrian to changing oceanographic conditions. The Ediacaran faunas are also significant because they predate the base of the Phanerozoic, historically placed at the first appearance of metazoans. This Precambrian–Phanerozoic boundary problem is under review (Cloud, 1976; Glaessner, 1979).

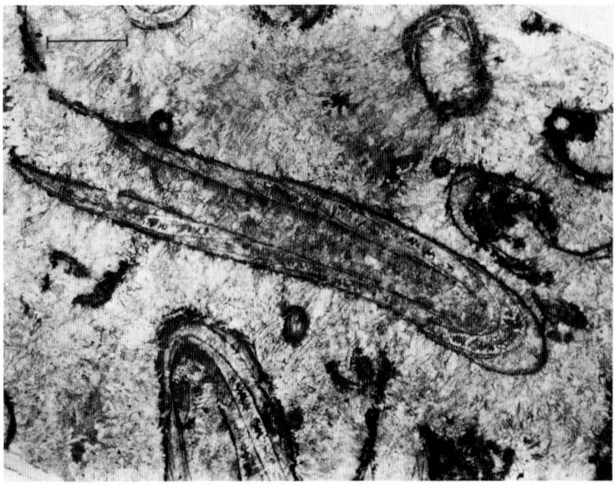

(A)

(B)

(C)

Figure 9-10. Nama metazoan fauna: (A) *Cloudina*, (B) *Rangea*, (C) *Cyclomedusa*, (D) *Phycodes* (trace fossil). [Courtesy of G.J.G. Germs.]

Fig. 9-10D

The Naukluft Nappe Complex

The spectacular thrust–nappe complex within the Naukluft Mountains on the northern margin of the Nama basin (Fig. 9-6) has, despite its small size, generated wide interest since it was first described by Korn and Martin (1959). Not only does it rank as a classic locality of thrust–nappe tectonics during the Pan African event but it also serves as a Rosetta stone for the understanding of Late Precambrian and Cambrian stratigraphy in Namibia. Critical evidence of age relations between the Nama Group and unusually well-preserved correlates of the Damaran Swakop Group are contained within the highly complex tectonostratigraphy of the Naukluft nappe pile.

The Naukluft nappe complex rests allochthonously on deformed Nama strata, separated from them by the "Unconformity Dolomite" (Korn and Martin, 1959). This thin, remarkably planar layer acted as a weak detachment zone and is underlain by tightly folded strata which are cut by imbricate thrust faults. Slaty cleavage is present in the autochthonous Nama basement for 15 km southeast of the nappe complex (Ahrendt *et al.*, 1978). Korn and Martin (1959) concluded that all components in the nappe pile were part of the Nama Group. Martin's (1974) discovery that at least part of the complex (the Dassie nappe) contains Damara rocks coincided with a detailed reinvestigation of the nappe complex by Hartnady (1974, 1976, 1978a, 1978c). Hartnady reported that much of the Naukluft complex consists of Damara rocks carried southeastward as thrust nappes for estimated distances of up to 78 km out of the rising axial zone (Khomas Ridge) of the Damaran orogen and onto the Kalahari craton where the Nama Group was accumulating (Hartnady, 1978d). Lower Nama sediments were deposited syntectonically and autochthonously incorporated within some of the nappes (Hartnady, 1978a).

Age relations between the Damara Supergroup and the Nama Group have long been speculative because the two successions are nowhere in contact outside the Naukluft Mountains. The Nama sediments have been tentatively regarded as younger than those in the Damara Geosyncline, mainly on account of the anorogenic tectonic setting of the Nama (e.g., Martin, 1965, p. 119, 1974; Haughton, 1969, p. 485). However, the accumulated evidence led Kröner (1971) and Hartnady (1974) to conclude that the Nama was a miogeoclinal shelf equivalent of the eugeosynclinal Swakop Group (Damara Supergroup). Germs (1972a) disagreed with this interpretation on paleontological grounds and because he opposed equating the poorly developed Schwarzrand (Nama) diamictites with the extensive and thick Chuos (Damara) diamictites as proposed by Kröner (1971). Hartnady (1976) showed that lower Nama sediments (Zebra River Formation) within the Naukluft nappe complex unconformably overlay erosionally truncated thrust faults in Damara sediments (Dassie Formation, renamed the Büllsport Formation by Hartnady, 1978a). It

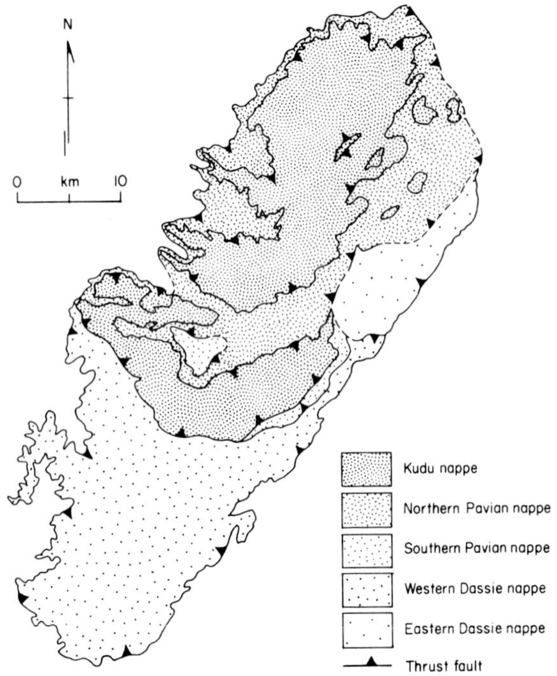

Figure 9-11. Map of the Naukluft nappe complex in the northwest Nama basin (Fig. 9-6). [After Hartnady (1978a).]

is therefore clear that the Nama Group is not a time equivalent of the eugeosynclinal Damaran Swakop Group but that the Nama began to accumulate syntectonically during the final stages of the Damaran orogeny. The upper Nama succession is most likely a late-syntectonic foredeep molasse deposit, probably equivalent in age and tectonic setting to the Mulden Group of probable Early Cambrian age on the northern side of the Damara Province (Kröner, 1971; Germs, 1972a; Kröner and Clauer, 1979).

The Naukluft nappe complex comprises five major south-verging thrust nappes (Fig. 9-11) which have been lithostratigraphically subdivided by Hartnady (1978a) into eight formations (Table 9-3). The Nama Group serves as an autochthonous foundation for the nappe complex and probable stratigraphic equivalents to it are also present within the complex as the Zebra River Formation. All other formations are correlated with the Damara Supergroup.

Angular unconformities at three main stratigraphic levels (Table 9-4) show that upper Damara and lower Nama formations were deposited during progressive southward translation

Table 9–3. Tectonic and Stratigraphic Units of the Naukluft Complex[a]

Tectonic Unit	Lithostratigraphic Unit
Kudu nappe	Klipbokrivier Formation
	? (erosional unconformity)
	Noab Formation
Northern Pavian nappe	Tsabisis Formation
	Blässkranz Formation
	Remhoogte Formation
Southern Pavian nappe	Zebra River Formation (?)
	?—?
	Aubschlucht Formation
Western Dassie nappe	Upper Zebra River Formation (?Nama Group)
	Lower Zebra River Formation (?Nama Group)
	Büllsport Formation
Eastern Dassie nappe	Zebra River Formation (?)
	Büllsport Formation

[a] ═══, tectonic boundary; ——, conformable contact; ∼∼, erosional unconformity. After Hartnady [1978a].

Table 9–4. Tectonostratigraphic History of the Naukluft Nappe Complex[a]

Deformational Episode	Depositional episode
	Deposition of dolomitic Noab and clastic Büllsport Formations in sabkha–tidal flat environment.
D0 Early (?tensional) fracturing of Noab Formation.	Deposition of largely clastic Remhoogte Formation (and Klipbokrivier Formation?) in shallow-marine environment.
D1 Deformation of Remhoogte Formation.	〰〰〰〰〰〰〰〰〰〰
	Deposition of olistostromic Blässkranz Formation breccias in shallow- to deep-marine environment, followed by volcaniclastics, dolomite, and purple shale of Tsabisis Formation in shallow-marine environment.
D2 Overthrust emplacement of dolomitic Kudu-Dassie nappe over Northern Pavian nappe (Remhoogte, Blässkranz, and Tsabisis Formations). Minimum cumulative displacement during D1 and D2 is 33 km toward azimuth 170° relative to Northern Pavian nappe.	Deposition of Aubschlucht Formation conglomerates and coarse clastics (in alluvial fan environment?).
	(probable major hiatus) 〰〰〰〰〰〰〰〰〰〰
	Deposition of lower Zebra River Formation in progressively deepening shallow-marine environment, transgressive across imbricated frontal zone of Dassie nappe.
D3 Emplacement of Kudu–Dassie nappe over Aubschlucht and Lower Zebra River Formations, and major imbrication and individualization of Kudu–Dassie–Northern Pavian nappe complex. Minimum displacement during D3 is 16 km toward azimuth 170° relative to frontal Dassie nappe.	〰〰〰〰〰〰〰〰〰〰
	Deposition of upper Zebra River Formation in shallow-marine environment.
D4 Folding and local imbrication of upper Zebra River Formation.	
D5 Formation of main basal décollement at sole of Naukluft complex and associated Unconformity Dolomite.	

[a] 〰, angular unconformity. After Hartnady [1978a].

and upward construction of the nappe pile (Hartnady, 1978a). According to Hartnady (1976), early pre-Nama deformation took place at deeper crustal levels (low-grade greenschist metamorphism) than the crustal levels of the Nama (burial metamorphism only) onto which the nappes were allochthonously emplaced during the late D5 episode (Table 9-4). The sole of the nappe complex cuts across the regional metamorphic zonation displayed by the underlying Nama (Ahrendt *et al.*, 1978). Because of their southward translation out of the Damara orogen, the Damaran rocks of the nappe complex largely escaped both the intense deformation and the postkinematic thermal metamorphism that affected the eugeosynclinal Damara Province (Hälbich, 1970; Hoffer, 1975; Kasch, 1978a, 1979a). During D5 the nappe complex as a whole was emplaced, possibly by riding on a lubricating sole represented by the "Unconformity Dolomite"; traction near the leading edge is thought to have induced the folding and imbricate thrusting in the underlying Nama Group (Hartnady, 1978a). The paucity of D5 strains within the nappe complex as a whole suggests that by this stage the thrust sheets had locked together and were acting as a single tectonic entity.

Work on the "Unconformity Dolomite" by Behr *et al.* (1981) suggests that this remarkable unit may have been derived from a Damaran evaporite sequence and emplaced as a hot intrusion. The rock type is an albite dolomite; its high content of sodium and boron silicates and its highly saline (40% NaCl) fluid inclusions suggest that it was derived from a carbonate–evaporite sequence; such a sequence has now been recognized in the Duruchaus facies of the Nosib Group some 70 km to the north (H. Martin, personal communication, 1980). Fluid-inclusion studies show that the "Unconformity Dolomite" was emplaced at temperatures 100–200°C higher than metamorphic temperatures affecting the allochthonous overburden and autochthonous basement, but at a crustal level 2 km higher. Behr *et al.* (1981) therefore concluded that the "Unconformity Dolomite" was emplaced as a suspension of fragmented carbonate–evaporite sediment and exotic material in hot brine at 495 Ma ago (Ahrendt *et al.*, 1978; Behr *et al.*, 1981; K/Ar white mica). Mylonitization associated with final nappe emplacement has affected the underlying rocks but not the "Unconformity Dolomite," indicating that the latter did not act as a lubricating layer in the final stages of thrusting.

The relationship between the Naukluft nappes and the Damaran orogeny is further discussed in Section 9.4.

Post-Nama Epeirogeny

Except for its fringes, the Nama platform sequence has been preserved in a virtually unmetamorphosed and undeformed state. The sequence has a very low regional dip to the east so that successively younger strata are exposed in this direction. The gentle homoclinal structure, which stands out vividly on multispectral Landsat imagery, is distorted by a broad north-northeast-plunging syncline–anticline pair to the east of Maltahöhe (Short *et al.*, 1976; their Plate 350). The generally flat-lying regional structure has been disrupted by folding on the southwestern and northwestern margins and by pervasive block faulting that controlled the uplift and consequent exposure of Namaqua and Gariep basement.

The Kuibis Subgroup northeast of Rosh Pinah mine (Fig. 9-6) has been deformed into upright folds with axial crenulation lineations that plunge gently northwest or southeast, in common with the underlying more deformed Numees Formation of the Gariep Group (McMillan, 1968). The northwest-trending folds can be traced northward into the area south of Aus village, where they extend up to 20 km away from the more strongly folded Gariep Province into the Nama foreland area (Jackson, 1976). This deformation clearly results from the final stages in the closure of the Gariep Geosyncline. Open folding of the upper Nama Nomtsas unconformity in the northwest near the Damara Geosyncline (Hartnady, 1978a) may be analogous.

In the south the Nama Group is laced by numerous northwest-trending faults parallel to the boundaries of the adjacent Gariep Province. A prominent example is that on the east side of the Rosh Pinah basement inlier with a throw estimated at 200–300 m (McMillan, 1968). Faults of this orientation extend northward to the Aus area where they dissect the Nama plateau into horst and graben slivers. That some of these are basement controlled is indicated by their concentration in the Nama cover over the Mid Proterozoic Kuckaus and Excelsior strike-slip shear belts, which also trend northwest (Jackson, 1976). Dip-slip reactivation along these deeply rooted shear belts continues today, as evidenced by the presence of Recent fault scarps along them (Jackson, 1976; Blignault, 1977). Northeast of Aus village, throws of 500 m are present along north-trending faults defining the Nama escarpment with no apparent basement control (Jackson, 1976).

Metamorphic grades in the Nama Group increase toward the Gariep and Damara Provinces, according to illite crystallinity indices in Schwarzrand shales (Germs, 1972a; Ahrendt *et al.*, 1978). The age of this metamorphism, which is synkinematic with respect to the folding and formation of slaty cleavage in the Nama Group and Naukluft nappes, is 530 Ma (Ahrendt *et al.*, 1978; K-Ar white micas). The K-Ar age of 495 Ma obtained by these authors from metapelites in the southern Damara Supergroup, the Nama Group, and the Naukluft nappe complex may mark the time of final emplacement of the Naukluft nappes as well as widespread uplift and cooling (Behr *et al.*, 1981).

9.3. The Malmesbury Geosyncline in the Western Saldanian Province

The Malmesbury Group of the southwestern Cape (Fig. 9-12) comprises a varied low-grade assemblage of poorly exposed pre-Cape inliers of sedimentary and metasedimentary rocks; they have been affected by tight upright folding along northwest-trending axes, with considerable displacement by shearing and faulting, during the Pan African orogeny. The term "Saldanian" has been proposed for the Late Precambrian tectonism affecting the Malmesbury Group and for the structural province defined by these rocks (Hartnady, 1969; Hartnady *et al.*, 1974); the term is extended here to cover the pre-Cape inliers farther east. Structural complexity and poor exposure obscure the facies and stratigraphic relations and conceal the extent of the lithostratigraphic units. The base of the sequence is not exposed.

In the absence of either fossil evidence or radiometric ages that can reasonably be interpreted as representing the time of formation of the Malmesbury Group, estimates of its age are based on its relationship to the late-syntectonic to posttectonic Cape granites. Dating of the Cape granites places a minimum age of 600 Ma for the start of the Saldanian orogeny (refer to section on "Syntectonic and Posttectonic Granitoids"). It is generally accepted that the Malmesbury succession postdates the Namaqua event, thus giving a maximum age of about 950 Ma (Burger and Coertze, 1973; Hartnady *et al.*, 1974). However, this cannot be proved because rocks indisputably belonging to the Malmesbury Group are nowhere in contact with rocks of the Namaqua Metamorphic Complex. In the Vanrhynsdorp area a pre-Nama metasedimentary succession, equated by Kröner (1968) with the Gariep Group, unconformably overlies the Namaqua gneisses. A correlation of these metasediments with the Malmesbury Group is also possible.

Two major zones of dislocation divide the Malmesbury Group into three discrete northwest-trending tectonic prisms: southwestern domain, central domain, and northeastern domain (Fig. 9-12). Each tectonic domain has been assigned an independent lithostratigraphy (Table 9-5). Stratigraphic relations within these domains are schematically shown in Figure 9-12A; correlations between domains are highly speculative, but it is possible that the principal graywacke-bearing units, the Tygerberg and Moorreesburg Formations, are equivalent (Hartnady *et al.*, 1974). Stratigraphic ordering has not been well established and the possibility of large-scale inversion is present. The present account draws heavily on the work of Hartnady *et al.* (1974), which is the only comparatively recent systematic review of the Malmesbury Group.

The western Malmesbury has a eugeosynclinal character, whereas the remainder has features of marine-shelf and alluvial environments of rapid deposition. The Malmesbury Geosyncline therefore appears to have formed adjacent to the continental Kalahari Province to the northeast.

Turbidite Sedimentation

Situated in the southwestern domain between the Atlantic coast and a zone of major faulting (Fig. 9-12B), the Tygerberg Formation is dominated by regular alternation of shale and fine-grained graywacke and subgraywacke which are parallel bedded over considerable distances (Table 9-5). The succession contains numerous upward coarsening and upward-fining cycles with bed thickness varying from a few centimeters to several meters. The more arenaceous beds are massive to laminated and commonly fine upward. Other sedimentary structures include cross-bedding, load casts, flame structures, and slump structures. Thin limestone and conglomerate beds are locally present. Altered volcanic rocks exposed along the coast include tuffs and brecciated flows of intermediate composition with amygdaloidal texture (Hartnady *et al.,* 1974).

This succession has been interpreted as the result of deposition by turbidity currents, possibly on a trench plain, with some volcanism (Hartnady *et al.*, 1974). The predominance of argillites and the characteristics of the graywackes suggest slow hemipelagic deposition interrupted by occasional turbidity currents between abyssal plains and lower fans (Mutti and Ricci Lucchi, 1972; Walker, 1970, 1978). Walker (1978) attributed parallel bedding of similar flysch deposits to deposition on a smooth, low-gradient basin plain,

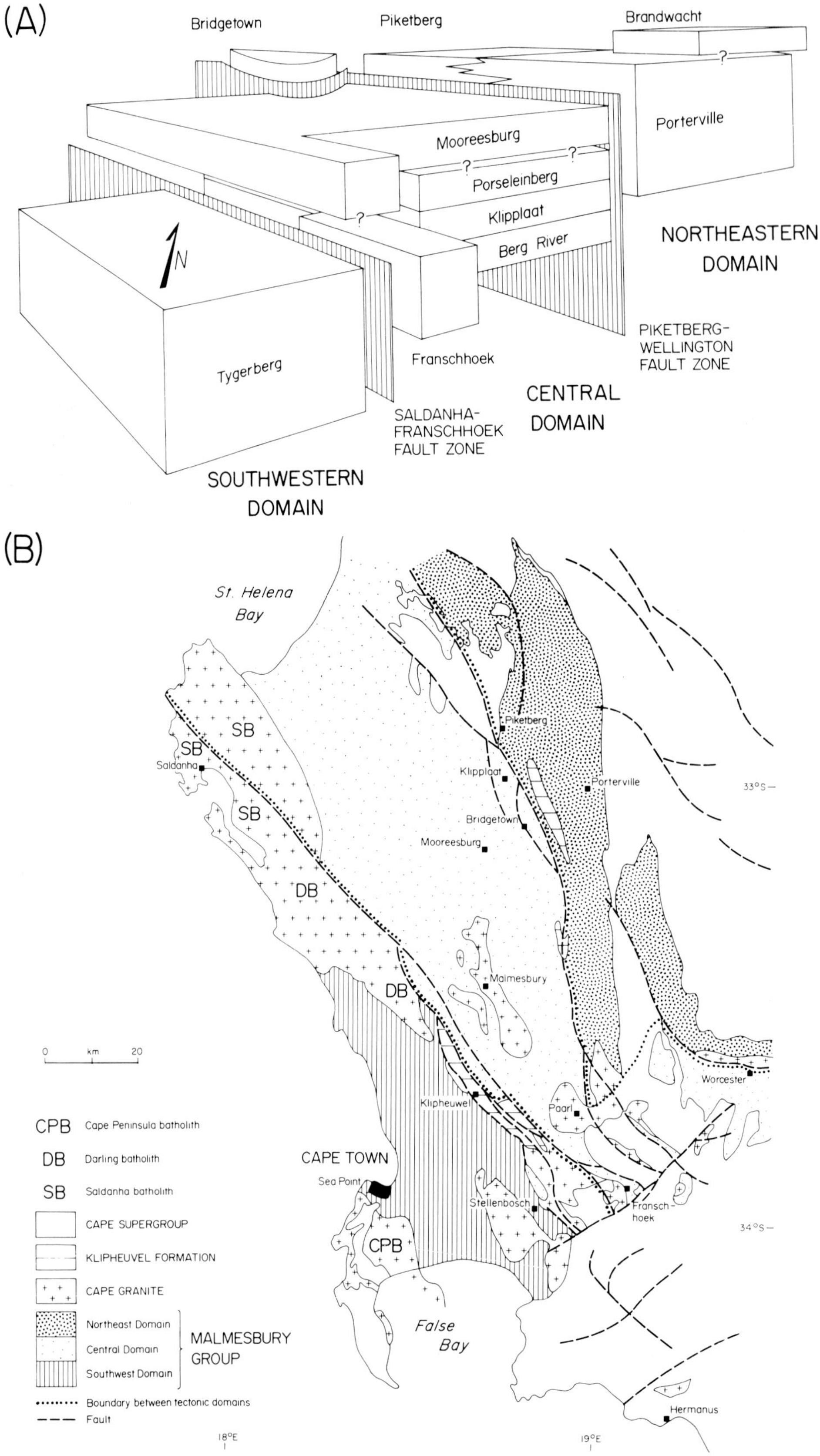

Figure 9-12. The Malmesbury Group, western Saldanian Province. (A) Block diagram to show stratigraphic relations between the three tectonic domains. (B) Geologic map of the Malmesbury Group and younger Cape granites. [Adapted from Hartnady *et al.* (1974).]

Table 9–5. Stratigraphy of the Malmesbury Group, Western Saldanian Province

Southwestern Tectonic Domain	Central Tectonic Domain	Northeastern Tectonic Domain
Tygerberg Formation: Alternating shale and fine-grained immature sandstone (regularly bedded; structures include graded bedding, cross-beds, load casts, slump structures); rare limestone and conglomerate. Intermediate volcanic rocks brecciated).	Franschhoek Formation: Pebble to boulder conglomerate, arkose, subgraywacke, quartzite, and slate (lenticular geometry). Bridgetown Formation: Basic rocks, chert, dolomite, limestone, shale, and phyllite. Swartland Subgroup Moorreesburg Formation: Irregular alternation of wacke and shale. Porseleinberg Formation: Well-laminated quartz–muscovite–sericite schist with interbedded phyllitic chlorite–muscovite schist. Klipplaat Formation: Quartz schist with mudclast breccia; mica schist, basally quartzitic and cherty; rare lenses of limestone. Berg River Formation: Alternating lenticular mica schist and quartz schist, immature wacke, thin limestone, quartzite.	Boland Subgroup Brandwacht Formation: Shale, graywacke with graded bedding, conglomerate; very conglomeratic basally; metabasites in upper part. Porterville Formation: Phyllitic shale and fine- to medium-grained graywacke in irregular units, conglomerate up to 90 m thick, limestone basally, quartzite, chert. Piketberg Formation: Coarse-grained feldspathic sandstone, graywacke, conglomerate, interbedded phyllite especially in south, rare feldspathic limestone.

[a]Compiled from De Villiers *et al.* (1964), Hartnady *et al.* (1974).

with bed thickness and grain size increasing proximally. The lithology of the Tygerberg Formation resembles that of other flysch successions such as the Cretaceous–Tertiary flysch of the northern Appenines (Mutti and Ricci Lucchi, 1972; Ricci Lucchi and Valmori, 1980) and the eastern Alps (Hesse, 1974, 1975).

Proximal (?) Deep-Sea Sedimentation

The Berg River Formation of the central domain contains an interlayered succession of lenticular mica schist and quartz schist, graywacke, and thin limestone in its lower parts (Table 9-5). Grain size and maturity increase upward. This sequence is superseded by quartz schist consisting of a mosaic of quartz grains with mud-clast breccias.

The lenticular bedding, indicative of more-channelized deposition, and the coarser grain sizes and conglomerates suggest more-proximal deposition than that inferred for the Tygerberg Formation. Nevertheless a eugeosynclinal origin is supported by other rock types similar to those in the Tygerberg succession. These include an alternating sequence of wackes and shales and the association of intrusive and extrusive basic rocks (Table 9-5). A suite of quartz–muscovite–biotite schists (Porseleinberg Formation), which are phyllitic at the lower and upper contacts, represents the most intensively deformed and metamorphosed part of the Malmesbury Group (De Villiers *et al.*, 1964; Hartnady *et al.*, 1974).

Fluvial Depositional Environments

The Franschhoek Formation in the central domain was assigned to the Malmesbury Group by Hartnady *et al.* (1974) although earlier authors (Du Toit, 1954; De Villiers *et al.*, 1964; Haughton, 1969) separated them. Several criteria suggest an origin during the later stages of geosynclinal infilling. Some of the conglomerates contain quartz–porphyry clasts ostensibly derived from the late-syntectonic Cape granites (Du Toit, 1954, p. 171). Furthermore the Franschhoek and Malmesbury deformational patterns are similar and both have been infolded with the granites (Haughton, 1969, p. 303). Although lithologically similar to the post-Malmesbury Klipheuwel Formation (Fig. 9-12B and Table 9-5), the Franschhoek metasediments have a much higher degree of deformation, a feature interpreted by Haughton (1969) to indicate their greater age; this does not necessarily follow, because of the known heterogeneity of Saldanian strain. The Klipheuwel Formation (De Villiers *et al.,* 1964) is a molasse-type deposit that has been attributed to prograding fluvial deposition on a basement with high relief (Rust, 1973; Vos and Tankard, 1981).

Deposition of the heterogeneous Franschhoek Formation was probably strongly controlled by the prevailing tectonic activity. Basin infilling was accomplished by dispersal systems that initially contributed small pebbles to boulders up to 30 cm in diameter. Conglomerates are most abundant near highly sheared and mylonitized granitoids (Hartnady *et al*., 1974), which raises the possibility that some of the "clasts" are porphyroclasts of tectonic origin. Superimposed is an upward-fining sequence of intertonguing arkose, subgraywacke, quartzite, mudstone, and slate; argillaceous varieties are more prominent toward the top (De Villiers *et al*., 1964; Haughton, 1969, p. 303; Hartnady *et al*., 1974).

Possible Shelf and Alluvial Environments in the Northeast Domain

A succession of shale and fine- to medium-grained graywacke, which display abrupt thickness and textural changes, dominates the Porterville Formation. Conglomeratic units up to 90 m thick, limestone, and chert are also represented (Table 9-5). A second unit along the northwest periphery, the Piketberg Formation, displays marked facies zonation. Metamorphosed conglomerate and subgraywacke merge southward with a succession of lenticular arkosic sandstone. The latter are interbedded with phyllite and thin limestone toward the south. Graywacke constitutes another major facies, the Brandwacht Formation, which is poorly exposed in the Worcester region in the east (Fig. 9-12). The graywacke varies in texture from pebbly and arenaceous to argillaceous; basal conglomerate is up to 100 m thick. Internally the graywacke is characterized by graded bedding (De Villiers *et al*., 1964; Hartnady *et al*., 1974). A large body of metabasite has been interpreted as metabasalt (Hartnady *et al*., 1974).

Available evidence is insufficient to further characterize the depositional environments. Hartnady *et al.* (1974) envisaged predominantly marine sedimentation within a geosynclinal setting. The lithologic types and the extensive development of graded bedding are comparable with some examples of ancient submarine fans (cf. Walker, 1978). Nevertheless parts of the Boland Subgroup, like the Franschhoek succession, may bear comparison with the Lower Paleozoic alluvial sediments of this region (Vos and Tankard, 1981). Stump (1976) believes that deposition occurred partly in marine-shelf and alluvial environments.

Saldanian Tectonism

The Saldanha–Franschhoek fault zone in the southwest and the Piketberg–Wellington fault zone in the northeast contain vertical shear belts and faults (Fig. 9-12B). These major discontinuities divide the relict geosyncline into two marginal tectonic zones of comparatively simple structure flanking a central zone of complex and less understood structure.

The southwestern and northeastern tectonic domains have several features in common (Hartnady *et al*., 1974). Bedding has commonly survived in both domains in the arenaceous rocks and is overprinted by strong D1 slaty cleavage in the argillites. The strata have been tightly folded into steeply inclined major D1 folds but stratigraphic inversion is rarely recognized. D2 deformation was weak and sporadic and took the form of minor folds, fracture cleavage, and dilation veins with quartz fibers. The sporadic development of these late structures and their association with sinistral simple shear indicate that they may be the result of the faulting and shearing described below.

The structural evolution of the northeastern domain in the extreme east around Worcester is more complex (Hartnady, 1969). The oldest structures are major, asymmetric south-facing folds, which may have been associated with thrusting. D2 shortening along northeast axes resulted in a regional foliation and superposed minor reclined folds on the major D1 fold limbs. The D2 strains appear to increase southward to the base of the Malmesbury Group, which is inferred to be a zone of décollement above the Cape granitoids (Hartnady *et al.*, 1974). The main foliation has been distorted by D3 folding and the formation of crenulation cleavage as conjugate sets in the hinge zones of D3 folds and sporadic kink bands elsewhere. A further phase of crenulation of subhorizontal planes also took place. Both types of crenulation may be the result of heterogeneous strain during the D2 fabric-forming event (Hartnady, 1969) or may have formed during a younger post-Cape orogeny (Hartnady *et al.*, 1974).

Polyphase deformation is most common in the Porseleinberg Formation of the central tectonic domain (Hartnady *et al.*, 1974). The oldest structure is a fine layering locally preserved in the spaces between younger crenulation cleavage; it has not been established whether this lamination represents primary stratification. Recumbent folding (D2) on northwest- or west-trending axes resulted in the form of a regional crenulation cleavage, axial intersection lineation, and intrafolial folds. Gentle flexural-slip folding during D3 deformed the regional foliation into upright periclinal folds trending north-northwest but produced no obvious cleavage. These elongated domes largely control the outcrop pattern of the various formations. Slight D4 shortening along the northwest axes resulted in sporadic formation of kink bands. In the overlying Moorreesburg Formation, D1 strains have not been recognized but D2 strains resulted in schistosity and a few intrafolial folds. Hartnady *et al.* (1974) suggested that the mica schists separating the Moorvesburg Formation from the underlying Porseleinberg Formation represented a tectonic contact rather than a true stratigraphic unit.

The Saldanha–Franschhoek and Piketberg–Wellington fault zones have clearly had a long history of spasmodic displacement (Hartnady *et al.*, 1974). Movement older than the Cape Supergroup is indicated by the truncation of some components both by younger members of the Cape granitoid suite and also by the unconformity at the base of the Cape Supergroup. Late-syntectonic emplacement of the Cape granitoids is indicated by the additional fact that shear belts cut through and border some of these intrusions (De Villiers *et al.*, 1964; Hartnady, 1969; Visser and Schoch, 1973). Earlier workers tended to confuse the porphyroclastic textures in the deformed granitoids with quartz porphyries, which are also present. Repeated reactivation of these fault zones is shown by the fact that many of their component faults are post-Klipheuwel and post-Cape in age. However, most of the shearing appears to have taken place before the Cape Supergroup was deposited at a time when the Malmesbury rocks were probably at a lower crustal level, although evidence is also cited for post-Cape shearing.

The crustal level at which the Malmesbury rocks were deformed and metamorphosed is presently unclear but there is no evidence for substantial burial. The regional metamorphic grade varies from nonmetamorphosed rocks that have undergone only diagenesis, through very low-grade rocks that have undergone burial metamorphism (Hartnady *et al.*, 1978) to low-grade rocks metamorphosed to greenschist facies (Bryson, 1975). Measurements of the b_0 spacing in white micas in metapelites from the Brandwacht Formation in the Worcester area led Hartnady *et al.* (1978) to infer a low P/T metamorphic facies series. Under an inferred high geotherm of 35°C/km, the Malmesbury rocks could have attained temperatures typical of greenschist facies merely by burial under the cratonic Cape and Karoo cover (Hartnady *et al.*, 1978). However, the associated tectonic fabrics indicate that metamorphism occurred during the pre-Cape Saldanian orogeny.

Syntectonic and Posttectonic Granitoids

During and following the Saldanian orogeny at least 25 high-level granitoid plutons intruded the geosynclinal Malmesbury sediments in the southwestern Cape at approximately the same time as similar granitoid plutons of the Kuboos–Bremen line were posttectonically intruding the Gariep Group in the Richtersveld. References to the voluminous literature on the Cape granites, as they are known, are given by Scholtz (1946) for the pre-1945 period and by Schoch *et al.* (1977) for the subsequent period.

The plutons are exposed as upfaulted or upfolded inliers in the Cape Fold Belt. Since the comprehensive study by Scholtz (1946), the Cape granites have been divided into two geographic groups. An eastern group, comprising a multiphase batholith at George (Fig. 9-13) and two small plutons of biotite granite elsewhere, is characterized by east–west elongation and penetrative tectonic fabrics. The remainder of the granites constitute a southwestern group characterized by elongation in a northwest direction and varying amounts of deformation (Fig. 9-12B). Both of these regional trends correspond to those in the Malmesbury country rocks and reflect their late-syntectonic emplacement. However, this apparent alignment is also caused by a similar fold trend in the Cape cover which controls the pattern of exposure of the basement granites.

The Cape granites can be divided into two groups on the basis of their major-element geochemistry (Schoch *et al.*, 1977). An older group is partly foliated and characterized by prominent trends of magmatic differentiation and by contamination from country rocks. Differentiation is manifest in a compositional change from hornblende–biotite granite to biotite granite and in a textural gradient from coarse grained and megacrystic to fine grained. Contamination is evident from two trends that are roughly proportional to the number of Malmesbury xenoliths per unit volume of granite: first, the appearance of cordierite and muscovite and, second, a negative correla-

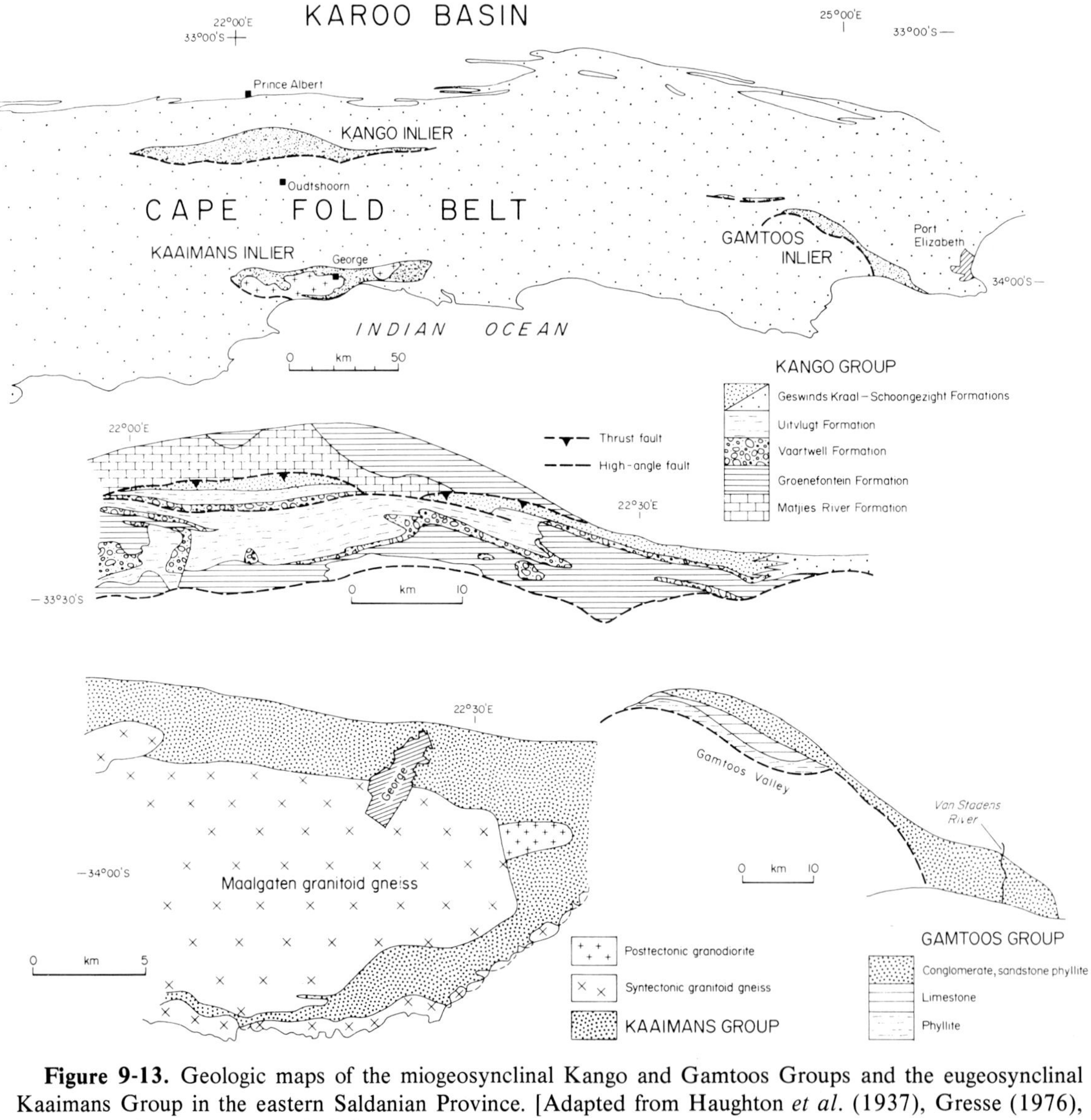

Figure 9-13. Geologic maps of the miogeosynclinal Kango and Gamtoos Groups and the eugeosynclinal Kaaimans Group in the eastern Saldanian Province. [Adapted from Haughton *et al.* (1937), Gresse (1976), Krynauw (1977), Le Roux (1977).]

tion between feldspar and biotite (Schoch *et al.*, 1977). A younger group, constituting only 5–10 percent of the suite is represented by leucogranites devoid of differentiation trends and with compositions close to that of minimum melts from biotite-free granites.

Schoch *et al.* (1977) interpret the older, generally coarse-grained megacrystic granitoids as having formed deep in the crust and subsequently ascended gradually, undergoing slow crystallization and differentiation and abundant contamination. Conversely, they conclude that the younger group cooled rapidly at a high crustal level, indicated by their common hypabyssal textures, as small plutons and sheets within the batholiths. The younger phases commonly form a "hood" over the main mass (Scholtz, 1946).

Large crystals of perthite within the megacrystic granites attain sizes of 10 cm in the Paarl batholith and 8 cm in the Cape Peninsula batholith (Fig. 9-12B). These giant crystals are frequently referred to as phenocrysts but at least some of them are porphyroblasts. The beautifully exposed Sea Point contact at Cape Town between megacrystic granite and Tygerberg slate of the Malmesbury Group drew the attention of Charles Darwin in 1836 and was subsequently declared a historic monument. Along this contact a migmatite zone has formed. At one side isolated, randomly oriented porphyroblasts of feldspar cut the slaty cleavage; toward the granite the slate is increasingly punctuated with aggregates of megacrysts and injected by roughly concordant sheets and veinlets of granitoid melt; this zone merges into one consisting largely of megacrystic granite with irregular xenoliths and schlieren of hornfelsed slate (Fig. 9-5D).

The granitoids are sparsely mineralized by wolframite, cassiterite, molybdenite, and tourmaline, generally in quartz veins or along northwest-trending fractures (Haughton, 1969, p. 309); scattered clots of tourmaline attain 30 cm in length (Gentle, 1977).

Field evidence for multiphase intrusion is abundant and this is supported by a range of Cambrian radiometric dates from 600 Ma to 500 Ma (Aldrich *et al.*, 1958; Nicolaysen, 1962; Allsopp and Kolbe, 1965; Burger and Coertze, 1973; Schoch *et al.*, 1975, Schoch and Burger, 1976). Some of these are mineral ages (e.g., the 600 ± 30 Ma Rb-Sr biotite age of Aldrich *et al.*, 1958) and may reflect subsequent Late Saldanian tectonism. On the other hand, the Pb-Pb isochron age of 522 ± 12 Ma for the Saldanha rhyolite porphyry (Schoch and Burger, 1976), 530 ± 15 Ma for the adjacent Bredenburg granite, and 600 ± 20 Ma for the Hoedjiespunt (part of Saldanha batholith), Darling, and Cape Peninsula granites (Schoch *et al.*, 1975) may reasonably be interpreted as primary ages of intrusion.

As described in the previous section, there is field evidence that Saldanian deformation took place both before and after intrusion of Cape granites. In many cases, however, it is difficult to separate the effects of Cambrian Saldanian orogeny from Permo–Triassic Cape orogeny unless, as in some cases, faults in the granites are erosionally truncated by the basal Cape unconformity. Tectonism of granite during the Cape orogeny, resulting in virtually complete isotopic overprinting, has been proved in the case of the largely offshore Groot Haelkraal granite. Gentle *et al.* (1978) obtained an Ar-Ar isochron of 248 ± 2 Ma on biotite from this granite, which underwent cataclastic recrystallization of biotite and quartz at this time.

In summary we may conclude that the regional fabric in at least part of the Malmesbury Group had already formed by the time that the Cape Peninsula granite was emplaced 600 Ma ago. Shearing along the major Saldanha–Franschhoek discontinuity continued after 500 Ma and affected the Saldanha batholith; but without radiometric age determinations of minerals from the mylonites, it is not possible to say whether this occurred during the Late Saldanian or the Cape orogeny.

9.4. Pre-Cape Basins in the Eastern Saldanian Province

In the southern Cape the Saldanian orogeny was accompanied by deposition of thick sequences of sediments in marginal and rifted basins. Exposure is restricted mainly to three elongate inliers in the Cape Fold Belt containing rocks known as the Kango, Gamtoos, and Kaaimans Groups (Fig. 9-13). The Kango and Gamtoos sedimentary successions have miogeosynclinal characteristics, whereas the Kaaimans Group is more eugeosynclinal. The sedimentary suites (Table 9–6) include terrigenous clastic and carbonate rock types that have been subjected to low-grade regional metamorphism. Multiple phases of deformation have sheared and folded the strata, locally overturning and thrusting them to the north and northeast along tectonic slides (Stocken, 1954; Gresse, 1976; Stump, 1976; Krynauw, 1977; Le Roux, 1977). These sedimentary assemblages are believed to be broadly contemporaneous with the Malmesbury Group of the southwestern Cape (Du Toit, 1954, p. 169; Haughton, 1969, p. 200; Stump, 1976). Le Roux (1977) suggested that the post-Kango Schoemans Poort Formation was deposited as a molasse succession in the closing stages of Saldanian orogeny, analogous to the Klipheuwel rocks of the southwestern Cape.

The Kango and Gamtoos inliers expose complex but genetically related suites of shelf and terrestrial sediments which shoal upward and appear to have formed in a protracted geotectonic cycle (Haughton *et al.*, 1937; Stocken, 1954; Rossouw *et al.*, 1964; Stump, 1976; Le Roux, 1977). The Kango Group apparently accumulated in a rifted east–west basin with tectonic uplift to the south (Le Roux, 1977). Basic dikes suggest an age of 782 Ma for the metamorphism (Hälbich, 1979a).

Kango Group

Carbonate-shelf and tidal-flat environments. Approximately 1000 m of shale and graywacke or subarkose are gradationally succeeded by a roughly equal thickness of oolitic limestone and dolomite (Table 9-6, unit 1). In the terrigenous clastic interval, shales and sandstones alternate gradationally, or erosively based sandstones form lenticular beds. In the carbonate-dominated interval there is a regular alternation of boudinaged limestone and phyllitic shale. Maximum metamorphic grade is described as lower greenschist (Le Roux, 1977) and the metamorphic facies series appears to be low P/T or medium P/T.

Le Roux (1977) inferred shallow-shelf, tidal-flat, and supratidal depositional environments, initially with steep eastward gradients dominated by fluvial processes. The limestones are believed to have accumulated on a shallow shelf and tidal flats. Diagnostic criteria include the purity and fine grain of the limestone, oolitic textures (Ginsburg and James, 1974), and the abundance of dolomite, which may indicate supratidal conditions (Shinn *et al.*, 1969). In modern shelf seas the oolith–nonskeletal grain associations are restricted mainly to tropical regions within 30° of the equator (Lees and Buller, 1972).

Deep-sea sedimentation. Sandstone and shale alternate rhythmically over a 2400-m interval (Table 9-6, unit 2). These sediments, together with 1500 m of Huis Rivier shale and graded quartz wackes in the west (unit 3), originated mainly from westward-flowing turbidity currents along the axis of a marginal cratonic basin (Le Roux, 1977; I.W. Hälbich, personal communication, 1980). Some characteristics shared with classic turbidites (Walker, 1978) include the regular bedding, shale layers above each sandstone unit, sole marks associated with erosive sandstone bases, graded bedding, ripple cross-lamination, and convolutions. Nonoolitic limestones occur only in the lower and upper parts of the succession as westward-tapering lenses, suggesting that the succession may have begun and ended with carbonate-shelf sedimentation (Le Roux, 1977). Coarse-grained lenticular sandstones near the top also thin to the west.

Braided alluvial-fan environments. This coarse-grained facies consists of thick eastward-tapering wedges of paraconglomerate containing a tectonite fabric with a maximum thickness of 1300 m in the western part of the basin (Table 9-6, unit 4). Clasts near the base, some of which are more than a meter in diameter, are matrix supported and consist of sandstone, shale, limestone, granite, and gneiss. However, most of the succession consists of well-rounded vein quartz and quartzite conglomerate in lenticular units up to 50 m long. An upward reduction in clast size is associated with increasingly common intercalations of cross-bedded sandstone (Le Roux, 1977).

These thick wedges of conglomerate have been interpreted as braided alluvial-fan deposits generated by tectonic rejuvenation and substantial relief (Le Roux, 1977). The paleoenvironment is believed to have included debris flows, generating poorly sorted, matrix-rich conglomerates on the steeper fan slopes, and channeled flow, which resulted in deposition of lenticular pebble conglomerates and distal cross-bedded sandstone. Eastward flow was directed down the plunging trough axis, resulting in downslope thinning of the conglomerate wedges and strongly unimodal cross-bed orientations (Le Roux, 1977).

Table 9-6. Pre-Cape Saldanian Basins of the Southern Coast[a]

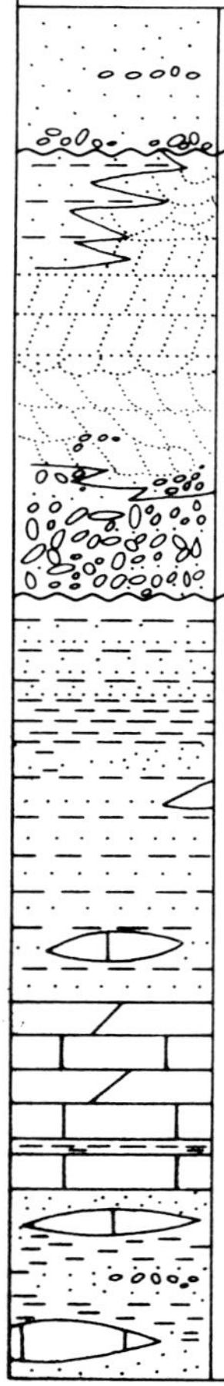

Kango Group (excluding unit 6)[b]	Gamtoos Group[c]	Kaaimans Group[d]
Unit 7: Schoemans Poort Formation, 600 m Subarkosic sandstone with bands and lenses of conglomerate		Unit 7: Homtini Formation Alternating slate, phyllite, graywacke; lenticular interbeds of coarse-grained sandstone and conglomerate
Unit 6: Gezwinds Kraal Fm (in W), 1500 m Schoongezight Fm (in E), 1100 m Pebbly sandstone grading westward into alternating sandstone and shale		
Unit 5: Uitvlugt Formation, 1000 m Subarkosic sandstone, cross-bedded, lenses of conglomerate and mudstone	Unit 5: Arkosic sandstone, medium to very coarse grained, cross-bedded, mudstone and limestone interbeds; discontinuous basal conglomerate; phyllitic upper parts	Unit 6: Victoria Bay Formation, 85 m Coarse-grained arkosic sandstone, lenticular conglomerate interbeds, occasional thin beds of marble
Unit 4: Vaartwell Formation, > 1000 m Conglomerate with sandstone interbeds		
Unit 3: Huis Rivier Fm, (in W), 1500 m Shale grading upward into graded quartz wacke with sole structures	Unit 4: (Kaan) Dark limestone (sparite), massive beds separated by thin beds of phyllite, quartzite, and calcareous sandstone	Unit 5: Soetkraal Formation, 300 m Phyllite and schist, locally rhythmically alternating; some hornfels and sandstone interbeds
Unit 2: Groenefontein Formation, 2400 m Alternating sandstone and shale, parallel bedded, sporadic limestone lenses	Unit 3: Phyllite and shale, and coarse-grained arkosic sandstone, thinly bedded; lenses of very coarse-grained granular sandstone and conglomerate	Unit 4: Skaapkop Formation, 260 m Quartzite, medium to coarse grained, phyllite medially. Phyllite (6m) basally Unit 3: Sandkraal Formation, 300 m Quartz-sericite schist, arenaceous, parallel bedded
Unit 1: Matjies River Formation, 2300 m 1300 m dolomitic limestone (oolitic) with shale, siltstone, and sandstone interbeds in lower part	Unit 2: (Kleinfontein) Dark dolomitic limestone (sparite), thin interbeds of calcareous and pyritic carbonaceous shale	Unit 2: Saasveld Formation, 600 m Andalusite-bearing biotite schist, hornfels, argillaceous mica schist
1000 m shale and graywacke, subordinate conglomerate lenses and limestone interbeds	Unit 1: Phyllite, with lenses of pyritic carbonaceous shale	Unit 1: Silver River Formation, 1300 m Chlorite-rich phyllonitic psammite and semipelite, some thinly bedded graywacke, rare limestone
(base unknown)	(base unknown)	(base unknown)

[a]Thicknesses approximate. Correlation of formations between groups and stratigraphic sequence within groups is speculative.
[b]Compiled from Stocken (1954); Rossouw *et al.*, (1964); Le Roux (1977); I.W. Hälbich (personal communication, 1980).
[c]Amm (1934); Haughton *et al.* (1937); Stump (1976).
[d]Potgieter (1950); Gresse (1976); Krynauw (1977).

Braided alluvial-plain environments. The conglomeratic alluvial-fan facies is gradationally succeeded by nearly 2000 m of fine- to coarse-grained pebbly arkose with shale and conglomerate intercalations (Table 9-6, units 5 and 6). Internally the sandstones are predominantly planar cross-bedded in sets up to 4 m thick, succeeded by smaller scale trough cross-bedding and asymmetric ripple forms. Le Roux (1977) found that this succession encompasses two different patterns of sedimentation. The lower 800 m (unit 5) contains eastward-tapering wedges of sandstone in which cross-bedding is also directed to the east. Only along the eastern margin of the inlier are azimuths more variable and sometimes bipolar. In contrast the rest of the succession (unit 6) displays westward fining, with the arkoses grading laterally into a thick upward-fining succession of interbedded sandstones and shales. Beds become thinner and less regular upward and westward.

A fluvial origin has been proposed on account of the composition and consistent unimodal paleocurrent pattern in each interval (Stocken, 1954; Le Roux, 1977). This succession grades into the alluvial-fan facies described previously, suggesting that the lower stratigraphic interval was deposited by eastward-flowing streams on a low-gradient sandy alluvial plain.

The upper part of the succession is also ascribed to braided fluvial deposition, except that tectonic activity succeeded in reversing the paleoslope so that stream flow was directed to the west (Stocken, 1954; Le Roux, 1977).

Molasse alluvial fans. The closing stages of Saldanian orogeny were marked by accumulation of thick molasse-type sequences of laterally restricted sandstones (Table 9-6, unit 7) that unconformably overlie older folded units (Le Roux, 1977). The Schoemans Poort Formation consists of massive feldspathic sandstone containing layers and lenses of matrix-supported conglomerate with clasts up to 30 cm in diameter. These characteristics can be attributed to deposition by periodic flood discharge on alluvial fans draining a southern highland (Le Roux, 1977). This succession is unconformably overlain by the Cape Supergroup and may be analogous to the Klipheuwel Formation.

Tectonic control. Late Precambrian and Cambrian miogeosynclinal sedimentation in the Kango basin reflects a distinct tectonic control. The provenance indicated by conglomerate clasts in the Vaartwell Formation is that of a granitoid–gneiss terrane to the west (Le Roux, 1977; Hälbich, 1979a). Le Roux (1977) envisaged the Kango basin as an elongate plunging trough open to the ocean in the southeast. Almost half of the exposed succession consists of intratectonic carbonates deposited in shelf and tidal-flat environments, followed by turbiditic flysch generated during tectonic deepening of the basin. Subsequent marine regression, emergence, and folding resulted in the angular unconformity at the base of the Vaartwell conglomerates (I.W. Hälbich, personal communication, 1980). The remainder of the succession comprises clastics deposited on braided alluvial fans and alluvial plains. These fluvial dispersal patterns bear the distinct imprint of episodic tectonic control. Flow was at first directed eastward down the trough axis but was subsequently reversed. The uppermost formation was deposited after considerable deformation, rifting, and erosion of older Kango deposits. Le Roux has suggested that the paleolatitudes were tropical to subtropical.

Basic dikes intruding the Kango Group have yielded a metamorphic age of 782 $\pm$ 6 Ma (Hälbich, 1979a; whole-rock K-Ar spectrum). Although the overall deformation intensity is low, northward-verging folding and thrusting took place under greenschist facies to give rise to an arcuate fold belt that is convex to the north (Le Roux, 1977). This suggests northward translation away from the more active part of the orogen. These structural trends are shared by the Cape cover and it is not possible to place an age on this tectonism (Le Roux, 1977).

Gamtoos Group

Pre-Cape rocks in the Gamtoos Valley (Fig. 9-13) contain one of South Africa's oldest mines, the Maitland Mine. Argentiferous galena, sulfides, and carbonates of copper associated with veins of quartz and calcite in the Kaan limestone have been known since 1792 (Haughton *et al.*, 1937). The stratigraphic succession is largely overturned to the northeast; it is characterized by mesoscopic folding and cleavage (Du Toit, 1954, p. 170; Stump, 1976) and is cut by basic sills and dikes. The upper part of the Gamtoos Group is believed to be conformable with the overlying Cape Supergroup (Amm, 1934; Haughton *et al.*, 1937; Stump, 1976).

Table 9-6 summarizes the lithology of the Gamtoos Group. This succession has not been divided into formal stratigraphic units, and thicknesses of its lenticular components are unknown. The lower part is dominated by two thick limestone formations with intervening phyllite. Lenses of quartzite, coarse-grained arkose, and pebble- to cobble-grade conglomerate occur within the phyllite. In contrast, the upper Gamtoos Group (Table 9-6, unit 6) is more arenaceous and consists of medium- to very coarse-grained arkose which is profusely cross-bedded. A discontinuous conglomerate with vein-quartz pebbles occurs basally, whereas mudstone and phyllite are common in the upper part of this group.

Stump (1976) attributed the thick Gamtoos limestones to shallow-shelf deposition. Increasing tectonism is correlated with accumulation of substantial thicknesses of coarser clastic material by fluvial and fan-delta processes. The stratigraphic, paleoenvironmental, and tectonic resemblance between the Gamtoos and Kango inliers is striking, suggesting that both are of the same elongate basin, the Gamtoos being the more distal. Both groups are dominated by limestone and phyllite in their lower parts and contain conglomerate and cross-bedded arkose and subarkose above. In the Kango Group thick conglomerates form eastward-tapering wedges equivalent to discontinuous conglomerates at the base of the uppermost Gamtoos interval. This eastward fining is supported by the paleocurrent distribution (Stocken, 1954).

An inverted sequence of metasedimentary rocks, constituting the Sardinia Bay Formation, forms the coastline west of Port Elizabeth and also in the Van Stadens Valley (Fig. 9-13) (Shone, 1979). Phyllite and quartz arenite predominate, with subordinate conglomerate. Sedimentary structures include cross-bedding in sets up to 1 m thick locally displaying a bimodal–bipolar foreset distribution, herringbone cross-bedding, flaser and lenticular bedding, and biogenic structures. The conglomerates contain well-rounded and imbricated clasts of vein quartz. Shone ascribed this succession to fluvial and tidal-shelf deposition.

Kaaimans Group

Of the three major inliers (Table 9-6, Fig. 9-13), the Kaaimans succession near George is the most argillaceous and uniform (Potgieter, 1950; Gresse, 1976). It is still doubtful whether or not the sequence is inverted (Krynauw and Gresse, 1980). Stump (1976) believed that the parallel-bedded phyllites, andalusite-bearing quartz–mica schists, and quartzites originated in deep water. Lithologically these rocks resemble the Tygerberg and Swartland turbidite suites of the southwestern Cape. Shallowing and emergence are envisaged for accumulation of the more arenaceous upper parts (units 6 and 7 in Table 9-6).

Granitoids with strong tectonite fabrics were syntectonically emplaced during D1 or D2 as southward dipping sheets parallel to the regional foliation, and these were followed by relatively undeformed granitoids (Gresse, 1976; Krynauw, 1977; Krynauw and Gresse, 1980). One of the former group, the megacrystic Maalgaten granite, yielded two Pb-Pb zircon ages of 504 Ma and 532 Ma (Gresse, 1976; Krynauw and Gresse, 1980) and whole-rock Rb-Sr ages ranging from 546 Ma to 521 Ma (Hälbich, 1979b), but in the absence of data on the initial $^{87}Sr/^{86}Sr$ ratios it is not known which metamorphic episode these ages reflect. Comparison of a U-Pb zircon age of 548 Ma from a granite near Wilderness with those obtained from the Cape granites suggests a primary age of 600 ± 20 Ma (Burger and Walraven, 1976).

Intrusion of the syntectonic granites at high levels (equivalent to 200–350 MPa water pressure) boosted the regional geotherm to metamorphose the Kaaimans Group to higher levels (locally to upper amphibolite facies, but generally to lower amphibolite facies) than the Kango or Gamtoos Groups. Intrusion was accompanied or followed by three episodes of low P/T greenschist-facies regional metamorphism and associated deformation (Gresse, 1976; Krynauw, 1977; Krynauw and Gresse, 1980). The first episode resulted in a southward-dipping schistosity and downdip lineation. During D2 the early syntectonic granitoids were refoliated by crenulation cleavage or fracture cleavage; F2 folds were apparently progressively rotated to north–south trends by later D2 deformation. D1 and D2 strains took place before deposition of the Cape cover (Chapter 10). Retrogression in post-Cape times was associated with east-trending major flexural-slip folding during D3 with conjugate minor folds, kink bands, and a shear belt. A final episode of deformation forming kink bands and shear belts has been designated D4. The first three deformation episodes indicate north–south shortening of the Kaaimans cover and the syntectonically intruding sheetlike granites (Krynauw and Gresse, 1980). These authors were uncertain of the cause of the D1 and D2 Pan African shortening, but they ascribed the post-Cape D3 deformation to A-subduction collision (Bally and Snelson, 1980) between the proto-African plate with a southern continent, as proposed by Rhodes (1974b) and Lock (1980) (Chapter 11).

9.5. The Damara Province: Keystone of the Pan African Framework

The Damara Province represents an integral part of the Pan African structural framework around the northwestern and southwestern margins of the Kalahari Province of southern Africa (Fig. 9-1). The Saldanian-Gariep geosynclinal chain extends some 1200 km northward along the Namibian coast into Angola; this extension is referred to as the coastal branch of the Damara Province. The coastal branch is divided into northern and southern parts by a massive oblique offshoot known as the intracontinental branch of the Damara Province (Fig. 9-14). The northern and southern arms of the coastal branch, together with the intracontinental branch, constitute a three-armed asymmetric orogenic junction with lithologic and structural continuity between each arm.

The intracontinental branch is 400–500 km wide, extending from the Atlantic coast across Namibia in a northeastward direction and disappears beneath the post-Paleozoic Kalahari basin in the continental interior. On the eastern side of the Kalahari basin the intracontinental branch may connect with other Pan African orogens, such as the Katangan and Zambezi Provinces. The intracontinental branch separates the Kalahari Province (>0.9 Ga) in the south from the Congo Province (>0.8 Ga) in the north (Figs. 9-1 and 9-14).

The Damara Province plays a pivotal role in understanding Late Precambrian sedimentation and tectonism on account of its massive size, its unique triple-junction shape, its thick geosynclinal succession, and its high degree of exposure. Attention focused on the Damara Province in recent years has yielded important glimpses of the style of orogeny at this stage of Earth's history, and a discussion of Late Precambrian tectonics would be incomplete without a summary of these findings. Here, the evidence for and against large differential movement between crustal plates in different parts of the Damara Province is examined and used as a basis for viewing the Pan African framework in broader perspective.

For the most part, the geosynclinal rocks known as the Damara Supergroup, which are preserved in the three arms of the orogen, appear to have been deposited on continental crust. This is evident from their shallow-water or terrestrial sedimentary affinities and from the presence of pre-Damara sialic basement exposed along the margins and as large inliers scattered throughout much of the Damara Province (Fig. 9-14). In many cases these basement inliers were elevated source areas in early Damara times. In the north the basement is represented by the supracrustal Epupa, Huab–Grootfontein, and Khoabendus Metamorphic Suites, by gneisses and granitoids such as the Franzfontein and Summas Intrusive Suites (1.8 Ga), and by the anorthositic Kunene Intrusive Suite (2.0 Ga) (Clifford *et al.*, 1962; Söhnge, 1964; Martin, 1965; Frets, 1969; Simpson, 1970; Köstlin, 1974; Burger *et al.*, 1976). In the central and southern areas largely allochthonous Damara metasediments rest on gneisses (1.9–1.0 Ga); the Geelkopf, Piksteel, and Gamsberg granitoids (1.7–1.0 Ga), and the supracrustal Abbabis, Mooirivier (?1.8 Ga), Elim, Marienhof (?1.6 Ga), Gaub Valley, Grauwater, Doornpoort, and Klein Aub Metamorphic Suites (De Kock, 1934; Martin, 1965; De Waal, 1966; Hälbich, 1970; Schalk, 1973; Hugo and Schalk, 1975; Watters, 1977, Malling, 1978; Jacob *et al.*, 1978). The basement rocks generally yield radiometric ages that suggest isotopic resetting during the time span of two periods of pre-Damara tectonism (Porada, 1979): these are the Eburnian (around 2000 Ma) and Kibaran (1400 to 900

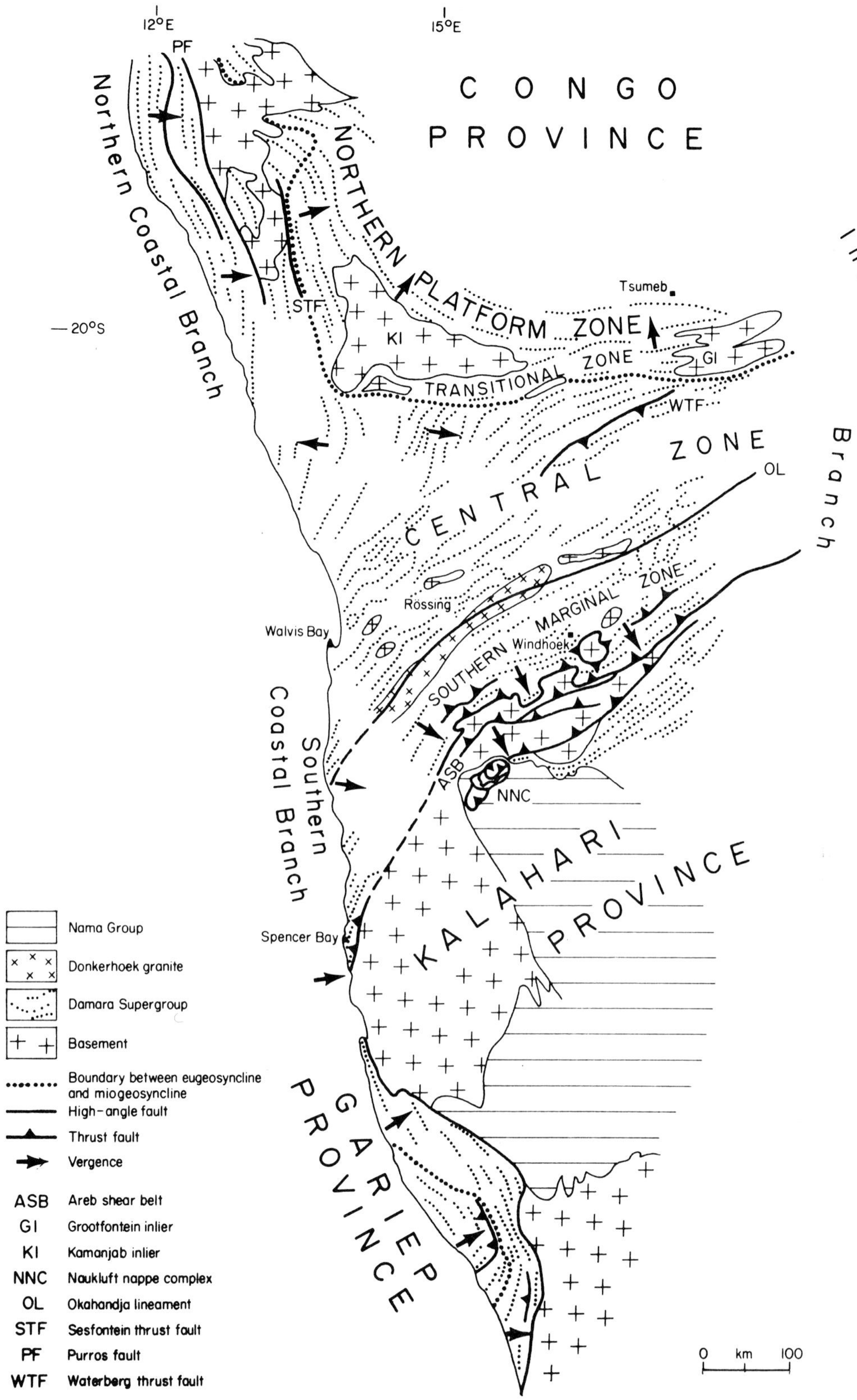

Figure 9-14. Tectonic map of the Damara and Gariep Provinces showing tectonic zones, eugeosyncline–miogeosyncline boundary, pre-Damara basement inliers, and southern marginal thrust zone. [Adapted from Porada (1979) with additional data from Martin and Porada (1977a), Hartnady (1979), R.M. Hedberg (1979).]

Ma), which may be correlated with synchronous tectonic cycles in South America, termed the Transamazonian and Uruacuano, respectively (Hurley *et al.*, 1967; Cordani *et al.*, 1973; Almeida *et al.*, 1976).

The Damara supracrustal rocks are largely of sedimentary origin. Despite the rarity of volcanic rocks, the bulk of this succession is commonly referred to as the eugeosynclinal Swakop Group (Martin, 1965); the miogeosynclinal Otavi Group, up to 7 km thick (R.M. Hedberg, 1979), lies to the north in the acute angle between the northern coastal branch and the core of the intracontinental branch (Fig. 9-14). Both facies are underlain by the predominantly coarse-clastic Nosib Group which represents the oldest part of the Damara succession (Figs. 9-15 and 9-16A). The maximum present thickness of the eugeosynclinal fill has been estimated at more than 10 km (Kröner, 1977) or 17 km (R.M. Hedberg, 1979). However, estimates such as these are not particularly meaningful where the rocks no longer lie in normal stratigraphic order. This is evident from the fact that gneissic basement units locally "conformably" overlie younger Damara cover because of imbricate thrusting and recumbent folding (Hartnady, 1978d).

Sedimentation in the Damara Province began 1000–900 Ma ago (Table 9-7) with infilling of fault-bounded troughs in rifted continental crust with up to 6 km of Nosib clastic sediments and minor volcanic rocks; each trough was approximately 50–70 km wide and 200 km long (Figs. 9-15 and 9-16A) (Martin, 1965; Martin and Porada, 1977a). Three parallel grabens trending northeast have been recognized in the intracontinental branch and one graben in the northern coastal branch, as well as analogous troughs in the Ribeira Province of Brazil, which is believed to have been contiguous during Late Precambrian and Gondwana times (Porada, 1979). Although they have been largely buried by younger sediments or tectonically obliterated, the positions of the ancient faults can be deduced from thickness

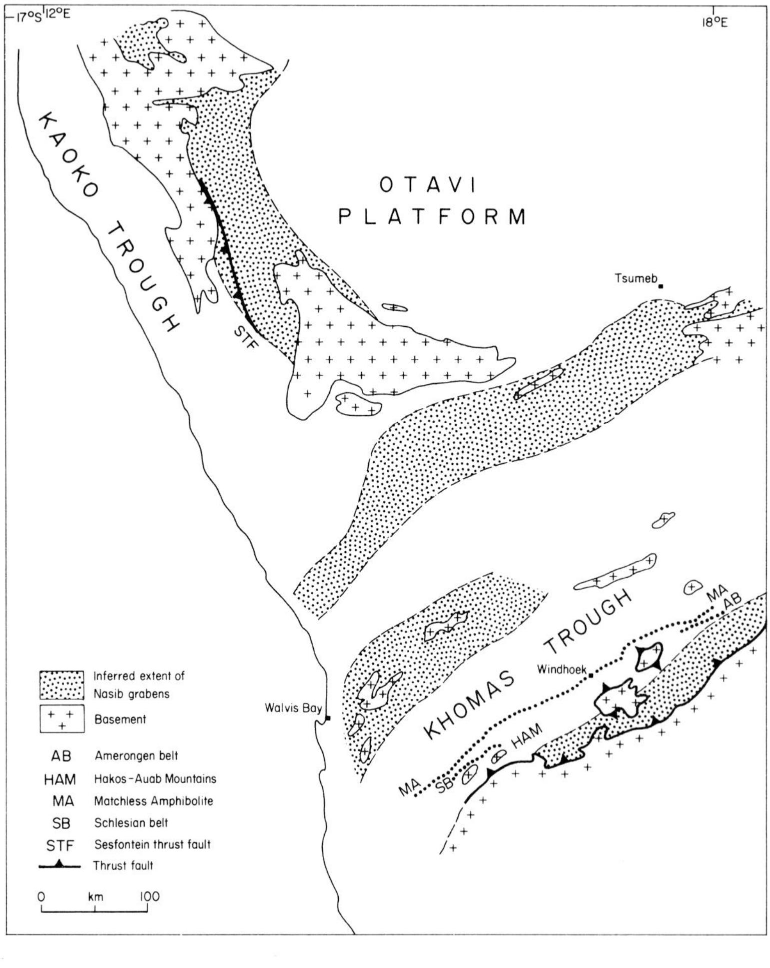

Figure 9-15. Geologic map showing inferred extent of early Nosib grabens in the Damara Province. [Adapted from Martin and Porada (1977a), Kasch (1979b), Porada (1979).]

Table 9–7. Rb-Sr Geochronologic Evolution of the Damara Province[a]

Ages (Ma)	Events
455–400	Uplift and cooling, closing of Rb-Sr isotopic systems
460	Emplacement of uraniferous Rössing alaskite
~510–480	Partial resetting of Rb-Sr isotopic systems D3 mantled gneiss domes, emplacement of late granitoids
~530–495	Nappe tectonics in Southern Marginal Zone and Nama foreland Activation of Okahandja lineament and emplacement of Donkerhoek granite
~560–540	M2 metamorphic peak, intrusion of post-D2 granitoids
570–540	D2, intrusion of syntectonic Salem granite
~660–570	Molasse-type deposition of Nama and Mulden Groups Uplift in center of orogen
~660–620	D1, M1: formation of regional fabric
777 ± 41	Intrusion of post-Otavi basic dikes
~830–760	Emplacement of Matchless amphibolite Deposition of Swakop and Otavi Groups
840 ± 12	Naauwpoort acid volcanism in upper Nosib Group
~1000–900	Start of Nosib rift sedimentation

[a]After Kröner (1980a).

changes in the Nosib and from the distribution of pre-Damara basement inliers (Figs. 9-15 and 9-16A). Arkosic arenites, which are locally conglomeratic, have been attributed to fluvial or shallow-marine infilling of the Nosib grabens (R.M. Hedberg, 1979). Marine incursions in Late Nosib times resulted in the formation of impure carbonates and anhydrite, now metamorphosed to calc–silicate rocks (Fig. 9-16A) (Martin, 1965; Nash, 1972; Martin and Porada, 1977a). Stratiform Cu and Fe sulfides and native Ag are present in the upper Nosib Group at the contact between continental red beds and marine rocks at Oamites mine in the southernmost graben and elsewhere in the Nosib Group (Lee and Glenister, 1976; Martin, 1978). Two of the three Nosib grabens contain rhyolites extruded during rifting, and one of these also contains alkaline and peralkaline volcanic, volcaniclastic, and intrusive rocks of considerable local thickness, known as the Naauwpoort Formation (Fig. 9-16A) (Frets, 1969; Miller, 1972, 1974a; Guj, 1974; Jacob, 1974; Blaine, 1977; Porada and Wittig, 1977). These volcanic rocks define a differentiation trend consistent with Carmichael *et al.* (1974) hypersthene-normative alkali–basalt series (Barnes and Sawyer, 1980). The association of Naauwpoort extrusive rocks with a large fault, the Summas fault, may reflect the structural control of volcanism (Miller, 1972). Thin layers of iron-formation are locally mineralized with Fe, Cu, and Zn sulfides of possible fumarolic origin (Blaine, 1977).

After the end of Nosib times, about 830 Ma ago (Table 9-7), two important trends controlled further deposition within the Damara Province. First, the area of deposition widened so that the original four narrow troughs coalesced and sediments overstepped the basement highs. Second, the miogeosyncline–eugeosyncline dichotomy began, so that a platform succession (the Otavi Group) accumulated on a gently subsiding foreland represented by the Congo craton while flysch-type clastic sediments (the Swakop Group) were deposited in a deepening trough south of the miogeosyncline (Fig. 9-16A).

The miogeosynclinal Otavi Group unconformably overlies the clastic Nosib Group and basement (Martin, 1965; Martin and Porada, 1977b; R.M. Hedberg, 1979). A thick platform succession of stromatolitic dolomitic limestone, with subordinate marl near the top, accumulated in a stable and largely shallow-marine environment 830–760 Ma ago (Kröner, 1980a). Mississippi Valley-type Pb-Zn mineralization is hosted by these rocks (Söhnge, 1964; Martin, 1978). Gentle folding and erosion, which cut an angular unconformity down to the basement in the west, was followed by molasse-type deposition of the Mulden Group in a continental environment, probably in a foredeep basin during the Early Cambrian around 550 Ma (Fig. 9-16A) (Martin, 1965; Frets, 1969; Guj, 1970, 1974; R.M. Hedberg, 1979; Kröner and Clauer, 1979). The Mulden Group comprises clastic sediments of basement provenance with carbonate intercalations. Deposition was unconformable within the eugeosyn-

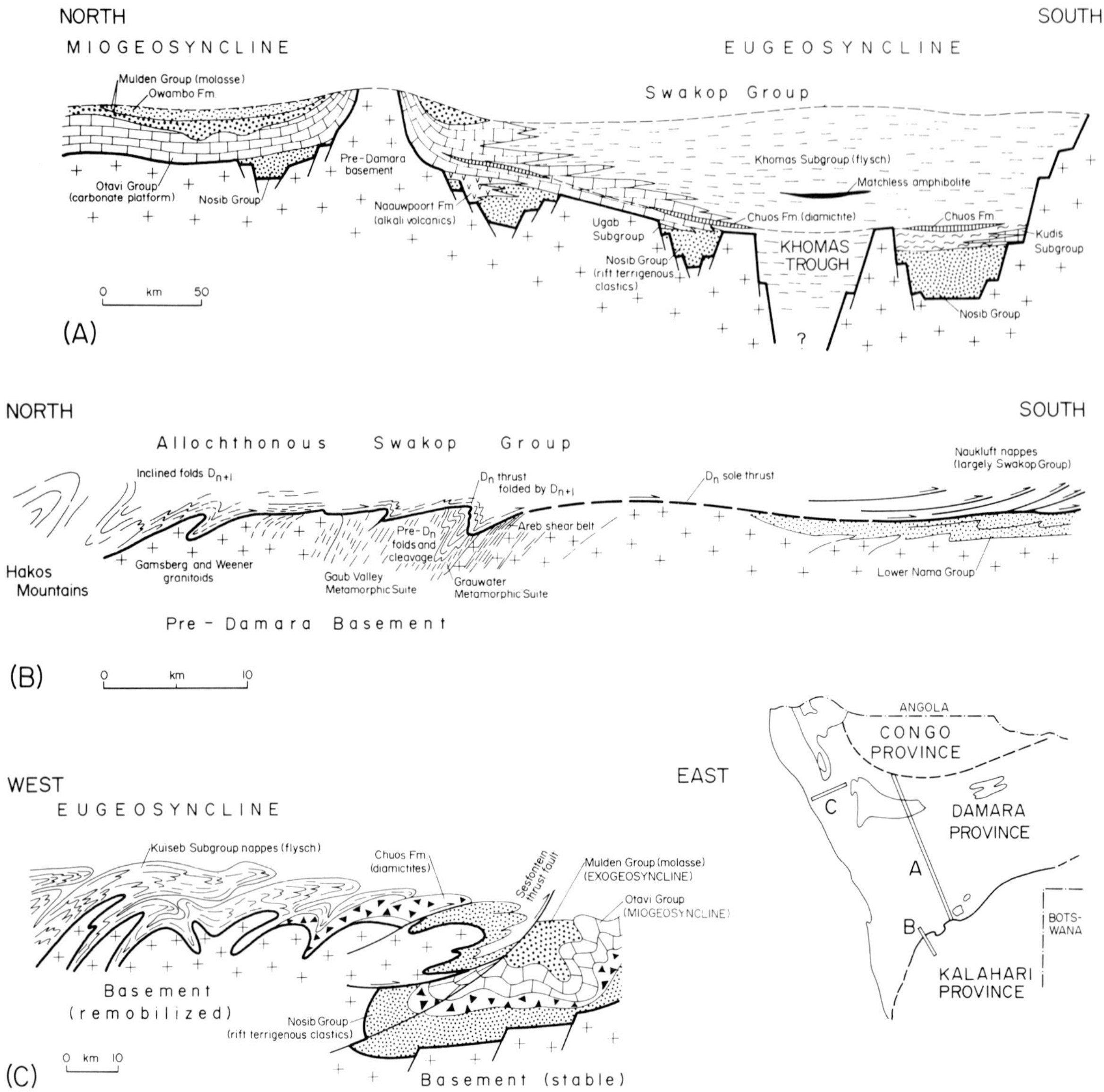

Figure 9-16. (A) Schematic stratigraphic cross section across the intracontinental branch of the Damara Province. [Adapted from Martin and Porada (1977a), R.M. Hedberg (1979), Porada (1979).] (B) Schematic structural cross section across the southern marginal thrust zone of the Damara Province, the Kalahari foreland and the Naukluft nappe complex. [After Bickle and Coward (1977).] (C) Schematic geologic cross section across the coastal branch of the Damara Province. [After Guj (1970), Porada (1979).]

cline to the west and paraconformable in the miogeosyncline (R.M. Hedberg, 1979). Sedimentation ended with accumulation of molasse-type fine-grained terrigenous clastics of the Owambo Formation from a source area probably to the south in the Damara orogen, which is inferred to have been of lower relief than in early Mulden times.

In the southern and central parts of the Damara Province, infilling of the Nosib grabens was followed by unconformable or paraconformable deposition of the Swakop Group in the deepening Khomas trough (Fig. 9-16A). Carbonates and terrigenous clastic sediments of the Kudis and Ugab Subgroups were sporadically deposited, in many places onlapping basement. The extremely thick Khomas Subgroup paraconformably overlies these sediments and consists largely of Kuiseb Formation mica schists representing a monotonous flysch-type succession of argillites overlying calcareous and quartzose turbidite sequences such as the Karibib Formation (Martin, 1965; Hälbich, 1970; Hoffer and Puhan, 1975; Porada, 1975, in preparation; Porada and Wittig, 1976).

Several diamictite units are present in the lower Swakop Group; the uppermost one, known as the Chuos Formation, can be traced across the intracontinental branch and into the northern coastal branch (Fig. 9-16A) (Gevers, 1931; Le Roux, 1941; Martin, 1965; Kröner and Rankama, 1972; Martin and Porada, 1977a; R.M. Hedberg, 1979). Because of its inferred glacial or glaciomarine origin, the Chuos Formation has acquired chronostratigraphic significance in almost all regional correlations, which is unfortunate in view of the fact that in places the diamictites are interbedded with turbidites and most of the inferred glaciogenic features can be explained by nonglacial mass-flow processes (Martin *et al.*, in preparation). Although glaciogenic material is present in the diamictites, its proportion is not yet clear. Volcanic rocks in the Khomas Subgroup are remarkably rare and are restricted to thin units of basic to intermediate lavas in the south. Best known of these basic units is the 300-km-long Matchless amphibolite, an enigmatic and strikingly continuous volcanic belt prominent on Landsat imagery as a narrow, dark band surrounded by the Khomas mica schists (Fig. 9-15) (Short *et al.*, 1976, their plate 349). These metalavas are possibly of ocean-floor type (Finnemore, 1978; Barnes and Sawyer, 1980; Hawkesworth *et al.*, 1981).

The asymmetry in sedimentary facies of the Damara Supergroup, which is expressed as a miogeosynclinal–eugeosynclinal pair, is paralleled by the strong structural and metamorphic asymmetry of the Damara Province. This asymmetry is best understood in terms of the contrasting geotectonic settings of the four tectonic zones, shown in Figure 9-14, that make up the intracontinental branch of the Damara Province (Martin and Porada, 1977a): the Northern Platform Zone (about 80 km wide), a Transitional Zone (about 40 km wide), the Central Zone (about 160 km wide), and the Southern Marginal Zone (about 150 km wide). The Southern Marginal Zone has been subdivided by Barnes and Sawyer (1980) into a Lineament Zone, a Trough Zone, and a Southern Margin Zone, the last-named zone extending southward to include parts of the Nama basin affected by Damara tectonism.

The Northern Platform Zone overlies the rigid and stabilized Congo craton and was therefore protected from orogeny. Tectonism was restricted to upright folding with slight northward vergence and weak metamorphism that affected the Nosib, Otavi, and Mulden Groups (Guj, 1970; Martin and Porada, 1977a; R.M. Hedberg, 1979). The Transitional Zone, which is marked by a string of large basement inliers, appears to have acted as a kind of hinge zone between the Otavi carbonate platform in the north and the subsiding trough of the Central Zone to the south (Figs. 9-14 and 9-16A). The Transitional Zone underwent greenschist-facies metamorphism and polyphase deformation, resulting in tight folds of variable axial trend (Frets, 1969; Martin and Porada, 1977a).

Between 660 and 460 Ma ago, the Central Zone underwent polyphase folding and voluminous (more than 50 percent of the area) intrusion of pretectonic, syntectonic, and posttectonic granitoids, of which the Salem Intrusive Suite is the most abundant; some granitoids may have been derived by partial melting of Khomas trough sediments and basement granitoid gneisses during high-temperature medium P/T metamorphism (Smith, 1965; Miller, 1974b, 1975; Jacob, 1974, 1978; Blaine, 1977; Martin and Porada, 1977a; Martin, 1978; Barnes and Sawyer, 1980). The initial Sr- and Nd-isotope ratios of posttectonic diorite and granites indicate that their magmas were derived both directly from the upper mantle and by melting of material with a previous crustal history (Hawkesworth *et al.*, 1981). In places sialic basement rose as mantled gneiss domes (Porada and Wittig, 1977; Jacob *et al.*, 1978; Barnes and Sawyer, 1980) or was sheared into the form of large recumbent sheath folds several kilometers in wavelength (Coward, 1981). Partial melting of Nosib metasediments and basement granitoid gneisses is postulated to have yielded uranium that was concentrated in alaskitic residual fractions to form the economically important Rössing uranium ores (Berning *et al.*, 1976; Jacob, 1978). Although this postulate has not been supported by studies of the uranium distribution in Nosib heavy-mineral concentrates, the Rössing alaskites were certainly derived by partial melting at high crustal levels, as indicated by unusually high initial $^{87}Sr/^{86}Sr$ (0.759) and low initial $^{143}Nd/^{144}Nd$ (0.5112) ratios (Hawkesworth *et al.*, 1981).

The Southern Marginal Zone preserves the deepest parts of the Khomas trough, in which great thicknesses of the Kuiseb metapelites have been tectonically thickened by recumbent folding and imbricate thrusting. The northern margin of this zone is marked by the Okahandja lineament (Miller, 1979), along which the Donkerhoek granite was intruded at about 520–500 Ma (Fig. 9-14) (Gevers, 1963; Faupel, 1973; Neiberding, 1976; Blaine, 1977; Sawyer, 1978; Miller, 1979).

The metamorphic grade decreases southward from the Central Zone, but the degree of deformation increases in the same direction (Martin and Porada, 1977a). In the Hakos–Auas Mountains intense strain fabrics, south- to southeast-verging overturned to recumbent early folds, and imbricate interslicing of Damara and basement by overthrusting characterize the tectonic style (Fig. 9-14) (Guj, 1969; Porada and Wittig, 1975; Kasch, 1976, 1978a; Hälbich, 1977a; Hartnady, 1978b, c, 1979). The degree of shortening in the northern part of this zone has been calculated at 50–80 percent (Sawyer, 1978). The southern margin of the Damara Province is marked by a southeast-verging thrust zone some 400 km long, which represents the culmination of horizontal tectonics in this area (Fig. 9-16B) (Bickle and Coward, 1977; Hartnady, 1978c, 1979; Kasch, 1976, 1979b). Allochthonous Damaran thrust nappes were emplaced as intensely deformed rocks on older undeformed granitoids and subsequently infolded with the basement.

The Naukluft nappes, consisting largely of Swakop Group metasediments with subordinate syntectonic Nama deposits, have been calculated to have traveled up to at least 78 km southeastward from their inferred root zone in the Hakos–Auas Mountains (Fig. 9-16B) (Hartnady, 1978a,d). However, equivalent rock types in the Damara Supergroup are not found in this area, so this inference is not well substantiated (H. Martin, personal communication, 1980). Suitable lithologies are confined to the southern margin of the Damara Province, where these Damaran rocks have been telescoped and intersliced with basement wedges. Nevertheless the presence of biotite in the Naukluft nappes 50 km southeast of the Damaran biotite isograd suggests that the nappes traveled at least this distance (Ahrendt *et al.*, 1978). Loading of the Kalahari craton by the southward-advancing Damaran nappes, which are partly preserved in the Naukluft Mountains, allowed transgression of the Late Nama Sea into the depressed foredeep at the toe of the thrust complex (Hartnady, 1979). During their translation, erosion of the overthrust nappes would have shed detritus into the depressed areas over which the nappes advanced. These syntectonic Nama deposits were later deformed as their original provenance overrode them (Fig. 9-16B) (Hartnady, 1978a, 1979).

The timing of orogenic activity within the Damara Province is summarized in Table 9-7. Most of the deformation and metamorphism took place during the Late Precambrian Damara orogeny, whereas the younger Pan African event *sensu stricto* is recorded mainly as Cambro–Ordovician mineral ages reflecting regional cooling and uplift of the orogen.

Speculation on the cause of the orogeny that formed the Damara Province, with its distinctive three-armed shape, is strongly constrained by two groups of evidence which, although not irreconcilable, lead to two contrasting hypotheses. An "ensialic school" holds that initial rifting along a triple junction was followed by aulacogen development, leading to ensialic orogeny along the intracontinental branch (e.g., Martin and Porada, 1977a; Kröner, 1977; Martin, 1978). This model must be reconciled with the marked sedimentological, structural, and metamorphic asymmetry of this branch which, together with the gigantic overthrusting of the southern margin, suggests continental collision along the presently intracontinental orogen (e.g., Hartnady, 1976, 1978c,d, 1979; Kasch, 1979b; Barnes and Sawyer, 1980). Evidence for these two viewpoints is discussed in the following three sections.

An Ensialic Model for the Intracontinental Branch

The terrestrial or shallow-marine deposition of the Nosib Group onto rifted continental crust accompanied by sporadic alkaline magmatism is strong evidence for the early history of the Damara Province being ensialic (Fig. 9-16A). It is likely that a three-armed continental rift system with at least eight individual grabens formed in a supercontinent including southern Africa and eastern South America some 1000–900 Ma ago, possibly in response to a rising mantle plume (Martin and Porada, 1977a; Porada, 1979). Plate divergence in the north–south arms progressed to opening of the proto-South Atlantic; subsequent closure was probably responsible for orogeny along the coastal branch of the Damara Province (Porada, 1979) and in the Gariep and Saldanian Provinces farther south. However, Martin and Porada (1977a) believed that plate divergence in the intracontinental branch of the Damara never proceeded to the stage of ocean opening but remained as a failed rift that evolved into an aulacogen rather than a true eugeosyncline (cf. Hoffman *et al.*, 1974). An asthenolith, or pillow of hot, mantle-derived material injected into the crust-mantle interface, was invoked as the cause of the initial rifting, the subsidence and coalescence of the original graben into a broad aulacogen, and the high-temperature tectonism that affected the basin (Fig. 9-17).

The original rifts were envisaged as keystone

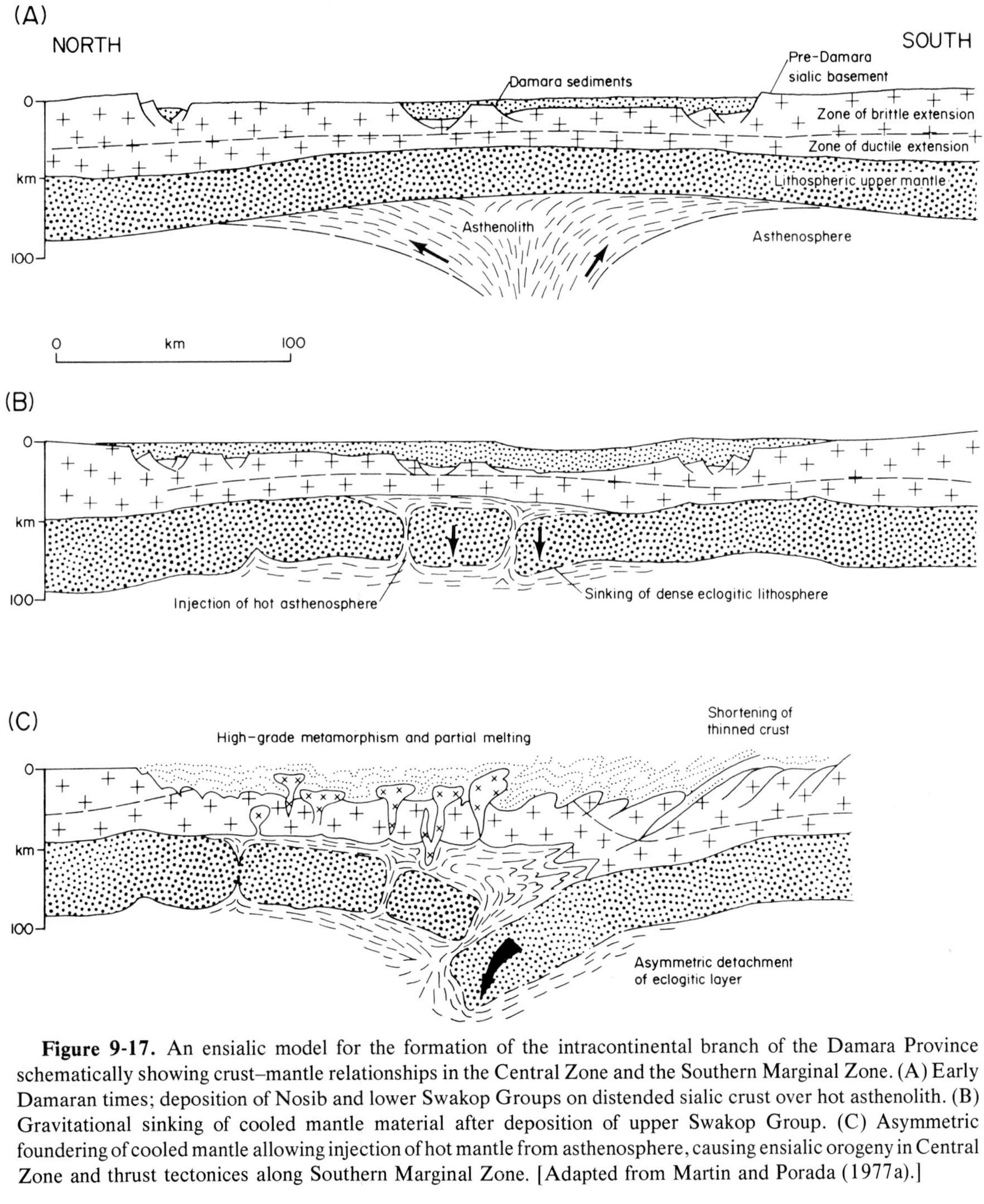

Figure 9-17. An ensialic model for the formation of the intracontinental branch of the Damara Province schematically showing crust–mantle relationships in the Central Zone and the Southern Marginal Zone. (A) Early Damaran times; deposition of Nosib and lower Swakop Groups on distended sialic crust over hot asthenolith. (B) Gravitational sinking of cooled mantle material after deposition of upper Swakop Group. (C) Asymmetric foundering of cooled mantle allowing injection of hot mantle from asthenosphere, causing ensialic orogeny in Central Zone and thrust tectonices along Southern Marginal Zone. [Adapted from Martin and Porada (1977a).]

grabens developing in the upper levels of stretched continental crust over the asthenolith cushion (Fig. 9-17A) (Martin and Porada, 1977a). At lower levels the crust would thin by more ductile extension, allowing the Mohorovičić discontinuity to rise. Evolution to an aulacogen, in the form of a broad downwarp filled by onlapping sediments with marginal shelf facies and distal accumulation of calc–turbidites, suggests that the hypothetical asthenolith was particularly massive. Cooling of the asthenolith would have converted it to cool, dense eclogitic mantle material overlying the less dense asthenosphere (Fig. 9-17B). The resulting density inversion was suggested as the cause of the deformation and metamorphism that followed (Martin and Porada, 1977a). Foundering of the dense, cooled asthenolith supposedly allowed injection of fresh, hot mantle material above it along the M discontinuity, leading to a rise of isotherms and consequent metamorphism and partial melting of crust (Fig. 9-17C). In order to explain why this process did not lead to renewed rifting and eventual ocean opening, Martin and Porada (1977a) were constrained to invoke a major change in the crustal stress regime from tensile to compressive.

This model accounts for all the major characteristics of the depositional and early tectonic history of the Damara Province, namely, the prevalence of continental basement; early rifting and infilling of the grabens by terrestrial and shallow-water clastic sediments; the almost ubiquitous presence of carbonates in the lower Swakop Group; the apparent continuity of possible chronostratigraphic units, such as the Chuos Formation across the orogen; and the relative scarcity of calc–alkaline volcanism. However, this ensialic–aulacogen model is less successful in explaining the later tectonic history, particularly the strong structural and metamorphic asymmetry, the presence of ocean-floor metabasalts and depleted upper-mantle ultrabasic rocks (Barnes and Sawyer, 1980), and the lack of a stable shelf to the south equivalent in age to the Otavi miogeosyncline. Martin and Porada (1977a) did not rule out limited continental convergence of a southern plate below a northern plate but felt that the strong southeast vergence of thrusting and folding was better explained by asymmetric sinking of the dense, cooled asthenolith into less-dense asthenosphere. Kröner (1977) ascribed the strong structural asymmetry to forceful rise of sialic material in the axial zone of the orogen, but in this case it is difficult to see why there was no orogenic spreading to the north equivalent to that in the Southern Marginal Zone.

A Continental Convergence Model

First proposed by Hartnady (1974) and Watters (1974), plate-tectonic models for the presently intracratonic branch of the Damara Province have received further support in recent years (Hartnady, 1978b, 1979; Blaine, 1977; Sawyer, 1978; Barnes, 1979; Kasch, 1979b; Barnes and Sawyer, 1980).

Watters (1974, 1976, 1977) compiled information on a 600–1000 km arc of calc–alkaline acid and basic volcanic and volcaniclastic rocks and granitoids which he termed the Rehoboth Magmatic Arc (Fig. 1-4). This arc lies parallel to the Damara Province but is about 200 km southeast of the inferred paleomargin of the eugeosyncline. This configuration, together with the calc–alkaline character of the igneous rocks, led Watters to propose that the arc was the surface expression of southward subduction of the Congo plate beneath the Kalahari plate. The age of the Rehoboth arc varies from 1360 Ma to 940 Ma (Watters, 1977) implying convergence and Damaran orogeny during this time span. It is difficult to reconcile the older ages with those of the Damara Supergroup (cf. Table 9-7). It is likely that the Rehoboth arc is the result of an earlier stage of major ocean closure during Kibaran (late Namaqua) tectonism because paleopoles for the Kalahari plate are closely similar to those of the Congo plate at this time (McElhinny and McWilliams, 1977; McWilliams *et al.*, 1981). The arcuate shape of the Rehoboth arc may be no more than apparent because of burial of these rocks on the inner side of the arc (Martin and Porada, 1977a).

Hartnady and others suggested that continental convergence along the Damara orogen is more likely to have been northwest facing, with the Kalahari plate being subducted beneath the leading Congo plate. The search for a suture zone was naturally directed toward the Southern Marginal Zone where the deep Khomas trough contains a thick infilling of intensely deformed flysch-like deposits. A large negative Bouguer gravity anomaly here is compatible with thickened continental crust; the very constant strike and steep dip of the intensely developed strain fabric further suggests that this area marks the site of collision (Hartnady, 1979). A prominent candidate for the actual suture zone was the strikingly persistent belt of volcanic rocks visible on Landsat imagery known as the Matchless amphibolite (Fig. 9-15). The chemistry of these sporadically pillowed intermediate to basic rocks indicates similarities with ocean-floor tholeiites (Finnemore, 1978 Sawyer, 1978; Barnes and Sawyer, 1980); this conclusion is also suggested by relatively high Sm/Nd and $^{143}Nd/^{144}Nd$ ratios (Hawkesworth *et al.*, 1981). These findings strengthen the view that these rocks represent an obducted sliver of ocean crust trapped between the converging Kalahari and Congo continents (Hartnady, 1976; Burke *et al.*, 1977). Geochemical (Finnemore, 1978), lithologic, structural, and ore-mineral evidence (Hartnady, 1979) suggest that the northward-dipping Matchless belt, which is associated with magnetite quartzites containing economic amounts of Besshi-type stratiform Cu and Zn ores (Goldberg, 1976; Sawkins, 1976), may be inverted and therefore youngs southward. Inversion of this type would be expected during ocean-floor obduction. Nevertheless the Matchless belt remains unconvincing as a suture zone (Kröner, 1977; Martin and Porada, 1977a; Martin, 1978; Hartnady, 1979) because (1) although shearing is common, there is no evidence of major thrust surfaces on either side of the basic rocks; (2) the amphibolites grade into or interfinger with the Khomas metagraywackes; (3) there are no significant stratigraphic differences in the metasediments on either side of the supposed suture zone; (4) there is no evidence for layered oceanic crust with sheeted dikes.

The suture zone is now postulated to be farther south than the Matchless Amphibolite, possibly along the sinuous trace of intermittently exposed Schlesian–Amerongen Alpine-type (Coleman, 1977) basic and ultrabasic rocks in a zone of overthrusting (Fig. 9-15) (Hartnady, 1978d, 1979; Sawyer, 1978; Barnes, 1979; Kasch 1979b; Barnes and Sawyer, 1980). These rocks were emplaced as "cold" solid bodies along D1 thrust faults and overprinted by D2 schistosity (Barnes and Sawyer, 1980).

Based on the presently understood sequence of deformation and metamorphism in the Southern Marginal Zone and known geothermal gradients in this area, it is possible to reconstruct in a simplified manner this hypothetical collision (Fig. 9-18).

Continental rifting in Nosib times evolved into ocean spreading in early Kuiseb times (Barnes and Sawyer, 1980). Prior to continental collision, and before 660 Ma, oceanic crust of the Kalahari plate is inferred to have been subducted northwestward beneath overriding continental crust of the Congo craton, which consisted of the Central Zone, with calc–alkaline Andean-type magmatism, and the Transitional and Northern Platform Zones of the Damara Province (Figs. 9-14 and 9-18A). By the classification of Irvine and Baraggar (1971) the voluminous Salem granitoids are almost entirely calc–alkaline (Barnes and Sawyer, 1980). Kasch (1979b) interpreted the Khomas trough as a forearc basin containing basement wedges intersliced with terrigenous forearc deposits and ophiolitic ocean crust. The northern boundary of the forearc basin was defined as the Okahandja lineament which is known to have had a large downthrow to the south at this stage of basin infilling (Fig. 9-14) (Blaine, 1977). Underthrusting of cold Kalahari oceanic crust caused depression of isotherms and gave rise to the steeply inclined isothermal surfaces that have been mapped by Kasch (1979a). No blueschist metamorphism (high P/T barometric type) has been proved, but the presence of kyanite and the b_0 values of micas in phyllites are compatible with medium or high P/T metamorphism. This M1 metamophism was syntectonic with respect to two periods of southeast-verging folds and thrusts, termed D1 and D3, as well as left-lateral D2 shearing with 15–20 km of estimated displacement (Kasch, 1978b, 1979a).

Zagros-type continental collision (Fig. 9-18B) is thought to have been responsible for the D1 and D2 folding and thrusting in the forearc basin, including tectonic emplacement of the Alpine-type ultrabasic rocks, and for the initiation of southward spreading of the Naukluft nappes over foreland basement units and the lower Nama shelf succession. The subduction rate presumably slowed on collision, but, owing to geodynamic momentum, isotherms would continue to rise to an M1 peak at about 650 Ma (Kasch, 1979b; Barnes and Sawyer, 1980).

If the bouyant Kalahari continent became locked into place beneath the Congo continent, the momentum of plate convergence would result in a new zone of underthrusting of continental crust south of the original suture (Fig. 9-18C). This would mark a progression from B-subduction (Benioff type) to A-subduction (Ampferer

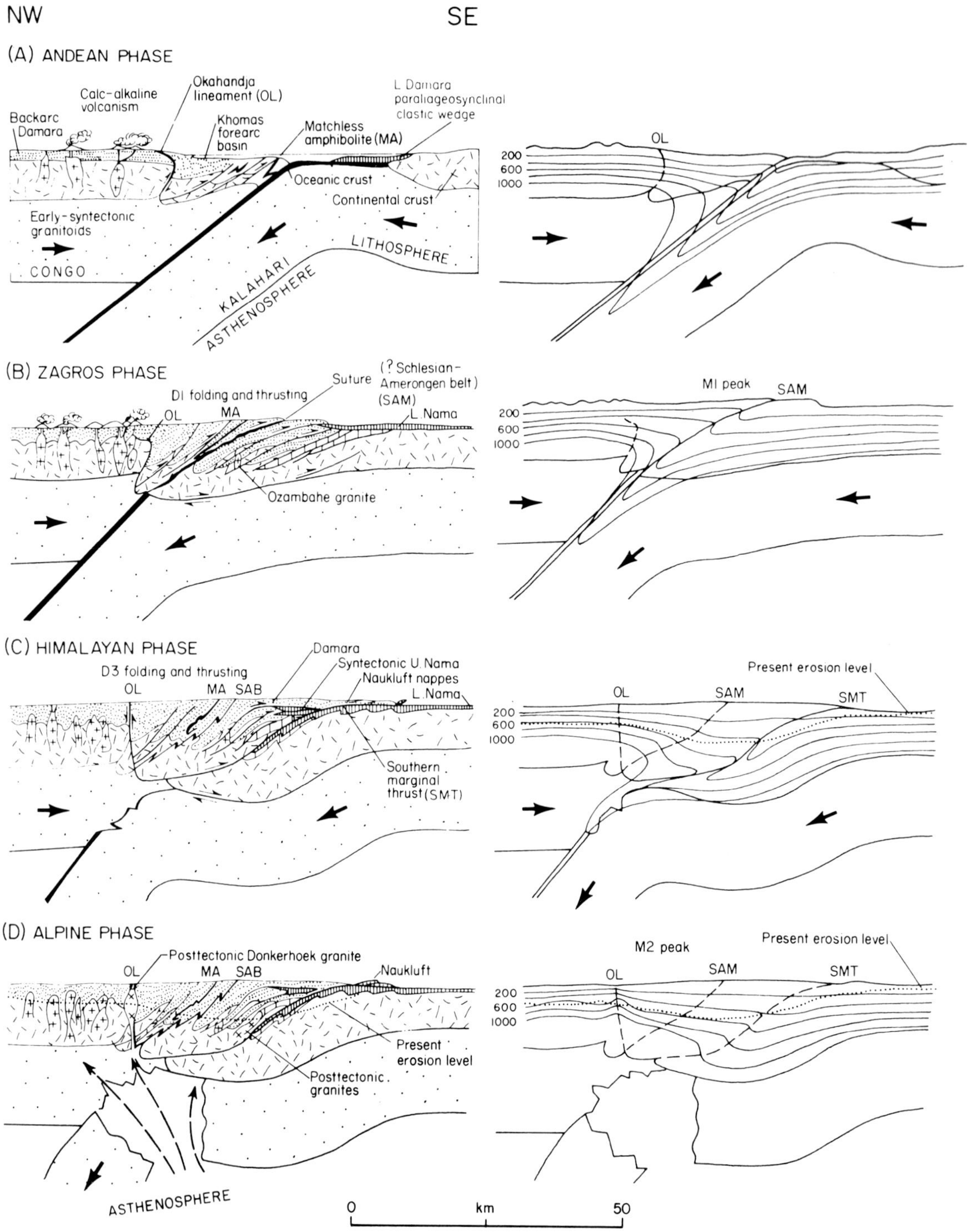

Figure 9-18. A model of continental collision for the intracontinental branch of the Damara (see text for explanation). Isotherms in right-hand column are based on geothermal gradients calculated from metamorphic parageneses and measured distances in the southern part of the Damara Province. [After Kasch (1979b, 1980).]

type) (Bally and Snelson, 1980). A Himalayan-type secondary collision zone, marked by the major thrust zone along the southern margin of the province, could absorb most of the strain of convergence by the growth of an imbricate pile of thrust sheets (Kasch, 1979b; Barnes and Sawyer, 1980). The observed reversal of dip slip down to the north along the Okahandja lineament at this stage (Blaine, 1977; Sawyer, 1978) can be accounted for during this late stage of convergence. Himalayan uplift around the original suture possibly provided the surface slope (cf. Elliott, 1976) necessary for the southeastward spreading of the Naukluft nappes and their syntectonic Nama deposits.

Detachment of oceanic crust and cessation of subduction would have elevated the isotherms because of injection of hot asthenosphere toward the base of the continental crust (Fig. 9-18D). Partial melting of this crust could account for the posttectonic Donkerhoek-type granitoids with their high initial $^{87}Sr/^{86}Sr$ ratios (Hawkesworth *et al.*, 1981) and for the posttectonic M2 metamorphic overprint at about 560 Ma (Kasch, 1979a, b). The geothermal gradient associated with M2 varied from 40-50°C/km in the Central Zone, then dropping sharply across the Okahandja lineament to 20–30°C/km in the Southern Marginal Zone (Barnes and Sawyer, 1980). Upwelling of hot asthenosphere and isostatic recovery of thickened continental crust in the collision zone combined to produce rapid Alpine-type uplift along the Southern Marginal Zone and Central Zone between 550 Ma and 530 Ma (Kasch, 1979b, 1980; Barnes and Sawyer, 1980). As the rate of uplift outpaced the rise of isothermal surfaces, the metamorphic grade would have retrogressed and orogeny ceased.

The extent of ocean spreading and separation of these continents before their hypothetical convergence and collision is a matter of much speculation. Hartnady (1979) noted that the proposed welding of the Kalahari and Congo continents provided a compelling and satisfying solution to a long-standing stratigraphic problem, namely the very dissimilar lithostratigraphy on either side of the Khomas trough (Gevers, 1934; Martin, 1965). This viewpoint implies that the two continents had an independent history before collision, in the manner of Asia and the Indian subcontinent. Radiometric ages from the Damara Province suggest an orogenic history of some 200 Ma from about 700 Ma to about 500 Ma with younger "cooling" ages (Table 9-7) (Hartnady, 1978b; Porada, 1979; Kröner, 1980a; Hawkesworth *et al.*, 1981). At this stage it is difficult to partition this spread of radiometric ages into the precollisional, collisional, and postcollisional phases proposed above. The large margin of error associated with earlier estimates of paleomagnetic pole positions for the Congo and Kalahari cratons (Piper, 1972; McElhinny *et al.*, 1974) allowed for some differential movement; but more-accurate estimates suggest that ocean opening, if it occurred at all, was extremely limited in extent (McElhinny and McWilliams, 1977; McWilliams *et al.*, 1981).

A Compromise Model

The continental collision model described above implies the closing of a broad expanse of oceanic crust. In order for an active plate margin to evolve through the various stages of collision from Andean to Himilayan, a large and forceful mantle convection system would have to have been operative. Subduction of this oceanic crust would entail formation of an extensive calc–alkaline volcanic arc in the Central Zone of the Damara Province (Fig. 9-18A,B). No such volcanic rocks are preserved, even in a metamorphic state, although the calc–alkaline nature of the extremely voluminous Salem granitoids in the Central Zone suggests that the roots of a broad magmatic arc remain. A problem with this interpretation is that most of the granitoids studied have high initial $^{87}Sr/^{86}Sr$ and low initial $^{143}Nd/^{144}Nd$ ratios (Hawkesworth *et al.*, 1981). This indicates an origin by remelting of pre-Damara basement or Damara metasediments with only limited addition of upper-mantle material. However, the diorite sampled by these authors from the Salem Suite and dated at ~550 Ma (Rb-Sr) has initial Sr- and Nd-isotope ratios that suggest upper-mantle derivation.

The absence of blueschist metamorphic assemblages and the inexplicable stratigraphic juxtaposition of ocean-floor basalts and terrigenous pelitic sediments in the Matchless belt militate somewhat against large-scale, rapid ocean closure. The interpretation of the Khomas Subgroup as forearc deposits is debatable: their Sm-Nd data suggest they represented a mixture of pre-Damara continental detritus and a young volcanic arc represented either by a Damara arc proposed by Kasch

(1980) or the 1300–1200 Ma Rehoboth Arc of Watters (1974, 1976, 1977).

Kröner (1980b) proposed what is essentially a compromise hypothesis between the nonconverging ensialic model and those models advocating conventional plate convergence. The early history of the orogen is envisaged as entirely ensialic with three rift grabens forming in thinned continental crust over a hot mantle plume, as proposed by Martin and Porada (1977a) (Fig. 9-19A). The aulacogen stage, in which the rift basins deepen, widen, and coalesce, is marked by two important processes, also advocated by Martin and Porada (1977a), near the base of the thinned sialic crust: (1) delamination of the dense subcrustal lithosphere by sinking; (2) concomitant intrusion of hot asthenosphere into the base of the crust as ocean-tholeiite intrusions and volcanic rocks (Fig. 9-19B). Subsequent compression leads to A-subduction of a flat plate of continental crust along the zone of delamination to produce thickened crust (Fig. 9-19C). As in the continental collision model, the suture at this stage is postulated to be along the southern boundary thrust at the margin of the Damara Province. Kröner's (1980b) model accommodates most of the features of the intracontinental branch of the Damara Province, namely the limited oceanic basalts, the paleomagnetic evidence against large-scale ocean opening, the paucity of calc–alkaline mantle-derived volcanism and blueschist metamorphism, the evidence for abundant crustal melting, and the considerable telescoping of the crust along the southern margin of the orogen as opposed to its gently deformed northern platform.

Continental Divergence and Convergence along the Coastal Branch of the Damara Province and the Gariep–Saldanian Provinces

The three Early Nosib grabens at the base of the Damara Supergroup in the intracontinental branch have their equivalent in the northern coastal branch west of the large Kamanjab basement inlier (Fig. 9-15) (Martin, 1965; Guj, 1970; Porada, 1979) and in the Stinkfontien trough in the Gariep Province farther south (Fig. 9-2). The north-northwest trend of the grabens in the northern coastal branch of the Damara Province and in the Gariep Province contrasts strongly with the east-northeast trend of the grabens in the intracontinental Damara Province; the grabens together define an early rift system developed over a triple-plate junction (Porada, 1979). Divergence of the Kalahari, Congo, and Brazilian plates may have been caused by a rising mantle plume beneath the triple junction (cf. Hoffman *et al.*, 1974) west of the Namibian coastline (Porada, 1979).

Bimodal igneous activity along the downwarped western margin of the Damara coastal graben is reflected in the Noerop acid lavas and the Khankhaib basic and ultrabasic intrusions in Namibia (Guj, 1970) and in the Chela Group of Angola (Kröner and Correia, 1980). Equivalent igneous activity took place in the intracratonic branch (Naauwpoort Formation) and in the Gariep Province (Stinkfontein Formation, Richtersveld Intrusive Suite, and Gannakouriep dike swarm).

Porada (1979) drew attention to analogous tectonic activity in the Brazilian plate, the third component of the triple junction. In the Late Precambrian Ribeira Province three grabens parallel to the coast are filled by Brasiliano cross-bedded clastic sediments overlain by carbonates and interfingering rhythmites and subordinate metabasites (Almeida *et al.*, 1973, 1976; Hasui *et al.*, 1975); the paleoslope was eastward. The graben-like nature of these accumulations is indicated by the fact that they are separated by zones of reworked basement overlain directly by late-syntectonic molasse-type deposits; these zones would therefore have been uplifted areas during early syntectonic stages (Porada, 1979).

Subsidence along the Damaran coastal branch during continental divergence was strongly asymmetric. East of the early graben lies the miogeosynclinal Otavi carbonate platform (Fig. 9-15). Nosib graben clastics are thin or absent west of the Sesfontein thrust fault (Guj, 1970), which marks the approximate western boundary of the early graben (Porada, 1979). In this area post-Nosib sediments directly overlie the pre-Damara basement, indicating that this was elevated in Nosib times (Fig. 9-16C). West of this elevated basement area metamorphosed turbidites and graywackes equivalent to the Kuiseb Formation occupy a deep eugeosynclinal trench, here termed the Kaoko trough, extending 350 km along the coast (Fig. 9-14); facies changes and thickness gradients are directed westward toward the Brazilian plate (Guj, 1970). The source area for the 4–5 km

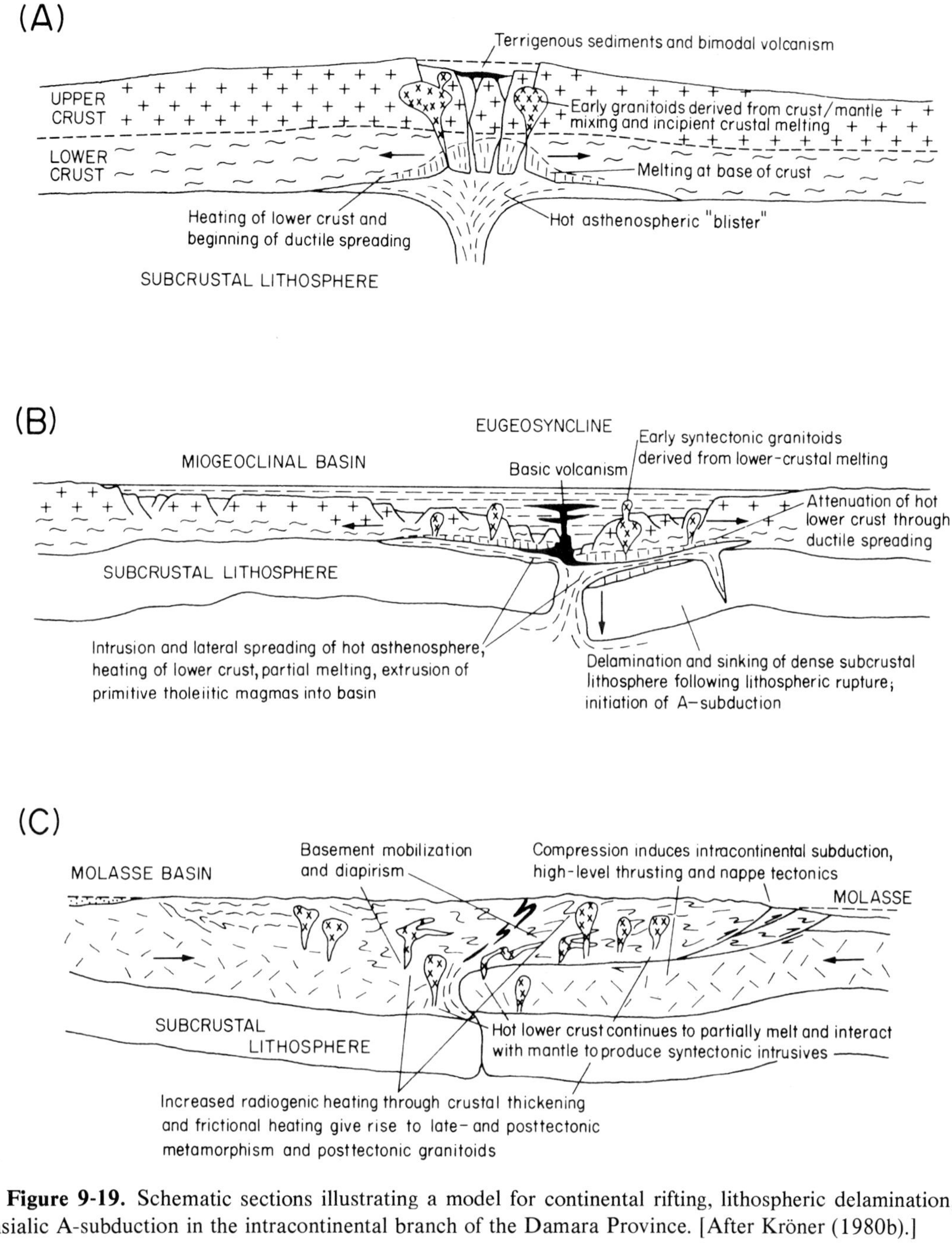

Figure 9-19. Schematic sections illustrating a model for continental rifting, lithospheric delamination, and ensialic A-subduction in the intracontinental branch of the Damara Province. [After Kröner (1980b).]

thickness of these flyschlike deposits is unlikely to have been to the east or northeast because this area was occupied by the Otavi carbonate platform and a narrow chain of uplifted basement blocks at this time (Fig. 9-15). Porada (1979) tentatively suggested source areas to the north, with infilling southward along the trough axis, or to the west from the Ribeira Province of Brazil.

From the evidence of abundant basic volcanism in the Grootderm Formation in the eugeosynclinal part of the Gariep Province, it may be concluded that opening of the proto-South Atlantic by divergence of the Brazilian plate away from Africa may have continued to the stage of ocean-floor spreading (Porada, 1979). Unfortunately this is conjectural because no possible remnants of ocean crust have yet been recognized in the coastal Damara Province. This crust would most likely be preserved in a suture zone, formed by later closure of the proto-ocean; the suture would now be hidden beneath Phanerozoic sediments on the present African continental shelf.

Hypothetical convergence and subsequent collision of the Brazilian plate with the Pan African plate is strikingly illustrated by the structural style of the coastal Damara Province (Fig. 9-16C). A strong eastward vergence of the folds and thrust faults led Porada (1979) to suggest that the Pan African plate was thrust beneath the leading Brazilian plate (cf. Dewey and Bird, 1970; Dewey and Burke, 1973). Basement-cored nappes of intensely deformed eugeosynclinal sediments were driven outward from the orogen, where they overrode a foredeep, or exogeosyncline, in the east filled by the molasse-type Mulden Group of Cambrian age (Fig. 9-16C) (Guj, 1970; R.M. Hedberg, 1979). Eastward-directed thrusting has also been recorded in the southern coastal branch of the Damara in a Nosib-type rift sequence called the Spencer Bay Formation (Kröner and Jackson, 1974) and in the eugeosynclinal part of the Gariep Province farther south (Fig. 9-14) (Martin, 1965; Kröner, 1975b).

In the coastal Damara the grade of regional metamorphism increases westward away from the foreland and toward the inferred offshore suture zone. In the northern coastal branch the grade rises westward from greenschist to upper amphibolite facies via a medium P/T facies series (Guj, 1970). The metamorphic grade in the Gariep Province rises westward from lower to upper greenschist facies, and possibly to high P/T blueschist facies (De Villiers and Söhnge, 1959; Kröner, 1975b).

In the leading Brazilian plate, erosion to deep crustal levels has exposed abundant syntectonic granitoids intruding Brasiliano metasediments metamorphosed to low- or medium-P/T facies series (Porada, 1979). Late-syntectonic molasse deposits and acid to intermediate volcanic rocks filled several intermontane grabens during uplift of the Ribeira Province (Almeida *et al.*, 1976; Porada, 1979). At this time shedding of molasse-type debris from the rising intracontinental branch of the Damara Province formed the Fish River Subgroup which represents the terminal phase of sedimentation in the Nama basin.

The Kalahari, Congo, and Brazilian plates remained fused to form components of the Gondwana supercontinent. This part of the supercontinent survived as an entity until its fragmentation in the Mesozoic (Chapter 12) despite an abortive episode of Early Paleozoic rifting along the grain of the Late Precambrian Saldanian Province that gave rise to the Cape Supergroup, discussed in Chapter 10.

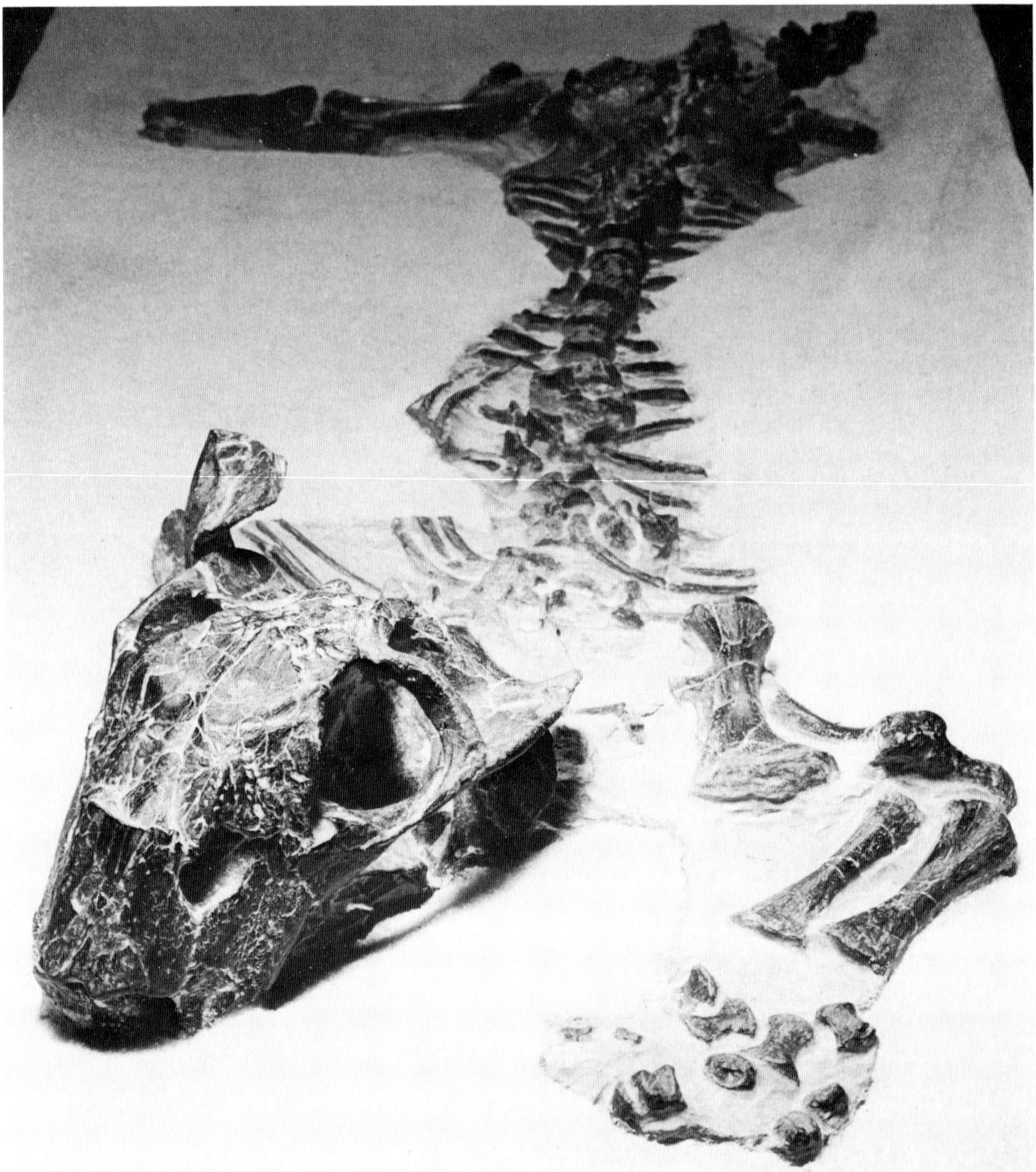

STAGE 4

The Gondwana Era

Paleozoic and Mesozoic basins have long provided compelling evidence for the former existence of the supercontinent Gondwana. The dicynodont therapsid, *Lystrosaurus*, preserved in great numbers in Africa, is also found in Europe, Asia, and Antarctica. A herbivorous mammal-like reptile, *Lystrosaurus* represents one of the first groups of tetrapods that were primary consumers. Lower Triassic Katberg Formation, specimen USNM 22945, Smithsonian Institution. [Courtesy of N. Hotton III, photographed by Chip Clark.]

Chapter 10

The Cape Trough: An Aborted Rift

The thermotectonic events at the end of the Precambrian that brought Malmesbury geosynclinal sedimentation to a close initiated a new style of sedimentation characteristic of the 8000-m-thick Lower Paleozoic Cape Supergroup. Deformation and low-grade metamorphism of Malmesbury formations were accompanied by emplacement of granite plutons between 600 and 500 Ma. Posttectonic rise of the Pan African orogens favored accumulation of molasse-type deposits under shallow-water and subaerial conditions. Both the Franschhoek Formation (Malmesbury Group) and the post-Malmesbury Klipheuwel Formation of the southern Cape are complex intertonguing mudstones, arkosic sandstones, and conglomerates of continental origin; the Franschhoek Formation is distinguished by more intense shearing and folding (Haughton, 1969, p. 304). Germs (1972a) equates the Klipheuwel Formation with the terminal Nama cycle. By Cambrian time the paleogeography of the southwestern Cape had evolved to a braided alluvial coastal plain bordering a shallow siliciclastic shelf (Vos and Tankard, 1981). Marine limestones accumulated in the most distal parts of the basin now represented by the Ellsworth and Transantarctic regions of Antarctica (Stump, 1976; Shergold *et al.*, 1976).

The oldest Klipheuwel strata of the southwestern Cape show complex deformation related to the final episode of the polyphase Pan African thermotectonic event (Rust, 1973; Hartnady *et al.*, 1974). They overlie deformed Malmesbury metasediments and Cape granites unconformably (Chapter 9) but have nevertheless been affected by folding, thrusting, and faulting (Visser, 1967; Haughton, 1969). The Klipheuwel Formation becomes progressively less deformed upward, and in its uppermost parts it is probably a facies equivalent of the basal Cape Supergroup.

The Klipheuwel Formation is up to 2000 m thick and forms a northward-thinning wedge. Sedimentary structures and lithic associations of intertonguing red-brown conglomerates, feldspathic sandstones, mudstones, and shales indicate a terrestrial origin. The conglomerates contain intrabasinal clasts of Malmesbury rock, granite, quartz porphyry, quartzite, and vein quartz and vary between clast and matrix supported. Angular clasts of Malmesbury graywacke up to 30 cm long occur at the base. The conglomerate facies thins southward (Visser, 1967), where it gives way to a sandstone succession containing conglomerate interbeds, shallow scour structures, southward-directed cross-bedding, interference ripples, and sand-filled desiccation cracks (Haughton, 1933; De Villiers *et al.*, 1964). Red mudstone is present as partings and as lenticular interbeds up to 30 cm thick; elsewhere it occurs as units up to 55 m thick with sandstone interbeds (Haughton, 1933; Visser, 1967).

The Klipheuwel succession is envisaged as a series of coalescing and stacked alluvial-fan and interlobe deposits of southward-flowing braided fluvial systems (Rust, 1973; Vos and Tankard, 1981). The coarseness and poor rounding of the conglomerates, their low degree of sorting, and the high proportion of clay matrix suggest limited distances of transportation and local debris flows in a block-faulted terrain such as commonly marks the culminating phase of orogenesis (cf. the Triassic rift basins of the Appalachians). Although the Klipheuwel Formation is in most places overlain unconformably by the Cape Supergroup, the two sequences are locally conformable. Considering its greater paleogeographic distribution in a reconstructed Gondwana (Du Toit, 1927) and the abundance of exotic clasts (Rust, 1973), the Piekenierskloof Formation of the basal Cape Supergroup probably represents the establishment of a more regional braided alluvial plain.

In the eastern Cape thick quartz arenites of the Cape Supergroup overlie metasedimentary rocks of the Kango, Gamtoos, and Kaaimans Groups (Chapter 9). These Late Precambrian rocks are conglomeratic and arkosic in their upper parts, and although they were metamorphosed to low or medium grade during the Pan African orogeny (Du Toit, 1954; Haughton, 1969), in places they are conformable with the overlying Peninsula Formation. Le Roux (1977) showed that the youngest of the Kango deposits may even be postorogenic and younger than the Klipheuwel Formation, so that Cambrian alluvial sedimentation was probably widespread in the early Cape basin.

Southward thickening of the Lower Paleozoic Cape succession represents nearly continuous sedimentation and indicates regional subsidence of a broad cratonic shelf and northward transgression of a marginal sea. Three lithostratigraphic groups of variable thickness total over 8000 m of sediment in the eastern Cape depocenter (Fig. 10-1), recording a progressive change from shallow-marine and terrestrial environments in the Table Mountain Group to destructive deltaic and shallow-marine shelf in the Bokkeveld and Witteberg Groups.

The most complete and continuous stratigraphic section is in the southwestern Cape. On the eastern seaboard (Transkei, Natal, and Zululand) deposition of the Natal Group took place in an elongate downwarp which, prior to continental breakup, was probably the failed rift of a triple junction (Tankard and Hobday, 1979). Because of Mesozoic breakup and subsequent erosion, the original extent of the Cape Supergroup is poorly known; however, the remarkably similar stratigraphy in parts of South America and the Falkland Islands was recognized by Du Toit (1927, 1937). Further evidence of the former contiguity of these land masses has been provided by Boucot *et al.* (1969) and Copper (1977).

10.1. Table Mountain Group: The Quartz Arenite Problem

Resting unconformably on Precambrian–Cambrian metasedimentary and granitic basement are up to 4000 m of quartz arenites, conglomerates, and mudstones which comprise the Early Ordovician to Early Devonian Table Mountain Group of the southern Cape (Table 10-1). The elongate depoaxis trended parallel to the present coast. Patterns of sedimentation and stacking of facies were probably accommodated by pronounced basement-controlled subsidence

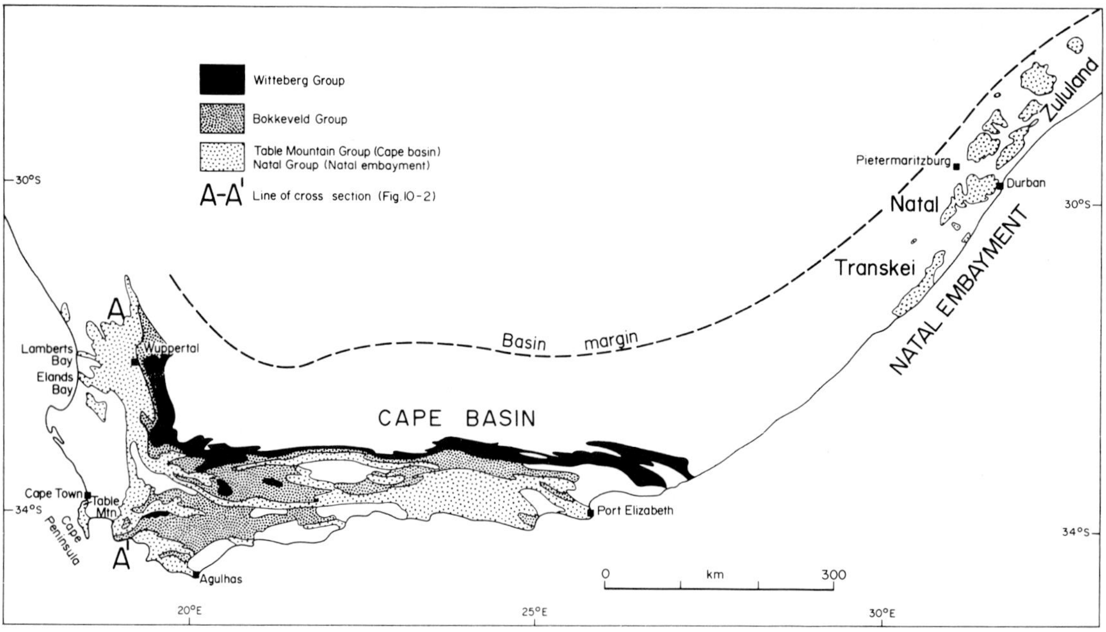

Figure 10-1. Geologic Map of the Cape Supergroup.

Table 10–1. Stratigraphy of the Table Mountain Group[a]

Western Cape			20°E	Eastern Cape		
Formation	Thickness (m)	Lithology	Formation	Thickness (m)	Lithology	Age
Upper						
			Baviaanskloof	150	Shale, mudstone, quartz arenite; marine invertebrates	Silurian–Siegenian (Early Devonian)
Nardouw	1100	Coarse-grained quartz arenite; trace fossils	Kouga	380	Quartz arenite	
			Tchando	280	Sandstone	
Cedarberg	140	Fine-grained sandstone, siltstone, and mudstone; marine invertebrates	Cedarberg	50	Shale, mudstone, fine-grained sandstone	Late Ashgillian (end Ordovician)
Pakhuis	120	Sandstone, conglomerate, diamictite				
Lower						
Peninsula	1800	Medium- to coarse-grained quartz arenite with quartz pebbles; trace fossils	Peninsula	2150	Medium- to coarse-grained quartz arenite with quartz pebbles; trace fossils	Early–Late Ordovician
Graafwater	440	Interbedded quartz arenite, siltstone, and mudstone; trace fossils				Early Ordovician
Piekenierskloof	800	Conglomerate and coarse-grained sandstone				Early Ordovician

[a]Sources: Cocks *et al.* (1970), Rust (1973, personal communication, 1980), Visser (1974), Johnson (1976), South African Committee for Stratigraphy (1980).

accompanied by faulting (Fig. 10-2) (Vos and Tankard, 1981; Tankard and Barwis, 1982). Variable thicknesses of the stratigraphic units are attributed to differential basinal subsidence and uneven rates of sedimentation. The succession is correlated lithologically with the Natal Group of the eastern margin of the subcontinent (Fig. 10-1). The stratigraphic nomenclature (Table 10-1, Fig. 10-2), here informally divided into lower and upper units, was established by Rust (1967, 1973).

Marine fossils occur sparsely in the Cedarberg Formation. The brachiopod fauna has several elements in common with the Late Ashgillian (end Ordovician) *Hirnantia* faunas of Morocco and South America (Wright, 1968; Cocks *et al.*, 1970); these *Hirnantia* faunas are generally associated with glaciogenic deposits (Whittington and Hughes, 1970). Age-diagnostic elements include a homalonotid trilobite and the brachiopod genera *Trematis, Eostropheodonta,* and *Plectothyrella*. Chitinozoans from the basal Cedarberg also indicate a Late Ordovician age (Cramer *et al.*, 1974). Near the top of the Table Mountain

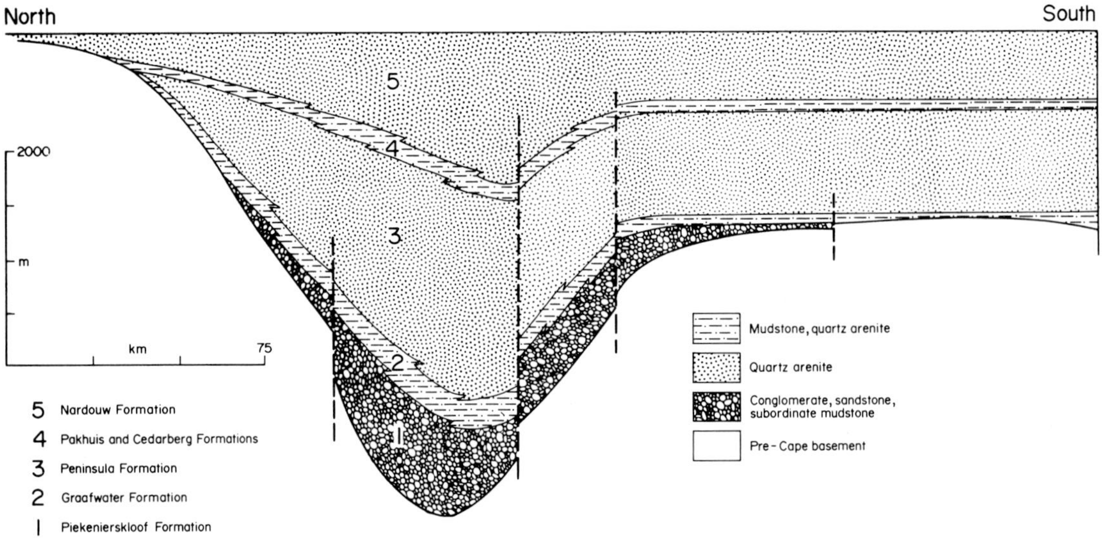

Figure 10-2. North–South stratigraphic cross section of the Table Mountain Group in the western Cape showing inferred contemporaneous faulting; line of section A–A′ in Figure 10-1. [Adapted from Rust (1967).]

Group in the eastern Cape, argillaceous intervals containing the brachiopod *Pleurothyrella* are probably of Siegenian (Early Devonian) age (Theron, 1972; A.J. Boucot, personal communication, 1981). Cocks *et al.* (1970, their table 1) have inferred a Late Cambrian to Early Ordovician age for the basal Table Mountain Group. With the Klipheuwel Formation, the Piekenierskloof Formation postdates the Cape granites (Rust, 1973; Hartnady *et al.*, 1974).

The succession is most spectacularly expressed in the Cape Fold Belt, where the degree of folding and shearing generally discourages detailed sedimentological studies. The following discussion emphasizes recent work in the western part of the basin (Rust, 1967, 1973, 1977, in press; Tankard and Hobday, 1977; Hobday and Tankard, 1978; Vos and Tankard, 1981). In the western Cape the Table Mountain Group lies unconformably on a remarkably planar granitic or metasediment platform that is broken only by minor faulting. This low relief permitted abrupt changes in sedimentation patterns resulting from minor changes in the balance between basinal subsidence and clastic influx.

Bedload Fluvial Complex

Previous studies of the Piekenierskloof Formation have shown a dominantly fluvial mode of deposition and distinct zonation of facies (Rust, 1967, 1973; Vos and Tankard, 1981). Coarse conglomerates with sandstone interbeds are well developed in the western apical region but recent geophysical surveys suggest that they may be considerably thinner than the 800 m originally estimated (I.C. Rust, personal communication, 1980). Distally the conglomerates merge with over 900 m of coarse-grained sandstones. The Piekenierskloof Formation is mostly flat lying or gently dipping and is commonly underlain by Klipheuwel rocks, the upper, relatively undeformed parts of which may represent distal facies equivalents of Piekenierskloof deposits.

Parts of the Klipheuwel succession consist of shallow basinal mudstones and fine-grained sandstones with crawling traces (Rust, 1967; Vos and Tankard, 1981). Elsewhere upward-coarsening intervals 50–80 m thick comprise siltstones, mudstones, and red sandstones arranged in 1–3 m thick sequences showing both upward-coarsening and erosively based, upward-fining patterns. Grain size and thickness of sedimentation sequences generally increase upward through the succession. Conglomerates near the top contain clasts up to 10 cm in diameter. A fan-delta environment is suggested for these deposits. The distal braided-stream facies contains multiple channel-fill sequences comprising basal conglomerate overlain by sandstone and mudstone. Sedimentary structures include planar and trough cross-bed types, with set thickness decreasing upward. Foresets dip mainly southeastward with considerable dispersion. The more tabular, gradationally based, argillaceous units that are prominent lower in the succession are overlain by multilateral sandstone units, reflecting the interaction between channelized flow and standing water.

Conglomerates and subordinate sandstones of the lower Piekenierskloof Formation generally overlie the distal braided facies with a deeply erosive contact (Fig. 10-3). In the Lamberts Bay area the red conglomerates and interbedded sandstones are thickest, thinning markedly to the southeast (Rust, 1967). In the apical regions clasts attain 40 cm in diameter and show a similar southeastward reduction in size. Rounded vein quartz is the dominant clast type in a diverse assemblage that includes quartzite, chert, jasper, acid and basic lavas, diorite, and gneiss.

Lenticular units of clast-supported, imbricated conglomerate up to 3 m thick decrease in thickness upward concomitantly with an increasing abundance of medium- to coarse-grained sandstone. Whereas paleoflow directions indicated by trough foresets, pebble alignment, and imbrication are broadly downstream, planar foresets are oriented at varying angles away from the conglomerate body, suggesting lateral accretion on the flanks of gravel bars.

Thick, medium- to coarse-grained sandstone replaces the conglomerate facies vertically and southward (Rust, 1967; Vos and Tankard, 1981). Channel-based sequences with a maximum thickness of 2.5 m consist mainly of coarse-grained pebbly sandstone with sporadic conglomerate lenses. Typically the basal erosion surfaces are overlain by conglomerate lag followed by sets of planar and trough cross-bedding with foresets directed uniformly to the southeast. Strike sections show side-filled channels 1 m deep and 5–15 m wide. Rare large-scale planar sets are directed into the channels, merging with smaller sets of trough cross-bedding. The lateral association of large-scale planar sets with trough cross-bedded channel fill has been ascribed to modification of braid-bar margins during falling-water stage (Collinson, 1970).

Thick sequences of plane-bedded to low-angle cross-bedded sandstone occur in the southern part of the basin. Graded laminae 0.5–2 cm thick occasionally preserve current lineation on bedding planes. Other structures include cross-bed intrasets and shallow southeastward-directed channel scours. Sheet flows associated with abnormal flooding of broad, low-gradient areas generate thick sheet sands with characteristics resembling those of this region (McKee *et al.*, 1967). Deposition was mainly under upper flow regime conditions with shallow channels and dunes forming during waning stages.

The variety of interrelated facies types and the fan-shaped paleocurrent distribution of the upper Klipheuwel and Piekenierskloof Formations reflect downstream changes in alluvial processes associated with decreasing gradients. High relief (Rust, 1973) was probably maintained along the northern margin of the basin (Fig. 10-2) by progressive subsidence relative to adjoining fault-bounded highlands. Downstream reduction from

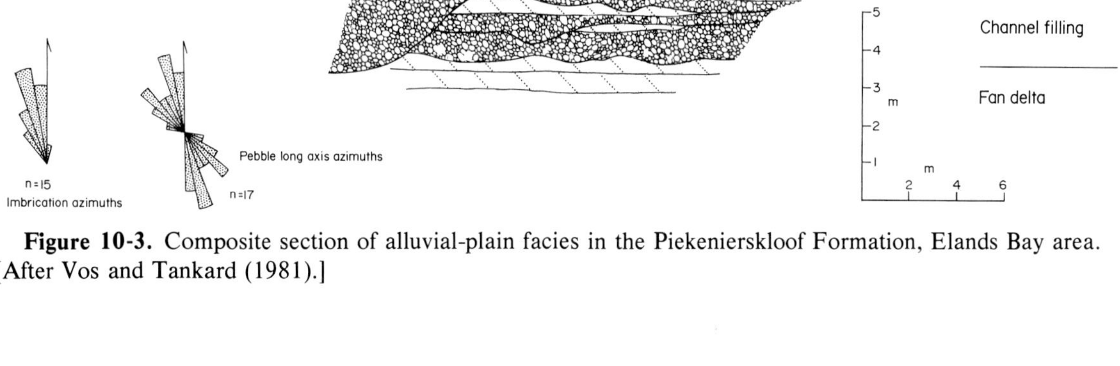

Figure 10-3. Composite section of alluvial-plain facies in the Piekenierskloof Formation, Elands Bay area. [After Vos and Tankard (1981).]

boulder-size sediments to sand was accompanied by a change from a dominance of longitudinal to transverse bars. Distally the overloaded streams debouched into a shallow sea forming a fan delta. Because of the absence of vegetation, low channel gradients, which commonly result in meandering river styles in modern rivers, characteristically supported braided patterns in the Early Paleozoic (Schumm, 1968a). Braided patterns were also favored by the low proportion of cohesive, fine-grained sediments. Flashy discharge resulting from unimpeded runoff in the vegetation-free Piekenierskloof landscape promoted substantial deposition of sheetflood sands.

Tidal-Flat Sedimentation

The Early Ordovician Graafwater Formation of the western Cape is believed to have originated in tidal-flat and shallow subtidal environments (Rust, 1973, 1977; Tankard and Hobday, 1977). It is characterized by thinly interbedded quartz arenites and red mudstones, with marine trace fossils, bipolar cross-bedding, and structures indicative of subaerial exposure. Progressive basinal subsidence and multiple stacking of facies facilitated maximum accumulation of more than 400 m of mud and sand in approximately equal proportion. Farther seaward vast quantities of quartz sand of the lower Peninsula Formation accumulated contemporaneously. Along the northern margin of the basin the Piekenierskloof and lower Graafwater Formations interfinger, indicating tidal reworking of the distal alluvial plain (Rust, 1967; Tankard and Hobday, 1977). Farther south the Piekenierskloof Formation is absent, and the Graafwater sequence lies directly on an even granite and metasediment platform.

Three lithologic types that characterize the Graafwater succession are commonly arranged in an upward-fining pattern. These are a quartz arenite facies, a mixed sandstone–mudstone facies and a mudstone facies.

The fine- to medium-grained quartz arenites form tabular and lenticular units separated by siltstone-draped erosion surfaces. Bed thickness and grain size tend to decrease upward within individual sequences. Trough cross-bedding is the dominant sedimentary structure although planar sets are also common. Overall foreset orientations are variable to bipolar, but within cosets they are strongly unidirectional. A southward to southwestward mode predominates, possibly reflecting an obliquely offshore trend relative to the Piekenierskloof paleogeography. However, weak landward-directed paleocurrents are associated with smaller scale sets, herringbone cross-bedding, and reactivation surfaces. Mud-clasts litter erosion surfaces and occur on foresets. Bedding-plane surfaces preserve a variety of ripple forms, including oscillation, ladderback (Fig. 10-4A), double-crested, and flat-topped varieties. Deformation and dewatering structures are common. Trace fossils are the only evidence of biological activity; *Skolithos* is the most common ichnogenus. Other traces include *Cruziana, Helminthoida, Platysolenites, Scolicia,* and large arthropod tracks (Fig. 10-4B) (Rust, 1967, 1973, 1977; Anderson, 1975a).

The quartz arenite facies is succeeded gradationally by a vertical alternation of thinly interbedded maroon-red mudstone and quartzose sandstone, with the clay content increasing upward. The quartz arenite units are rarely thicker than 20 cm and show small-scale solitary cross-bed sets and diverse ripple marks. Lenticular, flaser, and wavy bedding types characterize the finer units. Abundant sand-filled desiccation cracks (Fig. 10-4C) are evidence of periodic emergence.

Maroon-red mudstones are preserved in places, but generally they have had a low preservation potential. Sedimentary structures comprise mainly lenticular and flaser bedding, siltstone-filled desiccation cracks, and sporadic sandstone wedges, which have been compared with channel margin bars (Tankard and Hobday, 1977). Rare red-brown claystones contain abundant concretions.

The vertical distribution of the textures and primary sedimentary structures, together with evidence of shallowing (double-crested, ladderback, flat-topped ripples), exposure (desiccation cracks), and the presence of a marine biota, provide convincing evidence of tidal deposition when compared with modern prograding tidal flats. Bimodal–bipolar distribution of cross-bed azimuths, including herringbone and reactivation structures, is typical of tidal reversals. However, the absence of structures indicative of emergence within many sandstone sequences suggests subtidal deposition of much of the quartz arenite facies. Because bottom velocities and period of submergence decrease shoreward, tidal flats tend to be sandy at and below the mean low-tide level, grading landward into suspension-dominated mud flats.

Most modern tidal and estuarine flats are dissected by channels that transfer water through the tidal cycle. Tidal-creek point bars display upward-fining textural distributions similar to those produced by prograding tidal flats (Barwis, 1978). Evidence that many of the Graafwater sequences originated as tidal channels is provided by the lenticular–tabular sandstone units with erosive contacts and a lag of quartz pebbles and mud-clasts (Rust, 1977; Tankard and Hobday, 1977). Within cosets, paleoflow directions are generally similar and a southerly to southwesterly ebb component predominates. Although adjacent channel-fill units commonly display large differences in paleocurrent direction, northward-directed (flood-oriented) bedforms are most evident near the top of a channel sequence. Segregation of flood and ebb currents is common in estuarine environments, and velocity asymmetry of the tidal ellipse ensures that flood- and ebb-oriented bedforms occur at different elevations on tidal creek point bars (Barwis, 1978). The rock record tends to be biased toward ebb trends because the stratigraphically higher flood-oriented bedforms are commonly truncated. Clay-draped ripples and sand-filled desiccation cracks indicate short periods of submergence followed by lengthy periods of emergence.

The Graafwater Formation is transgressively overlain by a thick succession of quartz arenites, which have been compared with Holocene deposits in barrier island and tide-dominated shallow-shelf environments (Rust, 1973; Visser, 1974; Tankard and Hobday, 1977). The Graafwater is therefore visualized as representing an estuarine tidal environment fringing an alluvial plain (Piekenierskloof Formation), with a barrier complex (Peninsula Formation) forming the major strandline in the south. Tankard and Hobday (1977) inferred a lower mesotidal range (2–3 m) on the basis of upward-fining sequences which contained emergence runoff features and tidal-channel point-bar units.

Complete facies sequences are seldom encountered; the uppermost argillaceous members were usually truncated by transgressive erosion. However, retention of a substantial part of each sequence (Fig. 10-4D) is attributed to multiple, rapid, northward transgressions of the sea counterbalanced by substantial strike-fed sediment influx from a fluvial source. Stacking of numerous facies sequences has been caused by episodic subsidence. The red-brown, concretion-bearing claystone strongly resembles the supratidal flats of the Colorado River delta, where oxidizing conditions prevail (Thompson, 1975). Van Straaten (1959) has shown that in tropical regions desiccation cracks form in the midtide zone. The very large cracks of Chapmans Peak Drive suggest prolonged emergence between relatively infrequent episodes of inundation, such as may have been occasioned by storm-augmented or spring tides. Reddening is believed to be a diagenetic process favored by warm temperatures and alternating wet and dry conditions. Reconstructions by Smith *et al.* (1973) indicate an Early Ordovician paleolatitude for the Cape basin of about 30°.

It is possible that a sequence of metasediments comprising the Sardinia Bay Formation of the eastern Cape (p. 313) may be stratigraphically equivalent to the Graafwater Formation (I.C. Rust, personal communication, 1980). Certainly they represent the same depositional setting. The primary structures, including herringbone cross-bedding, small-scale textural alternations, and a variety of biogenic features, suggest deposition in a tidal environment (Shone, 1979).

High-Energy Barrier and Shelf Environments

The Peninsula Formation is a thick quartz arenite complex that transgressively overlies the backbarrier estuarine deposits of the Graafwater Formation in the Western Cape (Fig. 10-2); in the eastern part of the basin it rests on older units of metasedimentary and granitoid Precambrian–Cambrian basement. Representing one of the greatest quartz sand accumulations in the geologic record (Visser, 1974), the quartz arenites are more than 2000 m thick in the eastern Cape, and about 1800 m above the western basin axis. Today these rocks are topographically expressed in the 1000-km-long Cape Fold Belt and extend offshore onto the Agulhas Bank (Gentle, 1970).

Lithology and structures indicative of low-energy processes (argillaceous material or ripple cross-lamination) are rare, although Visser (1974) has noted rare lenses of black shale in the eastern Cape. Medium- to large-scale cross-bedding is the dominant sedimentary structure. Rust (1967) distinguishes two major facies tracts. The coarser and sporadically conglomeratic

(A)

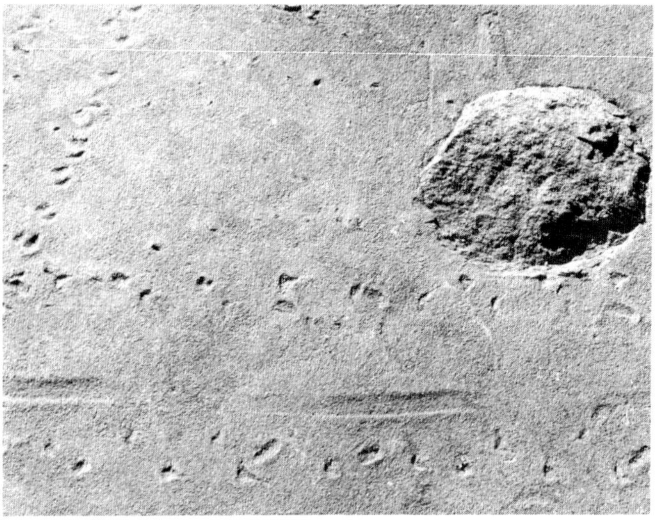

(B)

(C)

Figure 10-4. Graafwater Formation. (A) Ladderback ripples, (B) arthropod tracks, (C) sand-filled desiccation cracks and ball-and-pillow structures, (D) stacked tidal facies. Chapmans Peak Drive, Cape Peninsula.

Fig. 10-4 D.

northern margin of the basin shows a unimodal paleocurrent distribution reminiscent of the Piekenierskloof Formation. Throughout the remainder of the Peninsula Formation the quartz arenites are lithologically uniform, comprising medium- to coarse-grained sandstone with pebbles of well-rounded vein quartz, quartzite, slate, and intrabasinal mud-clasts. Biogenic structures include rare arthropod tracks, *Skolithos*, and truncated vertical tubes up to 120 cm long (Rust, 1967).

Despite the great thickness and areal extent of the Peninsula Formation, a barrier-beach and shallow-shelf model appears to apply throughout (Rust, 1967; Visser, 1974). The maturity of the sequence is ascribed to high-energy current and storm winnowing processes on a stable but slowly subsiding shelf. Even though tectonic subsidence varied from west to east there are believed to be few regional facies differences (Rust, 1967; Visser, 1974).

Excellent exposures of quartz arenite on the Cape Peninsula (Fig. 10-5) are free from the effects of deformation and shearing that inhibit detailed facies analysis elsewhere, particularly the eastern part of the Cape basin. Five sedimentary facies comprise distinctive assemblages of sedimentary structures (Hobday and Tankard, 1978). Small-scale channels and sheet sands of storm origin overlie backbarrier Graafwater deposits. Above these is a zone dominated by large, complex channel-fill sandstones followed by a zone of shelf sand bars. Renewed shoreline progradation and glaciogenic accumulation terminated deposition of the Peninsula Formation. Outcrops on the Cape Peninsula were used by Hobday and Tankard (1978) to expand on the shallow-shelf model established by Rust (1967, 1973) and Visser (1974).

The lower parts of the Peninsula Formation are characterized locally by 1- to 10-m-thick sequences of plane-bedded sandstones which dip gently (less than 4°) to the northwest. Distally the plane-beds tend to terminate in northwestward-dipping, high-angle, planar cross-bedding or are interlayered with small tidal-channel or tidal-flat facies of Graafwater lithology. In some areas the basal sheet sandstones are replaced entirely by small, multilateral channel-fill sandstones.

The relationship of the sheet sandstones to the sheltered tidal-flat deposits of the Graafwater Formation, together with their landward dip, suggest an origin by barrier overwash processes. Landward transition into planar cross-bedding is observed on modern washover fans, where it results from decreasing bed shear in the downdip direction. Washover fans are usually associated with storms that transport eroded beach and foredune material across the barrier crest. Lacking vegetation to stabilize foredunes, washover fans were probably more common on Early Paleozoic barriers such as those of the Peninsula Formation.

Another widespread facies in the lower Peninsula Formation consists of sandstones from 30 cm to 180 cm thick which are separated by erosive

Figure 10-5. Table Mountain with 750-m-thick quartz arenites of Peninsula Formation making up upper half. Cape Town in foreground.

contacts accentuated by a lag of quartz pebbles or mud-clasts. Trough and planar cross-beds are common. Within individual channels cross-bed azimuths tend to be unidirectional, although rare herringbone and reactivation structures are indicative of current reversals. Between adjacent or superimposed channel sandstones, however, the paleocurrent modes vary substantially, resulting typically in an overall multimodal foreset distribution. This facies is attributed to tidal-channel deposition in a sand-flat environment just landward of a transgressing barrier complex (Hobday and Tankard, 1978). Like their modern counterparts the predominance of southwestward-directed foresets indicates not only the greater preservation potential of the ebb bedforms which dominate the deeper parts of tidal channels but also shows an oblique orientation of the channels relative to the barrier coast (cf. Barwis, 1978).

A rare and poorly preserved facies consists of thin sequences of plane-bedded sandstones which dip gently seaward to the south and southeast. Internally the laminae are separated by heavy-mineral layers or meet with low-angle discordance. Possible antidune structures are developed and cross-bed intrasets show no consistent orientation. Beach foreshore deposition, where low-angle plane-bedding is generated by wave swash, is believed to account for this facies. Low-angle discordances represent delicate slope adjustments to varying wave energy conditions.

Channelized sandstone units up to 40 m thick are characterized by cross-bedding of variable scale and direction with a tendency for set thickness and grain size to decrease upward. Very large-scale, low-angle sets up to 2 m thick in the lower part of the sequence have a general southward or seaward orientation, whereas westward and rare northward dips become more prominent higher in the channel complex. Where preserved, plane-bedded beach sandstones overlie coarse material and terminate the sequence. Other structures include herringbone and reactivation structures and rare burrows.

These lenticular sandstones resemble the vertical sequences generated by modern tidal inlets as they migrate ahead of an accreting spit (Kumar and Sanders, 1974). The Ordovician inlets were maintained by strong tidal currents, as evidenced by large-sçale foreset reversals and reactivation surfaces, representing the daily exchange of water through the tidal cycle between the backbarrier and open marine environments. In modern inlets ebb and flood currents characteristically follow mutually evasive paths; ebb-oriented bedforms usually occupy the deeper central zone, and shoreward-directed bedforms occur along the inlet margins and in marginal flood channels. Some inlet sequences are overlain by seaward-dipping beach deposits formed in updrift spits. In the geologic record, channel-fill deposits are extensively eroded so that the ebb-dominated lower parts show a greater degree of preservation. Figure 10-6 shows a modern inlet on the coast of Massachusetts that is believed to resemble the inlets of Peninsula deposition.

Previous studies have suggested that the Peninsula Formation consists mainly of shallow shelf deposits dispersed by strong traction currents of variable direction (Rust, 1967; Visser, 1974). In the southwestern Cape thick (2–20 m), conformably based "bar" sequences display complex internal stratification. The large-scale master bedding, containing very large, low-angle foresets, is directed to the south. In many examples trough cross-bed intrasets dip obliquely along the strike of the major foresets. Periodic reworking is shown by rhythmic scour surfaces that truncate the large foreset bedding at very low angles and by sporadic herringbone structures.

Hobday and Tankard (1978) stressed the resemblance of these Peninsula sand bodies to the asymmetric sand ridges of the southern Bight of the North Sea (Houbolt, 1968). These modern ridges, with a maximum height of about 40 m, contain large-scale foresets and a secondary component of cross-bedding oblique to the long axis of the sand ridge, thus possibly explaining the complex internal geometry encountered in the Peninsula sand bar units. The rhythmic scour surfaces, like those on sand ridges of the North Sea, are possibly related to tidally induced fluctuations in current direction and intensity (Houbolt, 1968).

The substantial global rise in sea level through the Ordovician was balanced by the availability of sediment, derived by marine reworking of fluvially introduced sediment and by erosion of earlier deposits along the margin of the Cape basin; and locally caused landward overlapping of tidal flats by barrier washover sheet sands (Fig. 10-7). The wide areal extent and great thickness of the Peninsula sand bars imply a broad, gradually subsiding shallow shelf with strong tidal currents. Wide shelf seas, such as the North Sea and the China Sea, are characteristically tide dominated and develop major tidal sand ridges and bars (Off, 1963; Stride, 1963). Similar sand bodies accumulate under storm-dominated conditions and a lower tidal range, such as prevail off the eastern United States (Swift, 1975).

Storms are effective agents in the reworking cycle because shoaling storm-generated waves and currents can agitate the sea floor to greater

Figure 10-6. Modern example of some inferred Peninsula environments showing barrier island, inlet, and shallow backbarrier tidal channels. Essex Estuary, Massachusetts. [Courtesy of Jon Boothroyd.]

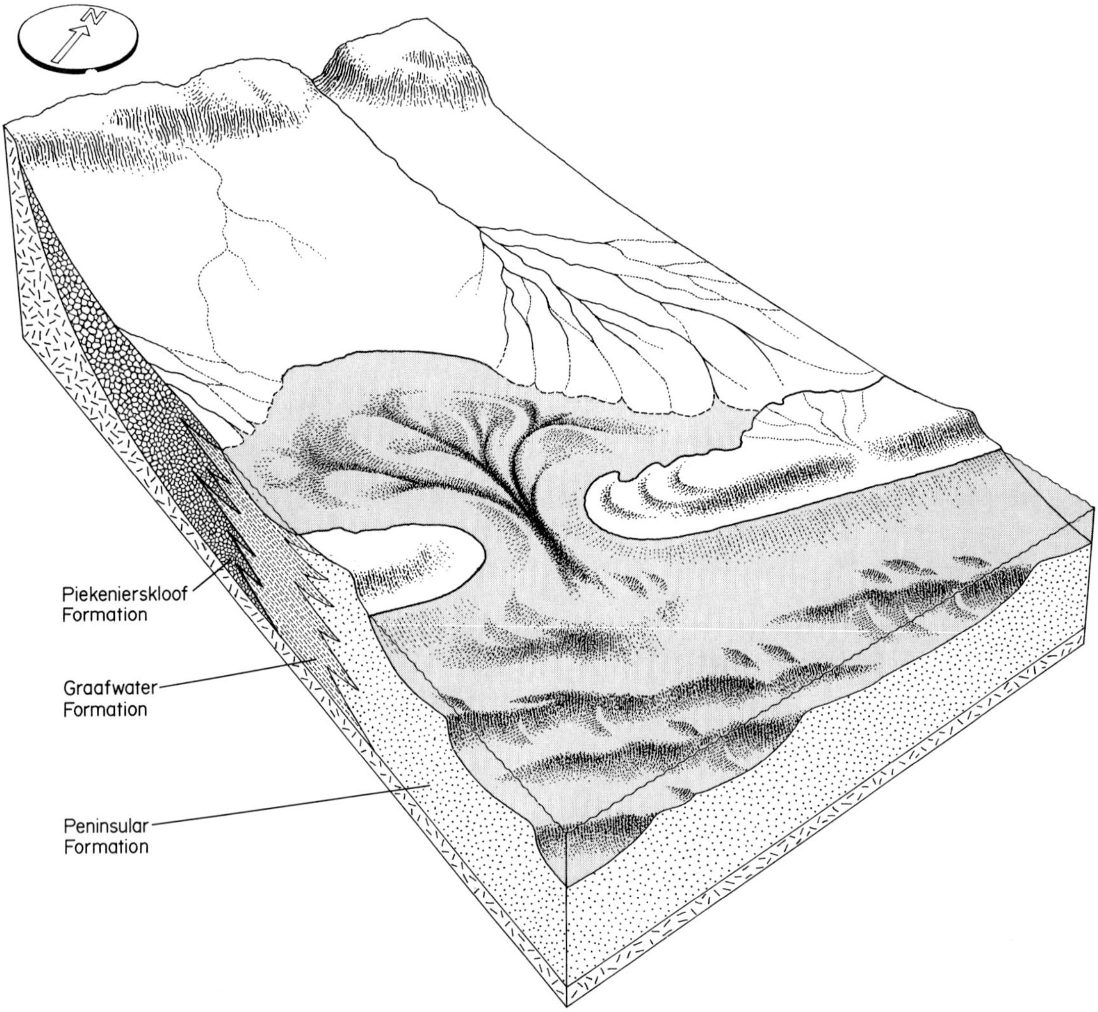

Figure 10-7. Schematic reconstruction of the depositional milieu of the lower Table Mountain Group including braided alluvial plain, fan delta, tidal flat and tidal estuary, barrier island, washover fan, tidal channel, inlet, and offshore tidal sand bar. [Drawn by Cedric Hunter.]

depths than fair-weather processes. Consequently ancient shallow-marine sediments commonly preserve a substantial storm imprint, as is characteristic of the Peninsula Formation. Washover features, channels incised into tidal sand bars, and seaward slumping of bedforms are possibly all products of storm surge and bottom return flow. Strongly ebbing storm currents are also believed to have generated units of coarse-grained winnowed lag with superimposed upper flow regime plane-beds.

The uppermost part of the Peninsula Formation is regressive, comprising medium- to coarse-grained sandstones and small-pebble conglomerates, which Rust (1967) has attributed to marginal reworking of glaciogenic deposits of the Pakhuis Formation. Deposition is believed to have occurred in tidal inlet, backbarrier sand shoal, and barrier washover environments (Hobday and Tankard, 1978). This Late Ashgillian regression is correlated with a worldwide glacio-eustatic fall in sea level (McKerrow, 1979).

Paleogeography of the Lower Table Mountain Group

Sedimentation within the basal Table Mountain Group occurred on a braided alluvial plain constructed by southeastward-flowing fluvial systems under moderately high runoff. Initial high relief was progressively reduced by erosion and aggradation. The low relief of the sub-Graafwater unconformity suggests planation by tidal and wave processes during the early transgression.

The principal components of the lower Table Mountain Group paleoenvironment were braided streams, tidal flats, barrier islands cut by well-developed tidal inlets, and tidal sand ridges on a shallow shelf (Fig. 10-7). The transition from coarse braided-stream to tidal-flat facies is abrupt, and the two show an interfingering relationship, except where the Piekenierskloof conglomerates underlie the Graafwater in the northern regions (Fig. 10-2). A fan-delta depositional setting is envisaged in the transitional zone. The development of coastal barriers was associated with reworking of sediment from braided alluvial deposits on a micro- to mesotidal coast influenced by strong tidal currents and moderate to high wave energy. Peninsula sand bodies were preserved as a result of substantial longshore sediment supply during landward migration of the facies belts. The maturity of the quartz arenites is believed to be the result of repeated reworking by marine processes on a broad, stable shelf.

Development of the Table Mountain succession can be interpreted in terms of worldwide eustatic sea level movements compounded by local tectonism. The earliest marine influence shown for the upper Klipheuwel-Piekenierskloof succession may be part of a major transgression resulting from Early Cambrian tectonism (Matthews and Cowie, 1979). Preservation of the Graafwater tidal deposits is attributed to a very rapid glacioeustatic transgression (1- to 2-Ma scale) in the Middle Ordovician (McKerrow, 1979). The thick Peninsula quartz arenites correspond to a prolonged period of basin and eustatic stability. Expansion of the Pakhuis ice sheets was associated with worldwide regression in Late Ashgillian times. Recent reconstructions by Smith *et al.* (1973) indicate an Early Ordovician paleolatitude for the Cape basin of about 30°. The Late Ashgillian *Hirnantia* faunas and associated glaciogenic deposits were close to 60° (Whittington and Hughes, 1970), implying a poleward drift of Gondwana averaging about 5 cm a year.

The Winterhoek Glaciation

During the Late Ordovician the Cape basin was on the periphery of an extensive Gondwana ice sheet centered in central Africa. Massive basal tillites with associated striated pavements and *roches moutonnées*, proglacially reworked tillites, and well-developed glaciolacustrine laminites characterize the Pakhuis Formation (Rust, 1967, 1969, 1973, in press). Final withdrawal of the ice sheet coincided with progressive development of thick transgressive mudstones that contain the "lagoonal" *Hirnantia* fauna of the Cedarberg Formation (Cocks *et al.*, 1970; Rust, 1973, in press). The composite Winterhoek succession thins from about 160 m in the west to about 50 m in the east (Table 10-1). In the western part of the basin Rust recognized two major expansions of ice in the glacial epoch, separated by erosion and reworking of tillites.

Glacial environments. Ordovician glaciation was not immediately accompanied by substantial lowering of sea level, and in the closing stages of Peninsula sedimentation glaciogenic material was reworked by coastal processes (Rust, 1967; Hobday and Tankard, 1978).

Regionally, tillite is closely associated with soft-sediment deformation of the underlying Peninsula Formation (Rust, 1967, 1973). Deformation structures include small-scale tight folds and larger scale folds, up to 100 m in amplitude, which are overfolded and refolded. The larger of these folds coincide with the depocenter of the Cape basin, and in the eastern Cape Province they are overfolded northward toward the depoaxis (Hälbich, 1979c). The tillite generally forms irregular erosional remnants occupying deformational troughs. It comprises nonstratified pebbly quartzose sandstone with exotic amethyst grains and finer grained sandstone with argillaceous matrix (Rust, 1967, 1973). Clasts are commonly polished, striated, and faceted and include several extrabasinal rock types, such as metamorphic rocks, chert, rhyolite, and granite (Rust, in press). Poorly bedded to cross-bedded, channeled tillite occurs locally. On top of Table Mountain rare wave ripples suggest reworking of glacial moraine in a low-energy, subaqueous environment (Hobday and Tankard, 1978).

Small-scale deformation structures resembling those of the basal Pakhuis, including tight and sheared folds, commonly result from subglacial stress (Edwards, 1978; Berthelsen, 1979). However, a similar interpretation for the large-scale folds (Haughton *et al.*, 1925; Blignault, 1970) is refuted by Rust (1973, in press), who attributes their geometries to glacial and tectonic distortion of load-cast structures. More recently, Hälbich (1979d) has ascribed similar structures in the eastern Cape to soft-sediment slumping toward the basin depoaxis.

Subglacial deposition is inferred for the glaciogenic material from its massive, unstratified texture, faceted and striated clasts, and wide distribution (cf. Edwards, 1978). Proglacial reworking, in part by marine processes, was important locally (Rust, 1973, in press).

Lenticular quartz arenites, rarely thicker than 3 m, in places overlie the tillites with erosive contact. A lag containing faceted pebbles is followed by cross-bedding including herringbone structures, indicative of deposition by traction currents. Asymmetrical current ripples commonly terminate the succession (Rust, 1967).

Daily and Cooper (1976) have cited evidence for permafrost conditions. On top of Table Mountain clastic dikes up to 10 cm wide and 1 m deep intrude the tillite from the overlying quartz arenite in a rectangular pattern resembling sand-wedge polygons. These polygons are believed to have formed by infilling of fissures created by contraction under extremely cold conditions. Daily and Cooper interpreted the intraformational folding at this locality as forming by loading after seasonal thawing of the underlying till.

Subglacial till was reworked by outwash stream and shore-zone processes in a cold, arid, periglacial climate. The tillite-shale association is attributed to clay sedimentation in ponds and lakes dammed by till.

Renewed expansion of the ice sheet resulted in widespread basal erosion and further deposition of thick glaciogenic sequences. These display marked north–south zonation of the terrestrial facies associations in the western Cape (Rust, 1967).

Along the northern margin of the basin the unconformity is extensively striated and retains other structures formed by subglacial stress, including small transverse step structures (Visser, 1962; Rust, 1967). Thin dropstone varvites occur locally above the unconformity. Maximum thickness of the tillite is about 120 m along the northern margin, but it thins rapidly toward the basin axis, averaging 20–30 m (Rust, in press). Local thickening of the deposits is related to depressions and *roches moutonnées* in the preglacial topography. The varicolored tillites contain matrix-supported pebble- to boulder-sized clasts. Erratics of chert and jasper are polished, faceted, and striated (Rust and Theron, 1964). Stratification is better developed upward, with increasing abundance of thin interbeds of dropstone laminites (Visser, 1962; Rust, 1967). These glaciogenic deposits grade into shales of the overlying Cedarberg Formation.

The southern facies equivalent is thinner, irregular in thickness, and generally more mature lithologically. Poorly stratified pebbly sandstones contain faceted and striated exotic clasts. The tillite grades northward (sourceward) into matrix-supported conglomeratic tillite and dropstone laminites, which were deposited on an uneven substrate. Younger tillite blankets the shales (Rust, 1967). Upward transition into the Cedarberg Formation is gradational or abrupt.

These rocks, excluding the laminites, strongly resemble modern subglacially deposited tills in terms of their poorly stratified diamictite texture, abundant striated and faceted clasts, and erosional, striated lower contacts. The thickness and sheetlike geometry implies "cold" glaciation (Boulton, 1970), and a wet-based regime is suggested by the abundance of polished, striated, and faceted clasts (Edwards, 1978). A general southward paleoslope (Rust, 1973, in press) is supported by the facies fabric and distribution, striations, and southward-increasing thickness. Glacial pavements and massive tillites are best developed in the north. Toward the south the tillites are progressively more quartzose and contain fewer and smaller clasts. Extrabasinal pre-Cape material predominates in the northern facies tract, but the ice sheet is believed to have eroded and redeposited large volumes of intrabasinal quartzose sands as it prograded. The laminites and shales are attributed to glaciolacustrine deposition in irregular proglacial environments marginal to waxing and waning ice sheets. Thick varved shale sequences (Rust, 1967) are rare but nevertheless indicate deposition in an extensive glacial lake (Ashley, 1975).

Postglacial Cedarberg transgression. Climatic amelioration and withdrawal of the Late Ashgillian ice sheets was accompanied by deposition of a persistent suite of argillaceous rocks along the east–west depoaxis. The lower shales are gradational with the interbedded tillite–laminite facies of the Pakhuis Formation. Rarely thicker than 40 m, the Cedarberg Formation consists mainly of thinly laminated shale with sporadic dropstone layers, some prominent varvites, and rare lenses of quartz arenite up to 10 m thick (Rust, 1967). In the western Cape mudstones coarsen upward into siltstones and fine-grained sandstones.

Up to 100 m of carbonaceous mudstones and fine-grained sandstones dominate the upper Cedarberg Formation. Thin-bedded, flasered siltstones and fine-grained sandstones grade upward into a 10-m-thick interval of bioturbated and fossiliferous mudstones and siltstones. The latter contain vertical burrows; crawling traces; fecal pellets; a low-diversity assemblage of invertebrate fossils, consisting principally of inarticulate and articulate brachiopods, fragmentary trilobites, and crinoids (Rust, 1967, 1973; Cocks *et al.*, 1970); and fossil algae (Plumstead, 1967). These mudstones grade vertically into a sequence of interbedded sandstones and mudstones that lithologically resemble the Graafwater Formation. Upward coarsening is reflected in the increasing thickness (15–30 cm) of the sandstone units and transition into the coarser grained Nardouw Formation (Rust, 1967).

Dropstone-laminites and fossiliferous mudstones in the Cedarberg Formation were deposited in proglacial and shallow-marine environments (Rust, 1967, 1973; Cocks *et al.*, 1970; Hobday and Tankard, 1978). Widespread distribution of these argillaceous sediments is attributed to isostatic depression in a tectonically active basin (Rust, 1973). The gradational association with the tillites and development of thick varvites suggest that deposition occurred initially in freshwater proglacial lakes similar to those in modern proglacial landscapes. Dropstones provide convincing evidence of ice rafting. These varvites are succeeded by nonvarved, laminated, fossiliferous mudstones that show a marked reduction in numbers of dropstones. The absence of varves is attributed to sediment mixing and flocculation in brackish or saline water. Fossil algae from these rocks imply shallow-water conditions (Plumstead, 1967). The brachiopod fauna (Cocks *et al.*, 1970) has several elements in common with the Ashgillian fauna of the northern hemisphere and with Early Silurian *Lingula–Eocoelia* communities (Ziegler *et al.*, 1968; Cocks, 1972). Cocks attributes the low diversity of the *Eostropheodonta–Orbiculoidea*–homalonotid community to a cold, shallow-water, benthic ecosystem.

The coarsening-upward interval of interbedded sandstones and mudstones has been compared with the tidally deposited Graafwater Formation (Rust, 1967). Rise of sea level is believed to have increased tidal circulation in the Cedarberg basin, resulting in traction sedimentation of coarser material.

Discussion. Ordovician ice sheets, centered in central Africa, spread out to South Africa, South America, West Africa, and North Africa (Edwards, 1978), covering an area of 200,000 km^2 (Rust, in press). Interpretation of the extensive Saharan tillites (Beuf *et al.*, 1971) is similar to that for the Pakhuis sequence, except that at least four major advances and retreats of the ice sheet are indicated by major unconformities, whereas in the Cape basin Rust (1973) recognized only two major expansions. The Saharan paleogeography was also characterized by striated pavements, *roches moutonnées*, and sourceward thinning of the massive tillites. Predominance of massive basal tillite is not unusual since this constitutes the most widespread on-land facies of the Pleistocene glaciation (Edwards, 1978).

Deglaciation was accompanied by glacioeustatic rise of sea level in the earliest Silurian (McKerrow, 1979) leading to marine flooding of an isostatically depressed basin. The brachiopods *Eostropheodonta* and *Orbiculoidea* and the homalonotid trilobite are common to the *Clarkeia* faunas of South America and South Africa, but the Cedarberg assemblage also has strong affinities with the *Hirnantia* faunas of the northern hemisphere (Wright, 1968; Cocks, 1972). These shallow-water, semirestricted *Hirnantia* (including *Clarkeia*) faunas and associated glaciogenic deposits of Gondwana (Fig. 10-8) were close to paleolatitude 60° (Whittington and Hughes, 1972).

Postglacial Regression

Sedimentary facies patterns established in the lower Table Mountain Group are largely repeated in the upper Cedarberg–Nardouw landscape. The Nardouw Formation comprises a 30- to 1100-m-thick westward-thickening wedge of quartz arenites that transgressively overlie Cedarberg deposits of inferred tidal origin. The Nardouw Formation is generally more fossiliferous than its Peninsula counterpart (Rust, 1967, 1973).

The lower 30–75 m of the Nardouw Formation consist of fine-grained sandstones, from 30 cm to 180 cm thick, separated by purple siltstone interbeds. Sedimentary structures include cross-

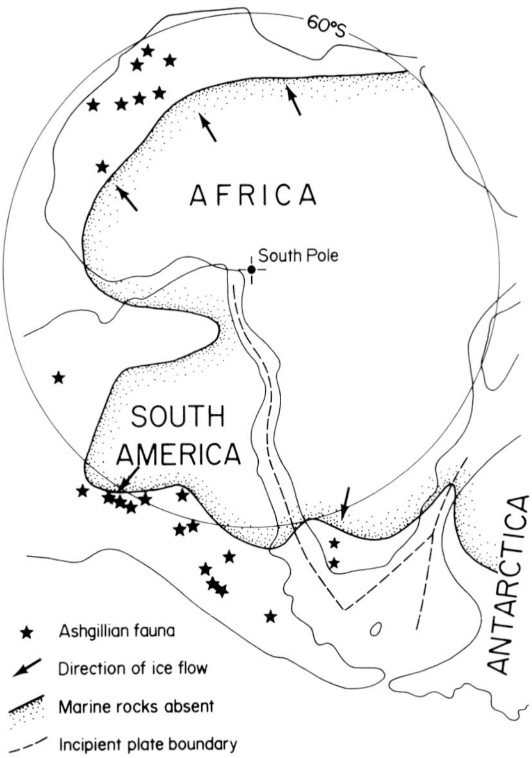

Figure 10-8. Ordovician reconstruction of Gondwana showing distribution of Ashgillian *Hirnantia* fauna, ice flow directions, and configurations of land area. [Adapted from Cocks (1972), Whittington and Hughes (1972).]

bedding, plane-bedding, and ripple cross-lamination. Crawling and feeding traces are common (Rust, 1967). Above this the Nardouw succession is dominated by thicker bedded (60–300 cm), medium- to coarse-grained quartz arenites similar to the Peninsula Formation. The lenticular sandstones are separated by erosive contacts that are commonly accentuated by pebble lag. Medium- to large-scale cross-bedding is ubiquitous, and upper flow regime plane-bedding is subordinate. Although foresets are mainly directed to the south, polymodal paleocurrent patterns and herringbone planar cross-bed sets show periodic current reversals. Horizontal and vertical trace fossils of marine affinity are exposed on bedding planes (Rust, 1967; Marchant, 1974).

Regional mapping demonstrates an overall northward increase in grain size, so that along the margin of the basin, units of very coarse-grained sandstone and conglomerate are up to 30 m thick with clasts of up to 25 cm in diameter. Quartzite and shale clasts are probably derived from pre-Cape strata (Rust, 1967).

The uppermost 90–200 m of the Nardouw Formation are less homogeneous, consisting of thinner bedded sandstones, and interbedded quartz arenites and carbonaceous mudstones (Rust, 1967; Theron, 1972); in the eastern Cape three separate formations constitute this stratigraphic interval (Table 10-1). Internally the sandstones are trough cross-bedded, with foresets generally overturned to the south. The carbonaceous mudstones are thoroughly bioturbated and contain a Siegenian *Pleurothyrella*-rich brachiopod fauna (Theron, 1972; A.J. Boucot, personal communication, 1981).

The lithologic characteristics and their vertical and regional zonation resemble the Piekenierskloof–Graafwater–Peninsula facies associations of the lower Table Mountain Group (Rust, 1967). Upward gradation from the Cedarberg to Nardouw Formation occurs through a succession of interbedded sandstones and mudstones with increasing thickness of the sandstone beds. This vertical pattern, together with the small-scale physical and biogenic structures, suggests increasing effectiveness of tidally dominated processes. Rust (1967, 1973) has compared the more massively bedded quartz arenites with the Peninsula Formation, suggesting that deposition occurred in shallow-shelf and shore-zone environments. Along the landward edge of the basin, coarse-grained sediments of alluvial plain origin were reworked by marine processes. The Table Mountain Group terminated with regression of the Nardouw coastline and deposition of fossiliferous sands and muds. Concomitant progradation of the Bokkeveld deltas initiated a new style of deposition (Theron, 1972; Tankard and Barwis, 1982).

10.2. The Natal Embayment

Of the three major subdivisions of the Cape Supergroup only the Table Mountain Group is believed to be represented in Natal and adjoining areas (Fig. 10-1), where it has been termed the Natal Group (South African Committee for Stratigraphy, 1980). The Natal Group attains a maximum thickness of approximately 1000 m. These sediments accumulated in an elongate downwarp, the Natal embayment (Visser, 1974), trending subparallel to the present coastline. This linear downwarp has the same general alignment as the Precambrian Mozambique Province to the north (Cox, 1970). The Natal embayment may therefore correspond to a zone of structural weakness that underwent Early Paleozoic crustal attenuation and rift subsidence but that aborted before reaching the spreading stage (Hobday and Von Brunn, 1979). This rift was twice reactivated during Dwyka and Ecca deposition (Late Paleozoic) and eventually was subject to Mesozoic tensional rifting north of Durban and strike-slip disruption in the south. Breakup probably occurred approximately along the axis of the Natal embayment (Matthews, 1970) so that only the western half of the Natal Group depository is now preserved along the eastern seaboard. Even there the record is fragmentary because of subsequent block faulting, seaward tilting, and erosion. The Natal Group of Natal and Transkei is separated from the Table Mountain Group of the southern Cape by a Karoo cover. Deep drilling has suggested the continuity of these strata in the subsurface (Visser, 1974).

The Natal embayment was characterized by a steep, rugged northern terminus in Zululand, with the axial gradient decreasing toward the southwest, where it opened into a major marine embayment (Fig. 10-9). The Natal Group lies uncon-

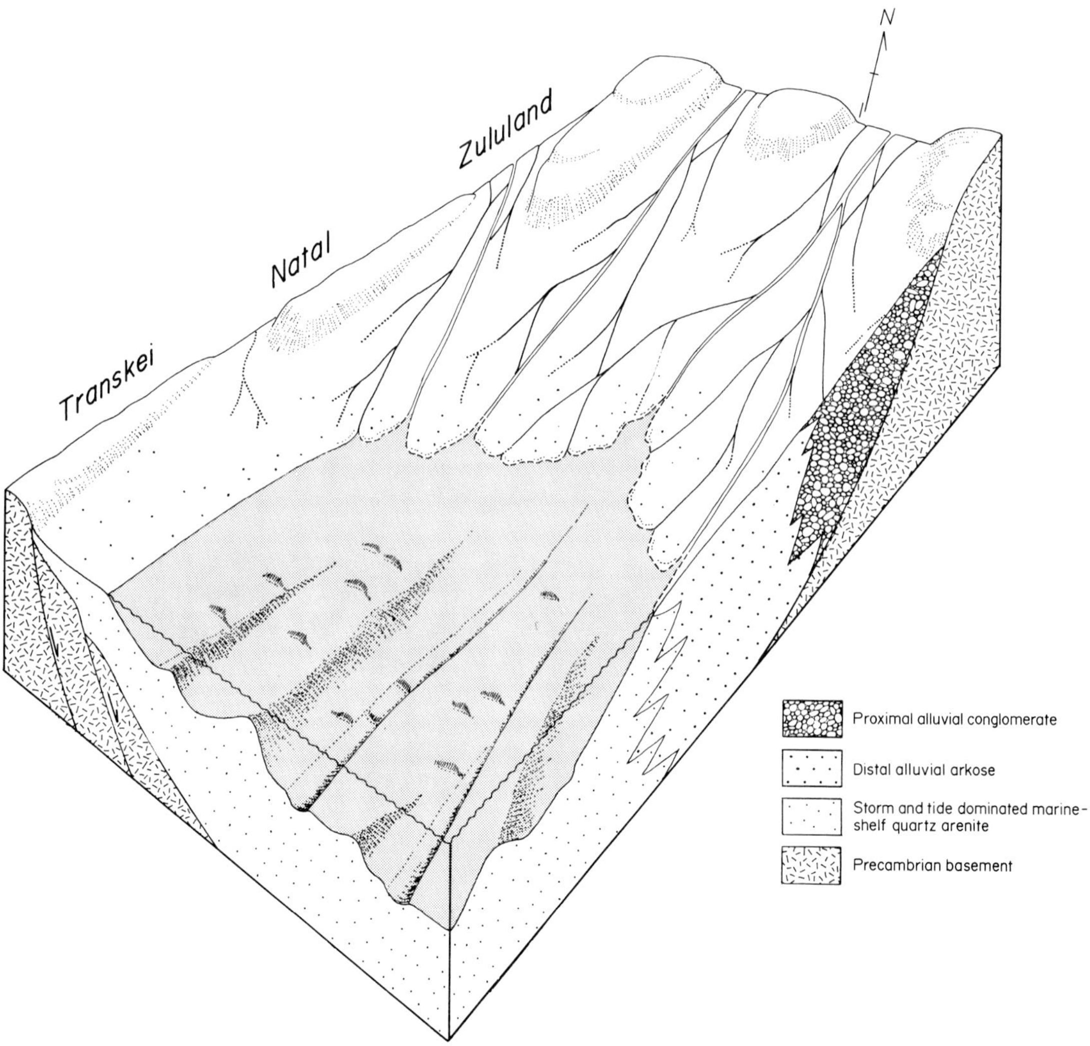

Figure 10-9. Schematic paleogeographic reconstruction of the Natal Group in the Natal embayment showing the relationship of alluvial arkoses to storm- and tide-dominated shelf quartz arenites. [Adapted from Hobday and Von Brunn (1979).]

formably on gneisses and granitoids of the Proterozoic Natal Province (Chapter 8) and Archean rocks of the Kaapvaal Province in the north (Chapter 2).

The deposits of the Natal embayment display well-defined vertical and longitudinal facies zonation (Kingsley, 1975; Hobday and Von Brunn, 1979). Coarse conglomerates occupy paleovalleys in the dissected apical regions, merging southward in the vicinity of the Tugela fault (Rhodes and Leith, 1967) into arkosic sandstones which overlie a gently undulating erosion surface. These arkosic sandstones extend northward over the coarse proximal conglomerates and grade upward into a thick but poorly preserved succession of fine-grained micaceous sandstones and siltstones. Intense penecontemporaneous deformation structures involving rock packages up to 15 m thick (Matthews, 1961) are characteristic of this fine-grained interval. In southern Natal the dip on the Precambrian basement steepens markedly, and uniform quartz arenites make up the entire expanded succession.

The proximal conglomerates of Zululand are up to 60 m thick and include boulders as large as 2 m in diameter near the base, decreasing upward to 25 cm or less near the top. Local paleorelief of 300 m is developed on a heterogeneous basement. Despite the variety of source rocks, only durable quartzites are preserved in abundance, attesting to vigorous weathering and considerable transport. Most boulders are well rounded and have abundant percussion marks. *In situ* abrasion was probably an important process. The conglomerate is predominantly clast supported and well imbricated. Coarse, pebbly sand settled into the gravel interstices during reduction in flow. Rare lenticular units up to 3 m thick show poor sorting and contain angular blocks; these may have been rapidly shed talus deposits that, for the most part, have been reworked by stream action.

The conglomerate facies thus accumulated in a humid intermontane environment without an effective vegetation cover. Intense physical and chemical weathering led to preferential preservation of only the most durable clasts. Initial aggradation within the lowest parts of the incised landscape was followed by the development of gravel bars as flow was dispersed over a wider area. These gravel accumulations may have resembled the sheetlike longitudinal bars of proximal outwash fans of northeast Gulf of Alaska (Boothroyd and Ashley, 1975). As gradients upstream continued to be reduced by aggradation, smaller particle size and increased bed relief were more characteristic of coarse midfans or gravelly braided streams with bars terminating in well-defined slipfaces. Bar migration gave rise to rare low-angle foresets interspersed with the more widespread plane-bedding resulting from aggradation. Conglomerate- or sand-filled scour channels and cross-bedded sand wedges are especially common toward the top of the conglomerate succession directly below the gradation into arkosic sandstone.

The arkoses attain a maximum thickness of 300–400 m and display the classic features of sandy braided-stream deposits. Large, solitary planar cross-bed sets record accretion on the flanks and downstream slipfaces of linguoid or transverse bars, giving a wide dispersion of current azimuths. Individual sets show pronounced changes in dip angle resulting from discharge fluctuations; the lower angle, tangentially based foresets in places contain smaller regressive ripples or reverse cross-lamination produced by strong flow separation (Hobday and Von Brunn, 1979). Exhumed sets show a typical bar morphology, with tops that are horizontal or slope gently upstream. The bars show varying degrees of dissection by small scour channels filled with cosets of trough cross-beds and, less commonly, by red siltstone.

Laterally persistent erosion surfaces accentuated by a veneer of argillaceous intraclasts in places define genetic packages up to 2 m thick. Each sequence contains one or more planar cross-bed sets, commonly truncated by cosets of smaller scale trough cross-bedding with siltstone at the top. The capping siltstone unit is more commonly eroded, however, and the clasts at the base of overlying sequences are generally the only record of the argillaceous top stratum.

The fine-grained, micaceous sediments at the top of the succession are 350–400 m thick. Channel patterns indicate deposition by narrower, steeper sided streams. This is typical of mixed-load streams, with a higher proportion of silt and clay providing more cohesive bank material (Schumm, 1960). Lateral accretion structures are probably of point-bar origin, but there is also evidence of large bars or islands of braided streams. Like many modern rivers at bankful discharge, a single, sinuous channel may have been present, whereas multiple channel courses divided around emergent bars during the dry season.

The widespread development of penecontemporaneous deformation structures may have been a consequence of abrupt changes in river stage, but the large scale and lateral persistence of deformed units suggest that earthquake tremors were the triggering mechanism. Matthews (1961) documented an eastward increase in the degree of deformation and showed that displacement was directed mainly northeastward toward the trough axis. These patterns suggest that contemporaneous downwarping of the Natal embayment was accompanied by seismic tremors, which may have induced instantaneous liquefaction. Abundant vertical dewatering structures are consistent with an earthquake origin.

The quartz arenites of southern Natal and Transkei were deposited in a shallow-shelf environment scoured by powerful tidal currents directed primarily perpendicular to the shore. Lenticular sandstones interpreted as tidal sandbars are preserved as channeled, truncated, stacked, and *en echelon* units with moderately to intensely burrowed upper surfaces and complex internal structures (Hobday and Mathew, 1974). Trace fossils include large *Rhizocorallium*, *Planolites*, *Skolithos*, and a variety of tracks and trails. Linear tidal channels of southwest trend show complicated patterns of scour and fill, with rare silty lenses and burrowed horizons. A macrotidal regime was probably favored by a funnel-shaped coastal configuration (Fig. 10-9). Oceanic storm swell and bottom return flow were possibly also important in remodeling the shoreline and inner shelf.

The remarkable degree of vertical accumulation of shallow-marine sands was presumably accomplished by slow but constant tectonic downwarping accompanied by a high rate of clastic sediment supply. Brief episodes of imbalance produced by accelerated subsidence caused two or three episodes of widespread marine transgression across the sandy alluvial plain. Intense reworking by marine processes left laterally persistent quartz arenite units up to 50 m thick with diverse internal structures of wave and current origin. Biological traces are rare, possibly because of constant physical reworking. These quartz arenite units, referred to as "orthoquartzite markers" by Rhodes and Leith (1967), are responsible for much of the plateau and scarp topography between Durban and Pietermaritzburg.

10.3. Paleogeographic Synthesis of the Table Mountain and Natal Groups

Regional synthesis of the tectonic–sedimentary framework needs to address the differing sedimentary styles of the Cape and Natal basins, the phenomenal thicknesses of quartzose sandstone, and the significance of these deposits in the Gondwana context.

The greatest volumes of sediment accumulated in the near-shore and shallow-shelf environments, where the character of these deposits was largely determined by the relative intensities of waves and tidal currents. Herein lies a fundamental difference between the embayed Natal depository and the relatively unconfined Cape basin. Amplification of the tidal wave in the funnel-shaped Natal embayment resulted in an alignment of bars and channels approximately parallel to the embayment axis (Figs. 10-9 and 10-10). Hayes (1975) discusses a similar phenomenon in the German North Sea Bight. Available evidence favors a low mesotidal regime in the southern Cape, where stunted barrier islands were separated by deep tidal inlets (Figs. 10-7 and 10-10). These differences in basin morphology, coupled with differences in tectonic behavior, may account for the broad lack of stratigraphic correspondence—specifically, the absence in the east of the threefold division in the Cape.

These sedimentological considerations show a logical relationship to inferred early rift patterns preceding Gondwana fragmentation. Reconstructions of the supercontinent show southern Africa and Antarctica juxtaposed, with the wedge-shaped Falkland Plateau intervening in the southwest (Francheteau and Le Pichon, 1972; De Wit, 1977). The aborted rifts along plate margins controlled the outline of the sedimentary basin. Pre-Mesozoic incipient plate boundaries between southern Africa, Antarctica, and the Falkland Plateau may have formed a triple junction (Fig. 10-10). This Early Paleozoic extensional tectonism was the first sign of impending breakup, but

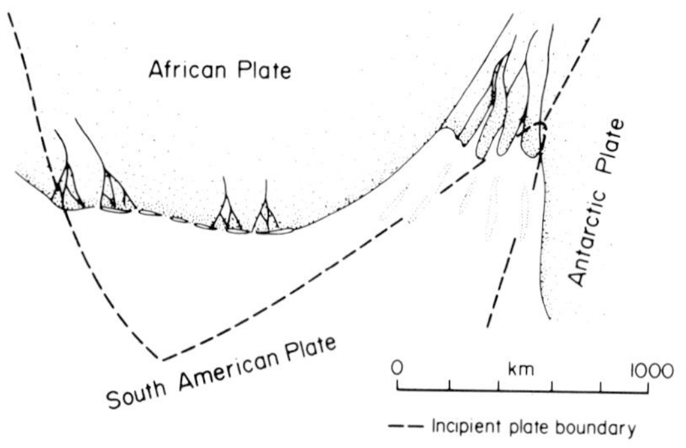

Figure 10-10. Predrift Gondwana with the Falkland Plateau intervening between southern Africa and Antarctica, defining a triple junction. Incipient plate boundaries shown as dashed lines; land area stippled. Shoreline geometry controlled the distribution of tidal facies, with micro- to mesotidal barrier systems dominating the western part of the region and broad open-mouthed estuaries in the Natal embayment.

rifting terminated before the spreading stage; it was reactivated in the Mesozoic.

The accumulation of 2000 m of shallow-marine, shore-zone, and alluvial sediments in relatively narrow depositories suggests that basin-margin fault displacement may have been accentuated by sediment loading. Thick quartz arenites such as those in the Peninsula and Nardouw Formations are thought to reflect long histories of reworking on a relatively stable but slowly subsiding shelf (Rust, 1973; Visser, 1974). Nio (1976) emphasizes the role of tidal currents in developing thick shelf sand accumulations such as those of the Table Mountain Group. Field (1980) has documented the accumulation of shelf sand on a transgressive, wave-dominated shelf. The average rate of subsidence and accumulation of the Cape basin quartz arenites is estimated as 3 cm per 1000 years (Rust, 1967); this is slow compared with sedimentation rates on modern barrier coasts, which are between 120 and 180 cm per 1000 years (Weimer, 1970). Provenance studies by Visser (1974) show that sands of the Cape basin were derived from a source of basement granitoids and gneisses.

In Paleozoic times the Falkland Plateau fitted against the southern and southeastern margin of South Africa (Fig. 10-10), juxtaposing metasedimentary and granitic rocks of the South American plate (Barker *et al.*, 1974) and continental crust of the Agulhas Bank (Simpson and Dingle, 1973), forming the broad and relatively shallow Malvinas shelf (Johnson and Boucot, 1973). There is no evidence for the existence of oceanic crust close to the Cape basin. The sedimentary rocks of the Cape basin are believed to have stratigraphic equivalents in the Andean basins and on the Falkland Islands (Du Toit, 1927, 1937, 1954, p. 262). Cocks (1972) has documented the widespread occurrence of the Ordovician–Silurian *Clarkeia* faunas, but their precise age relationships are still unresolved. In Argentina the Ordovician strata include 400-m-thick fossiliferous quartz arenites which, like the Peninsula Formation, terminate in glaciogenic deposits (Harrington, 1956). Deeper water argillaceous sediments and slump blocks up to 100 m thick and 3 km long occur along the western fringes of Bolivia and Argentina (Ahfeld, 1956; Harrington, 1956). Paleogeographic reconstructions (Fig. 10-10) show the Falkland (Malvinas) Islands situated in the outer parts of the Cape–Natal embayment. The 3000-m-thick Falkland succession has close affinities with the Cape Supergroup, including conglomerates, interbedded sandstones and mudstones, and prominent quartz arenites (Du Toit, 1927, 1937, 1954, p. 262; Greenway, 1972). Correlations with Antarctica are tenuous, although Visser (1974) suggests that metamorphosed geosynclinal rocks may be time equivalents.

10.4. Bokkeveld Group: Allocyclic Control Over Delta Progradation and Reworking

The Bokkeveld succession represents the most dynamic phase of Cape basin development. Large volumes of sediment were delivered by southward-flowing rivers. The eastern Cape area was subject to maximum subsidence (Table 10-2), accumulating more than 3200 m of poorly sorted sediments. Subsidence was slower in the western Cape, where at least 1500 m of sediment accumulated (Theron, 1972). The sediment packages are characteristically arranged in five or six superimposed upward-coarsening sequences (Figs. 10-11 and 10-12) which imply a history of tectonically controlled regressions and transgressions (Theron, 1972; Rust, 1973).

The Bokkeveld Group is best known for its diverse suite of invertebrates encompassed within

Table 10–2. Stratigraphy of the Bokkeveld Group[a]

Western Cape			20° E	Eastern Cape	
Formation	Thickness (m)	Lithology	Formation	Thickness (m)	Lithology
Bidouw Subgroup			Traka Subgroup		
Karoopoort	50	Siltstone and quartz arenite with shale interbeds	Sandpoort	400	Red shale, siltstone quartz arenite
Osberg	55	Feldspathic sandstone and quartz arenite			
Klipbokkop	170	Mudstone, graywacke, and lithic arenite	Adolphspoort		Siltstone, quartz arenite
Wuppertal	65	Quartz arenite, lithic arenite, siltstone			
Waboomberg	200	Siltstone, quartz arenite, shale, black shale near top	Karies	1200	Shale, siltstone, quartz arenite
Ceres Subgroup			Ceres Subgroup		
Boplaas	30	Quartz arenite, and lithic arenite	Boplaas	100	Feldspathic sandstone, quartz arenite, and mudstone
Tra-Tra	85	Mudstone, siltstone, subordinate sandstone	Tra-Tra	350	Mudstone, siltstone, sandstone
Hex River	100	Arkose, lithic arenite, and quartz arenite	Hex River	70	Lithic arenite, quartz arenite, siltstone, and mudstone
Voorstehoek	115	Siltstone, shale, fine-grained sandstone	Voorstehoek	300	Siltstone, shale, fine-grained sandstone
Gamka	135	Feldspathic sandstone, quartz arenite, mudstone	Gamka	200	Feldspathic sandstone, quartz arenite, mudstone
Gydo	160	Black shale and siltstone	Gydo	600	Shale, siltstone, fine-grained sandstone

[a]Marine invertebrates of the Malvinokaffric province and fish fossils are most common up to the Klipbokkop and Karies Formations; trace fossils and plant fossils are locally present throughout the stratigraphic column (J.N. Theron, personal communication, 1980). Arrows indicate direction of decreasing grain size. Sources: Theron (1972), South African Committee for Stratigraphy (1980).

the Malvinokaffric faunal province of western Gondwana (Clarke, 1913; Caster, 1952; Boucot *et al.*, 1969; Theron, 1972). An Early Emsian (late Early Devonian) age is suggested by the genera of Malvinokaffric brachiopods, particularly *Australospirifer*, *Scaphiocoelia*, *Pleurothyrella*, *Notiochonetes*, and *Australocoelia* (Boucot *et al.*, 1969). Also of Emsian age is the blastoid *Pachyblastus dicki* (Macurda, 1979). The trilobites are consistent with an Early Devonian age (Theron, 1972), but other taxonomic groups, including plant fossils (Plumstead, 1967, 1969; Chaloner *et al.*, 1980) and fish fossils (Chaloner *et al.*, 1980), are less diagnostic.

The stratigraphic framework (Table 10-2) established by Theron (1970, 1972) emphasizes the lateral variability of the Bokkeveld succession. Thick upward-coarsening sequences record progradation of lobate deltas along a coastline of moderately high marine energy (Theron, 1972;

Figure 10-11. Bokkeveld upward-coarsening sequences expressed in typical hogback topography of the Wuppertal area, western Cape.

Tankard and Barwis, 1982). These near-shore deposits grade southward into shelf mudstones. Major expansion of these fine-grained prodelta–shelf successions is characteristic of the eastern Cape area (Table 10-2), where maximum down-warping occurred, and of the extreme southern area (Theron, 1970, 1972).

Constructional Deltaic Facies

Centimeter-scale graded beds are present in the finer basal components of each facies sequence, with horizontal lamination and ripple cross-lamination in the upper siltstones (Fig. 10-12). The shelf deposits are extensively bioturbated and contain scattered casts and molds of trilobites, brachiopods, bivalves, and crinoid columnals. Bioclastic wackestones occur in lenses 10–15 cm thick. Large flow rolls and slumps reflect gravity mass movement on the oversteepened prodelta slope.

The marine faunas of the Bokkeveld and other Malvinokaffric basins are closely comparable. They appear to have inhabited shallow environments, *Australocoelia* possibly indicating depths shallower than 50 m. There is also evidence that the Parana (Brazil) counterparts also populated offshore areas in the vicinity of deltas (Copper, 1977). The thicknesses of the fossiliferous Bokkeveld strata are greater as a result of more active subsidence (Theron, 1972).

Bokkeveld distributary mouth bar deposits (Fig. 10-12) up to 100 m thick gradationally overlie the shelf and prodelta prism (Theron, 1972; Tankard and Barwis, 1982). Typically upward-coarsening, the silty lower 20–40 m contains small-scale, low-energy structures, such as ripple cross-lamination and thin cross-bed sets. Soft-sediment deformation structures are attributed to sediment loading and downslope gravity movement. Upward gradation to medium-grained lithic arenites is accomplished by thickening of cross-bed sets and decreasing abundance of argillaceous partings. Ripple bedforms show an upward change from symmetric to asymmetric forms. The upper parts of sandstone beds are characterized by straight, branching, and U-shaped burrows. Molds and casts of brachiopods and mollusks are less prolific than in the shelf and prodelta sediments, but transported plant debris litters bedding surfaces.

The overall textural distribution of these rocks and their vertical position is typical of seaward progradation of distributary mouth bar and distal bar complexes (Coleman and Gagliano, 1965; Wright, 1978). Suspension sedimentation and weak wave reworking characterize the distal parts. Progressive shoaling is suggested by current-produced structures higher in the sequence

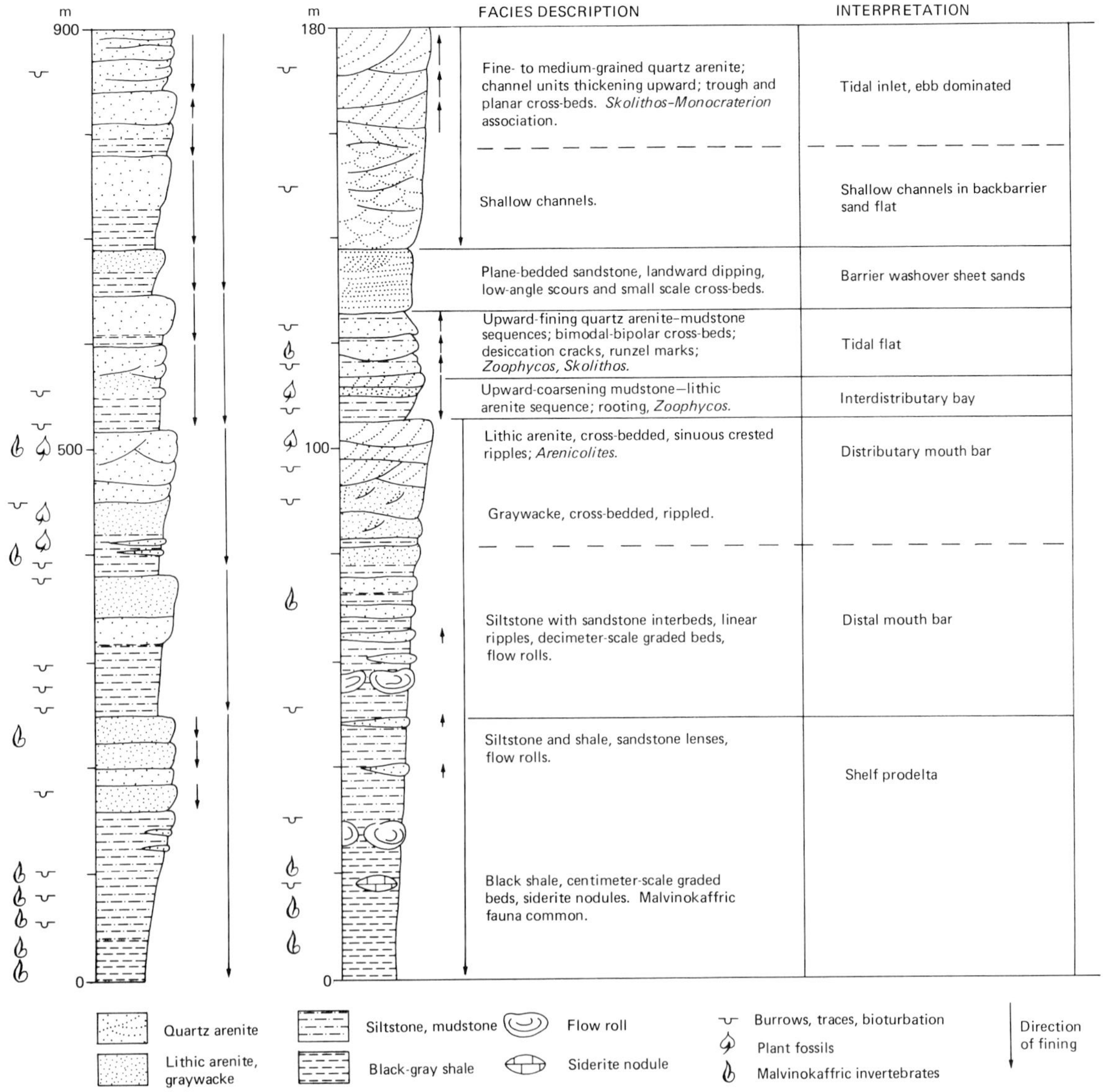

Figure 10-12. Bokkeveld stratigraphic section and detailed characteristics of an ideal genetic sequence. [After Tankard and Barwis (1982).]

and a tendency toward greater maturity and better sorting. The trace fossil assemblage indicates a dominance of suspension-feeding organisms.

Subordinate progradational delta sequences thinner than 10 m are locally superimposed on the thick mouth-bar deposits (Tankard and Barwis, 1982). These consist of dark mudstones that coarsen gradationally upward into lithic graywackes and lithic arenites. The argillaceous intervals contain isolated sandstone lenses and diagenetic siderite. The upper sandstones show graded beds, ripple-drift cross-lamination, and both interference and ladderback ripples. Highly carbonaceous mudstones and rooted zones occur rarely, with associated plant fossils, *Zoophycos*, and vertical burrows.

Studies of modern deltas have shown that the classic upward-coarsening bay-fill motif is the result of overbank flooding of fine sediment, crevassing, and encroachment of bay margins (Coleman and Gagliano, 1965). The gradational contact beneath the upper sandstones resulted because bedload material gradually prograded basinward over mixed bedload-suspended sediments of the shallow bay. Brackish circulation is suggested by a generally impoverished fauna and an abundance of *Zoophycos* spreiten burrows. In the Carboniferous coal measures of northern England, *Zoophycos* is also closely related to zones of rooting (Kennedy, 1975, p. 394). There is evidence that margins of the Bokkeveld bays were colonized by semiaquatic vegetation (cf. Plumstead, 1967).

An alluvial plain bounded the Bokkeveld basin in the north. Thick sequences of very coarse- to medium-grained feldspathic sandstone and conglomerate (Theron, 1972) are arranged in upward-fining units, with the coarser components overlying channel scours. Side-fill structures in the lenticular sandstones are further evidence of channelized flow. Periodic overbank sedimentation is inferred from argillaceous partings that contain plant fossils.

Marine-Reworked Deltaic Sands

Deltas commonly have a twofold history involving constructional and destructional episodes. During destructional episodes delta-plain and shallow delta-front sediments are redistributed by wave and tidal activity into mature barrier sand bodies (Fisher *et al.*, 1969; Saxena, 1976). Bokkeveld quartz arenites occupy a large outcrop area and terminate several of the upward-coarsening genetic sequences (Fig. 10-12). Rarely exceeding 70 m in thickness, these quartz arenites comprise two sedimentary facies: barrier washover sheets and active fill of tidal inlets and tidal channels (Tankard and Barwis, 1982).

Barrier washover sandstones in places overlap the bayfill sequences landward (Tankard and Barwis, 1982). Consisting mainly of fine-grained, plane-bedded sandstones with millimeter-scale laminae, sporadic low-angle scour surfaces, and washed-out ripples, they resemble modern storm-generated washover sheet sands that overlie lagoonal or bayfill muds (Andrews, 1970; Schwartz, 1975). Subordinate plane-bedded foreshore deposits contain numerous bedding-plane discontinuities and dip gently seaward.

Tidal inlet and tidal channel processes appear to have dominated deposition of the fine- to medium-grained quartz arenites in the Bokkeveld basin (Theron, 1972; Tankard and Barwis, 1982). The thick succession of stacked, lenticular sedimentation units displays an overall tendency toward increasing grain size and bed thickness, although the channel sandstones are individually upward fining. Cosets of trough and planar cross-bedding are common and show polymodal distribution of slipfaces with a southward, ebb-dominated mode. Other sedimentary structures include herringbone cross-bedding, reactivation surfaces, plane-bedding, and ripple cross-lamination. Deep side-filled channels are restricted to the uppermost parts of the quartz arenite facies. *Skolithos* burrows extend down from bedding surfaces; locally, *Monocraterion* overlies the *Skolithos* tubes where they are associated with small-scale cross-bedding.

Some modern tidal channel point bars have a *Skolithos–Monocraterion* type of association (Barwis and Goodwin, 1979). The thicker, coarser channel sandstones are believed to have originated as tidal-inlet fill (Tankard and Barwis, 1982). Most ancient tidal deposits, like those of the Bokkeveld, contain a high proportion of ebb-oriented bedforms. This is because of the greater preservation potential of the deeper channel areas where ebb currents predominate (Kumar and Sanders, 1974). In the Bokkeveld basin tidal channel and inlet deposits abruptly overlie delta-front lithologies, suggesting an origin by reworking and downdrift migration of inlets. Barrier transgression and landward overlap of channeled estuarine sand flats by deposits of the deeper and more energetic tidal inlets is inferred from the overall upward-coarsening pattern. Preservation of landward tidal-channel deposits beneath the eroding inlets implies extraordinarily rapid subsidence balanced by alongshore sediment supply. The reworked sandstones are generally succeeded by tidal facies.

Prograding tidal-flat environments have been invoked to explain some upward-fining sequences consisting of quartz arenite and mudstone (Theron, 1972; Tankard and Barwis, 1982). The lenticular quartz arenites are structured by cosets of trough and planar cross-bedding with bimodal–bipolar slipface orientations. Ripples are present on top of the sandstone units. A *Skolithos–Zoophycos* trace fossil association is common. The upper parts of these depositional sequences comprise a vertical alternation of thin sandstone and carbonaceous mudstone beds characterized by flaser and lenticular structures, overlain by fissile shales that are traversed by shallow channel fill. Periodic emergence is indicated by runzel marks, rill marks, sand-filled desiccation cracks, ladderback ripples, raindrop imprints, and rare lenses of gypsum (Theron, 1972; Tankard and Barwis, 1982). Bioturbation of the argillaceous material is extensive, and linguloid brachiopods are preserved in places. Tankard and Barwis (1982) have attributed the small channels to tidal drainage on microtidal or low-mesotidal flats.

Sands reworked by marine processes are locally overlain by argillaceous lagoonal deposits, with *Australospirifer* probably occupying an oyster niche (cf. Copper, 1977). The linguloid community, comprising homalonotid trilobites, infaunal bivalves, and orbiculoid and linguloid brachiopods, is believed to have populated a high intertidal zone (Isaacson, 1975). The decaying St. Bernard delta lobe of the Holocene Mississippi complex is an example of clean barrier sands separating lagoonal environments from shelf muds (Saxena, 1976).

Paleogeography of the Bokkeveld Group

The Bokkeveld Group is a southward-thickening sedimentary prism constructed largely by fluvial and shoal-water delta processes. Regional stratigraphic studies show that the thick delta-front sandstones are stacked above the depoaxis and that they taper southward into expanded prodelta-shelf facies (Theron, 1972). The characteristic Bokkeveld hogback topography (Fig. 10-11) permits individual sandstone units to be traced laterally and indicates lateral coalescence of reworked delta-front deposits.

Basinwide the succession contains five or six superimposed genetic packages (Fig. 10-12) each representing a major depositional episode (Tankard and Barwis, 1982). Stacking of successive deltaic wedges and restriction of the subaerial delta plain components to the proximal margin of the basin is attributed to tectonic subsidence. Episodes of marine transgression resulting from basin margin downwarping are reflected in the accumulation of destructive quartz arenites which erosively overlie the upward-coarsening progradational sequences (Fig. 10-12). The average rate of subsidence for the Bokkeveld Group was considerably more rapid than that for the Table Mountain Group.

Thin quartz arenites, possibly of storm origin, interrupt the lower two Bokkeveld sequences. However, quartz arenites become progressively thicker in the three upper sequences (Fig. 10-12), suggesting increasing effectiveness of marine reworking of delta-front sands.

The sandstones derived from reworked delta sands form upper boundary surfaces, constituting major regional stratigraphic markers. Theron (1970, 1972) has used the regional distribution of these stratigraphic markers for correlation between the western and eastern parts of the basin, which were frequently out of phase tectonically. Substantial thicknesses of fine-grained prodelta sediments in the south are attributed to continuous deposition while the bounding shelves were subjected to lengthy periods of reworking. Thick prodelta–shelf shales in the Karies Formation of the eastern Cape (Table 10-2) are attributed to localized tectonic deepening. An upward increase in concentrations of granitic tourmaline and brookite (Theron, 1972) is ascribed to progressive unroofing of cratonic basement in the drainage basin.

An Early Devonian epeiric sea drowned the southwestern margin of Gondwana (Fig. 10-13). Paleogeographic reconstructions show that the Bokkeveld infaunal bivalves and epifaunal brachiopods (Clarke, 1913; Reed, 1925; Boucot, 1971) are markedly endemic and characterize the Emsian Malvinokaffric province of southwestern Gondwana, including southern South America, Falkland Islands, South Africa, and Antarctica. Most of the fossils are concentrated on bedding surfaces in the lower Bokkeveld, where they are dominated in each instance by a few species. Boucot (1971) recognized four Malvinokaffric communities based on Bokkeveld collections: linguloid, mutationellid, *Notiochonetes*, and *Australospirifer*. Most of the Bokkeveld invertebrates are encompassed within a linguloid community or a mixed *Notiochonetes–Australospirifer* community (Boucot, 1971). A terebratuloid brachiopod association includes *Pleurothyrella, Derbyina,* and *Mutationella*. Mixing of taxa occurs where community boundaries meet.

The linguloid community (Fig. 10-14A) consists of infaunal nuculoid bivalves, orbiculoid and linguloid brachiopods, and homalonotid trilobites, which apparently occupied a lagoonal-intertidal flat niche (Boucot, 1971; Isaacson, 1975; Copper, 1977). Stratigraphically the tidal-flat–lagoon facies cap the progradational deltaic succession or are superimposed on the reworked sandstones.

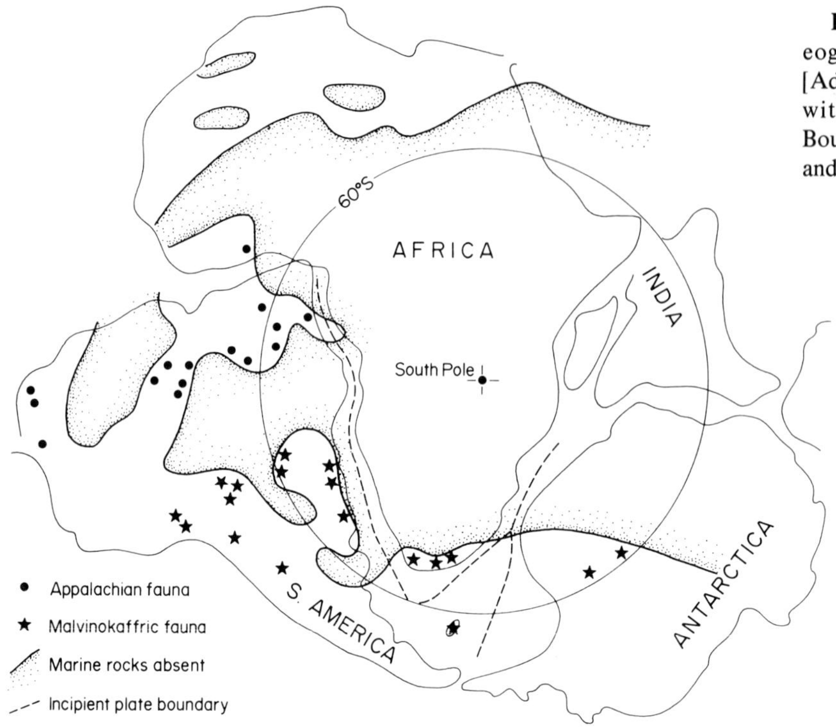

Figure 10-13. Emsian paleogeography of Gondwana. [Adapted from Copper (1977) with additional data from Boucot (1971) and Johnson and Boucot (1973).]

The mixed *Notiochonetes–Australospirifer* community (Fig. 10-14B) is more diverse and is believed to have occupied a slightly deeper habitat (Boucot, 1971; Isaacson, 1975). It is characterized by a rich epifaunal brachiopod fauna, particularly *Australospirifer*, *Notiochonetes*, *Derbyina*, and trilobites. *Australospirifer* colonies may have populated delta-front environments shallower than 50 m, locally forming "reefs" in front of linguloid mud flats (Copper, 1977).

It has been generally accepted that the composition of the Austral or Malvinokaffric faunas of Gondwana reflects a temperate climate (Clarke, 1913; Copper, 1977). Brittlestar-like echinoderms similar to *Encrinaster pontis* live today only in cold ecosystems. Further evidence of a cool but not severe climate is the scarcity of bryozoa, corals, or crinozoa. The taxa are not stunted and epifaunal brachiopods are common, suggesting that salinities were normal. As its name implies, the trilobite *Homalonotus colossus* was gigantic (attaining 50 cm in length according to Lake, 1904).

Freshwater fish fossils in upper Bokkeveld strata, representing the earliest record of fish in South Africa, are associated with a variety of lycopod plant remains (Chaloner *et al.*, 1980).The fishes include placoderms, an acanthodian, and the impression of a large egg case. These fossils have similarities with Devonian taxa from Australia, Antarctica, South America, and North America.

Potential changes in fluvial behavior stemmed from the first appearance of land plants in the Early Devonian. The earliest terrestrial plants in the Bokkeveld basin were the semiaquatic psilophytes, with the more advanced pterophytes (ferns) being associated with the upper Bokkeveld (Plumstead, 1967, 1969, 1977). This apparent stratigraphic ordering is believed to have paleoecological rather than evolutionary significance, considering the restricted time range of the Bokkeveld Group. Vegetation has a significant effect on fluvial and deltaic sedimentation by retarding erosion, controlling runoff and discharge, and affecting channel-bank cohesion (Schumm, 1968a). Pre-Devonian rivers tended to follow a braided pattern (Cotter, 1978). However, the earliest terrestrial vegetation was probably confined to the near-shore environment, suggesting that even in the Early Devonian, many deltas were probably characterized by braided distributary channels (Schumm, 1968a; Miall, 1977). Theron (1972) suggests that the Bokkeveld fluvial deposits reflect meandering channel deposition; this is supported by the presence of lateral accretion surfaces, probably of point-bar origin.

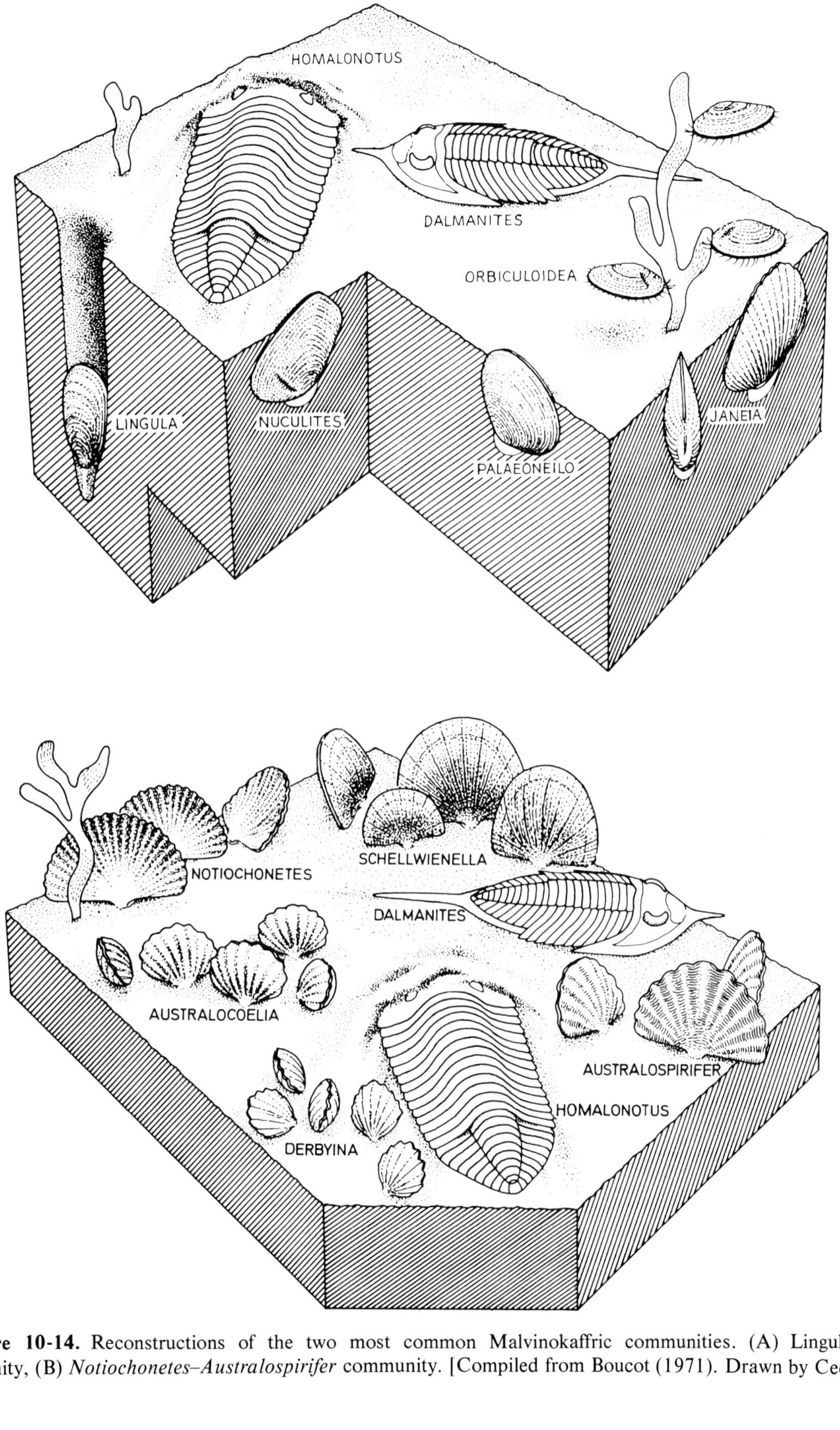

Figure 10-14. Reconstructions of the two most common Malvinokaffric communities. (A) Linguloid community, (B) *Notiochonetes–Australospirifer* community. [Compiled from Boucot (1971). Drawn by Cedric Hunter.]

Table 10–3A. Stratigraphy of the Lower Witteberg Group[a]

Western Cape			20°E	Eastern Cape		
Formation	Thickness (m)	Lithology	Formation	Thickness (m)	Lithology	Interpretation
Weltevrede Subgroup			Weltevrede Subgroup			
Witpoort	310	Quartz arenite with shale partings; plant fossils	Witpoort	850	Quartz arenite, mudstone partings; plant fossils (lycopods)	Barrier sands, tidal flats
Swartruggens	450	Siltstone and shale, thin sandstone interbeds; trace fossils (*Zoophycos*), plant fossils				Tidal flats
Blinkberg	135	Quartz arenite; plant fossils (lycopods)	Weltevrede	850	Shale, siltstone, thick quartz arenite; marine invertebrates, trace fossils (*Zoophycos*), plant fossils	Barrier sands of reworked delta
Wagen Drift	185	Shale and siltstone, interbedded sandstone; marine invertebrates, trace fossils (*Zoophycos*), plant fossils				Subaqueous delta

10.5. Witteberg Group: The Cape–Karoo Transition

The Witteberg Group represents a continuation of Bokkeveld sedimentary patterns in the lower part and encompasses glaciogenic components in the upper part (Du Toit, 1954, p. 261; Haughton, 1969, p. 341; Hiller and Dunlevey, 1978). Because of these affinities with both the Cape basin and the basal–Karoo Dwyka Formation, Witteberg boundaries have generated considerable debate among South African stratigraphers. More than 2000 m of terrigenous clastic sediments are represented in the Witteberg Group (Table 10-3), which derives its name from a prominent range of mountains in which white quartz arenites are impressively exposed. Initially the major locus of deposition remained the elongate Cape basin, with the western and eastern depocenters maintaining an element of tectonic independence (Rust, 1973).

Just as the lower and upper Witteberg strata display different stratigraphic affinities, so the associated fossils, scarce as they are, have distinct age ranges. The invertebrate fauna of the lower Witteberg, comprising mainly brachiopods, trilobites, and bivalves, is typically Malvinokaffric in character (Boucot *et al.*, 1969; Theron, 1970, 1972; House, 1979), but *Tropidoleptus* suggests a Givetian age (A.J. Boucot, personal communication, 1981). Marine fossils are absent from the upper Witteberg, but the biota includes plant and freshwater fish fossils. The plant fossils are more abundant and more advanced than the

Table 10–3B. Stratigraphy of the Upper Witteberg Group[a]

Western Cape			20°E	Eastern Cape		
Formation	Thickness (m)	Lithology	Formation	Thickness (m)	Lithology	Interpretation
			Kommadagga Subgroup			
			Dirkskraal	175	Feldspathic sandstone and lithic arenite	Fluvial
			Soutkloof	170	Mudstone, shale, varvite	Proglacial lacustrine
			Swartwaterspoort/Miller	100	Intertonguing sandstone and diamictite	Glaciofluvial Glaciogenic
Lake Mentz Subgroup			Lake Mentz Subgroup			
Waaipoort	70	Mudstone, rhythmite, graywacke, thin conglomerate; fish and plant fossils	Waaipoort	340	Gray shale, mudstone, feldspathic sandstone; fish and plant fossils	Glaciofluvial and lacustrine
Floriskraal	60	Feldspathic sandstone; plant fossils	Floriskraal	80	Shale, mudstone, quartz arenite; plant fossils	Fluvial and shore zone
Kweekvlei	130	Black, fissile shale	Kweekvlei	225	Shale, siltstone	Lacustrine (?), shelf (?)

[a]Sources: Theron (1962a, 1972), Loock (1967), Roussouw (1970), Johnson (1976), Hiller and Dunlevey (1978), South African Committee for Stratigraphy (1980), J.C. Loock and J.N. Theron (personal communication, 1980).

Bokkeveld flora and, by comparison with Argentinian taxa (Plumstead, 1967, 1969), suggest a Late Devonian age. Glaciogenic shales in the uppermost Witteberg and lower Dwyka successions contain paleoniscoid fishes, mainly *Cheirolepis* and *Mentzichthys* (Jubb, 1965), which are indicative of a Visean (Early Carboniferous) age (Gardiner, 1969). These rare bony fishes evolved in fresh water in the Middle Devonian (Young, 1962, p. 231).

The vertical distribution of distinctive lithologies and associations of physical and biogenic structures (Table 10-3) suggests that the complex history of alternating regressions and transgressions established in Bokkeveld time also controlled lower Witteberg sedimentation. Deltaic progradation and active basin subsidence allowed aggradation of thick delta-front, tidal-flat, interdistributary bay, and fluvial sediments (Theron, 1972; Johnson, 1976). Regionally the succession displays a marked southward and eastward decrease in grain size.

Lithofacies are similar to the progradational Bokkeveld facies. Genetic units are relatively sandy, as is characteristic of the deposits of deltas subject to powerful marine reworking. Marine-reworked deltas are typically cuspate in plan view when reworking occurs contemporaneously with fluvial input (Fisher *et al.*, 1969). Lobate deltas are subject to significant wave reworking only after abandonment. Elongate or birdfoot deltas may have developed following progradation to the shelf edge, where thick, mud-rich delta sequences are characteristic. Stages in Bokkeveld–Witteberg delta evolution may have been comparable to the Mississippi delta lobes, where thin but laterally extensive, sandy, lobate deltas were characteristic until the present delta lobe prograded to the shelf edge and assumed an elongate birdfoot form.

The tidal-flat facies consists of extensively bioturbated mudstones and sandstones, with desiccation cracks and rill marks reflecting periodic emergence. The marine fossil assemblage includes *Zoophycos* and other trace fossils, orthoceratid molds, articulate (chonetids) and inarticulate (*Lingula*) brachiopods, bivalves, and possible bryozoa (Theron, 1962a, 1972; Hiller and Dunlevey, 1978). Aquatic arachnoids, possibly related to this facies, are represented by a large eurypterid (*Hastimima*), one specimen of which was about 1.5 m long (B.W. Oelofsen, personal communication, 1980).

In the Witpoort Formation quartz arenites of coastal-barrier origin are up to 850 m thick (Theron, 1962a). They rest erosively on progradational deltaic sediments and tidal-flat lithosomes, suggesting that they were derived by transgressive reworking of delta lobes. In the western part of the Cape basin, Witteberg quartz arenites terminate the Karoopoort tidal facies of the Bokkeveld Group (Theron, 1972). These reworked sandstones comprise fine- to medium-grained quartz arenite lenses with numerous mudstone interbeds. Linear sand waves are developed and side-filled channels are accentuated by gravel lag. Cross-bedding displays an ebb-dominated, polymodal distribution. The lithologic associations and primary structures resemble those of the thick marginal marine facies of the Table Mountain Group (Theron, 1962a) and suggest that the Witteberg quartz arenites originated as barrier complexes seaward of tidal flats (Johnson, 1976). Evidence of channeled flow and traction currents of variable direction supports this model. The upper parts of the Witpoort sandstone are coarser grained and sporadically calcareous and contain lycopod stems (Theron, 1962a), suggesting a greater degree of fluvial influence.

Upper Witteberg strata are more argillaceous (Table 10-3), comprising a varied assemblage of black, fissile shales, micaceous mudstones, sandstones, and thin conglomerates or diamictites (Theron, 1962a; Loock, 1967; Roussouw, 1970; Rust *et al.*, 1970; Johnson, 1976). The sandstones are immature and cross-bedded, with rare trace fossils. Plant fragments are most abundantly preserved in the thick carbonaceous shales. Initiation of this distinctive upper Witteberg style of sedimentation may have resulted from basin tectonism at the end of Witpoort time (J.C. Loock, personal communication, 1980). There are insufficient data for a general paleogeographic reconstruction, but interpretations have included proglacial lacustrine to restricted shore-zone and shelf environments (Loock, 1967; Roussouw, 1970; Rust *et al.*, 1970; Johnson, 1976).

Vertical transition to Dwyka tillite takes place by interbedding of shales, varvites, and diamictites in the upper Witteberg (Du Toit, 1954, p. 259; Loock, 1967; Haughton, 1969, p. 341; Johnson, 1976). The known fossils are entirely nonmarine. Plant taxa are more advanced than the Bokkeveld flora and include *Protolepidodendron*, *Archaeosigillaria*, and *Platyphyllum* (Plumstead, 1967, 1969). Paleoniscoid fish, mainly *Cheirolepis* and *Mentzichthys*, are common to the upper Witteberg and lower Dwyka shales (Jubb, 1965; Gardiner, 1969; Rust, 1973). *Cheirolepis* was a rare air-breathing freshwater fish with sharp teeth and probably preyed on invertebrates (Young, 1962, p. 231), but the only evidence of this is enigmatic bivalve molds (Theron, 1962a). The mode of preservation of these fish suggests that they died of asphyxiation and freezing beneath packice (Theron, 1962b; Jubb, 1965; Gardiner, 1969).

By late Witteberg time the basin had diminished to only one third of its original extent (Roussouw, 1970; Rust, 1973). Although sediment infilling was important, the reduction in basin dimensions may have been partly tectonic. Loock (1967), Roussouw (1970), and Johnson (1976) described preDwyka deformation and folding of the Witteberg strata, the synsedimentary deformation being spectacular in places.

Marine faunas are restricted to the Bokkeveld-like lower units, their Malvinokaffric affinities suggesting an Emsian to Givetian age. There is no evidence that these marine faunas extend into the Middle Devonian (House, 1979); the Upper Devonian and Early Carboniferous horizons contain fish fossils and plant fossils resembling some Argentinian taxa (Plumstead, 1967, 1969). In South America the Devonian marine depositional cycle was in general regression by the Mid-Eifelian and was succeeded by nonmarine deposition. There are no Frasnian or Famennian marine faunas in South America (Boucot, 1971; Isaacson, 1977, 1978). The Upper Devonian Ponta Grossa Formation contains only plant fossils and was followed by Lower Carboniferous glaciogenic sedimentation (Copper, 1977).

Chapter 11

The Intracratonic Karoo Basin

A broad spectrum of paleoenvironments ranging from glacial through coarse bedload and mixed-load streams, deltaic, beach, proximal terrigenous shelf, and distal subsiding basin is preserved in the Karoo Supergroup. This sequence records a changing tectonic framework and the migration of Gondwana from polar to tropical latitudes over a span of 200 Ma. This Late Paleozoic to Mid-Mesozoic succession is best developed in the Karoo basin (Figs. 11-1 and 11-2), which covers half of South Africa. Subsidiary basins to the north include the southern terminus of the East African Rift system, represented by the north-trending Lebombo and the east-trending rifts in the northern Transvaal. In addition, there are the thin but widespread Karoo deposits of the Kalahari basin and Warmbad–Springbok Flats area. These southern African deposits formed the nucleus of the great Gondwana succession of the southern hemisphere continents and India, long recognized as irrefutable evidence of continental drift (Du Toit, 1937).

Permo–Carboniferous glaciation of an intensity and extent unparalleled in the Phanerozoic history of the Earth was a response to the migration of Gondwana over the South Pole. The polar position moved in a southeasterly direction across the southern African segment of the protocontinent (Creer, 1965). With the waning and final disappearance of the ice sheets, the vast, shallow Ecca sea was established, accumulating sediments from Early to Middle Permian time. Permo–Triassic Beaufort paleogeography was increasingly influenced by orogenic activity in the south. Encroaching fluviodeltaic tongues segmented the basin into a series of freshwater lakes of diminishing extent. Another major episode of northward fluvial influx in response to an orogenic pulse characterized deposition of the Triassic Molteno Formation. Oxidized alluvial sediments of the overlying Elliot Formation are followed by Clarens Formation deposits of a low-latitude desert environment of dune fields and playa lakes. This final phase of Karoo sedimentation was terminated by the voluminous Jurassic outpouring of basaltic magma.

Coals in the Dwyka–Ecca transition and Ecca Group of the northeastern Karoo basin and northern Transvaal basins are of paramount importance to South Africa. Coal liquefication processes, in operation since the 1950s, make these deposits economically crucial. The commercial potential of uranium deposits in the Beaufort Group of the southern and southwestern Karoo basin is currently being assessed; Karoo uranium may well increase in significance as other energy reserves decline worldwide and more nuclear generating facilities come into operation. Hydrocarbon exploration has been all but abandoned after only minor shows in the northern Ecca Group. Other Karoo minerals of potential value are heavy-mineral placers, molybdenum, and diamonds.

11.1. Glaciogene Dwyka Sedimentation

Moraine and Glaciolacustrine Facies

Over much of the Karoo basin, the Dwyka Formation rests with distinct unconformity on rocks of Early Paleozoic to Precambrian age, but in the vicinity of the subsiding Cape depoaxis the relationship with the underlying Witteberg is paraconformable. A considerable hiatus is represented by this basal discontinuity, possibly spanning much of the Carboniferous (Dunlevey and

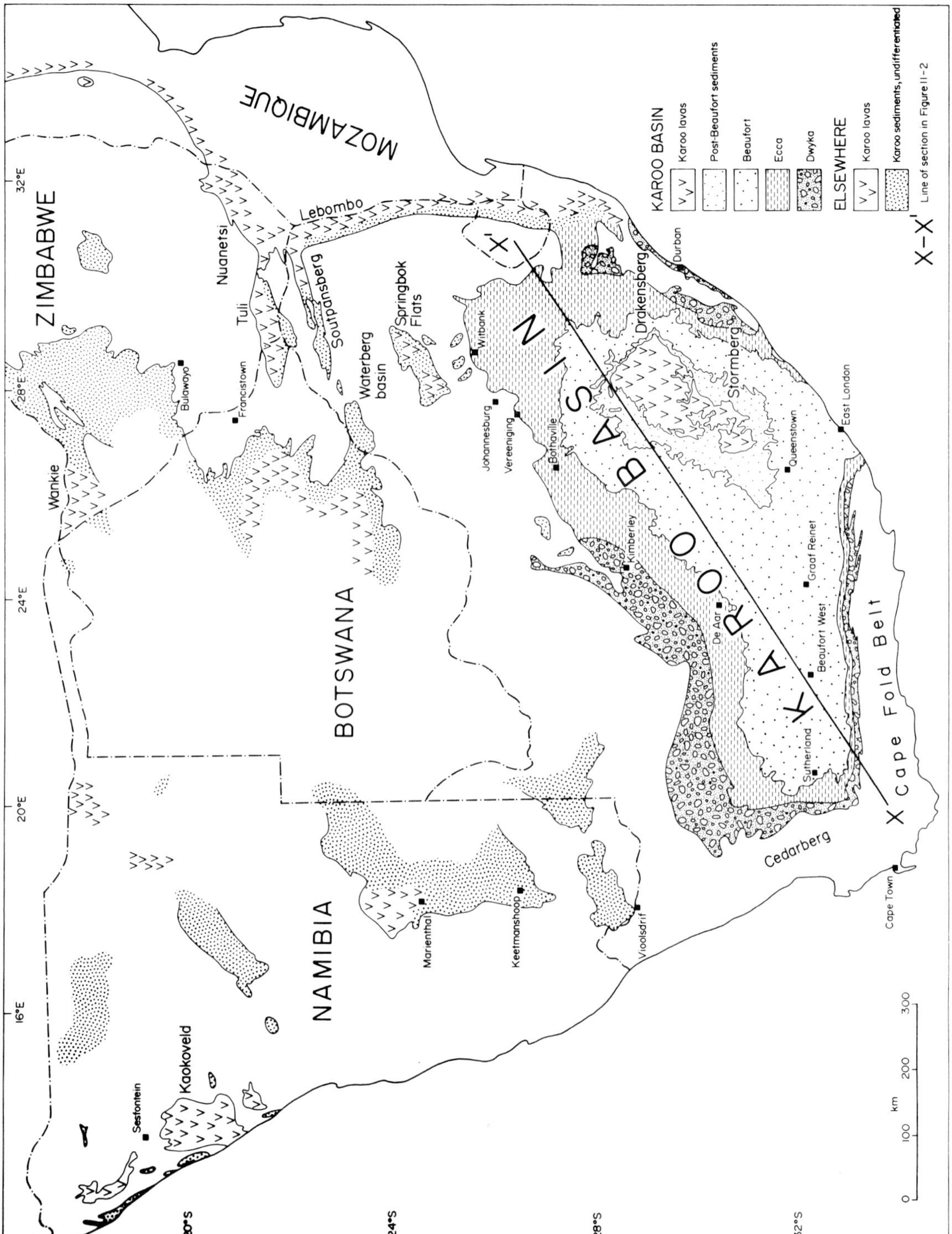

Figure 11-1. The Karoo basin, showing line of sections in Figure 11-2.

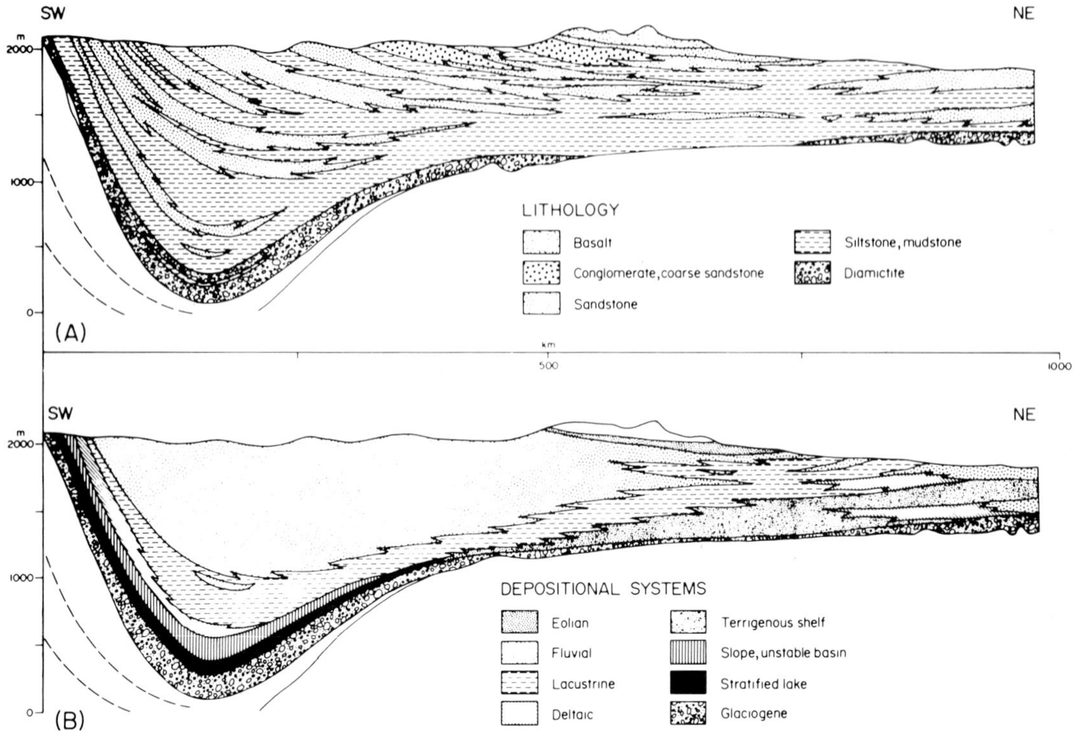

Figure 11-2. Northeast–southwest cross-sections through the Karoo basin showing (A) distribution of major lithologic units. [Adapted from Winter and Venter (1970)] (B) Major depositional systems.

Hiller, 1979). The Dwyka diamictite and associated deposits attain a maximum thickness in the south, thinning northward (Fig. 11-2) and wedging out over Precambrian basement in the northern Cape and Transvaal. Deposition on a highly dissected surface in the northern Orange Free State, southern Transvaal, and Natal has resulted in irregular thicknesses and complex facies relationships within the Dwyka; farther south a mature paleotopography is overlain by Dwyka of more uniform character. Following regional studies by Du Toit (1921) and Stratten (1967, 1968, 1970), who established the distribution and origin of major Dwyka facies, Von Brunn (1979) added a new dimension by relating the details of local facies variation to modern glacial environments.

The tillite (diamictite) comprises two distinct size modes with clasts in the pebble and boulder range supported by a silt and clay matrix. Toward the base of the Dwyka, the clasts commonly reflect local derivation, but higher in the succession clast lithology indicates transport from farther afield (Stratten, 1968). There is a general increase in maximum clast size toward the north, where Stratten observed boulders over 10 m in diameter. Small-pebble clasts tend to be subangular, but the degree of rounding improves with increasing size, so that large pebbles and boulders are commonly subrounded to rounded with moderate to high sphericity. Some display faceted and striated surfaces. Sand-size grains are angular, and in fresh tillite easily weathered minerals show little or no sign of alteration. Hamilton and Krinsley (1967) have observed by electron microscopy textures of glacial origin on grain surfaces, such as conchoidal fractures of variable size, subparallel and arcuate steps, and grinding indentations. The finest particles originated as rock flour, with almost complete absence of clay minerals apart from illite. This reflects negligible chemical weathering, probably in response to the dominance of physical erosive processes in a frigid climate, and to rapid burial.

The tillite is generally massive, but at some localities it displays crude horizontal bedding, particularly in the upper part of the succession. The clasts are not imbricated but show a weakly developed long-axis alignment parallel to the direction of ice flow as inferred from striations on subjacent glaciated pavements. This preferred orientation is observed in modern subglacial lodgement till which is overridden by ice (V. Von Brunn, personal communication, 1978) and is also produced by mass-flow processes where the clasts are supported in a viscous matrix. In either case, the resulting deposits typically contain matrix-supported clasts. On the northern flank of the Cape Fold Belt, where the Dwyka was involved in Mesozoic orogeny, the fabric is dominated by tectonic overprinting (Stratten, 1968).

In marked contrast are the relatively well-sorted, clast-supported conglomerates that occur interspersed with the tillite, particularly around the northern and western margins of the basin. Imbrication is usually well developed, and clast long axes are aligned transversely to the direction of ice flow. This is typical of coarse gravels deposited by traction currents, with the sandy matrix filtering into the interstices with waning flow.

Other lithologies within the tillite-dominated succession, both in the southern Karoo and around the northern periphery, are cross-laminated silts and carbonaceous shales. Considerable thicknesses of uniform siltstones display climbing ripples and small freeze–thaw deformation features. These fine-grained outwash and glaciolacustrine intercalations are most common high in the Dwyka succession, but in areas of elevated paleotopography they also occur near the base (V. Von Brunn, personal communication, 1978). Carbonaceous shales are typically laminated and in places display alternating light and dark laminae that are interpreted by some authors as varves. Graded, erosively based units were attributed by Anderson (1975b) to small-scale turbidity currents; others are thought to reflect differential settling from pulses of sediment-laden meltwater and lack the rhythmic alternation and small-scale textural gradation characteristic of glacial varves. The laminae were disturbed by rafted dropstones, slumping, ice puckering, gas escape, bioturbation, and downslope drainage of viscous mud. Trace fossils are abundant in the Dwyka shales of the northern Cape and Natal and record the activities of crustaceans, arthropods, and fish (Savage, 1970, 1971; Anderson, 1975 b).

The tillite accumulated largely as ground moraine and comprises basal lodgement debris plastered onto the substrate and molded by the overriding ice and ablation (supraglacial) till, which is thought to have been laid down during the final downwasting of debris-laden ice (V. Von Brunn, personal communication, 1978). Several authors have suggested that the Dwyka tillite may have been deposited almost entirely during the dwindling phase of glaciation, but Von Brunn (1977) has shown that there was significant moraine accumulation below advancing ice masses. According to Von Brunn, these subglacial lodgement tills can be distinguished by their very fine, dark matrix and the relatively low concentration of large inclusions. Pointed inclusions show a tendency to be aligned with the sharp end in the direction of ice flow. Ablation till tends to contain a greater assortment of clast sizes, the largest being several meters in diameter. The clasts although matrix supported, tend to be more closely spaced in lodgement deposits. A third variety of till, which is best developed in the upper part of the tillite succession, displays horizontal bedding planes and is thought to have been deposited subaqueously.

Several stages of glacial advance interspersed with interglacial retreat, such as typified the Pleistocene, are evident in various parts of the basin. In many areas, however, the stratigraphic evidence is highly complex and detailed study of sequences indicates that deposition was a response to composite events (V. Von Brunn, personal communication, 1980). The greatest degree of vertical ordering and lateral persistence of cycles is developed in the western and southern Karoo basin. In the west are four cycles consisting of massive tillite, attributed to deposition from grounded ice, overlain by bedded diamictite and both linear and sheetlike sandstones deposited largely from floating ice (Theron and Blignault, 1975). In the southern Karoo basin as many as nine cycles are superimposed. Each cycle is 60–100 m thick and is dominated by upward-fining diamictite overlain by glaciolacustrine shales (J.N.J. Visser, personal communication, 1979). There is evidence of accumulation both as terrestrial moraine and subaqueous terminal moraine. Crude stratification may have been produced by subaqueous gravity flow. There is abundant deformation resulting from slumping and loading. Visser *et al.* (1978) recognized two cycles in the Dwyka of the Kimberley-Britstown area of the northern Karoo basin; each cycle comprises a basal massive subglacial till and a stratified terminal zone deposited from floating ice. Associated mudstones reflect freshwater to brackish fluctuations. In the western Orange Free

State, the deposits of two major glacial phases are separated by interglacial shales.

Similar shale bodies occur sporadically at all levels in other parts of the basin but are most abundant in the north. Plant remains within the tillite (Plumstead, 1969) and miospores and cuticular material from intratillite shale bands (McLachlan and Anderson, 1973) indicate that the Karoo basin and surrounding areas were not uniformly glaciated but supported a plant cover in areas of local inactivity or recession.

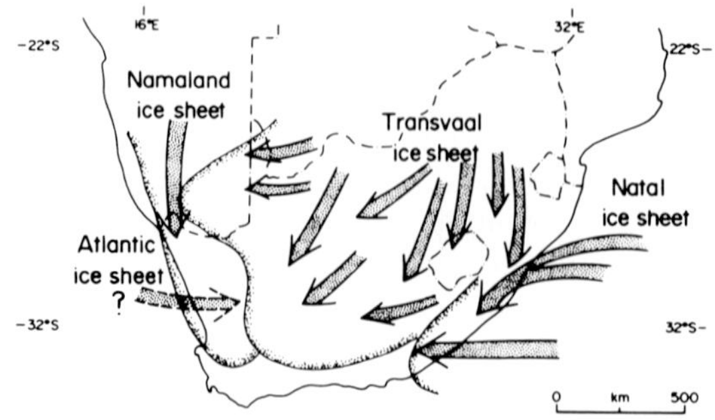

Figure 11-3. General patterns of ice movement from the major centers of glaciation. [Adapted from Stratten (1967, 1977), Crowell and Frakes (1972), and J.N.J. Visser (personal communication, 1980.]

In the northeast there is great difficulty in correlating these intercalations within the tillite, even between exposures only tens of kilometers apart, suggesting local extension of ice lobes accompanied by the development of ponds and outwash fans in adjacent areas (V. Von Brunn, personal communication, 1980). At some localities in the northern Karoo basin, black shales at the base of the Dwyka rest directly on Precambrian basement, whereas lodgement tillite is developed at the same level nearby. In this area there is a vertical repetition of glacial deposits overlain by fluvioglacial outwash, but lateral variations are pronounced. Von Brunn (personal communication, 1978) has shown that a typical cycle comprises lodgement and/or ablation till overlain by waterlain conglomerate, which in places shows major lateral accretion surfaces, followed by planar and trough cross-bedded sandstone, with cross-laminated siltstone at the top of the sequence. The upper deposits contain pods of tillite that depress the strata on which they rest; these were presumably liberated from isolated blocks of ice stranded in front of the retreating glacier. Also present are dropstones ranging from pea size to large boulders, indicating deposition from floating ice. Local evidence of intermittent ice flow and retreat comes from an intratillite pavement that was furrowed by the overriding ice and preserved beneath a blanket of glaciolacustrine siltstones deposited during the subsequent recession (Von Brunn, 1977). Abraded tillite clasts, clearly derived from Dwyka boulder clay, are common and point to deposition of moraine and its subsequent erosion as compacted or possibly frozen lumps.

The generally converging patterns of ice movement (Fig. 11-3) have been inferred from small-scale impressions in glaciated pavements (Fig. 11-4) and from larger geomorphic features. Around the basin margins are tillite-filled glacial

Figure 11-4. Sub-Dwyka striated pavement near Umzinto, Natal.

valleys whose orientation can be measured directly. Exhumed *roches moutonnées* with smoothly polished and striated upstream surfaces and jagged, glacially plucked downstream sides provide an accurate sense of ice movement. Excellent examples are present in northern Natal and at Nooitgedacht in the northeastern Cape. Boulder trains are occasionally preserved in the lee of such obstacles. Glacial grooves also record the direction of ice movement (Von Brunn, 1977), and where scribing pebbles are preserved at one end of the groove the sense of movement is indicated. In the western Cape, grooves are preserved in a paraglacial sand at the base of the Dwyka (Savage, 1972). Striations, friction cracks, crescentic gouges, lunate fractures, and bevelled boulders embedded in the pavement have been used to delineate movement of several ice sheets which originated over widely separated centers of maximum glaciation. The original concept of four glacial centers (Du Toit, 1921) has been expanded by Stratten (1968) with recognition of the Namaland, Griqualand, Atlantic, southern Cape, Transvaal, and Natal ice sheets. Evidence in Namibia and the western Cape of markedly different or reversed directions of ice movement (Stratten, 1967; 1968; Martin and Wilczewski, 1970) is explained by sequential episodes of glacial advance from different centers of ice accumulation as Gondwana drifted across the pole. Thus the earliest glaciation may have involved southward movement of the Namaland ice sheet and eastward movement of the Atlantic ice sheet; these were succeeded by the major glacial episode of the Transvaal ice sheet (Stratten, 1977). The record of the Griqualand and southern Cape ice sheets is less pronounced and has been questioned by some. Westward movement of the Natal ice sheet may have been contemporaneous with southward movement of the Transvaal ice sheet, with southwestward flow parallel to the present coastline of Natal resulting from convergence. Alternatively, according to Matthews (1970), this Natal lobe may have resulted from topographic deflection of a single ice sheet from the north down the axis of a trough inherited from the Early Paleozoic Natal embayment. Crowell and Frakes (1972) point out, however, that the record of ice flow on the Natal coast indicates a major contribution from the east.

As the area drifted out of polar latitudes, a general retreat of the ice sheets led to the establishment of a broad, shallow epicontinental basin. Hamilton and Krinsley (1967) suggest that depression of the crust by ice loading was not entirely reversed by subsequent rebound. Shales, which gradationally overlie the tillite, are best developed in the southwestern half of the Karoo basin, where they attain a thickness of 260 m. They are generally horizontally laminated with sporadic calcareous and ferruginous bands and phosphatic concretions. The shales are thin and sporadic in the east, and in the Transvaal where the shales are generally absent, the tillite was subject to prolonged subaerial exposure and reworking. Above the basal tillite of the western Kalahari basin are mudstones containing paleoniscid fish, gastropods, the bivalve *Eurydesma*, the cephalopods *Orthoceras* and *Eosianites*, brachiopods, echinoid spines, crinoid columnals, asteroids, radiolaria, and several genera of foraminifera (McLachlan and Anderson, 1973), together indicating marine conditions. In southern Namibia, a smaller but comparable assemblage of marine fossils occurs at the base of the shales which make up the upper two-thirds of the Dwyka succession. The discovery of marine fossils, including orthocerid cephalopods and brachiopods, in the Douglas-Blaauwkrantz area near Kimberley (McLachlan and Anderson, 1973) indicates that the marine incursion extended into the Karoo basin from the west.

Overlying the shales, which in the Kalahari basin contain thin coal seams in the upper part, is the thin (generally less than 30 m) but widespread "White Band," which was previously regarded as the uppermost unit of the Dwyka Formation but is now included at the base of the Ecca Group as the Whitehill Formation. In the west the Whitehill Formation consists of pyritic, carbonaceous shale with dolomitic and organic-rich layers and probable pseudomorphs after gypsum (Anderson *et al.*, 1977). Fossils include the fish *Palaeoniscus capensis*, the crustacean *Notocaris*, and the aquatic reptile *Mesosaurus*. Anderson *et al.* envisage deposition in anoxic bottom layers of a stratified lacustrine environment. The formerly contiguous Irati Formation in South America is an important oil producer, but hydrocarbon prospects in the Whitehill Formation have been severely limited by diagenesis.

The marine invasion that accompanied deglaciaton in the western half of the basin was at least in part contemporaneous with progradation of coarse clastic deposits in the northeast. The lower parts of the coal-bearing succession of the southeastern Transvaal contain abundant evidence of continued glacial influence, and scattered ice-rafted pebbles persist locally to near the top of the clastic wedge. Peripheral fluvioglacial systems prograded into a basin characterized by large areas of floating ice which was flushed from the diminishing icecap during the annual thaw. Precise delineation of the Dwyka–Ecca boundary is therefore difficult in areas where glacial influence, definitive of the Dwyka Formation, is reflected in the lower strata of major coalfields customarily regarded as classical Ecca deposits. Farther south there is a gradual vertical transition from fine-grained glaciolacustrine sediments to the dark, starved-basin mudstones of the Ecca.

Paraglacial Coal-forming Environments of the Northeastern Basin Margin: Dwyka–Ecca Transition

Deglaciation of the northeastern margins of the Karoo basin led to the accumulation of a complex suite of glaciolacustrine and glaciofluvial sediments in moraine-mantled depressions, with gradual onlap of intervening south-trending ridges. Diamictite was reworked by meltwater runoff, resulting in improved sorting in sheet gravels and small outwash fans. Gravitational instability of rapidly accumulating moraine, local impounding of meltwaters, and breaching of marine obstructions resulted in debris flows and deposition of poorly sorted conglomerates comprising subrounded, mud-supported granitic and felsic pebbles. These debris flow deposits are locally intercalated with current-reworked conglomerate layers and vary from sheetlike to lenticular depending on the topographic configuration of the floor at the time of deposition (Cairncross, 1979; LeBlanc-Smith and Eriksson, 1979).

Fan deltas developed along the irregular lacustrine shoreline and in depressions produced by ice stagnation and melting (LeBlanc-Smith and Eriksson, 1979). Progradation of these small, coarse-grained deltas is recorded in planar foreset units up to 5 m thick. Their bottomset deposits comprise ripple-drift cross-laminated fine-grained sandstones, siltstones, and mudstones. Associated graded units indicate frequent interruptions by pulses of gravity resedimentation followed by waning flow and increased fallout of the finer sediment grades. Outwash deltas are characterized by hyperpycnal processes whereby sediment-charged meltwaters are more dense than the lake water, resulting in density underflow. Associated massive lenticular sandstones contain deformation structures produced by excess pore fluid pressures brought about by rapid sedimentation during brief episodes of coarser grained subaqueous outwash. Farther basinward, deltaic bottomset deposits are composed of thin, graded silty sandstones and varved silty mudstones (rhythmites). Small deformation features reflect processes of loading, gravity slumping, and bottom current shear.

Updip environments comprised gravel and sand bedload streams of multiple-channel, low-sinuosity (braided) type. Bar migration deposited planar cross-beds in sets up to a meter thick. Exhumed transverse bars are preserved without erosion of thin mudstone drapes, suggesting that they may have been subjected to rapid freezing. Bar-front accretion sets are commonly overlain by upper flow regime plane-beds or cosets of small-scale cross-beds deposited by secondary bedform migration. These small bedforms advanced more rapidly than the underlying bar and avalanched down the bar slipface to give the characteristic graded foresets. Scour-based cosets of medium-scale trough cross-bedding originated from in-channel dune migration (LeBlanc-Smith and Eriksson, 1979). These aggrading alluvial sediments lapped onto glacial till or topographic highs.

Areas isolated from active sedimentation, such as over slightly elevated terrain or in inactive portions of the floodplain, developed a marshy cover of vegetation comparable to modern tundras. The resulting peats were widespread (the basal Number 1 Seam extends throughout the Witbank Coalfield); locally they developed almost directly on the basal diamictite (Fig. 11-5), whereas elsewhere they overlie the bedload stream deposits, with maximum development over abandoned channels marginal to the main fluvial depoaxes (Fig. 11-6). Maximum coal thicknesses of between 25 and 50 m accumulated above topographic depressions. Alternating vitrinite and inertinite-rich layers suggest fluctuating conditions of oxidation and reduction, probably related to seasonal water-table variations. The shallowly rooted, mossy vegetation and low intensity of chemical pedogenesis in the high-latitude climate may account for the absence of seat earths.

The typical Dwyka tillite to coal succession is

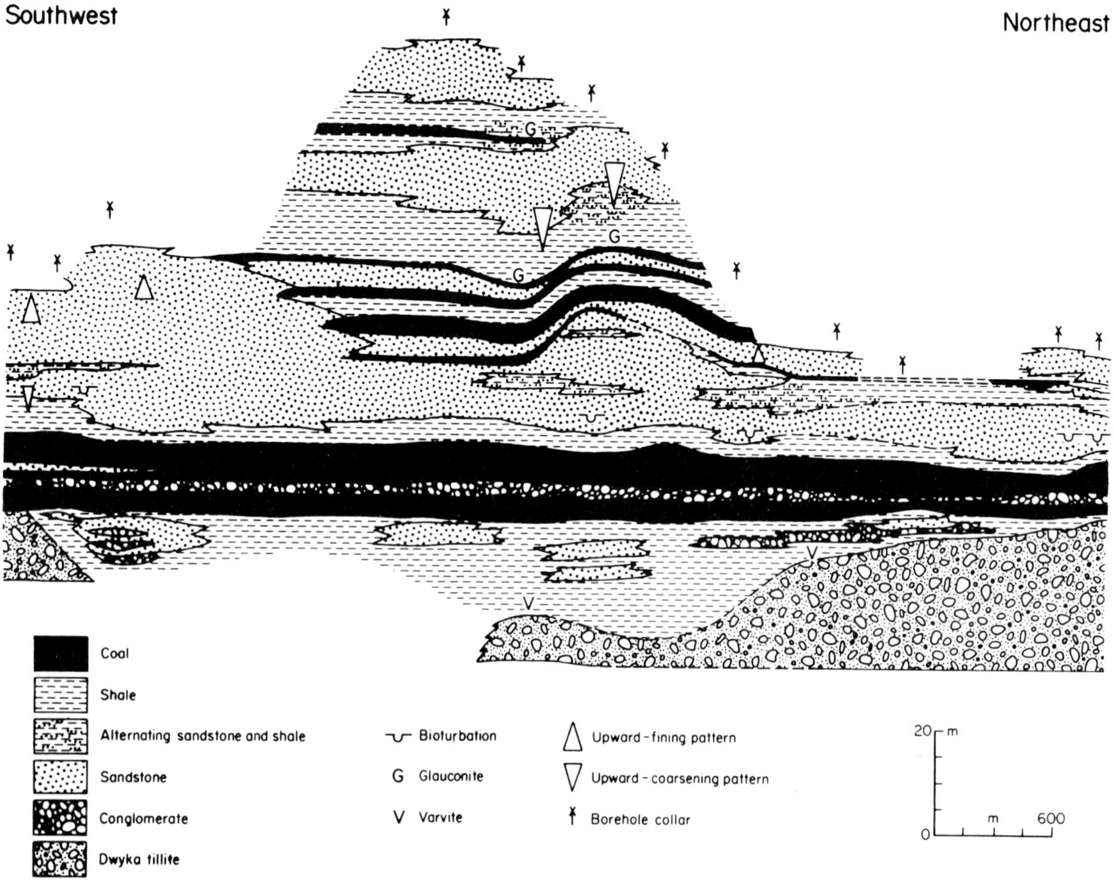

Figure 11-5. Seams of the Witbank coalfield at Van Dijks Drift Colliery. Lower seams, which locally impinge on Dwyka tillite, probably represent tundra-type peat environments and contrast with more normal fluviodeltaic seams 3, 4, and 5. [After Cairncross (1979).]

viewed by LeBlanc-Smith and Eriksson (1979) as representing a cycle of postglacial progradation, alluvial aggradation, and abandonment. Witbank seams, such as 1a, 1b, and 2, are related to the abandonment phase. Continued deposition by sluggish mixed-load streams resulted in *in situ* aggradation and the development of channel-fill sandstones of abnormally small width to thickness ratio (Cairncross, 1979).

11.2. Postglacial Epicontinental Ecca Basin

The Permian Ecca Group (Table 11-1) overlies the Dwyka succession gradationally and comprises predominantly clastic sediments deposited in an extensive landlocked basin experiencing only rare marine incursion. In this respect the Ecca basin resembled the Black Sea (Hart, 1964) but was shallower and had a larger surface area. The broad intracratonic depression inherited from the Dwyka ice cap may have been subject to further downwarping by sediment loading in Ecca depocenters. The tectonic framework was probably controlled to some extent by older structural elements. A broad, southward-deepening shelf possibly owed its stability to an underlying extension of the Kaapvaal Province. On the other hand the subsiding Karoo and Natal troughs, which accumulated sediment thicknesses of 3000 m and 1000 m, respectively, may have been located over belts of crustal thinning. These negative belts have been referred to as flysch troughs, but water depths were generally shallow, seldom exceeding 500 m even in the more active Karoo trough (Visser and Loock, 1978; Kingsley, 1979). The central Ecca basin was dominated by mud and silt

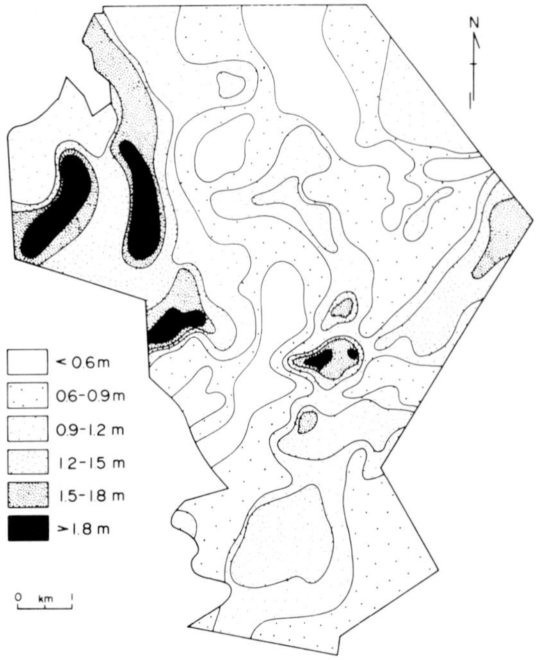

Figure 11-6. Isopach of No. 2A coal seam at Van Dijks Drift Colliery showing dip-oriented trends typical of a fluvial environment. Maximum thicknesses are developed over swales and abandoned channels. [After Cairncross (1979).]

accumulation. Visser *et al.* (1978) note an upward-coarsening tendency near the top of the succession. Arkosic fans developed around basement highs.

Initially, muds accumulated in an anoxic, starved basin, with coarse sediments restricted to the northern periphery. Because the Dwyka-Ecca boundary is diachronous, the marine invasion that accompanied deglaciation in the western half of the basin may have been at least in part contemporaneous with the major progradational phase of the Ecca in the northeast. Some workers are of the opinion that the Ecca basin was predominantly fresh to brackish (McLachlan, 1973), but the presence of marine acritarchs and other microfossils (Hart, 1964) and trace fossils that are customarily regarded as shallow marine forms (Cairncross, 1979; Stanistreet, 1979) possibly reflects the brief marine incursion documented in the Whitehill Formation to the west. Rare accounts of alleged marine mollusks have, for the most part, been discounted. A marine interpretation is further supported by glauconite that has replaced fecal pellets (Cadle and Hobday, 1977; Stanistreet, 1979). Substantial freshwater inflow that accompanied climatic warming, and a low rate of evaporation in a high-latitude situation, may have effectively diluted marine incursions, particularly within some of the large estuarine embayments along the irregular shoreline.

Deglaciated Basin Margin of the Southeastern Transvaal

In the southeastern Transvaal, the paraglacial milieu of glacial outwash fans, peat bogs, and moraine-dominated lakes was succeeded by shallow inundation and the development of small, elongate, and lobate deltas fed by high-gradient streams. LeBlanc-Smith (1979) has divided this 120-m-thick postglacial succession in the Witbank basin into the Coalville, Middelburg, and Van Dijks Drift formations. These terms remain in informal usage, and together with the Witbank formation, might better be regarded as subdivisions (members) of the Vryheid Formation.

Upward-coarsening deltaic sequences are thin and locally truncated by coarse fluvial channel sandstones, but delta-plain coals are particularly well developed. The destructive components of delta abandonment include organic-rich mudstones and glauconitic quartz sandstones.

Sedimentation patterns were strongly influenced by basement topography. Major channel trends were situated over glacially eroded valleys west of Witbank, south of Middelburg, and in the Van Dijks Drift area (LeBlanc-Smith, 1979). Precise delineation of channel geometry in the Van Dijks Drift area shows that although they were only 1200 m wide and sinuous, the channels were confined by a resilient mat of swamp vegetation (Cairncross, 1979). This rather unusual category of anastomosing channels underwent vertical accretion in a pattern similar to the modern channels in deglaciated intermontane valleys of Alberta, Canada (Smith and Smith, 1980).

Transgressive episodes flooded all but the residual ridges of basement granite and felsite that protruded as an archipelago chain. Recent studies by Stanistreet (1979) have revealed paleoenvironmental subtleties in the nature and orientation of trace fossils in the transgressive deposits. Both *Arenicolites* and *Rhizocorallium* show a preferred orientation in response to paleoslope and direction of wave approach; *Diplocraterion* has an arrangement of spreiten that indicates upward movement of the organism in an effort to stay a

Table 11–1. Stratigraphy of the Karoo Supergroup in the Karoo Basin[a]

Group	Formation			Biozone
Drakensberg (volcanics)				
(previously Stormberg)	Clarens			(dinosaurs)
	Elliot			
	Molteno			(*Dicroidium*)
Beaufort	*Southwest*	*Southeast*	*Northeast*	
Tarkastad Subgroup		Burgersdorp	Otterburn	Kannemeyeria–Diademodon
		Katberg	Belmont	Lystrosaurus–Thrinaxodon
Adelaide Subgroup		Balfour	Estcourt	Dicynodon lacerticeps–Whaitsia
	Teekloof			Aulacephalodon–Cistecephalus
		Middleton		Tropidostoma–Endothiodon
				Pristerognathus–Diictodon
	Abrahamskraal	Koonap		Dinocephalian
Ecca	Waterford		Volksrust	
	Fort Brown		Vryheid	
	Laingsburg	Ripon		
	Vischkuil		Pietermaritzburg	(*Glossopteris*)
	Collingham			
	Whitehill			
	Prince Albert			
	Dwyka			

[a]South African Committee for Stratigraphy (1980).

constant distance below the rapidly accreting sediment surface.

The degree of destructive wave reworking increases upward. LeBlanc-Smith (1979) proposed that the initial transgressive medium was fresh water, whereas the upper "marine" destructive phase fashioned linear glauconitic sand bodies up to 12 m thick and 6 km wide. Preservation of such thicknesses of relict barrier sands during transgression implies that substantial volumes of sand were supplied contemporaneously, probably by longshore processes. Furthermore, transgression was probably very rapid. This progressive increase in reservoir energy would have influenced delta morphology and depositional patterns. Early deltas were probably of small, elongate, and fan-delta types, depending on the volume and caliber of the fluvial sediments. Later, these would have evolved into lobate patterns more closely resembling deposits of the Lafourche delta of the Holocene Mississippi complex. Because of the microtidal range, probably comparable to the modern Gulf of Mexico, tidal processes were only important in shallow, open bays. Rippled sandstone surfaces reflecting variable currents and episodic emergence are a feature of some shallow sand shoal and bay-margin deposits in the northern Karoo basin.

Stable Shelf Delta Systems

Subsurface studies of over 1000 borehole logs from gold, coal, and hydrocarbon exploration (Van Vuuren and Cole, 1979) reveal a major buried paleoscarp that separates the dissected, till-veneered glacial highland area of Precambrian rocks in the north from a low, undulating plain underlain by granites and schists to the south. This southern shelf, with a northward-thinning mantle of Dwyka moraine, provided a relatively stable but gently subsiding platform for prograding delta systems. Because of the shallowness of the water and the low rate of subsidence, delta-front sands were spread out over a wide area to form extensive sheet sands (Fig. 11-7). Other factors influencing delta geometry were the large volumes of sediment supplied, including a high proportion of sandy bed load, and low to moderate reservoir energy.

The regressive fluviodeltaic wedge, which is a composite of as many as eight episodes of delta progradation and transgression (Hobday, 1973; Van Vuuren and Cole, 1979), constitutes the Vryheid Formation, previously known as the Middle Ecca. In the southeastern Transvaal, the Vryheid Formation rests directly on the Dwyka Formation or Precambrian basement, but to the south and southwest it is underlain by a thickening wedge of dark silty mudstones, the Pietermaritzburg Shale Formation, deposited on the postglacial platform. These terrigenous shelf deposits record the transition from sediment-starved polar basin to prodelta.

Although the basin waters may not have been truly marine at all times, they were more dense than inflowing fresh water, which extended out over the surface as a hypopycnal plane jet. Bedload was concentrated at the channel terminus, building up a mouth bar that constricted flow around its margins. This process was continued with successive channels, producing a bifurcating pattern of distributaries. With progradation, the subaerially emergent delta plain gained additional distributaries by crevassing. Suspended silt and clay settled offshore in the prodelta environment.

Episodes of delta progradation are recorded in upward-coarsening sequences comprising a repetitive vertical arrangement of lithologies, physical structures, and biogenic features. This preferred vertical arrangement is readily apparent in outcrop at localities such as Tugela Ferry, Natal (Hobday, 1973), where several upward-coarsening cycles are superimposed. Over a wider area this pattern of vertical ordering is confirmed by nonparametric statistical analysis (Hobday *et al.*,

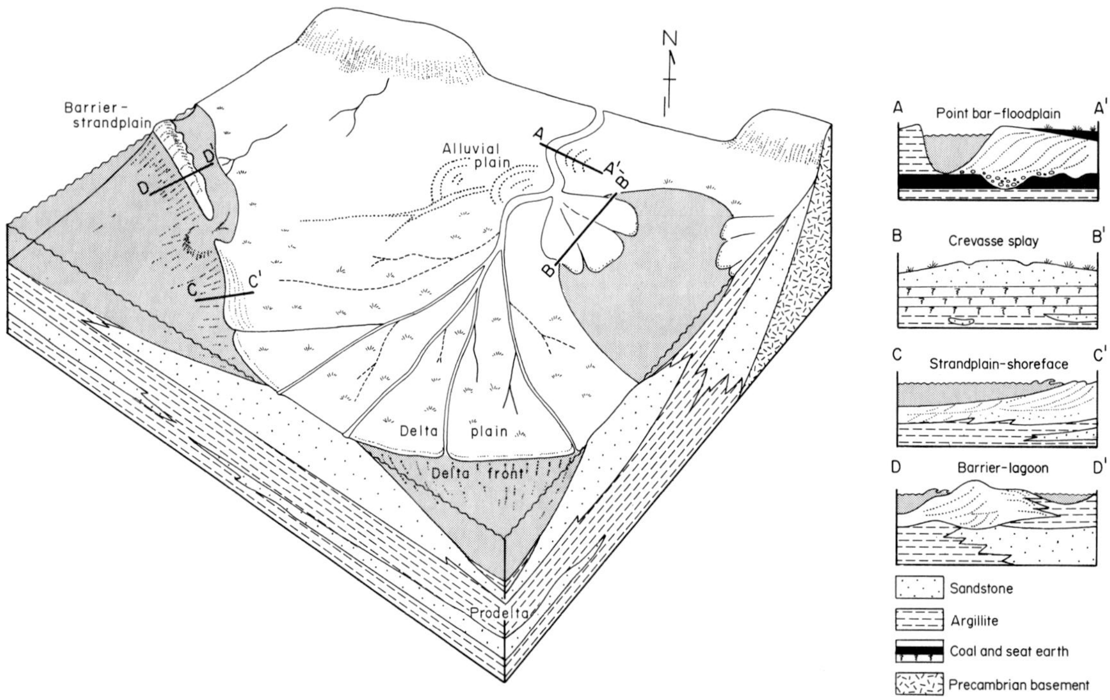

Figure 11-7. Schematic reconstruction of typical Vryheid Formation lobate delta fed by meandering, mixed-load channels between Precambrian residuals. A–A′ shows cross-section through channel-fill deposits; B–B′ is a cross-section through a crevasse splay subdelta that built over rooted levee and bay deposits; C–C′ is a cross-section through strand-plain-shoreface deposits marginal to the delta lobe; D–D′ represents barrier sand along the periphery of an abandoned lobe.

1975). Van Vuuren and Cole (1979) document the same regressive cycles in the subsurface of much of the northern Karoo basin.

Bioturbated or weakly laminated shelf mudstone and siltstone occur at the base of the ideal cycle and are best developed beneath the lowermost regressive deltaic sequence. This shelf-prodelta unit represents slow suspension sedimentation and reworking by sediment-ingesting organisms. Rare horizontal tracks and trails reflect the low physical energy preferences of the biota.

The overlying muddy siltstones and sandstones record the approach of the distal delta lobe. There is an irregular upward increase in thickness of the individual textural layers and in maximum grain size and proportion of sandstone, but a pattern of rhythmic layering on a scale of 2–10 cm is present throughout. The sandstone beds show current ripples of diverse trend, but with an offshore mode reflecting the influence of river currents. Rare oscillatory wave ripples, and the more common combined wave–current forms, generally strike subparallel or oblique to the shoreline trend. Downslope-directed flow rolls and slumps are common, reflecting the instability that is characteristic of the outer subaqueous portions of many modern deltas. Graded sandstone-siltstone units with occasional Bouma sequences signify gravity resedimentation or turbidity flow. Some of these rhythmic intercalations of sand may have been introduced by storms carrying out sand stored in distributary channels and mouth bars. Bedding-plane surfaces commonly show deposit-feeder traces, such as *Rhizocorallium*, together with trails such as *Scolicia* and the large U-burrow *Corophioides* (Hobday and Tavener-Smith, 1975).

Proximal distributary mouth bar deposits of the uppermost part of the deltaic cycle are medium- to very coarse-grained arkoses with pebbly lenses of vein quartz and feldspar. Trough cross-bedding, scour and fill structures, and irregular current rippled surfaces are the dominant structures. Paleocurrent patterns are mainly directed offshore toward the south or southwest but show considerable dispersion. Low-angle clinoforms record the successive profiles of delta progradation (Fig. 11-8).

The lower delta plain was traversed by numerous closely spaced, bifurcating channels 1–4 m deep, a pattern typical of deltas with a sandy sediment contribution that are building over relatively thin prodelta deposits. Distributary channel-fill deposits incise and intercalate with sandstones of the upper mouth bar, from which they are distinguished by their irregularly scoured, pebbly bases and by their internal geometry. Some of the channels are filled by thick, solitary sets of trough cross-bedding produced by downstream migration of major bedforms several meters high. Others contain vertically stacked sets of diminishing thickness, recording gradual reduction of flow, or show irregularly interspersed large-scale cross-bedded sandstone and clayey,

Figure 11-8. Gently inclined surfaces in the Vryheid Formation produced by distributary mouth-bar progradation over strongly burrowed, finer grained distal bar deposits, which in turn overlie prodelta mudstones (not shown). Clay layers caused small-scale diapiric deformation of the mouth bar sands. Zunguin tunnel outcrop near Vryheid, Natal.

rippled sets, indicating fluctuating discharge (Hobday, 1973; Hobday and Mathew, 1975).

Capping levee deposits are typically thin or absent in the lower delta plain but are prominent in the upper delta plain deposits where they comprise rooted siltstone with horizontal lamination and ripple-drift cross-lamination. Overbank deposition was largely restricted to shallow lakes and embayments, where 2–10 m thick, upward-coarsening bay-fill sequences are characteristic. These highly carbonaceous, bioturbated, sideritic mudstones and siltstones contain occasional poorly sorted crevasse splay sandstones, which are normally 10–60 cm thick. Persistent thin sandstone sheets produced by overbank flooding are present at the tops of some delta-plain mudstones and are usually rooted or associated with thin, impure coals which record the early development of delta-plain marshes.

The tops of delta cycles in Natal are riddled with the tubular burrow *Skolithos*. These dwelling burrows are up to 30 cm deep and were probably produced by wormlike organisms during the depositional hiatus following abandonment of the delta lobe. Delta abandonment was a response to the diminishing gradients as distributary channels extended their courses basinward. A modern analog is the Atchafalaya distributary which, without the intervention of the U.S. Army Corps of Engineers, would ultimately have led to total diversion of Mississippi River flow from the site of the modern birdfoot delta (Gould, 1970). The low levees of the Ecca streams would have been readily breached, thereby offering a significant gradient advantage. Where the sediment-laden waters debouched into an adjacent embayment a new delta lobe was initiated. In some cases the new system aborted soon after initiation of the crevasse subdelta. Examples in the Ecca are characterized by tongue-shaped sandstone units that are deeply scoured at their proximal ends and gradationally based toward their outer margins, where they interfinger with carbonaceous siltstones of a bay environment. All intermediate stages are developed between minor crevasse subdeltas less than 4 m thick and 500 m across, and major delta-front sandstones that can be traced over distances of 30 km or more. Maximum water depths, as indicated by the thickness of the deltaic cycles, were less than 100 m (Hobday, 1973, Hobday *et al*., 1975).

Successive sandstone units are commonly arranged in vertically offset or *en echelon* patterns, pointing to differential compaction. Of even greater importance were the effects of basement topography, which influenced depositional patterns at least to the middle of the Vryheid Formation (Van Vuuren and Cole, 1979). Development of coarse-grained multistoried sandstones was closely related to the presence of pre-Karoo valleys (Steyn, 1977).

Delta abandonment and inundation were accompanied by local reworking by waves and concentration of quartz and heavy-mineral sands. In lobate, river-dominated deltas of Ecca type, the destructive components of delta decay are typically restricted to the uppermost parts of the abandoned delta units. Reworked mature, glauconitic sands between 1 m and 5 m thick occur at the top of deltaic cycles in the southeastern Transvaal (Cadle and Hobday, 1977). These possibly represent the products of landward-migrating destructive barrier sands comparable to the Chandeleur barrier island chain (Louisiana, U.S.A.; Fig. 11-9), which developed following delta abandonment. Structureless carbonaceous mudstones that overlie the deltaic sequences farther updip represent the development of shallow bays and marshy lagoons. Deltaic deposits of the more rapidly subsiding areas, such as Tugela Ferry, show little or no abandonment reworking and are covered by thin, highly burrowed, organic-rich sediments that grade upward into shelf and prodelta deposits of the succeeding cycle. However, in the nearby Muden district, Ryan and Whitfield (1979) have documented a highly reworked deltaic sandstone unit. This indicates that local factors of shoreline orientation or reduced sediment supply may have been important. Furthermore, as conditions changed from net regressive to transgressive in the uppermost part of the Vryheid Formation, or as the delta prograded into deeper water, the significance of wave reworking probably increased.

Sandstones of the Vryheid Formation become progressively finer basinward, where they grade into a vertically homogeneous argillaceous prodelta and shelf succession. Classical regressive sequences from prodelta to lower delta plain are best developed in the medial to distal portions of the clastic wedge. Van Vuuren and Cole (1979) have shown that there is a relationship between the number of superimposed cycles and total thicknesses of the Vryheid Formation. Up to eight cycles are recognized in the areas of maximum

Figure 11-9. Wave-reworked, transgressive barrier sands along the margin of an abandoned delta lobe: the Chandeleur Islands of Louisiana, possible analog of the Ecca destructive deltaic sandstones. [Courtesy of H.H. Roberts.]

subsidence in northern Natal, whereas on the more stable platform only two cycles are developed.

As the proportion of sandstone to finer grained rocks increases updip, prodelta and bay facies thin at the expense of channel sandstones of the upper delta and alluvial plain (Hobday, 1978a; Mason and Tavener-Smith, 1978). Coals are typically well developed in this interval. The channels are up to 15 m deep, and in strongly regressive sequences they erode through the distributary mouth bar deposits. Lateral accretion structures, or epsilon cross-beds, indicate that the channels migrated by point-bar accretion. Channel-fill sequences conform to the established fine-grained meanderbelt model; a sporadic basal conglomerate contains both extrabasinal and locally derived clasts, together with occasional coalified tree trunks up to 30 m long aligned with the general southerly paleoslope. Overlying or replacing the basal conglomerate is commonly an interval of massive, coarse-grained arkosic sandstone with scattered pebbles and boulders, probably resulting from bank collapse. The overlying cross-bedded interval shows an upward reduction in set thickness and gradation into the fine-grained topstratum of ripple-drift cross-laminated levee siltstones and dark flood-basin mudstones with abundant *Glossopteris*, other plant debris, and sideritic concretions. Where coal is present it terminates the cycle.

Extensive point-bar accretion units permit reconstruction of the original cross-sectional dimensions of the channel. Individual accretion beds are assumed to have extended across two-thirds of the original channel width (Allen, 1965b), and the thickness of the unit containing the lateral accretion surfaces probably approximates to the channel depth at a meander bend. Accordingly, maximum channel depths of 10 m and widths of 120 m are obtained from outcrops in the eastern Transvaal coalfield (Hobday, 1978a, b). This is within the typical range of mixed-load fluvial channels (Schumm, 1972).

Strand-Plain Deposits

A major interval of strand-plain deposits is exposed along the Vaal River near Bothaville (Fig. 11-1). This southward-tapering sandstone tongue interfingers offshore with dark shale. In an updip direction it overlaps Precambrian lavas and metasediments. The sandstones are fine grained toward the base and distal extremities, coarsening upward in the thicker parts of the succession, where there are lenses and laminae of heavy-mineral placers, including ilmenite, zircon, garnet, rutile, monazite, chromite, staurolite, and tourmaline (Behr, 1965). The heavy minerals are fine grained and very well sorted, but intervening layers of quartz sand vary from medium to coarse grained and are less well sorted. Perhaps this was a consequence of textural inversion by short-lived, high-energy events that caused mixing of the lighter mineral grades.

On a regional scale the heavy minerals are concentrated in parallel northwest-trending belts (Behr, 1965) which are interpreted as strand-plain beach ridges. Bedding in the sandstone is inclined at a low angle toward the southwest and displays intersecting sets of current lineations, together with both normal and inverse textural grading. These attributes are characteristic of beach foreshore deposits. The beach prograded southwestward over deposits of the upper shoreface, which comprises erosively based cosets of offshore-directed trough cross-bedding displaying an upward decrease in set thickness. Deposition is thought to have been dominated by rip current processes that have eroded the upper shoreface and transported sand seaward. Lower shoreface sandstones are finer grained, with occasional coarse lenses and abundant micaceous, silty partings. Structures within the lower shoreface include horizontal lamination, ripple cross-lamination, and small-scale trough cross-beds. Bedding is partially disrupted by burrowing, with a large proportion of vertical tubular forms. Sedimentary processes of the lower shoreface probably involved a combination of wave surge and waning rip current flow (Vos and Hobday, 1977). The storm-dominated Bothaville deposits lack many of the features of beaches fashioned by fair-weather processes, such as landward-dipping ridge and runnel stratification. Offshore transport of sand by storm-related processes is also evidenced in the Durban area (Tavener-Smith, 1980). Sand tongues thinner than 2 m show erosive bases and intensely burrowed tops, attesting to their vigorous introduction and subsequent reestablishment of quiescent inner shelf conditions.

This development of a strand plain system on a delta-dominated shelf was probably caused by its marginal situation with respect to the major locus of delta development to the east and its position near the edge of the basement platform (Steyn, 1977). A lower degree of wave retardation by friction would have resulted in intense reworking of the limited volumes of detritus transported along strike by longshore drift from adjacent deltaic systems. There may have been some local contribution, including the heavy minerals, from the Precambrian rocks of the coastal hinterland.

Barrier and lagoonal deposits are also present near the base of the Vryheid Formation near Durban (Tavener-Smith, 1979a). A vertical succession reflecting storm-dominated shoreface, beach foreshore, rooted backbeach, and coastal marsh, records the development of a northeast-trending coastal barrier system fed by longshore currents from an active delta lobe situated to the southwest. An overlying complex of coastal lake, splay, fan delta, and fluvial channel-fill facies indicates the onset of more vigorous dip-fed sediment flux from the northwest.

Karoo Trough: An Intracratonic Flysch Basin

The east–west Karoo trough (Fig. 11-10) of the southern Karoo basin was established by continued post-Dwyka downwarping (Ryan, 1967). Maximum sediment accumulation of almost 2000 m of Ecca sediments occurred in the eastern Cape. Starved-basin mud accumulation prevailed initially, with density stratification in lakes providing a generally reduced substrate (Visser and Loock, 1978). Carbonate and phosphatic lenses, plant fragments, and sparse reptile remains are present. Rhythmically layered yellow tuffs are a feature of the Collingham Formation in places (Kingsley, 1979). These lacustrine conditions persisted in the western part of the trough during the initial phases of coarse influx in the east. The presence of the small reptile *Mesosaurus* is thought to reflect the shallow conditions of the basal Ecca; this was followed by an episode of rapid deepening.

In the eastern Cape, up to 1000 m of horizon-

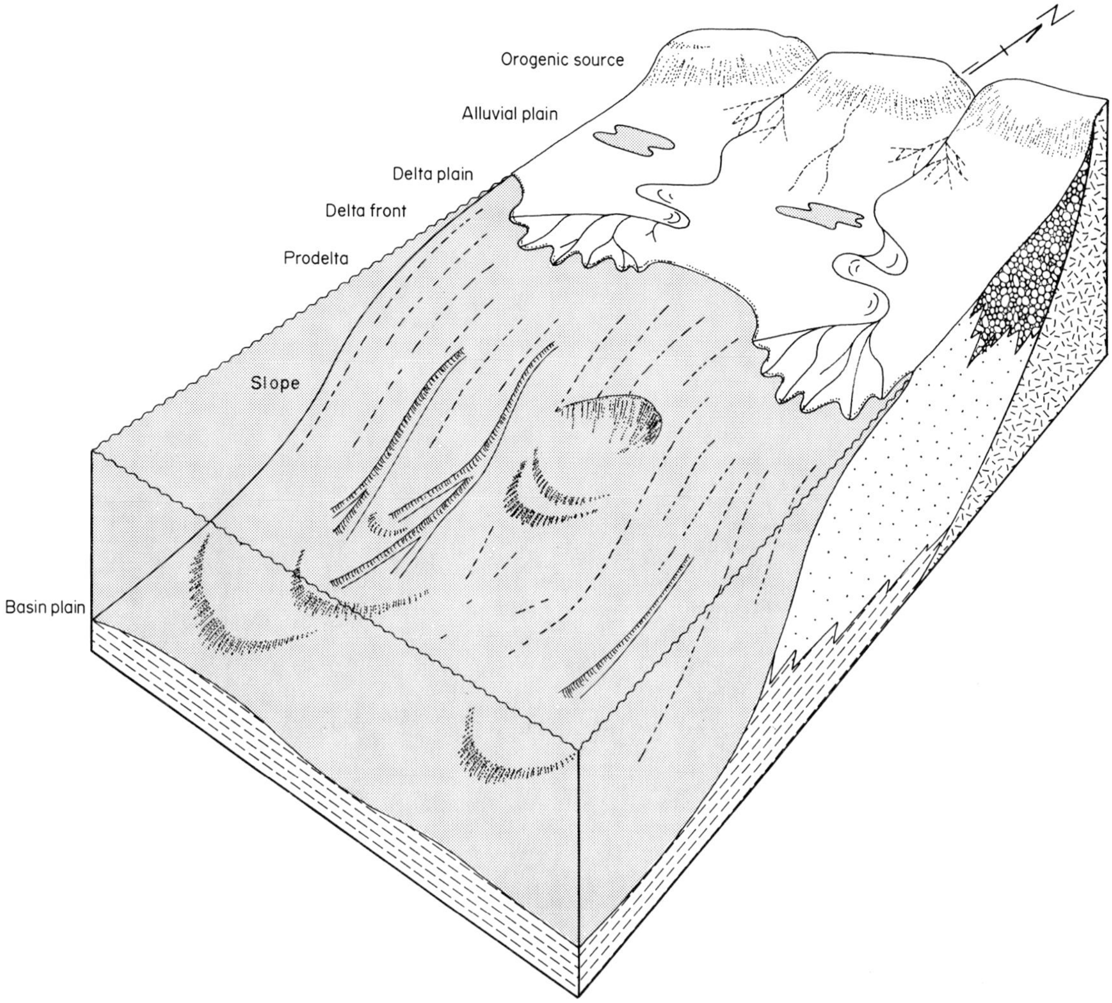

Figure 11-10. Schematic reconstruction of the east–west-trending Karoo trough subject to turbidity current influx and bordered to the south by alluvial and prograding deltaic systems fed by an elevated orogenic source. [Adapted from Kingsley (1979).]

tally interlayered muds and sands of the Ripon Formation accumulated in what has customarily been referred to as a flysch trough (Truswell and Ryan, 1969). At the base are banded gray and yellowish mudstones which coarsen upward by the incorporation of progressively thicker sandstone beds. The sandstone beds, of graywacke composition, are remarkably persistent along depositional strike but pinch out northward. Basal contacts are sharp, with well-developed sole marks. Graded bedding confirms deposition by turbidity currents (Kingsley, 1979). Also present are gravity slumps, load structures, flame structures, and convolutions. Dikes and sills resulted from sand injection of saturated muds during compactional loading (Truswell, 1972). Kingsley records the presence of *Helminthopsis* traces of deep-water type. Equivalent to the Ripon Formation is the thinner turbidite succession of the Laingsburg Formation to the west (Fig. 11-11).

The trough was clearly bounded to the south and southeast by elevated terrane. Uplift may have been related to continued activity in the Cape Fold Belt, which commenced during Witteberg deposition (Devonian), reaching a climax in the Triassic.

Water depth probably decreased with time as the subaqueous fans prograded northward. Sandstones higher in the Ripon Formation, although preserving characteristics of gravity flow, are considerably thicker and show a greater degree of channeling. This slope assemblage was fed by

Figure 11-11. Turbidites of the Laingsburg Formation. [Courtesy of V. Von Brunn.]

high-constructive delta systems approaching from the south. The overlying suite is made up largely of shelf muds with upward-coarsening prodelta and delta-front associations. These indications of shallowing are supported by trace fossils of the *Cruziana* community followed by *Skolithos* at the top (Visser and Loock, 1978; Kingsley, 1979).

The uppermost unit of the Ecca Group, the Waterford Formation, is coarsest in the west, where it has been interpreted as a low sinuosity fluvial channel facies by Turner (1978). Sheetlike sandstone units up to 25 m thick are overlain by thinner, fine-grained deposits. Upper flow regime plane-bedding predominates over cross-bedding. Such sequences commonly originate in "flashy" ephemeral streams subject to short-lived flood discharge. The development of plane-beds was also favored by the fine grain-size since sand waves and dunes do not form readily in sands below 0.1 mm mean size. Thus the plane-beds grade upward into ripple cross-lamination without the intervening phase of larger bedforms. Overbank fines at the top of the sequence show the typical association of desiccation cracks and calcareous concretions. In the east around Grahamstown, this interval is represented by dark mudstones and ripple-marked siltstones with numerous thin sandstones.

Fault-Bounded Ecca Basins

Elongate fault-bounded basins controlled Karoo sedimentation in adjoining states to the northeast. Southward continuation of this rift terrane (Rust, 1975) is manifest in the Lebombo monocline of Swaziland and Zululand, and the Soutpansberg trough of the northern Transvaal.

The narrow, linear outcrop of Ecca sediments that thickens southward along the Lebombo monocline demonstrates the effects of both contemporaneous and post-Karoo faulting. A study by Whateley (1979) shows that the Nongoma graben, a southward extension of the rift system, developed parallel to basement structures in response to crustal thinning. Rapid influx of arkosic sediments accompanied vertical displacement along the bounding faults. Allocyclic mechanisms are thus thought to have been responsible for the stacking of at least 15 cycles of delta progradation through a stratigraphic thickness of 750 m. Whateley records abundant prodelta turbidites, slumps, and other soft-sediment deformation characteristic of the unstable tectonic setting. Stable interludes of delta-plain sedimentation are represented by mixed-load channel fill, finer flood-basin facies, and laterally persistent coal seams.

The Natal trough to the south has a similar southwesterly trend to the Nongoma graben and may represent a further extension of this belt of incipient rifting. Like the Nongoma graben, the Natal trough shows convergent patterns of fluvial infilling (Christie and Tavener-Smith, 1979), with the basin deepening toward the south. This pattern confirms Ryan's (1967) concept of a highland sediment source to the east. Furthermore, it indicates westward migration of the negative belt relative to its Early Paleozoic position as the Natal embayment.

Other fault-bounded Ecca deposits have characteristics similar to the Nongoma graben fill, but in some areas they are substantially finer grained. Seemingly monotonous sandy siltstones show very slight upward-coarsening tendencies and represent low-energy infilling of 5-to 20-m-deep bays which may have occupied a position lateral to active delta lobes. Contacts between successive marginal bay-fill sequences are marked by thin root-disrupted intervals.

The intracratonic rift zone of the Soutpansberg trough (Fig. 11-1) was reactivated during Karoo times when sediments and lavas accumulated in block-faulted basins that connected with the Lebombo monocline in the east. In addition to predominantly vertical displacements, transcurrent faulting in the Soutpansberg trough may have accompanied initial breakup between Madagascar and Africa (Flores, 1970). Sediments of the Ecca Group include conglomerates, immature sandstones, mudstones, and coals. The depositional environment possibly included intermontane lakes, peat swamps, and humid alluvial fans.

This Permian fault basin tectonism is consistent with paleomagnetic evidence of relative movement among the constituent plates of Gondwana (Anderson and Schwyzer, 1977), transmitting stresses to the intraplate rift zones.

Karoo Coal

Nineteen Karoo coalfields, of which eighteen are in the Ecca Group and one in the Molteno Formation, contain some 81 billion tons of recoverable bituminous coal (D.A.M. Smith, 1978). The Ecca coalfields are distributed through five major coal basins, the Limpopo, Soutpansberg–Lebombo, Waterberg, Springbok Flats, and northern Karoo. The distribution of coal within these basins is strongly facies controlled, being largely restricted to paraglacial, upper delta-plain, strand-plain, and alluvial deposits (Hobday, 1978a,b; Cadle, 1979, Tavener-Smith 1979b). Rate of tectonic subsidence was another important factor. Tavener-Smith (1979b) notes that the stable platform area northwest of the Natal trough is characterized by economically important seams. In contrast, the less stable Natal trough is marked by numerous impersistent, thin seams in a considerably expanded succession of superimposed delta-front deposits capped by aggraded delta-plain facies up to 300 m thick. The coal-bearing succession at a particular locality is seldom thicker than 150 m and contains as many as five mineable seams, together with several subeconomic seams. LeBlanc-Smith (1979) regarded the four major coal-capped depositional sequences of the Witbank basin as representing major Gondwana coal cyclothems.

Distribution of the lower seams was dependent upon the postglacial paleotopography, peat development having been largely confined to topographic lowlands. The Witbank basin deposits accumulated north of an east-trending basement ridge. The uppermost seams are exposed in hillsides, whereas in northern Lesotho the deepest seams are 2500 m below the surface.

Coal seam characteristics are related to the depositional environment and to subsequent changes in rank caused by burial and diabase intrusion. The basal coals of the northern Karoo basin, such as the No. 1 and No. 2 seams of the Witbank basin, originated in a tundra-type peat bog environment dominated by mossy vegetation (Plumstead, 1969, p. 35; LeBlanc-Smith and Eriksson, 1979). Cairncross (1979) remarks that some organic debris was probably supplied directly from vegetated glaciers, a common feature of the present day. The No. 1 seam rests on a variety of footwall types, including tillite; in some areas, it overlies glacial outwash deposits. Debris flows were common in this unstable environment and locally interrupted peat accumulation. These coals of freshwater marsh origin have a characteristically low sulfur content.

Stream drainage became progressively more confined and assumed an anastomosing pattern. Sands accreted vertically in the channels, which were flanked by a resilient mat of vegetation subject to occasional overbank flooding and clay deposition. Thus, the No. 2 seam in some areas is split by elongate sands and more tabular silts and clays. Cairncross (1979) has shown that high ash values caused by inorganic contamination are associated with coals above and below the thicker developments of the shale parting; the lowest ash values occur beneath the channel-fill sands and where the shale parting is less than 1 m thick. High ash values in coals above the channel-fill sandstones are attributed to partial oxidation and increase in inherent ash because of the higher initial elevation of the peat. Careful mine planning is necessary where features such as anastomosing channel-fill or crevasse splay sands and bioturbated bay-fill shales may provide hazardous hanging-wall conditions (Cairncross, 1979).

An increased sulfur content in the higher No. 3, 4, and 5 seams of the Van Dijks Drift area (Cairncross, 1979) is consistent with the postulated marine influence. These coals, together with seams in Natal and the Orange Free State, originated in fluviodeltaic settings that were locally subject to marine inundation accompanying delta switching. Cadle (1979) recognized coals of lower delta-plain, upper delta-plain, and strandplain origin. Coals of the lower delta-plain environment are thin and have a low ash content, whereas thicker coals with a higher ash content are typical of the upper delta plain. Laterally persistent thin seams originated in strand-plain marshes. Cadle found that these strand-plain coals have the lowest ash content.

Beukes (1978b) noted that in contrast to the Witbank coalfield, which contains a large proportion of dull coal (durain) high in ash, lacustrine coals of the fault-bounded Limpopo, Soutpansberg–Lebombo, and Springbok Flats basins have a high vitrain content. This may reflect originally low Eh conditions.

The effects of diabase (dolerite) intrusion have been to introduce structural complexity by displacing coal seams and to bring about changes in the coal by temperatures of around 1000°C. Much coal was destroyed by burning, but in many areas the diabase had a beneficial effect, converting bituminous coal into metallurgical coking coal and anthracite. In addition to these thermally metamorphosed coals, the better bituminous coals from the Witbank coalfields and the straight coking coals of the Pafuri coalfield (Soutpansberg trough) are also considered high rank. The Vereeniging/Sasolburg coalfield (northern Karoo basin) produces the lowest rank coal, with 6–7 percent moisture content (D.A.M. Smith, 1978).

Coal is assuming increasing significance in subsidiary basins north of the main Karoo depository. Substantial seams are developed in the Waterberg basin and Soutpansberg trough. In the Wankie coalfield of Zimbabwe, a basal conglomeratic interval is overlain by a 10–90 m succession of dark mudstone and coal (Du Toit, 1954, p. 505). There is great potential for the discovery of additional coal resources in Zimbabwe.

Hydrocarbon Exploration in the Ecca Group

Although exploration by the Southern Oil Exploration Corporation (SOEKOR) between 1965 and 1977 did not produce any commercial hydrocarbon shows, it did reveal a close relationship between oil and gas occurrence and factors of sandstone geometry, regional and local structure, diagenetic trends, and diabase distribution.

Regional studies by Rowsell and De Swardt (1976) established a general southward increase in diagenesis. They indicate that although the Dwyka and Ecca shales of the southern Karoo basin initially had good source rock potential, high temperature associated with deep burial exceeded the oil-generating threshold where, at best, only dry gas could be generated. Shales in the central basin had a low organic content and there are no sandstone reservoirs. A lower level of diagenesis, together with the presence of sandstone reservoirs in the northern Karoo basin, accounts for the numerous small oil shows. Hydrocarbons were generated from kerogens of terrestrial and aquatic origin within shale source rocks in close proximity to sandstone reservoirs (Rowsell and Connan, 1979).

As summarized by Steyn (1977) and Rowsell and Connan (1979), oil and gas shows of the northern Karoo basin are concentrated above the southern and eastern margin of the Precambrian basement platform. Coarse sandstones located over paleovalleys and topographic highs constituted potential traps. Organic-rich shelf and prodelta clays were a good source for hydrocarbons which migrated updip into lower pressure sand, accumulating in traps along the platform edge and on the platform itself. Vryheid Formation shales in the north-central Karoo basin are within the oil-generation phase. Some oil was destroyed by the heat of diabase intrusion, the effects extending over a distance of approximately twice the thickness of the sill. Migration was probably complete prior to Jurassic volcanic intrusion, although some oil remigrated after intrusion.

Generally, however, the sandstones have low permeability values. This is a result of the initial poor sorting and further porosity reduction by diagenetic processes, particularly in the finer grained sandstones. As a result of this, as well as low oil-saturation values in some cases, it has not been possible to exploit the known hydrocarbon occurrences.

Some of the coarser grained fluvial sandstones in the north show late-stage secondary porosity (Rowsell and De Swardt, 1974), which may have developed after the main phase of hydrocarbon migration. This secondary porosity resulted from solution of certain mineral components, such as carbonates, during burial to an estimated depth of 3000 m, and would have led to a reduction in oil saturation. Rowsell and De Swardt suggested that leaching was accomplished either by the release of carbon dioxide as a result of coal diagenesis, or by the liberation of water during the conversion of montmorillonite to illite at depth. An intermediate stage of clay transformation is indeed represented in the mixed-layer illite–montmorillonite in shales of the northern Karoo basin. At the same time this released water could have provided an additional mechanism for oil migration. The organic geochemistry of the hydrocarbons indicates that in most cases the oil was not biodegraded or subsequently flushed by groundwater flow (Rowsell and Connan, 1979); low levels of oil saturation may be in part a result of poorly developed secondary migration.

11.3. The Beaufort Group: Fluvial Aggradation in a Foreland Basin

The Permian–Triassic Beaufort Group is exposed over an area of some 200,000 km^2 and attains a maximum thickness of 3000 m in the eastern Cape at the convergence of two linear downwarps trending east and north-northeast respectively. The east-trending Karoo trough depoaxis represents a northward shift from Ecca deposition. Fine-grained, varicolored sediments are characteristic and contain a distinctive suite of vertebrate remains that record the progression toward an increasingly terrestrial habitat. A succession of coarse regressive wedges developed in response to orogenic pulses in the south, tapering northward into the shrinking Ecca sea. Subordinate quantities of coarse sediment were supplied from locally elevated Precambrian terrane to the north and east. The boundary with the underlying Ecca Group is arbitrary and markedly diachronous and is most logically taken at the base of the lowermost regressive sandstone unit, a contact that varies with position in the basin. The presence of abundant reptilian remains and massive red and purple mudstone in the Beaufort has also been used to distinguish it from the underlying Ecca Group, particularly in the area west of longitude 25°E.

Renewed downwarping along an east–west trend may have been an isostatic response to the rising Cape Fold Belt, providing a classical foreland basin situation for the development of thick orogenic molasse-type wedges. These coarse-grained wedges are best developed in the eastern Cape and Transkei, but more detailed investigations have been conducted in connection with uranium exploration farther west, particularly around Beaufort West (Fig. 11-1). Episodes of maximum terrigenous influx were out of phase in different parts of the Karoo trough so that correspondence between stratigraphic subdivisions in the eastern Cape and the western part of the Karoo basin is not always apparent. However, Visser and Dukas (1979) have recently suggested correlation of thick tectonically related successions across the basin. Despite evidence of local marine conditions (Stapleton, 1974; Horowitz, 1976) and the presence of *Cruziana* produced by an enigmatic trilobite-like organism (Shone, 1978a), the prevailing opinion is that fluvial processes dominated Beaufort deposition, with progressive withdrawal of the shrinking lacustrine depository toward the east.

The importance of reptilian fossils in the Beaufort Group was first recognized by Bain in 1854. These fossils represent one of the best preserved ecological assemblages of premammalian terrestrial vertebrates in the world (Keyser and Smith, 1978) and afford a fascinating record of the evolutionary transition from reptile to mammal. Earlier stratigraphic breakdown into what became known as the Lower, Middle, and Upper "Stages" was based on a combination of lithologic and fossil vertebrate criteria (Rogers, 1905; Du Toit, 1954; Haughton, 1969). The current trend is toward the adoption of independent lithostratigraphic and biostratigraphic divisions, but with increasingly precise delineation of their spatial interrelationships.

The Adelaide Subgroup comprises predominantly greenish-gray and locally reddish mudstones, with a higher sand content and predominantly red mudstones in the overlying Tarkastad Subgroup. The boundary between these two subgroups is taken at the base of the widespread Katberg sandstone. More detailed subdivision of the succession in the eastern Cape by Johnson (1976) is shown in Table 11-1. A separate subdivision is used in the western Karoo basin. There is tentative correlation of the Koonap Formation with the Abrahamskraal Formation (Keyser and Smith, 1978). Visser and Dukas (1979) advocate extrapolation of the broader eastern Cape nomenclature to the area around Graaff-Reinet and to the west thereof.

As early as 1892 Seely recognized three biozones within the Beaufort Group. These were expanded by Broom (1909) into a sixfold zonation that was employed with minor modification (Watson, 1914) until 1970. Kitching (1970) recognized the Tapinocephalus, Cistecephalus, Daptocephalus, Lystrosaurus, and Cynognathus Zones, which he mapped basinwide. More recently, studies associated with uranium exploration in the western Karoo basin have led to local biostratigraphic revision by Keyser and Smith (1978). This new classification, based on a collection of 2500 fossils, mostly reptilian skulls, also takes cognizance of associated lithologic features and goes a long way toward providing a correlation of the eastern and western parts of the southern Beaufort outcrop belt. The biostratigraphic outline below draws mainly on the work of Keyser and Smith (1978) and Kitching (1970, 1977) as compiled by the South African Committee for Stratigraphy (1980).

The sparsely fossiliferous Dinocephalian Zone, equivalent to all but the upper part of the Tapinocephalus Zone and Abrahamskraal Formation (Table 11-1), is developed in the western part of the basin, pinching out eastward into the unfossiliferous Koonap Formation. The Dinocephalia were large herbivorous and carnivorous therapsids. The overlying Pristerognathus–Diictodon Zone, containing large numbers of small dicynodonts plus the carnivorous pristerognathids, is equivalent to the upper 300 m of the Tapinocephalus Zone. Its lower boundary coincides with a persistent green siliceous volcaniclastic ("chert") band. The Pristerognathus–Diictodon Zone corresponds with the Paalhuis Member at the top of the Abrahamskraal Formation, which contains the major uranium prospects.

The Tropidostoma and Cistecephalus Zones (Keyser and Smith, 1978) together make up the Cistecephalus Zone of Kitching (1970); these zones, now referred to the Tropidostoma–Endothiodon and Aulacephalodon–Cistecephalus Zones (South African Committee for Stratigraphy, 1980), are developed within the lower half of the Teekloof Formation in the west and the Middleton and lower Balfour Formations in the east. The Tropidostoma Zone is 240 m thick around Beaufort West. Large endothiodonts, dicynodonts, and gorgonopsians are present in addition to *Tropidostoma microtrema*. The Cistecephalus Zone contains the lowermost occurrences of the large herbivores *Oudenodon* and *Aulacephalodon* along with numerous dicynodonts and the zone fossil *Cistecephalus microrhinus*.

The Daptocephalus or Dicynodon lacerticeps–Whaitsia Zone is distinguished by the dicynodont *Daptocephalus leoniceps*. It is the lowermost of a series of concentric biostratigraphic zones encircling the Lesotho highlands (Kitching, 1970, 1977) and as such is represented in Transkei, Natal, and the eastern Orange Free State. A maximum thickness for the Daptocephalus Zone is 420 m near Graaff-Reinet.

The overlying Lystrosaurus–Thrinaxodon

Zone is highly fossiliferous and coincides approximately with the arenaceous Katberg Formation, a conspicuous unit in the eastern Cape, Transkei, and Natal that tapers northward from a maximum thickness of over 1000 m. The uppermost biostratigraphic subdivision of the Beaufort Group, the Cynognathus Zone, has been renamed the Kannemeyeria–Diademodon Zone (Table 11-1). Fossil fish are particularly numerous in this zone (Jubb and Gardiner, 1975). The Kannemeyeria–Diademodon Zone coincides with the upper two-thirds of the Burgersdorp Formation.

Although biostratigraphic units are conceptually quite different from chronostratigraphic units that are based on the inferred geologic time span represented, on a broad scale the Beaufort biostratigraphic zones may correspond to some degree with chronostratigraphic divisions. There is clearly a progression of Karoo vertebrates through the time of Beaufort deposition. Furthermore in some areas there is marked discordance between lithologic units and fossil zones; lithostratigraphic units generally ascend northward with respect to the biostratigraphic boundaries (Turner, 1978). On a shorter time span it is not possible to separate the effects of local environmental control. Many of the Karoo reptiles were specialized in their adaptation to a particular habitat, and these habitats shifted with time.

Sedimentation in the Western Karoo Trough

The episode of coarse influx represented by the Waterford Formation at the top of the Ecca Group was followed by a gradual reduction in fluvial energy (Fig. 11-12). Turner (1978) recognized a vertical succession from low-sinuosity channel facies (Waterford Formation) through high-sinuosity channel facies, to floodplain facies at the top. The high-sinuosity and channel facies correspond approximately to the Abrahamskraal and Teekloof Formations respectively.

Lowermost Beaufort deposits of the Abrahamskraal Formation in the extreme west near Sutherland comprise sandy bedload and mixed-load fluvial facies derived largely from the southwest. In the Beaufort West area, mudstone and siltstone predominate over sandstone. Channel and floodbasin components are recognized, with the latter making up approximately 80 percent of the succession (R.M.H. Smith, 1978). The sandstones are arkosic, with as much as 35 percent feldspar. Apart from quartz and orthoclase, the detrital matrix contains abundant volcanic rock fragments. Channel sandstones are 5–25 m thick and show the typical upward-fining motif and lateral accretion structures of high-sinuosity channels. Unequivocal evidence of a meandering origin is provided by exhumed channel sandstones showing arcuate ridge and swale topography (Kübler, 1977). Another remarkable aerial photograph (Fig. 11-13) of a sandstone paleosurface near Beaufort West illustrates a curved accretionary point bar, channel thalweg, and adjoining crevasse splay (Turner, 1979a).

The erosively based channel-fill sandstones incorporate a lag of mudstone intraclasts, reworked nodules, bone, and plant debris. Plane-beds and cross-beds dominate the lower point-bar deposits, with abundant internal scour discontinuities accentuated by silty drapes or a layer of mud-clasts. Sole marks, etch marks produced by rapidly falling water level, penecontemporaneous deformation, and sandstone dikes, occur locally (Turner, 1978, 1979a). Sandstones of the upper point bar are finer grained and silty, with horizontal lamination, ripples, and ripple-drift cross-lamination indicating rapid deposition from suspension where current flow was checked along the channel margin. Cross-bed azimuths are mainly toward the northeast and east; a northeasterly direction of elongation is discernible in many of the channel-fill units. Mud-filled abandoned channels are common.

Unique insight into the volumetrically dominant deposits of levee and interchannel subenvironments has been provided by R.M.H. Smith's studies (1978, 1979) of the spectacular three-dimensional paleosurfaces near Beaufort West. Smith distinguishes levee, crevasse splay, proximal flood basin, and distal flood basin subenvironments, each with a characteristic suite of vertebrate remains. Levee deposits are distinguished by interlaminated siltstone and mudstone, often as graded couplets, with desiccation cracks, raindrop impressions, leaf impressions, and pedogenic calcareous nodules. This pedogenic carbonate, possible syngenetic calichification (calcrete development; McPherson and Germs, 1979), gypsum pseudomorphs, and sedimentological evidence of flashy, seasonal discharge, confirm the semiarid or arid conditions postulated by Keyser (1966). Vegetation was largely confined to belts along the main channel courses and

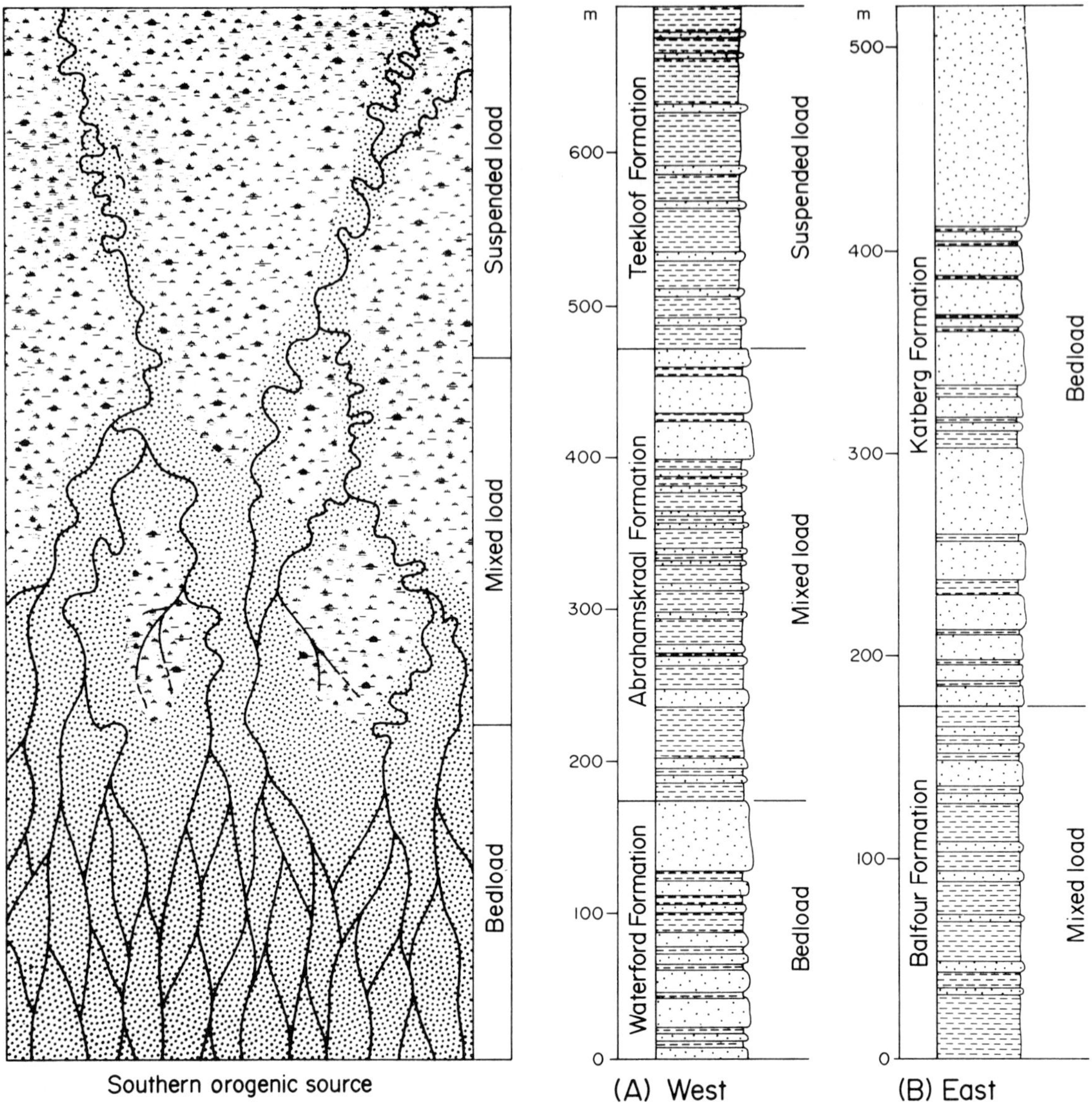

Figure 11-12. Comparative Beaufort successions of alluvial aggradation resulting from differential migration of alluvial facies tracts. (A) Vertical section through the southwestern Karoo reflecting progressive upward reduction in proportion of bedload to suspended load in response to decreasing fluvial energy. [Adapted from Turner, (1979a).] (B) The southeastern Karoo area, showing upward increase in bedload component reflecting basinward migration of high-energy proximal facies tracts.

provided a habitat for a diverse population of terrestrial, aquatic, and semiaquatic fauna (Keyser, 1970). R.M.H. Smith (1978) has incorporated taphonomic studies of the transported remains of these reptiles, amphibians, and fish in reconstructing the paleoenvironment. Concentrations of skulls and mandibles in the levee deposits represent a lag from which the other skeletal elements were winnowed by overbank flooding. The skulls are preserved in a position of maximum hydrodynamic stability, in an orientation that supports the interpretation of paleocurrents flowing laterally away from the trunk channels.

Crevasse splay sandstones show a scoured contact with underlying levee deposits, extending over adjacent floodbasin deposits as thin, uniform sheets. The crevasse sandstones are generally structureless internally, possibly attesting to their catastrophic introduction, but straight-crested and lunate bars with associated cross-bedding are

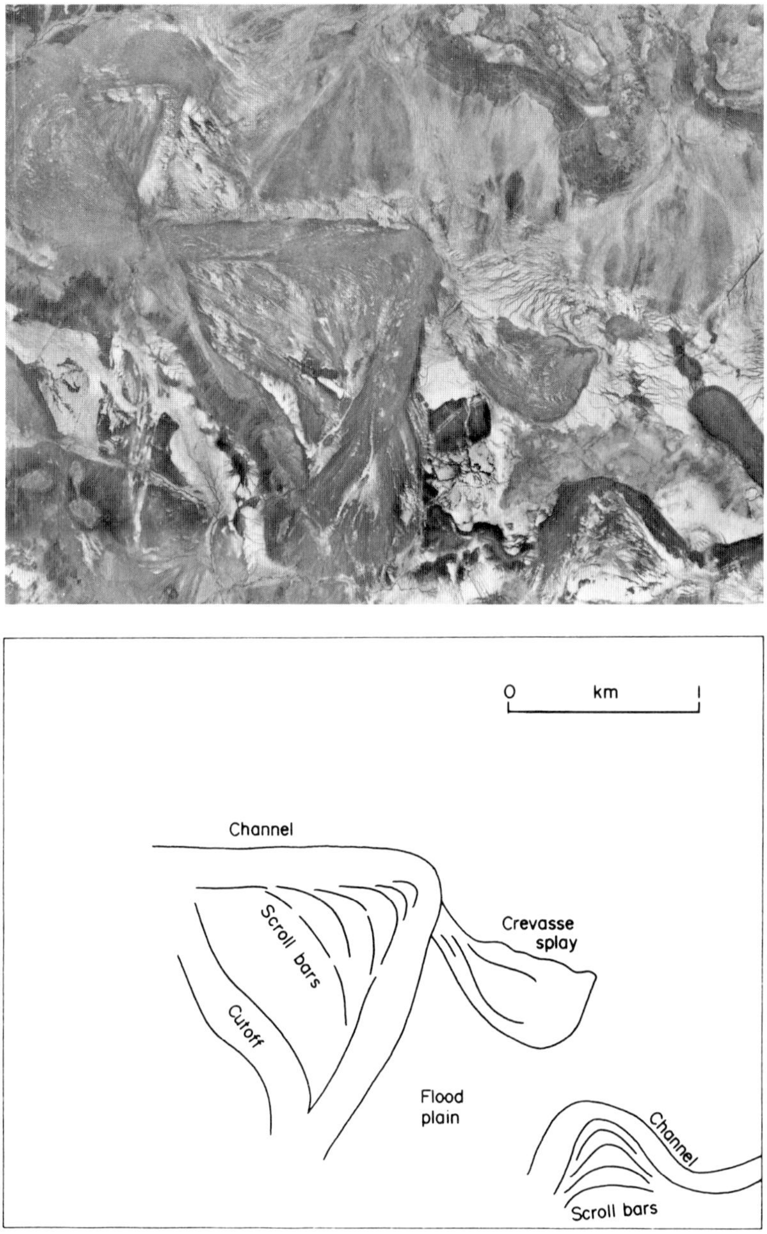

Figure 11-13. Exhumed Beaufort paleosurface showing sinuous channel, point bar with accretionary scroll bars, floodplain, and crevasse splay. [After Turner (1979a); reproduced under copyright authority 6636 of 18 July 1980 of the Government Printer, Republic of South Africa.]

developed locally. Some splay sheets thinner than 2 m persist laterally over distances of 10 km. Rapid incursion of sediment-laden floodwaters was followed by gradual recession of the ephemeral ponds and lakes. This slow drainage, coupled with the effects of wind stress on shallow water, left an extremely complex record of small bedforms (see Stear, 1978, his figs. 6–16), with some biological and deformational modification. Structures indicative of shallowing include terraces, flat-topped ripples, double-crested ripples, and complex interference ripple patterns such as ladderbacks. Strong wind action on a moist sand surface produced runzel marks and adhesion ripples. This combination of features is characteristic of tidal-flat environments, which are regarded as unlikely in the case of the Beaufort. Although marine conditions have been proposed by Stapleton (1974), among others, the remnant lake system to the east is likely to have been almost entirely landlocked and was too small for significant astronomical tides. Strong and variable local winds would have been capable of producing "wind tides" in shallow lakes; these effects may have been accentuated by irregular configuration and interconnection among semi-isolated water bodies. Similar wind-tide effects were interpreted in the Beaufort Group in Natal (Van Dijk *et al.*, 1978).

The upper surfaces of the crevasse splay sandstones display occasional algal mats, casts of gypsum needles and possible ice crystals, raindrop impressions, and chalcedony and calcite pseudomorphs after gypsum (Stear, 1978, 1979), a combination of features further suggestive of shallowing, emergence, and desiccation.

Interfluvial flood basin deposits of the Beaufort are divided by R.M.H. Smith (1978) into proximal and distal components on the basis of taphonomic attributes and distribution of nodules. Proximal flood basin deposits comprise varicolored mudstones and siltstones roughly equivalent in thickness to the main channel sandstones. Calcareous nodules are widespread but dispersed. R.M.H. Smith's studies indicate that bands of skulls extend over distances of kilometers. He reasons that carnivores such as pristerognathids and gorgonopsians were largely responsible for separation of the skull, which resisted transportation. In addition, lenticular beds of postcranial bones display considerable abrasion and desiccation, evidence of intermittent high-energy transport, and subaerial exposure. Associated nonmarine bivalves show imbrication of the disarticulated right valves. Other features of the proximal flood basin include small diapiric structures, slumps, faults and folds, sand-filled desiccation cracks, arthropod tracks and worm trails, large reptile burrows, plant impressions, and gypsum rosettes.

The distal flood basin is represented by gray to purple mudstone with subordinate siltstone. Calcareous nodules and fossils are rare. The warm, dry climate would have precluded the growth of dense vegetation in this backswamp environment and would have reduced the preservation potential of plant and animal material. Some of the more permanent ponds supported fish and semiaquatic herbivores (R.M.H. Smith, 1978).

Following brief episodes of flooding, these interfluvial lowlands were converted into playa lakes (Stear, 1979), producing diagenetic reddening, calichification, and precipitation of evaporites. Evaporites, however, are relatively poorly developed for a playa lake environment. This is possibly explained by a low initial content of dissolved salts or by subsequent dissolution or replacement of evaporite layers. McPherson and Germs (1979) propose several sources of calcium in Beaufort caliche, including concentration by plants from streams draining crystalline terrane, from *in situ* weathering, and from loessic dust. Caliche nodules show evidence of displacive growth.

Flood-basin silt and mud almost entirely dominated accumulation of the overlying 1000-m-thick Teekloof Formation (Fig. 11-12A). Sandstones are generally a minor component, but the persistence of a few trunk channel systems is evidenced by thicker local sandstone trends such as the Oukloof Member. Flood-stage deposition of muds over wide areas was followed by prolonged subaerial exposure under arid conditions. Vestiges of plant rootlets are recognized, but most organic material was destroyed by oxidation. These vast flood basins were probably comparable in scale to the modern Hwang-Ho Basin of China (Turner, 1979a), but the climatic setting may have borne closer resemblance to the head of the Gulf of California.

Uranium

Uranium investigations commenced in the southern Karoo basin in 1969, and since then a number of occurrences have been documented in the area of Beaufort exposure west of longitude 25°E. Initial exploration involved aerial radiometric surveys, geochemical sampling, geologic mapping, and facies analysis. Current methods mainly involve borehole sampling and gamma-ray logging.

Mineralization is strongly facies related, being confined mainly to the high-sinuosity channel sandstones of the Abrahamskraal Formation (Kübler, 1977; Turner, 1978), particularly in the vicinity of the Paalhuis Member or Pristerognathus–Diictodon Zone. Radioactivity is most concentrated in elongate or tabular pods rich in plant debris, particularly along the base and lateral margins of channels. The principal ore minerals are uraninite (UO_2) and complex silicates of coffinite type. Pyrite, arsenopyrite, molybdenum, and phosphorus are commonly associated with the uranium. Mineralized sandstones tend to be calcareous and contain a higher proportion of matrix (Anderson and Fraenkel, 1979). In addition to the recognizable uranium minerals there are urano-organic compounds. Surface radioactivity is commonly manifest in "koffieklip" (coffee stone), a dark-weathering product rich in calcite and iron and manganese oxides (Eddington and Harrison, 1979).

The source of uranium may have been intercalated volcaniclastics or basement granites in the southern highland source area. Greenish siliceous "chert" bands, which in some areas are particularly well developed in the immediate vicinity of uraniferous deposits, have been interpreted as altered volcanic tuffs (Martini, 1974). The tuffs are recrystallized to the extent that glass shards typical of volcanic ash are seldom recognizable. An alternative explanation for some of the "chert" bands is that they are altered evaporites (Rossouw, in R.M.H. Smith, 1978). Petrographic studies by Ho-Tun (1979) of calcareous sandstones associated with the uranium deposits have verified a significant sand-sized volcaniclastic component; this comprises shards with typically pointed ends and curved fractures, particles of microporphyritic lava of alkali trachyte composition, volcanic quartz, and tuffaceous ejectamenta. The zeolite laumontite occurs widely in the Beaufort Group (Fuller, 1970; Elliot and Watts, 1974) where it may represent slightly metamorphosed volcanic material. The anomalously high plagioclase content of the sandstones has also been cited as evidence of contemporaneous volcanism (Toens and Le Roux, 1978), which was a feature of formerly adjacent areas of Gondwana, such as Patagonia and western Antarctica. Toens and Le Roux (1978) favor a granitic and pegmatitic source, citing the great volume and close proximity of these rocks to the Beaufort uranium deposits. Indeed, the orthoclase content of the sandstones is commonly greater than 30 percent.

Whatever the uranium source, a favorable position within the groundwater flow system (i.e., an area of groundwater recharge) was a prerequisite. Furthermore, the very processes that liberate uranium in source rocks are commonly responsible for altering them beyond recognition. Leaching of the uranium from the parent rock was accomplished through weathering, pedogenesis, and freshwater flushing. Mobilized uranium entered the groundwater flow system only within updip areas of groundwater recharge. The dry climate favored maximum recharge through a thick, aerated phreatic zone (cf. Galloway *et al.*, 1979). The ground water was mildly reducing, carrying the uranium as uranyl carbonate complexes. This process differs from the Colorado or roll-front model, where oxidizing ground water leaves a distinct pattern of alteration updip from the ore front. The presence of zeolites and patterns of quartz overgrowth are cited in support of an alkaline groundwater chemistry (Turner, 1978).

The geometry of the isolated framework sand bodies provided only a moderately good conduit for groundwater flow, comparing unfavorably with humid alluvial-fan and bedload fluvial systems in this respect. The dip inherited from the initial paleoslope was probably enhanced by basin subsidence, improving the groundwater flow gradient. Freshwater tongues extended downdip along paths of maximum aquifer transmissivity provided by major fluvial sand trends. The coarse-grained basal portions of the channel sands were most permeable and consequently show the highest degree of mineralization. Two local controls over mineralization were the presence of intrinsic reductants, mainly concentrations of carbonaceous debris and pyrite, and permeability barriers provided by adjacent or interlayered argillaceous units. As a result the uranium occurs in isolated, stacked, or vertically offset lenses that are elongated with the paleochannel.

The primary uranium minerals, uraninite and coffinite, were probably precipitated by hydrogen sulfide generated by bacterial alteration of carbonaceous debris (Turner, 1978). The geochemical gradients would also have favored the precipitation of carbonate cement which sealed the uranium minerals from further alteration.

Although most of the ore bodies are peneconcordant, some cut discordantly across lithologic boundaries or extend along fracture zones. These probably represent modification and local remobilization of the primary deposits. Secondary minerals include carnotite, metazeunerite, metatorbernite, and metaautonite. Toens and Le Roux (1978) suggested that the ore solutions responsible for these modifications were of hydrothermal origin. Invasion by hot, chemically reactive waters from deeper levels may have been responsible for the precipitation of base metal sulfides associated with the uranium.

Toens and Le Roux (1978) pointed out that

northward from the main Beaufort West–Sutherland zone in the southern Karoo, uranium deposits ascend progressively higher in the succession until they occur in the overlying Molteno Formation in the northeastern Orange Free State.

Sedimentation in the Central Karoo Trough

A recent account of the Beaufort geology of the central part of the Karoo trough (Graaff-Reinet to De Aar) based partly on subsurface data, is provided by Visser and Loock (1979). They recognized three lithologic units within the Middleton Formation: a basal and an upper arenaceous zone separated by an argillaceous zone. The arenaceous units are interpreted as high-sinuosity stream deposits, and the argillaceous unit as floodplain deposits. These fluvial facies intertongue northward with deltaic deposits of the Ecca Group, dramatically illustrating the time-transgressive nature of the group boundary.

The Balfour Formation is represented by a basal unit of braided, bedload stream origin, overlain by, and intertonguing distally with, high-sinuosity stream deposits. A thick upper unit of flood-basin mudstones, siltstones, and sandstone lenses contains numerous carbonate concretions. This succession is interrupted in the south by a tongue of bedload-dominated stream deposits. Only 200 m of the Katberg sandstone is preserved in this area. Upward-fining fluvial cycles consist of basal clay clasts and bone fragments overlain by massive and plane-bedded arkosic sandstone followed by finer grained ripple cross-laminated sandstone.

The thick megacycles of the Middleton and Balfour Formations, comprising coarse bedload stream deposits at the base followed by finer grained fluvial deposits reflecting declining energy, are attributed by Visser and Loock (1979) to brief tectonic pulses. There were several distinct entry points into the basin, accounting for the spatially offset clastic lobes. The paleoslope was predominantly northeastward.

Eastern Karoo Trough

In the eastern Cape and Transkei, the Adelaide Subgroup is represented mainly by shallow lacustrine facies comprising horizontally laminated mudstones, wave- and current-rippled siltstones, and thin, discontinuous sandstone units. Reduced grayish colorations are more common than the reds that characterize equivalent strata to the west, but oxidized units are locally in evidence toward the middle and upper parts of the Adelaide Subgroup. In some areas this oxidative color change was a response to shallowing and emergence, as evidenced by desiccation features and plant roots; elsewhere the red coloration cuts discordantly across lithologic boundaries and was probably a diagenetic response to groundwater flow.

A thick, locally steeply dipping clastic succession attains 1000 m along the imposing cliff exposures of the Transkei coast (Hobday *et al.*, 1980). *Lystrosaurus* remains are sporadically present throughout. Although this succession equates for the most part to the Katberg sandstone, the lower few hundred meters are probably the coarser grained equivalent of the lacustrine facies of the Balfour Formation to the north and west. Isolated exposures of the Katberg sandstone are preserved in small down-faulted blocks and synclines toward East London (Hiller and Stavrakis, 1979) and in the eastern Cape interior.

Sandy fluvial facies aggraded rapidly in a tectonically downwarped foreland basin related to activity in the Cape Fold Belt and its eastward extension beyond the existing continental margin. This succession thins northward to 100 m in Natal and fingers out into red and gray siltstones and mudstones in the Orange Free State and southern Cape. Stavrakis (1980) shows that a major fine-grained unit, the Palingkloof shale, intervenes between the Balfour Formation and Katberg sandstone, thickening markedly northward. A broadly radiating paleocurrent pattern shows a dominantly northeasterly trend in Transkei, with a northerly mode around East London becoming more westerly beyond East London (Theron, 1975; Greenshields, 1978; Hiller and Stavrakis, 1979). The more proximal facies of this fluvial system were displaced by the subsequent breakup of Gondwana.

This preserved succession in Transkei comprises two distinct fluvial facies, mixed-load and bedload dominated respectively (Hobday, 1978a); these are complexly interlayered but show an irregular upward increase in the proportion of bedload facies. The Balfour Formation in the Transkei area comprises sandstones and siltstones in the ratio 3:1 with a small proportion of conglomerate. Vertical sequences are upward fining, with channel lag conglomerate consisting of extrabasinal pebbles of Cape Supergroup lithology, mudclasts, reworked concretions, and bone. This is followed by trough cross-beds in sets of upward-decreasing thickness, ripple cross-lamination, including ripple-drift, and root-penetrated concretionary siltstones at the top (Fig. 11-14). These cycles are comparable in many respects to fluvial facies of the Ecca Group but lack coal. Destruction of vegetation, evidenced only by root casts, may have resulted from a generally lower water table and more intense chemical breakdown in a warmer and drier climate. The prevailing reddish colors, desiccation cracks, and pedogenic concretions suggest a distinctly seasonal climate with high mean annual temperatures.

There were, nevertheless, semipermanent ponds and embayments, the largest of which were developed at the distal edge of the alluvial floodplain. Up to 15 m of silt and mud accumulated, with thin pyritic layers and almost complete skeletons of *Lystrosaurus*. This mammal-like reptile was probably a herbivorous inhabitant of the shallow, vegetated pools. The bay-fill deposits characteristically contain erratic sandy splays and small, ephemeral channel-mouth bars.

Stavrakis (1980) has a similar interpretation for the Balfour Formation of the eastern Cape, where he presents evidence of point-bar accretion and flood-basin aggradation. A sinuosity ratio of 2.2 and bankful discharge of 305 m^3/s are estimated for these channel deposits.

The overlying Katberg Formation represents the upstream equivalent of the mixed-load streams (Fig. 11-12) and, in the Transkei area, contains less than 10 percent sediment finer than sand grade. The average fully developed vertical succession comprises basal conglomerate followed by cross-bedded and plane-bedded sandstone, and siltstone; but the preservation potential of the upper units was low because of truncation by overlying sequences. Average preserved thicknesses are only 1.1 m per channel. Initial width to depth ratios were greater than 50, as is typical of channels with sandy banks. Channel patterns were braided and were dominated by transverse bars. Conglomerates, which accumulated in the deeper scour pools, contain granite and metasediment pebbles up to 12 cm in diameter, attesting to the high flow competence. Intrabasinal mudstone clasts range from small pebble size to angular slabs over 50 cm long. The latter show evidence of slumping from the thin argillaceous topstratum along the channel margin. Bone is fragmentary and highly abraded.

Commonly a single large-scale cross-bed set developed by braid-bar migration overlies the conglomerate layer and is discordantly overlain by conglomerate of the succeeding cycle. Foresets tend to be slightly sigmoidal, indicating rapid sedimentation on the bar front and surface; foresets generally merge upward into plane-beds without a break. Cosets of smaller scale cross-beds accumulated in channels and over the tops of some bars.

An unusual feature of the Katberg sandstone in this area of Transkei is the ubiquitous develop-

Figure 11-14. Channel-fill sequence in the Balfour Formation, Transkei coast, showing erosively based conglomeratic sandstone overlain by cross-bedded sandstone and ripple cross-laminated siltstone with pedogenic concretions.

Figure 11-15. Vertical water-escape structures and downcurrent-overturned foresets in large-scale planar crossbed set, Katberg Formation, Transkei.

ment of penecontemporaneous deformation features (Fig. 11-15). These range from simple overturned cross-beds to complex recumbent folds and microthrusts, with associated vertical water-escape pipes and sand volcanos. Deformation is frequently confined to individual sets, predating deposition of the next unit; stacked individually deformed sets persist through thicknesses of scores of meters. Deformation may have been a response to instantaneous liquefaction by seismic shocks from the tectonically active area to the south. On the other hand, similar features develop in some modern streams in response to rapid fluctuations in water level, although not in the abundance observed in the Transkei exposures of the Katberg sandstone.

A distal braided alluvial-fan setting is also envisaged for the Katberg sandstone in the eastern Cape (Hiller and Stavrakis, 1979; Stavrakis, 1980) and is compared to the modern sandy Platte River of Nebraska. Pebbles of granite, quartz porphyry, quartzite, and fossil wood are more numerous than in Transkei and indicate derivation from a granitoid terrane with a sedimentary cover that included rocks of Devonian or younger age. Maximum grain-size decreases markedly northward, and in the direction of Queenstown the Katberg Formation interfingers with shaly lithologies of the lower Burgersdorp Formation.

The Burgersdorp Formation of the eastern Cape and Transkei region attains a maximum thickness of 600 m (Du Toit, 1954, p. 294), thinning markedly northward. In many respects it resembles the stratigraphically lower flood basin deposits of the Teekloof Formation of the southwestern Karoo. Thick sequences of brownish-red mudstones predominate, with subordinate blue and green-gray bands and mottled patterns. Sandstones are widely spaced and are interpreted by Stavrakis (1980) to represent swiftly flowing ephemeral streams similar to the Katberg type.

Northeastern and Northern Karoo Basin

Broadly convergent or centripetal drainage patterns similar to the Ecca characterized the Beaufort Group of the eastern and northeastern Karoo basin (Theron, 1975). In Natal the lower 200–450 m are designated the Estcourt Formation, which is the stratigraphic equivalent of the upper Adelaide Subgroup to the south. The Estcourt Formation bears some resemblance to the Balfour Formation of the eastern Cape (Botha and Linström, 1978). It gradationally overlies the Volksrust shale of the upper Ecca, from which it is distinguished by the presence of thin sandstones. J.N.J. Visser (personal communication, 1980) has put forward a valid case for inclusion of the Estcourt Formation with the Ecca Group on the basis of its deltaic–lacustrine milieu; according to this view the Beaufort is exclusively fluvial in origin. The alternative approach, adopted here, is to regard the Dicynodon lacerticeps–Whaitsia (Daptocephalus) Zone, which in Natal conformably overlies the Ecca Volksrust Formation, as implying Beaufort affinities for that part of the Estcourt Formation corresponding with it. These stratigraphic uncertainties stem from the progressive northward migration of clastic wedges and the frequent lack of correspondence between lithostratigraphic and biostratigraphic units.

Gray siltstone and mudstone with subordinate lenticular sandstones reflect deposition in shallow lakes, small mouth bars, sinuous alluvial channels, and flood basins (Van Dijk *et al.*, 1978). A highly irregular shoreline of broad, open embayments and elongate or multilobate deltas fluctuated markedly in response to changes in the relative rates of subsidence and sedimentation. The main lake system, a segmented remnant of the Ecca sea, was probably less than 50 m deep and accumulated silts and clays. Graded varvelike lamination produced by differential settling and millimeter-scale color banding arising from periodic changes in water chemistry of the restricted basins may both reflect seasonal control. The lakes may have been temporarily stratified; dark gray coloration, finely divided plant debris, and disseminated pyrite in some strata indicate a generally anaerobic environment and possibly account for the low degree of bioturbation and the good preservation of occasional freshwater fish. Oxygenated conditions are evidenced by thin bioturbated units within dark silty mudstone successions. Coprolites and freshwater conchostraca are present locally.

These lacustrine deposits are generally sandy in their upper parts, with evidence of weak currents of variable direction, and short-period oscillatory waves. Arthropod tracks and mollusk trails are locally abundant, together with horizontal spreiten structures produced by sediment-feeding organisms. Sinuous traces were left by fish swimming over the muddy floor. Burrowing is particularly concentrated in 10 cm thick units containing bone fragments, fish scales, and coprolites, marking a brief hiatus. Wave ripples become larger upward and change from purely oscillatory, low-amplitude symmetrical forms to slightly asymmetrical combined wave-current forms followed by steeper, asymmetrical forms probably produced by translatory waves. Linear wave-ripple crestlines display a consistent north–south trend, parallel to the shoreline in this area and almost perpendicular to the westerly or northwesterly fluvial azimuths.

Distal mouth bar sandstones gradationally overlying the wave-rippled deposits are predominantly fine grained and microtrough cross-laminated. These are overlain by proximal mouth bar and distributary channel sandstones. The thickest, medial portions of the sandstone bodies represent the location of the main distributary channels. Here the sandstone is erosively based, substantially coarser grained, and arkosic; lenses of basal mud-clast conglomerate and large-scale cross-bedded units merge upward into ripple cross-lamination. The sandstone thins laterally from the axial channels, becomes conformably based, and coarsens upward slightly. The largest sandstone lenses in Natal are 20 m thick and are traced over distances of some 20 km before they interfinger with lacustrine or flood-basin silts. Varying degrees of fluvial reworking and modification were produced by channel incision during progradation of the relatively stable coastal plain. Channels were typically widely spaced and sinuous, but braided patterns may have been dominant locally. Similar sandstone geometries are present in the southeastern Transvaal and eastern Orange Free State, where the units are somewhat larger and are comparable to the delta lobes of the Vryheid Formation (J.R. Turner, 1977), which they resemble closely.

Flanking the distributary channels are thin, root-mottled levee deposits and laterally thickening siltstones and mudstones deposited in interdeltaic embayments. Characteristic of the levee toe and bay margins are lenticular, wavy, and flaser bedding; climbing ripples; flow rolls; desiccation cracks; and occasional sets of interference ripples (Van Dijk *et al.*, 1978). At some localities the subaqueous levee deposits show intense load convolution. Thin coals are developed in places along the bay margins, representing fringing marsh. The more distal, fine-grained bay-center deposits are burrowed, with some horizontal and wavy lamination preserved. A remarkably diverse and well-preserved suite of fossils includes *Glossopteris* leaves and horsetail rushes and rhizoids (Gordon-Gray *et al.*, 1976), together with a variety of insects; conchostraca; fish skeletons, impressions, and scattered scales; and the vertebrate *Daptocephalus*. Sandy crevasse splays range from irregular, debris-laden tongues to clean, wave- and current-reworked sheets.

Similar but thinner deposits originated in smaller lakes and backswamp basins on the fluvial–deltaic plain. Some flood-basin sequences of alternating calcareous and noncalcareous mudstones are ascribed to periodic influx and evaporation (Botha and Linström, 1978). These isolated ponds contained mosses and horsetails in abundance. Insect remains include large numbers of scattered wings, together with whole insects (Van Dijk *et al.*, 1978). Small vertical burrows are filled with pyrite.

In marked contrast to these generally fine-grained lacustrine and deltaic systems are the coarse arkosic aprons informally designated the Mooi River formation by Theron (1975). These coarse gravelly aprons extended across the floodplain from granitic highlands to the east and terminated in small fan deltas. The triggering mechanism for these interludes of coarse influx is uncertain. Perhaps climatic changes created a temporary drop in lake level and depleted the vegetation cover, thereby increasing the proportion of coarse bedload available; a more peaked discharge and slight increase in gradient would have provided an increase in stream competence. Alternatively an episode of tectonic warping and elevation of the crystalline margins may have been responsible. The main development of this coarse unit occurs fairly high in the succession, since Theron (1975) recorded the presence of fossils belonging to both the Daptocephalus and Lystrosaurus Zones in the underlying strata.

The upper part of the Estcourt Formation also contains a large proportion of red siltstone and mudstone and reflects a greater degree of fluvial dominance and prolonged subaerial exposure, diagenesis, and soil development. There were, nevertheless, fairly extensive water bodies up to 15 m deep. Their deposits include complete skeletons of the small therapsid *Thrinaxodon.*

The Belmont Formation of Natal represents the northern extension of the Katberg clastic wedge, thinning to less than 100 m in western Natal (Botha and Linström, 1978). The stacked bedload channel sandstones resemble the Transkei succession, but in Natal they are locally transected by thicker trunk channel units with significant proportions of finer sediments in their upper parts. Paleocurrents were mainly toward the north and northwest in the direction of interfingering with finer grained deposits.

The overlying Otterburn Formation is composed of red mudstone with subordinate lenticular channel sandstones. Associated with these channel deposits are locally reduced greenish mudstones with a higher content of organic matter (Botha and Linström, 1978). The Otterburn Formation is the stratigraphic equivalent of the considerably thicker Burgersdorp Formation of the eastern Cape. Together these correlative units constitute the upper Beaufort, an obsolete term for the innermost belt of predominantly argillaceous Beaufort strata that encircle Lesotho. Botha and Linström (1978) stress the difficulty of distinguishing between the Belmont and Otterburn Formations in the north and northwest where the Katberg–Belmont clastic wedge grades into red and purple mudstone. In these areas a more general designation as Tarkastad Subgroup (Table 11-1) is warranted.

11.4. Upper Karoo Sedimentation

The Molteno Molasse Wedge

The Upper Triassic Molteno Formation, although somewhat coarser grained, resembles the Katberg Formation of the Beaufort Group. It is an intracratonic, bedload-dominated fluvial wedge derived from a tectonically active source to the south and southeast (Rust, 1962, Turner, 1970). This orogenic source may represent renewed uplift of the eastern extension of the Cape Fold Belt or, as suggested by Turner (1979b), the tectonic pulse may have signaled initial rifting prior to withdrawal of the Falkland Plateau from a position adjacent to the present margin of southeast Africa. Sandstones of the Molteno Formation are distinguished by a characteristic sparkling appearance (Du Toit, 1954, p. 297) produced by crystal faces of secondary quartz overgrowths. Paleoenvironmental interpretation of the Molteno Formation is based largely on the work of B.R. Turner (1970, 1975, 1977, 1979b).

Each of three thick, composite sedimentary wedges contains one or more upward-fining sequences. These wedges die out northward into monotonous silty lacustrine and floodplain deposits; only the central unit, the Indwe Sandstone Member, extends to the northern basin margin. The Indwe sandstone represents the culminating phase of progressively more extensive episodes of fluvial progradation and was followed by a succession of sedimentary pulses of diminishing lateral extent (B.R. Turner, 1970, 1977).

The upward-fining sequences are dominated by coarse, pebbly arkosic sandstones which are overlain by fine-grained sandstone, shale, and coal. Boulders and pebbles of Witteberg quartzite are characteristic of the basal conglomerates (Du Toit, 1954, p. 565). A remarkable distinction between these cycles and comparable vertical sequences elsewhere in the Karoo basin is that they show great persistence along strike. The lenticular geometries of most of the Ecca and Beaufort sandstones are clearly a record of autocyclic control, involving abrupt shifting of depositional sites in response to local gradient advantage. The basinwide cycles of the Molteno Formation were a response to external control, probably tectonic (Fig. 11-16). Each episode of uplift in the fault-bounded southern highlands generated a new cycle of vigorous coarse-grained sediment influx followed by an episode of decreasing fluvial energy. The long axis of the depositional basin was north trending, and fluvial infilling proceeded longitudinally from the south.

An initial tectonic pulse gave rise to the lowermost of the three clastic wedges. High topographic gradients, rapid weathering in the source area, an ineffective vegetation cover, and high discharge combined to generate vast wet alluvial fans. Comparable modern systems are present south of the Himalayan Range and on the southeastern coast of Alaska, where glacial meltwater provides the bulk of the discharge. Estimates of fold amplitude in Cape Supergroup source rocks suggest that peaks may have extended above the snowline; thus as Turner (1979b) pointed out, coarse seasonal glaciogene clastic influx may have characterized the upper fan. Succeeding episodes of decelerated uplift and reduced gradients were accompanied by finer grained accumulation. The fans of the Indwe sandstone prograded across the entire basin from a source slightly east of the locus of initial influx. Peat swamps developed during the quiescent phase that followed. Some of these coals have been exploited commercially in the past. The nature of the lowland *Dicroidium* flora and annual growth rings in *Dadoxylon* point to a temperate climate with seasonal rainfall distribution. A few upland vegetation forms may represent rafted vegetation.

B.R. Turner (1977) has demonstrated that the fan surfaces were traversed by a network of low-sinuosity bedload channels, with occasional larger trunk channels. Gravel sheets were flushed out of the mountainous confines during the primary, tectonically controlled episodes of maximum flow competence. Sandy cosets of large-scale trough cross-beds were deposited by trains of dunes comparable to those of the braided Alaskan outwash fans. In addition, Turner recognized several distinct forms of planar cross-beds. Laterally persistent sets with steep, tabular, graded foresets were produced by avalanching on the lee face of channel bars. Isolated wedge-shaped sets up to 1.8 m thick originated by the migration of large transverse bars under fluctuating flow conditions. Occasional sigmoidal forms indicate rapid sedimentation on the bar front and upper bar surface. In the south many of the foresets were deformed by a combination of liquefaction by seismic shock and current drag (B.R. Turner, personal communication, 1980), thus closely resembling units in the Katberg sandstone of Transkei. Small-scale cosets were produced by accretion of minor channel bars with downstream margins oblique to the channel trend. Some of the large-scale planar cross-beds sets contain intrasets of smaller foresets produced by flow separation during high discharge, or by currents sweeping diagonally across a bar front. Using the empirical relationships established by Schumm (1968b), Turner (1979b) estimated that the major channels of the Indwe sandstone were up to 800 km long, with catchment areas of approximately 40,000 km^2.

The thin northern part of the Molteno Formation is represented only by the Indwe sandstone.

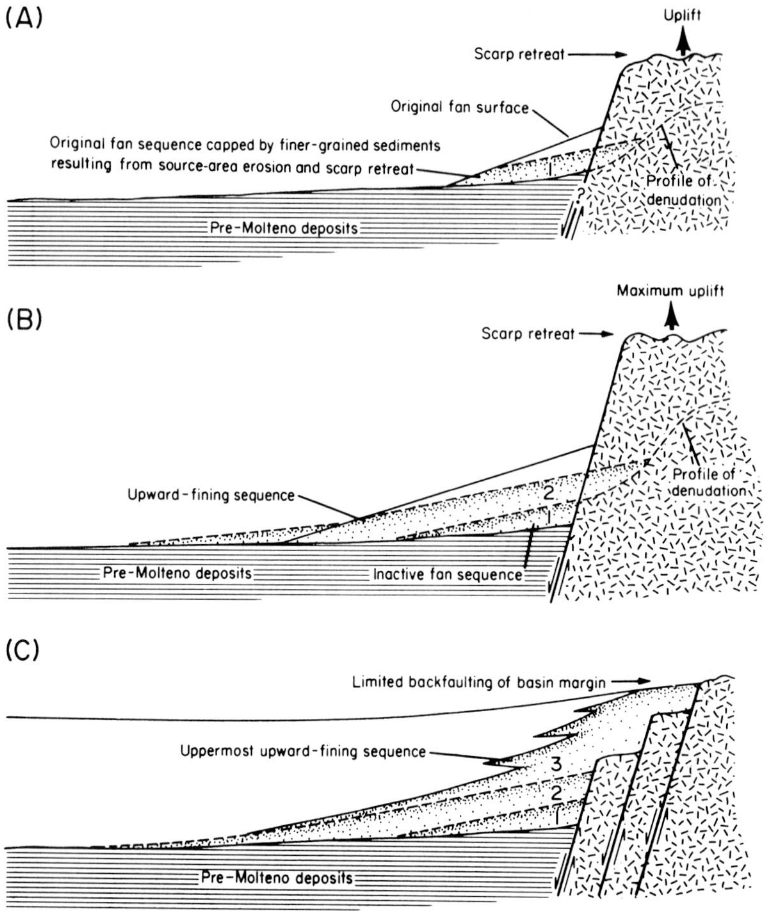

Figure 11-16. Turner's (in preparation) model for the evolution of laterally extensive upward-fining Molteno cycles in response to orogenic control.

Below this level, deposition of the Molteno cycles was contemporaneous with fine-grained sedimentation of the uppermost Beaufort Group, which represents the distal fluviolacustrine facies equivalent. Syngenetic uranium mineralization occurred in the northeastern Orange Free State. Above the Indwe sandstone, the Molteno tongues pinch out into fine-grained red beds of the Elliot Formation (Turner, 1975, his fig. 2). In the extreme north the Elliot Formation rests on the Beaufort with no apparent break.

Elliot Formation: Continental Red Beds

Fine-grained red beds and lenticular yellow sandstones make up the Elliot Formation, which follows conformably on the Molteno and represents a reduction in fluvial energy and proportion of bedload. The Elliot Formation attains a maximum thickness of about 500 m in the eastern Cape (Visser and Botha, 1980). Major trunk streams were widely spaced and were separated by vast areas of floodplain subject to seasonal inundation and prolonged subaerial exposure. Lacustrine facies are well developed in the northeastern Orange Free State, where vertebrate remains are sparse by comparison with the remainder of the formation (B.R. Turner, personal communication, 1980). A gradual climatic progression from warm and humid toward semiaridity (Du Toit, 1954, p. 566; Visser and Botha, 1980) was accompanied by a change in fluvial style, with the streams becoming broader, shallower, and more ephemeral. Paleocurrent patterns are mainly toward the north, northwest, and west (Visser and Botha, 1980). Medium to large dinosaurs characterize the lower half of the Elliot Formation, giving way upward to smaller, light-limbed forms in response to the need for greater mobility during progressive aridification (Haughton, 1924, 1953).

The contact with the underlying Molteno Formation is arbitrarily taken at the lowermost thick purple mudstone which, being part of a lenticular arrangement of lithofacies, varies in elevation (Botha, 1968; Visser and Botha, 1980). The thicker mudstone units attain 100 m and contain thin lenses and stringers of fine-grained sandstone. This association probably originated in sheltered lakes and embayments in the distal alluvial flats. In contrast to the thinner mudstones, these argillaceous units tend to be chemically reduced.

Tabular to lenticular bodies of fine- to medium-grained channel-fill sandstone typically have scoured bases with mudstone intraclasts and grade laterally into levee siltstones and flood-basin mudstones (Botha, 1968; Visser and Botha, 1980). Trough cross-bedding predominates and, in some cases, the thickness of the sets indicates migration of exceptionally large subaqueous dunes or transverse bars down the channel axis. Interfluvial mudstones vary from red or purple to green and gray, and contain desiccation cracks and vertebrate footprints (Du Toit, 1954, p. 298; Visser and Botha, 1980). Hematitic coloration characterizes the finest sediment grades, particularly in the upper half of the Elliot Formation. Thin greenish laminae enclosed within red mudstone suggest that oxidation may have been a product of early diagenesis. This is supported by associated concentrations of nodules, probably resulting from pedogenic processes in a markedly seasonal climate with a fluctuating water table. On the other hand, the pigmentation may have been derived from a red bed provenance. The uppermost part of the Elliot Formation contains evidence of eolian processes (Botha, 1968; Visser and Botha, 1980).

Uranium deposits in the eastern Orange Free State bear close resemblance to those in the Beaufort Group of the southern Karoo (B.R. Turner, personal communication, 1980).

Coarse, cross-bedded sandstones and conglomerates (Escarpment Grits) of northwestern Zimbabwe contain species of the *Dicroidium* flora, which have provided a basis for correlation with the Molteno–Elliot succession (Du Toit, 1954, p. 324). Diamonds are present in this unit (Du Toit, 1954, p. 325) and in northeastern Swaziland at approximately the same stratigraphic level (Hawthorne *et al.*, 1979). The Swaziland diamond placers were derived from a kimberlite diatreme 30 km to the west.

Clarens Formation: Midlatitude Arid Landscape

Progressive Late Triassic desiccation reached a maximum during deposition of the Clarens Formation (formerly Cave Sandstone). Eolian sand and associated playa lake, sheetflood, and ephemeral stream deposits occupy a curved north–south-trending belt through Lesotho, corresponding in general with the preceding fluvial depository of the Molteno and Elliot Formations. The Clarens Formation attains 300 m in the upper Orange River valley but is generally around 100 m thick (Du Toit, 1954, p. 300). Comparable strata are present in the Lebombo belt, northern Transvaal (Bushveld sandstone), Zimbabwe (Forest sandstone), and Namibia (Etjo sandstone). The Etjo sandstone, and possibly the northern Transvaal deposits, are regarded as somewhat younger (Bigarella and Van Eeden, 1971).

Beukes (1970) has divided the Clarens Formation into three units. The central unit represents true desert conditions in the north, grading southward into intercalated water-laid sediments. Paleowinds were from the west, redistributing fine-grained sands of the Elliot Formation and other units. Beukes distinguishes between large-scale cross-bedded units up to 10 m thick, formed by the migration of transverse dunes, and large trough sets up to 30 m thick, representing barchan dunes. After comparing eolian paleotransport directions to the major planetary wind systems of the present day, Bigarella and Van Eeden (1971) were of the opinion that markedly different circulation patterns could have been a consequence of the vast areal extent of Gondwana. The Lesotho area was approximately 50°S in Triassic time and may have been subject to midlatitude westerly winds similar to those blowing across the modern Gobi desert from the Siberian high--pressure cell in winter (Bigarella and Van Eeden, 1971). These authors suggest that the younger Etjo sandstone and correlatives postdate continental separation and were deposited by southwesterly winds originating in a South Atlantic high-pressure cell.

The lower and upper units reflect a less severe climate, with numerous shallow lake deposits, locally containing freshwater fish, crustaceans, remains of mobile, light-limbed dinosaurs, invertebrate traces, and dinosaur footprints. Desiccation cracks, syneresis structures, load casts, and evaporite pseudomorphs are present. Eolian sands were periodically saturated and subject to gravity flow, producing internally structureless and occasionally graded beds (Beukes, 1970). Local variations in this pattern have been observed by Pemberton (B.R. Turner, personal communication, 1980) who recognized a basal, diagenetically reddened unit overlain by eolian dune deposits, followed by massive sandstones deposited under conditions of increasing aridity.

Major channel-fill sandstones are commonly present toward the base of the Clarens Formation in the Natal Drakensberg, where a less arid paleoclimate is indicated. Some channel units are characteristically upward fining with root penetration at the the top (Van Dijk *et al.*, 1978); others comprise planar cross-bedded, medium- to coarse-grained sandstones deposited by converging wadi systems (P.G. Eriksson, 1979).

Playa lake deposits of the lower and upper units consist of interlaminated sand and clayey silt, frequently disrupted by burrowing or alternating episodes of wetting and drying accompanied by gypsum crystallization. Etch marks, rills, linear shrinkage cracks, runzel marks, and microripples delineate the fluctuating lake margins. At least 10 different types of vertebrate footprints are preserved; they reflect locomotion by bipedal hopping and by both quadrupedal and bipedal walking (Van Dijk *et al.*, 1978). The dinosaurs responsible probably ranged from the size of a sparrow to that of an ostrich (Ellenberger, 1970). Most playa lakes were shortlived and filled by sheetflooding, generating graded sandstone beds with a variety of solemarks, including flutes, grooves, and prod marks. Braided ephemeral stream deposits above show scour and fill structures, trough and planar cross-beds, and complex patterns of small current ripples.

Toward the close of Clarens deposition, the climate moderated somewhat and stream and sheetflood processes became more dominant (Fig. 11-17). Minor lava flows signaling the onset of Karoo volcanism are interlayered with the uppermost sedimentary units. P.G. Eriksson has

Figure 11-17. Upward-coarsening playa lake siltstones and sandstones of sheetflood and channel origin make up the Clarens Formation outcrops in the foreground, with erodible Elliot Formation red beds below. Basalts make up the upper part of the Drakensberg escarpment in the background. Giants Castle, Natal.

shown that the sandstones in Natal are, for the most part, remarkably immature. He proposes a marginal desert environment subject to high temperatures and rare, torrential discharge, with a source consisting in part of glacial deposits.

11.5. Cape Orogeny

The Cape Fold Belt is part of a persistent Gondwana orogenic belt, other dispersed remnants of which are presently preserved as far afield as Bolivia, Peru, Argentina, Antarctica, and eastern Australia (Du Toit, 1937; Lock, 1980). In the Cape, the Cedarberg range extends northward from syntaxis with the main east-trending fold belt. The latter comprises megaanticlines of Table Mountain Group and Witteberg quartz arenites expressed topographically in the Langeberg–Outeniqua and Witteberg–Swartberg mountains. Between these prominent ranges, and coinciding with a tract of erodible Bokkeveld strata in a broad synclinorium, are the strike-trending plains of the Little Karoo (Hälbich, 1977b).

Large-scale structures of the Cape Fold Belt include disharmonic, cascade, asymmetrical, and northward-overturned folds evidencing a dominance of gravitational deformation (Newton, 1973; De Swardt *et al.*, 1974). Immediately to the south of an area of disharmonic folding are high-angle thrust faults (Lock, 1980). Newton favored vertical tectonism as the driving mechanism for gravity-induced collapse of relatively unconsolidated sediments at shallow depth. Evidence cited in support of this model includes the overall fold geometry, sedimentological indications of northerly or easterly paleoslope, downslope overturning, downslope increase in fold intensity, the large proportion of ductile as opposed to brittle deformation, and the absence of volcanic activity, granite intrusion, and high-grade metamorphism. Downslope movement may have been enhanced by high pore-fluid pressures. According to Newton's reconstruction, syntaxis resulted from convergence of northward- and eastward-sliding masses.

The gravity-folding model has been critized on the grounds that the deformed strata may have been lithified prior to deformation (De Swardt, 1974), as much as 25 percent lateral shortening has been involved (Hälbich, 1977b), and high-angle thrusts indicate lateral compression (Lock, 1978a). Furthermore there are suggestions that the earlier Pan African structural trends may have had a major influence on the development of the fold belt (Lock, 1980). De Swardt *et al.* (1974) remarked that the coffin-shaped folds of the Cedarberg are fundamentally different in style and, as suggested by De Villiers (1944), could be earlier. Indeed, Lock (1980) is of the opinion that the Cedarberg and east-trending Cape ranges are genetically unrelated.

A conventional plate-tectonic model for the Cape Fold Belt was advocated by Rhodes (1974b) and was subsequently elaborated upon and refined by Lock (1978a, 1980). Rhodes envisaged accumulation of the Table Mountain and Bokkeveld Groups as a miogeoclinal prism extending onto an adjacent continental rise, implying an Atlantic-type trailing edge margin. An island-arc orogenic belt developed south of the present coastline as a result of Late Paleozoic subduction. This youthful, volcanically active mountain range supplied sediments of Witteberg and Karoo age. Cape folding probably reached a climax during the Early Triassic, deforming strata as young as middle Beaufort. Folding, according to Rhodes, was produced by underthrusting of continental lithosphere as a result of collision between the proto-African plate and a southern plate that included Patagonia and the Falkland Plateau; northward compression together with the buoyancy of the subducted continental material generated characteristic décollement structures such as disharmonic folds.

The plate tectonic model for the Cape Fold Belt received support from De Wit's (1977) suggestion that Andean-type subduction accounted for the entire Gondwanide orogeny. The apparently prohibitive distance of this postulated subduction zone, some 1000 km south of the Cape ranges, may be accounted for by Lock's (1980) flat-plate subduction model (Fig. 11-18). This mode of subduction requires very shallow dips for the descending oceanic plate. Lock suggests that this subducted plate was temporarily attached to the underside of the Gondwana plate in what is now Chile, Argentina, and the Falkland Plateau. Deformation of the Argentinian sierras and Cape Fold Belt occurred along the lines of detachment between the superimposed plates, thus concentrating the lateral forces along intraplate fold belts (Lock, 1980). Smithson *et al.* (1980) arrived at remarkably similar conclusions for the Laramide belt in Wyoming, based on deep seismic profiles.

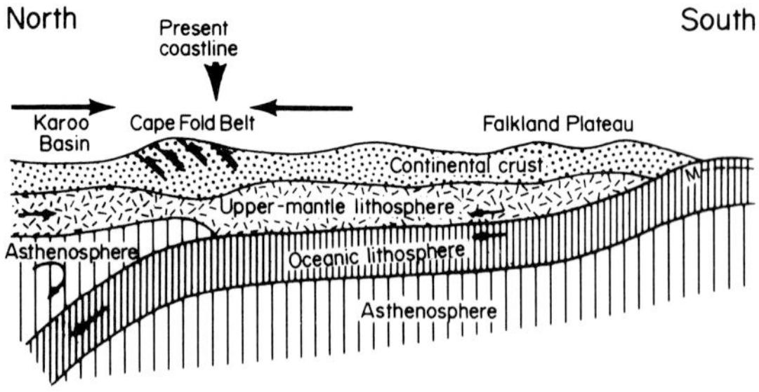

Figure 11-18. Flat-plate subduction model illustrating the development of the Cape Fold Belt along the line of separation of the descending oceanic plate from the upper plate. [After Lock (1980).]

11.6. Karoo Volcanism

A violent episode of Jurassic volcanism terminated Karoo deposition. Basaltic outpourings produced vast lavafields, of which the 140,000 km^2 Drakensberg plateau is an erosional remnant. Volcanism was a manifestation of Gondwana breakup but was initiated at least 20 Ma prior to fragmentation (Scrutton, 1973a). Indeed, older Karoo volcaniclastics are present in the Permian–Triassic Ecca and Beaufort Groups. In adjacent regions of Mozambique, such as the lower Zambezi valley, volcanism persisted intermittently into Cretaceous time (Woolley and Garson, 1970; Flores, 1973), long after the main outpourings accompanying continental separation.

Cox *et al.* (1967) showed that the Karoo volcanics represent two distinct geochemical provinces. A southern province comprises the basalts and diabases of the central Karoo (Stormberg Range of the eastern Cape; Drakensberg plateau of Lesotho) and Swaziland, with a normal tholeiitic chemistry. A northern province of basaltic volcanics in Zimbabwe and the northern Lebombo is characterized by values of K, Ti, P, Ba, and Zr that are abnormally high for tholeiitic rocks. The changes in magma composition between the southern and northern provinces were attributed by Woolley and Garson (1970) to increasing degrees of crustal fracturing. Similarly, Rhodes and Krohn (1972) suggested that differences between the Lesotho and northern Lebombo–Nuanetsi lavas resulted from derivation from different mantle levels.

On the basis of more recent geochemical studies, Duncan and Erlank (1979) recognized four major chemical provinces or magma types in the Karoo basaltic volcanics. These are compositionally distinct and characterize specific geographic regions with some overlap: (1) the tholeiitic central or Lesotho magma type of the Stormberg, Drakensberg, Springbok Flats, and Mariental (Namibia) lavafields and the widespread Karoo diabase intrusions (dolerites); (2) the Atlantic type, comprising the Kaokoveld lavas and correlatives in South America; (3) the southern Lebombo type, including lavas in Swaziland; and (4) the northern Lebombo type, which includes lavas in Zimbabwe.

According to Duncan and Erlank (1979) the magnesium–iron geochemistry of the magma types suggests that the compositional differences between the Atlantic type and other provinces may result from more advanced differentiation of the Atlantic-type magmas. In contrast, differences between the other provinces are such that they cannot readily be explained by different degrees of melting or differentiation. Similarly, contamination processes are unable to account for the differences, and the variations among the remaining three magma types are ascribed to mantle inhomogeneity, as are subordinate variations within the central type.

Duncan and Erlank (1979) observed that whereas basaltic magmas were generated over vast areas of Gondwana, acid volcanism was restricted to zones of complete crustal rupture and separation. They conjectured that both rifting and acid volcanism may have been symptomatic of locally elevated temperatures. Petrogenesis of the rhyolites presents a problem in view of the considerable volume of acid material found in the marginal areas and the almost total absence of intermediate rock types. Cox *et al.* (1965) proposed that the rhyolites represented primary acid magmas, although Sr-isotope studies led Manton (1968) to conclude that they had been derived from basic magmas. Similarly, Cleverly and Betton (1979) hypothesized that most of the rhyolites originated by fractionation, crystallization, or partial melting of basic Karoo rocks at the base of the crust.

Although rifting and volcanism are generally attributed to convective upwelling in the mantle (Cox, 1970), patterns of fracturing suggest strong but not total control by preexisting basement structures. For example, the Lebombo eruptive belt is aligned approximately parallel to both the Mozambique Province (Cox, 1970; Saggerson and Logan, 1970) and the zone of recurring

Paleozoic subsidence (Natal embayment and Natal trough) to the south. The Lesotho area was long subject to Karoo subsidence prior to volcanism, suggesting incipient crustal weakness (Rust, 1975).

The main volcanic episode extended between 200 and 190 Ma, with acid volcanism continuing until 140 Ma (Fitch and Miller, 1971; Flores, 1973). An initial explosive episode is evidenced by volcanic vents, diatremes, and tuffs. This explosive activity, with local caldera collapse, was confined to several distinct centers (Lock *et al.*, 1974). In southern Lesotho and the eastern Cape, Du Toit (1954, p. 301) described volcanic ash beds, brecciated lava flows, and agglomerates with fragments of solid ejectamenta of as much as a ton in the lower part of the volcanic succession. Lock *et al.* (1974) ascribed these diatremes and associated features to phreatic activity resulting from hot magma coming into contact with ground water, and the violent production of steam. In addition, Lock (1978b) provided evidence that eruption of ultrahigh-temperature lavas through crater lakes produced rapid quenching and generated mud flows. Clasts incorporated in the mudflow debris show a radial fracture pattern resulting from abrupt contraction.

The explosive phase was followed by extrusion of a monotonous succession of lava flows, of which the preserved thickness in the Drakensberg (Fig. 11-17) is 1400 m. The basal lava flows are irreglar in thickness, conforming with the dune and channel paleotopography of the top of the Clarens Formation. Pillow lavas developed in remnant playa lake systems (McCarthy, 1970). These early lava flows were in places covered by sand on which dinosaur tracks are preserved (Rust, 1975). The overlying tholeiitic flows of variable thickness are laterally persistent, with a massive central zone and amygdaloidal upper and lower boundaries. Upper surfaces show ropy or pàhoehoe structures and rare evidence of weathering prior to the succeeding flow. More basic, olivine-bearing lavas near the base of the succession may have evolved from picritic precursors (J.W. Bristow, personal communication, 1980). Eastern Cape (Stormberg) volcanics show greater variation, which has provided a basis for stratigraphic subdivision (Lock *et al.*, 1974).

Contemporaneous diabase intrusion affected almost the entire Karoo basin, with the exception of the thin northern succession and of the southern margin, which was compressed along the edge of the Cape Fold Belt. Textures range from glassy along chilled contacts to coarsely crystalline and gabbroic in appearance in the centers of the thickest sills. Compositional variants, such as the Effingham diabases (Frankel, 1969), are a product of contamination. Thick differentiated sheet intrusions, e.g., Insizwa in Transkei, show well-defined layering (Walker and Poldervaart, 1949), in general comprising an olivine-rich, ore-bearing basal zone, a central zone of gabbro and norite, and an acid roof zone (Maske, 1966).

Karoo volcanics of the Lebombo belt are considerably thicker than the succession centered on Lesotho, attaining at least 10,000 m. The Lebombo belt, which is downwarped to the east in a major north-trending monoclinal structure, extends from Zululand to the Limpopo River in the Transvaal. Basalts constitute the lower part and are overlain by rhyolites (Cleverly and Bristow, 1979). Monoclinal warping was in part contemporaneous with extrusion (Du Toit, 1929) but was accentuated during post-Karoo times. As a result the volcanic succession dips steeply eastward along the central limb of the monocline.

Rhyolite sheet flow units of the Lebombo monocline are typically 80–350 m thick and extend laterally for as much as 50 km (Bristow and Cleverly, 1979). As many as 30 superimposed sheet flows are present in Swaziland. Bristow and Cleverly recognized a vertical zonation within each flow unit. A basal tuffaceous zone incorporates weathered detritus from underlying flows, together with extraneous sedimentary and air-fall material. Above this is a thin microcrystalline or glassy rhyolite layer followed by the dominant massive, felsic, central zone, which grades upward into a highly contorted flow-banded zone capped by a breccia zone. These authors favor emplacement by ash-flow eruption, possibly involving degassed ignimbritic flows in which the primary features have been destroyed by intense welding remobilization and devitrification. The waning phase of Lebombo magmatism was marked by the emplacement of the Bumbeni complex of rhyolite flows and syenite intrusives cut by siliceous veins (Van Wyk, 1963; Bristow 1976). The complex has been dated at 133 $\pm$ 2 Ma (J.W. Bristow, personal communication, 1980; Rb-Sr).

North of the Limpopo River the monocline rotates into the northeast-trending Sabi monocline, which can be followed into the Lupata area of the East African Rift system. The Tuli and

Nuanetsi synclines, which also contain voluminous successions of Karoo volcanics, are located at the hinge formed by the two monoclines and extend along the Limpopo Province in Zimbabwe. With the exception of the Nuanetsi rhyolites the volcanic rocks of the Tuli, Nuanetsi, and Sabi areas are similar to those of the northern and central Lebombo (J.W. Bristow, personal communication, 1980). The Tuli syncline is located over the zonal structure of the Limpopo Province, whereas the Nuanetsi syncline is thought to have formed over the re-entrant produced by the converging Lebombo and Sabi monoclines (Cox *et al*., 1965).

Volcanic activity in Namibia has an age range of 183–121 Ma (Siedner and Miller, 1968; Gidskehang *et al*., 1975; Siedner and Mitchell, 1976). The first episode of diabase emplacement (183 Ma) and the extrusion of the Kalkrand (Hoachanas) basalts (168 Ma) were contemporaneous with Drakensberg magmatism. Younger flood basalts, andesites, and associated intrusions were related to rifting of western Gondwana (Chapter 12).

STAGE 5

After Gondwana

Modern climates and oceanic circulation are linked to the breakup of Gondwana and formation of a cold Southern Ocean. Neogene aridification of the western half of the subcontinent is exemplified by the smoking crest of a 250-m-high star dune on the precipitous southern margin of the Namib Sand Sea near Aus, Namibia.

Chapter 12

Fragmentation and Mesozoic Paleogeography

Fragmentation of Gondwana took place in two phases. Between 190 Ma and 160 Ma, separation followed a line down the eastern flank of the Mozambique Ridge (Dietz and Holden, 1970; Scrutton, 1973b; Kutina, 1975; Du Toit, 1976). This line of fracture paralleled the trend of the Lebombo rhyolites and tholeiitic basalts, a zone of crustal rupture that preceded drift by about 20 Ma (see Chapter 11). Volcanic activity also preceded the separation of South America from Africa and the initiation of the South Atlantic, which took place in the second phase of fragmentation between 135 Ma and 130 Ma (Siedner and Mitchell, 1976; K-Ar). The age of the oldest magnetic anomalies in the Cape Basin west of Cape Town are in close agreement with the Kaokoveld–Serra Geral volcanic activity (Larson and Ladd, 1973).

Originally the Falkland Plateau abutted the southern and southeastern margin of South Africa, intervening between the African and Antarctic plates (Fig. 12-1). Initial separation is believed to have occurred between Africa and Antarctica by the Late Jurassic (Simpson *et al.*, 1979). During the Upper Valanginian (Table 12-1) the Falkland Plateau remained with the South American plate and moved away from southern Africa along a transform margin with an initial 1000 km offset (Du Plessis, 1977; Rabinowitz and LaBrecque, 1979). Relative motion of the South American, African, and Antarctic plates (Fig. 12-1) is believed to have occurred at a triple junction, which had been in existence since the Early Paleozoic (Tankard and Hobday, 1979). LaBrecque and Hayes (1979) also inferred a triple junction from the presence of a magnetic anomaly bight (anomalies 33–34) about the Agulhas Plateau. The continental margins of southern Africa are therefore attributed to differing styles of fragmentation. The western and eastern margins are divergent structures formed by tensional rift faulting (Dingle and Scrutton, 1974). The southern to southeastern margin lies approximately along the trace of the Agulhas–Falkland fracture zone initiated by transform faulting (Francheteau and Le Pichon, 1972; LaBrecque and Hayes, 1979).

12.1. The Proto-Atlantic Margin

Upper Cretaceous sediments overstep Precambrian basement along the coastal fringe of Namibia, postdating extrusion of the Kaokoveld basalts; the ammonite *Proplacenticeras merenskyi* in limestones at Bogenfels (Fig. 12-2; Table 12-1) suggests a Cenomanian date for the transgression (Klinger, 1977). Other onshore exposures near Bogenfels are younger and apparently contemporaneous with offshore Orange basin deposits that contain the Middle Santonian to Late Campanian inoceramid *Sphenoceramus* cf. *schmidti*. Sedimentation is believed to have been continuous in the Orange basin from the Lower Aptian through the Maastrichtian, generating a seaward-thickening cone (Scrutton and Dingle, 1976). In the distal parts of the basin the succession fines upward from fossil wood-bearing carbonaceous shales, muddy sandstones, through gray and red shales to deep-water pelagic clays at the top (Deep Sea Drilling Project site 361; Siesser, 1977a). These are capped by Upper Cretaceous sandstones. Rapid influx of river-derived sediment was insured by the steep gradients maintained by repeated epeirogenic uplift, tilting, and down-faulting (Scrutton and Dingle, 1976).

Terrestrial sediments of Mesozoic age are less voluminous. Lignitic alluvium in the Henkries valley, just south of the Orange River, contains

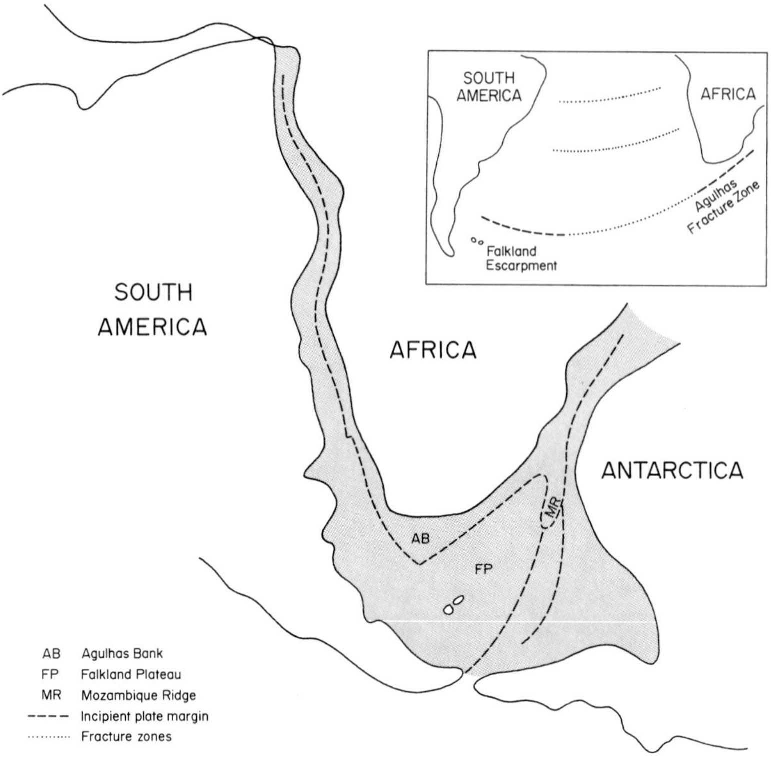

Figure 12-1. Pre-Cretaceous reconstruction showing the location of southern Africa in Gondwana and incipient plate margins. [Based on De Wit (1977), Norton and Sclater (1979).] Inset shows counterclockwise rotation of Africa relative to South America and the trace of the Falkland–Agulhas fracture zones. [After Rabinowitz and LaBrecque (1979).]

fossil wood and specimens of the Late Cretaceous iguanodont dinosaur *Kangnasaurus coetzeei*. Locally interbedded with the Kaokoveld lavas of Namibia is the Etjo sandstone deposited by westerly winds (Bigarella, 1970; Martin, 1973). Beyond the escarpment no definite Mesozoic sediments are recognized, although Du Toit (1954, p. 408) and Dixey (1955) have suggested that the basal parts of the terrestrial Kalahari basin may be of Cretaceous age.

12.2. Evolution of the Southern Continental Margin

Regional stress patterns produced by Early Cretaceous shearing along the southern continental margin caused tensional displacements on southward-dipping normal faults in the Cape Fold Belt (Lock, 1978a). Numerous small fault-controlled basins developed with an elongation parallel to the structural grain and subparallel to the present coast (Fig. 12-2). Structurally and stratigraphically these basins resemble the Triassic Newark basins of the Appalachian region (Lock *et al.*, 1975). The Algoa and Oudtshoorn basins are the largest of the onland depositories; the Outeniqua basin is a composite of several half-grabens below the continental shelf. Throws along faults such as the Worcester and Cango vary up to 6000 m and reflect long histories of activity.

Newark-type basins are half-grabens characterized by synsedimentary faulting and contemporaneous alkaline volcanism (Potter and Pettijohn, 1963, p. 244). In each of the southern Cape basins sedimentation was partially controlled by active boundary faulting, resulting in accumulation of conglomerates, finer grained terrigenous clastics, and rare lignite seams. Generally the strata dip toward the boundary faults and are dissected by numerous second-order faults (Fig. 12-3, inset). In the northern part of the Algoa basin Suurberg basalts and pyroclastics are interbedded with the Enon conglomerates. Late Jurassic–Cretaceous transgressions are related to continued evolution of the southern margin but failed to penetrate the landlocked basins.

The Algoa Basin

The Algoa basin is the largest of the onland Mesozoic basins, covering an area of 4000 km^2. It comprises a series of half-grabens which are fur-

Table 12-1. Correlation of Mesozoic Stratigraphy of Southern Africa[a]

System	Series	Stage	Age (Ma)	Igneous rock age distribution	Igneous activity	Algoa basin	Aqulhas Bank	Subsidiary basins	Zululand basin
CRETACEOUS	Upper	Maastrichtian		n = 143	Kimberlites			Needs Camp Igoda	St. Lucia (ZULULAND GROUP)
		Campanian							St. Lucia
		Santonian					Alphard	Mzamba	
		Coniacian							
		Turonian							
		Cenomanian						Bogenfels	Mzinene
	Lower	Albian	100						
		Aptian			Etendeka				
		Barremian					Sundays River and Kirkwood	Mgazana and Brenton	Makatini
		Hauterivian				Sundays River and Kirkwood (UITENHAGE GROUP)			
		Valanginian							
		Berriasian							
JURASSIC	Upper	Tithonian							
		Kimmeridgian						Enon	Basalt Acid volcanics Pyroclastics
		Oxfordian							
		Callovian	150			Enon	Enon	Robberg	?
		Bathonian			Kalkrand				Rhyolite Ash flows Tuffs
		Bajocian				?	?	?	
	Lower	Aalenian			Keetmanshoop; Karoo diabases; Drakensberg				?
		Toarcian							
		Pliensbachian							Basalt Andesite
		Sinemurian							
		Hettangian							
TR		Rhaetian	200						

[a]Sources: McLachlan and McMillan (1979), Winter (1979), H.C. Klinger (personal communication, 1977). Radiometric ages derived from Allsopp and Barrett (1975), Allsopp and Kramers (1977), Davis (1977), MacIntyre and Dawson (1977), Duncan *et al.* (1978), Siedner and Miller (1968), Moore (1980).

ther dislocated by normal block faulting. Two major depositories are bounded along their northern margins by the Coega and Suurberg faults respectively (Fig. 12-3). The thickest (4000 m) stratigraphic section in the northern basin is attributed to half-graben subsidence controlled by the Commando fault (Shone, 1976); extrusion of basaltic lava marked the early subsidence of this basin.

The stratigraphy established by Atherstone (1857) has recently been reviewed by Winter (1973, 1979), McLachlan and McMillan (1976, 1979), and Shone (1978b). Occurrence of the Suurberg volcanics within the succession has been the basis for a twofold subdivision into the Uitenhage and Suurberg Groups (Hill, 1972). The sedimentary rocks below the extrusive rocks are characterized by the presence of interbedded tuffs and smaller average clast size than the suite above. Field relationships suggest that extrusion of the basalts and intrusion of diabasic feeder dikes took place contemporaneously with deposition of the conglomerates. This interpretation is supported by the occurrence of pyroclastic rocks and cross-bedded tuffaceous sandstones.

The basalts within the basal Enon Formation have chemical affinities with the Drakensberg lavas and have been dated at 162 Ma (Du Toit, 1976; McLachlan and McMillan, 1976; K-Ar), suggesting that Enon deposition may have commenced in the Early or Middle Jurassic (Winter, 1979), although there is doubt about the age determination (Marsh *et al.*, 1979). Dingle and Scrutton (1974) believed that Enon sedimentation may have originated as synorogenic molasse in Middle Triassic intermontane valleys. Plant fos-

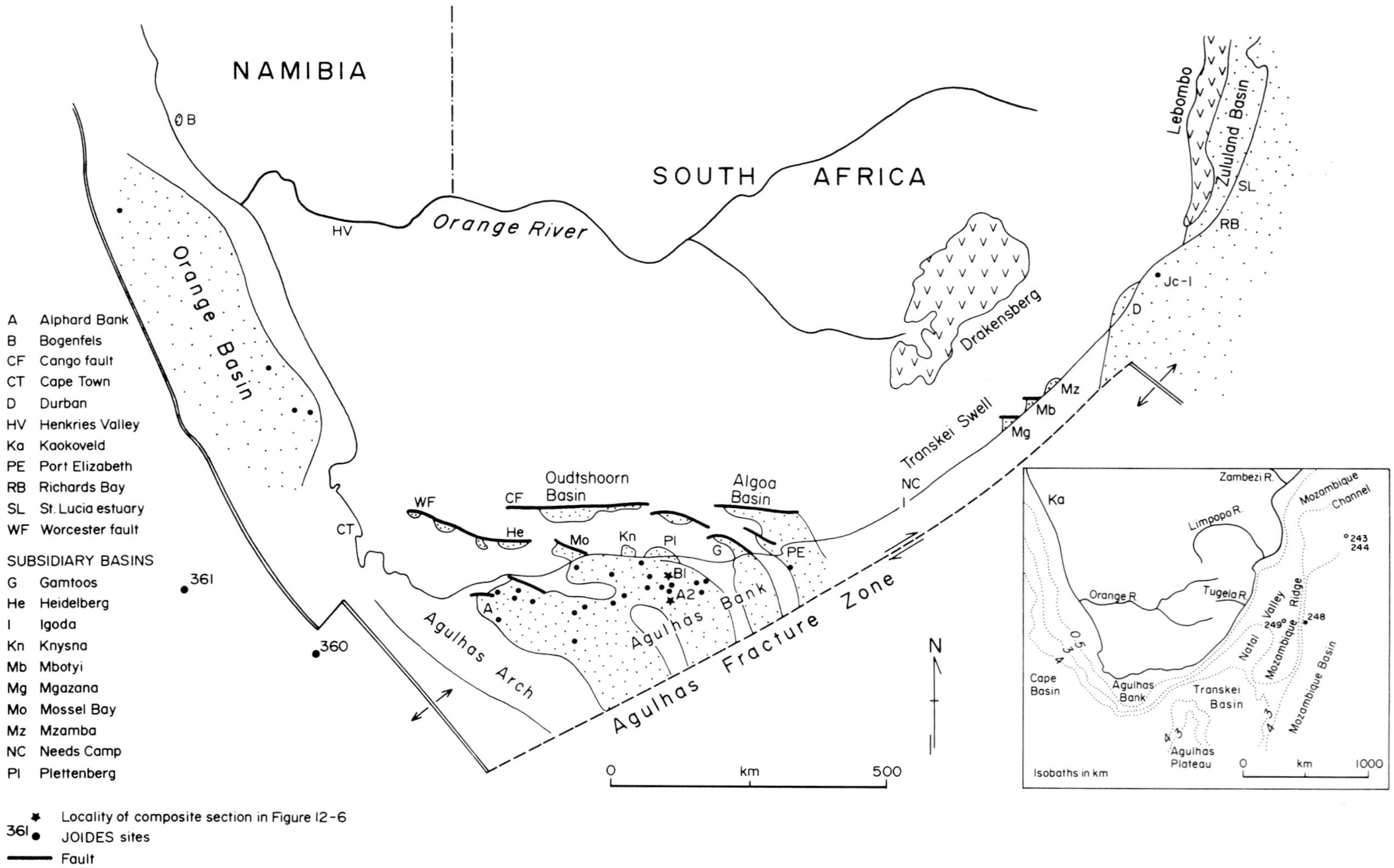

Figure 12-2. Locality map showing distribution of Mesozoic sedimentary basins. Inset shows major offshore physiographic features. The Outeniqua basin underlies the Agulhas Bank. [Sources: Dingle and Scrutton (1974), Scrutton and Dingle (1976), Du Toit (1979), McLachlan and McMillan (1979).]

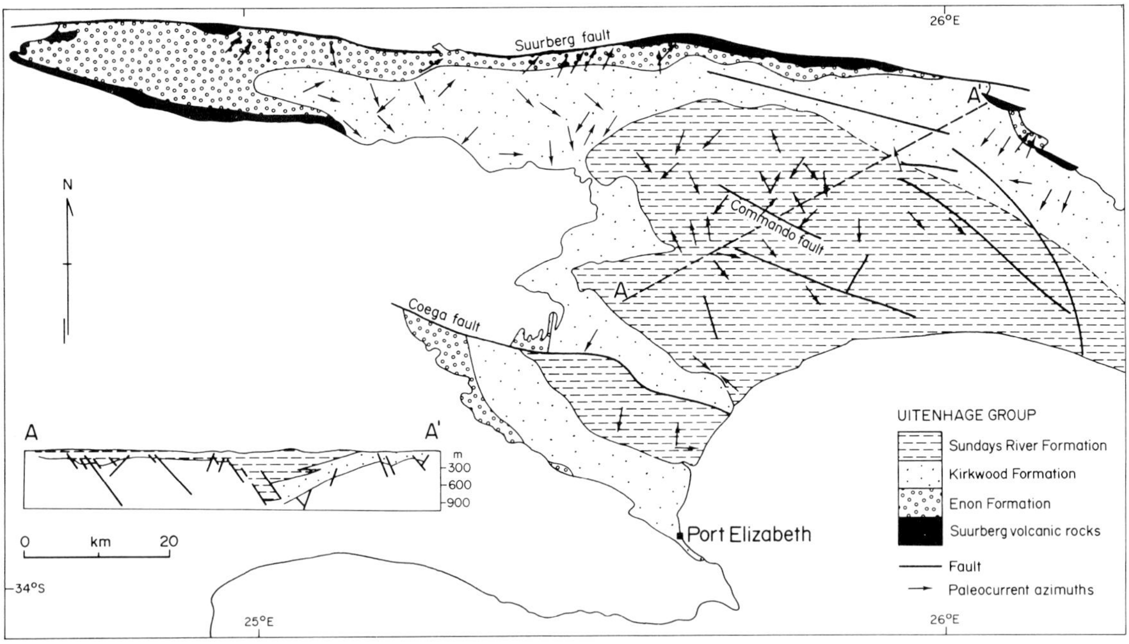

Figure 12-3. Geologic map of Algoa basin showing paleocurrent distribution. Inset based on seismic data. [Sources: Winter (1973), McLachlan and McMillan (1976), Lock *et al.* (1975).]

sils within the Kirkwood Formation are regarded as Lower Cretaceous (Seward, 1907; McLachlan and McMillan, 1976) rather than Upper Jurassic (Plumstead, 1966). Age-diagnostic ammonite and microfossil assemblages suggest a Late Valanginian to Hauterivian age range for marine sedimentation in the Kirkwood and Sundays River Formations (Spath, 1930; Beer, 1972; Cooper, 1973; Brenner and Oertli, 1976), although some Kirkwood faunas have latest Jurassic affinities (McLachlan and McMillan, 1979).

Stratigraphic reconstruction that takes into account lateral facies variations recognizes three main stratigraphic units, the Enon Conglomerate, Kirkwood, and Sundays River Formations (Table 12-1). The greatest thicknesses of Enon conglomerates, sandstones, and mudstones are associated with basement faulting. Basinward the Enon suite merges or interfingers with the finer grained Kirkwood Formation (the "variegated marls" and "wood beds" of Atherstone, 1857). The 2200-m-thick Kirkwood succession consists of interbedded pebbly sandstone and mudstone which contain foliage and wood fossils (Fig. 12-4) (Winter, 1973; Shone 1976). The most distal, argillaceous interval is the Sundays River Formation, which consists of up to 1860 m of impersistent sandstone, siltstone, and mudstone, the biota including plant debris and a diverse marine fauna.

Enon braided alluvial fans developed in response to active faulting along the margins of the basin. Resting unconformably on folded pre-Mesozoic rocks, the succession tapers northward and consists of poorly sorted conglomerate with subordinate sandstone, siltstone, and mudstone lenses. Large blocks of pre-Mesozoic rocks, some larger than 4 m in diameter, occur at the base near fault scarps (Rigassi and Dixon, 1972). In general, however, the clasts are smaller than 50 cm in diameter and are well rounded. Quartz arenite and other rock types identifiable with local exposures of Paleozoic and Precambrian rocks are the dominant clast lithologies.

Like many modern alluvial fans, these coarse conglomerates are associated with grabens. The greatest thicknesses and coarsest material are close to the bounding faults (Fig. 12-5). Debris-flow processes were common in the early stages of development of the Algoa basin, depositing poorly sorted matrix-supported conglomerates (Hill, 1972).

Apart from the initial debris-flow deposits, the Enon conglomerates are generally clast supported and the matrix silty; they occur in erosively based, poorly bedded units that are separated by red subarkosic sandstones in places. The sand-

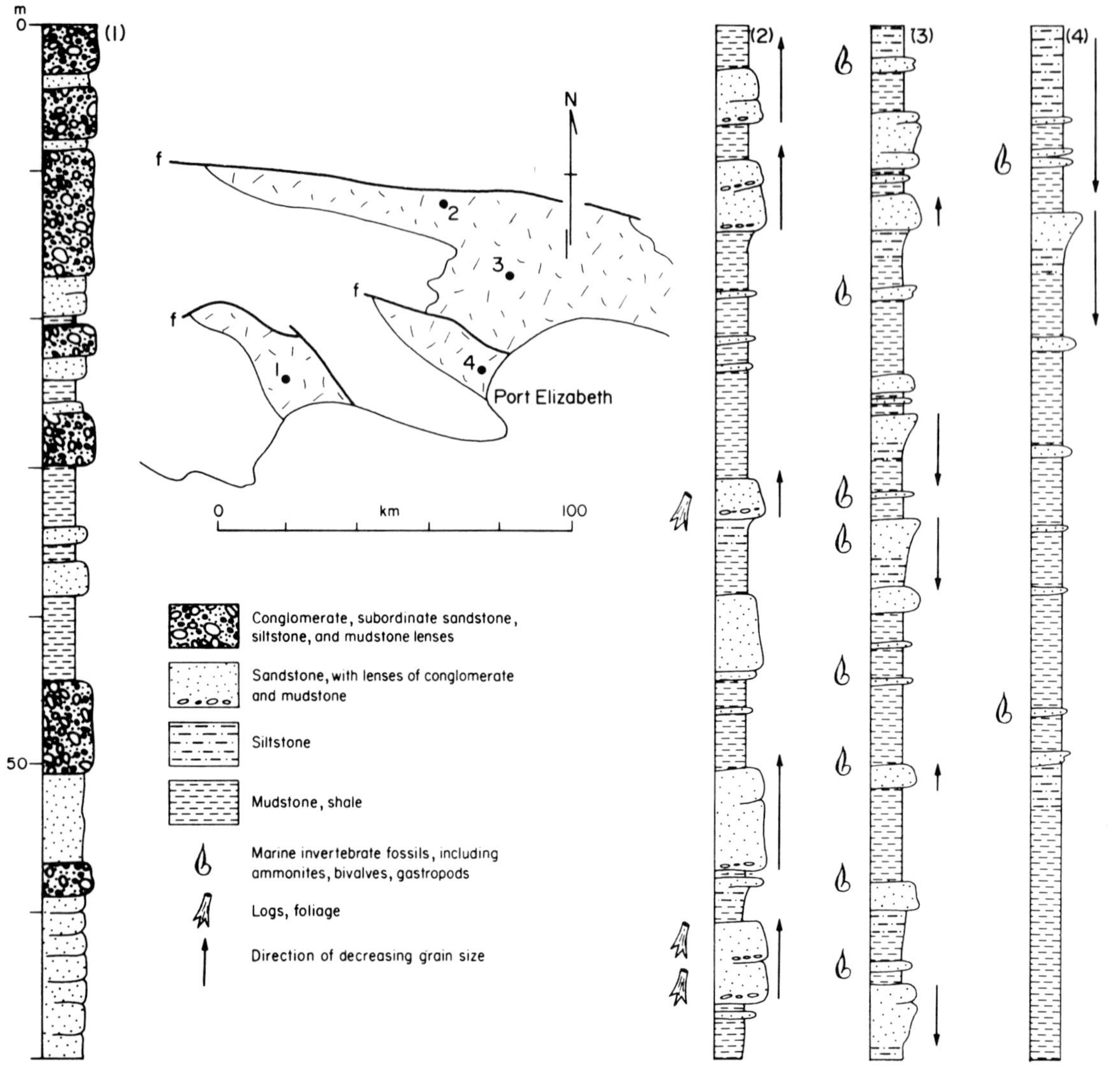

Figure 12-4. Stratigraphic sections: (1) Enon Formation in the Gamtoos basin; (2) and (3) Kirkwood Formation and (4) Sundays River Formation in the Algoa basin. [Sources: Shone (1976), personal communication, 1979).]

stones show planar and trough cross-bedding and less common plane-bedding. Paleocurrent data, including cross-bed azimuths and imbricated clasts, indicate deposition by low-sinuosity, northward-flowing streams (Fig. 12-3). Fossil wood occurs within the sediments, and wood and dinosaur-like fossils have been found in the tuffs and basalts (Haughton and Rogers, 1924; Hill, 1972; McLachlan and McMillan, 1976). Mudstone and other nonresistant clasts are rare in a suite dominated by quartz arenites, suggesting that the climate was humid. The crudely bedded and imbricated conglomerates are interpreted as the deposits of longitudinal bars in a braided-stream environment. Deposition during low-river stage is reflected in the lenses of finer grained material; thicker cross-bedded sandstones were probably deposited by migrating channel-floor dunes.

Alluvial-plain channel-fill processes contributed to accumulation of thick successions of lithic sandstones in the Enon and Sundays River Formations (Fig. 12-4). Lenticular units, rarely thicker than 2 m, are accentuated by bone-bearing pebble lag. Planar and trough cross-bedding indicate a southeasterly to southwesterly paleoflow system. The channel sandstones locally display a progressive vertical decrease in bedding thick-

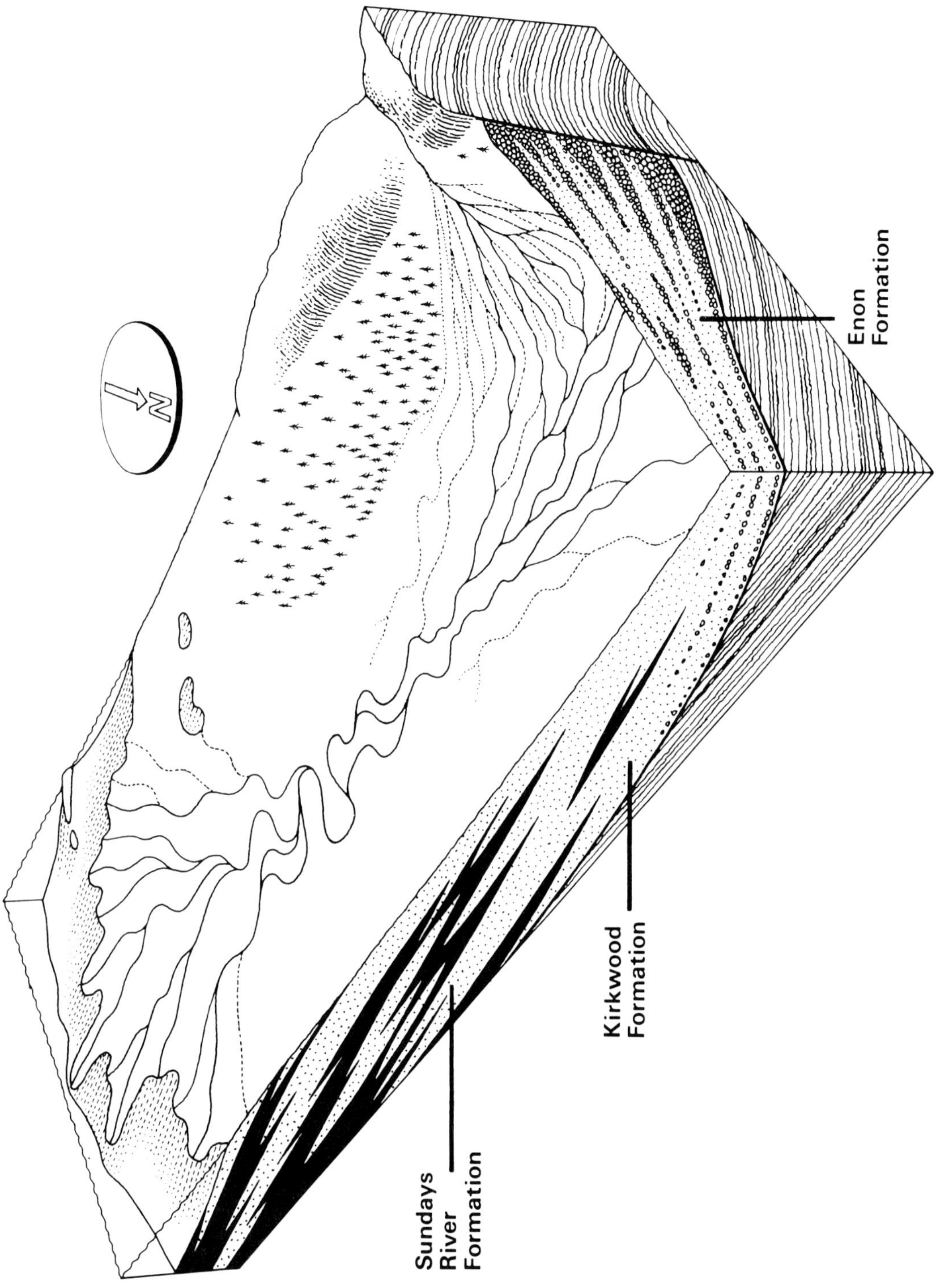

Figure 12-5. Schematic depositional model of Uitenhage Group in the Algoa basin indicating relationship to pre-Mesozoic basement.

ness; others show a tendency to become more thinly bedded only in their uppermost parts (Winter, 1973; Shone, 1976). Winter (1973) has inferred a fluvial point-bar origin for some of these motifs. Overlying many of the arenaceous units are rooted mudstones. Conglomerates are a minor component and form poorly bedded lenses and shallow channel-fill. Fossil wood (conifer and cycad) and fern fossils are abundant.

Paleocurrent patterns (Fig. 12-3) were significantly influenced by basin configuration. Northward-flowing streams that drained alluvial fans building away from valley-side fault scarps were deflected along the lower gradient sandy valley floor (Fig. 12-5).

Estuary and interdistributary bay environments are inferred from the thick development of bioturbated and rooted red and green mudstones in the Kirkwood and Sundays River Formations (Fig. 12-4). Sedimentary structures include flaser and wavy bedding. Gradationally or erosively based sandstone beds up to 2 m thick interrupt the argillaceous sequences and display an ebb-dominated bipolar paleoflow pattern (Shone, 1978b). Nonmarine fossils include plants and reptiles (*Algoasaurus*), and several invertebrates (e.g., *Unio*, *Modiola*) of brackish or freshwater affinity. Other infaunal and epifaunal invertebrate assemblages (*Astarte, Pecten, Perna, Ostrea,* trigoniids, *Turbo*, *Turritella*) are more characteristically marine (McLachlan and McMillan, 1976). Palynomorphs, ostracods, and arenaceous foraminifera occur in a carbonaceous shale (Colchester shale of Rigassi and Dixon, 1972).

Similar conditions of quiet-water deposition and bioturbation are found in the interdistributary bays and estuaries of modern depositional coasts. The mixed freshwater and marine fossil assemblages provide compelling evidence for a salinity gradient in these Algoa embayments. Traction processes were concentrated principally in tidal channels, depositing mud-clast and shell lag and causing channel-floor migration of dunes through the tidal cycle (Shone, 1978b). Bay-margin encroachment provided a platform for vegetation growth.

Progradational deltaic and shelf sedimentation is reflected in thick sequences of mudstone and siltstone in the Sundays River Formation which commonly display upward-coarsening tendencies through a 60–70 m interval (Fig. 12-4) (Winter, 1973; Shone, 1976, 1978b). Interlayered shell-bearing sandstones form sharp-based lenses. Sedimentary structures include flow rolls, large-scale slumps, and flaser and wavy bedding. Gradationally superimposed on these argillaceous rocks are fine- to medium-grained sandstones that are locally micritic. Channels up to 3 m deep are common, with scoured bases overlain by mud-clast conglomerate, shell lag and fossil wood. The sandstones are planar and trough cross-bedded and show an ebb-dominated polymodal slipface distribution (Shone, 1978b). Mollusk shells, horizontal and vertical trace fossils, and bioturbation are ubiquitous. The diversity and population density of the invertebrate faunas increase in a southeasterly direction; ammonites are most common in the distal parts of the basin.

The lithologic and biological characteristics of these upward-coarsening sequences suggest deposition by deltas subject to substantial marine reworking (Fig. 12-5). Outcrop and subsurface studies (M.J. Leith and S.R. Du Toit, personal communication, 1980) confirm these textural patterns. Depositional or erosional oversteepening subjected the subaqueous prodelta slopes to gravity mass movements.

In summary, deposition in the Algoa basin was accomplished by braided-fluvial and shoal-water deltaic processes in a Newark-type sedimentary basin (Fig. 12-5). Alluvial fans developed in response to active boundary faulting.

The Landlocked Oudtshoorn Basin

The Oudtshoorn basin (Fig. 12-2) is an east–west-trending fault-controlled basin, about 90 km long, which developed by north–south extension and collapse faulting (Kleywegt, 1971; Lock, 1978a). Enon rock types are entirely terrestrial in origin (Du Preez, 1944; Rust and Winter, 1979). Volcanic tuffs have been observed in the eastern part of the basin (Lock *et al.*, 1975). The sedimentary prism is more than 3500 m thick along the Cango fault (Rust and Winter, 1979).

Thick alluvial-fan conglomerates in the Oudtshoorn basin are transitional downslope into alluvial-plain sandstones and playa lake mudstones (Lock, 1978a; Rust and Winter, 1979). Clasts up to 2 m in diameter near the base include quartzite and mudstone derived from the Table Mountain and Bokkeveld Groups. Distal and upward decrease in grain size and bed thickness is accompanied by increasing clast roundness and maturity. Well-developed imbrication shows that flow was initially concentrated on northward-building alluvial fans, but diversion resulted in eastward migration of sand bedforms (Lock, 1978a).

Deposition of the upper parts of the Oudtshoorn succession was influenced by increasing aridity and subsidence along the northern boundary fault. Coarse, poorly bedded, matrix-supported conglomerates accumulated as southward-moving debris flows (Lock, 1978a). Clasts of Precambrian composition record progressive unroofing of the Kango Group with fault movement. Basinward these poorly sorted fan deposits are intimately associated with a monotonous succession of horizontally laminated mudstones with minor siltstones, sandstones, gypsum, and limestone lenses. The red coloration and evaporite formation point to a semiarid playa lake origin (Rust and Winter, 1979). Fossils include foliage, wood, rare lignite, and dinosaur teeth.

Several small faulted basins containing coarse terrigenous detritus occur east of the Oudtshoorn basin. Lock (1978a) suggested that, together with the Oudtshoorn basin, they formed a continuous, elongate, structurally controlled basin occupied by braided streams.

The Agulhas Bank

The Agulhas Bank is the triangular-shaped southern continental shelf which is underlain by Mesozoic and Cenozoic sediments of the Outeniqua basin (Fig. 12-2; Table 12-1); it is 240 km wide at 21°E (Du Toit, 1976). The Outeniqua basin (also referred to as the Agulhas basin by Simpson and Dingle, 1973, and the St. Francis basin by Emery *et al.*, 1975) is a composite feature comprising several smaller troughs separated by basement highs or arches. Large-scale normal faults mark the northern edge of each trough. The sedimentary prism is mainly underlain by continental crust and the arches appear to be southeasterly continuations of Cape Fold Belt anticlinoria. The Agulhas arch differs from the others in that its core probably comprises Cape granite and Malmesbury rock (Scrutton and Dingle, 1974).

Shearing of the Falkland Plateau from the African plate along the Agulhas fracture zone before the Valanginian (130 Ma) generated a marginal fracture ridge consisting of metamorphosed continental crust (Scrutton and Dingle, 1976; Du Toit, 1979). The Agulhas fracture zone separates the African continent from the deep ocean basin. It is characterized by a nearly linear positive magnetic signature (Simpson, 1968; Du Plessis and Simpson, 1974; Rabinowitz, 1976). Half-graben subsidence increased the depth of the separate basins still further. Recent studies show that Outeniqua sediments only overtopped the Agulhas marginal fracture ridge in the latest Cretaceous and Early Tertiary (Fig. 12-6) (Scrutton and Dingle, 1976).

Mesozoic sediment thicknesses range between 3000 and 5700 m in the half-grabens and grabens, thinning only over the horsts (Du Toit, 1976). Scrutton and Dingle (1976) suggest that high-density rocks in the lower crust beneath the Mesozoic cover isostatically compensate for the thick accumulation of low-density sediments.

The sedimentary history, established on the evidence of seismic and borehole data (Dingle, 1973a; Du Toit, 1976, 1979), shows similarities with the onshore Algoa basin. Enon-type conglomerates, sands, and muds were contributed to intermontane valleys by the fluvial systems draining the Cape mountainland as well as the elevated areas of the Falkland Plateau in the south (Dingle and Scrutton, 1974). By analogy with Algoa basin counterparts, Du Toit (1976, 1979) has suggested that deposition occurred on braided alluvial fans. Lateral motion of the Falkland Plateau in the Mesozoic removed the southern provenance and induced tensional faulting on the northern boundaries of the individual basins. The coarsest and thickest alluvial sediments accumulated against these boundary faults.

Conglomerates are volumetrically important along the proximal margin of the Outeniqua basin, but downslope and upward they merge with fine- to medium-grained sandstones and plant-bearing mudstones, which Du Toit (1976, 1979) attributes largely to fluvial deposition. This persistent textural gradient terminates in a 1300-m-thick interval of siltstone, mudstone, and shale which contains a sparse marine fauna.

Two major periods of delta progadation, separated by seismic horizon C (Valanginian unconformity), are reflected in the superimposed upward-coarsening sequences of glauconitic shale, siltstone, and sandstone. Shallowing and emergence are shown by the increasing dominance of bed-load sediments and lignites near the top (Du Toit, 1976, 1979). Electric-log patterns suggest that as much as 1400 m of shales and mudstones, containing disproportionately small amounts of sand-

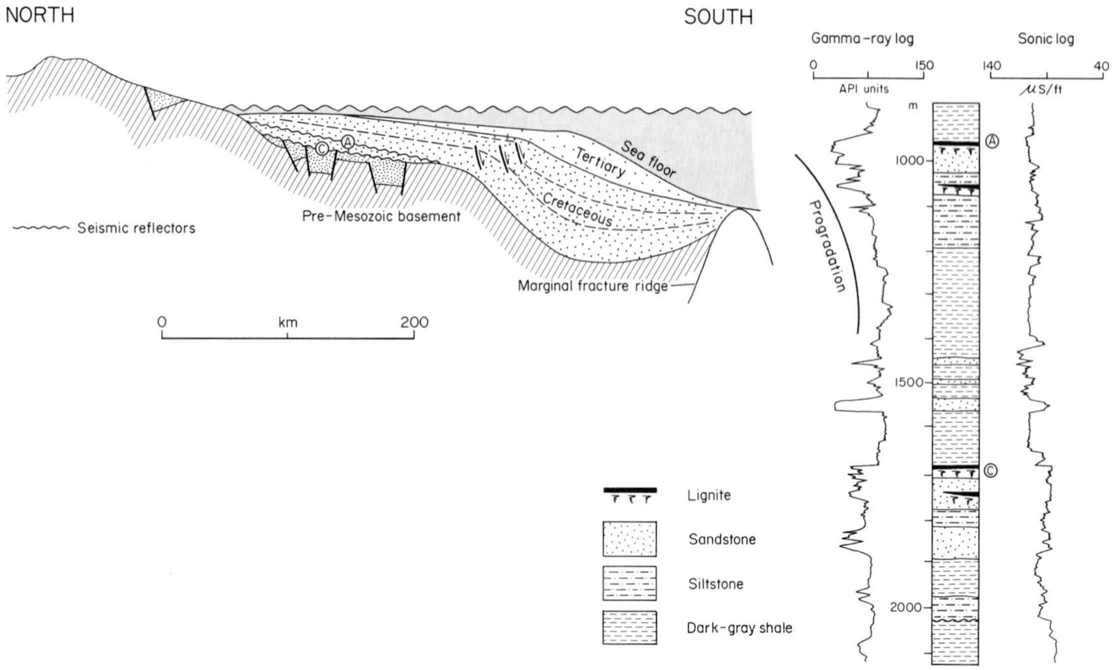

Figure 12-6. North–south stratigraphic section across the Agulhas Bank. The gamma–ray log shows that sedimentation between seismic reflector horizons C and A was initially argillaceous and interrupted by gravity-resedimented slope deposits, followed by delta progradation giving an upward-coarsening pattern. The gamma-ray and sonic logs are composites of two adjacent holes: that below seismic reflector C is from borehole Ga-A2, and that above from borehole Ga-B1 (see Fig. 12-2). Vertical scale refers to depth below sea level. [Source: Du Toit (1976).]

stone and limestone, overlie the deltaic facies. Du Toit suggests that these richly fossiliferous sediments were deposited on a tectonically deepened shelf.

Subsidiary Basins

Several smaller fault-controlled basins parallel the structural grain of the Cape Fold Belt (Fig. 12-2). The generation and preservation of thick conglomerate sequences resulted from substantial graben subsidence, amounting to 3030 m in the Gamtoos basin, 1460 m in the Mossel Bay basin, 2100 m in the Plettenberg basin, 1000 m in the Knysna basin, and 2130 m in the Heidelberg basin (Rigassi, 1970; McLachlan and McMillan, 1976). Close to the bounding faults red conglomerates with sandstone lenses reflect high-gradient braided alluvial-fan deposition. Grain size decreases basinward, where cross-bedded sandstones with interbedded conglomerates and mudstones are important. The mudstones are locally carbonaceous or lignitic and contain plant fossils. In the Gamtoos basin there are lignitic clays, resinous hydrocarbons, and conifer, cycad, and fern fossils (McLachlan and McMillan, 1976). Other taxa recognized in the Heidelberg basin include freshwater invertebrates, beetles, and fish scales. A Late Jurassic to Early Cretaceous age for the flora is probable (Seward, 1907; Martin, 1960).

Marine circulation influenced the Plettenberg and Knysna basins. Fluvial deposits in the former are interbedded with cross-bedded and ripple cross-laminated sandstones with tidal-bedding structures. Fossils of marine affinity are common, including coccoliths, asteroids, *Megatrigonia*, and rare ammonites of probable Early Cretaceous age (McLachlan and McMillan, 1976; Rust and Rogers, 1979). The terrestrial facies contains a diverse flora: Cycadophyta, Cycadales, *Pterophyllum*, *Ptilophyllum*, Pteridophyta ferns, and *Neocalamites* stems (Rust and Rogers, 1979).

A succession of rocks up to 20 m thick in the Knysna basin, referred to as the Brenton Formation, has been attributed to deposition in estuarine lagoonal and beach environments (McLachlan *et al.*, 1976a). The sedimentary suite includes carbonaceous shales, mudstones, cross-bedded sandstones, and conglomerates with a glauconitic sand matrix. Fossils are most common in the finer grained components; these include "*Trigonia*," *Nautilus*, *Belemnites*, ammonites, echinoids, foraminifera, ostracods, coccoliths, plesiosaur-like teeth, and dinoflagellates (McLachlan *et al*., 1976a). A Late Valanginian to Hauterivian age is inferred (Beer, 1972; Stapleton and Beer, 1977; McLachlan *et al*., 1976a), rather than Late Jurassic as suggested by Dingle and Klinger (1972).

Two shore-zone deposits of Late Cretaceous age are in close proximity at Needs Camp and Igoda (Fig. 12-2; Table 12-1). The bryozoa-dominated bioclastics of the former and the sands and shelly limestones of the latter share a fauna that is both diverse and abundant and of Late Campanian to Early Maastrichtian age (Chapman, 1916; McGowran and Moore, 1971; Klinger and Lock, 1978). This fauna includes corals, bryozoa, echinoderms, bivalves (*Inoceramus*, *Exogyra*, oysters), brachiopods, ammonites (*Baculites*, *Saghalinites*), ostracods, and foraminifera, with shark and plesiosaur teeth.

12.3. The Transkei Swell and the Zululand Basin

An anomalous region of positive relief in the southeastern part of the subcontinent, the Transkei swell (Simpson and Dingle, 1973), maintained faunal isolation of the Zululand and Algoa–Outeniqua basins into the Coniacian. In the south, crustal arching restricted Mesozoic sedimentation to three small depositories at Mzamba, Mbotyi, and Mgazana (Fig. 12-2).

The Mbotyi and Mgazana basins (Table 12-1) are both fault bounded and contain thick conglomeratic successions (McLachlan *et al*., 1976b). The largest clasts, some as much as 5 m in diameter, are associated with the boundary faults: Karoo diabase, sandstone, and mudstone are the dominant clast lithologies. In the Mgazana basin the distal conglomerates of possible fan-delta origin contain a diverse invertebrate fauna characterized by Late Valanginian belemnites and ammonites (McLachlan *et al*., 1976b).

Low seacliffs at Mzamba expose a suite of basal gravels and interbedded mudstones and bioclastic limestones above a platform of Lower Paleozoic rocks (Du Toit, 1954, p. 399; Haughton, 1969, p. 425). The gravel contains *Teredo*-bored logs, shark teeth, reptilian teeth, and turtle carapace fragments. Clast long axes indicate shoaling wave processes, which also left a record of symmetrical and asymmetrical ripples slightly oblique to the modern shoreline. Lithologically and paleontologically these outcrops strongly suggest deposition in a transgressive beach and shoreface environment. The bioclastic material is dominated by a diverse mollusk fauna, including abundant trigoniids, and coiled and uncoiled varieties of ammonites; echinoderms occur in the argillaceous horizons. Although the age is disputed, recent investigations by Klinger and Kennedy (1977) suggest a Middle Santonian to Early Campanian age.

Providing an important biostratigraphic link between the major Zululand depository and the Mzamba outcrop is a seaward-thickening wedge of Upper Cretaceous strata beneath Durban (Kennedy *et al*., 1973). The 100-m-thick succession consists of a basal boulder bed overlain by glauconitic mudstone, sandstone, and fossiliferous limestone. The biota is diverse, comprising mollusks, serpulids, corals, echinoderms, foraminifera, and plants.

Borehole Jc-1 off the Natal coast penetrated Upper Cretaceous through Danian and Tertiary clays and silts with thin limestone interbeds. This succession is interpreted as a shelf deposit reflecting substantial terrigenous influx from the ancestral Tugela River (Du Toit and Leith 1974).

Cretaceous sediments are developed extensively beneath the Zululand–Mozambique coastal plain (Fig. 12-7); this basin is 440 km at its widest in Mozambique. The oldest terrestrial sediments are thought to be related to Jurassic rifting (Flores, 1973; Dingle and Scrutton, 1974). Rifting and uplift was accompanied by north–south tensional faulting which generated seaward-stepping fault blocks. Strong relief resulted in an abundance of coarse clastic debris. Foundering of the basin during the opening of the Indian Ocean induced marine transgression from north and east. Fossil evidence indicates an interrupted marine succession from Barremian to Maastrichtian in the onshore basin (Kennedy and Klinger, 1972).

Unequal rates of basinal subsidence are reflected in the geometry of the sedimentary prism, which expands from 1000 m in thickness in the St. Lucia area (Kennedy and Klinger, 1975), to nearly 2500 m in the Zululand–Mozambique basin (Hales and Gough, 1962), and in excess of

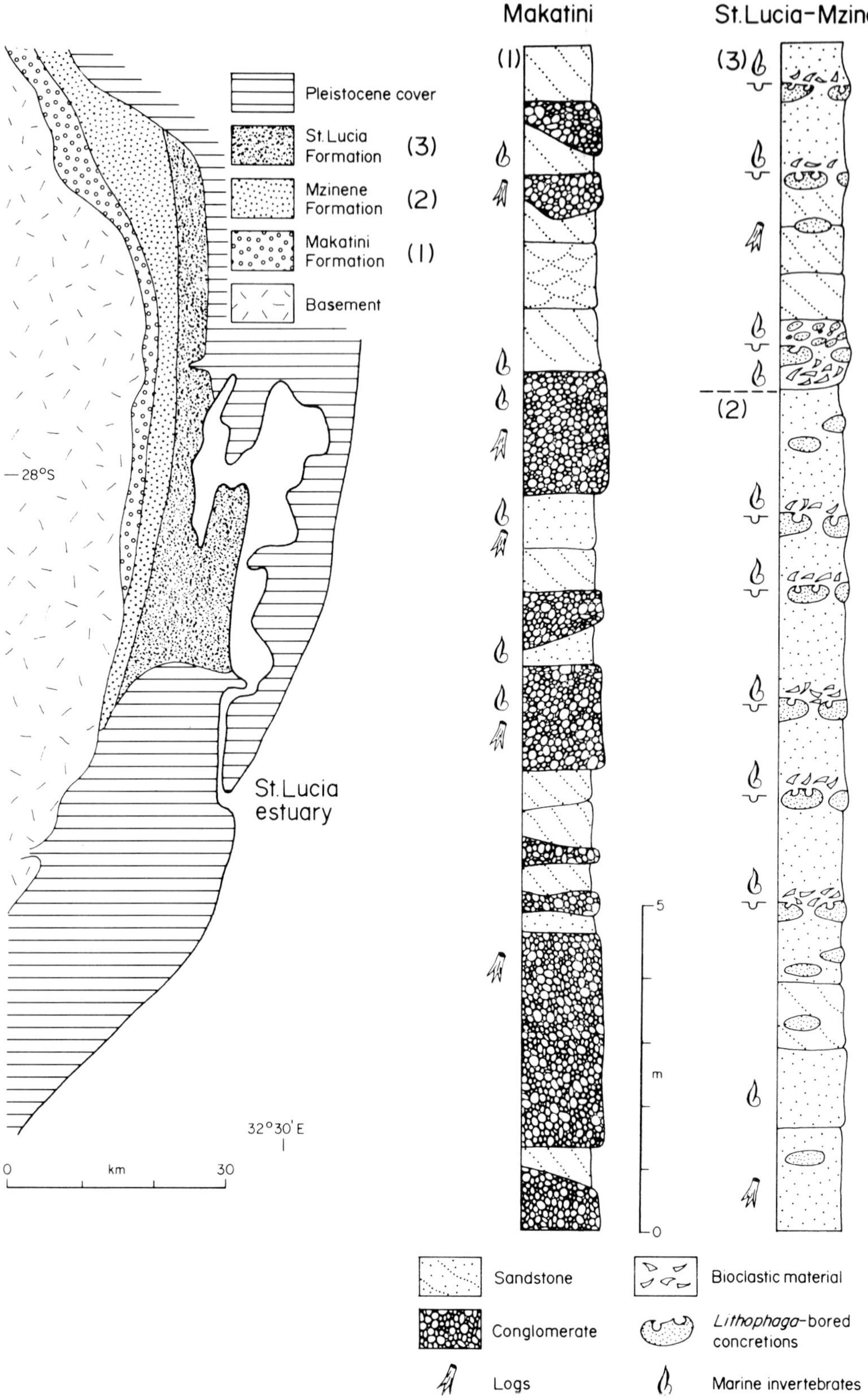

Figure 12-7. Composite stratigraphic sections in the Zululand basin showing (1) braided alluvial-plain and fan-delta facies of the Makatini Formation: (2) and (3) shallow-marine facies of the Mzinene Formation and St. Lucia Formation. [Sources: Kennedy and Klinger (1975), H.C. Klinger, (personal communication, 1977).]

4600 m in the offshore Zambezi basin (M.J. Leith, personal communication, 1980). The Mesozoic stratigraphy of the Zululand Group includes three formations (Table 12-1). Regional low-angle unconformities, which also mark biostratigraphic hiatuses, separate the Makatini, Mzinene, and St. Lucia Formations. The succession is best developed in the St. Lucia area (Fig. 12-7), where it overlies seaward-dipping Karoo lavas (Kennedy and Klinger, 1975).

The Makatini Formation, which consists of conglomerates, sandstones, and siltstones, is attributed to progradation of braided fluvial deposits into a shallow sea. The Mzinene and St. Lucia Formations consist of richly fossiliferous glauconitic siltstones and fine-grained sandstones characteristic of shallow-marine accumulation; they are separated from the Makatini Formation by an erosional or nondepositional hiatus marked by a bed of *Lithophaga*-bored concretions (Kennedy and Klinger, 1972). This paleontologically detectable break can be correlated with a major Aptian–Albian nonsequence in Madagascar (Collignon, 1963), suggesting that regional tectonism was related to rifting between eastern and western parts of Gondwana. A similar break, which has no correlative in Madagascar, separates the Mzinene and St. Lucia Formations. The slight angular unconformity is overlain by concretions containing Cenomanian fossils and a thin *Pterotrigonia* conglomerate with Coniacian *Proplacenticeras* (Fig. 12-7); Turonian rocks are absent (Kennedy and Klinger, 1972).

Alluvial-Plain Deposition

The thick development of conglomerates indicates a marked tectonic relief along the margin of the Zululand basin. Clast size rarely exceeds 80 cm in diameter (average 5–15 cm), the clasts comprising mainly volcanic and quartzitic rock types. Debris-flow processes are recognized on the basis of poorly sorted, massive, matrix-supported conglomerate texture. However, waterlaid deposits are more common and are typified by clast-supported conglomerates with minor cross-bedded intercalations of medium- to coarse-grained arkose. The clasts are better rounded than in the debris-flow deposits and are weakly imbricated. Generally the conglomerates form erosively based tabular and lenticular bodies with a predominantly massive appearance, although horizontal bedding and cross-bedding are not uncommon. Silicified logs up to 6 m long and 60 cm in diameter are common (H.C. Klinger, personal communication, 1977).

The horizontally bedded conglomerates possibly originated as longitudinal gravel bars. Under conditions of decreasing discharge, vertical accretion formed large-scale foreset slopes (Hein and Walker, 1977), followed by sand deposition as bar-margin sand wedges and by channel-floor dune migration. On modern outwash fans the upper surfaces of gravel bars are commonly littered with logs (Boothroyd, 1972).

Bed thickness and maximum clast dimensions decrease eastward toward a sandstone-dominated distal facies. The sandstones are commonly fine to medium grained and are characterized by eastward-directed trough and planar cross-bedding in sets up to 4 m thick (average, 30–40 cm) and plane-bedded sands and silts. Associated features include gravel lag, diffuse gravel sheets, mudstone intercalations, and lignite. *Teredo*-bored logs are abundant in the sandstones and in some thick conglomerate interbeds, and many are encrusted with oysters and serpulids. Fossils of shallow-marine affinity, including the bivalve *Gervillela*, trigoniids in life position, and anthropod burrows, are found in the upper parts of some sedimentation units. Bioclastic limestones and the large ammonite *Tropaeum* are characteristic of the distal facies (H.C. Klinger, personal communication, 1977).

A sandy alluvial floodplain origin is suggested for the sandstones on the basis of their interfingering relationship with the proximal conglomerate facies, their consistent easterly paleocurrent pattern, and the presence of plant fossils and lignite. In this context the high proportion of planar-cross bedding attributed to sandy transverse braid-bar deposition. The depositional setting may have resembled the braided distributary and tidal-flat complex of the Copper River fan delta of Alaska (Galloway, 1976).

Shallow Epeiric Seas

The glauconitic siltstones and fine-grained sandstones of the Zululand basin are richly fossiliferous and contain bored limestone concretions throughout (Fig. 12-7). Although structured mainly by plane-bedding, cross-bedding and cross-lamination are also common. The biota includes infaunal and epifaunal bivalves, gastropods, ammonites, nautiloids, echinoids, driftwood, and plant debris. *Teredo*-bored logs commonly lie among drifted shell lag on channel floors. The succession is characterized by a series of upward-fining sedimentation units which vary in thickness up to 3 m (Fig. 12-7). Each cycle begins with zone of *Lithophaga*-bored limestone concretions, which are commonly colonized by commensal serpulids and oysters, overlain by 20–30 cm of upward-fining shell lag in which shells are current aligned and imbricated. The overlying glauconitic sandstones and siltstones contain drifted mollusks and infaunal bivalves in life position (e.g., "*Trigonia*," "*Inoceramus*"). Bioturbation is extensive and large arthropod burrows are abundant (H.C. Klinger, personal communication, 1977).

Selwood (1970) has described small-scale sedimentary cycles in the British Lias which he ascribes to shore-zone sedimentation in an epeiric sea. In its early drift stage the Indian Ocean was probably similar. The invertebrate fossils suggest an intertidal to shallow-subtidal marine environment (e.g., *Teredo*, *Lithophaga*, *Trigonia*, *Ostrea*). A recent counterpart of the plane-bedded sandstones and siltstones is a flat, featureless, sandy intertidal and subtidal zone behind breaker bars along the margin of the Copper River fan delta (Galloway, 1976). The asymmetric cycles in the Zululand basin are attributed to fluctuations in the rate of sediment influx into a continuously subsiding basin. Westward transgression during periods of extremely low sediment influx was accomplished by erosional shore-zone retreat, the preserved record of which is a thin basal transgressive disconformity, or ravinement, with associated bored calcareous concretions. These concretions may have formed during periods of slow sedimentation that accompanied transgression (Young, 1957; Ryer, 1977). The dominant fine-grained part of each cycle resulted from renewed shoreline progradation. Although the benthic fauna was dominated by infaunal species, large shells of *Inoceramus*, which was adapted to living on soft substrates, and tests of ammonites provided hard surfaces for attachment of other epifaunal bivalves. During the Early Coniacian transgression the ammonite *Proplacenticeras* was far more abundant than other taxa; Cooper (1977) has attributed this population explosion to increased trophic resources.

12.4. Synthesis

Deep-sea drilling and geophysical investigations have shown the continental character of the Falkland Plateau (Barker *et al.*, 1976c; Rabinowitz *et al.*, 1976), thus imposing a reconstruction of Gondwana in which the wedge-shaped plateau fits between southern Africa and Antarctica in the southwest (Fig. 12-1). This arrangement is supported by the recent discovery of magnetic anomalies in the Mozambique basin and off Antarctica, which suggest that Dronning Maud Land was adjacent to southern Mozambique at the beginning of the Mesozoic (Bergh, 1977; Simpson *et al.*, 1979). However, the relationship between the Antarctic peninsula and the southern Andes and the paleoposition of Madagascar are still unresolved.

Disregarding reconstructions that overlap the Antarctic Peninsula and the Falkland Plateau, there is a choice of models that address the geologic similarities between Patagonia and the Antarctic Peninsula. Dalziel and Elliott (1973) have inferred oroclinal bending of a linear belt, a concept that has been modified by De Wit (1977) to include rotation of the Antarctic Peninsula so that it was continuous with the southern Andes. Because magnetic data argue against independent motion of the Antarctic Peninsula, Harrison *et al.* (1979) have proposed a reconstruction that places the Antarctic Peninsula adjacent to the Patagonian Andes. Prior to fragmentation the southwestern margin of Gondwana was apparently an arc–trench system (Suarez, 1976; De Wit, 1977; Harrison *et al.*, 1979).

The predrift position of Madagascar is also contentious and three interpretations have been proffered: a southerly position adjacent to Mozambique (Flores, 1970; Heirtzler, 1971; Tarling, 1972; Barron *et al.*, 1978); a northerly position adjacent to Kenya (Du Toit, 1937; Smith and Hallam, 1970; McElhinny *et al.*, 1976; Simpson *et al.*, 1979); and an unchanged position relative to Africa (Kent *et al.*, 1971; Darracott, 1974; Förster, 1975; Kutina, 1975). So far the evidence for Madagascar's paleoposition is equivocal and contradictory.

A southern paleoposition for Madagascar is unlikely because it lacks Waterberg Supergroup, Cape Supergroup, and Karoo volcanic rocks (Smith and Hallam, 1970). Recently, Simpson *et al.*, (1979) have identified east–west-trending

Mesozoic magnetic anomalies M2 through M22 in the northern Mozambique Basin, which are strong evidence against placing Madagascar against southeast Africa. Paleomagnetic results derived from Karoo-age rocks suggest that Kenya and Madagascar were neighbors until the Cretaceous (McElhinny *et al.*, 1976). Further support for this reconstruction is inferred from the north–south-trending ocean-floor topography (Bunce and Molnar, 1977; Segouffin, 1977; Simpson *et al.*, 1979). Yet the east African coastal plain contains a predrift history of marine deposition (Quennell *et al.*, 1956; Kent, 1973, 1974), possibly related to an early stage of rifting. The generally accepted model—which fits Madagascar against the Tanzania–Kenya bight—also requires transcurrent motion between India and Madagascar and separation of the lava fields of Mozambique and Madagascar (Kent, 1973; Barron *et al.*, 1978). However, seismic and subsurface data have been cited as evidence that the Mozambique Channel is underlain by continental crust (Kent *et al.*, 1971; Darracott, 1974; Förster, 1975).

The beginning of plate separation was marked by the formation of grabens and rapid accumulation of undated continental deposits and by intense faulting and magmatism. Magnetic anomalies observed in the Mozambique Basin and off Dronning Maud Land show that the earliest separation occurred between Africa and Antarctica by the Late Jurassic (Simpson *et al.*, 1979). Along its eastern edge the Mozambique Basin is probably separated from the Madagascar Basin by a plate boundary which separates different trends of fracture zones and magnetic anomalies (Bergh and Norton, 1976). The Indian Ocean only started opening between 190 Ma and 160 Ma by exploiting a rift along the eastern edge of the Mozambique Ridge (Dietz and Holden, 1970; Scrutton, 1973b). On this basis the Mozambique Channel had its origins in intermittent vertical tectonics which accompanied separation between east and west Gondwana (Kent, 1974; Kutina, 1975).

High basement relief and coarse fluvial deposition along the western margin of the Zululand basin are indicative of steep gradients and down-faulting related to rifting. However, opening of the Indian Ocean from the north (Heirtzler *et al.*, 1968) initiated a marine transgression that only reached Zululand latitudes in the Barremian. The depositional history was controlled by sporadic supply of large volumes of sediment, and tectonic instability of the newly formed continental margin. Late Cenomanian regression left a hiatus in the marginal sediments of Zululand and the southern Cape, but on the continental shelf (borehole Jc-1) sedimentation continued unabated through the entire Cretaceous and into the Danian. The stratigraphy of the Zululand Group, including the major Turonian hiatus, characterized a relatively small region and cannot be correlated precisely even with Madagascar.

Within a rift basin subjected to graben subsidence and tensional step faulting (Maud, 1961) it is difficult to separate eustatic effects from tectonic overprint. According to Sleep (1976), significant gaps in the geologic record could be produced by small sea-level fluctuations in a rapidly subsiding basin. Nevertheless, Kauffman (1973), Cooper (1977), and Vail *et al.* (1977) have recognized as many as 14 worldwide Cretaceous transgressive–regressive cycles, which they attributed to eustatic processes.

The Late Cretaceous transgression was distinctly diachronous along the eastern margin of the subcontinent, reaching the St. Lucia region in the Coniacian, the Richards Bay–Durban–Mzamba latitudes by the Middle Santonian (Klinger and Kennedy, 1977), and the Needs Camp–Igoda region in the Late Campanian to Maastrichtian (Klinger and Lock, 1978).

Subsidence and expansion of the Natal Valley (Fig. 12-2) was probably related to end of Jurassic faulting in coastal Natal as well as rifting and separation between Africa and the Falkland Plateau. Sediment discharged from the Limpopo and Tugela Rivers accumulated proximally as sediment cones and in the more distal parts of the Natal Valley as turbidites (Dingle and Scrutton, 1974). The post-Early Cretaceous dispersal patterns have been strongly influenced by the northeast–southwest topography of the Almirante Leite and Naudé ridges, and the larger, southward-plunging Mozambique Ridge (Dingle *et al.*, 1977).

Magnetic anomalies of Mesozoic age in the Cape Basin west of Cape Town date the opening between Africa and South America at between 135 Ma and 130 Ma (Larson and Ladd, 1973; Rabinowitz and LaBrecque, 1979). Recalculation of the early opening pole for the South Atlantic suggests that there was a large amount of extension in the southernmost regions (Rabinowitz and LaBrecque, 1979). During separation the Falkland Plateau acted as part of the South American plate and sheared westward along the Falkland–Agulhas fracture zone. The first widespread marine transgression affected the Outeniqua basin in the Late Valanginian. Thick lacustrine deposits seaward of Brazil, Gabon, and Angola show that the Torres Arch–Walvis Ridge complex was a barrier to marine circulation into that region until the Aptian (Rabinowitz and LaBrecque, 1979). A Late Aptian transgression is correlated with evaporite accumulation. Fully open-ocean circulation between the North and South Atlantic basins was not established until Late Albian to Early Turonian times (Reyment and Tait, 1972).

The evolution of Algoa basin stratigraphy was closely related to the history of continental rifting and graben subsidence. Deposition was accomplished by braided fluvial and shoal-water deltaic processes in a Newark-type sedimentary basin (Fig. 12-5). Alluvial fans, built mainly by northward-flowing streams, developed in response to boundary faulting. Southeasterly flow patterns within the finer grained fluvial facies are believed to be the combined result of flow diversion along the lower gradient alluvial plain, and later shift of dispersal centers as a result of decreasing activity along the boundary fault.

The first marine incursion into the Algoa basin possibly occurred by the Late Jurassic (McLachlan and McMillan, 1979) concurrently with separation between Africa and Antarctica (Simpson *et al.*, 1979); but it is unlikely that marine circulation was fully established before the Falkland Plateau moved westward in the Late Valanginian. Continental breakup established a coastal plain across the lower reaches of the braided fluvial system, after which the shoreline regressed by deltaic progradation (Fig. 12-5). The paleogeography probably consisted of an irregular shoreline, open embayments, and restricted lagoons on a shoal-water delta, while flow projected from distributary channels deposited subaqueous delta sediments onto a shelf shallower than 70 m. Open-marine conditions are indicated by the ammonites and many of the other mollusks. Stratigraphic variations can be explained in terms of the combined effects of tectonic subsidence, rates of sediment influx, and lateral displacements of sedimentary facies tracts.

Compared with the distinct and younger Cretaceous faunas of the Zululand basin, the Algoa invertebrate fauna was characterized by low population densities and low faunal diversity. Shallow-water infaunal bivalves predominate and gastropods and ammonites (mainly *Olcostephus* and *Bochianites*) constitute a minor part of the assemblage. Kitchin (1907) was unable to distinguish more than a single paleontologic stage within the Algoa basin.

Regional studies show that the offshore Outeniqua basin comprised several loci of deposition with characteristics similar to those of the onshore Algoa basin. Shearing away of the Falkland Plateau restricted the thickest and coarsest deposits to the vicinity of active boundary faults where fluvial processes were prevalent. Du Toit (1976, 1979) suggested that the alluvial-facies tract was fronted by an irregular delta plain that merged with a thicker subaqueous delta prism and shelf succession. Age determinations suggest that rifting on the Agulhas Bank was probably contemporaneous with Middle to Late Jurassic extrusion of Drakensberg basalts (Du Toit, 1979; McLachlan and McMillan, 1979) and separation between Africa and Antarctica (Simpson *et al.*, 1979). Syntectonic sedimentation characterized the rifting stage, at the end of which permanent marine circulation was established. The onset of separation between the African and South American plates is correlated with seismic reflector horizon C, which Du Toit interprets as a regional unconformity. Facies stacking is attributed to reactivation of normal faults and continued graben subsidence. By the end of the Cretaceous, shelf muds had draped the pre-Cenomanian topography and overtopped the Agulhas marginal fracture ridge to spill onto the upper slope (Fig. 12-6).

Upper Cretaceous deposition was brought to a close by a major worldwide eustatic regression attributed to subsidence of oceanic crust as a result of decreasing rates of sea-floor spreading (Flemming and Roberts, 1973; Hays and Pitman, 1973). Regression was accompanied by intrusion and extrusion of alkaline volcanic rocks of the Alphard Bank on the northeastern edge of the Agulhas arch. These Lower Tertiary (58 Ma) volcanics include tuffs, trachybasalt, aegirine-augite trachytes, and aegirine–augite phonolitic trachytes (Dingle and Gentle, 1972; K-Ar).

An anomalous region of positive relief in the southeastern part of the subcontinent, the Transkei swell (Simpson and Dingle, 1973), maintained faunal isolation between the Zululand and Algoa–Outeniqua basins into the Coniacian. It has been suggested that this basement high may have been continuous with the Ross orogen of Antarctica (Rigassi and Dixon, 1972). Marine Mesozoic outcrops on the Transkei swell are limited to small fault-bounded basins such as Mgazana and Mbotyi. Faunal links with the East African coast in the Coniacian are indicated by the occurrence of the ammonite *Yabeiceras* in the Alphard deposits of the Outeniqua basin (Klinger *et al.*, 1976).

A correlation of the Mesozoic stratigraphy of southern Africa is shown in Table 12-1. On available evidence it is difficult to separate eustatic processes from tectonic "overprint." A major marine regression that commenced in the Late Cenomanian and excluded Turonian rocks from the near-shore areas of all the basins was accompanied by regional uplift of the subcontinent (Scrutton and Dingle, 1976). Moore (1976) has related the main phase of kimberlite intrusion to this Turonian epeirogenesis and regression.

Siedner and Mitchell (1976; K-Ar) attributed the linear dike swarms along the coast of Namibia with an isochron age of 134 Ma to the earliest rift phase. There is some evidence that this volcanism is related to stresses that developed along northeast–southwest lineaments during opening of the South Atlantic (Larson and Ladd, 1973; Marsh, 1973; Scrutton, 1973b). Basalts, andesites, and capping rhyolites of the Etendeka plateau are believed to have originated simultaneously with effusion of the Serra Geral basalts of Brazil at 121 Ma (Siedner and Mitchell, 1976).

Radiometric dates distinguish three major episodes of alkaline volcanism since the breakup of Gondwana: Turonian, basal Tertiary, and Oligocene (Moore, 1976); these rocks are expanded upon in Chapter 13. In coastal Namibia a foyaite–nepheline syenite complex has been dated at 130 Ma (Marsh, 1975; K-Ar).

Another consequence of continental fragmentation is reflected in the Mesozoic paleoclimates. In the Late Triassic southern Africa was situated between south latitudes 45° and 60° (Dietz and Holden, 1970). Desert and semidesert conditions were universal and in late Karoo times the Stormberg environment was characterized by dune fields, playa lakes, and ephemeral streams (Van Dijk *et al.*, 1978). Bigarella (1970) has postulated an intensification of aridity through the Jurassic. With continental separation in the Cretaceous, shallow seas expanded, thus increasing shoreline length and the influence of marine conditions. The southern Cape and Zululand alluvial facies were deposited by fluvial processes with moderately high runoff and humid conditions. This is supported by a diverse and abundant fossil flora.

Chapter 13

Kimberlites and Associated Alkaline Magmatism

S. C. ERIKSSON

The substantial contribution of the alkaline igneous rocks to the South African economy is well known and far exceeds their abundance in the geologic record. Most notable are the diamondiferous kimberlites and the rocks of the Palabora Intrusive Suite containing copper, iron, phosphate, uranium, and vermiculite. Kimberlites are also significant because of their deep-seated origin, which provides direct evidence of the nature of the upper mantle and crust. This chapter examines the relationship of the kimberlites, carbonatites, olivine melilitites, and salic alkaline rocks in space and time.

Four major types of alkaline magmatism characterize the African subcontinent (Fig. 13-1). Kimberlites generally occur in isolation. Where they are associated with other types of alkaline magmatism it is still debated whether they represent true kimberlites (Mitchell, 1970, 1979). Carbonatites are a diverse suite and include a carbonatite-rich end member (e.g., Tweerivier), a carbonatite–ultrabasic silicate rock association (e.g., Palabora and Spitzkop), and a carbonatite-poor end member (e.g., Saltpetre Kop and Premier). Most of the alkaline volcanic rocks are basaltic, although they are commonly associated with kimberlite and carbonatite. Complexes containing a variety of plutonic and volcanic alkaline rocks, such as the Pilanesberg Province, also represent major centers of magmatic activity. The distribution of these alkaline igneous rocks is shown in Figure 13-1. Acidic alkaline rocks such as those associated with the Vredefort Dome, are discussed in Chapter 6.

Radiometric dating during the past decade indicates a long history of alkaline magmatism in southern Africa from Early Proterozoic to Recent (Table 13-1). Early Proterozoic alkaline magmatism formed the Palabora Intrusive Suite and is confined to the northern part of southern Africa. The Middle Proterozoic Pilanesberg magmatism terminated with carbonatite and kimberlite activity (Ferguson, 1973). Alkaline igneous activity was particularly common in Late Phanerozoic

Table 13–1. Examples of Alkaline Rocks of Various Ages

	References[a]
Early Proterozoic	
Palabora	1,2
Ultrabasic dikes, Kimberley	3
Middle Proterozoic	
Spitzkop	4
Leeuwfontein	5
Premier, National	6
Pilanesberg	7
Glenover	8
Late Proterozoic	
Beitbridge	9
Early Phanerozoic	
Dokolwayo	9
Mid-Phanerozoic	
Klein Spitzkop	10
Swartruggens, Okonjeje	11
Late Phanerozoic	
Roberts Victor, Finsch, Gross Brukkaros	11
Monastery, Bultfontein, Wesselton, Finsch	12
Zoutpan	13

[a] 1, Holmes and Cohen (1956); 2, Eriksson (1982); 3, Clement *et al.* (1979); 4, MacIntyre and Dawson (1971); 5, Oosthuysen and Burger (1964); 6, Allsopp and Kramers (1977); 7, Schreiner and Van Niekerk (1958); Van Niekerk (1962); 8, Verwoerd (1967); 9, Allsopp and Kramers (1979), in Hawthorne *et al.* (1979); 10, Hawthorne *et al.* (1979); 11, Allsopp and Barrett (1975); 12, Davis (1977); 13, Wagner (1922).

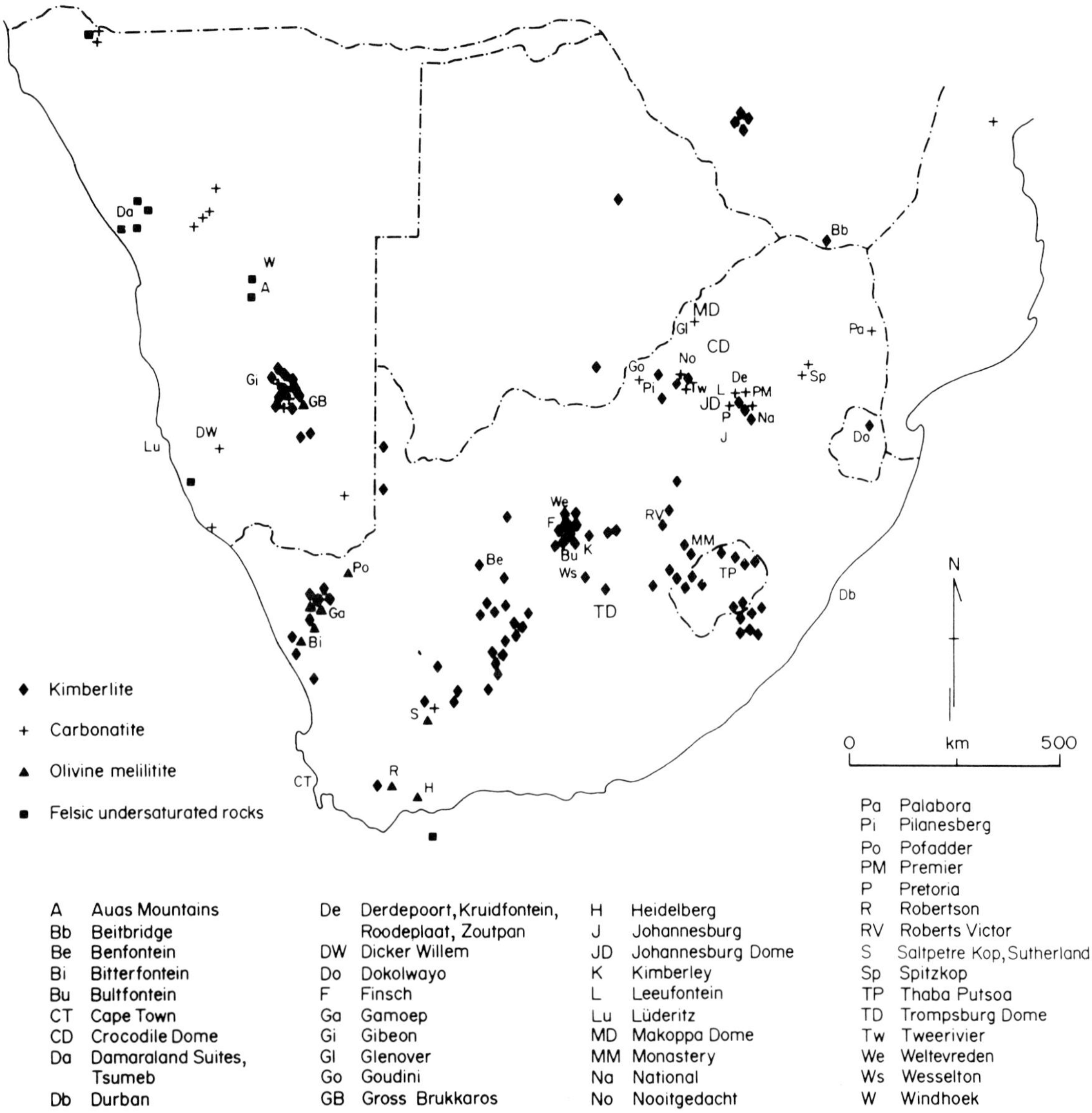

Figure 13-1. Distribution of alkaline igneous rocks in southern Africa.

time and is represented by four distinct episodes (Moore, 1976); the Early Cretaceous is represented by a western belt of subvolcanic intrusions and the Turonian, latest Cretaceous–Early Tertiary, and Oligocene by kimberlites, olivine melilitites, olivine nephelinites, phonolites, and trachytes (Table 12-1 and Fig. 14-6).

The various techniques used for dating kimberlites often give contradictory results. Multiple intrusions, late alteration, and noncognate samples may also cause rocks within a single body to give anomalous ages (Barrett, 1975; Hawthorne, 1975).

13.1. Carbonatites

Rocks containing a high proportion of carbonate minerals and associated with igneous activity are widely known as carbonatites. In southern Africa this usage is complicated where carbonatites have intruded metamorphosed sedimentary carbonates (Verwoerd, 1967). Nevertheless, the carbonate in carbonatite can commonly be distinguished from sedimentary carbonate sources on the basis of initial $^{87}Sr/^{86}Sr$ ratios (Powell *et al.*,1966). Carbonate in the Spitzkop intrusion, for example, has an initial $^{87}Sr/^{86}Sr$ ratio of 0.7023 ± 0.002 compared with a ratio of 0.7099 ± 0.002 for nearby dolomite in the Transvaal Supergroup. Where $^{87}Sr/^{86}Sr$ ratios are less discriminating, the magmatic and sedimentary carbonates should differ in their relative enrichment of rare earth elements, barium, and strontium (Gold, 1963; Bowden, 1968; Deans and Powell, 1968). Allsopp and Ferguson (1970) have characterized the Tsumeb pipe of Namibia in this way.

Verwoerd (1967) proposed an eightfold classification of southern African carbonatites, ranging from Na-rich carbonatitic volcanic breccias and tuffs, through feldspathoidal varieties, to the K-rich ultramafic associations. These carbonatites form dikes, ring dikes, and circular and irregular plugs, with an average diameter of about 1.6 km (Verwoerd, 1967). Only three intrusions are substantially larger; Nooitgedacht and Dicker Willem are both 3.5 km in diameter, and Tweerivier is 3.0 km in diameter. The circular shapes and scarcity of associated alkali silicate intrusions may reflect considerable erosive truncation of these carbonatite complexes (Jackson, 1976).

Besides the common association of carbonatite with plutonic rocks in suites such as Palabora and Spitzkop, there are also numerous examples of carbonatites in volcanic settings. These are most abundant north of Pretoria at Derdepoort, Goudini, Kruidfontein, Roodeplaat, and Zoutpan, in the western Cape at Saltpetre Kop, and in Namibia at Gross Brukkaros (Fig. 13-1).

The carbonatite in the Early Proterozoic Palabora Intrusive Suite has been intensively studied because large reserves of copper-bearing sulfides have prompted large-scale open-pit mining. The Palabora Suite is dominated by pyroxenites containing clinopyroxene, phlogopite, apatite, and microcline. The copper-bearing phoscorite and carbonatite are restricted to a small intrusive center. Plugs of syenite and pyroxenite, commonly brecciated and locally containing copper sulfides, are concentrated on the periphery of the complex; some of these plugs have the same paleomagnetic orientation as the pyroxenites of the main complex (Briden, 1976). Frick (1975b) has suggested that the syenites originated by differentiation of a trachytic magmatic line from a magma similar in composition to a shonkinite.

Cumulus processes were pervasive during Palabora magmatism, including intrusion of the carbonatite–phoscorite core (Eriksson, 1982). Typical "inch-scale" cumulus layering of the marginal micaceous pyroxenite is shown in Figure 13-2; this texture is ascribed to *in situ* crystal growth and accumulation (Hess, 1960; Morse, 1969, Irvine, 1974; Campbell, 1978). Other textures within the pyroxenites indicate coeval crystallization of pyroxene, phlogopite, and apatite, followed by breakup of the crystal aggregates by the intruding magma.

The mineral chemistry shows that the Palabora rocks crystallized at moderate temperatures and pressures from an ultrabasic magma similar in composition to many potassium-rich, ultrabasic volcanic rocks (Edgar *et al.*, 1976; Barton and Hamilton, 1978; Eriksson, 1982). Chemically, there is little to distinguish the phlogopites of phoscorite and carbonatite from those of the pyroxenites because both are high in Mg and Fe^{3+} and low in Al; this suggests that the carbonatite-bearing rocks are not highly differentiated fractions of the ultrabasic magma but are probably immiscible fractions separated early in the cooling history of the magma.

Evidence favors magmatic (Eriksson, 1982) rather than metasomatic processes (Brandt, 1948; Hanekom *et al.*, 1965) to explain the characteristic textures and large-scale rock relationships of the Palabora Intrusive Suite. Palabora magmas were derived from a mantle source similar in composition to that of kimberlite (Eriksson, 1982). Similar characteristics of mineralogy and texture are common in other carbonatites as well, although poor exposure generally precluded detailed studies. The Glenover carbonatite has abundant biotite pyroxenite and exhibits vertical zonation of minerals (Verwoerd, 1967). Closely akin to the Palabora Suite is the Schiel Suite of brecciated syenites, mica pyroxenite, and rare phoscorite (Mathias, 1974), although it is apparently much older (J.M. Barton, personal communication, 1979).

Several authors have emphasized the importance of metasomatism or fenitization in the formation of silicate rocks in alkaline complexes (McKie, 1966; Borodin and Pavlenko, 1974). Fenitization is characterized by alkali enrichment of country rock and the formation of new minerals, such as nepheline, mica, sodic amphiboles and pyroxenes, alkali feldspars, and apatite. Palabora fenite is recognized by the alteration of plagioclase to K-feldspar in the country rock. However, the major rock types are better interpreted in terms of magmatic crystallization. Other alkaline

Figure 13-2. "Inch-scale" cumulus layering in phlogopite–apatite–clinopyroxene rock of the Palabora Complex. Dark layers are philogopite; light layers are diopside and apatite. Located on northern margin of the micaceous pyroxenite zone.

complexes resulted from the addition of alkali elements from magma-derived fluids. Spitzkop's nepheline-bearing melteigite is believed to be the result of metasomatism of gabbroic theralite (Verwoerd, 1967).

Carbonate minerals are common in kimberlite, yet when they form carbonate rock in this association their classification as true carbonatite is still contentious (Mitchell, 1970, 1979; McMahon and Haggerty, 1979; Cox, 1980). The Premier kimberlite contains carbonate-rich dikes with carbon and oxygen isotopic ratios comparable to the carbonate fraction in the kimberlite (Deines and Gold, 1973). An interesting example of differentiation of a carbonate phase from kimberlite is the Benfontein sill (Dawson and Hawthorne, 1973), where ocelli of "carbonatite" occur in a kimberlitic groundmass. However, on the basis of oxide mineral composition, Boctor and Boyd (1979) argue that this carbonate rock is not true carbonatite. Carbonate is commonly present in nodules and megacrysts, although it is rare in mantle-derived peridotites (McGetchin and Besancon, 1973; Nixon and Boyd, 1973a).

13.2. Alkaline Complexes

Alkaline volcanic rocks are volumetrically insignificant in southern Africa, although sporadic pipelike intrusions are present (Fig. 13-1). There are no large lava flows to compare with those of the East African Rift system (Holmes, 1950) or Wyoming (Carmichael, 1967). These rocks comprise two main types; those of predominantly ultrabasic composition, and those in which the ultrabasic magmas are subordinate to more salic types.

Excluding the kimberlites, the ultrabasic plugs attract attention because their olivine melilitites may be genetically related to both kimberlite and carbonatite. These vents include Pofadder, Gamoep, Bitterfontein, Sutherland, Robertson, and Heidelberg (Fig. 13-1). The Sutherland olivine melilitite is a porphyritic rock with olivine phenocrysts set in a groundmass of melilite, phlogopite, olivine, augitic pyroxene, magnetite, perovskite, apatite, zeolite, monticellite, nepheline (McIver and Ferguson, 1979), and rare ilmenite (Gerrard, 1958). The mineral chemistry of the Gross Brukkaros and Saltpetre Kop rocks has distinct similarities with some kimberlites, particularly with respect to the high-Mg ilmenites (Janse, 1969, 1975; McIver and Ferguson, 1979). Derivation from a kimberlitic magma through fractionation, mineral reaction, and immiscibility has been suggested (Ferguson *et al.*, 1975; McIver and Ferguson, 1979).

The second type of alkaline intrusion is generally more salic, although not to the exclusion of ultrabasic material; it contains abundant feldspars, amphiboles, and silica-undersaturated minerals. The largest of these are the Lüderitz Igneous Province of Namibia and the Pilanesberg Igneous Province of the northwestern Transvaal

(Fig. 13-1). They are both multiple intrusions containing diverse rock types including alkali-oversaturated rocks (alkali rhyolites) and silica-saturated and -undersaturated varieties (phonolites and nepheline syenites) (Ferguson, 1973; Marsh, 1975, 1976). The Damaraland Igneous Province of Namibia contains similar assemblages of mixed acid and silica-undersaturated composition (Mathias, 1974).

13.3. Kimberlites

Because of the variety of compositions and textures it encompasses, kimberlite is a hybrid defying concise definition. According to Clement *et al.* (1977),

> Kimberlite is a volatile-rich, potassic, ultrabasic, igneous rock which has a distinctively inequigranular texture resulting from the presence of macrocrysts set in an essentially microporphyritic matrix. The matrix contains . . . primary . . . groundmass constituents of olivine and . . . phlogopite, calcite, serpentine, diopside, monticellite, apatite, spinels, perovskite and ilmenite. Other primary minerals may be present in accessory amounts.

The kimberlite assemblage includes rock fragments derived from mantle and crust, a variety of megacrysts, and a broad spectrum of groundmass compositions, which are related to the considerable depth of origin and the mechanism of emplacement. It has long been believed that these magmas were emplaced very rapidly through the continental crust (Wagner, 1914; Dawson, 1962; McGetchin and Besancon, 1973). The geometry of emplacement is variable and often complex; small diatremes and dikes predominate, but kimberlite sills are also known. Diatremes may form as offshoots from dikes in places (Dawson, 1967). Textures suggestive of igneous differentiation and cumulus processes occur in some sill kimberlites (Dawson and Hawthorne, 1971, 1973). Depending on the nature of the volatiles and the structure and force of intrusion, kimberlite magmatism may result in massive rock, intrusive breccias, or tuffs and tuff breccias. Cessation of volcanism may be followed by accumulation of epiclastic kimberlite by alluvial and debris-flow processes and by deposition of finer grained material in the crater lake (Hawthorne, 1975).

Inclusions in kimberlite include xenoliths, xenocrysts, and megacrysts of mantle and crustal origin, as well as megacrysts and autoliths of kimberlitic magma (Ferguson *et al.*, 1973; Danchin *et al.*, 1975; Gurney *et al.*, 1979). Recent studies have recognized the value of this material in yielding information about pressure, temperature, composition, and structure of the kimberlitic source area and the area it has sampled accidentally. Harte (1978) classified mantle-derived nodules into five categories: (1) peridotites and rare dunites (including harzburgites, garnet harzburgites, lherzolites, and garnet lherzolites); (2) rare garnet pyroxenites; (3) eclogites; (4) megacrysts or discrete nodules; and (5) amphibole-bearing and mica-rich nodules. Granolites and certain eclogites are believed to be derived from the deep crust.

Stability fields of minerals and various mineral assemblages, such as phlogopite (Modreski and Boettcher, 1973) and graphite–diamond (Bundy *et al.*, 1961), are gross indicators of P-T regimes, but P-T estimates based on compositions of coexisting phases are more precise. Several geobarometers are appropriate to kimberlite nodule assemblages. The Al solubility in orthopyroxene coexisting with garnet (MacGregor, 1974; Wood, 1974) and the Fe–Mg partitioning between garnet and clinopyroxene (Raheim and Green, 1974; Wood, 1976) are reliable pressure indicators depending on the compositional ranges over which they have been calibrated. Temperature estimates are based on the orthopyroxene–clinopyroxene pair (Davis and Boyd, 1966; Wood and Banno, 1973; Wells, 1977) and the garnet–clinopyroxene assemblage (Ellis and Green, 1979).

Garnet lherzolite nodules are common and contain the assemblage olivine, orthopyroxene, clinopyroxene, and garnet. Lherzolites are believed to have formed in a P-T range of from 900°C at 4000–6500 MPa to 1250°–1450°C at 7000 MPa (Harte, 1978) (1 kb = 100 MPa). Lherzolites from the Thaba Putsoa pipe in northern Lesotho fall into two categories with distinct textures, mineralogy, bulk composition, and mineral chemistry (Boyd and Nixon, 1973). The coarse-grained, granular lherzolites, with accessory phlogopite and chromite, and rare graphite formed at depths of 100–150 km. The second group, comprising "sheared" nodules with deformed textures, recrystallized olivine, and rare secondary phlogopite, had a deeper origin, 150–200 km. This bimodal distribution of textural and chemical parameters has since been widely recognized, separating the southern African kimberlite nodule suites into different depth ranges (MacGregor, 1975, 1979; Danchin and Boyd, 1976; Danchin, 1979). The data supports the P-T plot used by Boyd and Nixon (1975) to define a Mid-Cretaceous mantle geotherm (Fig. 13-3). Boyd and Nixon (1973) inferred an increase in the P/T gradient at high pressures but this has been refuted by Gurney *et al.* (1975) on the basis of data from Lesotho pipes; others question the significance of this inflection (Mercier and Carter, 1975; Harte, 1978). Both inflected and noninflected geotherms have been established on other continents (Boyd *et al.*, 1976; Boyd and Clement, 1977; Eggler and McCallum, 1976).

Xenoliths show that the upper mantle contains significantly higher Mg/(Mg + Fe) ratios than those predicted from models based on "pyrolite" and basaltic rocks (O'Hara *et al.*, 1975). Granular lherzolites from Lesotho kimberlites are depleted in Na, Ti, Fe, Ca, and Al relative to their sheared counterparts, and their Fe/(Fe + Mg) ratios are lower (<0.10 compared with >0.12) (Boyd and Nixon, 1973). Differences occur in mineral chemistry, with granular olivines possessing higher Mg/(Mg + Fe) ratios and smaller variations in composition than the sheared olivines. Mg/(Mg + Fe) ratios are also used to determine whether certain rocks are residual mantle [Mg/(Mg + Fe) = 0.92 − 0.95] (Boyd and Nixon, 1973, 1975) or cumulates [Mg/(Mg + Fe) = 0.90] (Boyd and Nixon, 1973). Anomalously high values for K_2O, TiO_2, and volatile components in xenoliths, compared to average mantle chemistry, are possibly the result of metasomatism.

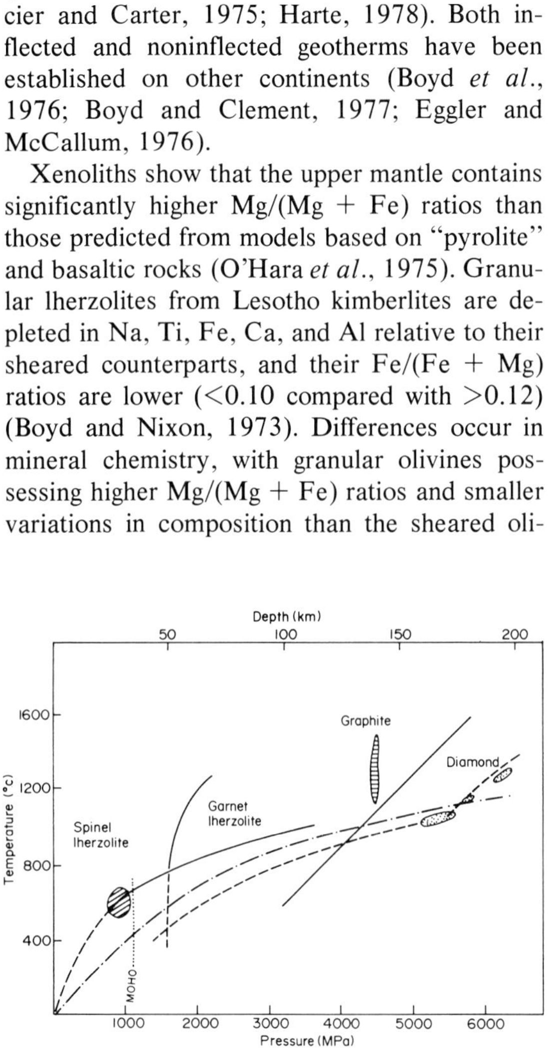

Figure 13-3. P-T regimes of some nodule and megacryst assemblages, showing their relationship to the mantle and shield geotherms.

Megacrysts are usually larger (1–20 cm) than the grains in xenoliths (<1 cm); they consist mainly of garnet, orthopyroxene, clinopyroxene, olivine, ilmenite, phlogopite, and rare zircon. Ilmenite is commonly associated with a silicate. In general, relative abundances of silicate megacrysts decrease in the order: garnet, diopside, enstatite, olivine; ilmenite may be associated with each.

The chemistry of these megacrysts is a useful indicator of the nature of the source area and fractionation trends during crystallization of the magma. Their mineral suites are also frequently used for exploration purposes. Megacrysts are divisible into a chrome-rich group and a rare chrome-poor group (Smith *et al.*, 1976). Some kimberlites, such as Weltevreden (Meyer *et al.*, 1979), contain both populations, whereas others, such as Monastery (Gurney *et al.*, 1979), contain only chrome-poor varieties.

Ilmenite is the most common megacryst in kimberlite. It is typically magnesium- rich and is also enriched in Cr, Ni, Co, Nb, Zr, Ta, and Hf, compared with ilmenites from other ultrabasic alkaline rocks (Gurney *et al.*, 1973; Mitchell, 1973, 1977), and impoverished in Cu, Zn, and Mn (Mitchell, 1977). Ilmenite forms discrete megacrysts or lamellar intergrowths or granular intergrowths with pyroxene and garnet. In association with silicate minerals the ilmenites, particularly the lamellar varieties, have compositions different from the discrete forms. It is debatable whether there is a simple relationship between the lamellar and the discrete types of ilmenite, and whether the ilmenite represents a phenocryst or a xenocryst of mantle origin. Both discrete ilmenites and lamellar intergrowths have been interpreted as high-pressure cumulates formed from a protokimberlitic magma (Mitchell, 1973; Wyatt *et al.*, 1975; Wyatt, 1977; Rawlinson and Dawson, 1979). Other interpretations are for the simultaneous crystallization of silicate and ilmenite under vapor-rich conditions (Mitchell, 1977), granular intergrowths as nonoriented intergrowths (Nixon and Boyd, 1973a), and Ti-metasomatism of garnet lherzolite mantle material (Harte and Gurney, 1975).

Besides this interest in mantle processes, there has also been a concerted effort to examine the behavior of the intruding kimberlite magma. Autoliths are fine-grained, nearly spherical bodies of rock, ranging in size from a few millimeters to several decimeters, which are believed to have crystallized from kimberlitic magma (Ferguson *et al.*, 1973; Danchin *et al.*, 1975). Autoliths are enriched in the incompatible elements Ti, P, and Mn, and impoverished in Mg compared to many kimberlites and may be derived through fractional crystallization of a kimberlite melt (Ferguson, 1973; Danchin *et al.*, 1975). Some autoliths have nucleated around granitic and basaltic fragments, proving that they crystallized at crustal pressures and temperatures. Certain textures, such as patches in nodules and kimberlite stringers cutting nodules, are believed to have formed during late-stage crystallization (Dawson and Smith, 1977). Groundmass minerals commonly have distinctive chemistry resulting from late-stage processes. Dawson and Smith (1977) cite the prevalence of late clinopyroxene which is distinguished from mantle-derived clinopyroxene by high Ca and low Al values (Figure 13-4). They postulated a temperature of formation of 400°–500°C at near-surface pressures. Elton and Ridley (1979) suggested that perovskite, magnetite, olivine, phlogopite, carbonate, apatite, and sulfides are produced when kimberlite crystallizes near the surface. These minerals also characterize dike kimberlites (Dawson and Hawthorne, 1973; Clement *et al.*, 1977).

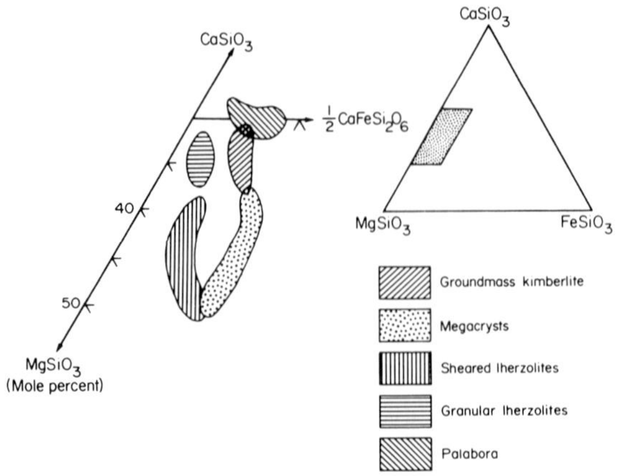

Figure 13-4. Plot of clinopyroxene compositions from various types of ultrabasic rocks in the system enstatite–ferrosalite–wollastonite. [Groundmass kimberlite from Dawson and Smith (1977); megacrysts, sheared lherzolites, and granular lherzolites from Boyd and Nixon (1973); Palabora from Eriksson (1982).]

13.4. Petrogenesis of Alkaline Rocks

Generation of Alkaline Magmas

It is now generally accepted that kimberlites, carbonatites, and other volatile-rich alkaline magmas are derived from the mantle, either directly by partial melting (Dawson, 1971; Mitchell and Crockett, 1971; Mysen and Boettcher, 1975), or through fractional crystallization of a mantle-derived magma (O'Hara and Yoder, 1967; MacGregor, 1979). The major constituent of the source magma is thought to be a four-phase peridotite consisting of olivine, orthopyroxene, clinopyroxene, and either spinel or garnet.

There is strong evidence that minerals and magmas derived from the mantle contain CO_2 and H_2O (Wyllie, 1979; Eggler, 1978). Experimental work shows that CO_2 produces liquids strongly depleted in SiO_2. Liquids of varying carbonatitic composition can be formed by varying the partial pressures of these volatile phases.

With constant composition, melting at different pressures results in magmas ranging in composition from quartz normative at 1500 MPa to leucite normative at pressures of 3500 MPa under vapor-absent conditions (Modreski and Boettcher, 1973). Under conditions of constant vapor composition and increasing pressure, Wendlandt (1977) produced carbonatitic to kimberlitic liquids. Eggler (1978) formed nephelinites from a similar composition at shallower pressures.

Under conditions of constant CO_2-rich vapor composition, the mineral assemblage is an important control on magma composition (Wyllie, 1979). Small degrees of partial melting of phlogopite peridotite produces a high-K_2O liquid with high Mg/Ca resembling leucite-normative magmas and kimberlites. In the absence of phlogopite, melilititic compositions are formed. The presence of dolomite ensures a low silica content in all these liquids.

Fractional crystallization may also determine the course of petrogenesis. For example, removal of an eclogite fraction from a picritic magma could produce kimberlitic compositions (O'Hara and Yoder, 1967; O'Hara, 1968). Green and Ringwood (1967b) have produced nephelinite from picrite.

The experimental and field evidence is sufficient to show that ascending liquids produced from various starting compositions under various P-T conditions are susceptible to modification to carbonatitic, kimberlitic, or melilititic magmas (cf. Wyllie, 1979). Eggler and Wendlandt (1979) found that liquids produced at 5000–6000 MPa could coexist with minerals of a particular megacryst assemblage, implying cogenetic origins for the megacryst and liquid from that depth. However, a fractional crystallization path at constant pressure over a wide temperature range was suggested for kimberlite megacrysts at Monastery (Gurney *et al.*, 1979). Fractional crystallization and olivine removal from a kimberlitic magma have also been suggested for the evolution of the melilitites of Saltpetre Kop and Sutherland (McIver and Ferguson, 1979) and Gross Brukkaros (Janse, 1975; Ferguson *et al.*, 1975). Liquid immiscibility between carbonate and silicate liquids at shallow depths has been demonstrated experimentally (Koster Van Groos and Wyllie, 1973). Textures supporting this process have also been observed in some dikelike kimberlites (Dawson and Hawthorne, 1973). The Mg-rich olivines and phlogopites in the Palabora carbonatite suggest that the carbonatite was derived as an immiscible product from an ultrabasic liquid rather than as the end product of crystal fractionation (Eriksson, 1982).

The chemistry of the clinopyroxenes from Palabora (Eriksson, 1982), from Benfontein (Dawson and Hawthorne, 1973), and from late-stage kimberlitic liquids (Clement *et al.*, 1979) is similar (Fig. 13-4), with high Ca, low Al and some Fe^{3+}. Nevertheless, there are pronounced differences in iron oxide mineralogy. Kimberlite iron oxides are invariably rich in Mg and Ti, whereas Palabora magnetites are poor in both elements (Hanekom *et al.*, 1965). There is still little consensus regarding the mineralogical evidence for a kimberlite–carbonatite relationship (Boctor and Boyd, 1979; Mitchell, 1979; McMahon and Haggerty, 1979; Eriksson, 1982).

The Nature of the Mantle

Tectonic activity and the source composition of kimberlite, and possibly alkaline magmas in general, may be reflected in the Mid-Cretaceous mantle geotherm established by Boyd and Nixon (1973). The inflection at depths between 150 km and 200 km (Fig. 13-3) could have resulted from compositional differences, degree of deformation, or phase changes (MacGregor, 1975). The bimodal distribution of textural and chemical properties of lherzolite nodules have been correlated with the two parts of the geotherm (Boyd and Nixon, 1973; MacGregor, 1975, 1979; Danchin and Boyd, 1976; Danchin, 1979). The less depleted, deformed nodules possibly had the deeper origin, corresponding to the low-velocity zone (Boyd, 1973). MacGregor (1975) compared the boundary between the two populations with the melting curve of vapor-saturated peridotite (Fig. 13-3) suggesting that the vapor phase or partial melts may have influenced deformation of the xenoliths. The Boyd–Nixon geotherm resembles the "shield geotherm" of Clark and Ringwood (1964), except for perturbation below 150 km. The inflection point may represent a physical boundary within the mantle. A similar geotherm based on garnet lherzolites from the Precambrian Premier kimberlite (Fig. 13-3) is indistinguishable from the Cretaceous geotherm, suggesting that the thermal evolution of southern Africa was virtually complete by 1115 Ma, and that subsequent events had no lasting effects on the geothermal gradient (Danchin, 1979).

The various mineral assemblages in the xenoliths reflect kimberlite source area "stratigraphy." Amphibole-bearing nodules are restricted to shallow depths in the garnet lherzolite P-T stability field (Erlank, 1976) and are not involved in melting in stable continental regions (Wyllie, 1979). Phlogopite is restricted largely to regions of relatively low pressure and temperature (Boyd, 1973; Boyd and Nixon, 1973), although it is present as syngenetic inclusions in some diamonds (Harris and Gurney, 1979). The low-velocity zone is either poorly defined or absent beneath continental shields, indicating little or no carbonate, phlogopite, or CO_2–H_2O vapor phases (Wyllie, 1979). Anomalously high concentrations of these components in the mantle may localize alkaline magmatism in cratonic areas. Isotopic studies of kimberlites and their xenoliths show that Pb-U heterogeneities have persisted in the mantle beneath southern Africa for at least 2.5 Ga (Kramers, 1977, 1979). Anomalous initial $^{87}Sr/^{86}Sr$ ratios of 0.7039–0.7100 for Palabora rocks may be indicative of a heterogeneous mantle (Eriksson, 1982). Metasomatism of the mantle may explain the enrichment of volatile and other elements, such as K, Ti, and P, in continental alkaline magmas (Lloyd and Bailey, 1975; Boettcher *et al.*, 1979; Wyllie, 1979).

The composition of some nodule assemblages reflects the nature of portions of the lower crust. Garnet granolites occur in kimberlites on the periphery of the Kaapvaal Province (Griffin *et al.*, 1979) or in the adjoining high-grade Namaqua Province (Fig. 1-4). "Griquaite," a type of eclogite, is restricted to kimberlites situated well within the Kaapvaal Province. Griffin *et al.* (1979) reconcile the nodule chemistry with a less siliceous lower crust than is generally believed (Heier, 1973).

The upper boundary of the low-velocity zone shallows toward the margins of the Kalahari Province from a maximum depth of 195 km in the central area around Kimberley; in Namibia it is 140 km below the surface, and in Lesotho 180 km (MacGregor, 1975) (Fig. 1-5). Diamondiferous kimberlites are generally older and concentrated within the craton, unlike the younger, diamond-free kimberlites, which show a preference for thinned margins of the Kalahari Province (Clifford, 1966, 1970; Dawson, 1971). The Dokolwayo diamond-bearing kimberlite in eastern Swaziland may be an exception, but its presence suggests that the eastern margin of the Kalahari Province extends at least to the Mozambique border (Hawthorne *et al.*, 1979) (Fig. 13-1).

Tectonic Control of Alkaline Magmatism

The distribution and timing of alkaline magmatism in southern Africa has generated considerable discussion. Kimberlites of Late Cretaceous age are particularly numerous and coincide with a reorganization of plate behavior that culminated in the breakup of Gondwana. Cox (1970) suggested that kimberlites resulted from reduced heat flow after the breakup of Gondwana when the lower degree of partial melting necessary for alkaline magmatism became prevalent. However, improved dating techniques have shown a greater spread of kimberlite ages than anticipated in this hypothesis (Allsopp and Kramers, 1977). Cratonic uplift, ascribed to retention of asthenospheric melt beneath the continental lithosphere (Sloss and Speed, 1974), and crustal structure are believed to be important controls of alkaline volcanism (Bailey, 1964). Less likely is the notion, based on deformation textures, that magmatism resulted from lateral displacement of a 200-km-thick continental plate over the low-velocity zone (Boyd, 1973; Nixon and Boyd, 1973); these textures probably originated during intrusion. Other mechanisms proposed to explain the heat distribution necessary for alkaline magmatism include mantle convective overturn (Dawson, 1970), diapiric upwelling of mantle material (Green and Gueguen, 1974), and mantle flow (Liu, 1977).

There appears to be a close relationship between major crustal flexures and emplacement of alkaline igneous rocks and carbonatites in southern Africa (Verwoerd, 1967; Ferguson, 1973; Cornelissen and Verwoerd, 1975; Pretorius, 1979). However, the details of the structural controls are still debated. Ferguson (1973) related Pilanesberg magmatism to a major northwest trending Early Proterozoic antiform which included the Johannesburg Dome, the Crocodile Dome, and the Makoppa Dome. With the exception of Glenover, all the Transvaal carbonatites appear to follow an east–west trend (Verwoerd, 1967). A third possible structural control of alkaline magmatism in the Transvaal is a north–northeast-trending abyssal fracture taken advantage of by the Late Archean Great Dyke of Zimbabwe, the Bushveld Complex, and the Trompsburg Intrusive Suite of the Orange Free State (Cousins, 1959). The general distribution of kimberlite diatremes is broadly coincident with a west-southwest-trending basement swell (Liu, 1977); Swaziland's Dokolwayo diatreme is on the axis of the Lebombo monocline. Alignment of alkaline intrusions between Angola and the southwestern Cape also implies some form of structural control on the western seaboard (Cornelissen and Verwoerd, 1975), possibly the result of continental extension of oceanic transform faults (Marsh, 1973).

Chapter 14

Changing Climates and Sea Levels: The Cenozoic Record

By the end of the Cretaceous the present coastal configuration of southern Africa was established, but the evolving Cenozoic environments continued to respond to external influences. The depositional trends reflect not only the relative rates of sediment supply and local subsidence, but also the major eustatic sea level movements brought about by volumetric changes in ocean basins. Glacioeustatic effects were pronounced only in the Late Cenozoic. Global climatic changes have also been imprinted on the sedimentology, chemistry, and biology of the Cenozoic basins.

Post-Cretaceous withdrawal of the marginal seas is believed to have been a consequence of the reduction in sea-floor spreading rates beginning about 85 Ma (Larson and Ladd, 1973). However, episodic spreading and sea-floor distortion continued to change the volumes of the ocean basins, and therefore sea levels, during the Tertiary (Rona, 1973). When the midoceanic ridges were active spreading centers they were in an expanded state; when inactive they contracted. These volumetric changes apparently resulted from the thermal behavior of mantle material. Response to inflation and contraction of the midocean ridge system caused substantial worldwide transgressions and regressions in the Late Cretaceous and Tertiary. A sea-level maximum in the Middle Eocene was terminated by ridge subsidence, beginning at 45 Ma, which initiated a long period of Oligocene–Miocene quiescence of the ridge system. Late Miocene transgression coincided with acceleration of spreading and ridge expansion at 10 Ma (Flemming and Roberts, 1973).

These eustatic effects were enhanced by epeirogenic uplift of the continental lithosphere, apparently related temporally to ocean-floor subsidence (Vogt *et al.*, 1971; Jacoby, 1972). This is thought to have resulted from coupling between continental and oceanic plates. Episodic uplift of this nature initiated widespread cycles of erosion (e.g., the pediplanation cycles of King, 1962).

Changing patterns of oceanic circulation and the climatic processes that led to Late Cenozoic aridification in southern Africa were intimately related to the history of high-latitude glaciation, particularly in Antarctica. Glaciation in Antarctica persisted through the Oligocene, but the ice sheet only grew to its present thickness between the early Middle Miocene and early Late Miocene (Shackleton and Kennett, 1975). By Late Miocene it had achieved its present invulnerability to climatic change and has existed in its present form since then.

Modern climate and oceanic circulation are linked to the Late Oligocene formation of the Circum-Antarctic Current when Australia and the South Tasman Rise finally separated from Antarctica (*cf.* Fig. 14-1) (Kennett *et al.*, 1975). This resulted in the divergence of high- and low-latitude oceanic temperatures in the Middle Miocene (Savin *et al.*, 1975). Widespread aridity in the western part of southern Africa is related to upwelling of cold water and the genesis of a cold Southern Ocean and so could not predate the Late Oligocene. Shackleton and Opdyke (1977) have shown from oxygen isotope and paleomagnetic studies of deep-sea cores that fluctuations from glacial to interglacial have characterized the climate for the past 3.2 Ma.

Since the Late Miocene the Antarctic ice sheet has contributed little to major sea-level changes, although a brief expansion of the ice sheet in the Pliocene did cause a temporary lowering of sea level (Shackleton and Kennett, 1975). In the Pleistocene the volume of water in the oceans has rarely exceeded present volume. High Pleistocene sea levels are attributed to the combined effect of alternating glaciation and deglaciation in the

Figure 14-1. Eocene paleogeographic setting of southern Africa before development of the Circum-Antarctic Current. [After McKenzie and Sclater (1971).]

northern hemisphere, volumetric changes in ocean basins, and local epeirogenesis (Tankard, 1976a). Because oceanic spreading rates are still very high, it is difficult to make precise inferences about glacioeustatic controls.

14.1. Tertiary Coastal Environments

The Tertiary depositional history of the coastal plain can be separated into several transgression–regression cycles. Each marine depositional complex is separated by a synchronous stratigraphic hiatus. Marine strata have their greatest inland extent where they rest on soft pre-Tertiary strata in the southern Cape and Zululand. The main depositories coincide with Mesozoic loci of deposition. From a discontinuous veneer beneath the coastal plain these Tertiary strata thicken seaward to more than 1 km below the outer shelf.

Paleogene Transgression and Regression

Although the Paleocene generally represents an important stratigraphic break because of worldwide withdrawal of marginal seas, sedimentation in the Zululand basin was continuous through the Maastrichtian and Danian, and the upward-fining succession and biofacies suggest deepening of the basin at Richards Bay. There the 25-m-thick sequence is richly fossiliferous and consists of interlaminated sandstone and siltstone with rare calcereous concretions. It has been dated as Danian by *Globioconusa daubjergensis* and *Globorotalia uncinata* Zones (Orr and Chapman, 1974; Maud and Orr, 1975). A midshelf depositional environment is envisaged because the fauna is dominated by planktonic foraminifera, and the subordinate benthic assemblage comprises mainly discorbid genera and very few shallow-water forms. Paleocene sediments also underlie the adjacent continental shelf (Siesser, 1977a). Beneath the southern Mozambique coastal plain contemporaneous deposition is shown by extensive glauconitic sands dated at

60 Ma (Flores, 1973). Paleocene deposits are unknown elsewhere on the southern African coastal margin.

Paleocene transgression reached its maximum extent in the Early Eocene, but only isolated remnants of the stratigraphic record have survived later transgressions (Siesser and Dingle, 1981). In Namibia Eocene marine sediments occur as pockets preserved in shallow basement depressions from 40 to 160 m above present sea level (Kaiser, 1926). At Buntfeldschuh (Fig. 14-2) a wave-cut platform is overlain by a transgressive veneer of gravel that contains fish teeth, followed by a thick regressive sequence of fine-grained sand (Martin, 1973). Siesser and Salmon (1979) have suggested an Early Eocene age. At Birbury in the eastern Cape Paleogene calcarenites overlie a seaward-sloping platform cut across Paleozoic strata at 200–300 m above present sea level. Assemblages of calcareous nannofossils and planktonic foraminifera indicate an Early Eocene age (Siesser and Miles, 1979). A similar age is inferred for the upper Needs Camp limestones.

Eocene depositional history was interrupted by a brief regressive episode in the Middle Eocene, although sediments of this age are preserved in Mozambique (Flores, 1973; Förster, 1975). Offshore a stratigraphic hiatus generally separates Lower and Upper Eocene rocks; however, some Middle Eocene strata have been penetrated by

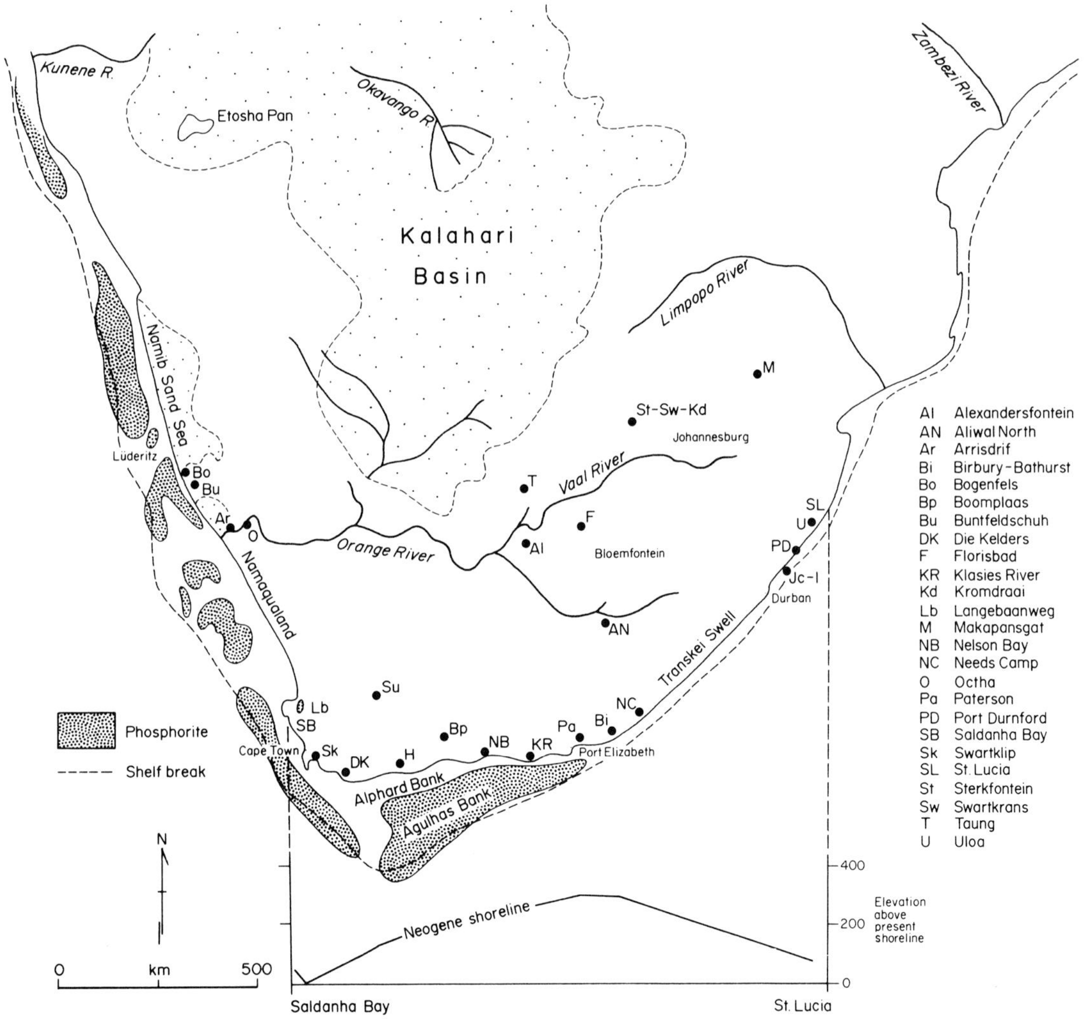

Figure 14-2. Locality map showing the distribution of shelf phosphorites [after Birch (1979b)] and the differential uplift of the Neogene shoreline along the southern and southeastern coastal margin.

oil-exploration boreholes (Du Toit and Leith, 1974; Siesser and Dingle, 1981).

The coastal margin was again flooded in the Late Eocene. A diverse marine fossil assemblage characterizes the calcareous deposits at Bogenfels in Namibia, including rare specimens of the nautiloid *Aturia lotzei* (Klinger, 1977). Ever since Böhm (1926) interpreted the mollusks as Middle to Late Eocene, the age of these strata has been disputed. However, recent nannofossil determinations have confirmed a Late Eocene age (Siesser, 1977b); phonolitic lavas that apparently overlie this regression complex have been dated at 36–38 Ma (Kröner, 1973b; K-Ar).

Available evidence indicates that the platform beneath the Zululand–Mozambique coastal plain is also a composite feature. In Zululand Late Eocene foraminifera have been reworked into Miocene shore-zone sediments (Frankel, 1968, 1972). Farther north, deposition of a 450-m-thick algal–oolitic limestone succession in southern Mozambique is attributed to continuous downwarping, although the Paleogene interval is nevertheless thinner than the Cretaceous sequence (Förster, 1975). These Upper Eocene deposits are characterized by lumachelles of larger shallow-marine foraminifera.

The Oligocene to Early Miocene interval was marked by a general withdrawal of the shelf sea and a paleoclimate less arid than present. In Namaqualand carbon-rich paludal sediments occupy depressions in weathered and deflated melilitite diatremes, some of the latter being dated at 38 Ma (Kröner, 1973b; K-Ar). The pollen and leaf flora, of probable Oligocene age (Estes, 1977), includes *Podocarpus*, *Gunnera*, "*Myrica*," *Ficus*, a proteaceous genus, and xerophilous-type leaves. This indicates an ecotonal area between temperate rainforest and sclerophyllous vegetation, with a relatively dry climate and summer rainfall (Estes, 1977; Axelrod and Raven, 1978). In contrast the present climate is semiarid with winter rainfall.

Vertebrate fossils in Namibia indicate that similar conditions prevailed there into the Miocene. These faunas resemble those of east Africa, which are believed to date to the Early or Middle Miocene (Stromer, 1926; Corvinus and Hendey, 1978). The Bogenfels assemblage, which occurs within fluvial sediments, indicates wooded grassland (antelopes and jumping hares) and riverine woodland (tragulids) (Hopwood, 1929). A second and somewhat younger assemblage at Arrisdrif in southern Namibia is found in gravel and coarse sand bars of a braided fluvial system. Rhinoceros and ruminant fossils suggest a wooded environment (Corvinus and Hendey, 1978).

Renewed Transgression: The Late Miocene–Pliocene Interval

Although not represented by thick sequences onshore, the Neogene transgression–regression complex is the most widespread of the Tertiary cycles in southern Africa. Seismic and borehole data show progressive seaward thickening (Du Toit and Leith, 1974; Du Toit, 1979). The lithofacies, their interrelationships, and their faunal content indicate deposition in shore-zone environments.

The Langebaanweg embayment. Neogene marine strata are rare in Namibia, but along the Namaqualand seaboard, phosphogenic deposits of probable Late Miocene age are sporadically preserved in basement depressions (Tankard, 1974a). However, in the Saldanha–Langebaanweg region of the southwestern Cape, eustatic transgression over Precambrian basement resulted in an embayed coastline (Fig. 14-3A), along which was deposited a complex suite of sediments (the Varswater Formation of Hendey, 1974). This stratigraphic interval is characterized by a diverse mammalian fossil fauna, economic grades of pelletal phosphorite, and extensive backbarrier peat deposits. The entire sedimentary succession is ascribed to an intermittent regression during which variable rates of detrital influx into a tectonically downwarped basin controlled the distribution of facies. Based on the occurrence of distinctive elephant, pig, and bovid fossils, a latest Miocene to Early Pliocene age has been suggested (Hendey, 1978a). However, some of the carnivores have close affinities with taxa from Montpellier, France, implying a more restricted 4.5–5.0 Ma age for the mammal-bearing units (Hendey, 1981). The Langebaanweg landscape is believed to have included backbarrier estuarine–peatswamp and near-shore environments (Figs. 14-3B and 14-4).

Backswamp environments are recognized on the basis of 1- to 3-m-thick beds of impure peat that overlie seat earths (Fig. 14-4). The lower peat contains a broad-leaf flora, including the

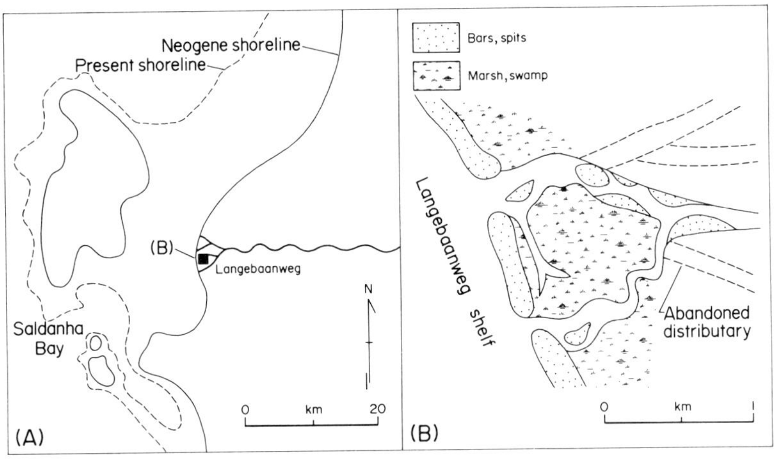

Figure 14-3. Neogene environments of the Saldanha region. (A) Coastal configuration [after Tankard and Rogers (1978)]; black square shows location of (B): (B) paleoenvironments of the Pliocene coastal plain at Langebaanweg.

mesic endemic *Curtisia dentata*. *Curtisia* is a temperate rainforest tree that cannot tolerate arid conditions (Tankard and Rogers, 1978). In contrast the upper peat contains a pollen flora that was better adapted to increasing aridity: grasses, reeds, bushes, Chenopodiaceae-type salt-flat bushes, and compositae. Restionaceae and *Cliffortia* are believed to be typical of the Cape Flora and indicate winter rainfall. The peats grade upward into carbonaceous and pyritic clays which form laterally extensive beds up to 6 m thick. This transitional relationship suggests gradual submergence of the peat swamps.

Phosphatic hardground intervenes between the upper two peat swamp cycles. The phosphorite consists of 1–1.5 m of quartz sand with shark teeth in a carbonaceous matrix cemented with microcrystalline carbonate–fluorapatite (Tankard, 1974b). Evidence of extensive bioturbation, the rigid substrate, the requirements of the molluscan fauna, and an overlying intraformational conglomerate suggest syndepositional cementation.

Generally phosphorite formation is believed to follow Kazakov's (1937) model of direct precipitation of marine apatite from upwelled phosphate-rich waters (Tankard, 1974b; Birch, 1977). The major locus of accumulation of the autochthonous phosphorites was a shallow basin adjacent to a belt of active upwelling (Fig. 14-5), a common characteristic of shallow marine phosphorite accumulations. The phosphatic hardground represents an intraformational disconformity.

Regressive estuarine deposits are superimposed on a thin transgressive interval comprising rounded phosphorite cobbles. Besides an abundance of shark teeth, the littoral taxa are dominated by an intertidal to shallow subtidal assemblage of mollusks characteristic of open coasts (Kensley, 1972).

The uppermost peat facies merges seaward with muddy sands which contain an invertebrate fossil assemblage indicative of a tidal flat milieu (Kensley, 1977). Beyond the northern margin of the peat swamp, deposition is believed to have occurred in an environment subjected to periodic fluctuation between freshwater and saline conditions. Although terrestrial vertebrates dominate the incredibly diverse fauna, aquatic species are not uncommon (seal, penguin, duck, fish). The principal large-fossil taxa include hyena, rhinoceros (*Ceratotherium*), elephant (*Mammuthus*), pig (*Nyanzachoerus*), bovids, giraffids, seal, whale, and birds (Hendey, 1976). Some of these estuarine bone accumulations show evidence of subaerial dispersion.

Shelf sedimentation is reflected in a 25- to 30-m-thick interval of pelletal phosphoritic sands (Fig. 14-4) in which P_2O_5 averages 15 percent (41 percent apatite). Bone and rip-up clasts are concentrated in the basal transgressive zone. Sedimentary structures in the sands indicate currents of variable direction and intensity. Together with the evidence provided by shallow-marine mollusk fossils and the absence of exposure-related structures, it appears that deposition took place in a tide-dominated shoreface environment; thick sets (3 m) of planar cross-bedding with nonerosive lower contacts suggest longshore migration of sand waves. Alternating episodes of bedform migration and relative quiescence are shown by the presence of bioturbated beds above the cross-bed megasets. Also supporting a marine-influenced near-shore environment are attached forms of foraminifera such as *Cibicides*

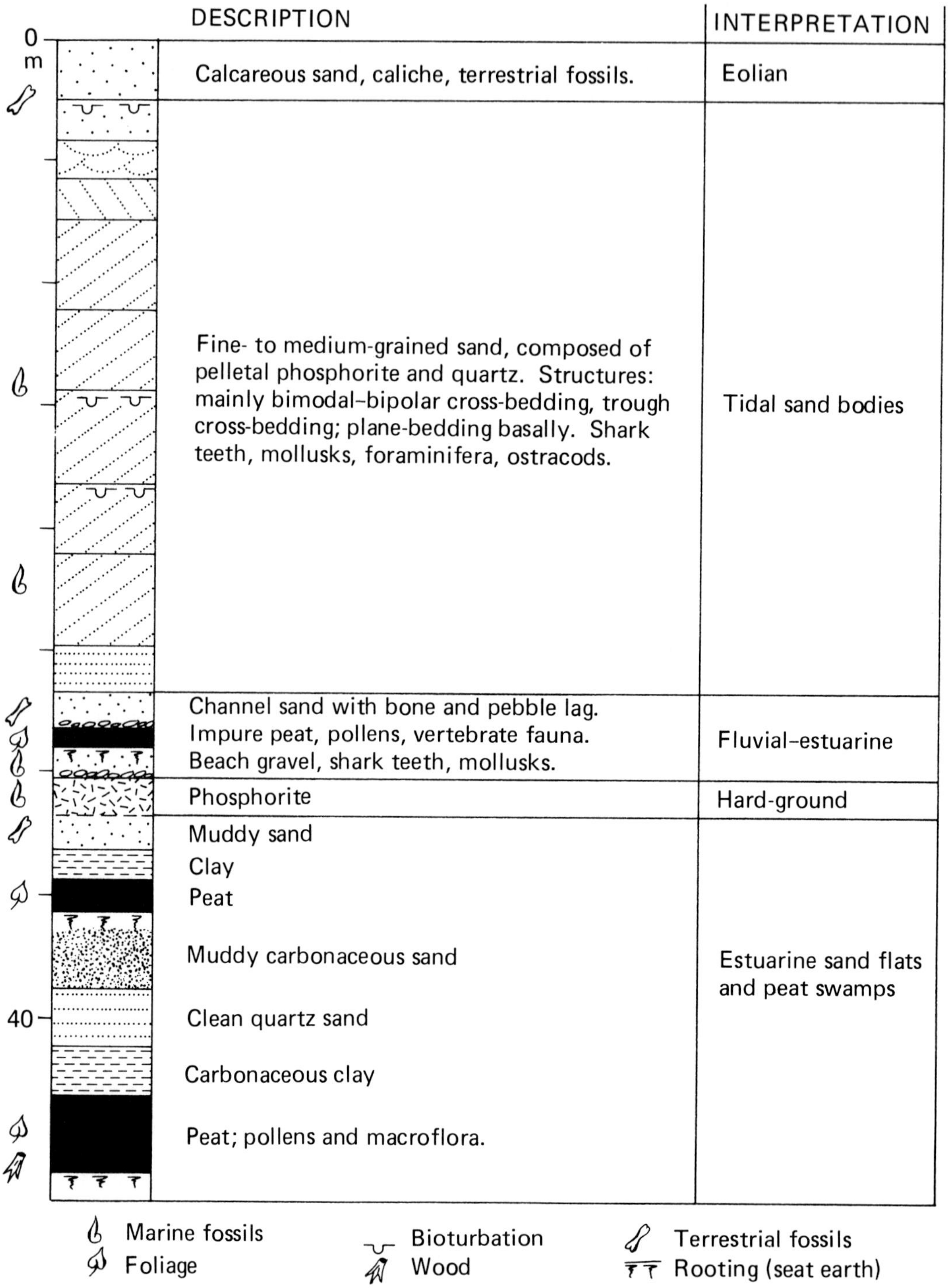

Figure 14-4. Stratigraphic column and interpretation of the Langebaanweg phosphogenic suite. [After Tankard (1975a).]

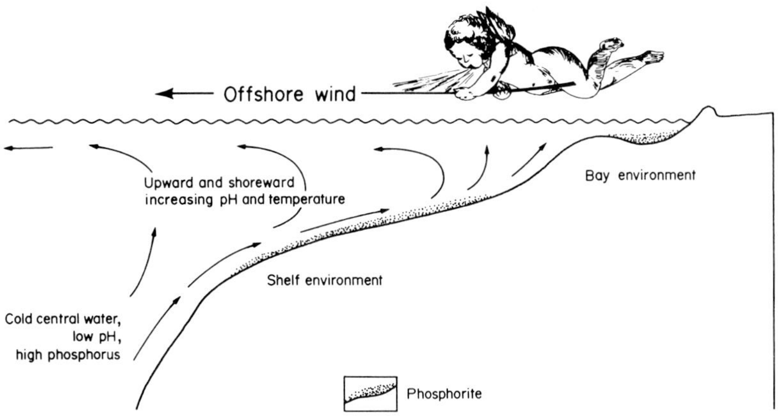

Figure 14-5. Model of phosphorite formation during Miocene sea-level high by inorganic precipitation and carbonate replacement on the continental shelf and Saldanha—Langebaanweg embayment. [After Tankard and Rogers (1978).]

lobatulus, *Planorbulina mediterranensis*, and *Rosalina* cf. *bradyii* (I.K. McMillan, personal communication, 1976). The sand-size pelletal phosphorite has petrologic affinities with the matrix of the phosphatic hardground from which it was probably eroded; fecal pellets and shell detritus form a minor component (Tankard, 1974a). Geochemical analyses (Table 14-1) have established the genetic relationship between the phosphorite pellets and the phosphatic hardground (Tankard, 1974c; Birch, 1977).

This coastal plain–shore-zone sedimentary prism in the Langebaanweg basin has considerable paleoclimatic significance. Gulbrandsen (1969) and Tooms *et al.* (1969) have found apatite precipitation to be temperature dependent; optimum conditions for worldwide phosphogenesis existed in warm seas in areas of upwelling. Although the mollusk taxa also indicate a warmer hydroclimate than today, pollen data and the large herbivore fossils support the suggestion of Axelrod and Raven (1978) that the Late Miocene–Pliocene environment in the Cape was characterized by sclerophyllous woodland and grassland trending toward winter rainfall (Tankard and Rogers, 1978). Long-necked giraffes (*Giraffa*) and other browsers, such as a sivathere, primitive proboscideans, and the boselaphine and tragelaphine antelope (*Mesembriportax* and *Tragelaphus*), indicate the presence of woodlands, whereas the high-crowned teeth of the rhinoceros (*Ceratotherium*), the horse (*Hipparion*), and alcelaphine antelopes are indicative of grasslands. Hendey (1973) visualized an environment characterized by riverine woodland flanked by grassland with seasonal rainfall.

Upper Pliocene to Lower Pleistocene fluvial deposits in this area (Baards) contain a hetero-

Table 14-1. Major-Element Geochemistry of Coastal Plain and Continental Shelf Phosphorite[a]

	Langebaanweg hardground n = 3	Langebaanweg phosphoritic sand 7	Agulhas shelf 21	Atlantic shelf 15
SiO_2	52.72	54.57	12.84	13.44
Al_2O_3	1.43	1.99	1.85	1.70
Fe_2O_3	1.00	1.33	8.24	5.55
MgO	0.99	1.45	1.35	1.29
CaO	22.37	20.40	37.29	39.27
Na_2O	0.25	0.43	0.67	0.98
K_2O	0.20	0.73	1.29	1.19
P_2O_5	16.68	14.66	16.18	17.83
F	1.37	1.21	2.10	2.46
CO_2	1.15	1.52	—	—
CaO/P_2O_5	1.34	1.39	2.31	2.23
F/P_2O_5	0.08	0.08	0.13	0.14

[a]Sources: Onland, Tankard (1975a); offshore, Birch (1977, 1979a)

geneous vertebrate fauna, the older assemblage of which has affinities with the early fauna of the Makapansgat and Sterkfontein hominid sites in the Transvaal. Most significantly, the Early Pliocene grazing rhinoceros (*Ceratotherium praecox*) was replaced in the Late Pliocene–Early Pleistocene by a browser (*Diceros bicornis*), suggesting progressive desiccation from grassland to a more xerophilous vegetation (Tankard and Rogers, 1978). However, as indicated by the *Hippopotamus*, surface water was still abundant.

Shorelines of the south coast. Along the southern coastal margin, marine and eolian calcarenites and limestones form a discontinuous veneer on a wave-cut platform that is 50 km wide in places. These coastal deposits are generally referred to as the Alexandria Formation, or, in the more westerly areas, the Bredasdorp Formation. They are 150 m at their thickest and comprise a sporadic basal conglomerate, shelly gravel, skeletal lime grainstones and packstones, and calcareous sandstones (Siesser, 1972). Near Bredasdorp (Fig. 14-2) basal conglomerates of unknown age are thick and silica-cemented and bear a striking resemblance to the fluvial Enon conglomerates (Wybergh, 1920).

The limestones developed in response to shorezone regression and preserve lithofacies characteristic of shoreface, beach, and east–west-trending foredune complexes which display extremely large foresets. In the Algoa region the Neogene sediments mantle a dissected seaward-sloping platform, the inland margin of which reaches 305 m above sea level. The littoral component is generally less than 10 m thick but nevertheless displays variations in lithology resulting from sea level fluctuation. Ruddock (1968, 1973) has profiled three distinct erosional platforms beneath the sediment cover and attributes these to Eocene, Miocene–Pliocene, and Late Pliocene transgressions coupled with seaward tilting. At most places the limestones are overlain by lagoonal silts and eolian sands.

These coastal facies are considered to be principally of Miocene to Pliocene age because of the close resemblance of the mollusks to northern hemisphere Neogene taxa (Newton, 1913; Du Toit, 1954, p. 439; Siesser and Dingle, 1981). This assemblage is ubiquitous on the south coast and is characterized by *Melapium patersonae*, *Pirenella stowi*, *Voluta africana*, *Cardium edgari*, *Notocallista schwarzi*, *Glycymeris borgesi*, *Isognomon gaudichaudi*, *Ostrea atherstonei*, and *Tivela baini*. Some exposures in the eastern Cape have also yielded Pliocene foraminifera (King, 1972).

The Neogene Natal–Zululand basin. Continuous emergence of the Transkei swell (p. 417) is shown by the lack of well-developed Tertiary marine strata. Farther north along the seaboard of Natal and Zululand, deposition took place mainly by shoreline regression and coastal dune formation (King, 1953, 1966; Frankel, 1960, 1966, 1968; Maud and Orr, 1975).

Shallow-marine deposits of the Uloa Formation transgressively overlie a seaward-sloping marine abrasion platform. The age of these deposits has long been disputed and estimates range from Early Miocene through Pliocene (King, 1953, 1966; Frankel, 1960, 1968; Davies, 1964). King (1953) envisaged two distinct depositional events, of Miocene and Pliocene ages, respectively, separated by an episode of erosion. More recent studies (Maud and Orr, 1975; Stapleton, 1977; Siesser and Miles, 1979) of the planktonic foraminifera and calcareous nannofossils suggest an uppermost Miocene to Lower Pliocene age for the entire sequence, probably ranging from foraminiferal Zone N16 to Zone N17 (9–6 Ma). On the basis of lithologic and paleontologic similarities and comparable terrace elevations and microfacies distributions, the coastal limestones of Zululand and Mozambique are believed to be closely related (Soares and Da Silva, 1970; Hobday, 1977).

Basal ferruginous gravel rests on evenly truncated Upper Cretaceous strata at elevations of 3–17 m above present sea level. Much of the gravel is composed of epigenetic concretions developed at the unconformity permeability barrier by groundwater migration. Phosphatic and pyritic nodules are common, some enclosing eroded Cretaceous ammonites. Reworked Eocene glauconitic sandstone clasts, foraminifera, shark teeth, and bone are present locally (Frankel, 1968).

At some localities a 20-cm ferruginous nodular layer resting disconformably on the Cretaceous is the only record of Miocene transgression (Hobday, 1965). More commonly this basal layer is overlain by a massive, shelly calcarenite or biosparrudite from which more than 100 species of macrofossils have been recovered (King, 1953). Thick-shelled mollusks predominate, with *Aequipecten uloa* most numerous. Deposition of this transgressive coquina was probably accomplished by rapid *in situ* production of shell debris, from which the finer fraction was winnowed by persistent currents and waves. Associated with this unit, particularly in its lower part, are subspherical to discoidal extrabasinal pebbles in distinct lithologic suites representing transport down different drainage systems, primarily the Mfolozi, Mkuze, and Pongola Rivers (Hobday, 1965).

Plane-bedded and locally cross-bedded calcarenites with scattered mollusks, mainly *Aequipecten uloa* and oysters, overlie the shelly limestones with a contact that varies between erosive and gradational. In the Lake St. Lucia area a 10-m-thick succession shows classical gradation from graded and multidirectionally cross-bedded bioclastic shoreface sands to seaward-dipping, plane-bedded foreshore sands with crustacean burrows (Hobday, 1965). Vestiges of high-angle eolian cross-bedding are preserved at the top of this regressive shoreline succession, which has largely been leached and diagenetically altered to structureless red sand (cf. McCarthy, 1967). Root disruption and fossil land snails are a feature of the subaerial strand-plain facies.

At a number of localities in Zululand and around Durban a thin boulder bed, representing the transgressive phase equivalent to the Uloa Formation, is present at the base of the regressive beach association (McCarthy, 1967).

Discussion

In southern Africa the end of the Cretaceous was marked by a general withdrawal of the sea. Locally subsiding areas remained and in the Zululand basin sedimentation persisted into the Danian. Throughout the remainder of the Tertiary, relative changes of sea level are indicated by cycles of coastal onlap which correlate with eustatic fluctuations (cf. Vail *et al.*, 1977). Between Namibia and Mozambique the coastal plain is underlain by rocks of the same general age, reflecting high sea levels in the Early Eocene, Late Eocene, and Late Miocene–Early Pliocene (Fig. 14-6). Stratigraphic hiatuses, representing periods of nondeposition, separate these transgressive pulses; the Oligocene hiatus is the most pronounced. The episodic nature of oceanic ridge activity is fundamental in explaining this coastal stratigraphy (Fig. 14-6).

Entrenched river drainages show that uplift was widespread in the Tertiary (King, 1967; Bond, 1979). In the Zululand basin the stratigraphy and the structure of the pre-Tertiary platform indicate little seaward tilting and a single Neogene transgressive episode (Maud and Orr, 1975; Stapleton, 1977). Repeated tilting characterized the Neogene history of the eastern Cape (Ruddock, 1968, 1973). Progressive uplift of the Transkei swell further deformed the Neogene shoreline. From elevations between 45 and 90 m in the western Cape and Zululand the Neogene shoreline is elevated to more than 300 m in the Bathurst region (Fig. 14-2) (Tankard, 1975a).

Sporadic volcanic activity persisted into the Tertiary (Fig. 14-6). Aegirine-augite trachytes and related plugs of the Alphard Bank on the southern continental shelf have been dated at 58 $\pm$ 2 Ma (Dingle and Gentle, 1972; K-Ar). Also included within this Alphard Bank igneous province are the olivine melilitite pipes in the Heidelberg region. Olivine melilitite volcanics in the Sutherland area also have a probable Early Tertiary age (Moore, 1976). Oligocene volcanic activity is shown by a date of 38 Ma on an olivine melilitite plug in Namaqualand, and by dates of 38 and 36 Ma on the Klinghardt phonolites of Namibia (Kröner, 1973b; K-Ar).

14.2. Tertiary Shelf Sedimentation

The shelf break off the western seaboard, varying between 200 and 500 m in depth, is one of the deepest in the world. Several relict shelf breaks characterize the outer shelf, where progradation and upbuilding have resulted from seaward tilting (Du Plessis *et al.*, 1972). Beneath the continental margin the basement descends gently oceanward and reaches a depth of 6–8 km; in places it is offset by large fracture zones (Scrutton and Dingle, 1974). In the Orange and Walvis basins a single seaward-building wedge of Cenozoic sediment with low seismic velocity spilled onto the oceanic crust. The low-velocity interval below the Atlantic shelf consists of up to 800 m of Paleogene glauconitic clays and limestones, and 300 m of Neogene glauconitic and phosphatic shelly limestones and sandstones of shallow-water origin (Dingle, 1973b; Siesser *et al.*, 1974). In two pockets, seaward of Saldanha and on the Lüderitz

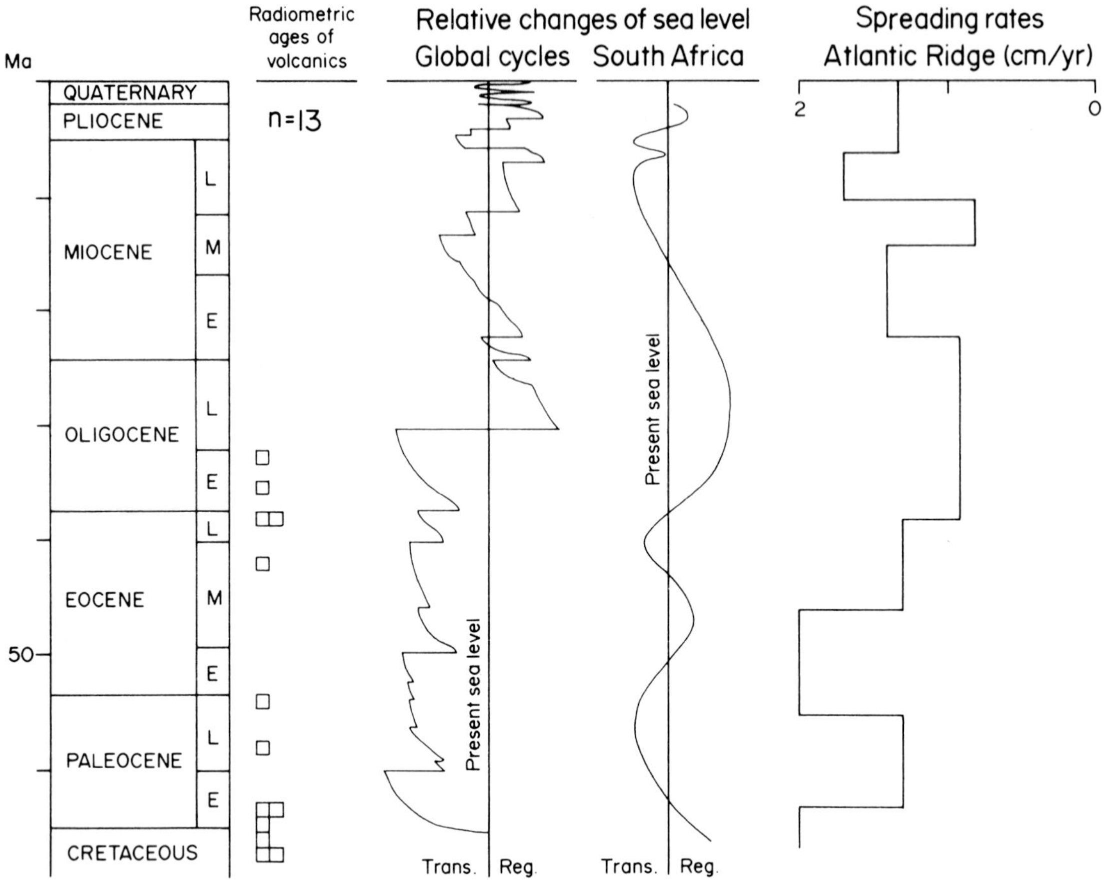

Figure 14-6. Comparison of South African sea-level history with global sea-level cycles [after Siesser and Dingle (1981), Vail *et al.* (1977)] and spreading history of the mid-Atlantic Ridge. [Sources: Phillips (1967), Larson and Pitman (1972), Latimore *et al*, (1974). Radiometric ages are derived from Dingle and Gentle (1972), Kröner (1973b), Davis (1977), Duncan *et al*, (1978), and McLachlan and McMillan (1979).]

Bank (Fig. 14-2), 96 percent of the shelf sediment is glauconitic. Numerous seismic reflectors punctuate the Tertiary succession, the most widespread of these being interpreted as a major Mid-Tertiary hiatus. Tertiary sediments are thinnest where they overlie the buoyant Cape pluton. In the south the continental shelf is traversed diagonally by the Cape Canyon (Simpson and Forder, 1968) which was possibly initiated by erosional regression at the end of the Eocene and enlarged by turbidity flows throughout the remainder of the Cenozoic (Siesser *et al.*, 1974).

The continental margin fronting the Indian Ocean has a shallower and better defined shelf break. Along the Agulhas Bank it varies between 120 m and 180 m in depth, and at 21°E the shelf is 240 km wide. East of the Agulhas Bank the shelf break shallows to 50 m and the shelf narrows to 15 km off the east coast (Siesser *et al.*, 1974). Surface irregularities are common in the vicinity of the Alphard Bank, where several basalt and trachyte plugs, dated at 58 Ma, are exposed (Dingle and Gentle, 1972; K-Ar). The low-velocity sedimentary prism contains up to 500 m of calcareous clays, glauconitic limestones, and sandstones and up to 400 m of foraminiferal–bryozoan limestones. The Agulhas Bank is a thick sediment wedge deposited over block-faulted continental basement (Siesser *et al.*, 1974). As on the west coast, a major seismic reflector records a Mid-Tertiary hiatus in the 1000-m-thick succession. For the most part sedimentation was concentrated on the outer shelf and slope where turbidity currents transported much of it into the deeper ocean basins.

A marked unconformity separates Paleogene and Neogene successions. The Paleogene interval includes gray-brown marls and calcareous silts with abundant intercalations of calcareous sandstone. Shelf sedimentation was apparently interrupted after Late Eocene time, and throughout the Oligocene and Early Miocene the upper Agulhas Bank was subjected to a period of erosion that elsewhere was marked by a biostratigraphic break (Dingle, 1973b; Du Toit, 1976). Seaward-plunging basement highs (Fig. 12-2) persisted as positive features through the Paleogene with the result that Neogene sediments directly overlie Mesozoic or basement rocks in their vicinity. The Neogene succession is dominated by micritic limestone in which bioclastic material varies up to 90 percent.

Phosphorite is one of the major rock types on the Agulhas Bank and the Atlantic shelf (Fig. 14-2) and is frequently associated with glauconite. First dredged from the Agulhas Bank by *H.M.S. Challenger* in 1873 (Murray and Renard, 1891), it forms slabs of phosphatized foraminiferal limestone and glauconitic–phosphatic breccia (Summerhayes, 1973). Iron-rich varieties include goethite in the cement, which otherwise comprises micrite and carbonate fluorapatite (francolite) (Parker and Siesser, 1972; Birch, 1979a). Shallow-water phosphorite is generally associated with glauconitic sands, but in deeper water glauconite is depleted and the proportion of planktonic foraminifera and other organic remains is increased (Dingle, 1974).

The main source of the phosphatizing solutions is the cold, nutrient-rich Central Water which moves toward the equator from the Southern Ocean at a depth of 500 m. Whereas the onshore phosphorites are enriched in Si, Al, and Fe derived from terrigenous materials, the offshore deposits are characterized by abundant calcite (Ca) and glauconite (K, Mg, Fe: Table 14-1) (Birch, 1977, 1979b). Furthermore the offshore phosphorites have formed by inorganic replacement of calcareous material, mainly calcareous mud, while their onshore counterparts originated by direct precipitation of apatite (Tankard, 1974b; Birch, 1977). Recent foraminiferal determinations show that phophatization probably commenced during the Middle Miocene and continued into the Pliocene but had ceased prior to the Quaternary (Siesser, 1978). Sand-size phosphoritic material has been related to reworking of the bedded Tertiary phosphorites and accumulation as a lag placer (Summerhayes, 1973; Birch, 1979c).

The shelf off the east coast is narrow and shallow and is underlain by a seaward-thickening sedimentary prism that records a substantial history of Tertiary sedimentation: Borehole Jc-1 penetrated a 1400 m succession (Du Toit and Leith, 1974). The Paleocene–Eocene interval is a monotonous accumulation of shelly mudstones that coarsen upward into glauconitic mudstones and limestones. A sequence representing a reduced rate of Oligocene sedimentation is followed by *Pecten*-rich bioclastics. Neogene sedimentation was terminated by erosional regression which generated an angular unconformity. Marine Oligocene conditions also persisted on the north side of the Zambezi delta (Flores, 1973). Nevertheless, the Oligocene was a time of widespread regression and marine deposits are mostly restricted to the deeper basins. Marginal regions were affected by emergence and erosion. A marked Oligocene hiatus on the Mozambique Ridge and the western edge of the Mozambique Basin (Deep Sea Drilling Project sites 249, 248) has been attributed to scouring caused by a shift in ocean currents following eustatic lowering (Dingle and Scrutton, 1974).

Large-scale slumping along the southern continental margin, possibly induced by seismic activity, involved strata as young as Pliocene. The largest of these submarine slumps is 750 km long and 106 km wide. Dingle (1977) attributed the magnitude of this feature to its association with the Agulhas fracture zone, a long and continuous shear margin.

14.3. Quaternary Transgressions and Regressions

Shoreline features of Quaternary age developed intermittently along the coastal plain and continental shelf in response to high-latitude glacial fluctuations (Maud, 1968; Carrington and Kensley, 1969; Butzer and Helgren, 1972). The Greenland and Antarctic ice caps, if melted, would raise sea level a maximum of 65 m, insufficient to account for Early Pleistocene levels above 100 m. Furthermore, available evidence argues against any substantial melting of these ice caps since the Tertiary, suggesting that most relict shoreline elevations are the combined result of glacio- and tectonoeustasism compounded by local epeirogenesis. Consequently elevation alone is no basis for shoreline correlation of any but the lowest (youngest), well preserved terraces. In the region of the Transkei swell epeirogenesis is inferred for very high marine terraces at 210–470 m above present sea level (Thompson, 1942; Davies, 1978).

Early and Middle Pleistocene Shorelines

Although geomorphic studies have shown that Pleistocene strandlines up to 100 m or more above present sea level are not uncommon along the southern and eastern seaboard of southern Africa, the most substantial record exists for the west coast. Carrington and Kensley (1969) have documented a series of transgressive–regressive complexes underlying the Namaqualand coastal plain, each of which is separated by an interval of terrestrial sedimentation: 75–90 m, 45–50 m, 29–34 m, 17–21 m, and the Late Quaternary levels at 7–8 m, 5 m, and 2 m above present sea level. These diamondiferous deposits overlie a granite–gneiss basement and occasional phosphogenic sediments of Neogene age. Basement dissection is associated with thick sequences of river channel fill.

The older and more elevated sequences contain a mollusk fauna of warm-water affinity (Haughton, 1931, 1969 p. 406). Most notable is the 45–50 m regressive complex, which consists of fine-grained sands overlying a thin transgressive gravel. The characteristic thermophilic taxa include *Striostrea margaritacea*, "*Chamelea*" *krigei*, *Donax haughtoni*, and *Fissurella robusta* (Carrington and Kensley, 1969). In contrast the younger 17–21 m regressive complex is coarser grained and contains a fauna more compatible with the present hydroclimate of the Atlantic coast. The zone fossil would undoubtedly be the more robust *Donax rogersi*. Progressive downwarping has lowered these Pleistocene strandlines in the south, and in the Saldanha region shore-zone deposits only 10 m above sea level are biostratigraphically equivalent to the 45–50 m regressive complex of the Namaqualand coast (Tankard, 1976b). The same trend is paralleled by the present shelf break, which is deeper west of Cape Town than off the Namaqualand coast.

Diamondiferous gravels are also associated with the mouths of the Buffels and Orange rivers, where repeated Pleistocene transgressions have reworked coarse fluvial sediments (Hallam, 1964). A general northward decrease in clast size, including diamonds, shows the dominance of the southwesterly ocean swell during the Pleistocene. Southwesterly wave approach also favored diamond concentration along the northern shores of bays. Terraces at the mouth of the Buffels River have elevations up to 90 m.

At the Orange River mouth the fluviomarine succession overlies an irregular Precambrian basement. The depositional history reflects alternating dominance by braided rivers during base-level lowering, and transgressive marine reworking. Throughout this period the source of the fluvial material was fixed by the deeply incised nature of the lower Orange valley. Most of the braid-plain conglomerates are unstratified or are horizontally bedded. However, planar cross-bedded conglomerates in sets up to 3 m thick are also common; they probably formed by the lateral accretion of longitudinal bars into deep channels. Inland of the Pleistocene shorelines the Orange River terraces are underlain by poorly bedded conglomeratic sequences up to 56 m thick, which Fowler (1976) attributed to rapid rates of sediment transport and accumulation.

The littoral lithosomes form more persistent units. Seaward-dipping foreshore accretion beds consist of sand and gravel (Fig. 14-7). Marine fossils and bioclastic material indicate a warm-water ecosystem for the shore-zone complexes above 25 m (Keyser, 1972). Zones of diamond accumulation are the result of transgressive shore-zone erosion of braid-plain deposits. The most valuable concentrations occur in the seaward beach gravel units and in their upper winnowed parts, and in bedrock depressions (Hallam, 1964; Keyser, 1972). Because of the long history of transportation and abrasion, the gems (average 1 carat) tend to be largest and of highest quality in the upper terraces. Mapping has shown the presence of six transgression–regression complexes with marked warping of the upper terraces (Martin, 1973; Fowler, 1976).

A unique occurrence of diamondiferous gravel occupies a bedrock depression at the foot of an abandoned waterfall at Octha (Fig. 14-2). This gravel contains extraordinarily high concentrations of diamonds; the largest industrial stone weighed 89 carats, and the largest gem, 73 carats (De Beers Consolidated Mines Ltd., 1976).

Strand-Plain Dune Ridges

Prominent ridges of predominantly eolian origin are a conspicuous feature of the Pleistocene coastal plain. Most are parallel or slightly oblique

Figure 14-7. Foreshore facies in the Pleistocene diamondiferous deposits at Oranjemund, Namibia.

to the present shoreline. Diagenetic processes have obliterated most of the primary sedimentary features and have imparted a reddish coloration.

Textural and stratification patterns nevertheless are discernible in some of the Zululand dune ridges mapped by Davies (1976). As many as six parallel cordons are developed. The base of the inner cordon is at 100 m above sea level where it abuts against Mesozoic volcanics. Successively younger ages of the seaward cordons are inferred from the less advanced stages of pedogenesis, higher relief, and archeological artifacts. A relatively youthful cordon along the western shore of Lake St. Lucia was dated archeologically by Davies as of Sangamon to Wisconsinan age.

Most ridges had a compound depositional history. An estuarine to shoreline boulder bed at the base of the innermost Zululand cordons gives way to scattered pebble layers in the lower parts of cordons of intermediate age. Rare pebble imbrication, in association with foresets of variable direction, indicates complex shore-zone circulation patterns (Hobday, 1965). The younger cordons immediately landward of the high Wisconsinan coastal dunes consist of homogeneous red sand throughout. These are generally regarded as eolian deposits, although evidence is sparse. Furthermore because of advanced diagenesis, these Pleistocene sands are identical in appearance to deeply weathered Neogene calcarenites (cf. McCarthy, 1967). The dune ridge extending up the western shore of Lake St. Lucia is clearly loessic in places (Hobday, 1965), with rare but unmistakable ventifacts. These eolian deposits overlie coarser, gravelly estuarine sands in characteristic upward-fining pattern. Because the basal gravels rest on a strike-persistent erosional platform, all of the dune ridges, apart from the youngest, are regarded as representing transgressive–regressive couplets. Alternatively they may represent successive stillstands during a general Cenozoic withdrawal of the sea from the coastal plain (McCarthy, 1967), difficult though this is to reconcile with established tectonoeustatic fluctuations.

Eolian dune ridges are also prominent in the eastern Cape. They are subparallel to the present coast and overlie a benched, wave-cut platform (Ruddock, 1973). Typically they overlie shorezone or estuarine boulder beds, or shell-bearing calcirudites. The eolian calcarenites and limestones are up to 150 m thick and reach over 350 m above present sea level in places. A discontinuous veneer of calcareous eolian material underlies the coastal plain westward, and in the southwestern Cape is characterized by the land snail *Trigonephrus* (Siesser, 1972; Tankard, 1976b). Deposition by southerly winds is indicated by consistently northward oriented cross-beds. *Terra rosa* sediments are locally associated with karst weathering. The dune complexes in the Cape have a Late Tertiary–Quaternary age range and are closely related to the transgressive–regressive history of the coastal plain. The youngest components are nearest the present shoreline and contain artifacts and Pleistocene vertebrate remains.

Relict Late Pleistocene dune ridges are also preserved on the inner continental shelf where they have been partly eroded by wave and current activity during the Flandrian transgression (McCarthy, 1967; Birch *et al*., 1978). The best known of these is the Aliwal Shoal which is situated 3–5 km offshore south of Durban. It is notorious as a shipping hazard. The Aliwal Shoal and Protea Banks are part of an extensive coastal dune complex that formed during a prolonged period of sea-level stillstand 68 m below present sea level. Locally this drowned foredune ridge has dammed the Quaternary nearshore sand wedge (Flemming, 1981).

Sangamon Sea Levels

A characteristic suite of three shorelines related to the last interglacial climatic optimum is now widely recognized (Hobday, 1975; Tankard, 1976a; Birch *et al*., 1978; Davies, 1978). The highest level at 6–8 m is the clearest record of Pleistocene transgression and is believed to correlate with the 120,000–130,000 year eustatic high, datum. Characterizing this Sangamon transgression is the distinctive Swartkops fauna whose presence shows epeirogenic upwarping to 15 m in the eastern Cape (Davies, 1978).

The last interglacial rise of sea level increased the size and number of estuaries and lagoons. By populating these protected embayments, tropical west African mollusks were able to extend their geographic ranges and maintain breeding populations in the western and southern Cape (Tankard, 1975b; Tankard and Rogers, 1978). Today the southern limit of these tropical species is at about 17°S, or the northern boundary of the Benguela Current (Table 14-2). This suggests a temperature about 5°C warmer than today in the hyperthermal estuaries and lagoons on the Atlantic coast. Although the fossils from open-coast habitats demonstrate the prevalence of upwelling north of the Cape Peninsula, contemporaneous evaporite formation is indicative of hot arid conditions. In the northern hemisphere a similar hydroclimate at that time is shown by the migration of Senegalese taxa into the Mediterranean (Butzer and Cuerda, 1962).

Highly fossiliferous limestones in False Bay, Lake St. Lucia, contain a diversity of thermophile corals, large reef mollusks, echinoderms, brachiopods, and open-marine microfossils that accumulated when False Bay opened directly to the sea during the last eustatic high. Reworked and *in situ* coral heads are preserved to a height of 3.5 m above present sea level.

At Port Durnford in Zululand, coastal erosion has exposed lagoonal deposits overlain by an allochthonous peat and destructive barrier sands beneath a coastal dune complex (Fig. 14-8) (Hobday and Orme, 1974; Hobday and Jackson, 1979). The oyster-bearing clays are believed to have resulted from infilling of a lagoon which

Table 14-2. Late Pleistocene Warm-Water Taxa Compared with their Modern Distribution[a]

Species	Modern latitudinal range
Nuculana bicuspidata	
Loripes liratula	
Felania diaphana	
Tellina madagascariensis	
Macoma tricostata	
Leporimetis hanleyi	
Venerupis dura	
Panopea glycymeris	

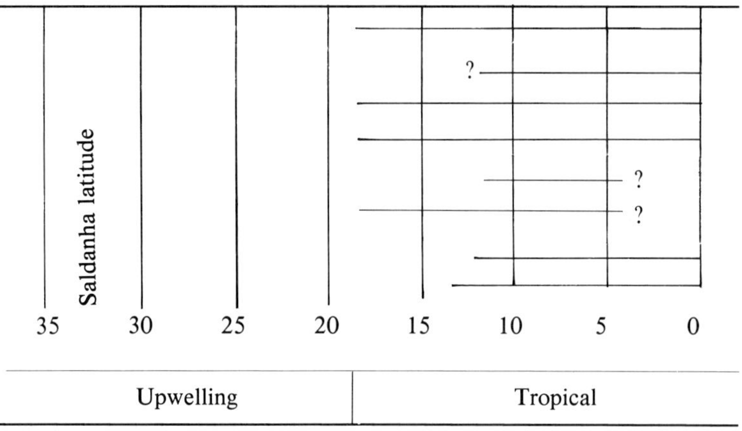

[a]Source: Tankard and Rogers (1978)

originated by flooding of a coastal dune field by the maximum Sangamon transgression. Some of the mollusks, such as *Corbula* and *Cardium* cf. *pseudolima*, are typical of tropical latitudes. These Port Durnford deposits have long been of interest because of their rich mammalian fauna, including rhinoceros, elephant, buffalo, hippopotamus, turtle, and crocodile remains, and the association of archeological artifacts (Anderson, 1907; McCarthy and Orr, 1978). Gradual barrier destruction commenced with renewed advance of the sea and overlap of the lagoon and peat swamp facies by washover sheet sands. Clay diapirism reflects loading of the lagoonal and swamp facies by the denser barrier and eolian sands (Jackson and Hobday, 1980). This was accompanied by gravity gliding along clay-rich décollement layers toward paleovalleys situated inland.

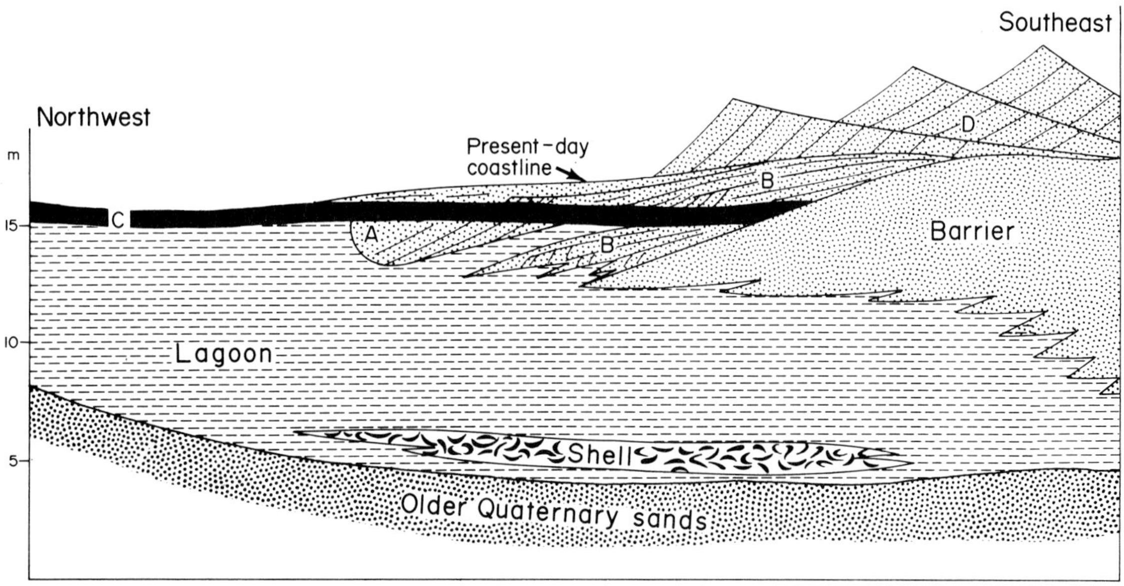

Figure 14-8. Schematic dip-oriented reconstruction of the Port Durnford barrier–lagoon complex showing relationships among (A) tidal channel, (B) washover, (C) peat, and (D) eolian dune subfacies. Scale is approximate. [After Hobday and Jackson (1979).]

The Late Pleistocene Hypothermal

During the Wisconsinan glacial advance, sea level oscillated between 20 m and 130 m below present datum, being most depressed during the glacial maximum at 17–18 ka (Tankard, 1976a). Retreat of the shorelines was accompanied by alluvial sedimentation as rivers extended their courses (Dingle and Rogers, 1972). Intensified winds and a sparse vegetation cover are reflected in the development of thick eolian complexes. At Swartklip in the western Cape the dune fields developed above washover sheet sands (Barwis and Tankard, in preparation). These eolianites show that the hypothermal (last glacial stage) wind pattern was similar to the present.

High Wisconsinan coastal dunes of Zululand and Mozambique attain heights of 200 m. These sands rest on a core of calcarenites and postdate development of the Sangamon terrace; they abut landward against older Pleistocene red sands. The rising dunes temporarily impounded the Lake St. Lucia system (Hobday, 1976) and sealed off Lake Sibaya from the sea.

The hypothermal climate of the subcontinent reflects the influence of global refrigeration, with temperatures as much as 10°C lower than today (Butzer, 1973a). Radiocarbon-dated pollen in the southwestern Cape records two invasions of temperate rainforest coincident with wet intervals between 33,000 and 45,000 years ago (Schalke, 1973). The rain forest, which included *Curtisia*, *Ilex*, and *Podocarpus*, displaced sclerophyllous vegetation. Relict populations of mesic endemics in Namaqualand also strongly suggest a moister hypothermal climate (Tankard and Rogers, 1978). Mammalian fossils also indicate a general expansion of grassland and forest vegetation in the southern Cape (Klein, 1974, 1975).

Van Zinderen Bakker (1976) estimated a 5° equatorward shift of the climatic belts during the hypothermal winter and expansion of the winter rainfall area to latitude 23°S. In Namaqualand these conditions resulted in a sequence of alternating lacustrine and eolian deposits and rejuvenation of coastal rivers (Coetzee, 1978; Tankard and Rogers, 1978). Excavations in coastal caves, such as Die Kelders, Klasies River, and Nelson Bay, show that they were briefly inhabited by Middle Stone Age hominids whose dietary remains reflect the considerable distance from the Wisconsinan shoreline (Klein, 1974; Tankard, 1976c). A major occupational hiatus has been attributed to the harsh climatic conditions which encouraged the inhabitants to seek a more protected environment away from the coastal plain. Excavation at the inland Boomplaas cave site provides proof of such occupation (Klein, 1975).

The Flandrian Transgression

After 17,000 years B.P. sea level recovered swiftly as the Wisconsinan ice sheets retreated, reaching its present position nearly 6000 years ago (Tankard, 1976a). Although transgression was rapid, up to 167 cm per century, frequent brief stillstands and even slight regressions interrupted its advance. The post-6000 year history is still debated, although the prevailing opinion favors a high about 5500 years ago (Flemming, 1977), probably as a result of hydroisostasy.

The shapes of most estuaries and lagoons along the coast were established by river erosion and dune development during the Wisconsinan lowering of sea level (Birch *et al*., 1978). These depressions were flooded by the Holocene transgression, producing large water bodies such as the Wilderness Lakes, Langebaan Lagoon, and Lake St. Lucia. Fossil oysters in Langebaan Lagoon have been dated at 6400 B.P. (Flemming, 1977). Langebaan Lagoon is a mixed siliciclastic–carbonate tidal environment (Fig. 14-9) in a hot, semiarid climate. Zululand lagoons have experienced 60 percent reduction in surface area, along with considerable shallowing, as a result of Holocene sedimentation (Orme, 1973). Farther north, flooding of Delagoa Bay on the Mozambique coast produced an interesting assemblage of coral reefs and tidal sand flats (Hobday, 1977).

Intrusion of the Agulhas Current over the outer shelf along the southeastern continental margin is shown by the southward migration of large sand waves (Fig. 14-10) observed on side-scan sonar in depths greater than 50 m (Flemming, 1978). These sharp-crested bedforms are superimposed on a biogenic basal conglomerate and have amplitudes up to 17 m and wavelengths as much as 700 m.

Generally the Indian and Atlantic shelves show a distinct shore-parallel zonation of sediment types (Siesser *et al*., 1974). Most of the continental slope and parts of the shelf edge along the western continental margin are draped with Holocene foraminiferal and coccolithophorid debris (Birch,1977). Seaward of the wave-dominated

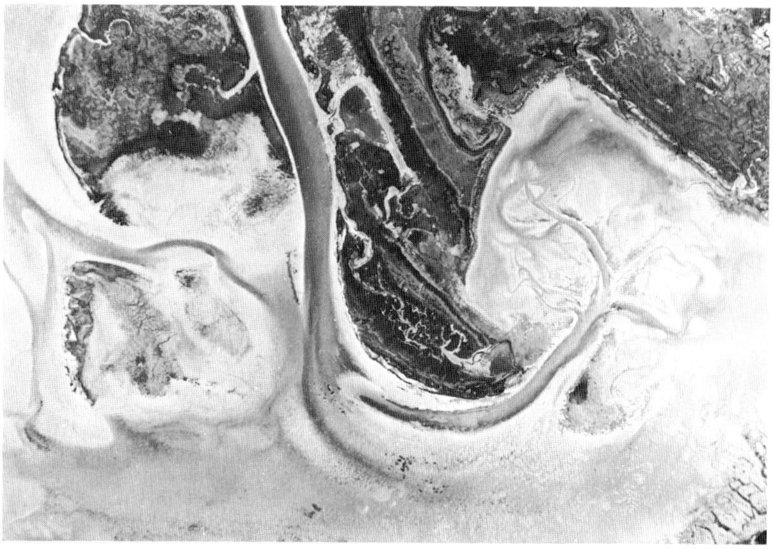

Figure 14-9. Langebaan Lagoon showing segregation of flood and ebb channels; channel termini are characterized by spillover lobes. [After Flemming (1977).]

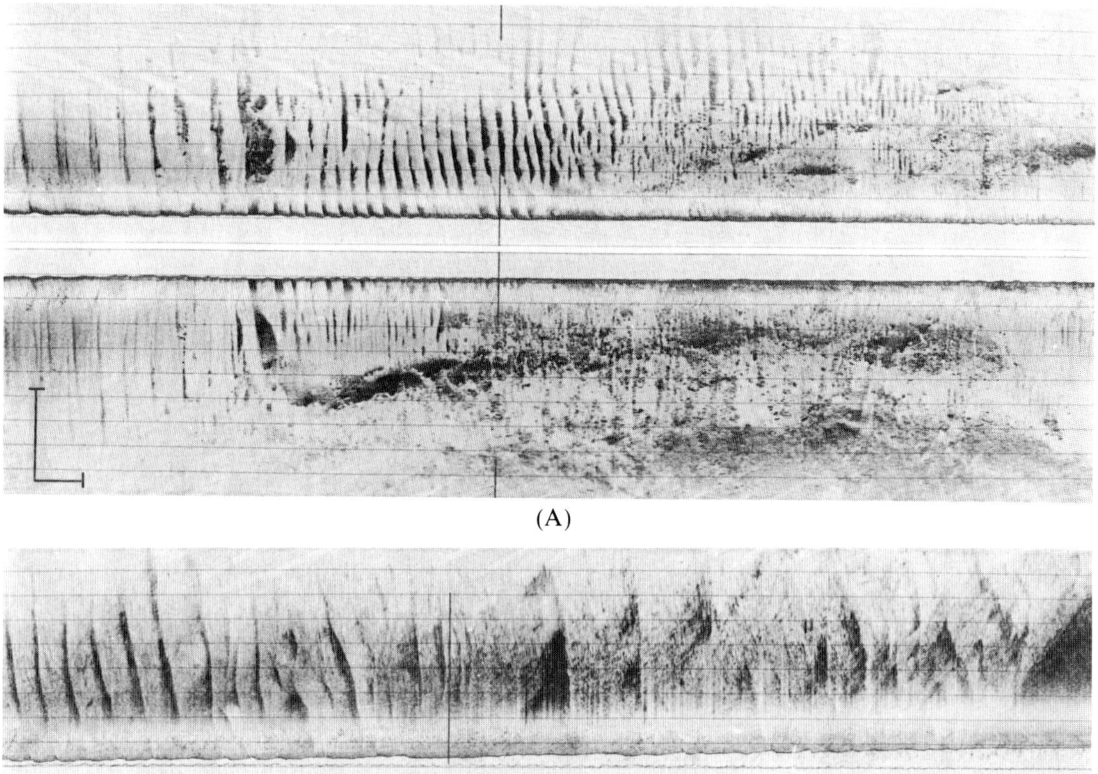

(A)

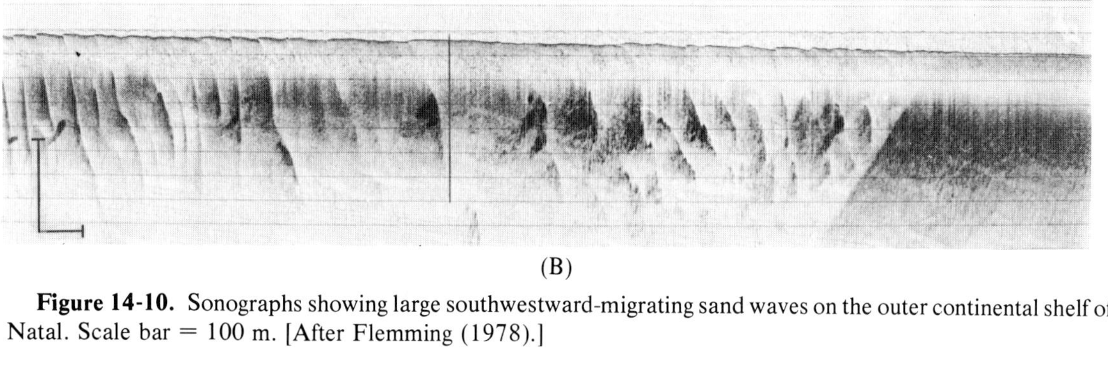

(B)

Figure 14-10. Sonographs showing large southwestward-migrating sand waves on the outer continental shelf of Natal. Scale bar = 100 m. [After Flemming (1978).]

Orange River delta the bedrock is buried beneath a subaqueous deltaic prism which is more than 60 m thick north of the mouth (Rogers, 1977).

Climatic amelioration in the Holocene is marked by reestablishment of the sclerophyllous vegetation in the western Cape and the reoccupation of coastal caves by Late Stone Age people.

14.4. The Interior Basin

To a considerable extent, epeirogenesis and widespread erosion characterized the Cenozoic history of the interior of South Africa. Toward the end of the Mesozoic the paleogeography of central southern Africa consisted of the internal drainage system of the Kalahari (Kgalagadi) basin and, farther south, the Orange–Vaal drainage basin (Fig. 14-2). Although deposition in the large Kalahari basin may have been initiated in the Cretaceous (Du Toit, 1954, p. 408), other deposits, including the hominid cave sites and river gravels, date back no further than the Pliocene.

The Kalahari Basin

The pre-Kalahari topography was molded by northward-flowing streams (Smit, 1977). Initially deposition was concentrated on ephemeral braided floodplains. Basal conglomerates are clast supported and show lithologic affinities with distant sources. However, the overlying conglomerates are more argillaceous and were probably derived from the basin margins. In places these alluvial gravels are thicker than 90 m over bedrock channels and are caliche cemented (Du Toit, 1954, p. 457). The remainder of the succession consists mainly of red shales, marls, and sandstones with duricrusts. Smit (1977) records a maximum thickness in excess of 180 m.

Caliche crusts are most conspicuous in the southern regions of the Kalahari basin and represent a series of large pan terraces. The presence of caliche and silcrete crusts, together with an abundance of dolomite, is indicative of an arid climate and saline conditions (Boocock and Van Straaten, 1962; Netterberg, 1969; Watts, 1980). The Etosha Pan in the western part of the Kalahari basin is a shallow ephemeral lake surrounded by ephemeral stream floodplains and is subjected to frequent desiccation (Gevers, 1930; Goudie, 1973, p. 127). Algal stromatolites in the Etosha sediments have been offered as further evidence of saline conditions (Martin and Wilczewski, 1972). The origin of the saline lakes has been attributed to Pliocene desiccation of the Kalahari–Etosha basin, although Lancaster (1979) has also noted stromatolites in saline Kalahari lakes which are dated at 17,000–15,000 B.P.

From its headwaters in Angola the Okavango River flows southward. In the northern part of the Kalahari it spreads out, forming distributary channels, lakes, and marshes in a delta-like complex. On the basis of microearthquake studies, Scholz *et al.* (1976) suggest that the Okavango delta may be related to a developing graben.

Since the Tertiary the Kalahari landscape has been characterized by eolian dune sand accumulation, punctuated periodically by cold, wet periods. These climatic oscillations are recorded in the dolomitic Ghaap escarpment along the southeastern margin of the Kalahari, where Butzer *et al.* (1978) have identified six major complexes of cryoclastic breccia and tufa accumulation. Late Quaternary pluvial conditions 18,000–11,000 years ago are correlated with high lake levels in the Makgadikgadi depression in the southern Kalahari (Heine, 1978).

The Vaal–Orange Drainage Basin

Much of the subcontinent is today drained by the Vaal and Orange rivers and their tributaries. Relict river sediments have been of interest ever since the first discovery of diamonds in the Vaal River gravels in the 1860s.

Diamondiferous gravels were discovered near Lichtenburg in the southwestern Transvaal in 1921. The clast- and matrix-supported gravels were deposited on high-gradient braided alluvial plains by southward-flowing tributaries of the Vaal River (Stratten, 1979).

Near its confluence with the Orange, the Vaal River and its tributaries have formed a complex succession of alluvial terraces (Partridge and Brink, 1967; Butzer *et al.*, 1973; Helgren, 1977). The oldest alluvial deposits are located 20–90 m above flood level. They consist of caliche-cemented cobble-grade conglomerates in units that rarely exceed 5 m in thickness. A Late Pliocene to Early Pleistocene age has been suggested on the basis of mammal remains and artifacts that are allegedly associated with these sediments (Cooke, 1963, 1967; Mason, 1967; Helgren, 1977). Most diagnostic is the Pliocene elephant *Mammuthus subplanifrons* (Maglio, 1970). A lithologically similar set of braided floodplain gravels and sand interbeds comprises the complex body of younger gravels. Rare fossils and artifacts suggest a Middle Pleistocene age. By Late Quaternary time the river system was transporting and depositing predominantly sand and silt grades.

South of the Vaal River the tributaries deposited finer grained sediments. Near Virginia in the Orange Free State, Butzer (1973b) has described a thick sequence of cross-bedded sands overlying a channel lag. Remains of an advanced *Mammuthus* in the lag gravel suggest a Pleistocene age (Hendey, 1978b). Farther east at Cornelia on the Upper Vaal, terrace sands and silts have been assigned to the Middle Pleistocene on the basis of faunal remains and artifacts (Butzer 1974).

The Late Pleistocene hypothermal climate of the Orange–Vaal basin was substantially cooler and wetter than today. Lake deposits at Alexandersfontein near Kimberley indicate water depths of 19 m and imply a 6°C temperature depression and doubling of the annual rainfall budget (Butzer *et al.*, 1973; Butzer, 1976). Palynological studies of the Florisbad and Aliwal North deposits confirm that these conditions persisted, and that grassland spread until about 12,000 years ago in the upper Orange River area (Coetzee, 1967).

The Australopithecine Cave Breccias

Since the first discovery of *Australopithecus* at Taung (Fig. 14-11) in the northern Cape in 1924, other hominid-bearing cave breccias have been found, including Sterkfontein, Swartkrans, Makapansgat, and Kromdraai in the Transvaal. Only comparatively recently has their known distribution been extended to east Africa with the

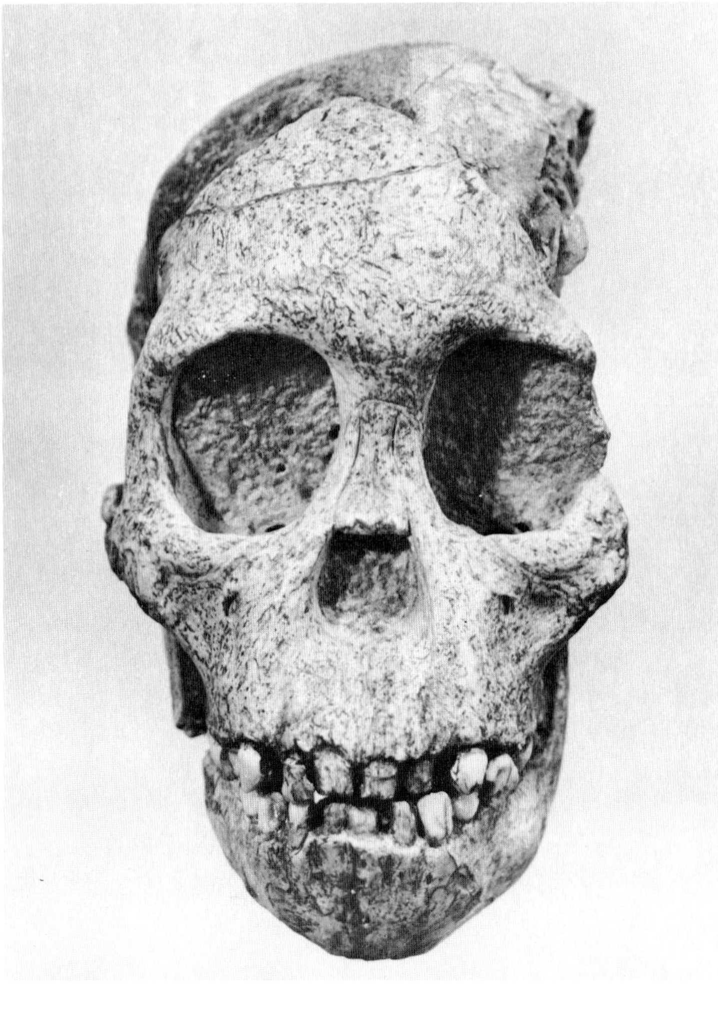

Figure 14-11. The Taung skull, *Australopithecus africanus*, frontal view. [Courtesy P.V. Tobias.]

discoveries by L.S.B. Leakey in Olduvai Gorge of Tanzania.

The australopithecine limestones of the Transvaal accumulated as cave fills (Brain, 1958; Butzer, 1971; Partridge, 1978). The caves were developed in dolomitic limestones of the Transvaal Supergroup as a result of karstic solution and enlargement by collapse in the vadose zone. Sedimentation, often contemporaneous with cave formation, reflects several different processes: brecciation of cavern walls, carbonate precipitation, accumulation of insoluble residues, and flushing of colluvial detritus into the caves. Most of the fossils occur in the carbonate-cemented colluvium and breccia (Butzer, 1971). The Sterkfontein limestones (Fig. 14-12) have yielded the greatest number of australopithecine specimens, as well as a diverse mammalian fauna, including *Homo*, in the upper levels.

Initially the Sterkfontein cavern had no direct opening to the outside so that sedimentation was slow and calcareous. Subsequent processes were dominated by rapid gravitative accretion of surface colluvium, roof rock, and bones, on a sedimentary cone beneath a continuously widening roof opening (Partridge, 1978). Detailed studies now show at least six sedimentary cycles, each separated by hiatuses that may be associated with calcite or flowstone accumulation, and periodic ponding and subaqueous deposition. A significant

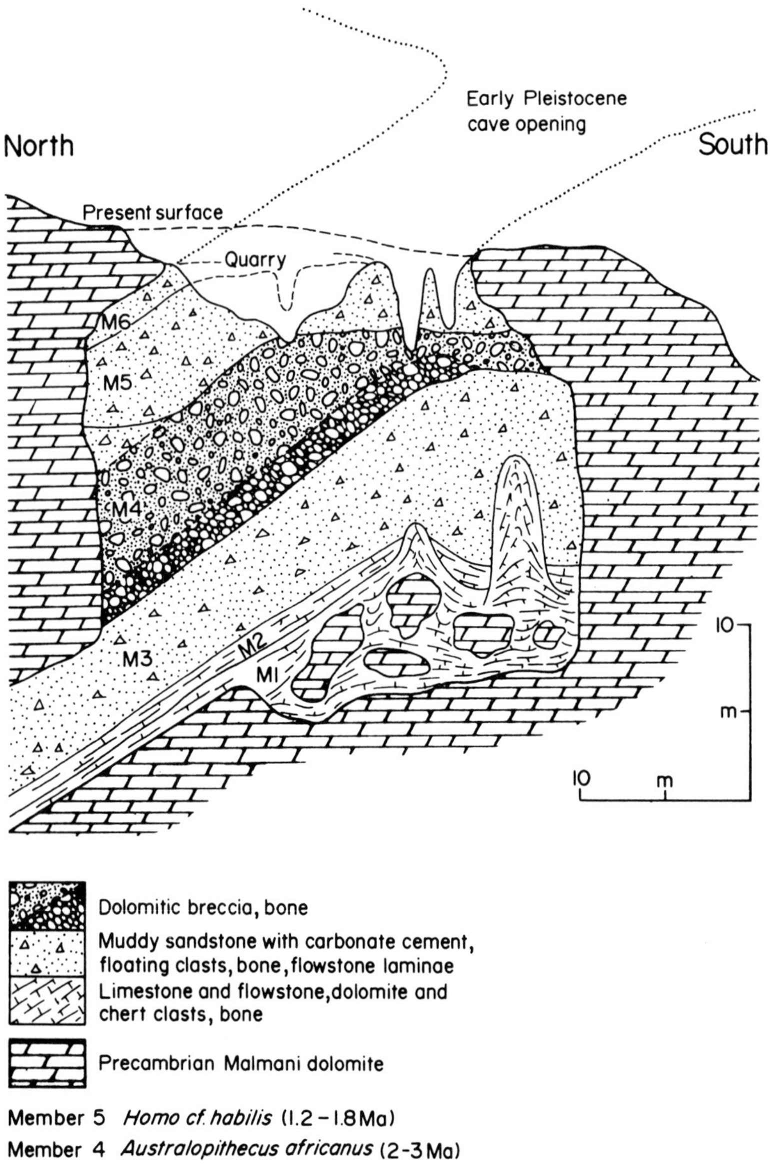

Figure 14-12. Stratigraphy of the australopithecine-bearing cave fill at Sterkfontein. [After Stiles and Partridge (1979).]

increase in ^{13}C values upward in the succession suggests progressively improved ventilation of the cave (Partridge and Talma, in press).

The nearby Swartkrans site also reflects substantial debris-cone accumulation, followed by extensive solution, channel erosion, and deposition (Butzer, 1976; Partridge, 1978). In the Makapansgat deposits some bone concentrations, including *Australopithecus* remains, have resulted where flow diversion around stalactites increased the flow strength.

Recent paleomagnetic results suggest that the australopithecine bone material at Makapansgat is about 3 Ma old (McFadden *et al.*, 1979). However, the Makapansgat faunal assemblage apparently predates the Sterkfontein taxa. Comparison with the australopithecine-bearing deposits of east Africa indicates an uppermost Pliocene age for the latter (Maglio, 1973; Vrba, 1975). The first appearance of *Homo* cf. *habilis* is associated with the Early Pleistocene sediments in the upper part of the Sterkfontein succession (Hughs and Tobias, 1977) (Fig. 14-12), which are thought to be about 1.5 Ma old (Stiles and Partridge, 1979). An Early to Middle Pleistocene age is suggested for the *Australopithecus africanus* strata at Taung (Butzer *et al.*, 1978).

The evidence from the sedimentary facies alone suggests accumulation under a relatively dry climate with little variation (Butzer, 1971). Isotopic studies confirm that although the climate fluctuated, these changes were never substantial (Partridge and Talma, in press). The composition of the fauna, dominated by bovids, is consistent with a savanna or wooded grassland environment (Vrba, 1976). Two species of *Australopithecus*, *robustus* and *africanus*, roamed the African savannas and were occasionally preyed on by leopards (Brain, 1958). Their dentition suggests that the species were separated by environmental and dietary preferences.

14.5. Cenozoic Biogeography and Climatic Evolution

Evolution of the Antarctic ice sheet and the pattern of Quaternary glaciation substantially influenced the Cenozoic environments of southern Africa. Antarctic glaciation dates from the Oligocene but only developed to its present extent by the Middle or Late Miocene. Modern climatic and oceanic circulation are also relatively youthful and find a common control in the Late Oligocene development of the Circum-Antarctic Current (Kennett *et al.*, 1975).

After the Mesozoic fragmentation of Gondwana the biota of the separate continents developed independently. Even in the Eocene and Oligocene, however, they were all characterized by a temperate to subtropical evergreen forest of Podocarpaceae, *Nothofagus*, Proteaceae, and ferns (Axelrod and Raven, 1978; Coetzee, 1978). During the Tertiary, tropical to subtropical palm and conifer forests alternated in the western Cape, but in the Late Miocene this flora was eliminated with development of the cold Benguela Current system and increasing aridity (Coetzee, 1978). A mosaic of sclerophyllous woodland, grassland, and scrub vegetation persisted through the Late Tertiary. During the Pliocene the trend toward a Mediterranean climate and evolution of a macchia flora in the southern and western Cape were firmly established (Tankard and Rogers, 1978). The terrestrial vertebrate faunas, limited by their dietary requirements, followed a parallel evolutionary pattern. Primitive bovids found in the Miocene sediments of Namibia indicate a moist wooded landscape. The earliest evidence of the modern semiarid environment and winter rainfall in the western Cape is Pliocene and coincides with a definite trend toward regionalism on the part of the mammal faunas (Coetzee, 1978; Tankard and Rogers, 1978). Full establishment of a macchia vegetation in the Cape Province is dramatically shown by the replacement of the Pliocene grazing rhinoceros (*Ceratotherium*) by a browzing rhinoceros (*Diceros*) in the Early Pleistocene.

Sedimentological, paleontological, and geochemical studies of cores from the Walvis Ridge abutment show that upwelling along the Atlantic coast was weakly developed by the early Late Miocene (Tankard and Rogers, 1978; Siesser, 1980). Warm surface waters and intermittent upwelling have been inferred for phosphogenesis. Marked intensification of upwelling and aridification occurred in the Late Miocene or Early Pliocene. Before the Late Miocene the mollusk taxa displayed a low degree of latitudinal zonation, suggesting that the coastal waters of southern Africa were undifferentiated at that time. The Zululand basin was also characterized by warmer water circulation in the Miocene (McLachlan and McMillan, 1979; Siesser and Miles, 1979).

The Quaternary climates of southern Africa were controlled by global climatic oscillations. However, only for the Late Quaternary are there sufficient data for a reasonably secure paleoclimatic reconstruction. Nevertheless, the model so derived still has significance in interpreting earlier Quaternary evidence.

Estuarine–lagoonal mollusk assemblages and evaporites attributed to the Sangamon sea-level high indicate a warm, dry climate along the Cape coastal margin. Conversely in the cold interval that followed there was a 5° equatorward shift of the climatic belts, allowing the region influenced by winter rainfall to expand into southern Namibia. Isotopic studies of speleothems in the Transvaal show that during this hypothermal the continental interior was 8–9°C colder than today (Talma *et al.*, 1974). Rejuvenation of ephemeral rivers on the Atlantic seaboard and filling of paleolakes in the Karoo and Kalahari were a direct consequence of increasing precipitation. Palynological studies have shown that during the cold, wet periods of the Upper Pleniglacial and Late Glacial, grassland also spread into these regions (Coetzee, 1978). Fossil taxa in coastal caves suggest that the coastal plain was covered with forest and bush, but the appearance of grazers at the time of the glacial maximum is evidence that the forest cover was in increasing competition with open grassland in the southern Cape (Klein, 1974; Coetzee, 1978; Tankard and Rogers, 1978). The sclerophylls only returned to the southern Cape with the postglacial expansion of the dry climate 12,000–9000 years ago.

With these climatic shifts the arid core was also displaced equatorward into Angola and the Congo basin. Available evidence suggests that the Namib desert system has existed since the end of the Pliocene, although Acheulian archeological remains (Korn and Martin, 1957) and increased river and lake activity (Selby, 1976; Selby *et al.*, 1979) indicate that southern Namibia has at times been less arid than at present. Early Pleistocene regressive complexes on the west coast are characterized by thermophilic mollusks (Tankard and Rogers, 1978).

In conclusion, the desert as a regional feature appears to be relatively youthful and has replaced a wooded grassland ecosystem. The desert taxa appear to have been derived by adaptation of woodland and savanna taxa to increasing drought. These endemic taxa, including the Tenebrionid beetles and the plant *Welwitschia*, are adapted to survive drought but do not necessarily require arid conditions.

References

Abbott, D., Ferguson, J. 1965. The Losberg intrusion, Fochville, Transvaal. Geol. Soc. S. Afr. Trans. 68:31–52.

Absolom, S. S. 1979. A sedimentological analysis of gold distribution in the Kimberley Reef at South Roodepoort Main Reef Areas Limited. Unpubl. M.Sc. thesis, Univ. Natal, Pietermaritzburg.

Adams, J., Zimpfer, G. L., McLane, C. F. 1978. Basin dynamics, channel processes, and placer formation: A model study. Econ. Geol. 73:416–426.

Ahfeld, F. 1956. Bolivia. *In* W. F. Jenks (ed.), Handbook of South American geology. Geol. Soc. Amer. Mem. 65:167–187.

Ahrendt, H., Hunziker, J. C., Weber, K. 1978. Age and degree of metamorphism and time of nappe emplacement along the southern margin of the Damara Orogen/Namibia (SW-Africa). Geol. Rundsch. 67:719–742.

Albat, H. M. 1979. Preliminary report on the geology of the Kliprand area in Namaqualand. Precambrian Res. Unit. Univ. Cape Town Ann. Rept. 16:47–57.

Aldrich, L.T., Wetherill, G. W., Davis, G. L., Tilton, G. R. 1958. Radioactive ages of micas from granitic rocks by Rb-Sr and K-Ar methods. Amer. Geophys. Union Trans. 39:1124–1134.

Allen, J. R. L. 1965a. A review of the origin and characteristics of Recent alluvial sediments. Sedimentology 5:91–191.

Allen, J. R. L. 1965b. The sedimentology and palaeogeography of the Old Red Sandstone of Anglesey, North Wales. Yorks. Geol. Soc. Proc. 35:139–185.

Allen, J. R. L. 1970. Studies in fluvial sedimentation: a comparison of fining-upward cyclothems with special reference to coarse-member composition and interpretation. Jour. Sediment. Petrol. 40:298–323.

Allsopp, H. L. 1964. Rubidium–Strontium ages from the western Transvaal. Nature 204:361–363.

Allsopp, H. L., Barrett, D. R. 1975. Rb-Sr age determinations on South African kimberlite pipes. *In* L. H. Ahrens, J. B. Dawson, A. R. Duncan, A. J. Erlank (eds.), Physics and chemistry of the earth, vol. 9, pp. 605–617. Oxford: Pergamon.

Allsopp, H. L., Ferguson, J. 1970. Measurements relating to the genesis of the Tsumeb Pipe, South West Africa. Earth Planet. Sci. Lett. 9:448–453.

Allsopp, H. L., Kolbe, P. 1965. Isotopic age determinations on the Cape granite and intruded Malmesbury sediments, Cape Peninsula, South Africa. Geochim. Cosmochim. Acta 29:1115–1130.

Allsopp, H. L., Kramers, J.D. 1977. Rb-Sr and U-Pb age determinations on southern African kimberlite pipes. 2nd Intl. Kimberlite Conf. Abstracts.

Allsopp, H. L., Welke, H. J. 1982. Age limits to the Witwatersrand Supergroup. Geol. Soc. S. Afr. Spec. Publ.

Allsopp, H. L. Erlank, A. J., Hornung, G. 1973. Rb-Sr age measurements on the Umkondo dolerites of Rhodesia. (Abstr.) Geol. Soc. S. Afr. Spec. Publ. No. 3, p. 217.

Allsopp, H. L., Köstlin, E. O., Welke, H. J., Burger, A. J., Kröner, A., Blignault, H. J. 1979. Rb-Sr and U-Pb geochronology of Late Precambrian–Early Palaeozoic igneous activity in the Richtersveld (South Africa) and southern South West Africa. Geol. Soc. S. Afr. Trans. 82:185–204.

Almeida, F. F. M. de, Amaral, G., Cordani, U. G., Kawachita, K. 1973. The Precambrian evolution of the South American cratonic margin south of the Amazon River. *In* A. E. M. Nairn, F. G. Stehli (eds). The ocean basins and margins. I. The South Atlantic, pp.441–446. New York:Plenum.

Almeida, F. F. M. de, Hasui, Y., De Brito Neves, B. B. 1976. The Upper Precambrian of South America. Inst. Geosci. Univ. Sao Paulo Boletim 7:45–80.

Amm, F. L. 1934. The pre-Cape rocks of the Uitenhage district. Geol. Soc. S. Afr. Trans. 37:69–86.

Anderson, A. M. 1975a. The "trilobite" trackways in the Table Mountain Group (Ordovician) of South Africa. Palaeont. Afr. 18:35–45.

Anderson, A. M. 1975b. Turbidites and arthropod trackways in the Dwyka glacial deposits (Early Permian) of southern Africa. Geol. Soc. S. Afr. Trans. 78:265–273.

Anderson, A. M., McLachlan, I. R., Oelofsen, B. W.

1977. The Lower Permian White Band and the Irati Formation. Geol. Soc. S. Afr. 17th Congr. Abstracts, 4–7.

Anderson, E. M. 1951. The dynamics of faulting. Edinburgh: Oliver and Boyd.

Anderson, J. M., Schwyzer, R. U. 1977. The biostratigraphy of the Permian and Triassic: Part 4. Palaeomagnetic evidence for large scale intra-Gondwana plate movements during the Carboniferous to Jurassic. Geol. Soc. S. Afr. Trans. 80:211–234.

Anderson, J. R., Fraenkel, H. C. 1979. The geology and mineralization of a uranium occurrence at Rietkuil farm, Beaufort West district, South Africa. Geol. Soc. S. Afr. 18th Congr. Abstracts 1:41–44.

Anderson, W. 1907. Third and final report of the Geological Survey of Natal and Zululand. Surveyor General's Dept., Natal Colony.

Andrews, P. B. 1970. Facies and genesis of a hurricane-washover fan, St. Joseph Island, Central Texas coast. Bur. Econ. Geol., Univ. Texas, Austin, Rept. Invest. 67.

Anhaeusser, C. R. 1966. A comparison of pebble and fold deformations in the Nelspruit granite contact aureole. Econ. Geol. Res. Unit Univ. Witwatersrand, Inf. Circ. 30.

Anhaeusser, C. R. 1972. The geology of the Jamestown Hills area of the Barberton Mountain Land, South Africa. Geol. Soc. S. Afr. Trans. 75: 225–255.

Anhaeusser, C. R. 1973. The geology and geochemistry of the Archaean granites and gneisses of the Johannesburg–Pretoria dome. Geol. Soc. S. Afr. Spec. Publ. 3: 361–385.

Anhaeusser, C. R. 1975. Precambian tectonic environments. Annl. Rev. Earth Planet. Sci. 3: 31–53.

Anhaeusser, C. R. 1976a. Archean metallogeny in southern Africa. Econ. Geol. 71: 16–43.

Anhaeusser, C. R. 1976b. The geology of the Sheba Hills area of the Barberton Mountain Land, South Africa, with particular reference to the Eureka Syncline. Geol. Soc. S. Afr. Trans. 79: 253–280.

Anhaeusser, C. R. 1977. Geological and geochemical investigations of the Roodekrans ultramafic complex and surrounding Archaean volcanic rocks. Geol. Soc. S. Afr. Trans. 80: 17–28.

Anhaeusser, C. R. 1978a. Rodingite occurrences in some Archaean ultramafic complexes in the Barberton Mountain Land, South Africa. Precambr. Res. 8: 49–76.

Anhaeusser, C. R. 1978b. The geological evolution of the primitive earth: evidence from the Barberton Mountain Land. *In* D. H. Tarling (ed.), Evolution of the earth's crust, pp. 71–106. London: Academic Press.

Anhaeusser, C. R. 1978c. The geology and geochemistry of the Muldersdrif ultramafic complex, and surrounding area, Krugersdorp district. Geol. Soc. S. Afr. Trans. 81: 193–203.

Anhaeusser, C. R. 1980. A geological investigation of the Archaean granite-greenstone terrane south of the Boesmanskop syenite pluton, Barberton Mountain Land. Geol. Soc. S. Afr. Trans. 83: 93–106.

Anhaeusser, C. R., and Robb, L. J. 1978. Regional and detailed field and geochemical studies of Archaean trondhjemitic gneisses, migmatites and greenstone xenoliths in the southern part of the Barberton Mountain Land, South Africa. Econ. Geol. Res. Unit Univ. Witwatersrand Inf. Circ. 143.

Anhaeusser, C. R., Mason, R., Viljoen, M. J., Viljoen, R. P. 1969. A reappraisal of some aspects of Precambrian shield geology. Geol. Soc. Amer. Bull. 80: 2175–2200.

Antrobus, E. S. A. 1982. The Klerksdorp goldfield. Geol. Soc. S. Afr. Spec. Publ. (in press).

Antrobus, E. S. A., Whiteside, H. C. M. 1964. The geology of certain mines in the East Rand. In S. H. Haughton (ed.). The geology of some ore deposits in southern Africa, vol. 1, pp. 125–160. Johannesburg: Geol. Soc. S. Afr.

Armstrong, G. C. 1966. A sedimentological study of the UK9 Kimberley Reefs in part of the East Rand. Unpubl. M. Sc. thesis, Univ. Witwatersrand, Johannesburg, South Africa.

Arndt, N. T., Naldrett, A. J., Pyke, D. R. 1977. Komatiitic and iron rich tholeiitic lavas of Munro Township, Northeast Ontario. Jour. Petrol. 18: 319–369.

Arth, J. G., Hanson, G. N. 1975. Geochemistry and origin of the Early Precambrian crust of northeastern Minnesota. Geochim. Cosmochim. Acta 39: 325–362.

Arth, J. G., Arndt, N. T., Naldrett, A. J. 1977. Genesis of Archean komatiites from Munro Township Ontario: trace element evidence. Geology 5: 590–594.

Arth, J. G., Barker, F., Peterman, Z.E., Friedman, I. 1978. Geochemistry of the gabbro-diorite-tonalite-trondhjemite suite of southwest Finland and its implications for the origin of tonalitic and trondhjemitic magmas. Jour. Petrol. 19: 289–316.

Ashley, G. M. 1975. Rhythmic sedimentation in glacial lake Hitchcock, Massachusetts-Connecticut. *In* A. V. Jopling, B. C. McDonald, (eds), Glaciofluvial and glaciolacustrine sedimentation. Soc. Econ. Paleont. Miner. Spec. Publ. No. 23, pp. 304–320.

Atherstone, W. G. 1857. Geology of Uitenhage. Eastern Province Monthly Mag. 1: 518–532.

Axelrod, D. I., Raven, P. H. 1978. Late Cretaceous and Tertiary vegetation history of Africa. *In* M. J. A. Werger, (ed.), Biogeography and ecology of southern Africa, pp. 77–130. The Hague: Junk.

Bahnemann, K. P. 1971. *In* E. R. Morrison, J. F. Wilson, (compilers), Excursion guidebook: symposium on the granites, gneisses and related rocks, pp. 18–22. Salisbury: Geol. Soc. S. Afr.

Bahnemann, K. P. 1972. A review of the structure, the stratigraphy and the metamorphism of the basement-rocks in the Messina District, northern Transvaal. Unpubl. D. Sc. thesis, Univ. Pretoria, South Africa.

Bahnemann, K. P. 1973. The origin of the Singelele granite gneiss near Messina, northern Transvaal. Geol. Soc. S. Afr. Spec. Publ. No. 3, pp. 235–244.

Bailey, D. K. 1964. Crustal warping, a possible tectonic control of alkaline magmatism. Jour. Geophys. Res. 69: 1103–1111.

Bally, A. W., Snelson, S. 1980. Realms of subsidence. *In* A. D. Miall, (ed), Facts and principles of world petroleum occurrence. Can. Soc. Petrol. Geol. Mem., No. 6, pp. 9–94.

Barker, O. B. 1976. The Soutpansberg trough (northern Transvaal)—An aulacogen. Discussion. Geol. Soc. S. Afr. Trans. 79:146.

Barker, O. B. 1979. A contribution to the geology of the Soutpansberg Group, Waterberg Supergroup, northern Transvaal. Unpubl. M.Sc. thesis, Univ. Witwatersrand, Johannesburg, South Africa.

Barker, P. F., Dalziel, I. W. D., Elliot, D. H., Von der Borch, C. C., Thompson, R. W., Pflaker, G., Tjalsma, R. C., Wise, S. W., Dinkelman, M. G., Gombos, A. M., Lonardi, A., Tarney, J. 1974. Southwestern Atlantic. Geotimes 19:16–18.

Barker, P. F., Dalziel, I. W. D., Dinkelman, M. G., Elliot, D. H., Gombos, A. M., Lonardi, A., Pflaker, G., Tarney, J., Thompson, R. W., Tjalsma, R. C., Von der Borch, C. C., Wise, S. W., Harris, W., Sliter, W. V. 1976c. Evolution of the southwestern Atlantic Ocean basin: Results of Leg 36, Deep Sea Drilling Project. Initial Reports of Deep Sea Drilling Project 36:993–1014. Washington, D.C.: U.S. Government Printing Office.

Barker, F., Friedman, I., Hunter, D. R., Gleason, J. D. 1976a. Oxygen isotopes of some trondhjemites, siliceous gneisses, and associated mafic rocks. Precambr. Res. 3: 547–557.

Barker, F., Millard, H. T., Reed, J. 1976b. Geochemistry of the Webb Canyon gneiss, Teton Range, Wyoming: An Archean bimodal rhyodacite-tholeiitic pair. Geol. Soc. Amer. Abstracts with Progr. 8:568–569.

Barnes, S.-J. 1979. Serpentinites in central South West Africa/Namibia—a reconnaissance study. Unpubl. M.Sc. thesis, Univ. Cape Town, South Africa.

Barnes, S.-J., Sawyer, E. W. 1980. An alternative model for the Damara Mobile Belt: Ocean crust subduction and continental convergence. Precambr. Res. 13:297–336.

Barrett, D. M., Jacobsen, J. B. E., McCarthy, T. S., Cawthorn, R.G. 1978. The structure of the Bushveld Complex south of Potgietersrus as revealed by a gravity survey. Geol. Soc. S. Afr. Trans. 81: 271–276.

Barrett, D. R. 1975. The genesis of kimberlites and associated rocks: strontium isotopic evidence. *In* L. H. Ahrens, J. B. Dawson, A. R. Duncan, A. J. Erlank (eds), Physics and chemistry of the earth, Vol. 9, pp.637–654. Oxford: Pergamon.

Barron, E. J., Harrison, C. G. A., Hay, W. W. 1978. A revised reconstruction of the southern continents. Eos 59:436–449.

Barton, J. M. 1979a. Crustal evolution clues. Nuclear Active 21:16–19.

Barton, J. M. 1979b. The chemical compositions, Rb-Sr isotopic systematics and tectonic setting of certain post-kinematic mafic igneous rocks, Limpopo Mobile Belt, southern Africa. Precambr. Res. 9: 57–80.

Barton, J. M., Hamilton, D. L. 1978. Water saturated melting relations to 5 kb of three Leucite Hills Lavas. Contr. Mineral. Petrol. 66:41–49.

Barton, J. M., Ryan, B. 1977. A review of the geochronologic framework of the Limpopo Mobile Belt. Botswana Geol. Surv. Bull. 12:183–200.

Barton, J. M., Fripp, R. E. P., Ryan, B. 1977. Rb/Sr ages and geological setting of ancient dykes in the Sand River area, Limpopo Mobile Belt, South Africa. Nature 267:487–490.

Barton, J. M., Ryan, B., Fripp, R. E. P. 1978. The relationship between Rb-Sr and U-Th-Pb whole-rock and zircon systems in the <3790 m.y. old Sand River gneisses, Limpopo Mobile Belt, southern Africa. *In* R.E. Zartman, (ed.), Short papers of the 4th Intl. Conf. Geochronology, Cosmochronology, Isotope Geology 1978, U.S. Geol. Surv. Open-File Rept. 78-701, pp. 27–28.

Barton, J. M., Fripp, R. E. P., Horrocks, P., McLean, N. 1979a. The geology, age and tectonic setting of the Messina Layered Intrusion, Limpopo Mobile Belt, southern Africa. Amer. Jour. Sci. 279:1108–1134.

Barton, J. M., Ryan, B., Fripp, R. E. P., Horrocks, P. 1979b. Effects of metamorphism on the Rb-Sr and U-Pb systematics of the Singelele and Bulai gneisses, Limpopo Mobile Belt, southern Africa. Geol. Soc. S. Afr. Trans. 82:259–269.

Barton, J. M., Hunter, D. R., Jackson, M. P. A., Wilson, A. C. 1980. Rb-Sr age and source of the Bimodal Suite of the Ancient Gneiss Complex, Swaziland. Nature 283:756–758.

Barwis, J. H. 1978. Sedimentology of some South Carolina tidal-creek point bars, and a comparison with their fluvial counterparts. *In* A.D. Miall, (ed.), Fluvial sedimentology. Can. Soc. Petrol. Geol. Mem. No.5, pp.129–160.

Barwis, J. H., Goodwin, P. W. 1979. Modern and ancient examples of annelid burrows as current direction indicators. Amer. Ass. Petrol. Geol. 1979 Annual meeting abstracts.

Barwis, J. H., Tankard, A. J. In Preparation. Pleistocene coastal environments and sea level history at Swartklip, False Bay, South Africa.

Beach, A., Coward, M. P., Graham, R. H. 1974. An interpretation of the structural evolution of the Laxford Front, north-west Scotland. Scottish Jour. Geol. 9:297–308.

Beer, E. M. 1972. Note. S. Afr. Geol. Surv. Ann. 9:25.

Behr, H. J., Ahrendt, H., Schmidt, A., Weber, K. 1981. Saline horizons acting as thrust planes along the southern margin of the Damara Orogen (Namibia/SW-Africa). Geol. Soc. London Spec. Publ. No. 9.: 167–172.

Behr, S. H. 1965. Heavy-mineral beach deposits in the Karoo System. S. Afr. Geol. Surv. Mem. No. 56, 116 pp.

Benedict, P. C., Wiid, D. de N., Cornelissen, A. K. 1964. Progress report on the geology of the O'okiep Copper District. *In* S. H. Haughton, (ed.), The geology of some ore deposits in southern Africa, vol.2, pp.238–314. Johannesburg: Geol. Soc. S. Afr.

Bennett, J. D. 1970. Craton-mobile belt relations with particular reference to the Mosetse-Matsitama area, north-eastern Botswana. Geol. Mag. 107: 113–123.

Bergh, H. W. 1977. Mesozoic sea floor off Dronning Maud Land, Antarctica. Nature 267:686–687.

Bergh, H. W., Norton, I. O. 1976. Prince Edward fracture zone and the evolution of the Mozambique Ridge. Jour. Geophys. Res. 81:5221–5239.

Berning, J., Cooke, R., Hiemstra, S. A., Hoffman, U. 1976. The Rössing uranium deposit, South West Africa. Econ. Geol. 71:351–368.

Berthelsen, A. 1979. Recumbent folds and boudinage structures formed by subglacial shear: an example of gravity tectonics. Geol. Mijnb. 58:253–260.

Bertrand, J. M. 1976. Granitoids and deformation sequence in the Goodhouse-Henkries area. A new interpretation of the relationship between rocks in the Vioolsdrif-Goodhouse area and the Namaqualand and Bushmanland gneisses. Precambr. Res. Unit Univ. Cape Town Ann. Rept. 13:61–70.

Bertrand-Sarfati, J., Eriksson, K. A. 1977. Columnar stromatolites from the Early Proterozoic Schmidtsdrif Formation, northern Cape Province, South Africa. Part 1: Systematic and diagnostic features. Palaeont. Afr. 20:1–26.

Beuf, S., Biju-Duval, B., De Charpal, O., Rognon, P., Gariel, O., Bennacef, A. 1971. Les Gres du Paleozoique inferieur au Sahara. Inst. Fr. Petrole, Coll. Sci. Tech. Petrole 18:1–464.

Beukes, G. J. 1973. 'n Geologiese ondersoek van die gebied suid van Warmbad, Suidwes-Africa, met spesiale verwysing na die metamorfmagmatiese assosiasies van die Voorkambriese gesteentes. Unpubl. D.Sc. thesis, Univ. Orange Free State, Bloemfontein.

Beukes, N. J. 1970. Stratigraphy and sedimentology of the Cave Sandstone Stage, Karroo System. 2nd Int. Gondwana Symposium. Proceedings and Papers, pp.321–342.

Beukes, N. J. 1973. Precambrian iron formations of southern Africa. Econ. Geol. 68:960–1004.

Beukes, N. J. 1977. Transition from siliciclastic to carbonate sedimentation near the base of the Transvaal Supergroup at Bothithong in the northern Cape Province, South Africa. Sediment. Geol. 18:201–222.

Beukes, N. J. 1978a. Die Karbonaatgesteentes en ysterformasies van die Ghaap-Groep van die Transvaal-Supergroep in Noord-Kaapland. Unpubl. DSc. thesis, Rand Afrikaans Univ., Johannesburg, South Africa.

Beukes, N. J. 1978b. An introduction to the sedimentology of coal deposits. Coal Processing vol.1, pp. 3/0–3/53. Johannesburg: Coal Processing Society, Rand Afrikaans University.

Bickle, M. J., Coward, M. P. 1977. A major thrust in the southern Damara belt, Namibia. Res. Inst. African Geol. Univ. Leeds Ann. Rept. 20:8–17.

Bickle, M. J., Martin, A., and Nisbet, E. G. 1975. Basaltic and peridotitic komatiites and stromatolites above a basal unconformity in the Belingwe greenstone belt, Rhodesia. Earth Planet. Sci. Lett. 27: 155–162.

Bickle, M. J., Ford, C. E., Nisbet, E. G. 1977. The petrogenesis of peridotitic komatiites: evidence from high-pressure melting experiments. Earth Planet. Sci. Lett. 37:97–106.

Biesheuval, K. 1970. An interpretation of a gravimetric survey of the area west of the Pilanesberg in the western Transvaal. Geol. Soc. S. Afr. Spec. Publ. No. 1, pp.266–282.

Bigarella, J. J. 1970. Continental drift and paleocurrent analysis. 2nd Intl. Gondwana Symposium. Proceedings and Papers, pp.73–97.

Bigarella, J. J., Van Eeden, O. R. 1971. Mesozoic palaeowind patterns and the problem of continental drift. Boletim Paranaense Geosci. 28/29:115–144.

Birch, G. F. 1977. Phosphorites from the Saldanha Bay region. Roy. Soc. S. Afr. Trans. 42:223–240.

Birch, G. F. 1979a. Phosphatic rocks on the western margin of South Africa. Jour. Sediment. Petrol. 49: 93–110.

Birch, G. F. 1979b. Phosphorite pellets and rock from the western continental margin and adjacent coastal terrace of South Africa. Mar. Geol. 33:91–116.

Birch, G. F. 1979c. The association of glauconite and apatite minerals in phosphatic rocks from the South African continental margin. Geol. Soc. S. Afr. Trans. 82:43–53.

Birch, G. F., Du Plessis, A., Willis, J. P. 1978. Offshore and onland geological and geophysical investigations in the Wilderness Lakes region. Geol. Soc. S. Afr. Trans. 81:339–352.

Bisschoff, A. A. 1969. The petrology of the igneous and metamorphic rocks in the Vredefort Dome and the adjoining parts of the Potchefstroom syncline. Unpubl. D.Sc. thesis, Univ. Pretoria, South Africa.

Bisschoff, A. A. 1972a. Tholeiitic intrusions in the Vredefort Dome. Geol. Soc. S. Afr. Trans. 75:23–30.

Bisschoff, A. A. 1972b. The dioritic rocks of the Vredefort Dome. Geol. Soc. S. Afr. Trans. 75:31–45.

Bisschoff, A. A. 1973. The petrology of some mafic and peralkaline intrusions in the Vredefort Dome. Geol. Soc. S. Afr. Trans. 76:27–52.

Blaine, J. L. 1977. Tectonic evolution of the Waldau Ridge structure and the Okahandja Lineament in part of the Central Damara Orogen, west of Okahandja, South West Africa. Precambrian Res. Unit Univ. Cape Town Bull. No.21.

Blignault, H. J. 1970. Contemporaneous glacial folding in the Table Mountain Group, Western Cape. Unpubl. M.Sc. thesis, Univ. Stellenbosch, South Africa.

Blignault, H. J. 1975. Towards a stratigraphic nomenclature for the Namaqua Metamorphic Complex and associated rocks. Precambrian Res. Unit Univ. Cape Town Ann. Rept. No.12, pp.41–42.

Blignault, H. J. 1977. Structural-metamorphic imprint on part of the Namaqua Mobile Belt in South West Africa. Precambrian Res. Unit Univ. Cape Town Bull. No.23.

Blignault, H. J., Jackson, M. P. A., Beukes, G. J., Toogood, D. J. 1974. The Namaqua Tectonic Province in South West Africa. Precambrian Res. Unit Univ. Cape Town Bull. No.15, pp.29–47.

Bluck, B. J., 1976. Sedimentation in some Scottish rivers of low sinuosity. Roy. Soc. Edinburgh Trans. 69:429–456.

Boctor, N. Z., Boyd, F. R. 1979. Oxide minerals in layered kimberlite-carbonate sills from Benfontein, South Africa. Carnegie Inst. Washington Yearbook 78:493–496.

Boettcher, A. L., O'Neil, J. R., Windom, K. E., Stewart, D. C., Wilshire, H. G. 1979. Metasomatism of the upper mantle and the genesis of kimberlites and alkali basalts. Second Intl. Kimberlite Conf. Proc. 2: 173–182.

Böhm, J. 1926. Über Tertiäre Versteinerungen von den Bogenfelser Diamantfeldern in die Diamantenwuste Südwestafrikas. *In* E. Kaiser, (ed.), Die Diamantenwuste Südwestafrikas, vol.2, pp.55–87. Berlin: Dietrich Reimer Verlag.

Bond, G .C. 1979. Evidence for some uplifts of large magnitude in continental platforms. Tectonophysics 61:285–305.

Boocock, C., Van Straaten, O. J. 1962. Notes on the geology and hydrogeology of the central Kalahari region, Bechuanaland Protectorate. Geol. Soc. S. Afr. Trans. 65:125–171.

Boothroyd, J. C. 1972. Coarse-grained sedimentation on a braided outwash fan, northeast Gulf of Alaska. Coastal Res. Division, Univ. South Carolina Tech. Rept. 6-CRD.

Boothroyd, J. C., Ashley, G. M. 1975. Processes, bar morphology and sedimentary structures on braided outwash fans, northeastern Gulf of Alaska. *In* A. V. Jopling, B. C. McDonald, (eds), Glaciofluvial and glaciolacustrine sedimentation. Soc. Econ. Paleont. Miner. Spec. Publ. No.23, pp.193–222.

Boothroyd, J. C., Nummedal, D. 1978. Proglacial braided outwash: a model for humid alluvial fan deposits. *In* A.D. Miall, (ed.), Fluvial sedimentology. Can. Soc. Petrol. Geol. Mem. No.5, pp.641–668.

Borchers, R. 1961. Exploration of the Witwatersrand System and its extensions. Geol. Soc. S. Afr. Proc. 64:67–98.

Borodin, L. S., Pavlenko, A. S. 1974. The role of metasomatic processes in the formation of alkaline rocks. *In* H. Sorenson, (ed.), The alkaline rocks, pp.515–534. London:Wiley.

Borradaile, G. J. 1978. Transected folds: A study illustrated with examples from Canada and Scotland. Geol. Soc. Amer. Bull. 89:481–493.

Botha, B. J. V. 1968. The stratigraphy of the Red Beds Stage, Karroo System, at Elliot, C.P. Geol. Soc. S. Afr. Trans. 71:101–113.

Botha, B. J. V., Grobler, N. J. 1979. Models for the geotectonic evolution of the Middle to Late Precambrian Namaqua Mobile Belt in eastern Namaqualand, South Africa. Precambrian Res. 10:21–41.

Botha, B. J. V., Linström, W. 1978. A note on the stratigraphy of the Beaufort Group in north-western Natal. Geol. Soc. S. Afr. Trans. 81:35–40.

Botha, B. J. V., Grobler, N. J., Linström, W., Smit, C. A. 1976. Stratigraphic correlation between the Kheis and Matsap formations and their relation to the Namaqualand Metamorphic Complex. Geol. Soc. S. Afr. Trans. 79:304–311.

Botha, B. J. V., Grobler, N. J., Linström, W., Smit, C. A. 1977. Major structural features of the area between the Langeberg Range and Kenhardt, northern Cape Province. Geol. Soc. S. Afr. Trans. 80:101–110.

Botha, B. J. V., Grobler, N. J., Burger, A. J. 1979. New U-Pb age measurements on the Koras Group, Cape Province and its significance as a time reference horizon in eastern Namaqualand. Geol. Soc. S. Afr. Trans. 82:1–5.

Boucot, A. J. 1971. Malvinokaffric Devonian marine community distribution and implications for Gondwana. Acad. Bras. Cienc. Anal. 43 (Suppl.):23–49.

Boucot, A. J., Johnson, J. G., Talent, J. A. 1969. Early Devonian brachiopod zoogeography. Geol.

Soc. Amer. Spec. Pap. No.199, pp.1–113.

Boulton, G. S. 1970. On the origin and transport of englacial debris in Svalbard glaciers. Jour. Glaciol. 7:391–412.

Bowden, P. 1968. Trace elements in carbonatites and limestones. Nature 219:716.

Boyd, F. R. 1973. The pyroxene geotherm. Geochim. Cosmochim. Acta 37:2533–2546.

Boyd, F. R. 1976. Inflected and noninflected geotherms. Carnegie Inst. Washington Yearbook 75: 521–523.

Boyd, F. R., Clement, C. R. 1977. Compositional zoning of olivines in kimberlite from the De Beers mine, Kimberley, South Africa. 2nd Intl. Kimberlite Conf. Abstracts.

Boyd, F. R., Nixon, P. H. 1973. Origin of the ilmenite-silicate nodules in kimberlites from Lesotho and South Africa. *In* P. H. Nixon, (ed.), Lesotho Kimberlites, pp. 254–268. Lesotho National Development Corporation.

Boyd, F. R., Nixon, P. H. 1975. Origins of the ultramafic nodules from some kimberlites of northern Lesotho and the Monastery Mine, South Africa. *In* L. H. Ahrens, J. B. Dawson, A. R. Duncan, A. J. Erlank (eds), Physics and chemistry of the earth vol.9, pp.431–454. Oxford: Pergamon.

Boyd, F. R., Fujii, T., Danchin, R. V. 1976. A noninflected geotherm for the Udachnaya kimberlite pipe, USSR. Carnegie Inst. Washington Yearbook 75:523–531.

Brain, C. K. 1958. The Transvaal ape-man-bearing cave deposits. Transvaal Mus. Mem. No.11.

Brandt, J. W. 1948. Die geologie van 'n gebied in noord-oos Transvaal met spesiale verwysing na die verspreiding en petrografie van die rotssoorte van die Palaborastallings-kompleks. Unpubl. D.Sc. thesis, Univ. Stellenbosch, South Africa.

Brenner, P., Oertli, H. J. 1976. Lower Cretaceous ostracodes (Valanginian to Hauterivian) from the Sundays River Formation, Algoa basin, South Africa. Centre Rech. Pau—S.N.P.A. Bull. 10:471–533.

Briden, J. 1976. Application of palaeomagnetism to Proterozoic tectonics. Roy. Soc. London Phil. Trans. A280:405–416.

Brink, W. C. 1950. The geology, structure and petrology of the Nuwerus area, Cape Province. Univ. Stellenbosch Ann. No.26A.

Bristow, J. W. 1976. The geology and geochemistry of the southern Lebombo. Unpubl. M.Sc. thesis, Univ. Natal, Durban.

Bristow, J. W., Cleverly, R. W. 1979. Volcanology of the Lebombo rhyolites. Geol. Soc. S. Afr. 18th Congr. Abstracts 1:60–63.

Brock, A., McFadden, P. L., Partridge, T. C. 1977. Some new palaeomagnetic results from Makapansgat. South African Society for Quaternary Research Conf., Pietermaritzburg, 1977 (Abstracts).

Brock, B. B., Pretorius, D. A. 1964. Rand Basin sedimentation and tectonics. *In* S. H. Haughton (ed.), The geology of some ore deposits in southern Africa, vol., 1, pp. 549–599. Johannesburg: Geol. Soc. S. Afr.

Brooks, J., Muir, M. D., Shaw, G. 1973. Chemistry and morphology of Precambrian microorganisms. Nature 244:215–217.

Broom, R. 1909. An attempt to determine the horizons of the fossil vertebrates of the Karroo. S. Afr. Mus. Ann. 7:285–289.

Brown, L. F., Fisher, W. L. 1977. Seismic stratigraphic interpretations of depositional systems: examples of Brazilian rift and pull-apart basins. *In* C. E. Payton, (ed.), Seismic stratigraphy—applications to hydrocarbon exploration. Amer. Ass. Petrol. Geol. Mem. 26:213–248.

Brynard, H. J., De Villiers, J. P. R., Viljoen, E. A. 1976. A mineralogical investigation of the Merensky Reef at the Western Platinum Mine, near Marikana, South Africa. Econ. Geol. 71:1299–1307.

Bryson, M. 1975. Geochemistry of the Brewelskloof mafic rocks. B.Sc. (Hons.) thesis, Univ. Cape Town, South Africa.

Buchanan, D. L. 1975. The petrography of the Bushveld Complex intersected by boreholes in the Bethal area. Geol. Soc. S. Afr. Trans. 78:335–348.

Bucher, W. H. 1956. Role of gravity in orogenesis. Geol. Soc. Amer. Bull. 67:1295–1318.

Buck, S. G. 1980. Stromatolite and ooid deposits within the fluvial and lacustrine sediments of the Precambrian Ventersdorp Supergroup of South Africa. Precambr. Res. 12:311–330.

Bunce, E. T., Molnar, P. 1977. Seismic reflection profiling and basement topography in the Somali basin: Possible fracture zones between Madagascar and Africa. Jour. Geophys. Res. 82:5305–5311.

Bundy, F. P., Bovenkerk, H. P., Strong, H. M., Wentorf, R. H. 1961. Diamond-graphite equilibrium line from growth and graphitisation of diamond. Jour. Chem. Phys. 35:383–391.

Burger, A. J. 1978. The geochronology of the Lüderitz "granites." Precambrian Res. Unit Univ. Cape Town Annl. Rept. 14/15:141–144.

Burger, A. J., Coertze, F. J. 1973. Radiometric age measurements on rocks from southern Africa to the end of 1971. S. Afr. Geol. Surv. Bull. No.58.

Burger, A. J., Potgieter, G. J. A. 1979. Note on U-Pb zircon ages from granitic rocks occurring near Prieskapoort, north-west Cape. Geol. Soc. S. Afr. Trans. 82:271–273.

Burger, A. J., Walraven, F. 1976. Summary of age determinations carried out during the period April 1975 to March 1976. S. Afr. Geol. Surv. Ann. 11: 323–329.

Burger, A. J., Walraven, F. 1979. Summary of age determinations carried out during the period April

1976 to March 1977; Summary of age determinations carried out during the period April 1977 to March 1978. S. Afr. Geol. Surv. Ann. 12:199–218.

Burger, A. J., Clifford, T. N., Miller, R. McG. 1976. Zircon U-Pb ages of the Franzfontein granitic suite, northern South West Africa. Precambr. Res. 3:415–431.

Burke, K. 1972. Longshore drift, submarine canyons, and submarine fans in development of Niger Delta. Amer. Ass. Petrol. Geol. Bull. 56:1975–1983.

Burke, K., Dewey, J. F., Kidd, W. S. F. 1977. World distribution of sutures—the sites of former oceans. Tectonophysics 40:69–100.

Burley, A. J., Evans, R. B., Gillingham, J. M., Masson Smith, D. 1970. Gravity anomalies in Swaziland. Swaziland Geol. Surv. Bull. 7:4–16.

Button, A. 1971. Early Proterozoic algal stromatolites of the Pretoria Group, Transvaal Sequence. Geol. Soc. S. Afr. Trans. 74:201–210.

Button, A. 1973a. The depositional history of the Wolkberg proto-basin, Transvaal. Geol. Soc. S. Afr. Trans. 76:15–25.

Button, A., 1973b. Algal Stromatolites in the Early Proterozoic Wolkberg Group, Transvaal Sequence. Jour. Sediment. Petrol. 43:160–167.

Button, A. 1973c. A regional study of the stratigraphy and development of the Transvaal basin in the eastern and northern Transvaal. Unpubl. Ph.D. thesis, Univ. Witwatersrand, Johannesburg, South Africa.

Button, A. 1973d. The stratigraphic history of the Malmani dolomite in the eastern and northeastern Transvaal. Geol. Soc. S. Afr. Trans. 76:230–247.

Button, A. 1974. Low-potash pillow basalts in the Pretoria Group, Transvaal Supergroup. Geol. Soc. S. Afr. Trans. 77:99–104.

Button, A. 1975. A palaeocurrent study of the Dwaal Heuvel Formation, Transvaal Supergroup. Geol. Soc. S. Afr. Trans. 78:173–183.

Button, A. 1976a. Transvaal and Hamersley basins—review of basin development and mineral deposits. Minerals Sci. Engng. 8:262–293.

Button, A. 1976b. Stratigraphy and relations of the Bushveld floor in the eastern Transvaal. Geol. Soc. S. Afr. Trans. 79:3–12.

Button, A. 1976c. Halite casts in the Umkondo System, southeastern Rhodesia. Geol. Soc. S. Afr. Trans. 79:177–178.

Button, A. 1977. Stratigraphic history of the Middle Proterozoic Umkondo Basin in the Chipinga area, southeastern Rhodesia. Econ. Geol. Res. Unit Univ. Witwatersrand, Inf. Circ. No.108.

Button, A. 1979a. Algal concentrations in the Sabi River, Rhodesia: Depositional model for Witwatersrand carbon. Econ. Geol. 74:1876–1889.

Button, A. 1979b. Upper Witwatersrand strata in the Heidelberg area. Econ. Geol. Res. Unit Univ. Witwatersrand, Ann. Rept. No.20.

Button, A. 1981. The Ventersdorp Supergroup. *In* D. R. Hunter (ed.), Precambrian of the southern hemisphere, pp. 520–527. Amsterdam:Elsevier.

Button, A., Tyler, N. 1981. The character and economic significance of Precambrian paleoweathering and erosion surfaces in southern Africa. Econ. Geol., 75th Anniv. Vol., 676–699.

Button, A., Vos. R. G. 1977. Subtidal and intertidal clastic and carbonate sedimentation in a macrotidal environment: An example from the Lower Proterozoic of South Africa. Sediment. Geol. 18:175–200.

Butzer, K. W. 1971. Another look at the australopithecine cave breccias of the Transvaal. Amer. Anthropol. 73:1197–1201.

Butzer, K. W. 1973a. Geology of Nelson Bay Cave, Robberg, South Africa. S. Afr. Archaeol. Bull. 28: 97–110.

Butzer, K. W. 1973b. On the geology of a Late Pliocene *Mammuthus* site, Virginia, Orange Free State. Natl. Mus. Bloemfontein Res. 2:386–393.

Butzer, K. W. 1974. Geology of the Cornelia beds, northeastern Orange Free State. Natl. Mus. Bloemfontein Mem. No.9, pp.7–32.

Butzer, K. W. 1976. Lithostratigraphy of the Swartkrans Formation. S. Afr. Jour. Sci. 72:136–141.

Butzer, K. W., Cuerda, J. 1962. Coastal stratigraphy of southern Mallorca and its implications for the Pleistocene chronology of the Mediterranean sea. Jour. Geol. 70:398–416.

Butzer, K. W., Helgren, D. M. 1972. Late Cenozoic evolution of the Cape coast between Knysna and Cape St. Francis, South Africa. Quat. Res. 2:143–169.

Butzer, K. W., Fock, G. J., Stuckenrath, R., Zilch, A. 1973. Palaeohydrology of a Pleistocene lake, Alexandersfontein, Kimberley, South Africa. Nature 243:328–330.

Butzer, K. W., Stuckenrath, R., Bruzewicz, A. J., Helgren, D. M. 1978. Late Cenozoic paleoclimates of the Gaap escarpment, Kalahari margin, South Africa. Quat. Res. 10:310–339.

Cadle, A. B. 1979. Depositional models from the Middle Ecca and their influence on coal accumulation in the eastern Transvaal. Geol. Soc. S. Afr. 18th Congr. Abstracts 1:77–78.

Cadle, A. B., Hobday, D. K. 1977. A subsurface investigation of the Middle Ecca and Lower Beaufort in the southeastern Transvaal and northern Natal. Geol. Soc. S. Afr. Trans. 80:111–115.

Cain, A. C. 1975. A preliminary review of the stratigraphic relationships and distribution of metamorphism in the northern part of the Natal-Namaquarides, South Africa. Geol. Rundsch. 64:192–216.

Cairncross, B. 1979. Depositional framework and control of coal distribution and quality, Van Dijks Drift area, northern Karoo basin. Unpubl. M.Sc. thesis, Univ. Natal, Pietermaritzburg, South Africa.

Camden-Smith, P. M. 1980. The sedimentology, geochemistry and diagenesis of West Rand Group sediments in the Heidelberg area, Transvaal. Unpubl. M.Sc. thesis, Univ. Cape Town, South Africa.

Cameron, E. N. 1970a. Compositions of certain coexisting phases in the eastern part of the Bushveld Complex. Geol. Soc. S. Afr. Spec. Publ. No.1, pp.46–58.

Cameron, E. N. 1970b. Discussion of paper by Ferguson and McCarthy. Geol. Soc. S. Afr. Spec. Publ. No.1, p.79.

Cameron, E. N. 1971. Problems of the eastern Bushveld Complex. Fortschr. Min. 48:86–108.

Cameron, E. N. 1977. Chromite in the central sector of the eastern Bushveld Complex, South Africa. Amer. Min. 62:1082–1096.

Cameron, E. N. 1978. The lower zone of the eastern Bushveld Complex in the Olifants River trough. Jour. Petrol. 19:437–462.

Cameron, E. N., Desborough, G. A. 1964. Origin of certain magnetite-bearing pegmatites in the eastern part of the Bushveld Complex, South Africa. Econ. Geol. 59:197–225.

Cameron, E. N., Desborough, G. A. 1969. Occurrence and characteristics of chromite deposits, eastern Bushveld Complex. Econ. Geol. Monogr. No.4, pp.23–40.

Campbell, I. H. 1978. Some problems with the cumulus theory. Lithos 11:311–323.

Cant, D. J. 1978. Development of a facies model for sandy braided river sedimentation: comparison of the South Saskatchewan River and Battery Point Formation. *In* A. D. Miall (ed.), Fluvial sedimentology. Can. Soc. Petrol. Geol. Mem. 5:627–639.

Carey, S. W. 1962. Folding. Alberta Soc. Petrol. Geol. Jour. 10:95–144.

Carmichael, I. S. E. 1967. The mineralogy and petrology of the volcanic rocks from the Leucite Hills, Wyoming. Contr. Mineral. Petrol. 15:24–66.

Carmichael, I. S. E., Turner, F. J., Verhoogen, J. 1974. Igneous petrology. New York: McGraw-Hill.

Carrington, A. J., Kensley, B. F. 1969. Pleistocene molluscs from the Namaqualand coast. S. Afr. Mus. Ann. 52:189–223.

Caster, K. E. 1952. Stratigraphic and paleontologic data relevant to the problem of Afro-American ligation during the Paleozoic and Mesozoic. Amer. Mus. Nat. Hist. Bull. 99:105–152.

Chaloner, W. G., Forey, P. L., Gardiner, B. G., Hill, A. J., Young, V. T. 1980. Devonian fish and plants from the Bokkeveld Series of South Africa. S. Afr. Mus. Ann. 81:127–157.

Chapman, F. 1916. Foraminifera and Ostracoda from the Upper Cretaceous Need's Camp, Buffalo River, Cape Province. S. Afr. Mus. Ann. 12:107–118.

Christie, A. D. M., Tavener-Smith, R. 1979. Middle Ecca sedimentary associations on the Natal coast north of Durban. Geol. Soc. S. Afr. 18th Congr. Abstract 1:69–70.

Clark, F. W. 1924. Data of geochemistry. U.S. Geol. Surv. Bull. No.770.

Clark, S. P., Ringwood, A. E. 1964. Density distribution and constitution of the mantle. Rev. Geophys. 2:35–88.

Clarke, J. M. 1913. Fosseis devonianos do Parana. Serv. Geol. Min. Brasil. Monogr. No.1.

Clement, C. R., Skinner, E. M., Scott, B. H. 1977. Kimberlite redefined. 2nd Intl. Kimberlite Conf. Abstracts.

Clement, C. R., Skinner, E. M., Hawthorne, J. B., Kleinjan, L., Allsopp, H. L. 1979. Precambrian ultramafic dykes with kimberlite affinities in the Kimberley area. 2nd Intl. Kimberlite Conf. Proc. 1: 101–110.

Cleverly, R. W., Betton, P. J. 1979. Petrogenesis of Lebombo rhyolites from Swaziland. Geol. Soc. S. Afr. 18th Congr. Abstracts part 1, pp.79–81.

Cleverly, R. W., Bristow, J. W. 1979. Revised volcanic stratigraphy of the Lebombo monocline. Geol. Soc. S. Afr. Trans. 82:227–230.

Cliff, R. A. 1975. Two granulite facies events near the northern margin of the Limpopo Belt? Res. Inst. African Geol. Univ. Leeds Annl. Rept. 19:68–69.

Clifford, T. N. 1966. Tectono-metallogenic units and metallogenic provinces of Africa. Earth Planet. Sci. Lett. 1:421–434.

Clifford, T. N. 1970. The structural framework of Africa. *In* T. N. Clifford, I. G. Gass (eds), African magmatism and tectonics, pp.1–26. Edinburgh: Oliver and Boyd.

Clifford, T. N., Nicolaysen, L. O., Burger, A. J. 1962. Petrology and age of the pre-Otavi basement granite at Franzfontein, northern South West Africa. Jour. Petrol. 3:244–279.

Clifford, T. N., Gronow, J., Rex, D. C., Burger, A. J. 1975a. Geochronological and petrogenetic studies of high-grade metamorphic rocks and intrusives in Namaqualand, South Africa. Jour. Petrol. 16:154–188.

Clifford, T. N., McIver, J. R., Stumpfl, E. F. 1975b. A sapphirine-cordierite-bronzite-phlogopite paragenesis from Namaqualand, South Africa. Mineralog. Mag. 40:347–356.

Clifton, H. E., Hunter, R. E., Phillips, R. L. 1971. Depositional structures and processes in the non-barred, high-energy nearshore. Jour. Sediment. Petrol. 41:651–670.

Cloud, P. E. 1972. A working model of the primitive earth. Amer. Jour. Sci. 272:537–548.

Cloud, P. E. 1976. Major features in crustal evolution. Geol Soc. S. Afr. Trans. 79(Annex):1–32.

Cloud, P. E., Licari, G. R. 1968. Microbiotas of the banded iron formations. Proc. Natl. Acad. Sci. U.S.A. 61:779–786.

Clubley-Armstrong, A. R. 1977. The geology of the Selonsrivier area, north of Middelburg, Transvaal, with special reference to the structure of the regions southeast of the Dennilton Dome. Unpubl. M. Sc. thesis, Univ. Pretoria, South Africa.

Cocks, L. R. M. 1972. The origin of the Silurian *Clarkeia* shelly fauna of South America, and its extension to West Africa. Palaeontology 15:623–630.

Cocks, L. R. M., Brunton, C. H. C., Rowell, A. J., Rust, I. C. 1970. The first Lower Palaeozoic fauna from South Africa. Geol. Soc. London Quart. Jour. 125:583–603.

Coertze, F. J. 1970. The geology of the western part of the Bushveld Igneous Complex. Geol. Soc. S. Afr. Spec. Publ. No.1, pp.5-22.

Coertze, F. J. 1974. The geology of the basic portion of the western part of the Bushveld Igneous Complex. S. Afr. Geol. Surv. Mem. No. 66.

Coertze, F. J. Burger, A. J., Walraven, F., Marlow, A. G., MacCaskie, D. R. 1978. Field relations and age determinations in the Bushveld Complex. Geol. Soc. S. Afr. Trans. 81:1–11.

Coetzee, C. B. 1941. The petrology of the Goodhouse-Pella area, Namaqualand, South Africa. Geol. Soc. S. Afr. Trans. 44:167–206.

Coetzee, C. B. 1942. Metamorphosed sediments from the Goodhouse-Pella area Namaqualand, South Africa. Roy. Soc. S. Afr. Trans. 29:91–112.

Coetzee, J. A. 1967. Pollen analytical studies in east and southern Africa. Palaeoecol. Afr. 3:1–146.

Coetzee, J. A. 1978. Late Cainozoic palaeoenvironments of southern Africa. *In* E. M. Van Zinderen Bakker (ed.), Antarctic glacial history and world palaeoenvironments, pp.115–127. Rotterdam: Balkema.

Coleman, J. M. 1969. Brahmaputra River channel processes and sedimentation. Sediment. Geol. 3:129–239.

Coleman, J. M., Gagliano, S. M. 1965. Sedimentary structures: Mississippi River deltaic plain. *In* G. V. Middleton, (ed.), Primary sedimentary structures and their hydrodynamic interpretation. Soc. Econ. Paleont. Mineral. Spec. Publ. No.12, pp.133–148.

Coleman, J. M., Wright, L. D. 1975. Modern river deltas: variability of processes and sand bodies. *In* M. L. Broussard (ed.), Deltas (2nd ed.), pp.99–150. Houston: Houston Geol. Soc.

Coleman, J. M., Wright, L. D. 1978. Sedimentation in an arid macrotidal alluvial river system and Ord River, Western Australia. Jour. Geol. 86:621–642.

Coleman, R. G. 1977. Ophiolites. Berlin: Springer-Verlag.

Collignon, M. 1963. Atlas des fossiles caracteristiques de Madagascar (Ammonites). Tananarive: Geol. Surv. Rep. Malagasy.

Collinson, J. D. 1970. Bedforms of the Tana River, Norway. Geogr. Annalr. 52-A:31–56.

Colliston, W. P. 1979. The stratigraphy of the Namaqualand Metamorphic Complex in Bushmanland. Geol. Soc. S. Afr. 18th Congr. Abstracts 1:85–105.

Condie, K. C., Harrison, N. M. 1976. Geochemistry of the Archean Bulawayan Group, Midlands greenstone belt, Rhodesia. Precambr. Res. 3:253–271.

Condie, K. C., Hunter, D. R. 1976. Trace element geochemistry of Archean granitic rocks from the Barberton region, South Africa. Earth Planet. Sci. Lett. 29:389–400.

Condie, K. C., Macke, J. E., Riemer, T. O. 1970. Petrology and geochemistry of Early Precambrian graywackes from the Fig Tree Group, South Africa. Geol. Soc. Amer. Bull. 81:2759–2776.

Cooke, H. B. S. 1963. Pleistocene mammal faunas of Africa, with particular reference to southern Africa. *In* F. C. Howell, F. Bourlière (eds.), African ecology and human evolution. Viking Fund Publ. Anthropol. 36:65–116.

Cooke, H. B. S. 1967. The Pleistocene sequence in South Africa and problems of correlation. *In* W. W. Bishop, J. D. Clark, (eds), Background to evolution in Africa, pp.175–184. Chicago: Chicago Univ. Press.

Coomer, P. G., Coward, M. P., Lintern, B. C. 1977. Stratigraphy, structure and geochronology of ore leads in the Matsitama schist belt of northern Botswana. Precambr. Res. 5:23–41.

Cooper, M. R. 1973. A revision of the Lower Cretaceous (Upper Valanginian) Cephalopoda from the Uitenhage Formation of South Africa. Unpubl. M.Sc. thesis, Univ. Natal, Durban.

Cooper, M. R. 1977. Eustasy during the Cretaceous: its implications and importance. Palaeogeogr. Paleoclim. Palaeoecol. 22:1–60.

Copper, P. 1977. Paleolatitudes in the Devonian of Brazil and the Frasnian-Famennian mass extinction. Palaeogeogr. Palaeoclim. Palaeoecol. 21: 165–207.

Cordani, U. G., Amaral, G., Kawashita, K. 1973. The Precambrian evolution of South America. Geol. Rundsch. 62:309–317.

Cornelissen, A. K., Verwoerd, W. J. 1975. The Bushmanland kimberlites and related rocks. *In* L. H. Ahrens, J. B. Dawson, A. R. Duncan, A. J. Erlank (eds.), Physics and chemistry of the earth 9:71–80. Oxford: Pergamon.

Cornell, D. H. 1975. Petrology of the Marydale metabasites. Unpubl. Ph.D. thesis, Cambridge Univ.

Cornell, D. H. 1977. A post-Transvaal age for the Marydale Formation, Kheis Group, southern Africa. Earth Planet. Sci. Lett. 37:117–123.

Cornell, D. H. 1978a. Petrologic studies at T'Kuip: evidence for metamorphism and metasomatic alterations of volcanic formations beneath the Transvaal volcanosedimentary pile. Geol. Soc. S. Afr. Trans. 81:261–270.

Cornell, D. H. 1978b. Age and metamorphism of the Copperton Formation, Prieska district. Geol. Soc. S. Afr. Spec. Publ. No.4, pp.223–234.

Cornell, D. H., Barton, E. S. 1979. Age of the Marydale Group banded iron formation by the Pb-Pb method. Geol. Soc. S. Afr. 18th Congr. Abstracts 1:111–114.

Corvinus, G., Hendey, Q. B. 1978. A new Miocene vertebrate locality at Arrisdrift in Namibia (South West Africa). N. Jb. Geol. Paläont. 1978 (4): 193–205.

Cotter, E. 1978. The evolution of fluvial style, with special reference to the Central Appalachian Paleozoic. *In* A. D. Miall (ed.), Fluvial sedimentology. Can. Soc. Petrol. Geol. Mem. No.5, pp.361–383.

Cousins, C. A. 1959. The structure of the mafic portion of the Bushveld Igneous Complex. Geol. Soc. S. Afr. Trans. 62:179–201.

Coward, M. P. 1976. Archaean deformation patterns in southern Africa. Roy. Soc. London Phil. Trans. A283:313–331.

Coward, M. P. 1981. Pan African gneiss domes, diapirs and sheath folds. Jour. Struct. Geol., 3:89–95.

Coward, M. P., James, P. R. 1974. The deformation patterns of two Archaean greenstone belts in Rhodesia and Botswana. Precambr. Res. 1:235–258.

Coward, M. P., Shackleton, R. M. 1975. Some aspects of the Archaean history of southern Rhodesia. Res. Inst. African Geol. Univ. Leeds Annl. Rept. 19: 55–58.

Coward, M. P., Graham, R. H., James, P. R., Wakefield, J. 1973. A structural interpretation of the northern margin of the Limpopo orogenic belt, South Africa. Roy. Soc. London Phil. Trans.A273: 487–491.

Coward, M. P., James, P. R., Wright, L. 1976a. Northern margin of the Limpopo Mobile Belt. Geol. Soc. Amer. Bull. 87:601–611.

Coward, M. P., Lintern, B. C., Wright, L. 1976b. The pre-cleavage deformation of the sediments and gneisses of the northern part of the Limpopo belt. *In* B. F. Windley, (ed.), The early history of the earth, pp.323–330. London: Wiley.

Cox, K. G. 1970. Tectonics and vulcanism of the Karroo Period and their bearing on the postulated fragmentation of Gondwanaland. *In* T. N. Clifford, I. G. Gass (eds.), African magmatism and tectonics, pp.211–235. Edinburgh: Oliver and Boyd.

Cox, K. G. 1978. Komatiites and other high magnesia lavas: some problems. Roy. Soc. London Phil. Trans. A288:599–609.

Cox, K. G. 1980. Kimberlite and carbonatite magmas. Nature 283:716–717.

Cox, K. G., Johnson, R. L., Monkman, L. J., Stillman, C. J., Vail, J. R., Wood, D. N. 1965. The geology of the Nuanetsi igneous province. Roy. Soc. London Phil. Trans. A257:71–218.

Cox, K. G., MacDonald, R., Hornung, G. 1967. Geochemical and petrographic provinces in the Karroo basalts of southern Africa. Amer. Mineral. 52: 1451–1474.

Cramer, F. H., Rust, I. C., Diez de Cramer, M. de C. R. 1974. Upper Ordovician chitinozoans from the Cedarberg Formation of South Africa. Preliminary note. Geol. Rundsch. 63:340–345.

Crampton, D. 1974. A note on the age of the Matsap Formation of the northern Cape Province. Geol. Soc. S. Afr. Trans. 77:71–72.

Creer, K. M. 1965. Palaeomagnetic data from the Gondwanic continents. Roy. Soc. London Phil. Trans. A258:27–40.

Crocker, I. T. 1979. Metallogenic aspects of the Bushveld granites: fluorite, tin and associated rare-metal-carbonate mineralization. Geol. Soc. S. Afr. Spec. Publ. No.5, pp.275–295.

Crockett, R. N. 1968. Brief explanation of the geology of the Shashe area (accompanying geological map, quarter degree sheet 2127A). Geol. Surv. Botswana Rept. (Unpubl.).

Crockett, R. N. 1971. The rocks of the Ventersdorp System of the Lobatsi and Ramotswa areas, Republic of Botswana: their possible origins and regional correlations. Geol. Soc. S. Afr. Trans. 74: 1–24.

Crook, K. A. W. 1974. Lithogenesis and geotectonics: the significance of compositional variations in flysch arenites (graywackes). Soc. Econ. Paleont. Miner. Spec. Publ. No.19, pp.304–310.

Crowell, J. C., Frakes, L. A. 1972. Late Paleozoic glaciation: Part V, Karoo basin, South Africa. Geol. Soc. Amer. Bull. 83:2887–2912.

Daily, B., Cooper, M. R. 1976. Clastic wedges and patterned ground in the Late Ordovician-Early Silurian tillites of South Africa. Sedimentology 23: 271–283.

Daly, R. A. 1928. The Bushveld Igneous Complex of the Transvaal. Geol. Soc. Amer. Bull. 55:125–145.

Dalziel, I. W. D., Elliot, D. H. 1973. The Scotia Arc and Antarctic margin. *In* A. E. M. Nairn, F. G. Stehli (eds.), The ocean basins and margins. I, The South Atlantic, pp.171–245. New York: Plenum.

Danchin, R. V. 1970. Aspects of the geochemistry of some selected South African fine-grained sediments. Unpubl. Ph.D. thesis, Univ. Cape Town, South Africa.

Danchin, R. V. 1979. Mineral and bulk chemistry of garnet lherzolite and garnet harzburgite xenoliths from the Premier Mine, South Africa. 2nd Intl. Kimberlite Conf. Proc. 2:104–126.

Danchin, R. V., Boyd, F. R. 1976. Ultramafic nodules from the Premier kimberlite pipe, South Africa. Carnegie Inst. Washington Yearbook 75:531–537.

Danchin, R. V., Ferguson, J., McIver, J. R., Nixon,

P. H. 1975. The composition of late stage kimberlite liquids as revealed by nucleated autoliths. *In* L. H. Ahrens, J. B. Dawson, A. R. Duncan, A. J. Erlank (eds.), Physics and chemistry of the earth, vol. 9, pp.235–246. Oxford: Pergamon.

Darracott, B. W. 1974. On the crustal structure and evolution of southeastern Africa and the adjacent Indian Ocean. Earth Planet. Sci. Lett. 24:282–290.

Darracott, B. W. 1975. The interpretation of the gravity-anomaly over the Barberton Mountain Land, South Africa. Geol. Soc. S. Afr. Trans. 78:123–128.

Davies, D. H. 1964. The Miocene shark fauna of the southern St. Lucia area. S. Afr. Assoc. Mar. Biol. Res. Invest. Rept. 10:1–16.

Davies, O. 1976. The older coastal dunes in Natal and Zululand and their relation to former shorelines. S. Afr. Mus. Ann. 71:19–32.

Davies, O. 1978. The influence of epeirogenetic movement on Quaternary shorelines on the eastern and southern coasts of South Africa. Palaeoecol. Afr. 10/11:149–154.

Davies, R. D., Allsopp, H. L. 1976. Strontium isotopic evidence relating to the evolution of the Lower Precambrian crust in Swaziland. Geology 4:553–556.

Davies, R. D., Allsopp, H. L., Erlank, A. J., Manton, W. I. 1970. Sr-isotopic studies on various layered mafic intrusions in southern Africa. Geol Soc. S. Afr. Spec. Publ. No.1, pp.576–593.

Davis, B. T. C., Boyd, F. R. 1966. The join $Mg_2Si_2O_6$-$CaMgSi_2O_6$ at 30 kilobars pressure and its application to pyroxenes from kimberlites. Jour. Geophys. Res. 71:3567–3576.

Davis, G. L. 1977. The ages and uranium contents of zircons from kimberlites and associated rocks. 2nd Intl. Kimberlite Conf. 1977, Abstracts.

Dawson, J. B. 1962. Basutoland kimberlites. Geol. Soc. Amer. Bull. 73:545–560.

Dawson, J. B. 1967. A review of the geology of kimberlites. *In* P. J. Wyllie (ed.), Ultramafic and related rocks, pp. 241–251. New York: Wiley.

Dawson, J. B. 1970. The structural setting of African kimberlite magmatism. *In* T. N. Clifford, I. G. Gass (eds.), African magmatism and tectonics, pp.321–335. Edinburgh: Oliver and Boyd.

Dawson, J. B. 1971. Advances in kimberlite geology. Earth Sci. Rev. 7:187–214.

Dawson, J. B., Hawthorne, J. B. 1971. Intrusion features of some hypabyssal South African kimberlites. Bull. Vulcanol. 34:740–757.

Dawson, J. B., Hawthorne, J. B. 1973. Magmatic sedimentation and carbonatitic differentiation in kimberlite sills at Benfontein, South Africa. Geol. Soc. London Jour. 129:61–85.

Dawson, J. B., Smith, J. V. 1977. Late-stage diopsides in kimberlite groundmass. 2nd Intl. Kimberlite Conf. 1977, Abstracts.

Deans, T., Powell, J. L. 1968. Trace elements and strontium isotopes in carbonatites, fluorites and limestones from India and Pakistan. Nature 218: 750–752.

De Beers Consolidated Mines Ltd. 1976. Diamonds. S. Afr. Geol. Surv. Handbook 7 (5th ed.), pp. 17–30.

De Bruiyn, H., Rhodes, R. C. 1975. A new variety of Bushveld granite in the Dennilton area, Transvaal. Geol. Soc. S. Afr. Trans. 78:89–95.

De Gasparis, A. A. A. 1967. Rb-Sr isotopic studies relating to problems of geochronology on the Nelspruit and Mpageni granites. Unpubl. M. Sc. thesis, Univ. Witwatersrand, Johannesburg, South Africa.

Deines, P., Gold, D. P. 1973. The isotopic composition of carbonatite and kimberlite carbonates and their bearing on the isotopic composition of deep-seated carbon. Geochim. Cosmochim. Acta 37: 1709–1733.

De Jager, D. H., Von Backström, J. W. 1961. The sillimanite deposits in Namaqualand near Pofadder. S. Afr. Geol. Surv. Bull. No. 33.

De Jager, F. S. J. 1964. The Witwatersrand system in the Springs-Nigel-Heidelberg section of the East Rand basin. *In* S. H. Haughton, (ed.), The geology of some ore deposits in southern Africa, vol. 1, pp. 161–190. Johannesburg: Geol. Soc. S. Afr.

De Kock, W. J. P. 1964. The geology and economic significance of the West Wits Line. *In* S. H. Haughton, (ed.), The geology of some ore deposits in southern Africa, vol. 1, pp. 323–386. Johannesburg: Geol. Soc. S. Afr.

De Kock, W. P. 1934. The geology of western Rehoboth: an explanation of Sheet F.33-W-3 (Rehoboth). South West Africa. Dept. Mines Mem. No. 1.

De Swardt, A. M. J. 1974. Note on orogenic style in the Cape Fold-Belt. Author's reply to discussion by A. R. Newton. Geol. Soc. S. Afr. Trans. 77:385–388.

De Swardt, A. M. J., Fletcher, O., Toschek, P. 1974. Note on orogenic style in the Cape Fold-Belt. Geol. Soc. S. Afr. Trans. 77:53–57.

De Villiers, J. 1944. A review of the Cape orogeny. Univ. Stellenbosch Ann. 22A:183–208.

De Villiers, J., Söhnge, P. G. 1959. The geology of the Richtersveld. S. Afr. Geol. Surv. Mem. No. 48.

De Villiers, J., Jansen, H., Mulder, M. P. 1964. Die geologie van die gebied tussen Worcester en Hermanus. S. Afr. Geol. Surv. Map Explanation Sheets 3319C and 3419A.

De Villiers, J. S. 1970. The structure and petrology of the mafic rocks of the Bushveld Complex south of Potgietersrust. Geol. Soc. S. Afr. Spec. Publ. No. 1, pp. 23–25.

De Villiers, P. R. 1970. The geology and mineralogy of the Kalahari manganese-field north of Sishen, Cape

Province. S. Afr. Geol. Surv. Mem. No. 59.

De Villiers, S. B. 1967. Aanvoerrigtings van sedimente in die Sisteme Loskop en Waterberg in Noord-Transvaal soos weerspiel deur kruislaagdheid. S. Afr. Geol. Surv. Ann. 6:63–67.

De Vries, W. C. P. 1968. Stratigraphy of the Waterberg System in the southern Waterberg area, northwestern Transvaal. S. Afr. Geol. Surv. Ann. 7:43–76.

De Waal, S. A. 1966. The Alberta Complex, a metamorphosed layered intrusion, north of Nauchas, South West Africa, the surrounding granites and repeated folding in the Damara System. Unpubl. D. Sc. thesis, Univ. Pretoria, South Africa.

De Waal, S. A. 1972. The Bushveld granites in the Zaaiplaats area. Geol. Soc. S. Afr. Trans. 75:135–147.

Dewey, J. F., Bird, J. M. 1970. Mountain belts and the new global tectonics. Jour. Geophys. Res. 75:2625–2647.

Dewey, J. F., Burke, K. 1973. Tibetan, Variscan and Precambrian basement reactivation: products of continental collision., Jour. Geol. 81:683–692.

De Wit, M. J. 1977. The evolution of the Scotia Arc as a key to the reconstruction of southwestern Gondwanaland. Tectonophysics 37:53–81.

De Wit, M. J., Stern, C. R. 1980. A 3500 Ma ophiolite complex from the Barberton greenstone belt, South Africa: Archean oceanic crust and its geotectonic implications. 2nd Intl. Archean Symp. Extended abstracts, pp. 85–87. Perth: Geol. Soc. Australia.

Dickinson, K. A., Berryhill, H. L., Holmes, C. W. 1972. Criteria for recognizing ancient barrier coastlines. Soc. Econ. Paleont. Mineral. Spec. Publ. No. 23, pp. 192–241.

Dickinson, W. R., Seely, D. R. 1979. Structure and stratigraphy of forearc regions. Amer. Ass. Petrol. Geol. Bull. 63:2–31.

Dietz, R. S., 1961. Vredefort ring-structure: Meteorite impact scar? Jour. Geol. 69:499–516.

Dietz, R. S., Holden, J. C., 1970. Reconstruction of Pangaea: breakup and dispersion of continents, Permian to present. Jour. Geophys. Res. 75:4939–4956.

Dimroth, E. 1977. Facies models 5. Models of physical sedimentation of iron formations. Geosci. Can. 4:23-30.

Dingle, R. V. 1973a. Post Palaeozoic stratigraphy of the eastern Agulhas Bank, South African continental margin. Mar. Geol. 15:1–23.

Dingle, R. V. 1973b. The geology of the continenta shelf between Lüderitz and Cape Town (Southwest Africa), with special reference to Tertiary strata. Geol. Soc. London Jour. 129:337–363.

Dingle, R. V. 1974. Agulhas Bank phosphorites: a review of 100 years of investigation. Geol. Soc. S. Afr. Trans. 77:261–264.

Dingle, R. V. 1977. The anatomy of a large submarine slump on a sheared continental margin (southeast Africa). S. Afr. Geol. Surv./Univ. Cape Town Joint Mar. Geol. Progr. Tech. Rept. 9, pp. 51–65.

Dingle, R. V., Gentle, R. I. 1972. Early Tertiary volcanic rocks on the Agulhas Bank, South African continental shelf. Geol. Mag. 109:127–136.

Dingle, R. V., Klinger, H. C. 1972. The stratigraphy and ostracod fauna of the Upper Jurassic sediments from Brenton, in the Knysna outlier, Cape Province. Roy. Soc. S. Afr. Trans. 40:279–298.

Dingle, R. V., Rogers, J. 1972. Pleistocene palaeogeography of the Agulhas Bank. Roy. Soc. S. Afr. Trans. 40:155–165.

Dingle, R. V., Scrutton, R. A. 1974. Continental breakup and the development of post-Paleozoic sedimentary basins around southern Africa. Geol. Soc. Amer. Bull. 85:1467–1474.

Dingle, R. V., Goodlad, S. W., Martin, A. K. 1977. Bathymetry and stratigraphy of the northern Natal Valley (SW Indian Ocean): A preliminary account. S. Afr. Geol. Surv./Univ. Cape Town Joint Mar. Geol. Progr. Tech. Rept. 9, pp. 66–74.

Dix, O. R. 1981. High-grade metamorphism and possible overturning of the Tugela Rand layered intrusion in southeast Africa. Geology 9:155–160.

Dixey, F. 1955. Some aspects of the geomorphology of central and southern Africa. Geol. Soc. S. Afr. Trans. 58 (Annex.).

Dixon, J. M. 1975. Finite strain and progressive deformation in models of diapiric structures. Tectonophysics 28:89–124.

Dörjes, J., Gadow, S., Reineck, H.-E., Singh, I. B. 1970. Sedimentologie und makrobenthos der Nordegründe und der Aubenjade (Nordsee). Senckenb. Marit. 2:31–59.

Douglas, R. J. W. 1980. On the age of rocks and Precambrian time scales. Geology 8:167–171.

Drake, L. D. 1972. Mechanisms of clast attrition in basal till. Geol. Soc. Amer. Bull. 83:2159–2166.

Drever, J. I. 1974. Geochemical model for the origin of Precambrian banded iron formations. Geol. Soc. Amer. Bull. 85:1099–1106.

Duncan, A. R., Erlank, A. J. 1979. Regional geochemistry of the Karoo volcanics. Geol. Soc. S. Afr. 18th Congr. Abstracts 1:128–131.

Duncan, A. R., Hargraves, R. B., Brey, G. P. 1978. Age, palaeomagnetism and chemistry of melilite basalts in southern Africa. Geol. Mag. 115:377–396.

Dunlevey, J. N., Hiller, N. 1979. The Witteberg-Dwyka contact in the south-western Cape. Geol. Soc. S. Afr. Trans. 82:251–256.

Duplan, L. 1978. A photogeological interpretation of the regional tectonics of an area east of the Richtersveld. Precamb. Res. Unit Univ. Cape Town Ann. Rept. 14/15:99–102.

Du Plessis, A. 1977. Seafloor spreading south of the

Agulhas Fracture zone. Nature 270:719–721.

Du Plessis, A., Simpson, E. S. W. 1974. Magnetic anomalies associated with the southeastern continental margin of South Africa. Mar. Geophys. Res. 2:99–110.

Du Plessis, A., Scrutton, R. A., Barnaby, A. M., Simpson, E. S. W. 1972. Shallow structures of the continental margin of southwestern Africa. Mar. Geol. 13:77–89.

Du Preez, J. W. 1944. Lithology, structure and mode of deposition of the Cretaceous deposits in the Oudtshoorn area. Univ. Stellenbosch Ann. 22A: 209–237.

Du Toit, A. L. 1921. The Carboniferous glaciation of South Africa. Geol. Soc. S. Afr. Trans. 24:188–227.

Du Toit, A. L. 1927. Geological comparison of South America with South Africa. Carnegie Inst. Washington Publ.

Du Toit, A. L. 1929. The volcanic belt of the Lebombo: A region of tension. Roy. Soc. S. Afr. Trans. 18:189–218.

Du Toit, A. L. 1937. Our wandering continents. London: Oliver and Boyd.

Du Toit, A. L. 1939. The geology of South Africa (2nd ed.). Edinburgh: Oliver and Boyd.

Du Toit, A. L. 1946. The geology of parts of Pondoland, East Griqualand and Natal. S. Afr. Geol. Surv. Map Explanation Sheet 119 (Port Shepstone).

Du Toit, A. L. 1954. The geology of South Africa (3rd ed.). Edinburgh: Oliver and Boyd.

Du Toit, M. C. 1965. A geological investigation and correlation of rocks belonging to the Koras Formation in the Gordonia and Kenhardt districts, northern Cape Province. Unpubl. M.Sc. thesis, Univ. Orange Free State, Bloemfontein.

Du Toit, M. C., Van Reenen, D. D. 1977. The southern margin of the Limpopo Mobile Belt, northern Transvaal, with special reference to metamorphism and structure. Botswana Geol. Surv. Bull. 12:83–97.

Du Toit, S. R. 1976. Mesozoic geology of the Agulhas Bank, South Africa. Unpubl. Ph.D. thesis, Univ. Cape Town.

Du Toit, S. R. 1979. The Mesozoic history of the Agulhas Bank in terms of the plate-tectonic theory. Geol. Soc. S. Afr. Spec. Publ. 6:197–203.

Du Toit, S. R., Leith, M. J. 1974. The J(c)-1 borehole on the continental shelf near Stanger, Natal. Geol. Soc. S. Afr. Trans. 77:247–252.

Eddington, S. M., Harrison, D. 1979. Ryst Kuil uranium deposit: A case history. Geol. Soc. S. Afr. 18th Congr. Abstracts 1:45–55.

Edgar, A. D., Green, D. H., Hibberson, W. O. 1976. Experimental petrology of a highly potassic magma. Jour. Petrol. 17:339–356.

Edwards, M. B. 1978. Glacial environments. *In* H. G. Reading (ed.), Sedimentary environments and facies, pp. 416–438. Oxford: Blackwell; New York: Elsevier.

Eggler, D. H. 1978. The effect of CO_2 upon partial melting of peridotite in the system Na_2O-CaO-Al_2O_3-MgO-SiO_2-CO_2 to 35 kb, with an analysis of melting of a peridotite-H_2O-CO_2 system. Amer. Jour. Sci. 278:64–94.

Eggler, D. H., McCallum, M. E. 1976. A geotherm for megacrysts in the Sloan kimberlite pipes, Colorado. Carnegie Inst. Washington Yearbook 75:631–636.

Eggler, D. H., Wendlandt, R. F. 1979. Experimental studies on the relationship between kimberlite magmas and partial melting of a peridotite. 2nd Intl. Kimberlite Conf. Proc. 1:330–338.

Ellenberger, P. 1970. Les niveaux paleontologiques de premiere apparition des mammiferes primordiaux en Afrique de sud et leur ichnologie. 2nd Intl. Gondwana Symposium. Proceedings and Papers, pp. 343–370.

Elliott, D. H. 1976. The motion of thrust sheets. Jour. Geophys. Res. 81:949–963.

Elliott, D. H., Watts, D. R. 1974. The nature and origin of volcaniclastic material in some Karroo and Beacon rocks. Geol. Soc. S. Afr. Trans. 77:109–111.

Ellis, D. J., Green, D. H. 1979. An experimental study of the effect of Ca upon garnet-clinopyroxene Fe-Mg exchange equilibria. Contr. Mineral Petrol. 71:13–22.

Elton, D., Ridley, W. I. 1979. The oxide and silicate mineral chemistry of a kimberlite from the Premier Mine: Implications for the evolution of kimberlitic magmas. 2nd Intl. Kimberlite Conf. Proc. 1:206–216.

Emery, K. O., Uchupi, E., Bowin, C. O., Phillips, T., Simpson, E. S. W. 1975. The continental margin off western Africa: Cape St Francis (South Africa) to Walvis Ridge (South-West Africa). Amer. Assoc. Petrol. Geol. Bull. 59:3–59.

Engel, A. E. J., Nagy, B., Nagy, L. A., Engel. C. G., Kremp, G. O. W., Drew, C. M. 1968. Alga-like forms in Onverwacht Series, South Africa: Oldest recognized life-like forms on earth. Science 161:1005–1008.

Eriksson, K. A. 1977a. Tidal deposits from the Archaean Moodies Group, Barberton Mountain Land, South Africa. Sediment. Geol. 18:257–281.

Eriksson, K. A. 1977b. Tidal flat and subtidal sedimentation in the Malmani dolomite, South Africa. Sediment. Geol. 18:223–244.

Eriksson, K. A. 1978. Alluvial and destructive beach facies from the Archaean Moodies Group, Barberton Mountain Land, South Africa and Swaziland. In A. D. Miall (ed.), Fluvial sedimentology. Can. Soc. Petrol. Geol. Mem. 5:287–311.

Eriksson, K. A. 1979. Marginal marine depositional processes from the Archaean Moodies Group,

Barberton Mountain Land, South Africa: Evidence and significance. Precambr. Res. 8:153–182.

Eriksson, K. A., 1980a. Hydrodynamic and palaeogeographic interpretation of turbidite deposits from the Archean Fig Tree Group of the Barberton Mountain Land, South Africa. Geol. Soc. Amer. Bull., 91:21–26.

Eriksson, K. A. 1980b. Transitional sedimentation styles in the Moodies and Fig Tree Groups, Barberton Mountain Land, South Africa: evidence favouring an Archaean continental margin. Precambr. Res. 12:141–160.

Eriksson, K. A., MacGregor, I. M. 1981. Precambrian palaeontology of southern Africa. *In* D. R. Hunter (ed.), Precambrian of the southern hemisphere, pp. 813–833. Amsterdam:Elsevier.

Eriksson, K. A., Truswell, J. F. 1974. Tidal flat associations from a Lower Proterozoic carbonate sequence in South Africa. Sedimentology 21:293–309.

Eriksson, K. A., Truswell, J. F. 1978. Geological processes and atmospheric evolution in the Precambrian. *In* D. H. Tarling (ed.), Evolution of the earth's crust, pp. 219–238. London: Academic Press.

Eriksson, K. A., Vos, R. G. 1979. A fluvial fan depositional model for Middle Proterozoic red beds from the Waterberg Group, South Africa. Precambr. Res. 9:169–188.

Eriksson, K. A., Truswell, J. F., Button, A. 1976. Palaeoenvironmental and geochemical models from a Lower Proterozoic succession in South Africa. *In* M. R. Walter (ed.), Developments in sedimentology. Stromatolites, pp. 635–643. Amsterdam: Elsevier.

Eriksson, K. A., Turner, B. R., Vos, R. G. 1981. Evidence of tidal processes from the lower part of the Witwatersrand Supergroup, South Africa. Sediment. Geol. 29., p. 309–325.

Eriksson, P. G. 1979. Mesozoic sheetflow and playa sediments of the Clarens Formation in the Kamberg area of the Natal Drakensberg. Geol. Soc. S. Afr. Trans. 82:257–258.

Eriksson, S. C. 1982. Aspects of the petrochemistry of the Palabora Igneous Complex, South Africa. Unpubl. Ph.D. thesis. Univ. Witwatersrand, Johannesburg.

Erlank, A. J. 1976. Upper mantle metasomatism as revealed by potassic richterite-bearing peridotite xenoliths from kimberlites. Eos 57:597.

Ermanovics, I., Key, R. M., Jones, M. T. 1978. The Palapye Group, central-eastern Botswana. Geol. Soc. S. Afr. Trans. 81:61–73.

Estes, R. 1977. Relationships of the South African fossil frog, *Eoxenopoides reuningi* (Anura, Pipidae). S. Afr. Mus. Ann. 73:49–80.

Eugster, H. P. 1969. Inorganic bedded cherts from the Magadi area, Kenya. Contr. Mineral. Petrol. 22:1–31.

Faupel, J. 1973. Geological and mineralogical investigation of the Donkerhoek granite, Karibib district, South West Africa (Preliminary Report). Neues Jahrb. Geol. Paläont. 1973:327–341.

Feather, C. E., Koen, G. M. 1975. The mineralogy of the Witwatersrand Reefs. Minerals Sci. Engng. 7:189–224.

Ferguson, J. 1969. Compositional variation in minerals from mafic rocks of the Bushveld Complex. Geol. Soc. S. Afr. Trans. 72:61–78.

Ferguson, J. 1973. The Pilanesberg alkaline province, southern Africa. Geol. Soc. S. Afr. Trans. 76:249–270.

Ferguson, J., Botha, E. 1963. Some aspects of igneous layering in the basic zone of the Bushveld Complex. Geol. Soc. S. Afr. Trans. 66:259–278.

Ferguson, J., McCarthy, T. S. 1970. Origin of an ultramafic pegmatoid in the eastern part of the Bushveld Complex. Geol. Soc. S. Afr. Spec. Publ. No. 1, 74–79.

Ferguson, J., Danchin, R. V., Nixon, P. H. 1973. Petrochemistry of kimberlite autoliths. *In* P. H. Nixon (ed.), Lesotho kimberlites, pp. 285–293. Lesotho National Development Corp.

Ferguson, J., Martin, H., Nicolaysen, L. O., Danchin, R. V. 1975. Gross Brukkaros: A kimberlite-carbonatite volcano. *In* L. H. Ahrens, J. B. Dawson, A. R. Duncan, A. J. Erlank (eds), Physics and chemistry of the earth, vol. 9, pp. 219–234. Oxford: Pergamon.

Field, M. E. 1980. Sand bodies on coastal plain shelves: Holocene record of the U.S. Atlantic inner shelf off Maryland. Jour. Sediment. Petrol. 50:505–528.

Finnemore, S. 1978. The geochemistry and origin of the Matchless amphibolite belt, Windhoek District, South West Africa. Geol. Soc. S. Afr. Spec. Publ. No. 4, pp. 433–447.

Fisher, W. L., Brown, L. F., Scott, A. J., McGowen, J. H. 1969. Delta systems in the exploration for oil and gas. Austin: Bur. Econ. Geol., University of Texas.

Fisk, H. N. 1944. Geological investigation of the alluvial valley of the lower Mississippi River. Vicksburg, Miss.: Mississippi River Commission.

Fitch, F. J., Miller, J. A. 1971. Potassium-argon radioages of Karroo volcanic rocks from Lesotho. Bull. Vulcanol. 35:64–84.

Flemming, B. W. 1977. Langebaan Lagoon - a mixed carbonate-siliclastic tidal environment in a semi-arid climate. Sediment. Geol. 18:61–95.

Flemming, B. W. 1978. Underwater sand dunes along the southeast African continental margin—observations and implications. Mar. Geol. 26:117–198.

Flemming, B. W. 1981. Factors controlling shelf sediment dispersal along the southeast African continental margin. Mar. Geol. (in press).

Flemming, N. C., Roberts, D. G. 1973. Tectono-eustatic changes in sea level and seafloor spreading. Nature 243:19–22.

Fleuty, M. J. 1964. The description of folds. Geol. Assoc. Proc. 75:461–492.

Flinn, D. 1965. On the symmetry principle and the deformation ellipsoid. Geol. Mag. 102:36–45.

Flores, G. 1970. Suggested origin of the Mozambique Channel. Geol. Soc. S. Afr. Trans. 73:1–16.

Flores, G. 1973. The Cretaceous and Tertiary sedimentary basins of Mozambique and Zululand. *In* G. Blant (ed.), Sedimentary basins of the African coast. Part 2, South and east coast, pp. 81–111. Paris: Association of African Geological Surveys.

Fockema, P. D. 1967. Crocidolite and associated rocks of the Kuruman area in the northern Cape Province. Unpubl. Ph.D. thesis, Univ. Witwatersrand, Johannesburg, South Africa.

Folk, R. L., Land, L. S. 1975. Mg/Ca ratio and salinity: two controls over crystallization of dolomite. Amer. Assoc. Petrol. Geol. Bull. 59:60–68.

Föster, R. 1975. The geological history of the sedimentary basin of southern Mozambique, and some aspects of the origin of the Mozambique Channel. Palaeogeogr. Palaeoclim. Palaeoecol. 17:267–287.

Fowler, J. A. 1976. The alluvial geology of the lower Orange River and adjacent coastal deposits, South West Africa. Unpubl. M.Sc. thesis, University of London, England.

Francheteau, J., Le Pichon, X. 1972. Marginal fracture zones as structural framework of continental margins in the South Atlantic Ocean. Amer. Assoc. Petrol. Geol. Bull. 56:991–1007.

Frankel, J. J. 1960. The geology along the Umfolosi River, south of Mtubatuba, Zululand. Reply to discussion. Geol. Soc. S. Africa. Trans. 63:256–263.

Frankel, J. J. 1966. Basal rocks of the Tertiary at Uloa, Zululand, South Africa. Geol. Mag. 103: 214–230.

Frankel, J. J. 1968. Tertiary sediments in the lower Umfolosi River valley, Zululand. Geol. Soc. S. Afr. Trans. 71:135–145.

Frankel, J. J. 1969. The distribution and origin of the Effingham rock type. Geol. Soc. Amer. Mem. 115:149–173.

Frankel, J. J. 1972. Shark teeth in the Tertiary of Zululand. Geol. Soc. S. Afr. Trans. 75:311–312.

French, B. M., Hargraves, R. B. 1971. Bushveld Igneous Complex, South Africa—absence of shock metamorphic effects in a preliminary search. Jour. Geol. 79:616–620.

Frets, D. C. 1969. Geology and structure of the Huab-Welwitschia area, southwest Africa. Precambrian Res. Unit Univ. Cape Town Bull. No. 5.

Frick, A. 1970. Sedimentological aspects of a portion of the southeastern Waterberg area. S. Afr. Geol. Surv. Ann. 8:69–74.

Frick, C. 1973. The sill phase and chill zone of the Bushveld Igneous Complex. Geol. Soc. S. Afr. Trans. 76:7–14.

Frick, C. 1975a. The geology and petrology of the Kafferskraal Igneous Complex. Geol. Soc. S. Afr. Trans. 78:11–23.

Frick, C. 1975b. The Phalaborwa Syenite Intrusion. Geol. Soc. S. Afr. Trans. 78:201–214.

Frick, C., Coetzee, C. B. 1974. The mineralogy and the petrology of the sillimanite deposits west of Pofadder, Namaqualand. Geol. Soc. S. Afr. Trans. 77:169–183.

Fripp, R. E. P. 1980. The ancient Sand River Gneisses, Limpopo Mobile Belt, South Africa. Intl. Archaean Symp. Proc. Perth 1980. (In press).

Fripp, R. E. P., Gay, N. C. 1972. Some structural aspects of the Hospital Hill Series on the north-central margin of the Witwatersrand basin. Geol. Soc. S. Afr. Trans. 75:187–196.

Fripp, R. E. P., Lilly, P. A., Barton, J. M. 1979. The structure and origin of the Singelele gneiss at the type locality near Messina, Limpopo Mobile Belt. Geol. Soc. S. Afr. Trans. 82:161–167.

Fripp, R. E. P., Van Nierop, D. A., Callow, M. J., Lilly, P. A., Du Plessis, L. U. 1980. Deformation in part of the Archaean Kaapvaal craton, South Africa. Precambr. Res. 13:241–251.

Fryer, B. J., Hutchinson, R. W. 1976. Generation of base metal deposits on the sea floor. Can. Jour. Earth Sci. 13:126–135.

Fuller, A. O. 1970. The occurrence of laumontite in strata of the Karroo System, South Africa. *In* 2nd Gondwana Symposium. Proceedings and Papers, pp. 455–456.

Fuller, A. O., Sprague, A. W., Waters, D., Willis, J. P., Camden-Smith, P. M. 1979. Chemistry and mineralogy of Archaean shales from the Witwatersrand Supergroup. Geol. Soc. S. Afr. 18th Congr. Abstracts 1:158.

Gaines, A. M. 1974. Protodolomite synthesis at 100°C and atmospheric pressure. Science 183: 518–520.

Galloway, W. E. 1976. Sediments and stratigraphic framework of the Copper River fan-delta, Alaska. Jour. Sediment. Petrol. 46:726–737.

Galloway, W. E., Kreitler, C. W., McGowen, J. H. 1979. Depositional and ground-water flow systems in the exploration for uranium. Bur. Econ. Geol., Univ. Texas Austin, Res. Colloquium notes.

Gardiner, B. G. 1969. New palaeoniscoid fish from the

Witteberg Series of South Africa. Zool. Jour. Linn. Soc. 48:423–452.

Gay, N. C. 1969. The analysis of strain in the Barberton Mountain Land, eastern Transvaal, using deformed pebbles. Jour. Geol. 77:377–396.

Gay, N. C., Jager, A. J. 1982. The influence of geological features on rock mechanics problems in Witwatersrand gold mines. Geol. Soc. S. Afr. Spec. Publ.

Gellatly, D. C. 1970. Cross-bedded tidal megaripples from King Sound. Sediment. Geol. 4:185–192.

Gentle, R. I. 1970. Pre-Quaternary geology of the continental shelf between Cape Infanta and Cape Town. S. Afr. Nat. Comm. Oceanogr. Res. Mar. Geol. Progr. Tech. Rept. No. 3.

Gentle, R. I. 1977. The Groot Haelkraal granite and its extension offshore. Geol. Soc. S. Afr. Trans. 80: 157–165.

Gentle, R. I., Hooker, P. J., Fitch, F. J., Miller, J. A. 1978. Evidence for Cape Fold Belt overprinting of the Groot Haelkraal granite during Upper Permian. Geol. Soc. S. Afr. Trans. 81:105–107.

Geringer, G. J. 1973. Die geologie van die Argeîese gesteentes en jonger formasies in die gebied wes van Upington met spesiale verwysing na die verskillende granietvoorkomste. Unpubl. D.Sc. thesis, University Orange Free State, Bloemfontein.

Geringer, G. J., Botha, B. J. V. 1977. The gneisses and regional structural pattern of the Namaqualand Mobile Belt in part of the Gordonia district, northwestern Cape. Geol. Soc. S. Afr. Trans. 80:93–95.

Germs, G. J. B. 1972a. The stratigraphy and palaeontology of the lower Nama Group, South West Africa. Precambrian Res. Unit Univ. Cape Town Bull. No. 12.

Germs, G. J. G. 1972b. New shelly fossils from the Nama Group, South West Africa. Amer. Jour. Sci. 272:752–761.

Germs, G. J. B. 1974. The Nama Group in South West Africa and its relationship to the Pan-African geosyncline. Jour. Geol. 82:301–317.

Gerrard, I. 1958. The olivine melilitites from the western Cape. Unpubl. M.Sc. thesis, Univ. Cape Town, South Africa.

Gevers, T. W. 1930. Terrestrer Dolomit in der Etosha-Pfanne, Südwestafrica. Cbl. Mineral. Geol. Paläont. Abtl. B:224–230.

Gevers, T. W. 1931. An ancient tillite in South West Africa. Geol. Soc. S. Afr. Trans. 34:1–17.

Gevers, T. W. 1934. The geology of the Windhoek district in South West Africa. Geol. Soc. S. Afr. Trans. 37:221–251.

Gevers, T. W. 1963. Geology along the northwestern margin of the Khomas Highlands between Otjimbingwe - Karabib and Okahandja, South West Africa. Geol. Soc. S. Afr. Trans. 66:199–258.

Gevers, T. W., Dunne, J. C. 1942. Charnockitic rocks near Port Edward in Alfred County, Natal. Geol. Soc. S. Afr. Trans. 45:183–214.

Gevers, T. W., Partridge, F. C., Joubert, G. K. 1937. The pegmatite area south of the Orange River in Namaqualand. S. Afr. Geol. Surv. Mem. No. 31.

Gidskehang, A., Creer, K. M., Mitchell, J. G. 1975. Palaeomagnetism and K-Ar ages of the South West African basalts and their bearing on the time of initial rifting of the South Atlantic Ocean. Geophys. Jour. Roy. Astr. Soc. 42:1–20.

Gijbels, R. H., Millard, H. T., Desborough, G. A., Bartel, A. J. 1974. Osmium, ruthenium, iridium, and uranium in silicates and chromite from the eastern Bushveld Complex, South Africa. Geochim. Cosmochim. Acta 38:319–337.

Ginsburg, R. N., James, N. P. 1974. Holocene carbonate sediments of continental shelves. *In* C. A. Burk, C. L. Drake (eds.), The geology of continental margins, pp. 137–155. Berlin: Springer.

Glaessner, M. F. 1959. Precambrian Coelenterata from Australia, Africa, and England. Nature 183: 1472–1473.

Glaessner, M. F. 1979. Precambrian biogeography and biostratigraphy. *In* R. A. Robison, C. Teichert (eds), Treatise on invertebrate paleontology. Part A Introduction, pp. A79–A118. Boulder, Colorado: Geol. Soc. Amer. and Univ. of Kansas.

Glikson, A. Y. 1976. Trace element geochemistry and origin of Early Precambrian acid igneous series, Barberton Mountain Land, Transvaal. Geochim. Cosmochim. Acta 40:1261–1280.

Gold, D. P. 1963. Average chemical composition of carbonatites. Econ. Geol. 58:83–91.

Goldberg, I. 1976. A preliminary account of the Otjihase copper deposit, South West Africa. Econ. Geol. 71:384–390.

Gole, C. V., Chitale, C. V. 1966. Inland delta building activity of the Kosi River. Amer. Soc. Civil. Engng. Proc., Jour. Hydr. Div. 92:111–126.

Goodwin, A. M. 1977. Archean volcanism in Superior Province, Canadian Shield. Geol. Assoc. Can. Spec. Pap. No. 11.

Gordon-Gray, K. D., Van Dijk, D. E., Lacey, W. S. 1976. Preliminary report on equisetalean plants from Lidgetton, Natal. Palaeont. Afr. 19:43–57.

Goudie, A. 1973. Duricrusts in tropical and subtropical landscapes. Oxford: Clarendon.

Gould, H. R. 1970. The Mississippi Delta complex. *In* J. P. Morgan (ed.), Deltaic sedimentation, modern and ancient. Soc. Econ. Paleont. Mineral. Spec. Publ. No. 15, pp. 3–30.

Graham, R. H. 1973. The southern margin of the Limpopo Mobile Belt, Transvaal, South Africa. Res. Inst. African Geol. Univ. Leeds Ann. Rept. 17:7–9.

Graham, R. H. 1974. A structural investigation of the southern part of the Limpopo belt and the adjacent

Kaapvaal craton, South Africa. Res. Inst. African Geol. Univ. Leeds Ann. Rept. 18:63–69.

Green, D. H. 1975. Genesis of Archaean peridotitic magmas and constraints on Archaean geothermal gradients and tectonics. Geology 3:15–18.

Green, D. H., Ringwood, A. E. 1967a. An experimental investigation of the gabbro to eclogite transformation and its petrological applications. Geochim. Cosmochim. Acta 31:767–833.

Green, D. H., Ringwood, A. E. 1967b. The genesis of basaltic magmas. Contr. Mineral. Petrol. 15:103–190.

Green, H. W. 1972. Metastable growth of coesite in highly strained quartz. Jour. Geophys. Res. 77: 2478–2482.

Green, H. W., Gueguen, Y. 1974. Origin of kimberlite pipes by diapiric upwelling in the upper mantle. Nature 249:617–620.

Greenman, L. 1966. The geology of area 2515C Lüderitz, South West Africa. Unpubl. M.Sc. thesis, Univ. Cape Town, South Africa.

Greenshields, H. D. 1978. A palaeoenvironmental analysis of the Beaufort Group along the Transkei and parts of the Ciskei coast. Unpubl. M.Sc. thesis, Univ. Natal, Pietermaritzburg.

Greenway, M. E. 1972. The geology of the Falkland Islands. Brit. Antarctic Surv., Sci. Rept. No. 76.

Gresse, P. G. 1976. The structure and metamorphism of the pre-Cape rocks in the coastal strip south of George. Unpubl. M.Sc. thesis, Univ. Stellenbosch, South Africa.

Griffin, W. L., Carswell, D. A., Nixon, P. H. 1979. Lower-crustal granulites and eclogites from Lesotho, southern Africa. 2nd Intl. Kimberlite Conf. Proc. 2:59–86.

Grobler, N. J. 1972. The geology of the Pietersburg greenstone belt. Unpubl. D.Sc. thesis, Univ. Orange Free State, Bloemfontein.

Grobler, N. J., Botha, B. J. V. 1976. Pillow-lavas and hyaloclastite in the Ongeluk Andesite Formation in a road-cutting west of Griquatown, South Africa. Geol. Soc. S. Afr. Trans. 79:53–57.

Grobler, N. J., Emslie, D. P. 1976. Stromatolitic limestone and chert in the Ventersdorp Supergroup at the T'Kuip hills area and surroundings, Britstown district, South Africa. Geol. Soc. S. Afr. Trans. 79:49–52.

Grobler, N. J., Botha, B. J. V., Smit, C. A. 1977. The tectonic setting of the Koras Group. Geol. Soc. S. Afr. Trans. 80:167–175.

Guj, P. 1969. Structural geology of the Auas Mountains, Windhoek district, South West Africa. S. Afr. Geol. Surv. Ann. 6:55–62.

Guj, P. 1970. The Damara Mobile Belt in the southwestern Kaokoveld, southwest Africa. Precambrian Res. Unit Univ. Cape Town Bull. No. 8.

Guj, P. 1974. A revision of the Damara stratigraphy along the southern margin of the Kamanjab inlier, South West Africa. Precambrian Res. Unit Univ. Cape Town Bull. No. 15, pp. 167–176.

Gulbrandsen, R. A. 1969. Physical and chemical factors in the formation of marine apatite. Econ. Geol. 64:365–382.

Gurney, J. J., Fesq, H. W., Kable, E. 1973. Clinopyroxene - ilmenite intergrowths from kimberlite: a reappraisal. *In* P. H. Nixon (ed.), Lesoth kimberlites, pp. 238–253. Lesotho National Development Corp.

Gurney, J. J., Haite, B., Cox, K. G. 1975. Mantle xenoliths in the Matsoku kimberlite pipe. *In* L. H. Ahrens, J. B. Dawson, A. R. Duncan, A. J. Erlank (eds.), Physics and chemistry of the earth, vol. 9, pp. 507–524. Oxford: Pergamon.

Gurney, J. J., Jakob, W. R. O., Dawson, J. B. 1979. Megacrysts from the Monastery kimberlite pipe, South Africa. 2nd Intl. Kimberlite Conf. 2:227–243.

Gustavson, T. C., Boothroyd, J. C., Ashley, G. M. 1975. Depositional sequences in glaciolacustrine deltas. *In* A. V. Jopling, B. C. McDonald (eds.), Glaciofluvial and glaciolacustrine sedimentation. Soc. Econ. Paleont. Mineral. Spec. Publ. No. 23, pp. 264–280.

Hälbich, I. W. 1970. The geology of the western Windhoek and Rehoboth districts: a stratigraphic-structural analysis of the Damara System. Unpubl. D.Sc thesis, Univ. Stellenbosch, South Africa.

Hälbich, I. W. 1977a. Structure and tectonics along the southern margin of the Damara Mobile Belt, South West Africa. Univ. Stellenbosch Ann. A2: 149–247.

Hälbich, I. W. 1977b. Fold profiles and tectonic shortening in the Cape Fold Belt. Geol. Soc. S. Afr. Trans. 80:253–265.

Hälbich, I. W. 1978. Minor structures in gneiss and the origin of steep structures in the Okiep Copper District. Geol. Soc. S. Afr. Spec. Publ. No. 4, pp. 297–322.

Hälbich, I. W. 1979a. The Cango Group in the Schoemanspoort. *In* I. C. Rust (ed.), Geokongres 79 Excursions Guide Book, pp. 44–52. Geol. Soc. S. Afr. 18th Congress.

Hälbich, I. W. 1979b. The George pluton and metamorphites. *In* I. C. Rust (ed.), Geokongres 79 Excursions Guide Book, pp. 69–81. Geol. Soc. S. Afr. 18th Congress.

Hälbich, I. W. 1979c. Table Mountain tectonics. *In* I. C. Rust (ed.), Geokongres 79 Excursions Guide Book, pp. 41–43. Geol. Soc. S. Afr. 18th Congress.

Hälbich, I. W. 1979d. Disharmonic folding, detachment and thrusting in the Cape Fold Belt. Geol. Soc. S. Afr. 18th Congr. Abstracts 2:167–169.

Hales, A. L., Gough, D. I. 1962. The gravity survey of the Republic of South Africa. Handbook 3. Part II.

Isostatic anomalies and crustal structure. Pretoria: Government Printer.

Hall, A. L. 1918. The geology of the Barberton gold mining district. S. Afr. Geol. Surv. Mem. No. 9.

Hall, A. L. 1932. The Bushveld Igneous Complex of the central Transvaal. S. Afr. Geol. Surv. Mem. No. 28.

Hall, A. L. 1938. Analyses of rocks, minerals, ores, coals, soils, and waters from southern Africa. S. Afr. Geol. Surv. Mem. No. 32.

Hallam, C. D. 1964. The geology of the coastal diamond deposits of southern Africa. *In* S. H. Haughton (ed.), Some ore deposits in southern Africa, vol. 2, pp. 671–728. Johannesburg: Geol. Soc. S. Afr.

Hallbauer, D. K. 1975. The plant origin of the Witwatersrand carbon. Minerals Sci. Engng. 7:111–131.

Hallbauer, D. K., Kable, E. J. D. 1979. Geochemical and fluid inclusion studies of quartz pebbles in Witwatersrand conglomerates and their relationship to gold mineralization. Geol. Soc. S. Afr. 18th Congr. Abstracts 1:176–186.

Hallbauer, D. K., Warmelo, K. T. 1974. Fossilized plants in thucholite from Precambrian rocks of the Witwatersrand, South Africa. Precambr. Res. 1:199–212.

Hallbauer, D. K., Jahns, H.M., Beltmann, H. A. 1977. Morphological and anatomical observations on some Precambrian plants from the Witwatersrand, South Africa. Geol. Rundsch. 66:477–491.

Hallberg, J. A., Williams, D. A. C. 1972. Archean mafic and ultramafic rock associations in the Eastern Goldfields region, Western Australia. Earth Planet. Sci. Lett. 15:191–200.

Hamilton, P. J. 1975. Isotope and trace element studies of some layered basic intrusions. Unpubl. Ph.D. thesis, Oxford Univ.

Hamilton, P. J. 1977. Sr isotope and trace element studies of the Great Dyke and Bushveld Mafic Phase and their relation to Early Proterozoic magma genesis in southern Africa. Jour. Petrol. 18:24–52.

Hamilton, P. J., Evensen, N. M., O'Nions, R. K., Smith, H. S., Erlank, A. J. 1979. Sm-Nd dating of Onverwacht Group volcanics, southern Africa. Nature 279:298–300.

Hamilton, W. 1970. Bushveld Complex—Product of impacts. Geol. Soc. S. Afr. Spec. Publ. No. 1, pp. 367–374.

Hamilton, W., Krinsley, D. 1967. Upper Paleozoic glacial deposits of South Africa and southern Australia. Geol. Soc. Amer. Bull. 78:783–800.

Hammerbeck, E. C. I., O'Okiep Copper Co., Messina (Transvaal) Development Co., Schoeman, J. J. 1976. Copper. S. Afr. Geol. Surv. Handbook 7:125–146.

Hanekom, H. J. 1966. The crocidolite deposits of the northern Cape Province. Unpubl. D.Sc. thesis, Univ. Pretoria, South Africa.

Hanekom, H. J., Van Staden, C. M. v. H., Smit, P. J., Pike, D. R. 1965. The geology of the Palabora Igneous Complex. S. Afr. Geol. Surv. Mem. No 54.

Hargraves, R. B. 1970. Palaeomagnetic evidence relevant to the origin of the Vredefort ring. Jour. Geol. 78:253–263.

Harmer, R. E. 1979. A tonalite of 'plagiogranite' composition from the Natal Mobile Belt. Geol. Soc. S. Afr. 18th Congr. Abstracts 1:188–192.

Harrington, H. J. 1956. Argentina. *In* W. F. Jenks (ed.), Handbook of South American geology. Geol. Soc. Amer. Mem. 65:129–166.

Harris, J. W., Gurney, J. J. 1979. Inclusions in diamond. *In* J. E. Field (ed.), The properties of diamond, pp. 555–591. London: Academic Press.

Harrison, C. G. A., Barron, E. J., Hay, W. W. 1979. Mesozoic evolution of the Antarctic Peninsula and the southern Andes. Geology 7:374–378.

Harrison, J. E., Peterman, Z. E. 1980. North American Commission on Stratigraphic Nomenclature Note 52–Preliminary proposal for a chronometric time scale for the Precambrian of the United States and Mexico. Geol. Soc. Amer. Bull. 91:377–380.

Hart, G. F. 1964. Where was the Lower Karroo Sea? Scient. S. Afr. 1:289–290.

Harte, B. 1978. Kimberlite nodules, upper mantle petrology, and geotherms. Roy. Soc. London Phil. Trans. A288:487–500.

Harte, B., Gurney, J. J. 1975. Evolution of clinopyroxene and garnet in an ecologite nodule from the Roberts Victor kimberlite pipe, South Africa. *In* L. H. Ahrens, J. B. Dawson, A. R. Duncan, A. J. Erlank (eds.), Physics and chemistry of the earth, vol. 9, pp. 367–388. Oxford: Pergamon.

Hartnady, C. J. H. 1969. Structural analysis of some pre-Cape formations in the Western Province. Precambrian Res. Unit Univ. Cape Town Bull. No. 6.

Hartnady, C. J. H. 1974. An ERTS-1 view of the southwestern part of the Damaran Mobile Belt. Precambrian Res. Unit. Univ. Cape Town Ann. Rept. 12:60.

Hartnady, C. J. H. 1976. The tectonic development of the Naukluft nappe complex. Precambrian Res. Unit Univ. Cape Town Ann. Rept. 13, 33–36.

Hartnady, C. J. H. 1978a. The stratigraphy and structure of the Naukluft nappe complex. Precambrian Res. Unit Univ. Cape Town Ann. Rept. 14/15:163–170.

Hartnady, C. J. H. 1978b. The age of orogenic episodes in the Pan-African mobile belts fringing the Kalahari craton: a review of new evidence. Precambrian Res. Unit Univ. Cape Town Ann. Rept. 14/15:151–162.

Hartnady, C. J. H. 1978c. The structural geology of the Naukluft nappe complex and its relationship to the Damara orogenic belt, SWA/Namibia. Unpubl. Ph.D. thesis, Univ. Cape Town, South Africa.

Hartnady, C. J. H. 1978d. Tectonic evolution of the southeastern part of the Hakos-Auas Mountains zone in the Damara orogenic belt. Precambrian Res. Unit Univ. Cape Town Ann. Rept. 14/15: 171–182.

Hartnady, C. J. H. 1979. Overthrust tectonics, stratigraphic problems and metallogenesis in the Khomas Ridge Province, Damara orogenic belt. Precambrian Res. Unit Univ. Cape Town Ann. Rept. 16:73–89.

Hartnady, C. J. H., Newton, A. R., Theron, J. N. 1974. The stratigraphy and structure of the Malmesbury Group in the southwestern Cape. Precambrian Res. Unit Univ. Cape Town Bull. 15:193–213.

Hartnady, C. J. H., Antrobus, B., Spector, D. 1978. Reconnaissance studies of regional metamorphism in the Malmesbury Group and the Nama Group of southern Namaqualand. Precambrian Res. Unit Univ. Cape Town Ann. Rept. 14/15:204–207.

Hasui, Y., Carneiro, C. D. R., Coimbra, A. W. 1975. The Ribeira folded belt. Rev. Bras. Geoscienc. 5:257–266.

Haughton, S. H. 1924. The fauna and stratigraphy of the Stormberg Series. S. Afr. Mus. Ann. 12:1–323.

Haughton, S. H. 1931.The Late Tertiary and Recent deposits of the west coast of South Africa. Geol. Soc. S. Afr. Trans. 34:19–57.

Haughton, S. H. 1933. The geology of Cape Town and adjoining country. S. Afr. Geol. Surv. Map Explanation Sheet No. 247.

Haughton, S. H. 1953. Gondwanaland and the distribution of early reptiles. Geol. Soc. S. Afr. Trans. 56 (Annex.).

Haughton, S. H. 1961. Review of probable Late-Precambrian glacial period in central and west Africa. Geol. Soc. S. Afr. Trans. 64:73–84.

Haughton, S. H. 1969. Geological history of southern Africa. Johannesburg: Geol. Soc. S. Afr.

Haughton, S. H., Rogers, A. W. 1924. The volcanic rocks south of Zuurberg. Roy. Soc. S. Afr. Trans. 11:235–249.

Haughton, S. H., Krige, L. J., Krige, A. V. 1925. On intraformational folding connected with the glacial bed in the Table Mountain Sandstone. Geol. Soc. S. Afr. Proc. 28:19–25.

Haughton, S. H., Frommurze, H. F., Visser, D. J. L. 1937. The geology of portion of the coastal belt near the Gamtoos Valley, Cape Province. S. Afr. Geol. Surv. Map Explanation Sheets 151N. and 151S. (Gamtoos River).

Hawkesworth, C. J., O'Nions, R. K. 1977. The petrogenesis of some Archaean volcanic rocks from southern Africa. Jour. Petrol. 18:487–520.

Hawkesworth, C. J., Moorbath, S., O'Nions, R. K., Wilson, J. F. 1975. Age relationships between greenstone belts and 'granites' in the Rhodesian Archaean craton. Earth Planet. Sci. Lett. 25:251–262.

Hawkesworth, C. J., Bickle, M. J., Gledhill, A. R., Wilson, J. F., Orpen, J. L. 1979. A 2.9-b.y. event in the Rhodesian Archean. Earth Planet. Sci. Lett. 43:285–297.

Hawkesworth, C. J., Kramers, J. D., Miller, R. McG. 1981. Old model Nd ages in Namibian Pan-African rocks. Nature 289:278–282.

Hawthorne, J. B. 1975. Model of a kimberlite pipe. *In* L. H. Ahrens, J. B. Dawson, A. R. Duncan, A. J. Erlank (eds), Physics and chemistry of the earth, vol. 9, pp. 1–15. Oxford: Pergamon.

Hawthorne, J. B., Carrington, A. J., Clement, C. R., Skinner, E. M. W. 1979. Geology of the Dokolwayo kimberlite and associated palaeo-alluvial diamond deposits. 2nd Intl. Kimberlite Conf. Proc. 1:59–70.

Hayes, M. O. 1975. Morphology of sand accumulation in estuaries: an introduction to the symposium. In L. E. Cronin (ed.), Estuarine research, vol. 2, pp. 3–22. New York: Academic Press.

Hayes, M. O., Anan, F. S., Bozeman, R. N. 1969. Sediment dispersal trends in the littoral zone: a problem in palaeogeographic reconstruction. *In* M. O. Hayes, (ed.), Coastal environments of northeastern Massachusetts and New Hampshire. Coastal Res. Group, Univ. Massachusetts Contr. No. 1, pp. 290–315.

Hays, J. D., Pitman, W. C. 1973. Lithospheric plate motion, sea level changes, and climatic and ecological consequences. Nature 246:18–22.

Hedberg, H. D. 1979. Progress and problems in stratigraphic classification. Geol. Soc. S. Afr. 18th Congr. Abstracts 1:1–3.

Hedberg, R. M. 1979. Stratigraphy of the Ovamboland basin, South West Africa. Precambrian Res. Unit Univ. Cape Town Bull. No. 24.

Heier, K. S. 1973. Geochemistry of granulite facies rocks and problems of their origin. Roy. Soc. Lond. Phil. Trans. A273:429–442.

Hein, F. J., Walker, R. G. 1977. Bar evolution and development of stratification in the gravelly, braided, Kicking Horse River, B. C. Can. Jour. Earth Sci. 14:562–570.

Heine, K. 1978. Radiocarbon chronology of Late Quaternary lakes in the Kalahari, southern Africa. Catena 5:145–149.

Heirtzler, J. R. 1971. The evolution of the southern oceans. *In* L. O. Quam (ed.), Research in the Antarctic, pp. 667–684. Washington, D.C.: Amer. Assoc. Adv. Sci.

Heirtzler, J. R., Dickson, G. O., Herron, E. M., Pitman, W. C., Le Pichon, X. 1968. Marine magnetic anomalies, geomagnetic field reversals, and motions of the ocean floor and continents. Jour. Geophys. Res. 73:2119–2136.

Helgren, D. M. 1977. Geological context of the Vaal River faunas. S. Afr. Jour. Sci. 73:303–307.

Henderson, J. B. 1975. Archean stromatolites in the northern Slave Province, Northwest Territories. Can. Jour. Earth Sci. 12:1619–1630.

Hendey, Q. B. 1973. Fossil occurrences at Langebaanweg, Cape Province. Nature 244:13–14.

Hendey, Q. B. 1974. The Late Cenozoic carnivora of the southwestern Cape Province. S. Afr. Mus. Ann. 63:1–369.

Hendey, Q. B. 1976. The Pliocene fossil occurrences in 'E' Quarry, Langebaanweg, South Africa. S. Afr. Mus. Ann. 69:215–247.

Hendey, Q. B. 1978a. Late Tertiary Mustelidae (Mammalia, Carnivora) from Langebaanweg, South Africa. S. Afr. Mus. Ann. 76:329–357.

Hendey, Q. B. 1978b. The age of the fossils from Baard's Quarry, Langebaanweg, South Africa. S. Afr. Mus. Ann. 75:1–24.

Hendey, Q. B. 1981. Palaeoecology of the Late Tertiary fossil occurrences in 'E' Quarry, Langebaanweg, South Africa, and a reinterpretation of their geological context. S. Afr. Mus. Ann. (In press).

Henley, R. W., Adams, J. 1979. On the evolution of giant gold placers. Inst. Min. Metall. Trans. 88:B41–B50.

Hepworth, J. V. 1973. A photogeological investigation of the Ancient Gneiss Complex and the Mozaan of Swaziland. Inst. Geol. Sciences (Overseas Division) London Rept. 20.

Herrmann, A. G., Blanchard, D. P., Haskin, L. A., Jacobs, J. W., Knake, D., Korotev, R. L., Brannon, J. C. 1976. Major, minor and trace element compositions of peridotitic and basaltic komatiites from the Precambrian crust of southern Africa. Contr. Mineral. Petrol. 59:1–12.

Hess, H. H. 1960. The Stillwater igneous complex: a comparative mineralogical study. Geol. Soc. Amer. Mem. No. 80.

Hesse, R. 1974. Long-distance continuity of turbidites: possible evidence for an Early Cretaceous trench-abyssal plain in the east Alps. Geol. Soc. Amer. Bull. 85:859–870.

Hesse, R. 1975. Turbiditic and non-turbiditic mudstone of Cretaceous flysch sections of the east Alps and other basins. Sedimentology 22:387–416.

Hewins, R. H. 1975. Pyroxene geothermometry of some granulite facies rocks. Contr. Mineral. Petrol. 50:205–209.

Hickman, M. H. 1974. 3500 m.y.-old granite in southern Africa. Nature 251:295–296.

Hickman, M. H. 1976. Geochronological investigations in the Limpopo Belt and parts of the adjacent Rhodesia craton. Unpubl. Ph.D. thesis, Univ. Leeds, England.

Hickman, M. H. 1978. Isotopic evidence for crustal reworking in the Rhodesian Archean craton, southern Africa. Geology 6:214–216.

Hickman, M. H., Wakefield, J. 1975. Tectonic implications of new geochronologic data from the Limpopo Belt at Pikwe, Botswana, southern Africa. Geol. Soc. Amer. Bull. 86:1468–1472.

Hill, R., Roeder, P. 1974. The crystallization of spinel from basaltic liquid as a function of oxygen fugacity. Jour. Geol. 82:709–729.

Hill, R. S. 1972. The geology of the northern Algoa basin, Port Elizabeth. Unpubl. M. Sc. thesis, Univ. Stellenbosch, South Africa.

Hiller, N., Dunlevey, J. N. 1978. The Bokkeveld-Witteberg boundary in the Montagu - Touws River area, Cape Province. Geol. Soc. S. Afr. Trans. 81:101–104.

Hiller, N., Stavrakis, N. 1979. Distal alluvial fan deposits in the Beaufort Group of the eastern Cape. Geol. Soc. S. Afr. 18th Congr. Abstract 2:113–120.

Hobbs, B. E., Means, W. D., Williams, P. F. 1976. An outline of structural geology. New York: Wiley.

Hobday, D. K. 1965. The geomorphology of the Lake St. Lucia area. Unpubl. M.Sc. thesis, Univ. Natal, Durban.

Hobday, D. K. 1973. Middle Ecca deltaic deposits in the Muden-Tugela Ferry area of Natal. Geol. Soc. S. Afr. Trans. 76:309–318.

Hobday, D. K. 1976. Quaternary sedimentation and development of the lagoonal complex, Lake St. Lucia, Zululand. S. Afr. Mus. Ann. 71:93–113.

Hobday, D. K. 1977. Late Quaternary sedimentary history of Inhaca Island, Mozambique. Geol. Soc. S. Afr. Trans. 80:183–191.

Hobday, D. K. 1978a. Fluvial deposits of the Ecca and Beaufort Groups in the eastern Karoo basin, South Africa. *In* A. D. Miall (ed.), Fluvial sedimentology. Can. Soc. Petrol. Geol. Mem. No. 5, pp. 413–429.

Hobday, D. K. 1978b. Palaeoenvironmental models in the eastern Karoo basin. Palaeont. Afr. 21:1–14.

Hobday, D. K., Jackson, M. P. A. 1979. Transgressive shorezone sedimentation and syndepositional deformation in the Pleistocene of Zululand, South Africa. Jour. Sediment. Petrol. 49:145–158.

Hobday, D. K., Mathew, D. 1974. Depositional environment of the Cape Supergroup in the Transkei. Geol. Soc. S. Afr. Trans. 77:223–227.

Hobday, D. K., Mathew, D. 1975. Late Paleozoic fluvial and deltaic deposits in the northeast Karoo basin. *In* M. S. L. Broussard (ed.), Deltas: Models for exploration, pp. 457–469. Houston: Houston Geol. Soc.

Hobday, D. K., Orme, A. R. 1974. The Port Durnford Formation. A major Pleistocene barrier lagoon complex along the Zululand coast. Geol. Soc. S. Afr. Trans. 77:141–149.

Hobday, D. K., Tankard, A. J. 1978. Transgressive barrier and shallow-shelf interpretation of the lower Paleozoic Peninsula Formation, South Africa.

Geol. Soc. Amer. Bull. 89:1733–1744.

Hobday, D. K., Tavener-Smith, R. 1975. Trace fossils in the Ecca Group of northern Natal and their paleoenvironmental significance. Palaeont. Afr. 18: 47–52.

Hobday, D. K., Von Brunn, V. 1979. Fluvial sedimentation and paleogeography of an Early Paleozoic failed rift, southeastern margin of Africa. Palaeogeogr. Palaeoclim. Palaeoecol. 28:169–184.

Hobday, D. K., Tavener-Smith, R., Mathew, D. 1975. Markov analysis and palaeoenvironments in the Ecca Group near Vryheid, Natal. Geol. Soc. S. Afr. Trans. 78:75–82.

Hobday, D. K., Tavener-Smith, R., Greenshields, H. D. 1980. A major fluviatile complex in the middle Beaufort of the Transkei, southern Africa. 4th Intl. Gondwana Symposium. Proceedings and Papers, pp. 522–562.

Hoffer, E. 1975. Die zeitliche Beziehung zwischen Tektogenese und Metamorphose: ein Beitrag zur Rekonstruktion der thermischen Geschichte des Damara-Orogens. Bericht des Sonderforschungsbereichs Univ. Gottingen 48: 170–183.

Hoffer, E., Puhan, D. 1975. Isoreaktionsgrade in den Metapeliten des Damara-Orogens. Bericht des Sonderforschungsbereichs Univ. Gottingen 48: 111–151.

Hoffman, P. 1973. Evolution of an Early Proterozoic continental margin: The Coronation geosyncline and associated aulacogens of the northwestern Canadian Shield. Roy. Soc. London Phil. Trans. A273:547–581.

Hoffman, P., Dewey, J. F., Burke, K. 1974. Aulacogens and their genetic relation to geosynclines with a Proterozoic example from Great Slave Lake, Canada. *In* R. H. Dott, R. H. Shaver, (eds.), Modern and ancient geosynclinal sedimentation. Soc. Econ. Paleont. Mineral. Spec. Publ. No. 19, pp. 38–55.

Hoffman, P. F. 1967. Algal stromatolites: Use in stratigraphic correlation and paleocurrent determination. Science 157:1043–1045.

Hofmeyr, P. K. 1971. The abundance and distribution of some trace elements in some selected South African shales. Unpubl. Ph.D. thesis, Univ. Cape Town, South Africa.

Holland, H. D. 1973. The oceans: A possible source of iron in iron-formations. Econ. Geol. 68:1169–1172.

Holmes, A. 1950. Petrogenesis of katungite and its associates. Amer. Mineral. 35:772–779.

Holmes, A., Cohen, L. 1956. Geochronologie africaine. Acad. Roy. Sci. Coloniales, Mem. in 8°, 5(1).

Hopwood, A. T. 1929. New and little-known mammals from the Miocene of Africa. Amer. Mus. Novit. 344:1–19.

Hor, A. K., Hutt, D. K., Smith, J. V., Wakefield, J., Windley, B. F. 1975. Petrochemistry and mineralogy of Early Precambrian anorthositic rocks of the Limpopo Mobile Belt, southern Africa. Lithos 8:297–310.

Horne, J. C., Ferm, J. C. 1976. Carboniferous depositional environments of the Pocahontas basin, eastern Kentucky and southern West Virginia. Field trip guide, Dept. Geology, Univ. South Carolina.

Horodyski, R. J., Vonder Haar, S. P. 1975. Recent calcareous stromatolites from Laguna Mormona (Baja California), Mexico. Jour. Sediment. Petrol. 45:894–906.

Horowitz, A. 1976. Environment of deposition and stratigraphy of the uranium-bearing strata around Beaufort West, South Africa. Pelindaba: S. Afr. Atomic Energy Board.

Horrocks, P. C. 1980. Ancient Archean supracrustal rocks from the Limpopo Mobile Belt. Nature 286: 596–599.

Ho-Tun, E. 1979. Volcaniclastic material in the lower Beaufort Group Karoo rocks. Geol. Soc. S. Afr. 18th Congr. Abstracts 1:197–199.

Houbolt, J. J. H. C. 1968. Recent sediments in the Southern Bight of the North Sea. Geol. Mijnbouw 47:245–273.

House, M. R. 1979. Devonian in the eastern hemisphere. *In* R. A. Robison, C. Teichert, (eds.), Treatise on invertebrate paleontology. A. Introduction, pp. A183–A217. Boulder: Geol. Soc. Amer. and Univ. Kansas.

Howard, J. D., Reineck, H.-E. 1972. Georgia coastal region, Sapelo Island, U.S.A.: Sedimentology and biology. IV. Physical and biogenic sedimentary structures of the nearshore shelf. Senckenb. Marit. 4:81–123.

Hubert, J. F., Reed, A. A. 1978. Red bed diagenesis in the East Berlin Formation, Newark Group, Connecticut Valley. Jour. Sediment. Petrol. 48:175–194.

Hughs, A. R., Tobias, P. V. 1977. A fossil skull probably of the genus *Homo* from Sterkfontein, Transvaal. Nature 265:310–312.

Hugo, P., Schalk, K. E. 1975. The isotopic ages of certain granites and acid lavas in the Rehoboth and Maltahöhe districts, South West Africa. S. Afr. Geol. Surv. Ann. 9:103–105.

Hunter, D. R. 1950. Usushwana Complex. Swaziland Geol. Surv. Annual Rept. for 1950; 40–41.

Hunter, D. R. 1968. The Precambrian terrain in Swaziland with particular reference to the granitic rocks. Unpubl. Ph.D. thesis, Univ. Witwatersrand, Johannesburg, South Africa.

Hunter, D. R. 1970. The Ancient Gneiss Complex in Swaziland. Geol. Soc. S. Afr. Trans. 73:107–150.

Hunter, D. R. 1973. Geochemistry of granitic and associated rocks in the Kaapvaal craton. Econ. Geol. Res. Unit Univ. Witwatersrand Inf. Circ. No. 81.

Hunter, D. R. 1974a. Crustal development in the Kaapvaal craton, I: The Archean. Precambr. Res. 1:259–294.

Hunter, D. R. 1974b. Crustal development in the Kaapvaal craton, II: The Proterozoic. Precambr. Res. 1:295–326.

Hunter, D. R. 1975. The regional setting of the Bushveld Complex. Johannesburg: Econ. Geol. Res. Unit Univ. Witwatersrand.

Hunter, D. R., Hamilton, P. J. 1978. The Bushveld Complex. *In* D. H. Tarling (ed.), Evolution of the earth's crust, pp. 107–173. London: Academic Press.

Hunter, D. R., Barker, F., Millard, H. T. 1978. The geochemical nature of the Archean Ancient Gneiss Complex and Granodiorite Suite, Swaziland: A preliminary study. Precambr. Res. 7:105–127.

Hutchison, R. I. 1975. The Witwatersrand system as a model of sedimentation in an intracratonic basin. Unpubl. D.Sc. thesis, Univ. Orange Free State, Bloemfontein.

Hutton, D. H. W. 1979. Tectonic slides: A review and reappraisal. Earth Sci. Reviews 15:151–172.

Hurley, P. M., Almeida, F. F. M. de, Melcher, G. C., Cordani, U. G., Rand, J. R., Kawashita, K., Vandoros, P., Pinson, W. H., Fairbairn, H. W. 1967. Test of continental drift by comparison of radiometric ages. Science 157:495–500.

Ianello, P. 1971. The Bushveld granites around Rooiberg, Transvaal, South Africa. Geol. Rundsch. 60:630–655.

Ingersoll, R. V. 1978. Submarine fan facies of the Upper Cretaceous Great Valley Sequence, northern and central California. Sediment. Geol. 21: 205–230.

Irvine, T. N. 1974. Petrology of the Duke Island ultramafic complex, southeastern Alaska. Geol. Soc. Amer. Mem. No. 138.

Irvine, T. N. 1975. Crystallization sequences in the Muskox intrusion. II. Origin of chromite layers and similar deposits of other magmatic ores. Geochim. Cosmochim. Acta 39:991–1020.

Irvine, T. N. 1977. Origin of chromitite layers in the Muskox intrusion and other stratiform intrusions: A new interpretation. Geology 5:273–277.

Irvine, T. N., Baragar, W. R. A. 1971. A guide to the chemical classification of the common volcanic rocks. Can. J. Earth. Sci., 8:523–548.

Irwin, M. L. 1965. General theory of epeiric clear water sedimentation. Amer. Assoc. Petrol. Geol. Bull. 49:445–459.

Isaacson, P. E. 1975. Evidence for a western extracontinental land source during the Devonian period in the central Andes. Geol. Soc. Amer. Bull. 86:39–46.

Isaacson, P. E. 1977. Devonian stratigraphy and brachiopod paleontology of Bolivia. Part A. Orthida and Strophomenida. Palaeontogr. Beitr. Naturgeschichte der Vorzeit No. A155.

Isaacson, P. E. 1978. Paleolatitudes in the Devonian of Brazil and the Frasnian-Famennian mass extinctions—comments. Palaeogeogr. Palaeoclim. Palaeoecol. 24:359–365.

Jack, A. M. 1980. The geology of western Namaqualand. Precambrian Res. Unit Univ. Cape Town Bull. 29.

Jackson, M. P. A. 1974a. Migmatization and metamorphism in a terrain of high-grade gneisses, Lüderitz District, South West Africa. Precambrian Res. Unit Univ. Cape Town Ann. Rept. 10/11:46–52.

Jackson, M. P. A. 1974b. Report on reconnaissance trip to Tschaukaib Mountains. Precambrian Res. Unit Univ. Cape Town Ann. Rept. 10/11:53–54.

Jackson, M. P. A. 1975. Field relations between key lithological units of the Namaqua Metamorphic Complex and the Naisib River Complex in the eastern Lüderitz District. Precambrian Res. Unit Univ. Cape Town Ann. Rept. 12:33–40.

Jackson, M. P. A. 1976. High-grade metamorphism and migmatization of the Namaqua Metamorphic Complex around Aus in the southern Namib desert, South West Africa. Precambrian Res. Unit Univ. Cape Town Bull. No. 18.

Jackson, M. P. A. 1979a. High-strain deformation history of the Ancient Gneiss Complex in the Mankayane area, Swaziland: a preliminary account. Geol. Soc. S. Afr. 18th Congr. Abstracts 1:210–216.

Jackson, M. P. A. 1979b. A major charnockite-granolite province in southwestern Africa. Geology 7: 22–26.

Jackson, M. P. A. 1980. Early-Precambrian gneiss tectonics in Swaziland and implications for greenstone-belt development. Geol. Soc. Amer. 1980. Annual Meeting Abstracts.

Jackson, M. P. A., Clarke, M. C. G. (In preparation). Structural geology of Inanje Hill, type area of the Archaean Mkhondo Valley Metamorphic Suite, Swaziland.

Jackson, M. P. A., Hobday, D. K. 1980. Gravity gliding and clay diapirism in a Pleistocene shoreline sequence in Zululand, South Africa. Amer. Jour. Sci. 280:333–362.

Jackson, M. P. A., Robertson, D. I. 1982. Regional implications of Early-Precambrian strains in the Onverwacht Group adjacent to the Lochiel granite, north-west Swaziland. Barberton Volume, S. African Geodynamics Program. Geol. Soc. S. Afr. Spec. Publ. (in press).

Jackson, M. P. A., Zelt, G. A. D. (In preparation). Crustal reworking and deformation of metabasite dikes, layered intrusions, and lavas in Namaqualand, South Africa.

Jackson, N. J., Ramsay, C. R. 1980. What is the "Pan-African"? A consensus is needed. Geology 8:210–211.

Jacob, R. E. 1974. Geology and metamorphic petrology of the southern margin of the Damara Orogen around Omitara, southwest Africa. Precambrian Res. Unit Univ. Cape Town Bull. No. 17.

Jacob, R. E. 1978. Granite genesis and associated mineralization in part of the central Damara belt. Geol. Soc. S. Afr. Spec. Publ. No. 4, pp. 417–432.

Jacob, R. E., Kröner, A., Burger, A. J. 1978. Areal extent and first U-Pb age of the Pre-Damaran Abbabis Complex in the central Damara belt of South West Africa (Namibia). Geol. Rundsch 67:706–718.

Jacoby, W. R. 1972. Plate theory, epeirogenesis and eustatic sea-level changes. Tectonophysics 15:187–196.

Jahn, B. M., Shih, C. 1974. On the age of the Onverwacht Group, Swaziland Sequence, South Africa. Geochim. Cosmochim. Acta 38:873–885.

Jahn, B. M., Shih, C. Y., Murthy, V. R. 1974. Trace element geochemistry of Archean volcanic rocks. Geochim. Cosmochim. Acta 38:611–627.

Jakeš, P., White, A. J. R. 1972. Major and trace element abundances in volcanic rocks or orogenic areas. Geol. Soc. Amer. Bull. 83:29–40.

James, P. R. 1975. A deformation study across the northern margin of the Limpopo Belt, Rhodesia. Unpubl. Ph.D. thesis, Univ. Leeds, England.

Janse, A. J. A. 1969. Gross Brukkaros, a probable carbonatite volcano in the Nama Plateau of South-West Africa. Geol. Soc. Amer. Bull. 83:573–586.

Janse, A. J. A. 1975. Kimberlite and related rocks from the Nama Plateau of South-west Africa. *In* L. H. Ahrens, J. B. Dawson, A. R. Duncan, A. J. Erlank (eds.), Physics and chemistry of the earth, vol. 9, pp. 81–94. Oxford: Pergamon.

Jansen, H. 1960. The geology of the Bitterfontein area, Cape Province. S. Afr. Geol. Surv. Map Explanation Sheet 253 (Bitterfontein).

Jansen, H. 1970. Volcanic rocks and associated sediments in the southern portion of the Waterberg basin. S. Afr. Geol. Surv. Ann. 8:53–61.

Jansen, H. 1975. The Soutpansberg trough (northern Transvaal)—An aulacogen. Geol. Soc. S. Afr. Trans. 78:129–136.

Jansen, H. 1976. The Waterberg and Soutpansberg Groups in the Blouberg area, northern Transvaal. Geol. Soc. S. Afr. Trans. 79:281–291.

Jansen, H. 1977. Limpopo belt - Transvaal craton relationships. Botswana Geol. Surv. Bull. 12:99–104.

Johnson, J. G., Boucot, A. J. 1973. Devonian brachiopods. *In* A. Hallam (ed.), Atlas of palaeobiogeography, pp. 89–96. Amsterdam: Elsevier.

Johnson, M. R. 1976. Stratigraphy and sedimentology of the Cape and Karoo sequences in the eastern Cape Province. Unpubl. Ph.D. thesis, Rhodes University, Grahamstown, South Africa.

Joliffe, A. W. 1965. Geology and iron ores of Steep Rock Lake. Econ. Geol. 68:913–1179.

Jones, H. A. 1973. Marine geology of the northwest Australian continental shelf. Bur. Min. Res. Geol. Geophys. Bull. No. 136.

Jopling, A. V. 1965. Laboratory study of sorting processes in cross-bedded deposits. Soc. Econ. Paleont. Mineral. Spec. Publ. No. 12, pp. 53–65.

Jopling, A. V., Walker, R. G. 1968. Morphology and origin of ripple-drift cross-lamination with examples from the Pleistocene of Massachusetts. Jour. Sediment. Petrol. 38:971–984.

Joubert, P. 1971. The regional tectonism of the gneisses of part of Namaqualand. Precambrian Res. Unit Univ. Cape Town Bull. No. 10.

Joubert, P. 1973. Geological maps of the Aggeneys and Pofadder area. Precambrian Res. Unit Univ. Cape Town (unpublished).

Joubert, P. 1974a. The gneisses of Namaqualand and their deformation. Geol. Soc. S. Afr. Trans. 77: 339–345.

Joubert, P. 1974b. Wrench-fault tectonics in the Namaqualand Metamorphic Complex. Precambrian Res. Unit Univ. Cape Town Bull. 15:17–25.

Joubert, P. 1976. The relationship between the Namaqualand Metamorphic Complex and the Kheis Group. S. Afr. Jour. Sci. 72:312–313.

Joubert, P., Kröner, A. 1972. The Stinkfontein Formation south of the Richtersveld. Geol. Soc. S. Afr. Trans. 75:47–54.

Jubb, R. A. 1965. A new palaeoniscid fish from the Witteberg Series (Lower Carboniferous) of South Africa. S. Afr. Mus. Ann. 48:267–272.

Jubb, R. A., Gardiner, B. G. 1975. A preliminary catalog of identifiable fossil fish material from South Africa. S. Afr. Mus. Ann. 67:381–440.

Kaiser, E. 1926. Die Diamantenwüste Südwestafrikas. 2 vol. Berlin: Dietrich Reimer.

Kartun, K. G., Moore, A. C. 1979. Geochemistry and genetic relationships of gabbroic and related rocks in the Tantalite Valley and Kumkum areas, south of Warmbad, S.W.A./Namibia. Geol. Soc. S. Afr. 18th Congr. Abstracts 1:223–226.

Kasch, K. W. 1976. A preliminary report on the geology of the southern margin of the Damara orogen around Omitara, South West Africa. Precambrian Res. Unit Univ. Cape Town Ann. Rept. 13:102–138.

Kasch, K. W. 1978a. Some structural and metamorphic aspects of the Damara orogenic belt around Omitara, S.W.A./Namibia. Precambrian Res. Unit Univ. Cape Town Ann. Rept. 14/15:194–203.

Kasch, K. W. 1978b. Recognition of a major shear zone in the southern Damara belt, S.W.A./Namibia. Geol. Soc. S. Afr. Trans. 81:205–209.

Kasch, K. W. 1979a. Metamorphic history of the southern Damara belt around Omitara (S.W.A./ Namibia). Precambrian Res. Unit Univ. Cape Town Ann. Rept. 16:90–100.

Kasch, K. W. 1979b. A continental collision model for the tectonothermal evolution of the (southerr Damara belt. Precambrian Res. Unit Univ. Cape Town Ann. Rept. 16:101–107.

Kasch, K. W. 1980. The structural geology, metamorphic petrology and tectono-thermal evolution of the southern Damara belt around Omitara, S.W.A./ Namibia. Unpubl. Ph.D. thesis, Univ. Cape Town.

Kauffman, E. G. 1973. Stratigraphic evidence of Cretaceous eustatic changes. Geol. Soc. Amer. Abstracts with programs 5:687.

Kazakov, A. V. 1937. The phosphorite facies and the genesis of phosphorites. Scient. Inst. Fertil. Insectofungic. Trans. 142:95–113.

Kennedy, W. J. 1975. Trace fossils in carbonate rocks. *In* R. W. Frey (ed.), The study of trace fossils, pp. 377–398. New York: Springer-Verlag.

Kennedy, W. J., Juignet, P. 1974. Carbonate banks and slump beds in the Upper Cretaceous (Upper Turonian-Santonian) of Haute Normandy, France. Sedimentology 21:1–42.

Kennedy, W. J., Klinger, H. C. 1972. Hiatus concretions and hardground horizons in the Cretaceous of Zululand. Palaeontology 15:539–549.

Kennedy, W. J., Klinger, H. C. 1975. Cretaceous faunas from Zululand and Natal, South Africa. Introduction, Stratigraphy. Brit. Mus. (Nat. Hist). Bull. Geol. No. 25.

Kennedy, W. J., Kauffman, E. G., Klinger, H. C. 1973. Upper Cretaceous invertebrate faunas from Durban, South Africa. Geol. Soc. S. Afr. Trans. 76:95–111.

Kennedy, W. Q. 1964. The structural differentiation of Africa in the Pan-African (± 500 m.y.) tectonic episode. Res. Inst. African Geol. Univ. Leeds Ann. Rept. 8:48–49.

Kennett, J. P., Houtz, R. E., Andrews, P. B., Edwards, A. R., Gostin, V. A., Hajos, M., Hampton, M., Jenkins, D. G., Margolis, S. V., Ovenshine, A. T., Perch-Nielsen, K. 1975. Cenozoic paleooceanography in the southwest Pacific Ocean, Antarctic glaciation, and the development of the circum-Antarctic Current. Initial reports of the Deep Sea Drilling Project, 29:1155–1169. Washington, D.C.: U.S. Govt. Printing Office.

Kensley, B. F. 1972. Pliocene marine invertebrates from Langebaanweg, Cape Province. S. Afr. Mus. Ann. 60:173–190.

Kensley, B. F. 1977. A second assemblage of Pliocene invertebrate fossils from Langebaanweg, Cape. S. Afr. Mus. Ann. 72:189–210.

Kent, L. E., Hugo, P. J. 1978. Aspects of the revised South African stratigraphic classification and a proposal for the chronostratigraphic subdivision of the Precambrian. Amer. Assoc. Petrol. Geol., Studies in Geology 6:367–379.

Kent, P. E. 1973. East African evidence of the paleoposition of Madagascar. *In* D. H. Tarling, S. K. Runcorn (eds.), Implications of continental drift to the earth sciences, vol. 2, pp. 873–875. London: Academic Press.

Kent, P. E. 1974. Continental margin of East Africa—A region of vertical movements. *In* C. A. Burk, C. L. Drake (eds.), The geology of continental margins, pp. 313–320. New York: Springer-Verlag.

Kent, P. E., Hunt, J. A., Johnstone, D. W. 1971. The geology and geophysics of coastal Tanzania. Inst. Geol. Sci. Geophys. Pap. No. 6.

Key, R. M. 1977. The geological history of the Limpopo Mobile Belt based on the field mapping of the Botswana Geological Survey. Botswana Geol. Surv. Bull. 12:41–59.

Key, R. M., Hutton, S. M. 1976. The tectonic generation of the Limpopo Mobile Belt, and a definition of its western extremity. Precambr. Res. 3:79–90.

Key, R. M., Litherland, M., Hepworth, J. V. 1976. The evolution of the Archaean crust of northeast Botswana. Precambr. Res. 3:375–413.

Keyser, A. W. 1966. Some indications of arid climate during the deposition of the Beaufort Series. S. Afr. Geol. Surv. Ann. 5:77–79.

Keyser, A. W. 1970. Some ecological aspects of the Cistecephalus Zone of the Beaufort Series of South Africa. 2nd Gondwana Symposium. Proceedings and Papers, pp. 687–689.

Keyser, A. W., Smith, R. M. H. 1978. Vertebrate zonation of the Beaufort Group with special reference to the western Karoo basin. S. Afr. Geol. Surv. Rept. No. 36.

Keyser, U. 1972. The occurrence of diamonds along the coast between the Orange River estuary and the Port Nolloth reserve. S. Afr. Geol. Surv. Bull. No. 54.

King, L. C. 1953. A marine Miocene fauna from Zululand. Geol. Soc. S. Afr. Trans. 56:59–91.

King, L. C. 1962. Morphology of the earth. Edinburgh: Oliver and Boyd.

King, L. C. 1966. An extensive marine Pliocene formation in Natal and Mozambique: Its geomorphological implications. Geol. Soc. S. Afr. Trans. 69:201–210.

King, L. C. 1967. South African scenery (3rd ed.). Edinburgh: Oliver and Boyd.

King, L. C. 1972. Pliocene marine fossils from the Alexandria Formation in the Paterson district, eastern Cape Province, and their geomorphic significance. Geol. Soc. S. Afr. Trans. 75:159–160.

Kingsley, C. S. 1975. A new stratigraphic classification implying a lithofacies change in the Table Mountain Sandstone in southern Natal. Geol. Soc. S. Afr. Trans. 78:43–55.

Kingsley, C. S. 1979. A composite turbidite - delta - fluvial model for Permian sedimentation in the Ecca and lower Beaufort of the eastern Cape Province: a new link with Gondwanaland. Geol. Soc. S. Afr. 18th Congr. Abstracts 1:96–112.

Kitchin, F. L. 1907. Notes on the invertebrate fauna of the Uitenhage Series in Cape Colony. Geol. Mag. 54:289–295.

Kitching, J. W. 1970. A short review of the Beaufort zoning in South Africa. 2nd Gondwana Symposium Proceedings and Papers, pp. 309–312.

Kitching, J. W. 1977. The distribution of Karroo vertebrate fauna. Bernard Price Inst. Palaeont. Res. Mem. No. 1.

Klein, G. de V. 1970. Depositional and dispersal dynamics of intertidal sand bars. Jour. Sediment. Petrol. 40:1095–1127.

Klein, G. de V. 1971. A sedimentary model for determining paleotidal range. Geol. Soc. Amer. Bull. 82:2585–2592.

Klein, G. de V. 1972. Determination of paleotidal range in clastic sedimentary rocks. Int. Geol. Congr. Proc. 24:397–405.

Klein, G. de V., Ryer, T. A. 1978. Tidal circulation patterns in Precambrian, Paleozoic, and Cretaceous epeiric and mioclinal shelf seas. Geol. Soc. Amer. Bull. 89:1050–1058.

Klein, R. G. 1974. Environment and subsistence of prehistoric man in the southern Cape Province, South Africa. World Archaeol. 5:249–284.

Klein, R. G. 1975. Palaeoanthropological implications of the nonarchaeological bone assemblage from Swartklip I, south-western Cape Province, South Africa. Quat. Res. 5:275–288.

Kleywegt, R. J. 1971. Interpretation of a gravity profile across the Oudtshoorn Cretaceous basin. S. Afr. Geol. Surv. Ann. 9:125–128.

Klinger, H. C. 1977. Cretaceous deposits near Bogenfels, South West Africa. S. Afr. Mus. Ann. 73:81–92.

Klinger, H. C., Kennedy, W. J. 1977. Upper Cretaceous ammonites from a borehole near Richards Bay, South Africa. S. Afr. Mus. Ann. 72:69–107.

Klinger, H. C., Lock, B. E. 1978. Upper Cretaceous sediments from the Igoda River mouth, East London, South Africa. S. Afr. Mus. Ann. 77:71–83.

Klinger, H. C., Kennedy, W. J., Siesser, W. G. 1976. *Yabeiceras* (Coniacian ammonite) from the Alphard Group off the southern Cape coast. S. Afr. Mus. Ann. 69:161–168.

Knetsch, G. 1937. Übersicht über die Geologie des Südlichen Lüderitzlandes. Geol. Rundsch. 28:208–223.

Knowles, A. G. 1967. A palaeocurrent study of the Ventersdorp Contact Reef at Western Deep Levels Limited on the Far West Rand. Unpubl. M.Sc. thesis, Univ. Witwatersrand, Johannesburg, South Africa.

Koeppel, V. 1978. Lead isotope studies of stratiform ore deposits of Namaqualand, northwest Cape Province, South Africa, and their implications on the age of the Bushmanland Supergroup. *In* R. E. Zartman (ed.), Short papers of the 4th Intl. Conf. Geochronology, Cosmochronology, Isotope Geology, 1978. U.S. Geol. Surv. Open-file Rept. 78-701, pp. 223–226.

Korn, H., Martin, H. 1957. The Pleistocene in South West Africa. *In* J.D. Clark, S. Cole (eds.), Third Pan-African congress on prehistory, Livingstone, 1955, pp. 14–22. London: Chatto and Windus.

Korn, H., Martin, H. 1959. Gravity tectonics in the Naukluft Mountains of South West Africa. Geol. Soc. Amer. Bull. 70:1047–1078.

Koster Van Groos, A., Wyllie, P. J. 1973. Liquid immiscibility in the join $NaAlSi_3O_8$-$CaAl_2Si_2O_8$-Na_2CO_3 - H_2O. Amer. Jour. Sci. 273:465–487.

Köstlin, E. C. 1974. The Kunene Basic Complex, northern South West Africa. Precambrian Res. Unit Univ. Cape Town Bull. 15:123–135.

Krahmann, R. 1936. The geophysical magnetometric investigations on west Witwatersrand areas between Randfontein and Potchefstroom, Transvaal. Geol. Soc. S. Afr. Trans. 39:1–44.

Kramers, J. D. 1977. Lead and strontium isotopes in Cretaceous kimberlites and mantle-derived xenoliths from southern Africa. Earth Planet. Sci. Lett. 34:419–431.

Kramers, J. D. 1979. Lead, uranium, strontium, potassium, and rubidium in inclusion-bearing diamonds and mantle derived xenoliths from southern Africa. Earth Planet. Sci. Lett. 42:58–70.

Krapez, B. 1980. A sedimentological study of the Ventersdorp Contact Reef at East Driefontein gold mine, Carletonville, South Africa. Unpubl. M.SC. thesis, Univ. Witwatersrand, Johannesburg, South Africa.

Krauskopf, K. B. 1967. Introduction to geochemistry. New York: McGraw-Hill.

Kröner, A. 1968. The gneiss-sediment relationships north-west of Vanrhynsdorp, Cape Province. Precambrian Res. Unit Univ. Cape Town Bull. No. 3.

Kröner, A. 1971. Late Precambrian correlation and the relationship between the Damara and Nama systems of South West Africa. Geol. Rundsch. 60: 1513–1523.

Kröner, A. 1972. Preliminary resume of the geology of the Gariep geosyncline in the Orange River area. Precambrian Res. Unit Univ. Cape Town Ann. Rept. 7/9:51–57.

Kröner, A. 1973a. Metasomatism and granitization in part of the Namaqualand granite-gneiss complex–A comparison between petrographic evidence and geochemical data. Geol. Soc. S. Afr. Spec. Publ. No. 3, pp. 411–417.

Kröner, A. 1973b. Comment on "Is the African plate stationery?" Nature 243:29.

Kröner, A. 1974a. The Gariep Group, Part I: Late-Precambrian formations in the western Richtersveld, northern Cape Province. Precambrian Res. Unit Univ. Cape Town Bull. No. 13.

Kröner, A. 1974b. Reports on research activities. Precambrian Res. Unit Univ. Cape Town Ann. Rept. 10/11:73–76.

Kröner, A. 1975a. Geochronology. Precambrian Res. Unit Univ. Cape Town Ann. Rept. 12:56–58.

Kröner, A. 1975b. Late Precambrian formations in the western Richtersveld, northern Cape Province. Roy. Soc. S. Afr. Trans. 41:375–433.

Kröner, A. 1976. Geochronology. Precambrian Res. Unit Univ. Cape Town Ann. Rept. 13:139–143.

Kröner, A. 1977. Precambrian mobile belts of southern and eastern Africa—Ancient sutures or sites of ensialic mobility? A case for crustal evolution towards plate tectonics. Tectonophysics 40:101–135.

Kröner, A. 1978a. The Namaqua Mobile Belt within the framework of Precambrian crustal evolution in southern Africa. Geol. Soc. S. Afr. Spec. Publ. No. 4, pp. 181–188.

Kröner, A. 1978b. Geochronology and some observations on the geology of the Pan-African orogenic belts. Precambrian Res. Unit Univ. Cape Town Ann. Rept. 14/15:208–210.

Kröner, A. 1979. Pan African plate tectonics and its repercussions on the crust of northeast Africa. Geol. Rundsch. 68:565–583.

Kröner, A. 1980a. Chronologic evolution of the Pan African Damara belt in Namibia, south western Africa. *In* H. Closs, *et al.* (eds.), Mobile earth: Final report of the geodynamics project, Federal Republic of Germany, pp. 221–224. Boppard: Harald Boldt Verlag.

Kröner, A. 1980b. Pan African crustal evolution. Episodes, International Union Geological Sciences, 1980, 2:3–8.

Kröner, A., Blignault, H. J. 1976. Towards a definition of some tectonic and igneous provinces in western South Africa and southern South West Africa. Geol. Soc. S. Afr. Trans. 79:232–238.

Kröner, A., Clauer, N. 1979. Isotopic dating of low-grade metamorphic shales in northern Namibia (South West Africa) and implications for the orogenic evolution of the Pan-African Damara belt. Precambr. Res. 10:59–72.

Kröner, A., Correia, H. 1980. Continuation of the Pan African Damara belt into Angola: A proposed correlation of the Chela Group in southern Angola with the Nosib Group in northern Namibia/S.W.A. Geol. Soc. S. Afr. Trans. 83:5–16.

Kröner, A., Germs, G. J. B. 1971. A re-interpretation of the Numees - Nama contact at Aussenkjer, South West Africa. Geol. Soc. S. Afr. Trans. 74:69–74.

Kröner, A., Jackson, M. P. A. 1974. Geological reconnaissance of the coast between Lüderitz and Marble Point, South West Africa. Precambrian Res. Unit Univ. Cape Town Bull. 15:79–103.

Kröner, A., Rankama, K. 1972. Late Precambrian glaciogenic sedimentary rocks in southern Africa: a compilation with definitions and correlations. Precambrian Res. Unit Univ. Cape Town Bull. No. 11.

Kröner, A., McWilliams, M. D., Germs, G. J. B., Reid, A. B., Schalk, K. E. L. 1981. Paleomagnetism of Late Precambrian to Early Paleozoic mixtite-bearing formations in Namibia (S.W.A.). Amer. Jour. Sci. (In press.)

Krynauw, J. R. 1977. The George granite pluton and its relationship to the Kaaimans Formation. Unpubl. M.Sc. thesis, Univ. Stellenbosch, South Africa.

Krynauw, J. R., Gresse, P. G. 1980. The Kaaimans Group in the George area, Cape Province: a model for the origin of deformation and metamorphism in the southern Cape Fold Belt. Geol. Soc. S. Afr. Trans. 83:23–38.

Krynine, P. O. 1950. Petrology, stratigraphy and origin of the Triassic sedimentary rocks of Connecticut. Conn. Geol. Nat. Hist. Surv. Bull. 73:175–194.

Kübler, M. 1977. The sedimentology and uranium mineralization of the Beaufort Group in the Beaufort West - Fraserburg - Merweville area, Cape Province. Unpubl. M.Sc. thesis, Univ. Witwatersrand, Johannesburg, South Africa.

Kumar, N., Sanders, J. E. 1974. Inlet sequence: a vertical succession of sedimentary structures and textures created by the lateral migration of tidal inlets. Sedimentology 21:491–532.

Kuschke, O. H. 1950. Granitic rocks of the Bushveld, northwest of Brits. Unpubl. M.Sc. thesis, Univ. Pretoria, South Africa.

Kutina, J. 1975. Tectonic development and metallogeny of Madagascar with reference to the fracture pattern of the Indian Ocean. Geol. Soc. Amer. Bull. 86:582–592.

La Berge, G. L. 1966. Altered pyroclastic rocks in South African iron formation. Econ. Geol. 61:572–581.

La Berge, G. L. 1973. Possible biological origin of Precambrian iron-formations. Econ. Geol. 68: 1098–1109.

LaBrecque, J. L., Hayes, D. E. 1979. Seafloor spreading history of the Agulhas basin. Earth Planet. Sci. Lett. 45:411–428.

Lake, P. 1904. The trilobites of the Bokkeveld Beds. S. Afr. Mus. Ann. 4:201–220.

Lamont, G. T. 1947. The geology of part of the Vanrhynsdorp division, Cape Province. Unpubl. Ph.D. thesis, Univ. Cape Town.

Lancaster, I. N. 1979. Evidence for a widespread Late Pleistocene humid period in the Kalahari. Nature, 279:145–146.

Larson, R. L., Ladd, J. 1973. Evidence for the opening of the South Atlantic in the Early Cretaceous. Nature 246:209–212.

Larson, R. L., Pitman, W. C. 1972. World-wide correlation of Mesozoic magnetic anomalies, and its implications. Geol. Soc. Amer. Bull. 83:3645–3662.

Latimore, R. K., Rona, P. A., De Wald, O. E. 1974. Magnetic anomaly sequence in the central North Atlantic. Jour. Geophys. Res. 79:1207–1209.

Latsky, R. 1942. The magmatic copper ores of Namaqualand. Geol. Soc. S. Afr. Trans. 45:109–150.

LeBlanc-Smith, G., 1979. High-constructive deltaic sedimentation and coal accumulation, Witbank basin. Geol. Soc. S. Afr. 18th Congr. Abstracts 2: 88–92.

LeBlanc-Smith, G., Eriksson, K. A. 1979. A fluvio-glacial and glaciolacustrine deltaic depositional model for Permo-Carboniferous coals of the north-eastern Karoo basin, South Africa. Palaeogeogr. Palaeoclim. Palaeoecol. 27:67–84.

Lee, J. E., Glenister, D. A. 1976. Stratiform sulfide mineralization at Oamites copper mine, South West Africa. Ecol. Geol. 71:369–383.

Lees, A., Buller, A. T. 1972. Modern temperate water and warm water shelf carbonate sediments contrasted. Mar. Geol. 13:1767–1773.

Lenthall, D. H. 1973. A proposed nomenclature system for the granophyres associated with the Bushveld Complex. Geol. Soc. S. Afr. Trans. 76:75–76.

Lenthall, D. H., Hunter, D. R. 1977. The geochemistry of the Bushveld granites in the Potgietersrus tin-field. Precambr. Res. 5:359–400.

Le Roux, H. D. 1941. A tillite in the Otavi Mountains, S.W.A. Geol. Soc. S. Afr. Trans. 44:207–218.

Le Roux, J. P. 1977. The stratigraphy, sedimentology and structure of the Cango Group north of Oudtshoorn, C. P. Unpubl. M.Sc. thesis, Univ. Stellenbosch, South Africa.

Liebenberg, C. J. 1960. The trace elements of rocks of the Bushveld Igneous Complex. Univ. Pretoria Publ. New Series No. 12.

Liebenberg, L. 1970. The sulphides in the layered sequence of the Bushveld Igneous Complex. Geol. Soc. S. Afr. Spec. Publ. No. 1, pp. 108–207.

Light, M. P. R., Watkeys, M. K. 1978. An outline of the Archaean and Early Proterozoic geological history of the region around Beitbridge. Rhodesia Geol. Surv. Ann. 3:35–40.

Light, M. P. R., Broderick, T. J., Watkeys, M. K. 1977. A preliminary report on the Central Zone of Limpopo Mobile Belt, Rhodesia. Botswana Geol. Surv. Bull. 12:61–73.

Lipson, R. D. 1979. Aspects of the geology of the Aggeneysberge, Namaqualand. Geol. Soc. S. Afr. 18th Congr. Abstracts 1:264–267.

Lipson, R. D., McCarthy, T. S. 1977. Geochemical correlation between some rock-types of the Namaqualand granite-gneiss complex. Geol. Soc. S. Afr. Trans. 80:177–181.

Lisle, R. J. 1977. The evaluation of Laxfordian deformation in the Carloway area, Isle of Lewis, Scotland. Tectonophysics 42:183–208.

Litherland, M. 1973. Uniformitarian approach to Archean "schist relics." Nature 242:125–127.

Liu, H. S. 1977. Convection pattern and stress system under the African plate. Phys. Earth Planet. Interiors 15:60–68.

Lloyd, F. E., Bailey, D. K. 1975. Light element metasomatism of the continental mantle: the evidence and the consequences. *In* L. H. Ahrens, J. B. Dawson, A. R. Duncan, A. J. Erlank (eds.), Physics and chemistry of the earth, vol. 9, pp. 389–416. Oxford: Pergamon.

Lock, B. E. 1978a. The Cape Fold Belt of South Africa: Tectonic control of sedimentation. Geol. Assoc. Proc. 89:263–281.

Lock, B. E. 1978b. Ultrahigh-temperature volcanic mudflows amongst the Drakensberg volcanic rocks: new criteria for their recognition. Geol. Soc. S. Afr. Trans. 81:55–59.

Lock, B. E. 1980. Flat-plate subduction and the Cape Fold Belt of South Africa. Geology 8:35–39.

Lock, B. E., Paverd, A. L., Broderick, T. J. 1974. Stratigraphy of the Karroo volcanic rocks of the Barkly East district. Geol. Soc. S. Afr. Trans. 77: 117–129.

Lock, B. E., Shone, R., Coates, A. T., Hatton, C. J. 1975. Mesozoic Newark-type sedimentary basins within the Cape Fold Belt of southern Africa. 9th Intl. Congr. Sedimentology, Nice, 1975.

Logan, B. W. 1961. Cryptozoon and associated stromatolites from the Recent, Shark Bay, Western Australia. Jour. Geol. 69: 517–533.

Logan, B. W., Davies, G. R., Read, J. F., Cebulski, D. E. 1970. Carbonate sedimentation and environments, Shark Bay, Western Australia. Amer. Assoc. Petrol. Geol. Mem. No. 13.

Logan, B. W., Hoffman, P., Gebelein, C. D. 1974. Algal mats, cryptalgal fabrics and structures, Hamelin Pool, Western Australia. Amer. Assoc. Petrol. Geol. Mem. 22:140–194.

Lombaard, A. F., Schreuder, F. J. G. 1978. Distribution pattern and general geological features of steep structures, megabreccias and basic rocks in the Okiep Copper District. Geol. Soc. S. Afr. Spec. Publ. No. 4, pp. 269–295.

Lombard, B. V. 1934. On the differentiation and relationships of rocks of the Bushveld Igneous Complex. Geol. Soc. S. Afr. Trans. 37:5–52.

Loock, J. C. 1967. The stratigraphy of the Witteberg-Dwyka contact beds. Unpubl. M.Sc. thesis, Univ. Stellenbosch, South Africa.

Lowe, D. R., Knauth, L. P. 1977. Sedimentology of the Onverwacht Group (3.4 billion years) Transvaal, South Africa, and its bearing on the characteristics and evolution of the early earth. Jour. Geol. 85:699–723.

Lowe, D. R., Knauth, L. P. 1978. The oldest marine carbonate ooids reinterpreted as volcanic accretionary lapilli, Onverwacht Group, South Africa. Jour. Sediment. Petrol. 48:709–722.

Lowenstam, H. A. 1980. What, if anything, happened at the transition from the Precambrian to the Phanerozoic? Precambr. Res. 11:89–91.

Macgregor, A. M. 1940. A pre-Cambrian algal limestone in Southern Rhodesia. Geol. Soc. S. Afr. Trans. 43:9–15.

MacGregor, I. D. 1974. The system $MgOAl_2O_3$-SiO_2: Solubility of Al_2O_3 in enstatite for spinel and garnet peridotite composition. Amer. Mineral. 59:110–119.

MacGregor, I. D. 1975. Petrologic and thermal structure of the upper mantle beneath South Africa in the Cretaceous. *In* L. H. Ahrens, J. B. Dawson, A. R. Duncan, A. J. Erlank (eds.), Physics and chemistry of the earth, vol. 9, pp. 455–466. Oxford: Pergamon.

MacGregor, I. D. 1979. Mafic and ultramafic xenoliths from the Kao kimberlite pipe. 2nd Intl. Kimberlite Conf. Proc. 2:156–172.

MacGregor, I. M., Truswell, J. F., Eriksson, K. A. 1974. Filamentous algae from the 2300 m.y. old Transvaal dolomite. Nature 247:538–540.

MacIntyre, R. M. Dawson, J. B. 1971. Radiometric dating of alkaline intrusions in southern Africa. (Abstract). Soc. Geol. Belgique Ann. 94:120.

Macurda, D. B. 1979. The Devonian blastoids of Bolivia. Jour. Paleont. 53:1361–1373.

Maglio, V. J. 1970. Early Elephantidae of Africa and a tentative correlation of African Plio-Pleistocene deposits. Nature 225:328–332.

Maglio, V. J. 1973. Origin and evolution of the Elephantidae. Amer. Phil. Soc. Trans. 63:1–149.

Malling, S. 1978. Some aspects of the lithostratigraphy and tectonometamorphic evolution in the Nauchas-Rehoboth area, S.W.A. (Namibia). Precambrian Res. Unit Univ. Cape Town Ann. Rept. 14/15: 183–193.

Mansfield, C. F. 1979. Possible biogenic origin for some sedimentary dolomite. Amer. Assoc. Petrol. Geol. Bull. 63:490 (Abstract).

Manton, W. I. 1968. The origin of associated basic and acid rocks in the Lebombo-Nuanetsi igneous province, southern Africa, as implied by strontium isotopes. Jour. Petrol. 9:23–39.

Marais, J. A. H., Packham, B. de V., Schreuder, F. J. G. 1975. The regional geology of the O'okiep Copper District. Geol. Soc. S. Afr. 16th Congr. Abstracts 1:88–89.

Marchant, J. W. 1974. Trace-fossils and tracks in the upper Table Mountain Group at Milner Peak, Cape Province. Geol. Soc. S. Afr. Trans. 77:369–370.

Marlow, A. G., Van der Merwe, M. J. 1977. The geology and potential economic significance of the Malope area, north-eastern Bushveld Complex. Geol. Soc. S. Afr. Trans. 80:117–123.

Marsh, J. S. 1973. Relationships between transform directions and alkaline igneous lineaments in Africa and South America. Earth Planet. Sci. Lett. 18: 317–323.

Marsh, J. S. 1975. The Lüderitz alkaline province, South West Africa. I. Descriptive petrology of the Granitberg foyaite complex. Geol. Soc. S. Afr. Trans. 78:215–224.

Marsh, J. S. 1976. The Lüderitz alkaline province, South West Africa. III. The Pomona and Drachenberg syenite complexes. Geol. Soc. S. Afr. Trans. 79:168–176.

Marsh, J. S., Lock, B. E., Fuchter, W. H. 1979. New chemical analyses of the Suurberg volcanic rocks and their significance in relation to Mesozoic volcanism in southern Africa. S. Afr. Jour. Sci. 75: 227–229.

Martin, A., Nisbet, E. G., and Bickle, M. J. 1980. Archaean stromatolites of the Belingwe Greenstone Belt, Zimbabwe (Rhodesia). Precambrian Res. 13:337–362.

Martin, A. R. H. 1960. A Mesozoic microflora from South Africa. Nature 186:95.

Martin, H. 1965. The Precambrian geology of South West Africa and Namaqualand. Cape Town: Precambrian Res. Unit. Univ. Cape Town.

Martin, H. 1973. The Atlantic margin of southern Africa between latitude 17° South and the Cape of Good Hope. *In* A. E. M. Nairn, F. G. Stehli (eds.), The ocean basins and margins I. The South Atlantic, pp. 277–300. New York: Plenum.

Martin, H. 1974. Damara rocks as nappes on the Naukluft Mountains, South West Africa. Precambrian Res. Unit Univ. Cape Town Bull. 15:153–165.

Martin, H. 1978. The mineralization of the ensialic Damara orogenic belt. Geol. Soc. S. Afr. Spec. Publ. No. 4, pp. 405–415.

Martin, H., Porada, H. 1977a. The intracratonic branch of the Damara orogen in South West Africa. Part I. Discussion of geodynamic models. Precambr. Res. 5:311–338.

Martin, H., Porada, H. 1977b. The intracratonic branch of the Damara orogen in South West Africa. Part II. Discussion of relationships with the Pan-African mobile belt system. Precambr. Res. 5:339–357.

Martin, H., Wilczewski, N. 1970. Palaeoecology, conditions of deposition, and the palaeogeography of the marine Dwyka beds of South West Africa. 2nd Gondwana Symposium. Proceedings and Papers, pp. 225–232.

Martin, H., Wilczewski, N. 1972. Algen-stromatolithen aus der Etoscha-Pfanne Südwest-Afrikas. N.

Jb. Geol. Paläont. Monatsh. 12:720–726.
Martin, H., Porada. H., Walliser, O. H. (in preparation). Damara mixtites and pebble-bearing schists of supposed glacial origin: Problems in interpretation.
Martini, J. E. J. 1974. The presence of ash beds and volcanic fragments in the graywackes of the Karroo System in the southern Cape Province (South Africa). Geol. Soc. S. Afr. Trans. 77:113–116.
Martini, J. E. J. 1978. Coesite and stishovite in the Vredefort Dome, South Africa. Nature 272:715–717.
Maske, S. 1966. The petrography of the Ingeli mountain range. Univ. Stellenbosch Ann. No. A41.
Mason, R. 1973. The Limpopo Mobile Belt - southern Africa. Roy. Soc. London Phil. Trans. A273:463–485.
Mason, R. J. 1967. The archaeology of the earliest superficial deposits in the lower Vaal basin near Holpan, Windsorten district. S. Afr. Geogr. Jour. 49:39–56.
Mason, T. R., Tavener-Smith, R. 1978. A fluviodeltaic Middle Ecca succession west of Empangeni, Zululand. Geol. Soc. S. Afr. Trans. 81:13–22.
Mason, T. R., Von Brunn, V. 1977. 3-Gyr-old stromatolites from South Africa. Nature 266:47–49.
Mathias, M. 1974. Alkaline rocks of southern Africa. In H. Sorenson (ed.), The alkaline rocks, pp. 189–202. London: Wiley.
Matthews, P. E. 1959. The metamorphism and tectonics of the pre-Cape formations in the post-Ntingwe thrust-belt, S.W. Zululand, Natal. Geol. Soc. S. Afr. Trans. 62:257–322.
Matthews, P. E. 1961. Slump structures in the Table Mountain series of Natal. Geol. Soc. S. Afr. Trans. 64:55–71.
Matthews, P. E. 1967. The pre-Karroo formations of the White Umfolozi inlier, northern Natal. Geol. Soc. S. Afr. Trans. 70:257–272.
Matthews, P. E. 1970. Paleorelief and Dwyka glaciation in the eastern region of South Africa. 2nd Gondwana Symposium. Proceedings and Papers, pp. 491–499.
Matthews, P. E. 1972. Possible Precambrian obduction and plate tectonics in southeastern Africa. Nature 240:37–39.
Matthews, P. E. 1981. Eastern or Natal sector of the Namaqua-Natal Mobile Belt in southern Africa. *In* D. R. Hunter (ed.), Precambrian of the southern hemisphere, pp. 705–715. Amsterdam: Elsevier.
Matthews, R. K. 1974. Dynamic stratigraphy. New York: Prentice-Hall.
Matthews, S. C., Cowie, J. W. 1979. Early Cambrian transgression. Geol. Soc. London Jour. 136:133–135.
Matthysen, J. L. 1953. 'n Nuwe stratigrafiese indeling van die Ventersdorp-Sisteem. Unpubl. M.Sc. thesis, Univ. Pretoria, South Africa.
Maud, R. R. 1961. A preliminary review of the structure of coastal Natal. Geol. Soc. S. Afr. Trans. 64: 247–256.
Maud, R. R. 1968. Quaternary geomorphology and soil formation in coastal Natal. Z. Geomorph. Suppl. 7:155–199.
Maud, R. R., Orr, W. N. 1975. Aspects of post-Karroo geology in the Richards Bay area. Geol. Soc. S. Afr. Trans. 78:101–109.
McBride, E. F., Shepherd, R. G., Crawley, R. A. 1975. Origin of parallel, near-horizontal laminae by migration of bedforms in a small flume. Jour. Sediment. Petrol. 45:132–139.
McCarthy, M. J. 1967. Stratigraphical and sedimentological evidence from the Durban region of major sea-level movements since the Late Tertiary. Geol. Soc. S. Afr. Trans. 70:135–165.
McCarthy, M. J. 1970. An occurrence of pillow lava in a basal flow of the Drakensberg volcanic stage, Ndedema Valley, Natal. 2nd Gondwana Symposium. Proceedings and Papers, pp. 433–439.
McCarthy, M. J., Orr, W. N. 1978. A rhinoceros tooth from basal strata of the Port Durnford Formation, Zululand, South Africa: The relocation of Anderson's mammalian fossil site. Natal. Mus. Ann. 23: 485–496.
McCarthy, T. S. 1976. Chemical interrelationships in a low-pressure granulite terrain in Namaqualand, South Africa, and their bearing on granite genesis and the composition of the lower crust. Geochim. Cosmochim. Acta 40:1057–1068.
McCarthy,T. S. 1978. A geochemical study of the gneisses of the Nababeep district, Namaqualand. Geol. Soc. S. Afr. Spec. Publ. No. 4, pp. 351–353.
McCarthy, T. S., Robb, L, J. 1977. On the relationship between cumulus mineralogy and trace and alkali chemistry in an Archaean granite from the Barberton region, South Africa. Econ. Geol. Res. Unit Univ. Witwatersrand Inf. Circ. No. 112.
McDaid, J. M. 1976. Preliminary report on the geology of the northern part of Diamond Area No. I, South West Africa. Precambrian Res. Unit Univ. Cape Town Ann. Rept. 13:89–96.
McDaid, J. M. 1978. The geology of the northern part of Diamond Area no. 1, South West Africa. Precambrian Res. Unit. Univ. Cape Town Ann. Rept. 14/15:124–140.
McDonald, J. A. 1967. Evolution of part of the lower critical zone, farm Ruighoek, western Bushveld. Jour. Petrol. 8:165–209.
McElhinny, M. W., McWilliams, M. O. 1977. Precambrian geodynamics—A palaeomagnetic view. Tectonophysics 40:137–159.
McElhinny, M. W., Gidding, J. W., Embleton, B. J. J. 1974. Palaeomagnetic results and Precambrian glaciations. Nature 248:557–561.
McElhinny, M. W., Embleton, B. J. J., Daly, L., Pozzi, J. .P. 1976. Paleomagnetic evidence for the

location of Madagascar in Gondwanaland. Geology 4:455–457.

McFadden, P. L., Brock, A., Partridge, T. C. 1979. Palaeomagnetism and the age of the Makapansgat hominid site. Earth Planet. Sci. Lett. 44:373–382.

McGetchin, T. R., Besancon, J. R. 1973. Carbonate inclusions in mantle-derived pyropes. Earth Planet. Sci. Lett. 18:408–410.

McGowen, J. H. 1970. Gum Hollow fan delta, Neuces Bay, Texas. Bur. Econ. Geol. Univ. Texas, Austin, Rept. Invest. No. 69.

McGowran, B., Moore, A. C. 1971. A reptilian tooth and Upper Cretaceous microfossils from the lower quarry at Needs Camp, South Africa. Geol. Soc. S. Afr. Trans. 74:103–105.

McGregor, V. R. 1973. The Early Precambrian gneisses of the Godthaab district, West Greenland. Roy. Soc. London Phil. Trans. A273:343–358.

McIver, J. R. 1966. Orthopyroxene-bearing granitic rocks from southern Natal. Geol. Soc. S. Afr. Trans. 69:99–117.

McIver, J. R., Ferguson, J. 1979. Kimberlitic, melilititic, trachytic and carbonatitic eruptives at Saltpetre Kop, Sutherland, South Africa. 2nd Intl. Kimberlite Conf. Proc. 1:111–128.

McIver, J. R., Lenthall, D. H. 1973. Mafic and ultramafic extrusives of the Onverwacht Group in terms of the system XO-YO-R_2O_3-ZO_2. Econ. Geol. Res. Unit Univ. Witwatersrand Inf. Circ. No. 80.

McIver, J. R., Cawthorn, R. G., Wyatt, B. A. 1981. The Ventersdorp Supergroup—The youngest komatiitic sequence in South Africa. *In* E. G. Nisbet, N. T. Arndt (eds.), Komatiites. London: Allan and Unwin.

McKee, E. D., Crosby, E. J., Berryhill, H. L. 1967. Flood deposits, Bijou Creek, Colorado, June 1965. Jour. Sediment. Petrol. 37:829–851.

McKenzie, D., Sclater, J. G. 1971. The evolution of the Indian Ocean since the Late Cretaceous. Geophys. Jour. Roy. Astr. Soc. 24:437–528.

McKerrow, W. S. 1979. Ordovician and Silurian changes in sea level. Geol. Soc. London Jour. 136:137–145.

McKie, D. 1966. Fenitization. *In* O. F. Tuttle, J. Gittins (eds.), Carbonatites, pp. 261–294. New York: Interscience.

McLachlan, I. R. 1973. Problematic microfossils from the lower Karroo beds in South Africa. Palaeont. Afr. 15:1–21.

McLachlan, I. R., Anderson, A. 1973. A review of the evidence for marine conditions in southern Africa during Dwyka times. Palaeont. Afr. 15:37–64.

McLachlin, I. R., McMillan, I. K. 1976. Review and stratigraphic significance of southern Cape Mesozoic palaeontology. Geol. Soc. S. Afr. Trans. 79:197–212.

McLachlan, I. R., McMillan, I. K. 1979. Microfaunal biostratigraphy, chronostratigraphy and history of Mesozoic and Cenozoic deposits on the coastal margin of South Africa. Geol. Soc. S. Afr. Spec. Publ. No. 6, pp. 161–181.

McLachlan, I. R., Brenner, P. W., McMillan, I. K. 1976a. The stratigraphy and micro-palaeontology of the Cretaceous Brenton Formation and the PB-A/1 well, near Knysna, Cape Province. Geol. Soc. S. Afr. Trans. 79:341–370.

McLachlan, I. R., McMillan, I. K., Brenner, P. W. 1976b. Micropalaeontological study of the Cretaceous beds at Mbotyi and Mngazana, Transkei, South Africa. Geol. Soc. S. Afr. Trans. 79:321–340.

McMahon, B. M., Haggerty, S. E. 1979. The Oka carbonatite complex: Magnetite compositions and the related role of titanium in pyrochlore. Second Intl. Kimberlite Conf. Proc. 1:382–392.

McMillan, M. D. 1968. The geology of the Witputs-Sendelingsdrif area. Precambrian Res. Unit Univ. Cape Town Bull. No. 4.

McPherson, J. G., Germs, G. J. B. 1979. Calcrete (caliche) in the Beaufort Group of the southern Karoo basin and its palaeoclimatic significance. Geol. Soc. S. Afr. 18th Congr. Abstracts 2:145–147.

McWilliams, M. O., McElhinny, M. W., Kröner, A. 1981. Pan African tectonics: evidence from paleomagnetism. Earth Planet. Sci. Lett. (In press.)

Mehnert, K. R. 1968. Migmatites and the origin of granitic rocks. Amsterdam: Elsevier.

Mercier, J. C., Carter, N. L. 1975. Pyroxene geotherms. Jour. Geophys. Res. 80:3349–3362.

Meyer, H. O., Tsoi, H., Gurney, J. J. 1979. A unique enstatite-megacryst with coexisting Cr-poor and Cr-rich garnet, Weltevreden Floors, South Africa. 2nd Intl. Kimberlite Conf. Proc. 1:279–291.

Miall, A. D. 1976. Proterozoic and Paleozoic geology of Banks Island, Arctic Canada. Geol. Surv. Can. Bull. No. 258.

Miall, A. D. 1977. A review of the braided-river depositional environment. Earth Sci. Rev. 13:1–62.

Miall, A. D. 1978. Lithofacies types and vertical profile models in braided river deposits: a summary. *In* A. D. Miall (ed.), Fluvial sedimentology. Can. Soc. Petrol. Geol. Mem. No. 5, pp. 597–604.

Middlemost, E. A. K. 1964. Petrology of the plutonic and dyke-rocks of the south-eastern Richtersveld. Geol. Soc. S. Afr. Trans. 67:227–261.

Middlemost, E. A. K. 1966. The genesis of the Stinkfontein Formation. Geol. Soc. S. Afr. Trans. 69:87–98.

Middlemost, E. A. K. 1967. Petrology of the Bremen granite-syenite complex, South West Africa. Geol. Soc. S. Afr. Trans. 70:117–134.

Middleton, G. V., Hampton, M. A. 1973. Sediment gravity flows: Mechanics of flow and deposition. *In* G. V. Middleton, A. H. Bouma (eds.), Turbidites and deep-water sedimentation, pp. 1–38. Soc. Econ. Paleont. Mineral. Short Course, Syllabus of Pacific Section.

Middleton, R. C. 1976. The geology of Prieska Copper Mines Limited. Econ. Geol. 71:328–350.

Miller, R. McG. 1972. The geology of a portion of southern Damaraland, southwest Africa, with particular reference to the petrogenesis of the Salem granite. Unpubl. Ph.D. thesis, Univ. Cape Town.

Miller, R. McG. 1974a. The stratigraphic significance of the Naauwpoort Formation of east central Damaraland, South West Africa. Geol. Soc. S. Afr. Trans. 77:363–367.

Miller, R. McG. 1974b. Some economic aspects of the crystallization of the Damara granites, South West Africa. Precambrian Res. Unit Univ. Cape Town Bull. 15:183–192.

Miller, R. McG. 1975. Experimental melting of the Khomas schist and Salem granite at 4 and 7 kilobars water pressure. Geol. Soc. S. Afr. 16th Congr. Abstracts 1:103–106.

Miller, R. McG. 1979. The Okahandja Lineament, a fundamental tectonic boundary in the Damara orogen of South West Africa/Namibia. Geol. Soc. S. Afr. Trans. 82:349–361.

Minnitt, R. C. A. 1975. The geology and geochemistry of the eastern portion of the Murchison Range. Geol. Soc. S. Afr. 16th Congr. Abstracts 1:107–108.

Minnitt, R. C. A. 1979. The geological setting of porphyry-type copper mineralization in the Haib River area, South West Africa. Unpubl. Ph.D thesis, Univ. Witwatersrand, Johannesburg, South Africa.

Minter, W. E. L. 1970. Gold distribution related to the sedimentology of a Precambrian Witwatersrand conglomerate, South Africa, as outlined by moving-average analysis. Econ. Geol. 65:963–969.

Minter, W. E. L. 1972. Sedimentology of the Vaal Reef in the Klerksdorp area. Unpubl. Ph.D thesis, Univ. Witwatersrand, Johannesburg, South Africa.

Minter, W. E. L. 1976. Detrital gold, uraninite and pyrite concentrations related to sedimentology in the Precambrian Vaal Reef Placer, Witwatersrand, South Africa. Econ. Geol. 71:157–175.

Minter, W. E. L. 1978. A sedimentological synthesis of placer gold, uranium and pyrite concentrations in Proterozoic Witwatersrand sediments. *In* A. D. Miall (ed.), Fluvial sedimentology. Can. Soc. Petrol. Geol. Mem. No. 5, pp. 801–829.

Minter, W. E. L. 1980. Debris flows and their value as climatic indicators. Econ. Geol. Res. Unit Univ. Witwatersrand, EGRUSEG Encaenia No. 8.

Minter, W. E. L. 1982. The geology of the West Wits Line. Geol. Soc. S. Afr. Spec. Publ. (In press).

Mitchell, R. H. 1970. Kimberlite and related rocks—A critical reappraisal. Jour. Geol. 78:686–704.

Mitchell, R. H. 1973. Magnesian ilmenite and its role in kimberlite petrogenesis. Jour. Geol. 81:301–311.

Mitchell, R. H. 1977. Geochemistry of magnesian ilmenites from kimberlites in South Africa and Lesotho. Lithos 10:29–37.

Mitchell, R. H. 1979. The alleged kimberlite-carbonatite relationship: additional contrary mineralogical evidence. Amer. Jour. Sci. 279:570–589.

Mitchell, R. H., Crockett, J. H. 1971. The isotopic composition of strontium in some South African kimberlites. Contr. Mineral. Petrol. 30:277–290.

Mitchum, R. M., Vail, P. R., Thompson, S. 1977. Seismic stratigraphy and global changes of sea level. Part 2. The depositional sequence as a basic unit for stratigraphic analysis. Amer. Ass. Petrol. Geol. Mem. 26:53–62.

Modreski, P. J., Boettcher, A. L. 1973. Phase relationships of phlogopite in the system K_2O-MgO-CaO-Al_2O_3-SiO_2-H_2O to 35 kilobars: A better model for micas in the interior of the earth. Amer. Jour. Sci. 273:385–414.

Mofjeld, H. O. 1976. Tidal currents. *In* D. J. Stanley, D. J. P. Swift (eds.), Marine sediment transport and environmental management, pp. 53–64. New York: Wiley.

Molyneux, T. G. 1970. The geology of the area in the vicinity of Magnet Heights, eastern Transvaal, with special reference to the magnetic iron ore. Geol. Soc. S. Afr. Spec. Publ. No. 1, pp. 228–241.

Molyneux, T. G. 1974. A geological investigation of the Bushveld Complex in Sekhukhuneland and part of the Steelpoort valley. Geol. Soc. S. Afr. Trans. 77:329–338.

Molyneux, T. G., Klinkert, P. S. 1978. A structural interpretation of part of the eastern mafic lobe of the Bushveld Complex and its surrounds. Geol. Soc. S. Afr. Trans. 81:359–368.

Monty, C. L. V. 1972. Recent algal stromatolitic deposits, Andros Island, Bahamas, preliminary report. Geol. Rundsch. 61:742–783.

Moody, J. D., Hill, M. J. 1956. Wrench-fault tectonics. Geol. Soc. Amer. Bull. 67:1207–1246.

Moorbath, S., Wilson, J. F., Cotterill, P. 1976. Early Archean age for the Sebakwian Group at Selukwe, Rhodesia. Nature 264:536–538.

Moorbath, S., Wilson, J. F., Goodwin, R., Humm, M. 1977. Further Rb-Sr age and isotope data on Early and Late Archaean rocks from the Rhodesian craton. Precambr. Res. 5:229–239.

Moore, A. C. 1975. The petrography and the regional setting of the Tantalite Valley Complex, South West Africa. Geol. Soc. S. Afr. Trans. 78:235–249.

Moore, A. C., Kartun, K. G., Waters, D. J. 1979. Metamorphic history of the aureole associated with the Tantalite Valley Complex, South West Africa/Namibia. Geol. Soc. S. Afr. Trans. 82:67–80.

Moore, A. E. 1976. Controls of post-Gondwanaland

alkaline volcanism in southern Africa. Earth Planet. Sci. Lett. 31:291–296.

Moore, A. E. 1980. The geochemistry of the olivine melilitites and related rocks of Namaqualand - Bushmanland. Unpubl. Ph.D. thesis, Univ. Cape Town, South Africa.

Moore, G. M., Peck, O.L. 1962. Accretionary lapilli in volcanic rocks of the western continental United States. Jour. Geol. 70:182–193.

Moore, J. M. 1977. The geology of Namiesberg, northern Cape. Precambrian Res. Unit Univ. Cape Town Bull. No. 20.

Morse, S. A. 1969. The Kiglapait layered intrusion, Labrador. Geol. Soc. Amer. Mem. No. 112.

Muir, M. D., Hall, D. O. 1974. Diverse microfossils in Precambrian Onverwacht Group rocks of South Africa. Nature 252:376–378.

Murray, J., Renard, A. F. 1891. Report on deep sea deposits based on the specimens collected during the voyage of *H. M. S. Challenger*. Rept. Voyage *Challenger* 1873–1876 (Deep Sea Deposits), pp. 391–400.

Mutti, E., Ricci Lucchi, F. 1972. Le torbiditi dell' Appennino settentrionale: Introduzione all'analisi di facies. Soc. Geol. Ital. Mem. 11:161–199. (English translation by T. H. Nielson, 1978. Int. Geol. Rev. 20, 125–166, AGI Reprint Series 3.)

Myers, J. S. 1976. Granitoid sheets, thrusting, and Archean crustal thickening in West Greenland. Geology 4:265–268.

Myers, J. S. 1978. Formation of banded gneisses by deformation of igneous rocks. Precambr. Res. 6: 43–64.

Mysen, B. O., Boettcher, A. L. 1975. Melting of a hydrous mantle, I. Phase relations of natural peridotite at high pressures and temperatures with controlled activities of water, carbon dioxide, and hydrogen. Jour. Petrol. 16:520–548.

Nagy B., Nagy, L. A. 1969. Early Precambrian Onverwacht microstructures: Possibly the oldest fossils on earth? Nature 223:1226–1229.

Nagy, L. A. 1978. New filamentous and crystous microfossils 2300 m.y. old from the Transvaal sequence. Jour. Paleont. 52:141–154.

Naldrett, A. J., Cabri, L. J. 1976. Ultramafic and related mafic rocks: Their classification and genesis with special reference to the concentration of nickel sulfides and platinum group elements. Econ. Geol. 71:1131–1158.

Nash, C. R. 1972. Primary anhydrite in Precambrian gneisses from the Swakopmund district, South West Africa. Contr. Mineral. Petrol. 36:27–32.

Nel, H. J. 1940. The basal rocks of the Bushveld Igneous Complex, north of Pretoria. Geol. Soc. S. Afr. Trans. 43:37–68.

Nel, W. A. 1968. The estimation and distribution of boron in fine-grained South African sediments. Unpubl. M.Sc. thesis, Univ. Cape Town, South Africa.

Netterberg, F. 1969. The interpretation of some basic calcrete types. S. Afr. Archaeol. Bull. 24:117–122.

Newton, A. R. 1973. A gravity-folding model for the Cape Fold-Belt. Geol. Soc. S. Afr. Trans. 76:145–152.

Newton, R. B. 1913. On some Kainozoic shells from South Africa. Albany Mus. Rec. 2:315–352.

Nicolaysen, L. O. 1962. Stratigraphic interpretation of age measurements in southern Africa. Geol. Soc. Amer. Buddington Vol., pp. 569–598.

Nicolaysen, L. O., Burger, A. J. 1965. Note on an extensive zone of 1000 million-year old metamorphic and igneous rocks in southern Africa. Sci. Terre 10:497–516.

Nicolaysen, L. O., Ferguson, J. 1981. Diapirs driven by high pore fluid pressure. Jour. Struct. Geology 3:89–95.

Nicolaysen, L. O., De Villiers, J. W. L. Burger, A. J., Strelow, F. W. E. 1958. New measurements relating to the absolute age of the Transvaal System and of the Bushveld Igneous Complex. Geol. Soc. S. Afr. Trans. 61:137–163.

Nicolaysen, L. O., Burger, A. J., Liebenberg, W. R. 1962. Evidence for the extreme age of certain minerals from the Dominion Reef conglomerates and the underlying granite in the western Transvaal. Geochim. Cosmochim. Acta 26:15–23.

Nicolaysen, L. O., Burger, A. J., Van Niekerk, C. B. 1963. The origin of the Vredefort Dome structure in the light of new isotopic data. (Abstract.) Intl. Union Geodesy Geophysics, 13th general assembly, Berkeley, Calif.

Nieberding, F. 1976. Die Grenze der zentralen Granitzone südwestlich Otjimbingwe (Karibib-District, Südwestafrika): Intrusions-verband, Tectonik, Petrographie. Gottinger Arb. Geol. Paläont. No. 19.

Nilsen, T. H., Abbott, P. L. 1979. Turbidite sedimentology of the Upper Cretaceous Point Loma and Cabrillo Formations, San Diego, California. *In* P. L. Abbott (ed.), Geological excursions in the southern California area, pp. 139–166. San Diego State Univ.

Nio, S. D. 1976. Marine transgressions as a factor in the formation of sandwave complexes. Geol. Mijnb. 55:18–40.

Nixon, P. H., Boyd, F. R. 1973a The discrete nodule (megacryst) association in kimberlites from northern Lesotho. *In* P. H. Nixon (ed.), Lesotho kimberlites, pp. 67–75. Maseru: Lesotho National Development Corp.

Nixon, P. H., Boyd, F. R. 1973b. Petrogenesis of the granular and sheared ultrabasic nodule suite in kimberlite. *In* P. H. Nixon (ed.), Lesotho kimberlites, pp. 48–56. Maseru: Lesotho National Development Corp.

Nockolds, S. R. 1954. Average chemical composition of some igneous rocks. Geol. Soc. Amer. Bull. 65: 1007–1032.

Norton, I. O., Sclater, J. S. 1979. A model for the evolution of the Indian Ocean and the breakup of Gondwanaland. Jour. Geophys. Res. 84: 6803–6830.

Odell, J. 1975. Explanation of the geological map of the country around Bangala dam. Rhodesia Geol. Surv. Short Rept. No. 42.

O'Driscoll, E. S. 1964. Interference patterns from inclined shear fold systems. Can. Petrol. Geol. Bull.12:279–10.

Oehler, D. Z., Schopf, J. W., Kvenvolden, K. A. 1972. Carbon isotope studies of organic matter in Precambrian rocks. Science 175: 1246–1248.

Off, T. 1963. Rhythmic linear sand bodies caused by tidal currents. Amer. Assoc. Petrol. Geol. Bull. 47:324–341.

O'Hara, M. J. 1968. The bearing of phase equilibria studies in synthetic and natural systems on the origin and evolution of basic and ultrabasic rocks. Earth Sci. Rev. 4:69–133.

O'Hara, M. J., Yoder, H. S. 1967. Formation and fractionation of basic magmas at high pressures. Scot. Jour. Geol. 3:67–117.

O'Hara, M. J., Saunders, M. J., Mercy, E. L. P. 1975. Garnet peridotite, primary ultrabasic magma and eclogite: interpretation of upper mantle processes in kimberlite. *In* L. H. Ahrens, J. B. Dawson, A. R. Duncan, A. J. Erlank (eds.), Physics and chemistry of the earth vol. 9, pp. 571–604. Oxford: Pergamon.

O'Nions, R. K., Evensen, N. M., Hamilton, P. J., Carter, S. R. 1978. Melting of the mantle past and present: isotopic and trace element evidence. Roy. Soc. London Phil. Trans. A288:547–559.

Oosthuysen, E. J. 1970. The geochronology of a suite of rocks from the granitic terrain surrounding the Barberton Mountain Land. Unpubl. Ph.D. thesis, Univ. Witwatersrand, Johannesburg, South Africa.

Oosthuysen, E. J., Burger, A. J. 1964. Radiometric dating of intrusives associated with the Waterberg System. S. Afr. Geol. Surv. Ann. 3:87–106.

Orme, A. R. 1973. Barrier and lagoon systems along the Zululand coast, South Africa. *In* D. R. Coates (ed.), Coastal geomorphology, pp. 181–217. New York: N.Y. State Univ. Press.

Orpen, J. L. 1976. Discussion of "Some aspects of the Archean History of Southern Rhodesia" by M. P. Coward and R. M. Shackleton. Res. Inst. Africa Geol. Univ. Leeds Ann. Rept. 20:28–9.

Orpen, J. L., Wilson, J. F. 1981. Stromatolites at ~3,500 Myr and a greenstone-granite unconformity in the Zimbabwean Archaean. Nature 291: 218–220.

Orr, W. N., Chapman, J. R. 1974. Danian (Palaeocene) marine rocks at Richards Bay, South Africa. S. Afr. Jour. Sci. 70:247–249.

Otto, J. D. T. 1973. The geology and petrology of Marble Delta. Unpubl. D.Sc. thesis, Univ. Stellenbosch, South Africa.

Page, D. C., Watson, M. D. 1976. The Pb-Zn deposit of Rosh Pinah Mine, South West Africa. Econ. Geol. 71:306–327.

Papenfus, J. A. 1957. The geology of the Government Gold Mining Areas and New State Areas, with particular reference to channel sediments and the origin of minerals of economic importance. Unpubl. Ph.D. thesis, Univ. South Africa, Pretoria.

Papenfus, J. A. 1964. The Black Reef Series within the Witwatersrand basin with special reference to its occurrence at Government Gold Mining Areas. *In* S. H. Haughton (ed.), The geology of some ore deposits in southern Africa, Vol. 1, pp. 191–218. Johannesburg: Geol. Soc. S. Afr.

Paraguassu, A. B. 1972. Experimental silicification of sandstone. Geol. Soc. Amer. Bull. 83:2853–2858.

Park, R. G., Cresswell, D. 1973. The dykes of the Laxfordian belts. *In* R. G. Park, J. Tarney (eds.), The Early Precambrian of Scotland and related rocks of Greenland, pp. 119–130. Newcastle: Univ. Keele, Dept. Geology.

Parker, R. J. Siesser, W. G. 1972. Petrology and origin of some phosphorites from the South African continental margin. Jour. Sediment. Petrol. 42:434–440.

Partridge, T. C. 1978. Re-appraisal of lithostratigraphy of Sterkfontein hominid site. Nature 275:282–287.

Partridge, T. C., Brink, A. B. A. 1967. Gravels and terraces of the lower Vaal River basin. S. Afr. Geogr. Jour. 49:21–38.

Partridge, T. C., Talma, A. S. (In press.) Carbon and oxygen isotopic measurements at Makapansgat and Sterkfontein, and their palaeoenvironmental significance. Ninth Cong. UISPP Proc. Nice.

Pearce, T. H., Gorman, B. E., Birkett, T. C. 1977. The TiO_2-K_2O-P_2O_5 diagram: A method of discriminating between oceanic and non-oceanic basalts. Earth Planet. Sci. Lett. 24:419–426.

Pearce, T. H., Gorman, B. E., Birkett, T. C. 1977. The relationship between major element chemistry and tectonic environment of basic and intermediate volcanic rocks. Earth Planet. Sci. Lett. 36:121–132.

Petri, S., Suguio, K. 1969. Sobre os metasedimentos do grupo Açungui do extremo sul do Estado de Sao Paulo. Univ. Sao Paulo, Convenio USP/DAEE.

Phillips, J. D. 1967. Magnetic anomalies over the mid-Atlantic ridge near 27°N. Science 175:920–923.

Piper, J. D. A. 1972. Palaeomagnetic studies of Precambrian rocks of Africa. Res. Inst. African Geol. Univ. Leeds Ann. Rept. 16:41–47.

Piper, J. D. A. 1974. Proterozoic crustal distribution, mobile belts and apparent polar movements. Nature 251:381–384.

Pitcher, W. S. 1978. The anatomy of a batholith. Geol. Soc. London Jour. 135:157–182.

Plumstead, E. P. 1966. Recent palaeobotanical advances and problems in Africa. Symposium on floristics and stratigraphy of Gondwanaland 1964, pp. 1–12. Birbal Sahni Institute, Lucknow, India.

Plumstead, E. P. 1967. A general review of the Devonian fossil plants found in the Cape System of South Africa. Palaeont. Afr. 10:1–83.

Plumstead, E. P. 1969. Three thousand million years of plant life in Africa. Geol. Soc. S. Afr. Trans. 72 (Annex.).

Plumstead, E. P. 1977. A new phytostratigraphical Devonian zone in southern Africa which includes the first record of *Zosterophyllum*. Geol. Soc. S. Afr. Trans. 80:267–277.

Poldervaart, A. 1966. Archaean charnockitic adamellite phacoliths in the Keimoes-Kakamas region, Cape Province, South Africa. Geol. Soc. S. Afr. Trans. 69:139–154.

Poldervaart, A., Von Backström, J. W. 1949. A study of an area at Kakamas (Cape Province). Geol. Soc. S. Afr. Trans. 52:433–495.

Porada, H. 1975. Stratigraphie und Fazies des südlichen Damara-Belts. Bericht des Sonderforschungsbereichs Univ. Göttingen 48:40–50.

Porada, H. 1979. The Damara-Ribeira orogen of the Pan-African Brasiliano cycle in Namibia (Southwest Africa) and Brazil as interpreted in terms of continental collison. Tectonophysics 57:237–265.

Porada, H. (In preparation). Turbidites in the Damara orogen.

Porada H., Wittig, R. 1975. Zur Tektonik des südlichen Damara-Belts. Bericht des Sonderforschungsbereichs, Univ. Göttingen 48:51–70.

Porada, H., Wittig, R. 1976. Das Causib-Turbiditbecken am Südrand des Damara-Orogens, Südwest-Afrika. Geol. Rundsch. 65:1002–1019.

Porada. H., Wittig, R. 1977. Die Entwicklung der basalen Damara-Folge am Abbabis Inlier (Südwest-Afrika) und Uberlegungen zur Geodynamik dieses Bereichs. N. Jb. Geol. Paläont. Monatsh. 8:475–502.

Potgieter, C. T. 1950. The structure and petrology of the George granite plutons and the invaded pre-Cape sedimentary rocks. Univ. Stellenbosch Ann. A26:323–412.

Potter, P. E., Pettijohn, F. J. 1963. Paleocurrents and basin analysis. Berlin: Springer-Verlag.

Powell, J. L., Hurley, P. M., Fairbairn, H. W. 1966. The strontium isotopic composition and origin of carbonatites. *In* O. F. Tuttle, J. Gittins (eds.), Carbonatites, pp. 365–378. New York: Interscience.

Pretorius, D. A. 1964. The geology of the South Rand Goldfield. *In* S. H. Haughton, (ed.), The geology of some ore deposits in southern Africa, vol. 1, pp. 219–282. Johannesburg: Geol. Soc. S. Afr.

Pretorius, D. A. 1974. The structural boundary between the Kaapvaal and Sonama crustal provinces. Econ. Geol. Res. Unit Univ. Witwatersrand Inf. Circ. No. 88.

Pretorius, D. A. 1976. The nature of the Witwatersrand gold-uranium deposits. *In* K. H. Wolf (Ed.), Handbook of strata-bound and stratiform ore deposits, 7:29–88. Amsterdam: Elsevier.

Pretorius, D. A. 1979. The crustal architecture of southern Africa. Geol. Soc. S. Afr. Trans. 82 (Annex).

Pretorius, D. A. 1981. The Witwatersrand Supergroup. *In* D. R. Hunter (ed.), Precambrian of the southern hemiphere, pp. 511–520. Amsterdam: Elsevier.

Price, R. A., Douglas, R. J. W. 1972. Variations in tectonic styles in Canada. Geol. Assoc. Can. Spec. Paper No. 11.

Prins, P. 1972. A thermal aureole around the Homeep East mafic body, Namaqualand. Geol. Soc. S. Afr. Trans. 75:165–175.

Quennell, A. M., McKinlay, A. C. M., Aitken, W. G. 1956. Summary of the geology of Tanganyika. Part I. Introduction and stratigraphy. Tanganyika Geol. Surv. Mem. No. 1.

Rabinowitz, P. D. 1976. A geophysical study of the continental margin of southern Africa. Geol. Soc. Amer. Bull. 87:1643–1653.

Rabinowitz, P. D., LaBrecque, J. L. 1979. The Mesozoic South Atlantic Ocean and evolution of its continental margins. Jour. Geophys. Res. 84:5973–6002.

Rabinowitz, P. D., Cande, S. C., LaBrecque, J. L. 1976. The Falkland escarpment and Agulhas fracture zone: the boundary between oceanic and continental basement at conjugate continental margins. Acad. Brasil. Cienc. Ann. 48:241–251.

Räheim, A., Green, D. H. 1974. Experimental determination of the temperature and pressure dependence of the Fe-Mg partition coefficient for coexisting garnet and clinopyroxene. Contr. Mineral. Petrol. 48:179–203.

Ramsay, J. G. 1963. Structural investigations in the Barberton Mountain Land, eastern Transvaal. Geol. Soc. S. Afr. Trans. 66:352–398.

Ramsay, J. G. 1967. Folding and fracturing of rocks. New York: McGraw-Hill.

Ramsay, J. G., Graham, R. H. 1970. Strain variation in shear belts. Can. Jour. Earth Sci. 7:786–813.

Rankama, K. 1970. Proterozoic, Archaean and other weeds in the Precambrian rock garden. Bull. Geol. Soc. Finland 42:211–222.

Rawlinson, P. J., Dawson, J. B. 1979. A quench pyroxene-ilmenite xenolith from kimberlite: Implications for pyroxene-ilmenite intergrowths. 2nd Intl. Kimberlite Conf. Proc. 2:292–299.

Read, H. H. 1951. Metamorphism and granitization. Geol. Soc. S. Afr. Trans. 54 (Annex.).

Redfield, A. C. 1958. The influence of continental shelf on the tides of the Atlantic coast of the U.S.A. Jour. Mar. Res. 17:432–448.

Reed, F. R. C. 1925. Revision of the fauna of the Bokkeveld Beds. S. Afr. Mus. Ann. 22:27–225.

Reeves, C. V., Hutchins, D. G. 1975. New data on

crustal structures in central southern Africa. Nature 254:408–410.

Reid, D. L. 1977. Geochemistry of Precambrian igneous rocks in the lower Orange River region. Precambrian Res. Unit Univ. Cape Town Bull. No. 22.

Reid, D. L. 1979a. Petrogenesis of calc-alkaline metalavas in the mid-Proterozoic Haib Volcanic Subgroup, lower Orange River region. Geol. Soc. S. Afr. Trans. 82:109–131.

Reid, D. L. 1979b. Total rock Rb-Sr and U-Th-Pb isotopic study of Precambrian metavolcanic rocks in the lower Orange River region, southern Africa. Earth Planet. Sci. Lett. 42:368–378.

Reid, D. L. 1979c. Age relationships within the Mid-Proterozoic Vioolsdrif batholith, lower Orange River region. Geol. Soc. S. Afr. Trans. 82:305–311.

Reid, D. L. 1979d. Basic-ultrabasic complexes within the Vioolsdrif batholith, lower Orange River region. Geol. Soc. S. Afr. 18th Congr. Abstracts 1:286–289.

Reid, D. L. 1979e. The Late Precambrian Gannakouriep mafic dyke swarm, lower Orange River region. Geol. Soc. S. Afr. 18th Congr. Abstracts 1:282–285.

Reid, D. L., Moore, A. C., Duncan, A. R. 1979. Petrology and geochemistry of the Nouzees intrusion, NW Cape Province. Geol. Soc. S. Afr. 18th Congr. Abstracts 1:290–296.

Reimer, T. O. 1972. Diagenetic reactions in Early Precambrian graywackes of the Barberton Mountain Land (South Africa). Sediment. Geol. 7:263–282.

Reineck, H.-E. 1967. Layered sediments of tidal flats, beaches and shelf of the North Sea. *In* Lauff, G. H. (ed.), Estuaries. Amer. Assoc. Adv. Sci. Publ. 83: 191–206.

Reineck, H.-E. 1972. Tidal flats. *In* Rigby, J. K., Hamblin, W. K. (eds.), Recognition of ancient sedimentary environments. Soc. Econ. Paleont. Mineral. Spec. Publ. No. 16, pp. 146–156.

Reineck, H.-E., Singh, I. B. 1973. Depositional sedimentary environments. New York: Springer-Verlag.

Reyment, R. A., Tait, E. A. 1972. Biostratigraphical dating of the early history of the South Atlantic Ocean. Roy. Soc. London Phil. Trans. B264:55–95.

Rhodes, R. C. 1974a. Petrochemical characteristics of the Bushveld granite and Rooiberg felsite. Geol. Soc. S. Afr. Trans. 77:93–98.

Rhodes, R. C. 1974b. Discussion of paper by A. R. Newton: A gravity folding model of the Cape Fold-Belt. Geol. Soc. S. Afr. Trans. 77:207–209.

Rhodes, R. C., 1975. Bushveld granophyre in the Stavoren tin district, Transvaal. Geol. Soc. S. Afr. Trans. 78:71–74.

Rhodes, R. C., Du Plessis, M. D. 1976. Notes on some stratigraphic relations in the Rooiberg felsite. Geol. Soc. S. Afr. Trans. 79:183–185.

Rhodes, R. C., Krohn, D. H. 1972. Tectonic control over regional geochemical variations in the Karroo basaltic province of southern Africa. Geol. Soc. S. Afr. Trans. 75:11–21.

Rhodes, R. C., Leith, M. J. 1967. Lithostratigraphic zones in the Table Mountain Series of Natal. Geol. Soc. S. Afr. Trans. 70:15–28.

Rhodes, R. C., Leith, M. J. 1971. Granitic rocks in the basement complex around Mapumulo, Natal, in the light of zircon studies. Geol. Soc. S. Afr. Trans. 74: 57–68.

Ricci Lucchi, F., Valmori, E. 1980. Basin-wide turbidites in a Miocene, over-supplied deep-sea plain: a geometrical analysis. Sedimentology 27:241–270.

Rigassi, D. A. 1970. Cretaceous of the Cape Province, Republic of South Africa. Unpubl. report, Petroconsultants S.A., Geneva, Switzerland.

Rigassi, D. A., Dixon, G. 1972. Cretaceous of the Cape Province, Republic of South Africa. Ibadan Univ. Conf. on Afr. Geol., December 1970, pp. 513–523.

Ritter, U. 1976. Preliminary report on the geology of the north-eastern Richtersveld. Precambrian Res. Unit Univ. Cape Town Ann. Rept. 13:42–48.

Ritter, U. 1978. The rocks of the south-eastern Richtersveld between Eksteenfontein and Klein Helskloof: their relationships to Gariep metamorphism and tectonism and to the Namaqualand metamorphic complex. Precambrian Res. Unit Univ. Cape Town Ann. Rept. 14/15:83–98.

Ritter, U. 1980. The Precambrian evolution of the eastern Richtersveld. Prevambrian Res. Unit Univ. Cape Town Bull. 26.

Robb, L. J. 1978. A general geological description of the Archaean granitic terrane between Nelspruit and Bushbuckridge, eastern Transvaal. Geol. Soc. S. Afr. Trans. 81:331–338.

Roberts, J. L. 1970. The intrusion of magma into brittle rocks. *In* G. Newall, N. Rast (eds.), Mechanisms of igneous intrusion. Liverpool Geol. Soc. Geol. Jour. Spec. Issue No. 2, pp. 187–338.

Robertson, I. D. M. 1968. Granulite metamorphism of the Basement Complex in the Limpopo Metamorphic Zone. Geol. Soc. S. Afr. Trans. 71 (Annex.): 125–133.

Robertson, I. D. M. 1973a. The geology of the country around Mount Towla, Gwanda district. Rhodesia Geol. Surv. Bull. No. 68.

Robertson, I. D. M. 1973b. Potash granites of the southern edge of the Rhodesian craton and the northern granulite zone of the Limpopo Mobile Belt. Geol. Soc. S. Afr. Spec. Publ. 3:265–276.

Robertson, I. D. M. 1977. Some granulite facies metasediments of the Rhodesian part of the North Marginal Zone of the Limpopo Mobile Belt. Botswana Geol. Surv. Bull. 12:157–176.

Robertson, I. D. M., Du Toit, M. C. 1981. The Limpopo Belt. *In* D. R. Hunter (ed.), Precambrian of the southern hemisphere, pp. 641–671. Amsterdam: Elsevier.

Robertson, I. D. M., Van Breemen, O. 1970. The southern satellite dykes of the Great Dyke, Rhodesia. Geol. Soc. S. Afr. Spec. Publ. No. 1, pp. 621–644.

Robey, J. 1979. Probable lower crust metamorphics from the Sonama-Kaapvaal craton junction - evidence from kimberlites. Geol. Soc. S. Afr. 18th Congr. Abstracts. 1:301.

Rogers, A. W. 1904. Geological survey of the northwestern part of Van Rhyn's Dorp. Cape of Good Hope Geological Commission Ann. Rept. 9:11–46.

Rogers, A. W. 1905. An introduction to the geology of the Cape Colony. London: Longmans.

Rogers, A. W. 1910. The Kheis Series. Geol. Soc. S. Afr. Trans. 13:93–104.

Rogers, A. W. 1915. The geology of part of Namaqualand. Geol. Soc. S. Afr. Trans. 18:72–101.

Rogers, A. W. 1916. The nature of the copper deposits of Little Namaqualand. Geol. Soc. S. Afr. Proc. 18:21–35.

Rogers, A. W., Du Toit, A. L. 1908. Report on the geology of parts of Prieska, Hay, Britstown, Carnarvon and Victoria West. Cape of Good Hope Geological Commission Ann. Rept. 13:7–109.

Rogers, A. W., Du Toit, A. L. 1909. Report on the geology of parts of Kenhardt, Prieska, Victoria West and Carnarvon. Cape of Good Hope Geological Commission Ann. Rept. 14:8–108.

Rogers, J. 1977. Sedimentation on the continental margin off the Orange River and the Namib Desert. S. Afr. Geol. Surv./Univ. Cape Town Joint Mar. Geoscience Group Bull. 7:1–148.

Rona, P. A. 1973. Relations between rates of sediment accumulation on continental shelves, sea-floor spreading, and eustasy inferred from central north Atlantic. Geol. Soc. Amer. Bull. 84:2851–2872.

Roscoe, S. M. 1969. Huronian rocks and uraniferous conglomerates in the Canadian Shield. Geol. Surv. Can. Pap. No. 68-40.

Roussouw, P. J. 1970. The Witteberg-Dwyka contact in the Willowmore and Steytlerville districts. 2nd Intl. Gondwana Symposium. Proceedings and Papers, pp. 205–207.

Roussouw, P. J., Meyer, E. I., Mulder, M. P., Stocken, C. G. 1964. The geology of the Swartberge, the Cango Valley and the environs of Prince Albert, C.P. S. Afr. Geol. Surv. Map Explanation Sheets 3321B (Gamkapoort) and 3322A (Prince Albert).

Roux, A. T. 1967. The application of geophysics to gold exploration in South Africa. Can. Geol. Surv. Econ. Geol. Rept 26.

Rowsell, D. M., Connan, J. 1979. Oil generation, migration and preservation in the Middle Ecca sequence near Dannhauser and Wakkerstroom. Geol. Soc. S. Afr. Spec. Publ. No. 6, pp. 131–150.

Rowsell, D. M., De Swardt, A. M. J. 1974. Secondary leaching porosity in Middle Ecca sandstones. Geol. Soc. S. Afr. Trans. 77:131–140.

Rowsell, D. M., De Swardt, A. M. J. 1976. Diagenesis in Cape and Karoo sediments, South Africa, and its bearing on their hydrocarbon potential. Geol. Soc. S. Afr. Trans. 79:81–153.

Rozendaal, A. 1978. The Gamsberg zinc deposit, Namaqualand. Geol. Soc. S. Afr. Spec. Publ. No. 4, pp. 235–265.

Rozendaal, A. 1979. The Gamsberg zinc deposit: A banded stratiform base metal sulphide ore deposit. Geol. Soc. S. Afr. 18th Congr. Abstracts 1:323–328.

Ruddock, A. 1968. Cainozoic sea-levels and diastrophism in a region bordering Algoa Bay. Geol. Soc. S. Afr. Trans. 71:209–233.

Ruddock, A. 1973. The Tertiary limestones of the southern coastal regions of Cape Province, South Africa. *In* G. Blant (ed.), Sedimentary basins of the African coasts. Part 2. South and East Coasts, pp. 49–62. Paris: Assoc. African Geol. Surveys.

Rundle, C. C., Snelling, N. J. 1977. The geochronology of uraniferous minerals in the Witwatersrand triad; an interpretation of new U-Pb data on rock and minerals from the Dominion Reef, Witwatersrand and Ventersdorp Supergroups. Roy. Soc. London Phil. Trans. A286:567–583.

Rust, B. R. 1972. Structures and processes in a braided river. Sedimentology 18:221–245.

Rust, I. C. 1962. On the sedimentation of the Molteno sandstones in the vicinity of Molteno, C. P. Univ. Stellenbosch Ann. 37:165–234.

Rust, I. C. 1967. On the sedimentation of the Table Mountain Group in the western Cape Province. Unpubl. D.Sc. thesis, Univ. Stellenbosch, South Africa.

Rust, I. C. 1969. The western Cape some 450 million years ago. S. Afr. Jour. Geogr. 3:351–358.

Rust, I. C. 1973. The evolution of the Paleozoic Cape basin, southern margin of Africa. *In* A. E. M. Nairn, F. G. Stehli (eds.), The ocean basins and margins, I. The South Atlantic, pp. 247–276. New York: Plenum.

Rust, I. C. 1975. Tectonic and sedimentary framework of Gondwana basins in southern Africa. *In* K. S. W. Campbell (ed.), Gondwana Geology, pp. 537–564. Canberra: Australian National Univ. Press.

Rust, I. C. 1977. Evidence of shallow marine and tidal sedimentation in the Ordovician Graafwater Formation, Cape Province, South Africa. Sediment. Geol. 18:123–133.

Rust, I. C. (In press). Lower Paleozoic Pakhuis tillite, South Africa. *In* W. B. Harland, M. J. Hambrey, (eds.), Earth's pre-Pleistocene glacial record. IGCP Project 38—Pre-Pleistocene tillites. Cambridge: Cambridge Univ. Press.

Rust, I. C., Rogers, J. 1979. Robberg and environs of Plettenberg Bay. *In* I. C. Rust (ed.), Excursions guide book, Geokongres 79, pp. 91–100. Johannesburg: Geol. Soc. S. Afr.

Rust, I. C., Theron, J. N. 1964. Some aspects of the Table Mountain Series near Vanrhynsdorp. Geol. Soc. S. Afr. Trans. 67:133–137.

Rust, I. C., Winter, H. de la R. 1979. The Cretaceous Oudtshoorn basin. *In* I. C. Rust (ed.), Excursions guidebook, Geokongres 79, pp. 53–57. Johannesburg: Geol. Soc. S. Afr.

Rust, I. C., Theron, J. N., Loock, J. C. 1970. Trans-Karoo excursion: Ceres to Oudtshoorn. 2nd Intl. Gondwana Symposium, Guide book 3:21–31.

Ryan, P. J. 1967. Stratigraphic and paleocurrent analysis of the Ecca Series and lowermost Beaufort beds in the Karroo basin of South Africa. Unpubl. Ph.D. thesis, Univ. Witwatersrand, Johannesburg, South Africa.

Ryan, P. J., Whitfield, G. G. 1979. Basinal analysis of the Ecca and lowermost Beaufort beds and associated coal, uranium, and heavy mineral beach sand occurrences. Geol. Soc. S. Afr. Spec. Publ. No. 6, pp. 97–101.

Ryer, T. A. 1977. Patterns of Cretaceous shallow-marine sedimentation, Coalville and Rockport areas, Utah. Geol. Soc. Amer. Bull. 88:177–188.

Saager, R., Köppel, V. 1976. Lead isotopes and trace elements from sulfides of Archean greenstone belts in South Africa—A contribution to the knowledge of the oldest known mineralizations. Econ. Geol. 71:44–57.

Saggerson, E. P., Logan, C. T. 1970. Distribution controls of layered and differentiated mafic intrusions in the Lebombo volcanic sub-province. Geol. Soc. S. Afr. Spec. Publ. No. 1, pp. 721–733.

Saggerson, E. P., Turner, L. M. 1976. A review of the distribution of metamorphism in the ancient Rhodesian craton. Precambr. Res. 3:1–53.

Sassi, F. P., Scolari, A. 1974. The b_0 value of the potassic white micas as a barometric indicator in low grade metamorphism in pelitic schists. Contr. Mineral. Petrol. 45:143–152.

Savage, N. M. 1970. A preliminary note on arthropod trace fossils from the Dwyka Series in Natal. 2nd Gondwana Symposium. Proceedings and Papers, pp. 627–636.

Savage, N. M. 1971. A varvite ichnocoenosis from the Dwyka Series of Natal. Lethaia 4:217–233.

Savage, N. M. 1972. Soft-sediment glacial grooving of Dwyka age in South Africa. Jour. Sediment. Petrol. 42:307–308.

Savin, S. M., Douglas, R. G., Stehli, F. G. 1975. Tertiary marine paleotemperatures. Geol. Soc. Amer. Bull. 86:1499–1510.

Sawkins, F. J. 1976. Massive sulphide deposits in relation to geotectonics. Geol. Soc. Can. Spec. Pap. No. 14, pp. 221–240.

Sawyer, E. W. 1978. Damaran structural and metamorphic geology of an area south-east of Walvis Bay, South West Africa/Namibia. Unpubl. M.Sc. thesis, Univ. Cape Town, South Africa.

Saxena, R. S. 1976. Sand bodies and sedimentary environments of the modern Mississippi delta—An excellent model for exploration in deltaic sandstone reservoirs. Egypt General Petroleum Corporation Exploration Seminar, Cairo, 1976.

Schalk, K. E. L. 1973. Some Late Precambrian formations in central South West Africa. S. Afr. Geol. Surv. Ann. 8:29–47.

Schalke, H. J. W. G. 1973. The Upper Quaternary of the Cape Flats area (Cape Province, South Africa). Scripta Geol. 15:1–57.

Schermerhorn, L. J. G. 1974. Late Precambrian mixtites: Glacial and/or nonglacial? Amer. Jour. Sci. 274:673–824.

Schmidt, E. R. 1952. The structure and composition of the Merensky Reef and associated rocks of the Rustenburg platinum mine. Geol. Soc. S. Afr. Trans. 55:233–279.

Schoch, A. E., Burger, A. J. 1976. U-Pb age of the Saldanha quartz porphyry. Geol. Soc. S. Afr. Trans. 79:239–241.

Schoch, A. E., Leygonie, F. E., Burger, A. J. 1975. U-Pb ages for Cape granites from the Saldanha batholith: a preliminary report. Geol. Soc. S. Afr. Trans. 78:97–100.

Schoch, A. E., Leterrier, J., La Roche, H. De. 1977. Major element geochemical trends in the Cape granites. Geol. Soc. S. Afr. Trans. 80:197–209.

Scholtz, D. L. 1946. On the younger Precambrian granite plutons of the Cape Province. Geol. Soc. S. Afr. Proc. 49:35–82.

Scholz, C. H., Koczynski, T. A., Hutchins, D. G. 1976. Evidence for incipient rifting in southern Africa. Geophys. Jour. Roy. Astr. Soc. 44:135–144.

Schopf, J. W. 1976. Evidence of Archean life: a brief appraisal. *In* B. W. Windley (ed.), The early history of the earth, pp. 589–593. New York: Wiley.

Schopf, J. W., Oehler, D. Z., Horodyski, R. J., Kvenvolden, K. A. 1971. Biogenicity and the significance of the oldest known stromatolites. Jour. Paleont. 45: 477–485.

Schreiner, G. D. L., Van Niekerk, C. B. 1958. The age of a Pilanesberg dyke from the Central Witwatersrand. Geol. Soc. S. Afr. Trans. 61:198–199.

Schreyer, W., Abraham, K. 1978. Symplectitic cordierite-orthopyroxene-garnet assemblages as products of contact metamorphism of pre-existing basement granulites in the Vredefort structure, South Africa, and their relations to pseudotachylite. Contr. Mineral. Petrol. 68:53–72.

Schultz, R. 1978. Origin of the so-called charnockitic

adamellite porphyry from the Upington geotraverse. Precambrian Res. Unit Univ. Cape Town Ann. Rept. 14/15:48–82.

Schumm, S. A. 1960. The effect of sediment type on the shape and stratification of some modern fluvial deposits. Amer. Jour. Sci. 258:177–184.

Schumm, S. A. 1968a. Speculations concerning paleohydrologic controls of terrestrial sedimentation. Geol. Soc. Amer. Bull. 79:1573–1588.

Schumm, S. A. 1968b. River adjustment to altered hydraulic regimen—Murrumbidgee River and paleochannels. U.S. Geol. Surv. Prof. Pap. No. 598.

Schumm, S. A. 1972. Fluvial paleochannels. *In* J. K. Rigby, W. K. Hamblin (eds.), Recognition of ancient sedimentary environments. Soc. Econ. Paleont. Mineral. Spec. Publ. No. 16, pp. 98–107.

Schutte, I. C. 1972. The main pegmatites of the area between Steinkopf, Vioolsdrif and Goodhouse. S. Afr. Geol. Surv. Mem. No. 60.

Schwartz, R. K. 1975. Nature and genesis of some washover deposits. U.S. Army Corps of Engineers, Coastal Engineering Research Centr. Tech. Memo. No. 61.

Schwellnus, C. M. 1941. The Nama tillite in the Klein-Kharas Mountains, South West Africa. Geol. Soc. S. Afr. Trans. 44:19–33.

Schwellnus, J. S. I. 1956. The basal portion of the Bushveld Igneous Complex and the adjoining metamorphosed sediments in the north-eastern Transvaal. Unpubl. D.Sc. thesis, Univ. Pretoria, South Africa.

Schwellnus, J. S. I., Engelbrecht, L. N. J., Coertze, F. J., Russell, M. C., Malherbe, J. J., Van Rooyen, D. P., Cooke, R. 1962. The geology of the Olifants River area. S. Afr. Geol. Surv. Map Explanation Sheets 2429B and 2430A.

Schwerdtner, W. M., Sutcliffe, R. H., Tröeng, B. 1978. Patterns of total strain in the crestal region of immature diapirs. Can. Jour. Earth Sci. 15: 1437–1447.

Scrutton, R. A. 1973a. The age relationship of igneous activity and continental breakup. Geol. Mag. 110: 227–234.

Scrutton, R. A. 1973b. Structures and evolution of the sea floor south of South Africa. Earth Planet. Sci. Lett. 19:250–256.

Scrutton, R. A., Dingle, R. V. 1974. Basement control over sedimentation on the continental margin west of southern Africa. Geol. Soc. S. Afr. Trans. 77:253–260.

Scrutton, R. A., Dingle, R. V. 1976. Observations on the processes of sedimentary basin formation at the margins of southern Africa. Tectonophysics 36:143–156.

Segouffin, J. 1977. Carte bathymétrique du canal de Mozambique. Rapport d'activité du Laboratoire de Géophysique marine de Physique du Globe de Paris, pour la période Mars 1976-Février 1977.

Selby, M. J. 1976. Some thoughts on the geomorphology of the central Namib Desert. Desert Ecol. Res. Unit Bull. 1:5–6.

Selby, M. J., Hendy, C. H., Seely, M. K. 1979. A Late Quaternary lake in the central Namib Desert, southern Africa, and some implications. Palaeogeogr. Palaeoclim. Palaeoecol. 26:37–41.

Sellwood, B. W. 1970. The relation of trace fossils to small scale sedimentary cycles in the British Lias. *In* T. P. Crimes, J. C. Harper (eds.), Trace fossils. Jour. Geol. Spec. Issue 3:489–504.

Seward, A. C. 1907. Notes of fossil plants from South Africa. Geol. Mag. 54:481–487.

Shackleton, N. J., Kennett, J. P. 1975. Palaeotemperature history of the Cenozoic and the initiation of Antarctic glaciation: oxygen and carbon isotope analysis in D. S. D. P. sites 279, 277, and 281. Initial Reports of the Deep Sea Drilling Project No. 29, pp. 743–755.

Shackleton, N. J., Opdyke, N. D. 1977. Oxygen isotope and palaeomagnetic evidence for early northern hemisphere glaciation. Nature 270:216–219.

Shackleton, R. M. 1956. Downward-facing structures of the Highland Border. Geol. Soc. London Jour. 113:361–392.

Shand, S. J. 1916. The pseudotachylite of Parys (Orange Free State) and its relation to "trapshotten gneiss" and "fluity crush-rock." Geol. Soc. London Quart. Jour. 72:198–217.

Sharpe, M. R., and Snyman, J. A. 1980. A model for the emplacement of the eastern compartment of the Bushveld Complex. Tectonophysics 65:85–110.

Shergold, J. H., Cooper, R. A., MacKinnon, D. I., Yochelson, E. C. 1976. Late Cambrian Brachiopoda, Mollusca, and Trilobita from northern Victoria Land, Antarctica. Palaeontology 19:247–291.

Shinn, E. A. 1968. Practical significance of birdseye structures in carbonate rocks. Jour. Sediment. Petrol. 38:215–223.

Shinn, E. A. 1973a. Carbonate coastal accretion in an area of longshore transport, north-eastern Quatar. *In* B. H. Purser (ed.), The Persian Gulf, pp. 211–223. New York: Springer-Verlag.

Shinn, E. A. 1973b. Sedimentary accretion along the leeward, S. E. coast of Quatar Peninsula, Persian Gulf. *In* B. H. Purser (ed.), The Persian Gulf, pp. 199–209. New York: Springer-Verlag.

Shinn, E. A., Lloyd, R. M., Ginsburg, R. N. 1969. Anatomy of a modern carbonate tidal flat, Andros Island, Bahamas. Jour. Sediment. Petrol. 39:1202–1228.

Shone, R. W. 1976. The sedimentology of the Mesozoic Algoa basin. Unpubl. M.Sc. thesis, Univ. Port Elizabeth, South Africa.

Shone, R. W. 1978a. Giant *Cruziana* from the Beaufort Group. Geol. Soc. S. Afr. Trans. 81:327–329.

Shone, R. W. 1978b. A case for lateral gradation between the Kirkwood and Sundays River Formations, Algoa basin. Geol. Soc. S. Afr. Trans. 81: 319–326.

Shone, R. W. 1979. Sardinia Bay excursion. *In* I. C. Rust (ed.), Excursion guidebook, Geol. Soc. S. Afr. Geokongres 79, pp. 173–180.

Short, N. M., Lowman, P. D., Freden, S. C. 1976. Mission to Earth: Landsat views the world. Washington, D.C.: Natl. Aeronaut. Space Admin. Scientific and Technical Office.

Sibson, R. H. 1975. Generation of pseudotachylyte by ancient seismic faulting. Geophys. Jour. Roy. Astr. Soc. 43:775–794.

Siedner, G., Miller, J. A. 1968. K-Ar determinations on basaltic rocks from South West Africa and their bearing on continental drift. Earth Planet. Sci. Lett. 4:451–458.

Siedner, G., Mitchell, J. G. 1976. Episodic Mesozoic volcanism in Namibia and Brazil: a K-Ar isochron study bearing on the opening of the South Atlantic. Earth Planet. Sci. Lett. 30:292–303.

Siesser, W. G. 1972. Petrology of the Cainozoic coastal limestones of the Cape Province, South Africa. Geol. Soc. S. Afr. Trans. 75:177–185.

Siesser, W. G. 1977a. Biostratigraphy and micropalaeontology of continental margin samples. S. Afr. Geol. Surv./Univ. Cape Town Joint Mar. Geol. Progr. Tech. Rept. 9:108–117.

Siesser, W. G. 1977b. Upper Eocene age of marine sediments at Bogenfels, South West Africa, based on calcareous nannofossils. S. Afr. Geol. Surv. Bull. 60:73–74.

Siesser, W. G. 1978. Age of phosphorites on the South African continental margin. Mar. Geol. 26:M17–M28.

Siesser, W. G. 1980. Late Miocene origin of the Benguela upwelling system off northern Namibia. Science 208:283–285.

Siesser, W. G., Dingle, R. V. 1981. Tertiary sea-level movements around southern Africa. Jour. Geol. 89: 523–536.

Siesser, W. G., Miles, G. A. 1979. Calcareous nannofossils and planktic foraminifers in the Tertiary limestones, Natal and eastern Cape, South Africa. S. Afr. Mus. Ann. 79:139–158.

Siesser, W. G., Salmon, D. 1979. Eocene marine sediments in the Sperrgebiet, South West Africa. S. Afr. Mus. Ann. 79:9–34.

Siesser, W. G., Scrutton, R. A., Simpson, E. S. W. 1974. Atlantic and Indian Ocean margins of southern Africa. *In* C. A. Burk, C. L. Drake (eds.), The geology of continental margins, pp. 641–654. New York: Springer-Verlag.

Simpson, C. 1978. The structure of the rim synclinorium of the Vredefort Dome. Geol. Soc. S. Afr. Trans. 81:115–121.

Simpson, E. S. W. 1968. Marine geology: progress and problems. Geol. Soc. S. Afr. Trans. 71:97–111.

Simpson, E. S. W. 1970. The anorthosite of southern Angola: A review of present data. *In* T. N. Clifford, I. G. Gass (eds.), African magmatism and tectonics, pp. 89–96. Edinburgh: Oliver and Boyd.

Simpson, E. S. W., Dingle, R. V. 1973. Offshore sedimentary basins on the southeastern continental margin of South Africa. *In* G. Blant (ed.), Sedimentary basins of the African coast. Part 2. South and east coast, pp. 63–67. Paris: Assoc. African Geol. Surveys.

Simpson, E. S. W., Forder, E. 1968. The Cape submarine canyon. Part I. Bathymetry. S. Afr. Div. Sea Fisheries Bull. 5:35–38.

Simpson, E. S. W., Tregidga, J. A. 1956. The Archaean rocks of the Marble Delta district, Natal. Geol. Soc. S. Afr. Trans. 59:237–258.

Simpson, E. S. W., Sclater, J. G., Parsons, B., Norton, I., Meinke, L. 1979. Mesozoic magnetic lineations in the Mozambique basin. Earth Planet. Sci. Lett. 43:260–264.

Sims, P. K. 1980. Subdivision of the Proterozoic and Archean eon: Reconstructions and suggestions by the International Subcommission on Precambrian Stratigraphy. Precambr. Res. 13:379–380.

Singer, S. F. 1970. Origin of the moon by capture and its consequences. Amer. Geophys. Union Trans. 51:637–641.

Sleep, N. H. 1976. Platform subsidence mechanisms and "eustatic" sea-level changes. Tectonophysics 36:45–56.

Sloss, L. L., Speed, R. C. 1974. Relationships of cratonic and continental margin tectonic episodes. *In* W. R. Dickinson (ed.) Tectonics and sedimentation. Soc. Econ. Paleont. Mineral. Spec. Publ. No. 22, pp. 98–119.

Smit, P. J. 1973. The Waterberg System in the northern Cape. Geol. Soc. S. Afr. 15th Congr. Abstracts 1:47–48.

Smit, P. J. 1977. Die geohidrologie in die opvanggebied van die Moloporivier in die noordelike Kalahari. Unpubl. Ph.D. thesis, Univ. Orange Free State, Bloemfontein.

Smit, P. J., Hales, A. L., Gough, D. J. 1962. The gravity survey of the Republic of South Africa. S. Afr. Geol. Surv. Handbook No. 3.

Smith, A. G., Hallam, A. 1970. The fit of the southern continents. Nature 225:139–144.

Smith, A. G., Briden, J. C., Drewry, G. E. 1973. Phanerozoic world maps. Systematics Assoc. Publ. No. 9.

Smith, C. B., McCallum, M. E., Eggler, D. H. 1976. Clinopyroxene-ilmenite intergrowths from the Iron Mountain kimberlite district, Wyoming. Carnegie Inst. Washington Yearbook 75:542–546.

Smith, D. A. M. 1965. The geology of the area around the Khan and Swakop Rivers in South West Africa. S. Afr. Geol. Surv. Mem. (South West Africa series) No. 3.

Smith, D. A. M. 1978. The location and nature of South Africa's coal resources. *In* Coal Processing, vol. 1, pp. 5/0–5/17. Johannesburg: South African Coal Processing Soc., Rand Afrikaans University.

Smith, D. G., Smith, N. D. 1980. Sedimentation in anastomosed river systems: Examples from alluvial valleys near Banff, Alberta. Jour. Sediment. Petrol. 50:157–164.

Smith, N. D. 1971a. Transverse bars and braiding in the lower Platte River, Nebraska. Geol. Soc. Amer. Bull. 81:2993–3014.

Smith, N. D. 1971b. Pseudo-planar stratification produced by very low amplitude sand waves. Jour. Sediment. Petrol. 41:69–73.

Smith, N. D., Minter, W. E. L. 1980. Sedimentological controls of gold and uranium in two Witwatersrand paleoplacers. Ecol. Geol. 75:1–14.

Smith, R. M. H. 1978. A study of the sedimentology and taphonomy of floodplain deposits of the lower Beaufort (Adelaide Subgroup) strata near Beaufort West, Cape Province. S. Afr. Geol. Surv. Ann. 12:37–68.

Smith, R. M. H. 1979. Sedimentology and taphonomy of lower Beaufort floodplain deposits. Geol. Soc. S. Afr. 18th Congr. Abstracts 2:136–144.

Smithson, S. B., Oliver, J. E., Kaufman, S., Brown, L. D. 1980. The Laramide orogeny: evidence from COCORP deep crustal seismic profiles in the Wind River Mountains, Wyoming. Tectonophysics 62: 165–189.

Snethlage, R., Von Gruenewaldt, G. 1977. Oxygen fugacity and its bearing on the origin of chromitite layers in the Bushveld Complex. *In* D. D. Klemm, H. J. Schneider (eds.), Time- and strata-bound ore deposits, pp. 353–370. Berlin, Heidelberg: Springer-Verlag.

Soares, A. F., Da Silva, G. H. 1970. Contribuição para o estudo da geologia do Maputo. Revta Ciênc. Geol. Lourenço Marques A3:1–85.

Sohl, N. F. 1977. Note 45—Application for amendment concerning terminology for igneous and high-grade metamorphic rocks. Amer. Assoc. Petrol. Geol. Bull. 61:248–251.

Söhnge, P. G. 1946. The geology of the Messina copper mines and surrounding country. S. Afr. Geol. Surv. Mem. No. 40.

Söhnge, P. G. 1964. The geology of the Tsumeb mine. Geol. Soc. S. Afr. Spec. Publ. No. 2, pp. 367–382.

Söhnge, P. G. 1980. Fifth meeting of Sub-Commission on Precambrian stratigraphy. Geol. Soc. S. Afr. Quart. News Bull. March 1980, pp. 13–15.

Söhnge, P. G., De Villiers, J. 1948. The Kuboos pluton and its associated line of intrusives. Geol. Soc. S. Afr. Trans. 51:1–31.

Söhnge, P. G., Le Roex, H. D., Nel, H. J. 1948. The geology of the country around Messina. S. Afr. Geol.. Surv. Map Explanation Sheet No. 46.

South African Committee for Stratigraphy. 1980. Stratigraphy of South Africa. Part I. Lithostratigraphy of the Republic of South Africa, South West Africa/Namibia, and the Republics of Bophuthatswana, Transkei and Venda. S. Afr. Geol. Surv. Handbook No. 8.

Spath, L. F. 1930. On the Cephalopoda of the Uitenhage beds. S. Afr. Mus. Ann. 28:131–157.

Stanley, S. M. 1976. Fossil data and the Precambrian-Cambrian evolutionary transition. Amer. Jour. Sci. 276:56–76.

Stanistreet, I. G. 1979. Trace fossils as sedimentological and palaeoenvironmental indices in the Ecca Group of the Transvaal. Geol. Soc. S. Afr. 18th Congr. Abstracts 2:84–87.

Stapleton, R. P. 1974. Dinoflagellates from the Lower Beaufort Stage. S. Afr. Geol. Surv. Ann. 10:87–89.

Stapleton, R. P. 1977. Planktonic Foraminifera and the age of the Uloa *Pecten* bed. S. Afr. Geol. Surv. Bull. 60:11–15.

Stapleton, R. P., Beer, E. M. 1977. Micropalaeontological age determination for the Brenton beds. S. Afr. Geol. Surv. Bull. 60:1–9.

Stavrakis, N. 1980. Sedimentation of the Katberg sandstone and adjacent formations in the southeastern Karoo basin. Geol. Soc. S. Afr. Trans. 83, (in press).

Stear, W. M. 1977. The stratigraphy and sedimentation of the Pretoria Group at Rooiberg, Transvaal. Geol. Soc. S. Afr. Trans. 80:53–65.

Stear, W. M. 1978. Sedimentary structures related to fluctuating hydrodynamic conditions in flood plain deposits of the Beaufort Group near Beaufort West, Cape. Geol. Soc. S. Afr. Trans. 81:393–399.

Stear, W. M. 1979. Channel sandstone and bar morphology of the lower Beaufort Group uranium district near Beaufort West, Cape. Geol. Soc. S. Afr. 18th Congr. Abstracts 2:132–135.

Steyn, J. G. D. 1962. The mineralogy of the more important tin-bearing pipes on Stavoren. S. Afr. Geol. Surv. Mem. No. 51.

Steyn, L. S. 1964. The sedimentology and gold distribution pattern of the Livingstone Reefs of the West Rand. Unpubl. M.Sc. thesis, Univ. Witwatersrand, Johannesburg, South Africa.

Steyn, L. S. 1977. Regional controls of oil and gas accumulations in the northern part of the Karoo basin. Geol. Soc. S. Afr. 17th Congr. Abstracts 1: 99–102.

Stiles, D. N., Partridge, T. C. 1979. Results of recent archaeological and palaeoenvironmental studies at the Sterkfontein extension site. S. Afr. Jour. Sci. 75:346–352.

Stocken, C. G. 1954. The pre-Cape rocks of the central Cango. Unpubl. Ph.D. thesis, Univ. Cape Town, South Africa.

Stockwell, C. H. 1969. Tectonic map of Canada. Can. Geol. Surv. Comm. Tectonic Map Canada.

Stow, G. W. 1874. Geological notes upon Griqualand West. Geol. Soc. London Quart. Jour. 30:581–680.

Stowe, C. W. 1973. The older tonalite gneiss complex in the Selukwe area, Rhodesia. Geol. Soc. S. Afr. Spec. Publ. No. 3, pp. 85–95.

Stowe, C. W. 1979. The Upington Geotraverse National Geodynamics Project—1979. Progress report and outline of macrostructure. Precambrian Res. Unit Univ. Cape Town Ann. Rept. 16:20–34.

Stratten, T. 1967. A preliminary report on a direction study of the Dwyka tillite in the Karroo basin of South Africa. *In* Gondwana stratigraphy, vol. 2, pp. 741–762. Proc. 1st I.U.G.S. Gondwana Symposium, Buenos Aires.

Stratten, T. 1968. The Dwyka glaciation and its relationship to the pre-Karroo surface. Unpubl. Ph.D. thesis, Univ. Witwatersrand, Johannesburg, South Africa.

Stratten, T. 1970. Tectonic framework of sedimentation during the Dwyka Period in South Africa. 2nd Int. Gondwana Symposium. Proceedings and Papers, pp. 483–490.

Stratten, T. 1977. Conflicting directions of ice flow in the western Cape Province and southern South West Africa. Geol. Soc. S. Afr. Trans. 80:79–86.

Stratten, T. 1979. The origin of the diamondiferous alluvial gravels in the southwestern Transvaal. Geol. Soc. S. Afr. Spec. Publ. No. 6, pp. 219–228.

Strauss, C. A. 1941. The geology, copper-ore deposits and groundwater hydrology of the area around Springbok and O'okiep, Namaqualand. Unpubl. D.Sc. thesis, Univ. Stellenbosch, South Africa.

Strauss. C. A. 1947. Granitization and rheomorphism associated with the Bushveld granite near Leeuwpoort tin mine. Geol. Soc. S. Afr. Trans. 50:161–170.

Strauss, C. A. 1954. The geology and mineral deposits of the Potgietersrus tin-fields. S. Afr. Geol. Surv. Mem. No. 46.

Strauss, C. A. 1964. The iron ore deposits in the Sishen area, Cape Province. *In* S. H. Haughton (ed.), The geology of some ore deposits in southern Africa, vol. 2, pp. 393–403. Johannesburg: Geol. Soc. S. Afr.

Strauss, C. A., Truter, F. C. 1944. The Bushveld granites in the Zaaiplaats tin mining area. Geol. Soc. S. Afr. Trans. 47:47–88.

Stride, A. H. 1963. Current swept floors near the southern half of Great Britain. Geol. Soc. London Quart. Jour. 119:175–199.

Stromer, E. 1926. Reste land- und süsswasser bewohnender Wirbeltiere aus den Diamantenfelden Deutsch-Südwestafrikas. *In* E. Kaiser (ed.), Die Diamanternwüste Südwestafrikas, pp. 107–153. Berlin: Reimer.

Strydom, P. M. 1951. The arenaceous and rudaceous beds in the Jeppestown Series on Brakpan Mines Ltd., and S. A. Land and Exploration Co. Ltd. Unpubl. M.Sc. thesis, Univ. Witwatersrand, Johannesburg, South Africa.

Stump, E. 1976. On the Late Precambrian-Early Paleozoic metavolcanic and metasedimentary rocks of the Queen Maud Mountains, Antarctica and a comparison with rocks of similar age from southern Africa. Inst. Polar Studies Ohio State Univ. Rept. No. 62.

Stumpfl, E. F. 1978. Variations in silicate, oxide and sulphide mineralogy in the Okiep Copper District. Geol. Soc. S. Afr. Spec. Publ. No. 4, pp. 349–350.

Suarez, M. 1976. Plate tectonic model for the southern Antarctic Peninsula and its relation to southern Andes. Geology 4:211–214.

Summerhayes, C. P. 1973. Distribution, origin and economic potential of phosphatic sediments from the Agulhas Bank, South Africa. Geol. Soc. S. Afr. Trans. 76:271–277.

Sun, S. S., Nesbitt, R. W. 1978. Petrogenesis of Archaean ultrabasic and basic volcanics: Evidence from rare earths. Contr. Mineral Petrol. 65:301–325.

Swift, D. J. P. 1975. Tidal sand ridges and shoal retreat massifs. Mar. Geol. 18:105–134.

Sydora, L. J., Jones, F. W., Lambert, R. St J. 1980. Model calculations of the thermal fields of subducting lithospheric slabs and partial melting. Tectonophysics 62:233–249.

Talbot, C. J. 1973. The geology of the eastern end of the Zambesi orogenic belt, Rhodesia. Geol. Soc. S. Afr. Trans. 76:113–133.

Talbot, C. J. 1979. Infrastructural migmatitic upwelling in East Greenland interpreted as thermal convective structures. Precambr. Res. 8:77–93.

Talma, A. S., Vogel, J. C., Partridge, T. C. 1974. Isotopic contents of some Transvaal speleothems and their palaeoclimatic significance. S. Afr. Jour. Sci. 70:135–140.

Tankard, A. J. 1974a. Varswater Formation of the Langebaanweg-Saldanha area, Cape Province. Geol. Soc. S. Afr. Trans. 77:265–283.

Tankard, A. J. 1974b. Petrology and origin of the phosphorite and aluminium phosphate rock of the Langebaanweg-Saldanha area, southwestern Cape Province. S. Afr. Mus. Ann. 65:217–249.

Tankard, A. J. 1974c. Chemical composition of the phosphorites from the Langebaanweg-Saldanha area, Cape Province. Geol. Soc. S. Afr. Trans. 77: 185–190.

Tankard, A. J. 1975a. The Late Cenozoic history and palaeoenvironments of the coastal margin of the south-western Cape Province, South Africa. Unpubl. Ph.D. thesis, Rhodes Univ. Grahamstown, South Africa.

Tankard, A. J. 1975b. Thermally anomalous Late Pleistocene molluscs from the south-western Cape

Province, South Africa. S. Afr. Mus. Ann. 69:17–45.

Tankard, A. J. 1976a. Cenozoic sea-level changes: A discussion. S. Afr. Mus. Ann. 71:1–17.

Tankard, A. J. 1976b. The Late Pleistocene history and coastal morphology of the Ysterfontein-Elands Bay area, Cape Province. S. Afr. Mus. Ann. 69: 73–119.

Tankard, A. J. 1976c. Stratigraphy of a coastal cave and its palaeoclimatic significance. Palaeoecol. Afr. 9:151–159.

Tankard, A. J., Barwis, J. H. 1982. Wave-dominated deltaic sedimentation in the Devonian Bokkeveld basin of South Africa. Jour. Sediment. Petrol.

Tankard, A. J., Hobday, D. K. 1977. Tide-dominated back-barrier sedimentation, Early Ordovician Cape basin, Cape Peninsula, South Africa. Sediment. Geol. 18:135–159.

Tankard, A. J., Hobday, D. K. 1979. Sedimentation patterns in Early Paleozoic tensional failed arms: The Table Mountain Group of southern Africa. Geol. Soc. Amer. 1979 Congr. Abstract.

Tankard, A. J., Rogers, J. 1978. Late Cenozoic palaeoenvironments on the west coast of southern Africa. Jour. Biogeogr. 5:319–337.

Tarling, D. H. 1972. Another Gondwanaland. Nature 238:92–93.

Tarney, J. 1973. The Scourie dyke suite and the nature of the Inverian event in Assynt. *In* R. G. Park, J. Tarney (eds.), The Early Precambrian of Scotland and related rocks of Greenland, pp. 105–118. Newcastle: Univ. Keele, Dept. Geol.

Tarney, J., Dalziel, I. W. D., De Wit, M. J. 1976. Marginal basin 'Rocas Verdes' complex from south Chile: a model for Archaean greenstone belt formation. *In* B. F. Windley (ed.), The early history of the earth, pp. 131–146. London: Wiley.

Tavener-Smith, R. 1979a. Prograding Middle Ecca coastal associations from the Effingham Quarries near Durban. Geol. Soc. S. Afr. 18th Congr. Abstracts 2:61–68.

Tavener-Smith, R. 1979b. Regional control on coal seam occurrence in the north-east Karoo basin. S. Afr. Jour. Sci. 75:353–355.

Tavener-Smith, R. 1980. The *Skolithos*-bearing sandstone at Newlands Estate, Durban: A suspected shallow water, off-shore sand tongue in the Lower Ecca Group (Pietermaritzburg Formation). Geol. Soc. S. Afr. Trans. 83:87–91.

Taylor, H. P., Mordeckai, M. 1975. Oxygen and hydrogen isotope studies of 2.6 - 3.4 b.y. old granites from the Barberton Mountain Land, and the Rhodesian craton. Geol. Soc. Amer. Abstracts with Progr. 7:1293.

Taylor, S. R., Hallberg, J. A. 1977. Rare earth elements in the Marda calc-alkaline suite: an Archaean geochemical analogue of Andean-type volcanism. Geochim. Cosmochim. Acta 41:1125–1129.

Theart, H. F. J. 1980. The geology of the Precambrian terrane in parts of western Namaqualand. Precambrian Res. Unit Univ. Cape Town Bull. 30.

Theron, J. C. 1975. Sedimentological evidence for the extension of the African continent southwards during the Late Permian - Early Triassic times. *In* K. S. W. Campbell (ed.), Gondwana geology, pp. 61–71. Canberra: Australian National Univ. Press.

Theron, J. N. 1962a. An analysis of the Cape folding in the district of Willowmore, Cape Province. Univ. Stellenbosch Ann. A37:247–419.

Theron, J. N. 1962b. Occurrence of fish remains in the Witteberg/Dwyka transition zone. S. Afr. Geol. Surv. Ann. 1:263–266.

Theron, J. N. 1970. A stratigraphical study of the Bokkeveld Group (Series). 2nd Intl. Gondwana Symposium. Proceedings and Papers, pp. 197–204.

Theron, J. N. 1972. The stratigraphy and sedimentation of the Bokkeveld Group. Unpubl. D.Sc. thesis, Univ. Stellenbosch, South Africa.

Theron, J. N., Blignault, H. J. 1975. A model for sedimentation of the Dwyka glacials in the southwestern Cape. *In* K. S. W. Campbell (ed.), Gondwana geology, pp. 347–356. Canberra: Australian National Univ. Press.

Thomas, D. 1977. A study and evaluation of the Ventersdorp Contact Reef at Vaal Reefs Exploration and Mining Company Limited, Klerksdorp, South Africa. Unpubl. M.Sc. thesis, Royal School of Mines, Univ. London.

Thompson, A. O. 1942. Coastal terraces of eastern Pondoland. Geol. Soc. S. Afr. Trans. 45:37–53.

Thompson, R. W. 1975. Tidal flat sediments of the Colorado River delta, northwestern Gulf of California. *In* R. N. Ginsburg (ed.), Tidal deposits, pp. 57–65. Berlin: Springer-Verlag.

Tickel, S. J. 1975. Braided river deposits in the Waterberg Supergroup. Geol. Soc. S. Afr. Trans. 78:83–88.

Toens, P. D., Le Roux, J. P. 1978. The Permo-Triassic uranium deposits of southern Africa within the Gondwana framework. S. Afr. Atomic Energy Board Rept. VB-579.

Toogood, D. J. 1975. Tectonic interpretation of the Namaqualand Mobile Belt in southeastern South West Africa. Precambrian Res. Unit Univ. Cape Town Ann. Rept. 12:27–32.

Toogood, D. J. 1976. Structural and metamorphic evolution of a gneiss terrane in the Namaqua belt near Onseepkans, South West Africa. Precambrian Res. Unit Univ. Cape Town Bull. No. 19.

Tooms, J. S., Summerhayes, C. P., Cronan, D. S. 1969. Geochemistry of marine phosphate and manganese deposits. Oceanogr. Mar. Biol. 7:48–100.

Truswell, J. F. 1972. Sandstone sheets and related intrusions from Coffee Bay, Transkei, South

Africa. Jour. Sediment. Petrol. 42:578–583.

Truswell, J. F. 1977. The geological evolution of South Africa. Cape Town: Purnell.

Truswell, J. F., Eriksson, K. A. 1973. Stromatolitic associations and their palaeoenvironmental significance: a re-appraisal of a Lower Proterozoic locality from the northern Cape Province, South Africa. Sediment. Geol. 10:1–23.

Truswell, J. F., Eriksson, K. A. 1975. A palaeoenvironmental interpretation of the Early Proterozoic Malmani dolomite from Zwartkops, South Africa. Precambr. Res. 2:277–303.

Truswell, J. F., Ryan, P. J. 1969. A flysch facies in the Lower Ecca Group of the southern Karroo and a portion of the Transkei. Geol. Soc. S. Afr. Trans. 72:151–157.

Truter, F. C. 1955. Modern concepts of the Bushveld Igneous Complex. C.C.T.A. Southern Regional Committee, Compte rendu, Salisbury meeting, pp. 77–92.

Truter, F. C. 1962. Discussion of paper by C. A. Cousins. Geol. Soc. S. Afr. Trans. 65:33–35.

Tucker, R. F. 1980. The sedimentology and mineralogy of the Composite Reef on Cooke Section, Randfontein Estates gold mine, Witwatersrand, South Africa. Unpubl. M.Sc. thesis, Univ. Witwatersrand, Johannesburg, South Africa.

Turner, B. R. 1970. Facies analysis of the Molteno sedimentary cycle. 2nd. Intl. Gondwana Symposium. Proceedings and Papers, pp. 313–319.

Turner, B. R. 1975. Statistical appraisal of Molteno (Triassic) sedimentary cycles from the upper part of the Karroo (Gondwana) System in South Africa. Jour. Sediment. Petrol. 45:95–104.

Turner, B. R. 1977. Fluviatile cross-bedding patterns in the Upper Triassic Molteno Formation of the Karoo (Gondwana) Supergroup in South Africa and Lesotho. Geol. Soc. S. Afr. Trans. 80:241–252.

Turner, B. R. 1978. Sedimentary patterns of uranium mineralization in the Beaufort Group of the southern Karoo (Gondwana) basin, South Africa. *In* A. D. Miall (ed.), Fluvial sedimentology. Can. Soc. Petrol. Geol. Mem. No. 5, pp. 831–848.

Turner, B. R. (ed.) 1979a. Geology of the uraniferous Beaufort Group near Beaufort West. Geol. Soc. S. Afr. 18th Congr. Guidebook.

Turner, B. R. 1979b. Palaeohydraulics of an Upper Triassic braided river system in the main Karoo basin, South Africa. Geol. Soc. S. Afr. 18th Congr. Abstracts 2:167.

Turner, B. R. In preparation.

Turner, J. R. 1977. A sedimentological study of the lower Beaufort Group. Unpubl. M.Sc. thesis, Univ. Natal, Pietermaritzburg.

Tuyezov, I. K. 1967. Transition stages of ancient platforms in U.S.S.R. Intl. Geol. Rev. 9:584–597.

Tweedie, E. B. 1978. History, geology, and value distribution of the Evander Goldfield, eastern Transvaal, South Africa. 11th Commonw. Min. Metall. Congr. Hong Kong Proc., 489–496.

Tyler, N. 1978a. Stratigraphy, origin and correlation of the Kanye Volcanic Group in the west-central Transvaal. Econ. Geol. Res. Unit Univ. Witwatersrand Inf. Circ. No. 130.

Tyler, N. 1978b. A stratigraphic analysis of the pre-Chuniespoort Group strata around the Makoppa Dome, west-central Transvaal. Unpubl. M.Sc. thesis, Univ. Witwatersrand, Johannesburg, South Africa.

Tyler, N. 1979a. Stratigraphy, geochemistry, and correlation of the Ventersdrop Supergroup in the Derdepoort area, west-central Transvaal. Geol. Soc. S. Afr. Trans. 82:133–147.

Tyler, N. 1979b. The stratigraphy of the Early-Proterozoic Buffalo Springs Group in the Thabazimbi area, west-central Transvaal. Geol. Soc. S. Afr. Trans. 82:215–226.

Tyler, N. 1979c. Early-Proterozoic meandering stream sedimentation in the Buffalo Springs Group of the Transvaal Supergroup. Econ. Geol. Res. Unit Univ. Witwatersrand Inf. Circ. No. 121.

Tyler, N. 1979d. Depositional sedimentary environments within the Black Reef Quartzite of the west-central Transvaal. Econ. Geol. Res. Unit Univ. Witwatersrand Inf. Circ. No. 132.

Vail, J. R., Dodson, M. H. 1969. Geochronology of Rhodesia. Geol. Soc. S. Afr. Trans. 72:79–113.

Vail, P. R., Mitchum, R. M., Thompson, S. 1977. Seismic stratigraphy and global changes of sea level, part 4: Global cycles of relative changes of sea level. *In* C. E. Payton (ed.), Seismic stratigraphy—Applications to hydrocarbon exploration. Amer. Assoc. Petrol. Geol. Mem. No. 26, pp. 83–97.

Vajner, V. 1974a. Note on the Precambrian tectonic history of parts of the Namaqua front and foreland in the northern Cape. Precambrian Res. Unit Univ. Cape Town Ann. Rept. 10/11:12–23.

Vajner, V. 1974b. Crustal evolution of the Namaqua Mobile Belt and its foreland in parts of the northern Cape. Precambrian Res. Unit Univ. Cape Town Bull. No. 15, pp. 1–15.

Vajner, V. 1974c. The tectonic development of the Namaqua Mobile Belt and its foreland in parts of the northern Cape. Precambrian Res. Unit Univ. Cape Town Bull. No. 14.

Vajner, V. 1975. Preliminary report on the geology and structure of parts of the Namaqua foreland in the Upington area. Precambrian Res. Unit Univ. Cape Town Ann. Rept. 12:11–19.

Vajner, V. 1976. Progress report on the statistical evaluation of structural data from the Namaqua foreland, northern Cape. Precambrian Res. Unit Univ. Cape Town Ann. Rept. 13:7–18.

Vajner, V. 1978. The Wilgenhoutdrif Formation. Precambrian Res. Unit Univ. Cape Town Ann. Rept. 14/15:13–25.

Vajner, V., Jackson, M. P. A. 1974. Brief report on a field trip to the area between Kakamas, Kenhardt, Marydale and Upington. Precambrian Res. Unit Univ. Cape Town Ann. Rept. 10/11:56–57.

Van Beeck, J. L., Koster, E. A. 1972. Fluvial and estuarine sediments exposed along the Oude Maas. Sedimentology 19:237–256.

Van Bever Donker, J. M. 1980. Structural and metamorphic evolution of an area around Kakamas and Keimoes, Cape Province, South Africa. Precambrian Res. Unit Univ. Cape Town Bull. No. 28.

Van Biljon, S. 1939. The Kuboos batholith in Namaqualand, South Africa. Geol. Soc. S. Afr. Trans. 42:123–219.

Van Breeman, O., Dodson, M. H. 1972. Metamorphic chronology of the Limpopo belt, southern Africa. Geol. Soc. Amer. Bull. 83:2005–2018.

Van den Bergh, J. J. 1946. Petrofabric analysis of the Bushveld gabbro from Bon Accord. Geol. Soc. S. Afr. Trans. 49:156–203.

Van der Merwe, M. J. 1976. The layered sequence of the Potgietersrus limb of the Bushveld Complex. Econ. Geol. 71:1337–1351.

Van Dijk, D. E., Hobday, D. K., Tankard, A. J. 1978. Permo-Triassic lacustrine deposits in the eastern Karoo basin, Natal, South Africa. *In* A. Matter, M. E. Tucker (eds.), Modern and ancient lake sediments. Intl. Assoc. Sedimentologists Spec. Publ. No. 2, pp. 229–235.

Van Eeden, O. R., Visser, H. N., Van Zyl, J. S., Coertze, F. J., Wessels, J. T. 1955. The geology of the eastern Soutpansberg and the Lowveld to the north. S. Afr. Geol. Surv. Map Explanation Sheet 42.

Van Eeden, O. R., De Wet, N. P., Strauss, C. A. 1963. The geology of the area around Schweizer-Reneke. S. Afr. Geol. Surv. Map Explanation Sheets 2724B (Pudmore) and 2725A (Schweizer-Reneke).

Van Houten, F. B. 1968. Iron oxides in red beds. Geol. Soc. Amer. Bull. 79:399–416.

Van Niekerk, C. B. 1962. The age of the Gemspost dyke from the Venterspost Gold Mine. Geol. Soc. S. Afr. Trans. 65:105–111.

Van Niekerk, C. B. 1968. The suitability of extrusive rocks for U-Pb radiometric dating. Unpubl. Ph.D. thesis, Univ. Cape Town, South Africa.

Van Niekerk, C. B. 1969. The uranium-lead isotopic dating of South African acid lavas. Volcan. Bull. 32:481–498.

Van Niekerk, C. B., Burger, A. J. 1964. The age of the Ventersdorp System. S. Afr. Geol. Surv. Ann. 3:75–86.

Van Niekerk, C. B., Burger, A. J. 1969. Lead isotopic data relating to the age of Dominion Reef lava. Geol. Soc. S. Afr. Trans. 72:37–45.

Van Niekerk, C. B., Burger, A. J. 1978a. The age of the Moodies conglomerate boulders. Geol. Soc. S. Afr. Spec. Publ. No. 4, pp. 99–106.

Van Niekerk, C. B., Burger, A. J. 1978b. A new age for the Ventersdorp acidic lavas. Geol. Soc. S. Afr. Trans. 81:155–163.

Van Reenen, D. D. 1979. Phase relations in cordierite-garnet-hypersthene bearing assemblages as an indication of changing metamorphic conditions during a second granulite event in the Limpopo metamorphic complex south of the Soutpansberg. Geol. Soc. S. Afr. 18th Congr. Abstracts 1:307–399.

Van Reenen, D. D., Du Toit, M. C. 1977. Mineral reactions and the timing of metamorphic events in the Limpopo metamorphic complex south of the Soutpansberg. Botswana Geol. Surv. Bull. 12:107–127.

Van Rooyen, D. P. 1950. The metamorphic rocks constituting the floor of the Bushveld Igneous Complex, north of Potgietersrus. Geol. Soc. S. Afr. Trans. 53:65–72.

Van Straaten, L. M. J. U. 1959. Minor structures of some Recent littoral and neritic sediments. Geol. Mijnb. 21:197–216.

Van Straaten, L. M. J. U., Kuenen, Ph. H. 1957. Accumulation of fine-grained sediments in the Dutch Wadden Sea. Geol. Mijnb. 19:329–354.

Van Vuuren, C. J., Cole, D. I. 1979. The stratigraphy and depositional environments of the Ecca Group in the northern part of the Karoo basin. Geol. Soc. S. Afr. Spec. Publ. No. 6, pp. 103–111.

Van Wyk, W. L. 1963. Groundwater studies in northern Natal, Zululand and surrounding areas. S. Afr. Geol. Surv. Mem. No. 52.

Van Zinderen Bakker, E. M. 1976. The evolution of Late-Cenozoic palaeoclimates of southern Africa. Palaeoecol. Afr. 9:160–202.

Van Zyl, C. Z. 1979. The Upington Geotraverse - a progress report. Precambrian Res. Unit Univ. Cape Town Ann. Rept. 16:10–19.

Van Zyl, D. 1967. The geology of the O'okiep Copper Mine, Namaqualand. Univ. Stellenbosch Ann. A42.

Van Zyl, D. 1978. A petrological approach towards the ore-bearing potentialities of the Okiep basic intrusives in Namaqualand. Geol. Soc. S. Afr. Spec. Publ. No. 4, pp. 323–329.

Vermaak, C. F. 1970. The geology of the lower portion of the Bushveld Complex and its relationship to the floor rocks in the area west of the Pilanesberg, western Transvaal. Geol. Soc. S. Afr. Spec. Publ. No. 1, pp. 242–262.

Vermaak, C. F. 1976a. The Merensky Reef - thoughts on its environment and genesis. Econ. Geol. 71:1270–1298.

Vermaak, C. F. 1976b. The nickel pipes of Vlakfontein and vicinity, western Transvaal. Econ. Geol. 71:261–286.

Vermaak, C. F., Hendriks, P. 1976. A review of the mineralogy of the Merensky Reef with specific reference to new data on the previous metal mineralogy. Econ. Geol. 71:1244–1269.

Verwoerd, W. J. 1963. Die geologiese struktuur van die Krokodilrivierfragment. Geol. Soc. S. Afr. Trans. 66:49–66.

Verwoerd, W. J. 1967. The carbonatites of South Africa and South West Africa. S. Afr. Geol. Surv. Handbook No. 6.

Viljoen, M. J., Viljoen, R. P. 1969a. The geology and geochemistry of the Lower Ultramafic Unit of the Onverwacht Group and a proposed new class of igneous rocks. Geol. Soc. S. Afr. Spec. Publ. No. 2, pp. 55–85.

Viljoen, M. J., Viljoen, R. P. 1969b. The geochemical evolution of the granitic rocks of the Barberton region. Geol. Soc. S. Afr. Spec. Publ. No. 2, pp. 189–219.

Viljoen, M. J., Viljoen, R. P. 1970. Archaean volcanicity and continental evolution in the Barberton region, Transvaal. *In* T. N. Clifford, I. G. Gass (eds.), African magmatism and tectonics, pp. 27–49. Edinburgh: Oliver and Boyd.

Viljoen, M. J., Van Vuuren, C., Pearton, T. N., Minnitt, R., Muff, R., Cilliers, P. 1978. The regional geological setting of mineralization in the Murchison Range with particular reference to antimony. Geol. Soc. S. Afr. Spec. Publ. No. 4, pp. 55–76.

Viljoen, M. J., Viljoen, R. P., Pearton, T. N. 1981. The nature and distribution of Archaean komatiites in South Africa. *In* E. G. Nisbet, N. T. Arndt (eds.), Komatiites. London: Allan and Unwin.

Viljoen, R. P., Viljoen, M. J. 1969. The geological and geochemical significance of the upper formations of the Onverwacht Group. Geol. Soc. S. Afr. Spec. Publ. No. 2, pp. 113–151.

Visher, G. S. 1976. A Pennsylvanian interdistributary tidal flat deposit. *In* R. N. Ginsburg (ed.), Tidal deposits, pp. 179–185. New York: Springer-Verlag.

Visser, D. J. L. 1957. The structural evolution of the Union. Geol. Soc. S. Afr. Proc. 60:xiii–1.

Visser, H. N. 1967. Verspreiding en korrelasie van die Formasie Klipheuwel in die Swartland en die Sandveld. S. Afr. Geol. Surv. Ann. 6:31–38.

Visser, H. N., Schoch, A. E. 1973. The geology and mineral resources of the Saldanha Bay area. S. Afr. Geol. Surv. Mem. No. 63.

Visser, J. N. J. 1962. Die voorkoms en oorsprong van die tillietband in die Serie Tafelberg. Unpubl. M.Sc. thesis, Univ. Orange Free State, Bloemfontein, South Africa.

Visser, J. N. J. 1971. The deposition of the Griquatown glacial member in the Transvaal Supergroup. Geol. Soc. S. Afr. Trans 74:187–199.

Visser, J. N. J. 1974. The Table Mountain Group: a study in the deposition of quartz arenites on a stable shelf. Geol. Soc. S. Afr. Trans. 77:229–237.

Visser, J. N. J., Botha, B. J. V. 1980. Meander channel, point bar, crevasse splay and aeolian deposits from the Elliot Formation in Barkly Pass, north-eastern Cape. Geol. Soc. S. Afr. Trans. 83: 55–62.

Visser, J. N. J., Dukas, B. A. 1979. Upward-fining fluviatile megacycles in the Beaufort Group north of Graaff-Reinet, Cape Province. Geol. Soc. S. Afr. Trans. 82:149–154.

Visser, J. N. J., Loock, J. C. 1978. Water depth in the main Karoo basin, South Africa, during Ecca (Permian) sedimentation. Geol. Soc. S. Afr. Trans. 81:185–191.

Visser, J. N. J., Loock, J. C. 1979. The Beaufort Group between Graaff-Reinet and De Aar, Cape Province. Geol. Soc. S. Afr. 18th Congr. Abstracts 2:158–166.

Visser, J. N. J., Loock, J. C., Van der Merwe, J., Joubert, C. W., Potgieter, C. D., McLaren, C. H., Potgieter, G. J. A., Van der Westhuizen, W. A., Nel, L., Lemmer, W. M. 1978. The Dwyka Formation and Ecca Group, Karoo sequence, in the northern Karoo basin, Kimberley-Britstown area. S. Afr. Geol. Surv. Ann. 12:143–176.

Vogt, P. R., Johnson, G. L., Holcombe, T. L., Gilg, J. C., Avery, O. E. 1971. Episodes of seafloor spreading recorded by the North Atlantic basement. Tectonophysics 12:211–234.

Von Backström, J. W. 1952. The Dominion Reef and Witwatersrand Systems between Wolmaransstad and Ottosdal, Transvaal. Geol. Soc. S. Afr. Trans. 55:53–71.

Von Backström, J. W. 1964. The geology of an area around Keimoes, Cape Province, with special reference to phacoliths of charnockitic adamellite-porphyry. S. Afr. Geol. Surv. Mem. No. 53.

Von Backström, J. W. 1967. The geology and mineral deposits of the Riemvasmaak area, northwest Cape Province. S. Afr. Geol. Surv. Ann. 6:43–51.

Von Backström, J. W. 1976. Pegmatite deposits. S. Afr. Geol. Surv. Handbook 7:183–197.

Von Backström, J. W., De Villiers, J. 1972. The geology along the Orange River valley between Onseepkans and the Richtersveld. S. Afr. Geol. Surv. Map Explanation Sheets 2817D (Vioolsdrif), 2818C and D (Goodhouse), 2819C (Onseepkans).

Von Brunn, V. 1969. Igneous rocks of the Nagatis and Sinclair Formations north-east of Lüderitz, South West Africa. Precambrian Res. Unit Univ. Cape Town bull. No. 7.

Von Brunn, V. 1974. Tidalites of the Pongola Supergroup (Early Precambrian) in the Swart-Mfolozi area, north Natal. Precambrian Res. Unit Univ. Cape Town Bull. 15:107–122.

Von Brunn, V. 1977. A furrowed intratillite pavement in the Dwyka Group of northern Natal. Geol. Soc.

S. Afr. Trans. 80:125–130.

Von Brunn, V. 1979. Sedimentary structures in the upper Dwyka deposits of northern Natal. Geol. Soc. S. Afr. 18th Congr. Abstracts 2:60.

Von Brunn, V., Hobday, D. K. 1976. Early Precambrian tidal sedimentation in the Pongola Supergroup of South Africa. Jour. Sediment. Petrol. 46: 670–679.

Von Brunn, V., Mason, T. R. 1977. Siliciclastic-carbonate tidal deposits from the 3000 m.y. old Pongola Supergroup, South Africa. Sediment. Geol. 18:245–255.

Von Gruenewaldt, G. 1968. The Rooiberg felsite north of Middelburg and its relation to the layered sequence of the Bushveld Complex. Geol. Soc. S. Afr. Trans. 71:153–172.

Von Gruenewaldt, G. 1972. The origin of the roof-rocks of the Bushveld Complex between Tauteshoogte and Pardekop in the eastern Transvaal. Geol. Soc. S. Afr. Trans. 75:121–134.

Von Gruenewaldt, G. 1973. The main and upper zones of the Bushveld Complex in the Roossenekal area, eastern Transvaal. Geol. Soc. S. Afr. Trans. 76:207–227.

Von Gruenewaldt, G. 1976. Sulfides in the upper zone of the eastern Bushveld Complex. Econ. Geol. 71:1324–1336.

Von Gruenewaldt, G. 1977. Mineral resources of the Bushveld Complex. Minerals Sci. Engng. 9:83–95.

Vos, R. G. 1975. An alluvial plain and lacustrine model for the Precambrian Witwatersrand deposits of South Africa. Jour. Sediment. Petrol. 45:480–493.

Vos, R. G., Eriksson, K. A. 1977. An embayment model for tidal and wave-swash deposits occurring within a fluvially-dominated Proterzoic sequence in South Africa. Sediment. Geol. 18:161–173.

Vos, R. G., Hobday, D. K. 1977. Storm beach deposits of the Upper Paleozoic Ecca Group of South Africa. Sediment. Geol. 19:217–232.

Vos, R. G., Tankard, A. J. (1981). Braided fluvial sedimentation in the Lower Paleozoic Cape Basin, South Africa. Sediment. Geol. 29: 171–193.

Vrba, E. S. 1975. Some evidence of chronology and palaeoecology of Sterkfontein, Swartkrans and Kromdraai from the fossil Bovidae. Nature 254:301–304.

Vrba, E. S. 1976. The fossil Bovidae of Sterkfontein, Swartkrans and Kromdraai. Transvaal Mus. Mem. No. 21.

Wagener, G. F. 1972. Sedimentary tectonics of the East Rand basin with special reference to the Brakpan Formation. Unpubl. M.Sc. thesis, Univ. Stellenbosch, South Africa.

Wager, L. R., Brown, G. M. 1967. Layered igneous rocks. London: Oliver and Boyd.

Wagner, P. 1914. The diamond fields of southern Africa. Johannesburg: *The Transvaal Leader*.

Wagner, P. A. 1922. The Pretoria salt-pan, a soda caldera. S. Afr. Geol. Surv. Mem. No. 20.

Wakefield, J. 1974. The geology of the Pikwe-Cu province, eastern Botswana. Unpubl. Ph.D. thesis, Univ. Leeds, England.

Wakefield, J. 1976. The structural and metamorphic evolution of the Pikwe Ni-Cu sulfide deposit, Selebi-Pikwe, eastern Botswana. Econ. Geol. 71: 988–1005.

Wakefield, J. 1977a. The structural and metamorphic history of the Pikwe-area—evidence for post-Great Dyke deformation in the Limpopo Mobile Belt. Botswana Geol. Surv. Bull. No. 12, pp. 141–156.

Wakefield, J. 1977b. Mylonitization in the Lethakane shear zone, eastern Botswana. Geol. Soc. London Jour. 133:263–275.

Walker, F., Poldervaart, A. 1949. Karroo dolerites of the Union of South Africa. Geol. Soc. Amer. Bull. 60:591–706.

Walker, R. G. 1970. Review of the geometry and facies organization of turbidites and turbidite-bearing basins. *In* J. Lajoie (ed.), Flysch sedimentology in North America. Geol. Assoc. Can. Spec. Pap. 7, pp. 219–251.

Walker, R. G. 1978. Deep-water sandstone facies and ancient submarine fans: Models for exploration for stratigraphic traps. Amer. Assoc. Petrol. Geol. Bull. 62:932–966.

Walker, T. R. 1976. Diagenetic origin of continental red beds. *In* H. Falke (ed.), The continental Permian in Central, West and South Europe, pp. 240–282. Dordrecht (Holland): Reidel.

Walraven, F. 1976. Notes on the late-stage history of the western Bushveld Complex. Geol. Soc. S. Afr. Trans. 79:13–21.

Walraven, F., Darracott, B. W. 1976. Quantitative interpretation of a gravity profile across the western Bushveld Complex. Geol. Soc. S. Afr. Trans. 79:22–26.

Walter, M. R., 1972a. Stromatolites and the biostratigraphy of the Australian Precambrian and Cambrian. Palaeont. Assoc. London Spec. Pap. No. 11.

Walter, M. R. 1972b. A hot spring analog for the depositional environment of Precambrian iron formations of the Lake Superior region. Econ. Geol. 67:965–980.

Ward, J. H. W. 1974. The Vioolsdrif pegmatite belt. Precambrian Res. Unit Univ. Cape Town Ann. Rept. 10/11:38–42.

Watchorn, M. B. 1978. Sedimentology of the Mozaan Group in the south-eastern Transvaal and northern Natal. Unpubl. M.Sc. thesis, Univ. Natal, South Africa.

Watchorn, M. A. 1980a. Fluvial and tidal sedimentation in the 3000 Ma Mozaan basin, South Africa. Precambr. Res. 13:27–42.

Watchorn, M. B. 1980b. The stratigraphic and sedimentological development of the Witwatersrand

West Rand basin in the Klerksdorp area. Econ. Geol. Res. Unit Univ. Witwatersrand, Inf. Circ. 148.

Waters, D. J., McDaid, J. 1979. Phase relationships of some high grade metapelites from Lüderitz district, S.W.A./Namibia. Geol. Soc. S. Afr. 18th Congr. Abstracts 1:437–441.

Watson, D. M. S. 1914. The zones of the Beaufort beds of the Karroo System in South Africa. Geol. Mag. 1:203–208.

Watters, B. R. 1974. Stratigraphy, igneous petrology and evolution of the Sinclair Group in southern South West Africa. Precambrian Res. Unit Univ. Cape Town Bull. No. 16.

Watters, B. R. 1976. Possible Late Precambrian subduction zone in South West Africa. Nature 259: 471–473.

Watters, B. R. 1977. The Sinclair Group: Definition and regional correlation. Geol. Soc. S. Afr. Trans. 80:9–16.

Watts, N. L. 1980. Quaternary pedogenic calcretes from the Kalahari (southern Africa): mineralogy, genesis, and diagenesis. Sedimentology 27:661–686.

Weimer, R. J. 1970. Rates of deltaic sedimentation and intrabasin deformation, Upper Cretaceous of Rocky Mountain Region. *In* J. P. Morgan, R. H. Shaver (eds.), Deltaic sedimentation, modern and ancient. Soc. Econ. Paleont. Mineral. Spec. Publ. No. 15, pp. 270–292.

Welke, H. J., Burger, A. J., Corner, B., Kröner, A., Blignault, H. J. 1979. U-Pb and Rb-Sr age determinations on Middle Proterozoic rocks from the lower Orange River area, southwestern Africa. Geol. Soc. S. Afr. Trans. 82:205–214.

Wells, P. 1977. Pyroxene thermometry in simple and complex systems. Contr. Mineral. Petrol. 62:129–139.

Wendlandt, R. F. 1977. The system K_2O-MgO-Al_2O_3-SiO_2-H_2O-CO_2: phase relations of the joins Ph-H_2O-CO_2 and Ph-Eh-Mag at high temperatures and high pressures and applications to the genesis of alkalic magmas. 2nd Intl. Kimberlite Conf. Extended Abstracts.

Whateley, M. K. G. 1979. Deltaic and fluvial deposits of the Ecca Group (Late Carboniferous-Permian), Nongoma graben, Zululand. Geol. Soc. S. Afr. 18th Congr. Abstracts 2:71–76.

Whiteside, H. C. M. 1970. Volcanic rocks of the Witwatersrand triad. *In* T. N. Clifford, I. G. Gass, (eds.), African magmatism and tectonics, pp. 78–87. Edinburgh: Oliver and Boyd.

Whitford, D. J., Arndt, N. T. 1978. Rare earth element abundances in a thick, layered komatiite flow from Ontario, Canada. Earth Planet. Sci. Lett. 41:188–196.

Whitten, E. H. T. 1966. Structural geology of folded rocks. Chicago: Rand McNally.

Whittington, H. B., Hughes, C. P. 1970. Ordovician geography and faunal provinces deduced from trilobite distribution. Roy. Soc. London Phil. Trans. B263:235–278.

Wilcox, R. E., Harding, T. P., Seely, D. R. 1973. Basic wrench tectonics. Amer. Assoc. Petrol. Geol. Bull. 57:74–96.

Willemse, J. 1959. The floor of the Bushveld Igneous Complex and its relationships with special reference to the eastern Transvaal. Geol. Soc. S. Afr. Proc. 62:xxi–lxxx.

Willemse, J. 1964. A brief outline of the geology of the Bushveld Complex. *In* S. H. Haughton (ed.), The geology of some ore deposits of southern Africa, vol. 2, pp. 91–128. Johannesburg: Geol. Soc. S. Afr.

Willemse, J. 1969. The geology of the Bushveld Igneous Complex: The largest repository of magmatic ore deposits in the world. Econ. Geol. Monogr. 4:1–22.

Williams, D. A. C. 1972. Archaean ultramafic and mafic and associated rocks, Mt. Monger, Western Australia. Geol. Soc. Aust. Jour. 19:163–188.

Williams, D. A. C., Furnell, R. G. 1979. Reassessment of part of the Barberton type area, South Africa. Precambr. Res. 9:325–347.

Williams, P. F., Rust, B. R. 1969. The sedimentology of a braided river. Jour. Sediment. Petrol. 39:649–679.

Wilson, J. F., Bickle, M. J., Hawkesworth, C. J., Martin, A., Nisbet, E. G., Orpen, J. L. 1978. Granite-greenstone terrains of the Rhodesian Archaean craton. Nature 271:23–27.

Windley, B. F. 1973. Archaean anorthosites: A review with the Fiskenaesset Complex, West Greenland, as a model for interpretation. Geol. Soc. S. Afr. Spec. Publ. No. 3, pp. 319–332.

Winkler, H. G. F. 1974. Petrogenesis of metamorphic rocks (3rd ed.). New York: Springer-Verlag.

Winter, H. de la R. 1963. Algal structures in sediments of the Ventersdorp System. Geol. Soc. S. Afr. Trans. 66:115–128.

Winter, H. de la R. 1965. The stratigraphy of the Ventersdorp System in the Bothaville district and adjoining areas. Unpubl. Ph.D. thesis, Univ. Witwatersrand, Johannesburg, South Africa.

Winter, H. de la R. 1973. Geology of the Algoa basin, South Africa. *In* G. Blant (ed.), Sedimentary basins of the African coast. Part 2. South and east coast, p. 17–48. Paris: Assoc. African Geol. Surveys.

Winter, H. de la R. 1976. A lithostratigraphic classification of the Ventersdorp succession. Geol. Soc. S. Afr. Trans. 79:31–48.

Winter, H. de la R. 1979. Application of basic principles of stratigraphy to the Jurassic-Cretaceous interval in southern Africa. Geol. Soc. S. Afr. Spec. Publ. No. 6, pp. 183–196.

Winter, H. de la R., Venter, J. J. 1970. Lithostratigraphic correlation of recent deep boreholes in the Karoo-Cape sequence. 2nd Gondwana Symposium. Proceedings and Papers, pp. 395–408.

Wise, D. U. 1969. Origin of the moon from the earth: Some new mechanisms and comparisons. Jour. Geophys. Res. 74:6035–6045.

Wood, B. J. 1974. The solubility of alumina in orthopyroxene co-existing with garnet. Contr. Mineral. Petrol. 46:1–15.

Wood, B. J. 1976. The partitioning of iron and magnesium between garnet and clinopyroxene. Carnegie Inst. Washington Yearbook 75:571–573.

Wood, B. J., Banno, S. 1973. Garnet-orthopyroxene and orthopyroxene - clinopyroxene relationships in simple and complex systems. Contr. Mineral. Petrol. 42:100–124.

Woolley, A. R., Garson, M. S. 1970. Petrochemical and tectonic relationship of the Malawi carbonatite–alkaline province and the Lupata - Lebombo volcanics. *In* T. N. Clifford, I. G. Gass (eds.), African magmatism and tectonics, pp. 237–262. Edinburgh: Oliver and Boyd.

Worst, B. G. 1962. The geology of the Buhwa iron ore deposits and adjoining country: Belingwe district. Southern Rhodesia Geol. Surv. Bull. No. 53.

Wright, A. D. 1968. A westward extension of the Upper Ashgillian *Hirnantia* fauna. Lethaia 1:353–367.

Wright, L. 1973. Preliminary report on the structures present in the central zone of the Limpopo belt, eastern Botswana. Res. Inst. African Geol. Univ. Leeds Ann. Rept. 17:16–17.

Wright, L. 1974. The structure of the Limpopo belt in eastern Botswana. Res. Inst. African Geol. Univ. Leeds Ann. Rept. 18:50–53.

Wright, L. 1975. A deformation study of the Gwanda schist belt and its implications for the formation of the belt. Res. Inst. African Geol. Univ. Leeds Ann. Rept. 19:59–65.

Wright, L. D. 1978. River deltas. *In* R. A. Davis (ed.), Coastal sedimentary environments, pp. 5–68. New York: Springer-Verlag.

Wyatt, B. A. 1976. The geology and geochemistry of the Klipriviersberg volcanics, Ventersdorp Supergroup, south of Johannesburg. Unpubl. M.Sc. thesis, Univ. Witwatersrand, Johannesburg, South Africa.

Wyatt, B. A. 1977. The melting and crystallization behaviour of a natural clinopyroxene–ilmenite intergrowth. Contr. Mineral. Petrol. 61:1–9.

Wyatt, B. A., McAllister, R. H., Boyd, F. R., Ohashi, Y. 1975. An experimentally produced clinopyroxene-ilmenite intergrowth. Carnegie Inst. Washington Yearbook 74:536–539.

Wybergh, W. 1920. The coastal limestone of the Cape Province. Geol. Soc. S. Afr. Trans. 22:46–67.

Wyllie, P. J. 1979. Kimberlite magmas from the system peridotite -CO_2-H_2O. 2nd Intl. Kimberlite Conf. Proc. 1:319–329.

Wyllie, P. J., Huang, W. L. 1975. Influence of mantle CO_2 in the generation of carbonatites and kimberlites. Nature 257:297–299.

Yoder, H. S. 1976. Generation of basaltic magma. Washington: National Adacemy of Sciences.

Young, J. Z. 1962. The life of vertebrates (2nd ed.). Oxford: Clarendon.

Young, R. G. 1957. Late Cretaceous cyclic deposits, Book Cliffs, eastern Utah. Amer. Assoc. Petrol. Geol. Bull. 44:156–194.

Ziegler, A. M., Cocks, L. R. M., Bambach, R. J. 1968. The composition and structure of Lower Silurian marine communities. Lethaia 1:1–27.

Zelt, G. A. D. 1980a. A geotraverse across Namaqualand, South Africa: The petrology, geochemistry, and structural relations of a Proterozoic high-grade metamorphic terrain. Unpubl. Ph.D thesis, Univ. Natal, Pietermaritzburg.

Zelt, G. A. D. 1980b. Granulite-facies metamorphism in Namaqualand, South Africa. Precambr. Res. 13:253–274.

Zumberge, J. E., Sigleo, A. C. Nagy, B. 1978. Molecular and elemental analysis of the carbonaceous matter in the gold and uranium bearing Vaal Reef carbon seams, Witwatersrand sequence. Minerals Sci. Engng. 10:223–246.

Index

A

C

D

G

H

I

J

K

N

O

P

T

U

V